ASTRONOMICAL DATA ANALYSIS SOFTWARE AND SYSTEMS VI

A SERIES OF BOOKS ON RECENT DEVELOPMENTS IN ASTRONOMY AND ASTROPHYSICS

Managing Editor, D. Harold McNamara
Production Manager, Elizabeth S. Holloman

Printed by BookCrafters, Inc.

First published 1997

Library of Congress Catalog Card Number: 97-73839
ISBN 1-886733-45-7

Please contact proper address for information on:

PUBLISHING:
Managing Editor
PO Box 24463
Brigham Young University
Provo, UT 84602-4463
801-378-2298

pasp@astro.byu.edu
Fax: 801-378-2265

ORDERING BOOKS:
Astronomical Society of the Pacific
CONFERENCE SERIES
390 Ashton Avenue
San Francisco, CA 94112 - 1722 USA
415-337-2624

catalog@aspsky.org
Fax: 415-337-5205

A SERIES OF BOOKS ON RECENT DEVELOPMENTS IN ASTRONOMY AND ASTROPHYSICS

Vol. 1-Progress and Opportunities in Southern Hemisphere Optical Astronomy: The CTIO 25th Anniversary Symposium
ed. V. M. Blanco and M. M. Phillips ISBN 0-937707-18-X

Vol. 2-Proceedings of a Workshop on Optical Surveys for Quasars
ed. P. S. Osmer, A. C. Porter, R. F. Green, and C. B. Foltz ISBN 0-937707-19-8

Vol. 3-Fiber Optics in Astronomy
ed. S. C. Barden ISBN 0-937707-20-1

Vol. 4-The Extragalactic Distance Scale: Proceedings of the ASP 100th Anniversary Symposium
ed. S. van den Bergh and C. J. Pritchet ISBN 0-937707-21-X

Vol. 5-The Minnesota Lectures on Clusters of Galaxies and Large-Scale Structure
ed. J. M. Dickey ISBN 0-937707-22-8

Vol. 6-Synthesis Imaging in Radio Astronomy: A Collection of Lectures from the Third NRAO Synthesis Imaging Summer School
ed. R. A. Perley, F. R. Schwab, and A. H. Bridle ISBN 0-937707-23-6

Vol. 7-Properties of Hot Luminous Stars: Boulder-Munich Workshop
ed. C. D. Garmany ISBN 0-937707-24-4

Vol. 8-CCDs in Astronomy
ed. G. H. Jacoby ISBN 0-937707-25-2

Vol. 9-Cool Stars, Stellar Systems, and the Sun. Sixth Cambridge Workshop
ed. G. Wallerstein ISBN 0-937707-27-9

Vol. 10-The Evolution of the Universe of Galaxies. The Edwin Hubble Centennial Symposium
ed. R. G. Kron ISBN 0-937707-28-7

Vol. 11-Confrontation Between Stellar Pulsation and Evolution
ed. C. Cacciari and G. Clementini ISBN 0-937707-30-9

Vol. 12-The Evolution of the Interstellar Medium
ed. L. Blitz ISBN 0-937707-31-7

Vol. 13-The Formation and Evolution of Star Clusters
ed. K. Janes ISBN 0-937707-32-5

Vol. 14-Astrophysics with Infrared Arrays
ed. R. Elston ISBN 0-937707-33-3

Vol. 15-Large-Scale Structures and Peculiar Motions in the Universe
ed. D. W. Latham and L. A. N. da Costa ISBN 0-937707-34-1

Vol. 16-Atoms, Ions and Molecules: New Results in Spectral Line Astrophysics
ed. A. D. Haschick and P. T. P. Ho ISBN 0-937707-35-X

Vol. 17-Light Pollution, Radio Interference, and Space Debris
ed. D. L. Crawford ISBN 0-937707-36-8

Vol. 18-The Interpretation of Modern Synthesis Observations of Spiral Galaxies
ed. N. Duric and P. C. Crane ISBN 0-937707-37-6

Vol. 19-Radio Interferometry: Theory, Techniques, and Application, IAU Colloquium 131
ed. T. J. Cornwell and R. A. Perley ISBN 0-937707-38-4

Vol. 20-Frontiers of Stellar Evolution, celebrating the 50th Anniversary of McDonald Observatory
ed. D. L. Lambert ISBN 0-937707-39-2

Vol. 21-The Space Distribution of Quasars
ed. D. Crampton ISBN 0-937707-40-6

Vol. 22-Nonisotropic and Variable Outflows from Stars
ed. L. Drissen, C. Leitherer, and A. Nota ISBN 0-937707-41-4

Vol. 23-Astronomical CCD Observing and Reduction Techniques
ed. S. B. Howell ISBN 0-937707-42-4

Vol. 24-Cosmology and Large-Scale Structure in the Universe
ed. R. R. de Carvalho ISBN 0-937707-43-0

Vol. 25-Astronomical Data Analysis Software and Systems I
ed. D. M. Worrall, C. Biemesderfer, and J. Barnes ISBN 0-937707-44-9

Vol. 26-Cool Stars, Stellar Systems, and the Sun, Seventh Cambridge Workshop
ed. M. S. Giampapa and J. A. Bookbinder ISBN 0-937707-45-7

Vol. 27-The Solar Cycle
ed. K. L. Harvey ISBN 0-937707-46-5

Vol. 28-Automated Telescopes for Photometry and Imaging
ed. S. J. Adelman, R. J. Dukes, Jr., and C. J. Adelman ISBN 0-937707-47-3

Vol. 29-Workshop on Cataclysmic Variable Stars
ed. N. Vogt ISBN 0-937707-48-1

Vol. 30-Variable Stars and Galaxies, in honor of M. S. Feast on his retirement
ed. B. Warner ISBN 0-937707-49-X

Vol. 31-Relationships Between Active Galactic Nuclei and Starburst Galaxies
ed. A. V. Filippenko ISBN 0-937707-50-3

Vol. 32-Complementary Approaches to Double and Multiple Star Research, IAU Collouquium 135
ed. H. A. McAlister and W. I. Hartkopf ISBN 0-937707-51-1

Vol. 33-Research Amateur Astronomy
ed. S. J. Edberg ISBN 0-937707-52-X

Vol. 34-Robotic Telescopes in the 1990s
ed. A. V. Filippenko ISBN 0-937707-53-8

Vol. 35-Massive Stars: Their Lives in the Interstellar Medium
ed. J. P. Cassinelli and E. B. Churchwell ISBN 0-937707-54-6

Vol. 36-Planets and Pulsars
ed. J. A. Phillips, S. E. Thorsett, and S. R. Kulkarni ISBN 0-937707-55-4

Vol. 37-Fiber Optics in Astronomy II
ed. P. M. Gray ISBN 0-937707-56-2

Vol. 38-New Frontiers in Binary Star Research
ed. K. C. Leung and I. S. Nha ISBN 0-937707-57-0

Vol. 39-The Minnesota Lectures on the Structure and Dynamics of the Milky Way
ed. Roberta M. Humphreys ISBN 0-937707-58-9

Vol. 40-Inside the Stars, IAU Colloquium 137
ed. Werner W. Weiss and Annie Baglin ISBN 0-937707-59-7

Vol. 41-Astronomical Infrared Spectroscopy: Future Observational Directions
ed. Sun Kwok ISBN 0-937707-60-0

Vol. 42-GONG 1992: Seismic Investigation of the Sun and Stars
ed. Timothy M. Brown ISBN 0-937707-61-9

Vol. 43-Sky Surveys: Protostars to Protogalaxies
ed. B. T. Soifer ISBN 0-937707-62-7

Vol. 44-Peculiar Versus Normal Phenomena in A-Type and Related Stars
ed. M. M. Dworetsky, F. Castelli, and R. Faraggiana ISBN 0-937707-63-5

Vol. 45-Luminous High-Latitude Stars
ed. D. D. Sasselov ISBN 0-937707-64-3

Vol. 46-The Magnetic and Velocity Fields of Solar Active Regions, IAU Colloquium 141
ed. H. Zirin, G. Ai, and H. Wang ISBN 0-937707-65-1

Vol. 47-Third Decinnial US-USSR Conference on SETI
ed. G. Seth Shostak ISBN 0-937707-66-X

Vol. 48-The Globular Cluster-Galaxy Connection
ed. Graeme H. Smith and Jean P. Brodie ISBN 0-937707-67-8

Vol. 49-Galaxy Evolution: The Milky Way Perspective
ed. Steven R. Majewski ISBN 0-937707-68-6

Vol. 50-Structure and Dynamics of Globular Clusters
ed. S. G. Djorgovski and G. Meylan ISBN 0-937707-69-4

Vol. 51-Observational Cosmology
ed. G. Chincarini, A. Iovino, T. Maccacaro, and D. Maccagni ISBN 0-937707-70-8

Vol. 52-Astronomical Data Analysis Software and Systems II
ed. R. J. Hanisch, J. V. Brissenden, and Jeannette Barnes ISBN 0-937707-71-6

Vol. 53-Blue Stragglers
ed. Rex A. Saffer ISBN 0-937707-72-4

Vol. 54-The First Stromlo Symposium: The Physics of Active Galaxies
ed. Geoffrey V. Bicknell, Michael A. Dopita, and Peter J. Quinn ISBN 0-937707-73-2

Vol. 55-Optical Astronomy from the Earth and Moon
ed. Diane M. Pyper and Ronald J. Angione ISBN 0-937707-74-0

Vol. 56-Interacting Binary Stars
ed. Allen W. Shafter ISBN 0-937707-75-9

Vol. 57-Stellar and Circumstellar Astrophysics
ed. George Wallerstein and Alberto Noriega-Crespo ISBN 0-937707-76-7

Vol. 58-The First Symposium on the Infrared Cirrus and Diffuse Interstellar Clouds
ed. Roc M. Cutri and William B. Latter ISBN 0-937707-77-5

Vol. 59-Astronomy with Millimeter and Submillimeter Wave Interferometry
ed. M. Ishiguro and Wm. J. Welch ISBN 0-937707-78-3

Vol. 60-The MK Process at 50 Years: A Powerful Tool for Astrophysical Insight
ed. C. J. Corbally, R. O. Gray, and R. F. Garrison ISBN 0-937707-79-1

Vol. 61-Astronomical Data Analysis Software and Systems III
ed. Dennis R. Crabtree, R. J. Hanisch, and Jeannette Barnes ISBN 0-937707-80-5

Vol. 62-The Nature and Evolutionary Status of Herbig Ae / Be Stars
ed. P. S. Thé, M. R. Pérez, and E. P. J. van den Heuvel ISBN 0-937707-81-3

Vol. 63-Seventy-Five Years of Hirayama Asteroid Families: The role of Collisions in the Solar System History
ed. R. Binzel, Y. Kozai, and T. Hirayama ISBN 0-937707-82-1

Vol. 64-Cool Stars, Stellar Systems, and the Sun, Eighth Cambridge Workshop
ed. Jean-Pierre Caillault ISBN 0-937707-83-X

Vol. 65-Clouds, Cores, and Low Mass Stars
ed. Dan P. Clemens and Richard Barvainis ISBN 0-937707-84-8

Vol. 66- Physics of the Gaseous and Stellar Disks of the Galaxy
ed. Ivan R. King ISBN 0-937707-85-6

Vol. 67-Unveiling Large-Scale Structures Behind the Milky Way
ed. C. Balkowski and R. C. Kraan-Korteweg ISBN 0-937707-86-4

Vol. 68-Solar Active Region Evolution: Comparing Models with Observations
ed. K. S. Balasubramaniam and George W. Simon ISBN 0-937707-87-2

Vol. 69-Reverberation Mapping of the Broad-Line Region in Active Galactic Nuclei
ed. P. M. Gondhalekar, K. Horne, and B. M. Peterson ISBN 0-937707-88-0

Vol. 70-Groups of Galaxies
ed. Otto G. Richter and Kirk Borne ISBN 0-937707-89-9

Vol. 71-Tridimensional Optical Spectroscopic Methods in Astrophysics
ed. G. Comte and M. Marcelin ISBN 0-937707-90-2

Vol. 72-Millisecond Pulsars—A Decade of Surprise, ed. A. A. Fruchter
M. Tavani, and D. C. Backer ISBN 0-937707-91-0

Vol. 73-Airborne Astronomy Symposium on the Galactic Ecosystem: From Gas to Stars to Dust
ed. M. R. Haas, J. A. Davidson, and E. F. Erickson ISBN 0-937707-92-9

Vol. 74-Progress in the Search for Extraterrestrial Life,
ed. G. Seth Shostak ISBN 0-937707-93-7

Vol. 75-Multi-Feed Systems for Radio Telescopes
ed. D. T. Emerson and J. M. Payne ISBN 0-937707-94-5

Vol. 76-GONG '94: Helio- and Astero-Seismology from the Earth and Space
ed. Roger K. Ulrich, Edward J. Rhodes, Jr., and Werner Däppen ISBN 0-937707-95-3

Vol. 77-Astronomical Data Analysis Software and Systems IV
ed. R. A. Shaw, H. E. Payne, and J. J. E. Hayes ISBN 0-937707-96-1

Vol. 78-Astrophysical Applications of Powerful New Databases
ed. S. J. Adelman and W. L. Wiese ISBN 0-937707-97-X

Vol. 79-Robotic Telescopes: Current Capabilities, Present Developments, and Future Prospects for Automated Astronomy
ed. Gregory W. Henry and Joel A. Eaton ISBN 0-937707-98-8

Vol. 80-The Physics of the Interstellar Medium and Intergalactic Medium
ed. A. Ferrara, C. F. McKee, C. Heiles, and P. R. Shapiro ISBN 0-937707-99-6

Vol. 81-Laboratory and Astronomical High Resolution Spectra
ed. A. J. Sauval, R. Blomme, and N. Grevesse ISBN 1-886733-01-5

Vol. 82-Very Long Baseline Interferometry and the VLBA,
ed. J. A. Zensus, P. K. Diamond, and P. J. Napier ISBN 1-886733-02-3

Vol. 83-Astrophysical Applications of Stellar Pulsation, IAU Colloquium 155
ed. R. S. Stobie and P. A. Whitelock ISBN 1-886733-03-1

Vol. 84-The Future Utilisation of Schmidt Telescopes, IAU Colloquium 148
ed. Jessica Chapman, Russell Cannon, Sandra Harrison, and Bambang Hidayat ISBN 1-886733-05-8

Vol. 85-Cape Workshop on Magnetic Cataclysmic Variables
ed. D. A. H. Buckley and B. Warner ISBN 1-886733-06-6

Vol. 86-Fresh Views of Elliptical Galaxies
ed. Alberto Buzzoni, Alvio Renzini, and Alfonso Serrano ISBN 1-886733-07-4

Vol. 87-New Observing Modes for the Next Century
ed. Todd Boroson, John Davies, and Ian Robson ISBN 1-886733-08-2

Vol. 88-Clusters, Lensing, and the Future of the Universe
ed. Virginia Trimble and Andreas Reisenegger ISBN 1-886733-09-0

Vol. 89-Astronomy Education: Current Developments, Future Coordination
ed. John R. Percy ISBN 1-886733-10-4

Vol. 90-The Origins, Evolution, and Destinies of Binary Stars in Clusters
ed. E. F. Milone and J. -C. Mermilliod ISBN 1-886733-11-2

Vol. 91-Barred Galaxies, IAU Colloquium 157
ed. R. Buta, D. A. Crocker, and B. G. Elmegreen ISBN 1-886733-12-0

Vol. 92-Formation of the Galactic Halo--Inside and Out
ed. H. L. Morrison and A. Sarajedini ISBN 1-886733-13-9

Vol. 93-Radio Emission from the Stars and the Sun
ed. A. R. Taylor and J. M. Paredes ISBN 1-886733-14-7

Vol. 94-Mapping, Measuring, and Modelling the Universe
ed. Peter Coles, Vicent Martinez, and Maria-Jesus Pons-Borderia ISBN 1-886733-15-5

Vol. 95-Solar Drivers of Interplanetary and Terrestrial Disturbances
ed. K.S. Balasubramaniam, S. L. Keil, and R. N. Smartt ISBN 1-886733-16-3

Vol. 96-Hydrogen-Deficient Stars
ed. C. S. Jeffery and U. Heber ISBN 1-886733-17-1

Vol. 97-Polarimetry of the Interstellar Medium
ed. W. G. Roberge and D. C. B. Whittet ISBN 1-886733-18-X

Vol. 98-From Stars to Galaxies: The Impact of Stellar Physics on Galaxy Evolution
ed. Claus Leitherer, Uta Fritze-von Alvensleben, and John Huchra ISBN 1-886733-19-8

Vol. 99-Cosmic Abundances
ed. Stephen S. Holt and George Sonneborn ISBN 1-886733-20-1

Vol. 100-Energy Transport in Radio Galaxies and Quasars
ed. P. E. Hardee, A. H. Bridle, and J. A. Zensus ISBN 1-886733-21-X

Vol. 101-Astronomical Data Analysis Software and Systems V
ed. George H. Jacoby and Jeannette Barnes ISSN 1080-7926

Vol. 102-The Galactic Center, 4th ESO/CTIO Workshop
ed. Roland Gredel ISBN 1-886733-22-8

Vol. 103-The Physics of Liners in View of Recent Observations
ed. M. Eracleous, A. Koratkar, C. Leitherer, and L. Ho ISBN 1-886733-23-6

Vol. 104-Physics, Chemistry, and Dynamics of Interplanetary Dust, IAU Colloquium 150
ed. Bo Å. S. Gustafson and Martha S. Hanner ISBN 1-886733-24-4

Vol. 105-Pulsars: Problems and Progress, IAU Colloquium 160
ed. M. Bailes, S. Johnston, and M.A. Walker ISBN 1-886733-25-2

Vol. 106-Minnesota Lectures on Extragalactic Neutral Hydrogen
ed. Evan D. Skillman ISBN 1-886733-26-0

Vol. 107-Completing the Inventory of the Solar System
ed. Terrence W. Rettig and Joseph Hahn ISBN 1-886733-27-9

Vol. 108-Model Atmospheres and Spectrum Synthesis
ed. S. J. Adelman, F. Kupka, and W. W. Weiss ISBN 1-886733-28-7

Vol. 109-Cool Stars, Stellar Systems, and the Sun, Ninth Cambridge Workshop
ed. Roberto Pallavicini and Andrea K. Dupree ISBN 1-886733-29-5

Vol. 110-Blazar Continuum Variability
ed. H. R. Miller, J. R. Webb, and J. C. Noble ISBN 1-886733-30-9

Vol. 111-Magnetic Reconnection in the Solar Atmosphere
ed. R. D. Bentley and J. T. Mariska ISBN 1-886733-31-7

Vol. 112-Galactic Chemodynamics 4: The History of the Milky Way System and Its Satellite System
ed. A. Burkert, D. H. Hartmann, and S. A. Majewski ISBN 1-886733-32-5

Vol. 113-Emission Lines in Active Galaxies: New Methods and Techniques, IAU Colloquium 159
ed. B. M. Peterson, F.-Z. Cheng, and A. S. Wilson ISBN 1-886733-33-3

Vol. 114-Young Galaxies and QSO Absorption-Line Systems
ed. Sueli M. Viegas, Ruth Gruenwald, and Reinaldo R. de Carvalho ISBN 1-886733-34-1

Vol. 115-Galactic and Cluster Cooling Flows
ed. Noam Soker ISBN 1-886733-35-X

Vol. 116-The Second Stromlo Symposium: The Nature of Elliptical Galaxies
ed. M. Arnaboldi, G. S. Da Costa, and P. Saha ISBN 1-886733-36-8

Vol. 117-Dark and Visible Matter in Galaxies
ed. Massimo Persic and Paolo Salucci ISBN 1-866733-37-6

Vol. 118-First Advances in Solar Physics Euroconference: Advances in the Physics of Sunspots
ed. B. Schmieder, J. C. del Toro Iniesta, and M. Vázquez ISBN 1-886733-38-4

Vol. 119-Planets Beyond the Solar System and the Next Generation of Space Missions
ed. David R. Soderblom ISBN 1-886733-39-2

Vol. 120-Luminous Blue Variables: Massive Stars in Transition
ed. Antonella Nota and Henny J. G. L. M. Lamers ISBN 1-886733-40-6

Vol. 121-Accretion Phenomena and Related Outflows
ed. D. T. Wickramasinghe, G. V. Bicknell and L. Ferrario ISBN 1-886733-41-4

Vol. 122-From Stardust to Planetesimals
ed. Yvonne J. Pendleton and A. G. G. M. Tielens ISBN 1-886733-42-2

Vol. 123-Computational Astrophysics
ed. David A. Clarke and Michael J. West ISBN 1-886733-43-0

Vol. 124-Diffuse Infrared Radiation and the IRTS
ed. Haruyuki Okuda, Toshio Matsumoto, and Thomas L. Roellig ISBN 1-886733-44-9

Vol. 125-Astronomical Data Analysis Softward and Systems VI
ed. Gareth Hunt and H. E. Payne ISBN 1-886733-45-7

Inquiries concerning these volumes should be directed to the:
Astronomical Society of the Pacific
CONFERENCE SERIES
390 Ashton Avenue
San Francisco, CA 94112-1722 USA
415-337-2126
catalog@aspsky.org
Fax: 415-337-5205

ASTRONOMICAL SOCIETY OF THE PACIFIC
CONFERENCE SERIES

Volume 125

ASTRONOMICAL DATA ANALYSIS SOFTWARE AND SYSTEMS VI

Meeting held at Charlottesville, Virginia
22-25 September 1996

Edited by
Gareth Hunt and H. E. Payne

Contents

Part 2. Science Software Applications

Part 3. Algorithms

Part 4. Modeling

Part 5. FITS—Flexible Image Transport System

Part 6. Data Archives

Part 7. Database Applications

Part 8. Proposal Processing

Part 9. Real-Time Systems

Part 10. Instrument-Specific Software

Part 11. AXAF

Preface

This volume contains the papers presented at the sixth annual conference on Astronomical Data Analysis Software and Systems (ADASS VI), which was held at the Omni Charlottesville Hotel in Charlottesville, Virginia, 22–26 September 1996, hosted by the National Radio Astronomy Observatory (NRAO) and the University of Virginia Departments of Astronomy and Computer Science. There were 249 registered participants at the meeting, with 81 people representing 17 countries outside the United States and Canada. So, while the overall attendance was slightly lower than for the previous two meetings, the "overseas" participation was higher, making this the most international meeting of the series so far—this is sure to change dramatically at next year's meeting to be held in Germany.

The ADASS VI program included 8 invited speakers and 31 contributed oral papers. In addition, 81 poster papers and 11 computer demonstrations were available throughout the meeting. This year's conference also featured 11 BoF (Birds of a Feather) sessions on special topics including FITS, Tcl/Tk, the Astrophysical Data System (ADS), AIPS and AIPS++, IDL, IRAF, Interactive Data Analysis Environments, Linux in Astronomy, Irregular Time Series, and a second discussion of the Future of Astronomical Data Systems. This volume contains 117 contributions, consisting of 34 papers presented orally (including 5 of the invited papers), 77 poster papers, 3 demonstration summaries, and 3 BoF reports.

The Sunday before the conference saw a Java tutorial, presented by Harry Foxwell of Sun Microsystems, Inc., the two sessions of which framed a lovely brunch buffet in the Omni hotel. Others took a bus tour to the NRAO's Green Bank, West Virginia site, which included a thrilling climb on the enormous Green Bank Telescope, then, as now, under construction. The Thursday after the conference was devoted to the IRAF Developers Workshop, as has become standard at ADASS conferences. The weather cooperated magnificently to enhance the historic setting of the conference banquet, which was held outdoors at Ash Lawn-Highland, the estate of President James Monroe. The conference site, located on Charlottesville's downtown pedestrian mall, brought within easy walking distance a number of fine restaurants, coffee shops, bakeries, ice cream parlors, and, of course, pubs, all enveloped in fine early autumn weather. One of your editors (H.E.P.) fondly remembers giving a couple of visitors their first look at President Thomas Jefferson's magnificent Monticello on a postcard-perfect late Sunday afternoon.

The conference was sponsored by Associated Universities, Inc., the European Southern Observatory, the Gemini 8m Telescopes Project, the Infrared Processing and Analysis Center, the National Aeronautics and Space Administration, the National Center for Supercomputing Applications, the National Optical Astronomy Observatory, the National Radio Astronomy Observatory, the National Research Council of Canada, the National Science Foundation, the Smithsonian Astrophysical Observatory, the Space Telescope Science Institute, and the University of Virginia. Corporate sponsors for the conference included GE Fanuc Automation, Sprint Communications, and Sun Microsystems, Inc. We are very grateful to both the sponsoring institutions and the corporate sponsors for their generous support.

A conference of this size cannot possibly succeed without the efforts and dedication of a large number of people; we are indebted to them all. The conference Program Organizing Committee (POC) was comprised of: Rudi Albrecht (ST-ECF/ESO), Roger Brissenden (SAO), Tim Cornwell (NRAO), Dennis Crabtree (DAO/CADC), Bob Hanisch (chair; STScI), Gareth Hunt (NRAO), George Jacoby (NOAO), Barry Madore (IPAC), Jonathan McDowell (SAO), Jan Noordam (NFRA), Dick Shaw (STScI), Karen Strom (UMass), and Doug Tody (NOAO).

The Local Organizing Committee (LOC) was chaired by Richard Simon (NRAO). Other members were Alan Batson (UVa), David Brown (NRAO), Gareth Hunt (NRAO), Pat Murphy (NRAO), Bob O'Connell (UVa), Gene Runion (NRAO), Charles Tolbert (UVa), Jeff Uphoff (NRAO), and Carolyn White (NRAO). The LOC would like to express its thanks to Gayle Dodson, Phyllis Jackson, Joanne Nance, Bill Porter, Warren Richardson, Fred Schwab, Amy Shepherd, David Simon, Desiree Simon, Pat Smiley, and Dot Tarleton for invaluable help in making the conference a success.

ADASS VII will be held in Sonthofen, Bavaria, 14–17 September 1997. This will be hosted by the Space Telescope European Coordinating Facility and the European Southern Observatory. Send e-mail to adass@eso.org, or see the Web page at http://ecf.hq.eso.org/adass/adass97.html.

After many years of dedicated service, Bob Hanisch stepped down from the POC at the conclusion of this conference. He has been the chair of the POC of this conference series since its inception, and his guiding hand has been behind all of the conferences to date. We are sure that we speak for the whole community served by this conference series when we thank him profusely for his efforts. We look forward to seeing him and hearing from him as a participant in future conferences.

Gareth Hunt
National Radio Astronomy Observatory

Harry E. Payne
Space Telescope Science Institute

May 1997

Cover Illustration: This figure, from the paper by Gary Ferland, Kirk Korista, and Dima Verner (page 213), summarizes the results of thousands of complete spectral synthesis calculations performed with the Cloudy program. The equivalent width of C IV 1549Å line is plotted as a function of the ionizing photon flux and the gas density. These calculations show that this line—one of the strongest quasar emission lines—is produced efficiently only over a narrow range of physical conditions. While the physical conditions corresponding to the peak of this plot had been taken as "standard" quasar conditions, these calculations argue that this is an observational selection effect.

Participant List

Mark Abbott, Center for EUV Astrophysics, University of California, 2150 Kittredge Street, Berkeley, CA 94704, USA (mabbott@cea.berkeley.edu)

Alberto Accomazzi, Smithsonian Astrophysical Observatory, 60 Garden Street, Cambridge, MA 02138, USA (alberto@cfa.harvard.edu)

Michail Agafonov, Radiophysical Research Institute, 25 B. Pecherskaya St., Nizhny Novgorod, 603600, Russia (agfn@nirfi.nnov.su)

Miguel Albrecht, European Southern Observatory, Karl-Schwarzschild-Straße 2, D-85748 Garching bei München, Germany (malbrech@eso.org)

Rudolf Albrecht, Space Telescope - European Coordinating Facility/European Space Agency, Karl-Schwarzschild-Straße 2, D-85748 Garching bei München, Germany (ralbrech@eso.org)

Anastasia Alexov, Smithsonian Astrophysical Observatory, 60 Garden Street, Cambridge, MA 02138, USA (asquared@head-cfa.harvard.edu)

Steve Allen, Lick Observatory, University of California, Santa Cruz, CA 95064, USA (sla@ucolick.org)

Eric Anterrieu, Centre National de la Recherche Scientifique, 14 Avenue E. Belin, F-31400 Toulouse, France (anterieu@obs-mip.fr)

David Armoza, Naval Research Laboratory, 4555 Overlook Avenue, SW, Washington, DC 20375, USA (armoza@ait.nrl.navy.mil)

Drew Asson, Space Telescope Science Institute, 3700 San Martin Drive, Baltimore, MD 21218, USA (asson@stsci.edu)

Pascal Ballester, European Southern Observatory, Karl-Schwarzschild-Straße 2, D-85748 Garching bei München, Germany (pballest@eso.org)

Raymond Bambery, Jet Propulsion Laboratory, MS 168-414, 4800 Oak Grove Drive, Pasadena, CA 91001, USA (raymond.j.bambery@jpl.nasa.gov)

Klaus Banse, European Southern Observatory, Karl-Schwarzschild-Straße 2, D-85748 Garching bei München, Germany (kbanse@eso.org)

Jeannette Barnes, National Optical Astronomy Observatory, 950 N. Cherry, Tucson, AZ 85726-6732, USA (jbarnes@noao.edu)

Paul Barrett, Code 668.1, Goddard Space Flight Center, Greenbelt, MD 20771, USA (barrett@compass.gsfc.nasa.gov)

Tony Beasley, National Radio Astronomy Observatory, P. O. Box 0, Socorro, NM 87801, USA (tbeasley@nrao.edu)

Stephane Beland, National Radio Astronomy Observatory, P.O. Box 0, Socorro, NM 87801, USA (sbeland@nrao.edu)

David Bell, National Optical Astronomy Observatories, P.O. Box 26732, Tucson, AZ 85726, USA (dbell@noao.edu)

Christine Boyer, Space Telescope Science Institute, 3700 San Martin Drive, Baltimore, MD 21218, USA (heller@stsci.edu)

Willem Brouw, Australia Telescope National Facility, P. O. Box 76, Epping, NSW 2121, Australia (wbrouw@atnf.csiro.au)

Lawrence Brown, Code 664, Goddard Space Flight Center, Greenbelt, MD 20771, USA (elwin@redshift.gsfc.nasa.gov)

Howard Bushouse, Space Telescope Science Institute, 3700 San Martin Drive, Baltimore, MD 21218, USA (bushouse@stsci.edu)

Ivo Busko, Space Telescope Science Institute, 3700 San Martin Drive, Baltimore, MD 21218, USA (busko@stsci.edu)

Paul Butler, Department of Astronomy, University of California, Berkeley, CA 94720, USA (paul@further.berkeley.edu)

S. Josephine Chan, Instituto de Astrofisica de Canarias, C/VIa Lactea s/n, La Laguna, Tenerife E-38200, Spain (sjchan@iac.es)

Alberto Chavan, European Southern Observatory, Karl-Schwarzschild-Straße 2, D-85748 Garching bei München, Germany (amchavan@eso.org)

Judy Chen, MS 81, Smithsonian Astrophysical Observatory, 60 Garden Street, Cambridge, MA 02138, USA (chen@head-cfa.harvard.edu)

Yong Chen, P. O. Box 918-3, Beijing, China (cheny@astrosv1.ihep.ac.cn)

Yoshihiro Chikada, National Astronomical Observatory, 2-21-1 Osawa, Mitaka, Tokyo 181, Japan (chikada@optik.mtk.nao.ac.jp)

Teck Choo, Space Telescope Science Institute, 3700 San Martin Drive, Baltimore, MD 21218, USA (choo@stsci.edu)

De Clarke, University of California Oberatories/Lick Observatory, 1156 High Street, Santa Cruz, CA 95064, USA (de@ucolick.org)

Thomas Comeau, Space Telescope Science Institute, 3700 San Martin Drive, Baltimore, MD 21218, USA (tcomeau@stsci.edu)

Sarah Conger, Hughes STX/Naval Research Lab, Building 209, Code 7604, 4555 Overlook Ave., SW, Washington, DC 20375-5352, USA (conger@mustang.nrl.navy.mil)

Al Conrad, Keck Observatory, 65-1120 Mamiahoa Highway, Kamuela, HI 96743, USA (aconrad@keck.hawaii.edu)

Maureen Conroy, Smithsonian Astrophysical Observatory, 60 Garden Street, Cambridge, MA 02138, USA (mo@cfa.harvard.edu)

Michael Corcoran, Code 660.2, USRA/Goddard Space Flight Center, Greenbelt, MD 20771, USA (corcoran@barnegat.gsfc.nasa.gov)

Mark Cornell, McDonald Observatory, Rm. 15.308, University of Texas, Austin, TX 78712, USA (cornell@puck.as.utexas.edu)

Tim Cornwell, National Radio Astronomy Observatory, P.O. Box 0, Socorro, NM 87801, USA (tcornwell@nrao.edu)

Richard Crutcher, 103 Astronomy Building, University of Illinois, 1002 West Green Street, Urbana, IL 61801, USA (crutcher@uiuc.edu)

Philip Daly, UKIRT, 660 N. Aohuku Place, Hilo, HI 96720, USA (pnd@jach.hawaii.edu)

Lindsey Davis, National Optical Astronomy Observatory, P.O. Box 26732, Tucson, AZ 85726-6732, USA (davis@noao.edu)

Cor de Vries, Space Research Organization, Sorbonnelaan 2, 35O4 CA Utrecht, Netherlands (c.devries@sron.ruu.nl)

Jean-Pierre De Cuyper, Royal Observatory of Belgium, Ringlaan 3, B1180 Ukkel, Belgium (jean-pierre.decuyper@oma.be)

Janet DePonte, Harvard-Smithsonian Center for Astrophysics, 60 Garden Street, Cambridge, MA 02138, USA (janet@cfa.harvard.edu)

Ketan Desai, National Radio Astronomy Observatory, P. O. Box 0, Socorro, NM 87801, USA (kdesai@nrao.edu)

Erik Deul, Sterrewacht Leiden, Niels Bohrweg 2, 2300 RA Leiden, Netherlands (deul@strw.leidenuniv.nl)

Stephen Doe, Smithsonian Astrophysical Observatory/AXAF Science Center, MS 81, 60 Garden Street, Cambridge, MA 02138, USA (sdoe@head-cfa.harvard.edu)

Rodger Doxsey, Space Telescope Science Institute, 3700 San Martin Drive, Baltimore, MD 21218, USA (doxsey@stsci.edu)

Margo Duesterhaus, Code 664 Bldg. T2/79, Goddard Space Flight Center, Greenbelt, MD 20771, USA (duesterhaus@lheavx.gsfc.nasa.gov)

Jonathan Dunfee, Code 664 Bldg. T-2 Rm. 62, Goddard Space Flight Center, Greenbelt, MD 20771, USA (dunfee@rosserv.gsfc.nasa.gov)

Kimberly DuPrie, Smithsonian Astrophysical Observatory, MS 81, 60 Garden Street, Cambridge, MA 02138, USA (kduprie@cfa.harvard.edu)

Daniel Durand, Dominion Astrophysical Observatory, Canadian Astronomy Data Centre, 5071 W. Saanich Road, Victoria, BC V8X 4M6, Canada (durand@dao.nrc.ca)

Frossie Economou, Joint Astronomy Centre, 660 North Aohoku Place, University Park, Hilo, HI 96720, USA (frossie@jach.hawaii.edu)

Daniel Egret, Centre de Donées astronomiques de Strasbourg, 11 Rue De L'Université, 67000 Strasbourg, France (egret@astro.u-strasbg.fr)

Guenther Eichhorn, Harvard-Smithsonian Center for Astrophysics, 60 Garden Street, Cambridge, MA 02138, USA (gei@cfa.harvard.edu)

Jonathan Eisenhamer, Space Telescope Science Institute, 3700 San Martin Drive, Baltimore, MD 21218, USA (eisenham@stsci.edu)

Nicholas Elias, US Naval Observatory/Naval Research Laboratory Optical Interferometer Project, 3450 Massachusetts Avenue NW, Washington, DC 20392-5420, USA (nme@fornax.usno.navy.mil)

Brian Elza, MS 664, Goddard Space Flight Center, Greenbelt, MD 20771, USA (elza@rosserv.gsfc.nasa.gov)

Asghar Esfandiari, Naval Research Laboratory, Code 7660, 4555 Overlook Avenue, SW, Washington, DC 20375, USA (esfandiari@susim.nrl.navy.mil)

Maria Falvella, Agenzia Spaziale Italiana, Via Di Villa Patrizi 13, 00187 Roma, Italy (falvella@asirom.rm.asi.it)

Allen Farris, Space Telescope Science Institute, 3700 San Martin Drive, Baltimore, MD 21218, USA (farris@stsci.edu)

Gary Ferland, Department of Physics and Astronomy, University of Kentucky, Lexington, KY 40506, USA (gary@asta.pa.uky.edu)

Michael Fitzpatrick, National Optical Astronomy Observatory, P.O Box 26732, Tucson, AZ 85726-6732, USA (fitz@noao.edu)

Harry Foxwell, Sun Microsystems, Suite 400, 2650 Park Tower Drive, Vienna, VA 22180-7306 USA (hfoxwell@sundc.east.sun.com)

Marian Freuh, McDonald Observatory, University of Texas, P.O. Box 1337, Fort Davis, TX 79734, USA (frueh@astro.as.utexas.edu)

Jim Fullton, Center for Networked Information Discovery and Retrieval, 3021 Cornwallis Road, Research Triangle Park, NC 27709, USA (jim.fullton@cnidr.org)

Carolos Gabriel, Villafranca del Castillo Tracking Station, Aptdo. 50727, 28080 Madrid, Spain (cgabriel@iso.vilspa.esa.es)

Niall Gaffney, Department of Astronomy, RLM 15.308, University of Texas, Austin, TX 78712, USA (niall@rhea.as.utexas.edu)

Severin Gaudet, Dominion Astrophysical Observatory, Canadian Astronomy Data Center, 5071 W. Saanich Road, Victoria, BC V8X 4M6, Canada (gaudet@dao.nrc.ca)

Françoise Genova, Centre de Donées astronomiques de Strasbourg, 11 rue de L'Université, 67000 Strasbourg, France (genova@cdsxb6.u-strasbg.fr)

Barry Giles, Code 662, Goddard Space Flight Center, Greenbelt, MD 20771, USA (barry@pcasun3.gsfc.nasa.gov)

Kim Gillies, National Optical Astronomy Observatories, P.O. Box 26732, Tucson, AZ 85726, USA (gillies@noao.edu)

Brian Glendenning, National Radio Astronomy Observatory, PO Box 0, Socorro, NM 87801, USA (bglenden@nrao.edu)

Raymond Gonzalez, Cornell University/Arecibo Observatory, P. O. Box 995, Arecibo, PR 00613, USA (rgonzale@naic.edu)

Perry Greenfield, Space Telescope Science Institute, 3700 San Martin Drive, Baltimore, MD 21218, USA (perry@stsci.edu)

Preben Grosbøl, European Southern Observatory, Karl-Schwarzschild-Straße 2, D-85748 Garching bei München, Germany (pgrosbol@eso.org)

Lee Groves, National Optical Astronomy Observatory, PO Box 26732, Tucson, AZ 85726, USA (groves@noao.edu)

Nancy Hamilton, Department of Defense, 930 Penobscot Harbour, Pasadena, MD 21122, USA (nancyh@romulus.ncsc.mil)

Robert Hanisch, Space Telescope Science Institute, 3700 San Martin Drive, Baltimore, MD 21218, USA (hanisch@stsci.edu)

F. Rick Harnden, Jr., NASA Headquarters, #2226, 900 N. Stafford St., Arlington, VA 22203, USA (frh@hq.nasa.gov)

Joe Harrington, Code 693, Goddard Space Flight Center, Greenbelt, MD 20771, USA (jh@tecate.gsfc.nasa.gov)

Daniel Harris, Harvard-Smithsonian Center for Astrophysics, 60 Garden Street, Cambridge, MA 02138, USA (harris@cfa.harvard.edu)

Jill Harrison, Space Telescope Science Institute, 3700 San Martin Drive, Baltimore, MD 21218, USA (jharrison@stsci.edu)

R. Lee Hawkins, Wellesley College, Department of Astronomy, Whitin Observatory, Wellesley, MA 02181, USA (lhawkins@wellesley.edu)

Helen He, Smithsonian Astrophysical Observatory, 60 Garden Street, Cambridge, MA 02138, USA (hhe@cfa.harvard.edu)

Jose Herrero, Smithsonian Astrophysical Observatory, 60 Garden Street, Cambridge, MA 02138, USA (jherrero@cfa.harvard.edu)

Lloyd Higgs, Dominion Radio Astronomy Observatory, Box 248, Penticton, BC V2A 6K3, Canada (lah@drao.nrc.ca)

Kerry Hilldrup, Code 668, Goddard Space Flight Center, Greenbelt, MD 20771, USA (khilldru@xenolith.gsfc.nasa.gov)

Dean Hinshaw, Code 664, Goddard Space Flight Center, Greenbelt, MD 20771, USA (dah@lheamail.gsfc.nasa.gov)

Robert Hjellming, National Radio Astronomy Observatory, P.O. Box 0, Socorro, NM 87801-0379, USA (rhjellmi@nrao.edu)

Philip Hodge, Space Telescope Science Institute, 3700 San Martin Drive, Baltimore, MD 21218, USA (hodge@stsci.edu)

Richard Hook, European Southern Observatory, Karl-Schwarzschild-Straße 2, D-85748 Garching bei München, Germany (rhook@eso.org)

Keith Horne, School of Physics and Astronomy, University of St. Andrews, North Haugh, St. Andrews, Fife KY16 9SS, Scotland (kdh1@st-and.ac.uk)

Allan Hornstrup, Danish Space Research Institute, Juliane Maries Vej 30, DK-2100 Copenhagen 0, Denmark (allan@dsri.dk)

Zhenping Huang, N130 Bldg. 28, Goddard Space Flight Center, Greenbelt, MD 20771, USA (huang@ssvs.gsfc.nasa.gov)

David Huenemoerder, 37-667, Center for Space Research, Massachusetts Institute of Technology, Cambridge, MA 02139, USA (dph@space.mit.edu)

Edwin Huizinga, Space Telescope Science Institute, 3700 San Martin Drive, Baltimore, MD 21218, USA (huizinga@stsci.edu)

Gareth Hunt, National Radio Astronomy Observatory, 520 Edgemont Road, Charlottesville, VA 22903, USA (ghunt@nrao.edu)

Eric Huygen, K.U. Leuven, Celestijnenlaan 200B, Heverlee, 3001, Belgium (rik@ster.kuleuven.ac.be)

Peter Hyde, Space Telescope Science Institute, 3700 San Martin Drive, Baltimore, MD 21218, USA (phyde@stsci.edu)

Yashide Ishihara, Fujitsu Limited, 1-9-3 Nakase, Chiba, Chiba 291, Japan (ishi@ssd.se.fujitsu.co.jp)

Peter Jackson, Code 935.0, Goddard Space Flight Center, Greenbelt, MD 20771, USA (pjackson@leaf.gsfc.nasa.gov)

Tim Jenness, Joint Astronomy Centre, 660 North a'ohoku Place, University Park, Hilo, HI 96720, USA (timj@jach.hawaii.edu)

Gareth Jones, University of Wales College. Department of Physics/Astronomy, College of Cardiff, Cardiff, South Glamorgan CF2 3YB, United Kingdom (gareth.jones@astro.cf.ac.uk)

William Joye, Harvard-Smithsonian Center for Astrophysics, 60 Garden Street, Cambridge. MA 02138, USA (wjoye@cfa.harvard.edu)

Edward Kaita, Code 923, Goddard Space Flight Center, Greenbelt, MD 20771, USA (kaita@astroboy.gsfc.nasa.gov)

Margarita Karovska. Harvard-Smithsonian Center for Astrophysics, 60 Garden Street, Cambridge, MA 02138, USA (karovska@cfa.harvard.edu)

Phillip Keegstra. Hughes STX-COBE, PO Box 3242, Laurel, MD 20709. USA (keegstra@clark.net)

Athol Kemball, National Radio Astronomy Observatory, PO Box 0, Socorro, NM 87801, USA (akemball@nrao.edu)

Hyun-Goo Kim, San 36-1 Whaam-dong, Yoosung-ku, Taejon, 305-348, Korea (hgkim@hanul.issa.re.kr)

Young-Soo Kim, Dept. Physics and Astronomy, University College London, London, WC1E 6BT, United Kingdom (ysk@star.ucl.ac.uk)

Tim Kimball, Space Telescope Science Institute, 3700 San Martin Drive, Baltimore, MD 21218, USA (kimball@stsci.edu)

Edward King, Australia Telescope National Facility c/o COSSA, GPO Box 3023, Canberra, ACT 2601, Australia (eking@atnf.csiro.au)

Leonid Kogan, National Radio Astronomy Observatory, PO Box 0, Socorro, NM 87801, USA (lkogan@aoc.nrao.edu)

Sergey Konovalov, Institute of Solar-Terrestrial Physics, Lermontova Str. 126, Irkutsk, 664033, Russia (root@sitmis.irkutsk.su)

George Kosugi, National Astronomical Observatory, Osawa 2-21-1, Mitaka, Tokyo 181, Japan (george@optik.mtk.nao.ac.jp)

Reinhold Kroll, Instituto Astrofisico de Canarias, c/ Via Lactea s/n, La Laguna, TF 38200, Spain (kroll@lac.es)

Uwe Lammets, Space Science Department, European Space Agency/ESTEC, P.O. Box 299, 2200 AG Noordwijk, Netherlands (uwe.lammers@astro.estec.esa.nl)

Wayne Landsman, Code 681, Hughes/STX, Greenbelt, MD 20771, USA (landsman@stars.gsfc.nasa.gov)

Glen Langston, National Radio Astronomy Observatory, P.O. Box 2, Green Bank, WV 24944, USA (glangsto@nrao.edu)

André Lannes, Observatoire Midi-Pyrénées, 14 Avenue E. Belin, F-31400 Toulouse, France (lannes@obs-mip.fr)

Youngung Lee, Korea Astronomy Observatory, Whaam-dong San 36-1, Yusung-ku, Taejon, 305-348, Korea (yulee@hanul.issa.re.kr)

Zoltan Levay, Space Telescope Science Institute, 3700 San Martin Drive, Baltimore, MD 21218, USA (levay@stsci.edu)

James Lewis, Royal Greenwich Observatory, Madingley Road, Cambridge, CB3 0EZ, United Kingdom (jrl@ast.cam.ac.uk)

Peter Linde, Lund Observatory, Box 43, S-22100 Lund, Sweden (peter@astro.lu.se)

Harvey Liszt, National Radio Astronomy Observatory, 520 Edgemont Road, Charlottesville, VA 2293, USA (hliszt@nrao.edu)

Rick Lively, National Radio Astronomy Observatory, PO Box 0, Socorro, NM 87801, USA (rlively@nrao.edu)

Antoine Llebaria, Laboratoire d'Astronomie Spatiale/Centre National de la Recherche, Scientifique, Traverse du Siphon, 13 Arr, BP 8 13376 Marseille, France (antoine@astrsp-mrs.fr)

Antonio Longinotti, European Southern Observatory, Karl-Shcwarzschild Str. 2, D-85748 Garching bei München, Germany (alongino@eso.org)

Dyer Lytle, University of Arizona, 3319-1 E. Presidio, Tucson, AZ 85721, USA (dlytle@as.arizona.edu)

Allan MacKay, TRW/Smithsonian Astrophysical Observatory, 60 Garden Street, Cambridge, MA 02138, USA (amackay@cfa.harvard.edu)

Jun Maekawa, 1-7-5-103 Higashi-Izumi, Komae, Tokyo 201, Japan (maekawa@nro.nao.ac.jp)

Oleg Malkov, Institute of Astronomy, Russian Academy of Sciences 48 Pyatnitskaya st, Moscow, 109017, Russia (malkov@inasan.rssi.ru)

Eric Mandel, Smithsonian Astrophysical Observatory, 60 Garden Street, Cambridge, MA 02138, USA (eric@cfa.harvard.edu)

Pierre Maréchal, Observatoire Midi-Pyrénées, 14 Avenue E. Belin, F-31400 Toulouse, France (marechal@obs-mip.fr)

Jean-Michel Martin, Observatory de Paris, Place J. J. Janssen, F-92190 Meudon, France (jmmartin@obspm.fr)

Ralph Martin, Royal Greenwich Observatory, Madingley Road, Cambridge, CB3 0EZ, United Kingdom (ralf@ast.cam.ac.uk)

Jonathan McDowell, Smithsonian Astrophysical Observatory, 60 Garden Street, Cambridge, MA 02138, USA (jcm@urania.harvard.edu)

Douglas McElroy, Smithsonian Astrophysical Observatory/AXAF Science Center, 60 Garden Street, Cambridge, MA 02138, USA (dmcelroy@cfa.harvard.edu)

Thomas McGlynn, 660.2, Goddard Space Flight Center, Greenbelt, MD 20771, USA (tam@silk.gsfc.nasa.gov)

Brian McIlwrath, Rutherford Appleton Laboratory, Starlink, Chilton Didcot, Oxfordshire, OX11 0QX, United Kingdom (bkm@star.rl.ac.uk)

Brian McLean, Space Telescope Science Institute, 3700 San Martin Drive, Baltimore, MD 21218, USA (mclean@stsci.edu)

Alberto Micol, European Southern Observatory, Karl-Schwarzschild-Straße 2, D-85748 Garching bei München, Germany (amicol@eso.org)

Glenn Miller, Space Telescope Science Institute, 3700 San Martin Drive, Baltimore, MD 21218, USA (miller@stsci.edu)

David Mills, National Optical Astronomy Observatories, PO Box 26732, Tucson, AZ 85726, USA (dmills@noao.edu)

Douglas Mink, Smithsonian Astrophysical Observatory, 60 Garden Street, Cambridge, MA 02138, USA (dmink@cfa.harvard.edu)

Yoshihiko Mizumoto, National Astronomical Observatory, Osawa 2-21-1, Mitaka, Tokyo 181, Japan (mizumoto@optik.mtk.nao.ac.jp)

Jinger Mo, Space Telescope Science Institute, 3700 San Martin Drive, Baltimore, MD 21218, USA (jinger@stsci.edu)

Patrick Murphy, National Radio Astronomy Observatory, 520 Edgemont Road, Charlottesville, VA 22903, USA (pmurphy@nrao.edu)

Dan Nguyen, Smithsonian Astrophysical Observatory, 60 Garden Street, Cambridge, MA 02138, USA (dnguyen@head-cfa.harvard.edu)

Natalie Nigro, Space Telescope Science Institute, 3700 San Martin Drive, Baltimore, MD 21218, USA (nigro@stsci.edu)

Jan Noordam, Netherlands Foundation for Research in Astronomy, P.O. Box 2, 7990 AA Dwingeloo, Netherlands (jnoordam@nfra.nl)

Junichi Noumaru, National Astronomical Observatory, Osawa 2-21-1, Mitaka, Tokyo 181, Japan (noumaru@optik.mtk.nao.ac.jp)

Robert O'Connell, University of Virginia, PO Box 3818, University Station, Charlottesville, VA 22903, USA (rwo@virginia.edu)

Earl O'Neil, Kitt Peak National Observatory, P. O. Box 26732, Tucson, AZ 85726, USA (oneil@noao.edu)

Oliver Oberdorf, Smithsonian Astrophysical Observatory, MS 81, 60 Garden Street, Cambridge, MA 02135, USA (oly@head-cfa.harvard.edu)

Victor Oknyanskij, Sternberg State Astronomical Institute, Universitetskij Prospekt 13, Moscow, 119899, Russia (oknyan@lnfm1.sai.msu.su)

Eric Olson, Center for EUV Astrophysics, University of California, 2150 Kittredge Street, Berkeley, CA 94720, USA (ericco@cea.berkeley.edu)

Tom Oosterloo, Australia Telescope National Facility, P.O.Box 76, Epping, NSW 2121, Australia (toosterl@atnf.csiro.au)

Stephan Ott, European Space Agency, Villafranca del Castillo Satellite Tracking Station, 28080 Madrid, Spain (sott@iso.vilspa.esa.es)

Clive Page, University of Leicester, Physics Department, Leicester, LE1 7RH, United Kingdom (cgp@star.le.ac.uk)

Scott Paswaters, Naval Research Laboratory, Code 7660, 4555 Overlook Avenue, SW, Washington, DC 20375, USA (scott@argus.nrl.navy.mil)

Harry Payne, Space Telescope Science Institute, 3700 San Martin Drive, Baltimore, MD 21218, USA (payne@stsci.edu)

Robert Payne, National Radio Astronomy Observatory, P.O. Box 2, Green Bank, WV 24944, USA (rpayne@nrao.edu)

William Pence, Code 662, Goddard Space Flight Center, Greenbelt, MD 20771, USA (pence@gsfc.nasa.gov)

Michele Peron, European Southern Observatory, Karl-Schwarzschild-Straße 2, D-85748 Garching bei München, Germany (mperon@eso.org)

Benoît Pirenne, European Southern Observatory/Space Telescope - European Coordinating Facility, Karl-Schwarzschild-Straße 2, D-85748 Garching bei München, Germany (bpirenne@eso.org)

Richard Pisarski, Code 631, Goddard Space Flight Center, Greenbelt, MD 20771, USA (rlp@ros5.gsfc.nasa.gov)

Raymond Plante, National Center for Supercomputing Applications/University of Illinois, 5211 Bechman Institute, Drawer 25, 405 N. Mathews Avenue, Urbana, IL 61801, USA (rplante@ncsa.uiuc.edu)

Andrea Prestwich, Smithsonian Astrophysical Observatory, 60 Garden Street, Cambridge, MA 02138, USA (prestwich@cfa.harvard.edu)

Mauro Pucillo, Trieste Astronomical Observatory, Via Tiepolo 11, P. O. Box Succ. 5, I-36434 Trieste, Italy (pucillo@ts.astro.it)

Peter Quinn, European Southern Observatory, Karl-Schwarzschild-Straße 2, D-85748 Garching bei München, Germany (pjq@eso.org)

Walter Rauh, Max-Planck-Institut für Astronomie, Königstuhl 17, D-69117 Heidelberg, Germany (rauh@mpia-hd.mpg.de)

Joel Richon, Space Telescope Science Institute, 3700 San Martin Drive, Baltimore, MD 21218, USA (richon@stsci.edu)

Doug Roberts, University of Illinois, 405 N. Mathews Ave., Urbana, IL 61820, USA (droberts@lai.ncsa.uiuc.edu)

Michael Rosa, Space Telescope - European Coordinating Facility, Karl-Schwardschild-Str. 2, D-85748 Garching bei München, Germany (mrosa@eso.org)

Jim Rose, Space Telescope Science Institute, 3700 San Martin Drive, Baltimore, MD 21218, USA (rose@stsci.edu)

Arnold Rots, Code 660.2, Goddard Space Flight Center, Greenbelt, MD 20771, USA (arots@xebec.gsfc.nasa.gov)

Lee Rottler, University of California Obseratories/Lick Observatory, 1156 High Street, Santa Cruz, CA 95060, USA (rottler@ucolick.org)

Bert Rust, Bldg. 820, Rm 365, National Institute of Standards and Technology, Gaithersburg, MD 20899, USA (bwr@cam.nist.gov)

Mary Ryan, Space Telescope Science Institute, 3700 San Martin Drive, Baltimore, MD 21218, USA (mryan@stsci.edu)

Roberto Samson, Space Telescope Science Institute, 3700 San Martin Drive, Baltimore, MD 21218, USA (samson@stsci.edu)

Darrell Schiebel, National Radio Astronomy Observatory, 520 Edgemont Road, Charlottesville, VA 22903, USA (drs@nrao.edu)

Juergen Schmidt, Max-Planck-Institut für Radioastronomie, Auf dem Hugel 69, D-53121 Bonn, Germany (jsm@mpifr-bonn.mpg.de)

Eric Schulman, National Radio Astronomy Observatory, 520 Edgemont Road, Charlottesville, VA 22903, USA (eschulma@nrao.edu)

Fred Schwab, National Radio Astronomy Observatory, 520 Edgemont Road, Charlottesville, VA 22903, USA (fschwab@nrao.edu)

Steve Scott, California Institute of Technology/Owens Valley Radio Observatory, PO Box 968, Big Pine, CA 93513, USA (scott@ovro.caltech.edu)

Robert Seaman, National Optical Astronomy Observatories, P.O. Box 26732, Tucson, AZ 85726, USA (seaman@noao.edu)

William Sebok, University of Maryland, Astronomy Department, College Park, MD 20742, USA (wls@astro.umd.edu)

Paul Shannon, National Radio Astronomy Observatory, 520 Edgemont Rd, Charlottesville, VA 22903, USA (pshannon@nrao.edu)

Richard Shaw, Space Telescope Science Institute, 3700 San Martin Drive, Baltimore, MD 21218, USA (shaw@stsci.edu)

Martin Shepherd, MS 105-24, California Institute of Technology, Pasadena, CA 91125, USA (mcs@astro.caltech.edu)

Patrick Shopbell, MS 105-24, Dept of Astronomy, California Institute of Technology, Pasadena, CA 91125, USA (pls@astro.caltech.edu)

Richard Simon, National Radio Astronomy Observatory, 520 Edgemont Road, Charlottesville, VA 22903, USA (rsimon@nrao.edu)

Riccardo Smareglia, Astronomical Observatory of Trieste, Via Tiepolo 11, 34131 Trieste, Italy (smareglia@oat.ts.astro.it)

Oleg Smirnov, Institute of Astronomy, Russian Academy of Sciences 48 Pyatnitskaya St, Moscow, 109017, Russia (oms@inasan.rssi.ru)

Eric Smith, Code 681, Goddard Space Flight Center, Greenbelt, MD 20771, USA (esmith@hubble.gsfc.nasa.gov)

William Snyder, Naval Research Laboratory/BDC, Code 7604, Washington, DC 20375, USA (snyder@bdcv8.nrl.navy.mil)

David Stern, Research Systems, Inc., Suite 203, 2995 Wilderness Place, Boulder, CO 80027, USA (dave@rsinc.com)

Ivan Stern, Smithsonian Astrophysical Observatory, 60 Garden Street, Cambridge, MA 02138, USA (istern@head-cfa.harvard.edu)

Elizabeth Stobie, Steward Observatory, University of Arizona, 933 N. Cherry Avenue, Tucson, AZ 85721, USA (bstobie@as.arizona.edu)

Karen Strom, Astronomy Program GRC-517, Univeristy of Massachusetts, Amherst, MA 01003, USA (kstrom@hanksville.phast.umass.edu)

Wai-Ming Tai, 5 Merrion Square, Dublin 2, Dublin, Ireland (wai@cp.dias.ie)

Tanya Taidakova, Crimean Astrophysical Observatory, Simeiz, Yalta, Crimea 334242, Ukraine (tat@cat.crimea.ua)

Tadafumi Takata, National Astronomical Observatory, Osawa 2-21-1, Mitaka, Tokyo 181, Japan (takata@optik.mtk.nao.ac.jp)

Ian Taylor, University of Wales, Department of Physics/Astronomy, College of Cardiff, Cardiff, South Glamorgan CF2 3YB, United Kingdom (ian.taylor@astro.cf.ac.uk)

Peter Teuben, Astronomy Department, University of Maryland, College Park, MD 20742, USA (teuben@astro.umd.edu)

James Theiler, MS D436, Los Alamos National Laboratory, Los Alamos, NM 87545, USA (jt@lanl.gov)

Remo Tilanus, Joint Astronomy Centre, 660 N. Aohoku Place, University Park, Hilo, HI 96720, USA (rpt@jach.hawaii.edu)

Doug Tody, National Optical Astronomy Observatories, P.O. Box 26732, Tucson, AZ 85726, USA (tody@noao.edu)

Susan Tokarz, Smithsonian Astrophysical Observatory, 60 Garden Street, Cambridge, MA 02138, USA (tokarz@cfa.harvard.edu)

Jay Travisano, Space Telescope Science Institute, 3700 San Martin Drive, Baltimore, MD 21218, USA (jay@stsci.edu)

Francesca Tribioli, Osservatorio Astrofisico di Arecetri, Largo E. Fermi 5, I-50127 Firenze, Italy (tribioli@arcetri.astro.it)

Michael Tripicco, Hughes STX Corp./NASA HEASARC, Mail Code 664.0, Goddard Space Flight Center, Greenbelt, MD 20771, USA (miket@rosserv.gsfc.nasa.gov)

Takahiro Tsutsumi, Nobeyama Radio Observatory, Minamisaku Nagano 384-13, Japan (tsutsumi@nro.nao.ac.jp)

Patricia Tyler, Code 660.2, Goddard Space Flight Center, Greenbelt, MD 20771, USA (tyler@lheavx.gsfc.nasa.gov)

Frank Valdes, National Optical Astronomy Observatory/CCS, P.O. Box 26732, Tucson, AZ 85726, USA (fvaldes@noao.edu)

Robert Vallance, Birmingham University, School of Physics and Space Research, Edgbaston Park Road, Birmingham, B15 2TT, United Kingdom (rjv@star.sr.bham.ac.uk)

Ger Van Diepen, Netherlands Foundation for Radio Astronomy, PO Box 2, 7990 AA Dwingeloo, Netherlands (gvd@nfra.nl)

Bart Vandenbussche, European Space Agency-ISO Science Operations Centre, Appartado de Correos 50727, 28080 Madrid, Spain (bart@iso.vilspa.esa.es)

Sergei Vasilyev, SOLERC, P. O. Box 59, Kharkov, 310052, Ukraine (vvs@land.kharkov.ua)

Irina Vavilova, Observatornaya Str. 3, Kiev, 254053, Ukraine (vavilova@rcrm.freenet.kiev.ua)

Oleg Verkhodanov, Special Astrophysical Observatory, Nizhnij Arkkhys, Carachaj-Cherkessia, 357147, Russia (vo@sao.ru)

Veniamin Vityazev, Astronomy Department, St. Petersburg University, Petrodvorets, Bibliotechnaya pl. 2, St. Petersburg, 198904, Russia (vityazev@venvi.usr.pu.ru)

Stephen Voels, Astrophysics Data Facility, Code 631, Goddard Space Flight Center, Greenbelt, MD 20771, USA (voels@cannon.gsfc.nasa.gov)

Martin Vogelaar, Kapteyn Astronomical Institute, Landleven 12, 9747 AD Groningen, Netherlands (vogelaar@astro.rug.nl)

Shane Walker, National Optical Astronomy Observatories, PO Box 26732, Tucson, AZ 85726, USA (shane@noao.edu)

Patrick Wallace, Rutherford Appleton Laboratory, Chilton Didcot, Oxfordshire, QX11 0QX, United Kingdom (ptw@star.rl.ac.uk)

Dennis Wang, 14120 Parke Long Ct. 130, Chantilly, VA 22021, USA (wang@ares.nrl.navy.mil)

Zhong Wang, Smithsonian Astrophysical Observatory, MS-66, 60 Garden Street, Cambridge, MA 02138, USA (zwang@cfa.harvard.edu)

Michael Ward, McDonald Observatory, P.O. Box 1337, Fort Davis, TX 79734, USA (ward@astro.as.utexas.edu)

Rein Warmels, European Southern Observatory, Karl-Schwarzschild-Straße 2, D-85748 Garching bei München, Germany (rwarmels@eso.org)

Archibald Warnock III, A/WWW Enterprises, 6652 Hawkeye Run, Columbia, MD 21044, USA (warnock@clark.net)

Don Wells, National Radio Astronomy Observatory, 520 Edgemont Road, Charlottesville, VA 22903, USA (dwells@nrao.edu)

Carolyn White, National Radio Astronomy Observatory, 520 Edgemont Road, Charlottesville, VA 22903, USA (cwhite@nrao.edu)

Andreas Wicenec, Institute for Astronomy and Astrophysics, Waldhaeuserstr 64, D-72076 Tübingen, Germany (wicenec@astro.uni-tuebingen.de)

Ramon Williamson II, Space Telescope Science Institute, 3700 San Martin Drive, Baltimore, MD 21218, USA (williamson@stsci.edu)

Michael Wise, MIT Center for Research, Rm. 37-644, 70 Vassar Street, Cambridge, MA 02139-4307, USA (wise@space.mit.edu)

Jennifer Wiseman, National Radio Astronomy Observatory, 520 Edgemont Road, Charlottesville, VA 22903, USA (jwiseman@nrao.edu)

Jim Wright, Gemini 8-m, 450 N Cherry Ave., Tucson, AZ 85719, USA (jwright@gemini.edu)

Nailong Wu, York University, Department of Physics and Astronomy, 4700 Keele St., North York, Ontario M3J 1P3, Canada (nwu@sal.phys.yorku.ca)

Xiuqin Wu, MS 100-22, California Institute of Technology, Pasadena, CA 91125, USA (xiuqin@ipac.caltech.edu)

Michitoshi Yoshida, Okayama Astrophysical Observatory, Kamogata-cho, Asakuchi-gun, Okayama 719-02, Japan (yoshida@oao.nao.ac.jp)

Peter Young, Australian National University, Private Bag Weston Creek Post Office, Weston, ACT 2611, Australia (pjy@mso.anu.edu.au)

Bob Zacher, Smithsonian Astrophysical Observatory, MS 83, 60 Garden Street, Cambridge, MA 02138, USA (zacher@cfa.harvard.edu)

Nelson Zarate, National Optical Astronomy Observatory, 950 N. Cherry, Tucson, AZ 85726, USA (nzarate@noao.edu)

Anton Zensus, National Radio Astronomy Observatory, 520 Edgemont Road, Charlottesville, VA 22903, USA (azensus@nrao.edu)

Houri Ziaeepour, European Southern Observatory, Karl-Schwarzschild-Straße 2, D-85748 Garching bei München, Germany (hziaeepo@eso.org)

Panagoula Zografou, Smithsonian Astrophysical Observatory, 60 Garden Street, Cambridge, MA 02138, USA (pz@head-cfa.harvard.edu)

Sixth Annual Conference on Astronomical Data Analysis Software and Systems
September 22–25, 1996, Charlottesville, VA

Part 1. Software Systems

Astronomical Data Analysis Software and Systems VI
ASP Conference Series, Vol. 125, 1997
Gareth Hunt and H. E. Payne, eds.

A Home-Grown But Widely-Distributed Data Analysis System

H. S. Liszt

National Radio Astronomy Observatory, 520 Edgemont Road, Charlottesville VA, 22903-2475

Abstract. I stopped doing research during the five years 1986–1990, and crafted a series of programs for doing spectral line and image analysis on DOS-based IBM PCs. These programs, most notably drawspec and hazel, are now fairly widely distributed. With only a modicum of hard evidence, I attribute this to their ease of use, small hardware demands, lack of cost, and high level of support. Recognizing that my experience is unusual, and perhaps even of marginal relevance to the design of new programs by teams of expert professional programmers, I have attempted to distill from it some words of advice for the attendees of ADASS VI.

1. Introduction

After accepting the invitation to address the meeting, I began to wonder whether I could actually contribute to a gathering of professional astronomical programmers, my own experiences in programming being so bizarre and particular. For my talk I did not try to discuss in detail the analysis system from which the title was drawn, but rather focussed on how programming and computing look to me now that I am supposed to be able to tell others how to go about them.

At the end of 1985, NRAO turned off the IBM mainframe which, with its predecessors, had been the backbone of all Observatory computing. Although each internal computing review had based its recommendations on the continuing presence of such a machine—how else to preserve the vast library of PL/1 code we had amassed in support of single dish observing in Green Bank and Tucson?—synthesis computing needs had overtaken such concerns. The IBM was turned off, and its code libraries were abandoned without compunction or provision.

For pure numerics, I had always programmed in FORTRAN, but most of my effort was in a set of custom, PL/1, spectral line data-handling routines for VLA and single dish data, mostly to gain the use of record structures and the NRAO code libraries. Even if my library of data analysis routines was not rendered exactly obsolete—we had bought a PL/1 compiler for our VAX/780—the VAX was next to impossible to use. Bogged down by AIPS, addressed over internal serial links, with only a tiny fraction of a megabyte allotted per personal account, even editing a file was next to impossible. Still less disk space was available on our shiny new AIPS-machine, a Convex C-1.

Over Christmas vacation, just before the termination of the IBM, I read some data off a round tape on the VAX and ported it to the AT&T 6300 PC/XT-clone which the Observatory Director, Hein Hvatum, had ordered placed on my desk some months before. I had complained pointedly that each of our secretaries had a PC with more disk space (20 MB) than was alloted for the entire scientific staff, and Hein, in essence, dared me to come up with a reason why this should not continue to be the case. Having been warned away from C by one of the professional programmers in our building ("it's so ugly"), I had happily adopted Turbo Pascal as my PL/1 successor.

Having earlier figured out the 6300's non-standard 640×400 graphics scheme, and with my own nascent library of fast graphics routines, I devised a binary structure to hold some spectral line scans and proceeded to write a program to call up and draw on-screen spectra in histogram form. The speed with which this occurred, compared to all my past experience, was breathtaking and utterly seductive: in a few seconds I was made to understand that I had more computing power on my desk than was available at the NRAO to the entire community of spectral line observers, of which I was a member. The source file was drawspec.pas, a name which persists to this day. One minor change to the binary file format was made in 1986. In the early 1990s I added to drawspec the native ability to read FITS images and cubes.

Between 1986 and 1991, I all but abandoned my research career and coded a series of programs which deal with spectral line and image analysis. This was a passionate and costly exercise (I estimate that I spent over $10,000 in current dollars on two machines and sundry software) carried out largely in my basement in what otherwise would have been all my spare time. When I came to my senses in 1991 and began to rebuild my research career, a whole generation of graduate students was unfamiliar to me.

The drawspec program, comprised of 24,000 lines of code (with another few thousand lines of i86 and i87 assembler thrown in for good measure), is supplemented by scanmstr.exe, a 9,000 line utility program which imports and exports various other data formats and reproduces most of the functionality of the code libraries abandoned by NRAO in 1985. There is also 5,500-line hazel.exe (Harvey's AZ-EL program) which is used to view the sky and plan observing. Additionally, I wrote for myself a simple curve-plotting package (4,000 lines) which does the line and point graphics not available in drawspec. The first three of these programs are documented, supported, and distributed by me as a serious sideline. Anyone can access them from my home page.

I would say that the essential elements of building this family of programs began on the system side with a good development environment. Turbo Pascal was developing rapidly but always usefully, and extraordinary insights (and much dross) were available on the BBSs one could prowl. On my side, I think the most important elements were good I/O routines; a good test environment (the programs themselves!) consisting of a GUI and a consistent pattern of interaction, with the user (usually myself) relying on menus to take the place of explicit language; and good on-screen and PostScript graphics (post-1990). Once a useful programming idiom had been mastered, exhaustion, or perhaps stupefaction, became the ultimate bound on my productivity. My efforts would not have survived difficulties as great as those which seem to have been encountered by AIPS++ in the C++ environments they have used.

It is not my intention to proselytize for drawspec or to discuss its capabilities, because 1996 is hardly the time to advertise a DOS-based program written in an archaic language. I have kept the program current (relative to its competitors) and it is used all around the world by a community which, I must admit, is often largely unknown to me. When Sandy Sandquist noticed the abstract of this talk on the Web, he sent me a copy of Sandquist & Hagström (1995) which points out that drawspec has been adopted as part of the standard undergraduate astronomy curriculum in Sweden. My programs are also used in Mongolia, Brazil, Texas, and the Netherlands.

2. Imagery and Dialogue

Recently I had occasion to look for a rental car in a small lot of such cars. The tag on my keys gave the make, model and license number, as well as the color, written as "Rust." This seemed an unlikely and singularly infelicitous choice; are we to imagine ourselves actually selecting "rust" from a palette of car colors? Or asking a store clerk for rust-colored touchup paint at some time in the future? This same color on my kitchen refrigerator is called "Harvest Gold."

The computer industry is rife with its own barely-noticed versions of this phenomenon. Compaq recently unveiled its line of "Armada" laptops, ignoring that the Spanish Armada had been a fleet of singularly heavy, unwieldy vehicles which (according to English-language text books) was roundly defeated by a more mobile assortment of much smaller and therefore more useful craft. The logo for Java, today's language *du jour*, is a cup of coffee which exists only because of the evanescent arrangement of its own vapors. Vapors? Would I, as a computer manager, not be instantly forgiven for believing I was being mocked by anyone who suggested basing a new project on a technology whose very symbol was "vapor?"

Not paying attention to the use of inappropriate ordinary language isn't terribly costly in the examples I've given, merely absurd. But there is a misuse of language which has had a cost. I believe it is brought on by too much raw contact with objects (in the sense of OOP programming) and I call it speaking in tongues, reminiscent as it is of the snake-handling in some local religious ceremonies that leads to unusual verbal outbursts (which are misconstrued by other true believers as being divinely inspired).

Simply put, there is a lamentable tendency among OO-programmers to attempt to grant false dignity to their activities by renaming everything. In the AIPS++ project, I objected to the desktop being called the "arena" (since arenas are used locally only for cock and dog fights), and to the command log being called a "transparency tool" (I found this usage opaque). There seem to be no such things as "data" or "files," just measurementSets, Tables, Iterators, *usw*. Simple questions which astronomers know how to ask, like "What will I have to do to read a scan from the file containing last night's data" are practically guaranteed to elicit answers which contain not a single intelligible noun or verb phrase.

The inevitable result of this is to inspire terror in the hearts of users who have struggled to gain whatever meager purchase they might have on their computer literacy. The terms "object" and "object-oriented" inspire genuine, instinctive, all-consuming, visceral fear and loathing in almost any astronomer not fully engaged in programming. The problems of incapacitated companies like Borland seem to have been assimilated into the collective unconscious. This firm, as you may remember, went into the drink when it tried to use OO techniques in its own product development, instead of trying to make money by foisting OOP tools on slow-witted consumers (this is a common failing of dope dealers, too, of course). When there was left not a single Borland employee who could speak intelligibly to a banker or a spouse, the firm withered like a community of Shakers.

3. Programs, Not Toolkits

In my own work, it first seemed reasonable to distribute not programs but code or toolkits; I would empower those like-minded individuals who would enjoy, every bit as much as I had, the numerous overwhelmingly tedious and exacting tasks which go into handling real data or writing laser printer drivers. I would be the venerated leader of a band of *sympatico* individuals who admired the elegance and practicality of my handiwork. My particular binary data format, every odd byte of it, would become second nature for a generation of Turbo Pascal-programming, data-reducing astronomers.

This loony fantasy evaporated when I realized that ensuring correct, consistent, and productive use of my code could best be effected by providing a working program. Explaining the proper use of the code in toolkit form was not itself an easy chore, and the best example of how to use it would in fact be—what else?—a program. Mastering complexities and resolving ambiguities seemed preferable to exposing them. And the vast majority of any market for my offline data analysis programs were to be the very souls whom I would forever have disenfranchised by asking each of them to roll their very own.

4. Algorithm and Interface

Aperture synthesis projects, in which large quantities of data are merged to produce a single product like a map or cube, are compute-intensive, so that the ratio of time spent in the interface to that spent waiting for CPU and I/O to finish is typically small. Synthesis work, embodied by our AIPS package, is all algorithm and no interface. This is not to belittle the AIPS user interface, which for all its seeming clunkiness has proved very sturdy and highly serviceable. The 15-year span over which it has been used to reduce so much data is equal to the interval between Bill Monroe forming the Bluegrass Boys and writing "Blue Moon of Kentucky" and Elvis Presley doing his first commercial recording, of just that tune.

For the sort of chores which my programs tackle, the reverse is often true; large numbers of small, highly interactive steps with quick turnaround time are employed; this single dish processing is all interface and no algorithm. Of course these unequivocal statements are limits; one of my programs' failings is that they often deal with datasets as collections, not entities, and are sometimes insufficiently macroscopic. Other programs, like AIPS, are perhaps insufficiently granular.

5. Don't Write Coy Programs

Even in the most supposedly highly interactive programs, there are often elements of coyness and a certain Alphonse/Gaston quality about whose turn it actually is to go first. You bring up some program which is supposed to run interactively and it stares back blankly at you waiting for input. If the program or one of its subroutines has an easily encapsulated goal, should they not examine their own condition and suggest to you what should happen next to achieve it? If this point seems obscure, imagine trying to hold a dialog or complete a project with people who behaved the way most interactive programs do.

There is another aspect to the coy behaviour of many programs; it is when they know something the user doesn't, and so oblige the user to make a choice between begging and proceeding in ignorance. Making a request of a program has two elements of learned behaviour (how to ask and what to ask for), each of which is likely to require a separate traversal of the manual. If there are pieces of information or derived quantities which are important to the conduct of a session with the program, a way should be found to make them available without demanding that the user beg for them.

Yet another aspect of coyness concerns data formats. As the author of an offline data analysis program employing a binary data file format which no telescope produces, and so lacking captive datasets and/or clientele, my mantra is "data is where you find it." Making FITS-reading second nature for drawspec enlarged its scope greatly. Importing data shouldn't be like arranging a marriage between two people of differing, equally stubbornly-held beliefs ("I'll only marry you if you convert to 1-byte values"). If a dataset has coordinates, they should be honored. If it has values, their precision should not be degraded. If large quantities of interesting data exist in a common format (like FITS), that format should be read with a minimum of fuss.

6. Don't be Afraid to Distribute Crap

Any useful analysis program soon begins to grow horizontally and vertically as the result of a wholly natural and predictable process. Once users have made the investment of getting data into an application, they will require that it do more and more for them. Why? Because the only alternative would be to make the *additional* effort to export from that package and import into yet another, at which point the whole futile cycle would only have begun again.

As an example, consider that AIPS started out as a package to image calibrated synthesis data, and eventually subsumed all the calibration chores and the task of producing publication-quality maps. The low esthetic quality of many of these maps certainly diminishes the quality of the publications which agree to print them unadorned, but AIPS' maps have faithful axes which users really appreciate. The continuing refusal of optical astronomers to demand coordinates on their images is a mysterious phenomenon to radio astronomers.

I have discovered that there is a natural cradle-to-grave cycle which must be supported when dealing with data:

C alibration or the answers will be wrong,
R eduction or there will be no answers,
A nalysis or there will be no insight, and
P resentation or there will be no readers.

This is easy to remember *via* its acronym. I am proud to support and ship programs which are so full of it.

7. Why Your Product Should Really Have a Manual, and Why the Manual Must Be Kept Secret

Astronomers hate to read manuals under all circumstances. The idea that one would sit down with a manual prior to using even the most complex device is, at best, anachronistic. Users take umbrage at the mere suggestion that they would be proffered a product which needed any level of understanding or insight whatsoever.

Indubitably, the same product, distributed with a thinner or a thicker manual, will seem easier to use if provided only with the former. In the limit, a product with no manual is deemed entirely intuitive, and one with a thick, detailed tome appears so daunting as to appear intractable. Since documentation is expensive to produce and distribute, the choice is a seeming no-brainer: dump the idea of a user's manual. Borland even carried this idea into the realm of computer languages, all but refusing to supply reference manuals for the last versions of their Object Pascal dialect.

Why then do I counsel the provision of a user's manual? Why, to explain to the programmer what he was thinking! There is nothing more embarrassing than forgetting how one's own code is supposed to work, perhaps assuming it is broken when the debugger shows that it performs as designed: nothing is more baffling than trying to figure out why now it works one way when it should be another.

Of course distributing the manual is another matter. Once the manual exists, it exposes not only the flaws and limitations of your product, but also the inability of its author to write a manual.

8. Lying

Somewhere in every program there are code fragments which are destined to return an incorrect or inappropriate response to the user; often they are well-known and seemingly inviolate. For a period of perhaps 5 years in the early 1980s there was a bug in the plotting of spectra from the standard NRAO single dish program such that the x-axis labeling was in error by one-half of one channel; any x-value measured from our hardcopies was wrong by this amount. Although the error was known to the staff and was evident in the hardcopy, there was no mechanism by which a bug fix could be mandated. The staff knew about it, most users did not, and there is no way of knowing how many slightly incorrect values are entombed in the literature on this account.

I have been mortified when, in the past, I discovered (and I have personally found more bugs in my code than my users have communicated to me) that my code was producing visibly wrong results of varying severity and consequence. This is equivalent to lying and lying is quite simply antithetical to the conduct of research. This is the reason why a bug fix always takes priority over any other work I might want to do. But do I spend my days roaming my code looking for errors? No more than I might spend them checking my own old papers in the ApJ for typos. I simply fix bugs as they become apparent. Never complain, never explain, just slipstream.

9. Programming Hazardous to Your Health

I was standing in line at a receptie (that's Nederlands) after a promotie (Ph.D. orals) at Leiden University, when the person behind me, a well-respected non-Dutch astronomer whom I had never met, introduced himself and told me that many times he had wanted to strangle me! For various reasons he had had to read the native drawspec binary format on 32-bit and 64-bit UNIX workstations, and my native format is guaranteed to have an odd number of bytes per scan.

Of course I would not make this mistake now. But in 1985, when I programmed the way astronomers usually do (without really knowing what they are doing), with files having been as exotic as they were in the FORTRAN/mainframe era when there was almost NO permanent disk storage for data, it never occurred to me that the 1-byte granularity of DOS, a 2-bit operating system, was unusual.

10. The Future

I am always asked whether I have ported my programs to the MS-Windows environment (this was the only question I was asked after my talk). The answer is no, partly because I don't own a computer which can run the 32-bit versions of Windows or Turbo Pascal and partly because I am still resting up from producing the DOS version. It's now the turn of AIPS++ to produce a single dish package for the GBT, whose Project Scientist I've been for the last year. I can't wait to see it.

Acknowledgments. I thank the ADASS VI organizers, who either didn't know what they were getting into when they invited me to speak at their meeting, or didn't care. Neologisms like Compaq, Borland, Turbo Pascal, and MS-Windows are the property of their respective owners. The National Radio Astronomy Observatory is a facility of the National Science Foundation, operated by Associated Universities, Inc. under a cooperative agreement.

References

Sandquist, Aa., & Hagström, M. 1995, Highlights in Astronomy, 10, 172

Astronomical Data Analysis Software and Systems VI
ASP Conference Series, Vol. 125, 1997
Gareth Hunt and H. E. Payne, eds.

The Design and Implementation of Synthesis Calibration and Imaging in AIPS++

Tim Cornwell

National Radio Astronomy Observatory, PO Box 0, Socorro, NM 87801, E-Mail: tcornwel@nrao.edu

Mark Wieringa

Australia Telescope National Facility, Locked Bag 194, Narrabri, NSW 2390, Australia, E-Mail: mwiering@atnf.csiro.au

Abstract. Calibration of and imaging from data measured by synthesis arrays is accomplished in AIPS++ using a very general formulation of interferometry due to Hamaker, Bregman, & Sault (1996). We have designed and implemented a set of C++ classes to provide very flexible capabilities for processing synthesis data from a wide range of nominally different synthesis arrays. We describe the design, the design process, and some details of the implementation. We also comment on the use of AIPS++ for this type of work.

1. Introduction

The processing of data from synthesis arrays has become increasingly sophisticated in recent years, so much so that the final performance of such telescopes is inextricably wedded to the performance of the algorithms used to process the data. Furthermore, some telescopes, such as the proposed Millimeter Array and the Square Kilometer Array, are hard to envisage in operation without the necessary processing algorithms. The drawback to this development is that as the algorithms have increased in sophistication, the difficulty of implementing them in existing packages has increased correspondingly. This difficulty is of course one of the motivations for the AIPS++ Project, in which object-oriented techniques have been adopted with the goal of simplifying development and maintenance of processing algorithms.

AIPS++ is a package for (principally radio) astronomy processing, written in C++ by a collaboration formed between a number of organizations. Those now actively involved are:

- Australia Telescope National Facility,
- Berkeley-Illinois-Maryland Array,
- National Radio Astronomy Observatory, which hosts the Project Center in Socorro, and
- Netherlands Foundation for Research in Astronomy.

AIPS++ is scheduled to undergo a Beta release in early 1997, followed by a limited public release about six months later. More information on AIPS++ may be obtained from the AIPS++ On-Line Documentation[1] page.

The user specifications for AIPS++ were written in late 1991, based upon contributions from a large number of astronomers from the AIPS++ consortium. It is fair to say that the specifications cover all known and some speculative processing algorithms. As well as providing specific functionality in the form of applications, it was also desired that the AIPS++ system provide a means for long-term development of entirely new processing algorithms. The goals for the support of synthesis processing in AIPS++ can be summarized as follows:

- allow all important and, hopefully, most extant calibration and imaging schemes to be coded in AIPS++,
- allow observatories to develop custom calibration and imaging packages,
- allow a programmer to add new calibration and deconvolution algorithms as easily and naturally as possible,
- allow future instruments (such as the NRAO MMA and the SKAI) to be supported in AIPS++,
- support standard observing modes: mosaicked, spectral-line, multi-feed, polarization observations with synthesis arrays and single dishes,
- allow calibration and imaging algorithms to be used as part of higher level algorithms, and
- achieve performance goals that will allow the calibration and imaging system to be used for standard production processing.

Such open-ended flexibility is very demanding of the design of the system, and requires that the abstraction of synthesis telescopes be as general as possible. Thus, ambition drives the level of abstraction higher and higher. However, too high a level of abstraction is difficult to work with and so some compromise must be found. Early attempts to find the correct level of abstraction were not entirely successful. One of the authors advocated an approach based upon the linear mathematics common to all imaging telescopes. A better approach, and one that the Project eventually adopted, is to base the abstraction on the physics of synthesis arrays. In this paper, we present the details of the physical model for synthesis arrays that was adopted. We describe how this was translated into an object-oriented design, and how that design was implemented in AIPS++. We also show some examples of synthesis data processed using this framework. First, we must briefly review synthesis calibration and imaging.

2. Review of Synthesis Calibration and Imaging

Interferometric arrays measure the Fourier transform of the Sky Brightness:

---

[1] http://aips2.nrao.edu/aips++/docs/html/aips++.html

$$V(u,v) = \int I(x,y) e^{j2\pi(u \cdot x + v \cdot y)} dx dy \quad (1)$$

In practice, this transform is sampled only at a limited number of discrete points, and it is typically corrupted by antenna-based calibration errors:

$$V_{i,j}(u,v) = g_i g_j^* \int I(x,y) e^{j2\pi(u \cdot x + v \cdot y)} dx dy \quad (2)$$

Calibration and imaging in radio interferometry corresponds to solving for the calibration parameters g and the sky brightness I. There are many special cases where these equations must be modified:

- Wide fields: which cannot be handled as a simple 2-D transform,
- Wide fields: where power receptivity patterns limit the field of view, imposing a need to mosaic, also single dish observations,
- Wide fields: with bandwidth and time-averaging smearing, and
- Polarized radiation: which introduces more terms.

3. The Hamaker-Bregman-Sault Measurement Equation

The simple two-dimensional Fourier transform has been generalized in many different ways. Ultimately, it can be derived from the law for the propagation of mutual coherence in optics and the van Cittert-Zernike equation. Hamaker, Bregman, & Sault (1996) came across one very useful generalization by considering the polarization properties of synthesis arrays.

The required changes to the above equations must take into account the following:

- Each antenna measures two polarizations: either Right and Left Hand Circular or X and Y Linear.
- Typically, the correlator forms all cross correlations, e.g., RR, RL, LR, and LL.
- One often wishes to image Stokes I, Q, U, and V.
- For high quality imaging, it is usually necessary to calibrate, and even self-calibrate, gains and leakages (e.g., R to L and vice versa).

The Hamaker-Bregman-Sault Measurement Equation does take these into account. It is expressed in direct products of 2×2 matrices (Jones Matrices) representing the antenna-based calibration effects, and 4 vectors representing the measured coherence and sky brightness:

$$\vec{V}_{\mathsf{ij}} = X_{\mathsf{ij}} \left(\mathsf{M}_{\mathsf{ij}} \left[\mathsf{J}^{\mathsf{vis}}{}_{\mathsf{i}} \otimes \mathsf{J}^{\mathsf{vis}}{}_{\mathsf{j}}{}^{*} \right] \sum_k \left[\mathsf{J}^{\mathsf{sky}}{}_{\mathsf{i}}(\vec{\rho}_k) \otimes \mathsf{J}^{\mathsf{sky}}{}_{\mathsf{j}}(\vec{\rho}_k)^{*} \right] \mathsf{S}\, \vec{I}_k + \vec{A}_{\mathsf{ij}} \right) \quad (3)$$

where:

$\vec{V}$ is the 4 vector of coherences, e.g., (RR, RL, LR, LL),

X is the non-linear correlator response function,

M represents correlator-based gain errors,

$\mathsf{J}^{\mathsf{vis}}$ represents antenna-based and non-direction dependent gain effects,

$\mathsf{J}^{\mathsf{sky}}$ represents antenna-based and direction dependent gain effects (*including the Fourier transform phase factors*),

S is a 4×4 matrix to convert Stokes I, Q, U, V to polarization, e.g., RR, RL, LR, LL,

$\vec{I}_k$ is the 4 vector of Stokes I, Q, U, V for pixel k, and

$\vec{A}$ is an additive error per coherence sample.

Not surprisingly, this equation does specialize to the usual Fourier transform equation between sky brightness, and furthermore, by the correct choice of the components, such as the $\mathsf{J}^{\mathsf{sky}}$, one can derive other imaging modes, such as mosaicking.

For this equation to be useful in handling general problems of calibration and imaging, we need to be able to do the following things:

- evaluate predicted coherences $\vec{V}_{\mathsf{ij}}$ if given all terms on right hand side,
- given observed coherences, construct an image of the Sky Brightness, and
- solve for other terms on right hand side given observed coherences and a model for the sky brightness.

Before proceeding further, it is a good idea to split this one equation into separate Sky and Vis Equations:

$$\vec{V}_{\mathsf{ij}}^{Sky} = \sum_k \left[\mathsf{J}^{\mathsf{sky}}{}_{\mathsf{i}}\,(\vec{\rho}_k) \otimes \mathsf{J}^{\mathsf{sky}}{}_{\mathsf{j}}\,(\vec{\rho}_k)^*\right] \mathsf{S}\, \vec{I}_k \tag{4}$$

$$\vec{V}_{\mathsf{ij}} = X_{\mathsf{ij}}\left(\mathsf{M}_{\mathsf{ij}}\left[\mathsf{J}^{\mathsf{vis}}{}_{\mathsf{i}} \otimes \mathsf{J}^{\mathsf{vis}}{}_{\mathsf{j}}{}^*\right]\vec{V}_{\mathsf{ij}}^{sky} + \vec{A}_{\mathsf{ij}}\right) \tag{5}$$

We now also have to make some choice about the class of algorithms that we wish to support. For the moment, we have chosen to limit the main support to those algorithms that can be related to least-squares methods. Perhaps surprisingly, this accommodates many different calibration and imaging algorithms.

For simplicity, consider χ^2 for one time interval:

$$\chi^2 = \sum_{\mathsf{ij}} \Delta\vec{V}_{\mathsf{ij}}^{*T} \Lambda_{\mathsf{ij}} \Delta\vec{V}_{\mathsf{ij}} \tag{6}$$

A large class of algorithms, including Maximum Entropy, optimize some combination of χ^2 and another term. For these, one requires the derivatives of χ^2 with respect to the unknown pixels:

$$\frac{\partial \chi^2}{\partial \vec{I}_k} = -2\,\Re\left[\sum_{\mathsf{ij}} \left[\mathsf{J_i}\left(\vec{\rho}_k\right) \otimes \mathsf{J_j}\left(\vec{\rho}_k\right)^*\right]^{*T} \Lambda_{\mathsf{ij}}\, \Delta\vec{V}_{\mathsf{ij}}\right] \tag{7}$$

$$\frac{\partial^2 \chi^2}{\partial \vec{I}_k \partial \vec{I}_k^T} = 2\,\Re\left[\sum_{\mathsf{ij}} \mathsf{S}^{*T}\left[\mathsf{J_i}\left(\vec{\rho}_k\right) \otimes \mathsf{J_j}\left(\vec{\rho}_k\right)^*\right]^{*T} \Lambda_{\mathsf{ij}} \left[\mathsf{J_i}\left(\vec{\rho}_k\right) \otimes \mathsf{J_j}\left(\vec{\rho}_k\right)^*\right] \mathsf{S}\right] \tag{8}$$

Another whole class of algorithms, represented by CLEAN, require a residual image of some sort. A little thought convinces one that a *generalized* residual image can be defined as:

$$\vec{I}_k^D = -\left[\frac{\partial^2 \chi^2}{\partial \vec{I}_k \partial \vec{I}_k^T}\right]^{-1} \frac{\partial \chi^2}{\partial \vec{I}_k} \Big|_{\vec{I}_k=0} \tag{9}$$

The utility of this generalized residual image is that it may be used in many iterative-update type algorithms in which a trial estimate image is updated from the residual image.

Note that the corresponding PSF may well not be shift-invariant, and so either some approximation must be used or the algorithm must accommodate a shift-variant PSF.

Many *calibration* algorithms require derivatives of χ^2 with respect to the various gain matrices. As an example, consider the case where only one term is required:

$$\frac{\partial \chi^2}{\partial \mathsf{G}_{\mathsf{i},p,q}} = -2 \sum_{\mathsf{j}} \left[U_{p,q} \otimes \mathsf{G_j}^*\right]^{*T} \Lambda_{\mathsf{ij}}\, \Delta\vec{V}_{\mathsf{ij}} \tag{10}$$

$$\frac{\partial^2 \chi^2}{\partial \mathsf{G}_{\mathsf{i},p,q} \partial \mathsf{G}_{\mathsf{i},p,q}{}^{*T}} = 2 \sum_{\mathsf{j}} \left[U_{p,q} \otimes \mathsf{G_j}^*\right]^{*T} \Lambda_{\mathsf{ij}} \left[U_{p,q} \otimes \mathsf{G_j}^*\right] \tag{11}$$

where the matrix $U_{p,q}$ is unity for element (p,q) and zero otherwise.

These gradients may be used in a least squares solution for the calibration parameters.

In an object-oriented system, one would want to provide services to evaluate these gradients (and the original equations) for quite general forms of the components of the equations. The interface of the components need only be specified as far as it concerns the calculation of these quantities. Otherwise, the behavior of the components is unspecified, and may be varied to suit different contexts.

A consumer of these services has to be responsible for the model or trial estimates but not for evaluating χ^2 and associated gradients.

4. C++ Classes

The support for the Hamaker-Bregman-Sault Measurement Equation is provided by a number of C++ classes:

SkyEquation and VisEquation are concrete classes responsible for evaluating these equations, χ^2, and the gradients of χ^2,

VisSet is a concrete class that provides coherence data to the **SkyEquation** and **VisEquation** classes and stores the results of prediction and correction,

SkyModel supplies a set of images such as $\vec{I}$ to the **SkyEquation** class,

SkyJones supplies sky-plane-based calibration effects to the **SkyEquation** class by multiplying a given image by matrices, such as $\mathsf{J}^{\mathsf{sky}}$,

FTCoh and IFTCoh perform forward and inverse Fourier transforms,

VisJones applies visibility-plane-based calibration effects via either 2×2 or 4×4 matrices such as $\mathsf{J}^{\mathsf{vis}}$,

MIfr applies interferometer-based gain effects **VisEquation** via a 4×4 matrix M,

ACoh applies interferometer-based offsets via a 4 vector $\vec{A}$, and

XCorr applies a non-linear correlator function via a function X to a 4 vector.

The *MeasurementComponents* SkyModel, SkyJones, VisJones, MIfr, ACoh, and XCorr can solve for themselves given a specific SkyEquation or VisEquation and VisSet. Programmers write special versions of these MeasurementComponents for particular circumstances. As a concrete example, consider the calibration of a phase screen across a compact array such as the proposed Millimeter Array. In conventional self-calibration procedures, one would solve for the phases of all 40 antennas. A more sensible and robust approach is to solve only for a limited number of parameters describing a phase screen across the array. This could be simply a tilted screen or perhaps a few curvature terms, but the main point is that since the number of degrees of freedom would be limited, the robustness and sensitivity would both improve. In the AIPS++ approach, the programmer merely has to develop a MeasurementComponent that uses the VisEquation services to get gradients of χ^2 for the relevant Jones matrices, and converts this gradient information into update information for parameters of the phase screen. The interface to the VisEquation is thus fixed, but the internal behavior of the component is unspecified.

A similar approach works for the SkyModel. Here all the gradients are calculated in terms of a discrete pixellated images, but any given SkyModel may use these gradients as desired, perhaps to drive the solution for the parameters of Gaussian components.

Simulation is straightforwardly accommodated by providing MeasurementComponents that generate, rather than solve for, instrumental effects.

5. Applications Level Code

The design of the applications level code has been through a number of revisions. We initially used a monolithic, command-line program with lots of switches to control behavior—the prime advantage of this approach is simplicity: one could exercise all of the code this way. For the end-user, it has one prime drawback: the behavior of the application is controlled by a number of switches and it is hard to understand and document. Despite these shortcomings, it is still the best way to write a test program.

With this in mind, we changed the design to be based upon objects that could be directly invoked from the AIPS++ CLI, Glish, using the AIPS++ Distributed Object mechanism. By decomposing the code from the monolithic applications, we derived objects with different responsibilities, all coded in terms of the underlying C++ classes. The five user-level objects were:

calibrater which reads, writes, interpolates, applies, and solves for calibration information, e.g., to correct a MeasurementSet for parallactic angle:

```
calibrater:=imager.calibrater();
calibrater.initialize(calibrater, '3C273XC1.MS');
calibrater.set(calibrater, 'P', 30.0);
calibrater.correct(calibrater);
calibrater.write(calibrater);
```

flagger which flags data using a number of methods,

imagemaker which makes an empty template image,

imagesolver which solves for an image (i.e., deconvolves using, e.g., CLEAN or Dan Brigg's NNLS method), and

weighter which applies weighting to a dataset.

In the final revision of the application code, we watched what people did and then built standard sequences of object invocations into CLI functions. This is the `imager` package in AIPS++. A typical example of the use of the `imager` would be as follows:

```
include "imager.g"           # Include definition of imager and di
di.initialize("3C273XC1")    # Initialize inputs
di.getfits()                 # Get data from the FITS file
di.show()                    # Show the inputs
di.cell:=0.7                 # Cellsize
di.makeimage()               # Make an empty template image
di.P.t:=60.0                 # Time scale for parallactic angle
di.G.t:=300.0                # Time scale for antenna gain changes
di.D.t:=3600.0               # Time scale for polarization leakage changes
di.initcal()                 # Set up calibration objects
di.beam:=[2.5, 2.5, 0]       # Beam size
di.clean.niter:=10000        # Number of clean iterations
di.deconv()                  # Perform clean deconvolution
di.dis()                     # Display the image
for (selfcal in 1:5)            # Perform 5 loops of....
  {
     di.scal()               # Perform one selfcal iteration
     di.plotres()            # Plot residual visibility
     di.dis()                # Display the image
  }
```

The object `di` is the default instance of the class imager. Our next planned step is to build a GUI interface that invokes these functions and integrates display and plotting.

6. Comments on Design, Implementation, and Developing in AIPS++

We performed a detailed analysis of the Hamaker-Bregman-Sault formalism before starting the design, and read a lot. But in the end, we did not adopt any formal methodology or diagramming tools. We did lots of prototype coding, all in C++, although next time, we would probably do at least the early stages in Glish. We split responsibilities for development in different areas, and reconciled code divergences, by hand, every few days. The overall breakup into classes was settled fairly early on and changed relatively little. The assignment of responsibilities to classes changed a lot as we looked for and found a natural split of the evaluation of gradients of χ^2. This probably represented most of the experimentation that was performed. Symmetry in the design between the Jones matrices and the `SkyModel` was the most powerful organizing principle that we came upon. Enforcing uniformity of interface across objects was also important. Finally, we came to the split between `SkyEquation` and `VisEquation` quite late on. While the split was obvious, the reasons for making the design split became more pressing as we got deeper into the details and began to think in more detail about, e.g., Single Dish processing.

Clearly this is not a simple waterfall design process, in which specification leads to analysis, design, implementation, testing, and deployment. Instead it is probably better described in terms of the spiral model whereby one revisits each of the above stages a number of different times, each time having progressed from the last time. The design is still evolving, somewhat, as we try to optimize performance, especially on spectral line datasets. We think that such a spiral design process is mandated by the complexity of the problem that we were trying to solve.

We wrote special classes for matrices with few elements and special symmetries (SquareMatrix) and vectors with few elements (RigidVector). It was necessary to drop into FORTRAN for some inner loops, e.g., FFTs, gridding, and CLEAN loops.

Our conclusions on developing within AIPS++ are as follows. AIPS++ is (currently) very complex but is well-featured. It seems that C++ is a net win but the buy-in is large (6–12 months for competence to do the type of design work presented here). There are many new concepts that one must learn (e.g., iterators) that simply cannot be avoided. Compilers are an obstacle: *Gnu* is the standard compiler and is the best of the lot. Finally, we found the loose collection of AIPS++ tools united by Glish to be wonderful for development.

Acknowledgments. We thank Brian Glendenning and Ger van Diepen for advice and help in working inside AIPS++, and Jan Noordam and Johan Hamaker for discussions.

References

Hamaker, J. P., Bregman, J. D., & Sault, R. J. 1996, A&AS, 117, 137

Astronomical Data Analysis Software and Systems VI
ASP Conference Series, Vol. 125, 1997
Gareth Hunt and H. E. Payne, eds.

The Grid Signal Processing System

I. J. Taylor[1]

Department of Physics and Astronomy, University of Wales, College of Cardiff, PO BOX 913, Cardiff, Wales, UK,
E-mail: Ian.Taylor@astro.cf.ac.uk

B. F. Schutz[2]

Albert Einstein Institute, Max Planck Institute for Gravitational Physics, Schlaatzweg 1, Potsdam, Germany.
E-mail: schutz@aei-potsdam.mpg.de

Abstract. In this paper, we present an interactive signal-processing environment called Grid. Grid enables users to create programs by graphically connecting a desired set of tools. It is written in Java and therefore can run on virtually any computer platform. Grid is being developed to provide a quick-look data analysis system for the GEO 600 gravitational wave detector, but we have found its use is much more general with currently many collaborators using it within their own specific fields.

1. Introduction

Signal-processing systems are fast becoming an essential tool within the signal-processing community. This is primarily due to the need for constructing large complex signal-processing algorithms which would take many hours of work to construct using conventional programming languages. Here, we present such a signal-processing system, called Grid. Grid is a graphical interactive *multi-threaded* signal-processing environment which allows the creation of complex arrays of algorithms (called units within Grid) by simply choosing the desired units from a selection of toolboxes then graphically wiring them together.

2. An Overview

When Grid is run three windows are displayed. A *ToolBox* window, a *Main-Grid* window and a *Dustbin window.* Briefly, the ToolBox window shows the various tools available within Grid, the MainGrid window allows algorithms to be connected together and the Dustbin window allows unwanted units to be discarded.

---

[1] A post doctoral researcher at Cardiff who has been developing Grid since January 1996. For more information, see http://www.astro.cf.ac.uk/pub/Ian.Taylor/Grid.

[2] Professor Schutz is a Director of the Albert Einstein Institute

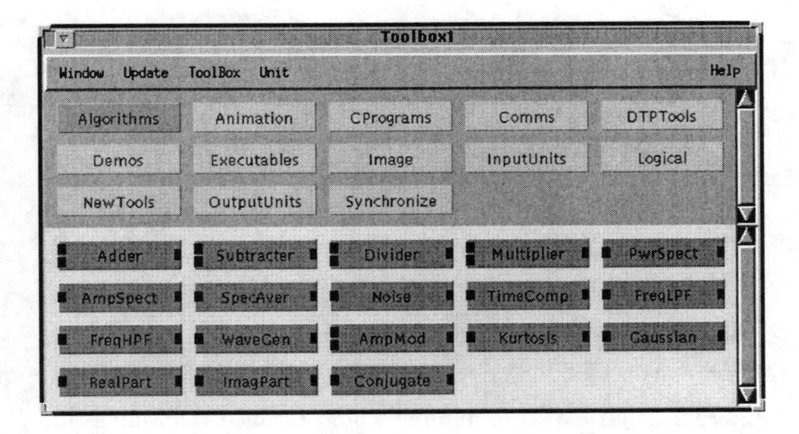

Figure 1. Grid's ToolBox window consists of a section which lists the available toolboxes and another showing its associated tools.

2.1. Distributed ToolBoxes

Figure 1 shows the *ToolBox* window which is divided into two sections. The top section shows the available toolboxes and the bottom section shows the tools which are contained within the selected toolbox. When Grid is run, it scans the toolbox paths (specified by the *TOOLBOXES* environment variable) and shows the detected toolboxes on the upper part of the *Toolbox* window. Toolboxes (and the associated tools) can be stored on a local server or distributed throughout several network servers. The network addresses are specified in the *TOOLBOXES* and the standard Java *CLASSPATH* environment variables. Grid uses its own *class loader* which makes the loading in of classes via the Internet or a local server totally transparent.

2.2. Programming Within Grid

The MainGrid window (see Figure 2) gives a typical example of an algorithm constructed within Grid. Within Grid, instead of re-coding computer algorithms when the specific connectivity needs to be altered slightly, we simply wire the algorithm in different way. Units are created by *dragging* them from the ToolBox window to the desired position in the MainGrid window and then connected together by dragging from a socket on the right of a sending unit to the socket on the left of a receiving unit. Once the desired connectivity is in place, the algorithm can be started by clicking on the *start* button and run in a *single step* fashion (i.e., one step at a time) or continuously, i.e., where an algorithm may be applied to continuous data (for example, in analysing the output from a gravitational wave detector or when animating the formation of star clusters, etc.). Each unit is run as a separate *thread* and therefore automatically load-balanced by the specific operating system.

The algorithm, shown in Figure 2 implements a simple algorithm which compares two signals. The basis of the signal is formed by contaminating a sine wave (of 2 kHz) with Gaussian noise (5 times its amplitude) and transforming

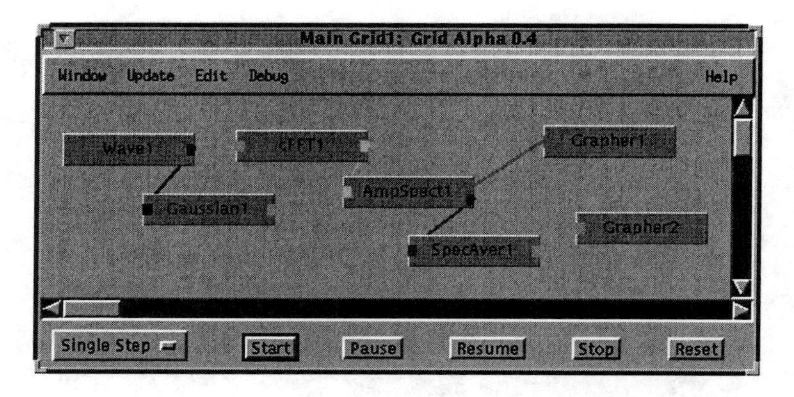

Figure 2. Grid's programming window, called the MainGrid.

this into the frequency domain by using an FFT.[3] The output of the Spectrum is then split into two, with one copy being sent straight to the first Grapher and the other sent via a spectral averaging unit to the second Grapher. The spectral averager receives a specified number of spectra (in our case 20) and outputs their average. The signals are displayed using two Graphers (but we could have just as easily have used one). As many concurrent signals as the user wants can be displayed on the same Grapher, each having its own drawing colour and line style). The result of the second Grapher is given in Figure 3. The first Grapher simply shows each new incoming signal.

Once the signal is displayed, it can be investigated further by using one of the Grapher's various zooming facilities. Zooming can be controlled via a zoom window which allows specific ranges to be set or by simply using the mouse to *drag* throughout the image. For example, by holding the control key down and dragging down the image is zoomed in vertically and by dragging across from left to right zoomed in horizontally. The reverse operations allow zooming out. Also once zoomed in, by holding the shift key and the control key down the mouse can be used to move around the particular area you are interested in. We also have another powerful zooming function which literally allows the user to drag to the position of interest and the image will zoom in accordingly.

3. For What Platforms is Grid Available?

Grid is written mostly in Java (e.g., there are about 33,000 lines of Java source code/documentation and about 500 lines of C code). A previous version was written in C++ using InterViews (Taylor & Schutz 1995) but was abandoned earlier in the year.[4] Grid is now in its fourth *alpha* testing stage with many

[3]the FFT is written in C for efficiency and dynamically linked into Java at run time

[4]We are grateful to Justin Shuttleworth who implemented the original design of Grid as the ideas presented in this version provided a strong foundation for the current design of Grid's graphical user interface

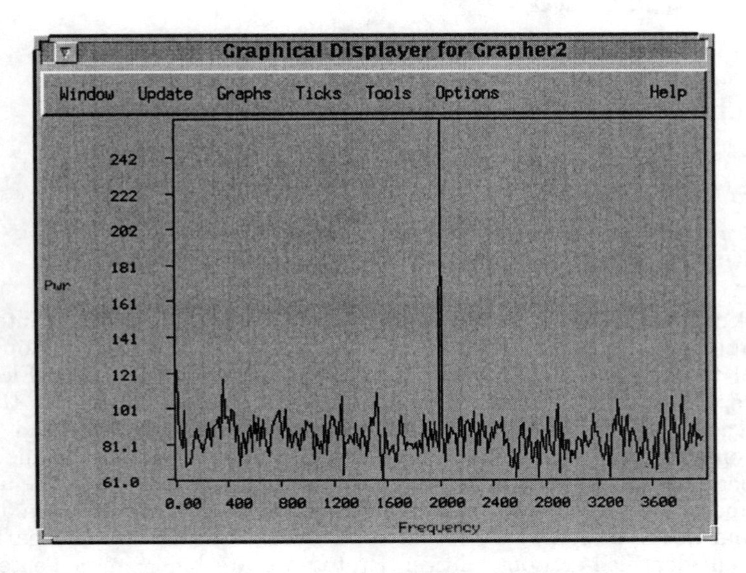

Figure 3. The Grapher can simultaneously display as many signals as required each with its own colour and line style but here we show just one display of a 2 kHz sine wave with Gaussian noise added.

more features being planned for the coming year. Grid is being made available via the World Wide Web.[5,6]

4. Current Applications of Grid

Although Grid is being developed to provide a quick-look data analysis system for the GEO 600 gravitational wave detector (expected to be built by mid-1999) its use is much more general. For example, we have created toolboxes for animation, image viewing, and certain desk-top publishing tasks. Collaborators are already working on units which can manipulate FITS images, apply a variety of signal and image processing algorithms to a variety of sources, and provide multi-media teaching aids; there is also a project to construct a musical composition system. In the future, we anticipate that Grid will be applied to many other varied subject areas.

References

Taylor, I. J. & Schutz, B. F. 1995. in Proceedings of the International Computer Music Conference 95, ed. R. Bidlack (San Francisco, CA: ICMA), 371

---

[5] http://www.astro.cf.ac.uk/pub/Ian.Taylor/Grid

[6] We do not discount the possibility of a version of Grid being made commercially available but none-the-less we will always provide a free version for download.

Astronomical Data Analysis Software and Systems VI
ASP Conference Series, Vol. 125, 1997
Gareth Hunt and H. E. Payne, eds.

The VLT Data Flow Concept

P. Grosbøl and M. Peron

European Southern Observatory,
Karl-Schwarzschild-Str. 2, D-85748 Garching, Germany

Abstract. The size and complexity of the VLT require an integrated approach to the handling of science data associated with the operation of the observatory. In this paper, we present the VLT Data Flow concept: a strategy to fulfill the scientific requirements outlined in the VLT Science Operation Plan—namely, ensure optimal efficiency and a high, constant, and predictable data quality. The VLT Data Flow model has been designed following an object-oriented methodology. In the analysis phase, we were able to get a clear understanding of the problem domain, and in the design phase to partition the system into smaller parts with well identified responsibilities. Prototypes are being implemented, and will be tested on the New Technology Telescope (NTT). They will allow us to clarify the astronomical requirements, verify the system architecture, and determine whether new technologies, such as distributed object environments, are appropriate.

1. Introduction

Space experiments normally include software components for performing full calibration and reduction of the acquired data. This is less common for ground based observatories, due to the higher cost involved. However, when size and complexity become too great, even ground based observatories cannot be operated reliably without a global, consistent concept.

In this paper, we argue that the VLT exceeds the limit for facilities which can be managed in an *ad hoc* manner. An important part of the operational concept of an observatory is the flow of science data from the initial proposal to the reduced observations. The concept of a VLT Data Flow System and its architecture are discussed, including first considerations for their realization.

2. Why is a Data Flow Concept Needed?

The VLT is a very large, complex facility which consists not only of four 8 m telescopes, but also of several smaller auxiliary telescopes for interferometry. Two additional facts complicate the operation further, namely: *a)* that the four telescopes have to be considered both as individual units and in any combination, since a combined focus is foreseen, and *b)* that several sites (e.g., Paranal, Chile and ESO Headquarters, Germany) will share operational duties.

Major ground based astronomical observatories have an operational lifetime which exceeds 25 years. A facility like the VLT is likely to have at least this life-time to fully amortize the investment. This corresponds to several generations of electronics, computer hardware, and software. Thus, it must be expected that most of the control hardware and software components will be exchanged several times during the operational period.

The large investment made in the VLT by the ESO community places a large responsibility on ESO to operate it in the most efficient way. This can only be achieved by analyzing the system dependencies, and establishing a well defined work-flow which takes them in to account. The distributed nature of the operation makes this more difficult, and requires a global view based on basic astronomical requirements.

Finally, the observatory should not only work efficiently in a technical sense, but astronomers must also be able to use it in an optimal way. It will be difficult for normal users to develop a comprehensive overview of the operation, due to the relatively short observing time allocated to them. It is therefore important that a clear concept for the astronomical use is presented, so that observations can be specified in a simple but exact way, with predictable results.

3. What Will a Data Flow System Give?

The Data Flow System provides a top-down view of the flow of science data in the VLT environment, with emphasis on astronomical usage (Baade 1995). Several simple, but general concepts were derived by focusing on the scientific requirements.

The architecture, based on the astronomical tasks to be performed by the individual subsystems, makes the design more robust to changes over time (Peron et al. 1996). Clear responsibilities of the different components, together small interfaces between subsystems, improves the maintainability. Upgrading or replacement of components is also facilitated by the global view and its modularity.

4. General Concepts

An important issue for the Data Flow is to present a simple framework of concepts which both are easy for the standard user to grasp, and which hold the system together. The astronomer's view is as follows:

- Astronomical programs for the VLT are prepared and submitted using equipment description and instrument simulators. They are then evaluated in terms of scientific merits and technical feasibility. This functionality is provided by the *Program Handling* subsystem, which supplies the downstream flow with accepted observing programs. In addition, a long term schedule is established to ensure that the correct equipment and resources are in place when needed.
- For each program, the detailed description of the observations required to meet the science objectives must be specified. This is done by breaking the full program up in to *ObservingBlocks*, which are self-contained units. Each *ObservingBlock* specifies both a target and a set of exposures with a given instrument configuration. The astronomer is guaranteed that exposures inside an *ObservingBlock* are executed as an atomic unit in the

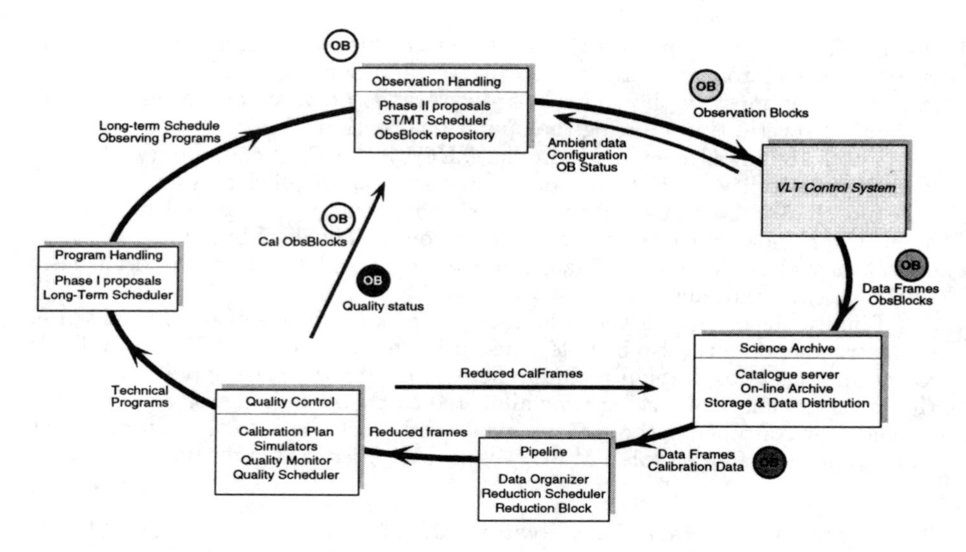

Figure 1. VLT Data Flow subsystem breakdown.

exact sequence specified. In service mode, a *Short Term Scheduler* provides to the ESO staff an ordered list of *ObservationBlocks* computed to take full advantage of weather conditions and instrumental configurations. The *Observation Handling* (Chavan & Albrecht 1997) provides functionality for handling *ObservationBlocks*, including possible scheduling, and forwards them for actual execution to the *VLT control* system.

- The external Data Flow view of the *VLT control system* is very simple, although internally it is extremely complex. From the outside, the *Observation Handling* subsystem gets environmental data and configuration parameters for instruments, while it provides the *VLT control system* with an optimized sequence of *ObservingBlocks* for execution. The resulting observations are then sent to the *Science Archive* subsystem, which keeps track of new data, log information, etc.
- The *Pipeline* subsystem receives the new data frames through the *Science Archive.* After classifying an incoming frame, a *ReductionBlock* is constructed by combining the raw data frame with necessary calibrations and an appropriate *Reduction Recipe*, specifying the actual reduction procedure. *ReductionBlocks*, like *ObservingBlocks*, are self-contained units which detail the standard pipeline reduction, independent of a given reduction system. The reduced data are made available to the user, and forwarded to *Quality Control* for validation.
- Finally, the *Quality Control* subsystem monitors the performance of the instruments by analyzing the reduced data from the *Pipeline* (Ballester et al. 1997). It generates calibration data and thereby establishes the quality level that can be achieved. Instrument simulators are maintained to reflect the current performance, for use by observers and the *Program Handling* subsystem. Calibration *ObservingBlocks* are also submitted to

Observation Handling according to a Calibration Plan, to ensure that specifications are met.

The main Data Flow subsystems are shown in Figure 1, where the general flow of science data is indicated. All along the flow, information is added to the *ObservingBlocks* so that a full record of events, from the original specification to the reduced data, is kept.

5. How Will it be Realized?

The analysis of the astronomical requirements was done by a mixed group of astronomers and software engineers to ensure a common view. An object-oriented methodology was applied since it provides a good mapping between the real world and software entities. The specific methodology, `OMT`, also gives a simple graphical representation of designs, which proved very useful in the discussions. An object-oriented approach also gives better modularity by strong encapsulation. The responsibilities of each component (i.e., class/object) are well defined.

The design concepts will be verified by prototyping on the NTT at La Silla. These implementations will test the general flow of information between subsystems and be used to analyze the merits of different choices. A baseline version will then be prepared for the VLT. At the same time, new technologies (e.g., OMG/CORBA and Java) will be evaluated.

6. Conclusions

A Data Flow concept is mandatory for the VLT due to its size and complexity. It provides the end user a simple view based on general concepts, like *ObservingBlocks* and *ReductionBlock*, which will yield a more homogeneous user interface and thereby help to increase efficiency. The tight *Quality Control* feedback loop will ensure that problems are detected early and corrected rapidly. The use of an object-oriented methodology has given a cleaner and more traceable system design. Finally, testing prototypes on the NTT is an important step in understanding the system behavior in a live, operational environment and in establishing a stable baseline for the VLT.

Acknowledgments. The VLT Data Flow concept presented here is based on work and discussions in the VLT On-line Data Flow Working Group and the Data Flow Project Team in ESO. We would like to thank all people involved in this process.

References

Baade, D. 1995, `VLT-PLA-ESO-10000-0441`, ESO

Ballester, P., Banse, K. & Grosbøl, P. 1997, this volume, 415

Chavan, A. M. & Albrecht, M. A. 1997, this volume, 367

Peron, M., Grosbøl, P. & van den Berg, R. J. G. 1996, in ASP Conf. Ser., Vol. 87, New Observing Modes for the Next Century, ed. T. Boroson, J. Davies, & I. Robson (San Francisco, ASP), 183

Astronomical Data Analysis Software and Systems VI
ASP Conference Series, Vol. 125, 1997
Gareth Hunt and H. E. Payne, eds.

Experience with a Software Quality Process

Richard A. Shaw and Perry Greenfield

Space Telescope Science Institute, Baltimore, MD 21218

Abstract. We describe the methodology and virtues of software design and source code inspections as vital elements of a quality management process. We also recount the experience of the Science Software Group at ST ScI with these methods, which were formalized and implemented during the past year.

1. Introduction

To hear W. Wayt Gibbs (1994) tell it, the software crisis identified more than 25 years ago is alive and well in the mid-1990s. Citing well known projects with spectacular cost and schedule overruns, Gibbs points out that losses of hundreds of millions or billions of dollars on failed software development projects are distressingly common in industry. The statistics are sobering indeed: something like 25% of all large software development projects are outright failures, and another 50% either do not implement all the promised functionality or are not used. According to Gibbs, "the average software development project overshoots its schedule by half." Another study of large distributed systems showed that over half exceeded their cost estimates, two-thirds exceeded their development schedules, and almost nine out of ten involved substantial redesigns. The problem, argues Gibbs, is that a solid engineering methodology has yet to be fully developed for what is still in large measure the art of computer programming.

Perhaps such grim statistics do not hold for software development efforts in astronomy. Yet many participants in the ADASS conferences have probably been involved in some software project that, for whatever reason, could have turned out a little better: perhaps the project took longer than anticipated, or the planned functionality was never fully achieved, or the result was plagued with bugs.

The importance of computers and software to astronomy has grown dramatically during the past few decades. Yet the specialized needs of astronomers are still, in large measure, addressed with custom-built software systems and applications. Generally, this software must be built and maintained with rather modest (by industry standards) resources. So it is important to identify the best methods of software engineering in order to minimize the risks associated with software development, while increasing productivity and product quality.

Our experience in the Science Software Group at ST ScI is that the cost of correcting bugs in a large software system that is distributed to the community is rarely less than one-half staff day, and can consume several days in some cases. This surprisingly large cost includes the effort to reproduce a problem, isolate the bug, devise and test a fix, install the fix in the configured system, document the changes (in the source and user documentation), and advertise the fix and likely user impact in the release notes and on-line news postings. Clearly, any process that identifies defects before code is released to the community is worth investigating.

2. Quality and the Software Development Process

There is compelling evidence that a studied, methodical emphasis on software quality is the best means to contain development costs and minimize schedule delays, particularly for larger projects. As a result, a number of software companies have, over the last two decades, instituted technical review processes. The benefits of a technical review process flow from a few key aspects of the software development process. These include the observation that all software developers are blind to certain kinds of defects in their own code, and that other developers—with different blind spots—can discover them with only modest effort. Secondly, the sooner software defects are discovered, the cheaper and simpler they are to fix. There is no single, correct method for assuring defect-free software, but it does take a serious commitment on the part of management for a software quality process to succeed.

Software development can be broken down into several distinct activities, including: problem definition, requirements analysis, high-level design, detailed design, construction, integration, unit testing, system testing, maintenance, and enhancements. Although our focus in this paper is on the design and construction phases, an effective software quality effort is characterized by a focus on identifying and correcting defects at all stages of development. A successful software quality effort must also be adaptable to local circumstances and therefore have a feedback mechanism. Most importantly, though, the process must produce results that are quantifiable, and the effects of local adaptations must also be quantifiable. Without that, it will never really be known whether or how well the process is working, and whether it is worth the effort.

3. The Review Process

According to various studies (see McConnell 1993 and references therein) the single most effective means to identify and correct code defects is the review process. This paper focuses on design and code reviews, although formal reviews can be applied to any stage of software development. The most formal variety of reviews—inspections—typically catch about 60% of the defects present in software (Jones 1986). While these rates are generally much better than that achieved by simply testing the end product, the kinds of errors found with each method are often quite different. In this sense, design and code reviews complement, rather than replace, testing as a tool for assuring quality software. A significant difference between inspections and testing, though, is that defects are identified and corrected in one step. Testing simply identifies a defect—it reveals nothing about how to fix it. Formal design and code inspections provide an effective means to catch defects early, where they are easiest and cheapest to fix.

Reviews can actually take a number of forms, including management reviews, walkthroughs, code reading, and inspections. The first of these is usually a poor means of discovering problems, though it may serve other useful purposes. Code readings or walkthroughs are more effective, and may be more appropriate for very small programming groups, but these methods are usually less effective at uncovering defects than formal inspections (McConnell 1993). Design and code inspections, being the most formal kind of review, involve preparation by the participants, a meeting, and a formal inspection report. There are three distinct roles in the inspection process:

The Moderator handles the review logistics and ensures that the review materials and reviewers are prepared before allowing the inspection to proceed. She keeps the reviews focused on detecting (rather than solving) problems, and records the defects found by the reviewers. The moderator limits the review meetings to 1.5–2 hours, and enforces proper review etiquette. She prepares an inspection report listing all defects, plus various statistics (e.g., time spent by the reviewers in preparing for the meeting, the number and type of defects found, etc.). It takes some training and experience to be a good moderator.

The Author notifies the moderator that a review should be scheduled, and distributes paper copies of the design or code to be reviewed. The author answers the reviewers' questions during the review meeting, and afterward documents how each identified defect was corrected or addressed. It is the author's responsibility to partition larger projects into segments that can each be reviewed within the time constraints for each review.

The Reviewers (usually two) must be technically competent and have no managerial role over the author. They review the design or code in advance, though it is best to limit preparation to roughly 2 hours per review meeting. (If they need more time, it is usually a sign that the review product should have been partitioned into smaller parts.) During the meeting, the reviewers should stick to technical issues and practice proper review etiquette. Most programmers can be successful reviewers with little additional training.

A significant point is that management must *not* participate directly in the reviews. For the process to succeed, the authors must not perceive the review as an open forum on their abilities or performance. The purpose of the reviews is to improve product quality, and the presence of management at the review undermines that purpose. Managers should demonstrate commitment to the process by not letting schedule pressures shortcut or otherwise cripple the review policy. The point is to distribute responsibility for the product to the review team, not just the author, and to reward the quality of the review and the product as corrected by the review. Management should, however, correct problems with the review process, such as poor or irresponsible behavior on the part of the participants. Management should see that the recommendations of the review panel are addressed, and should override the panel only under extraordinary circumstances.

It is not possible in a short paper to cover all the issues relating to software inspections. See McConnell (1993), Freedman & Weinberg (1990), and Humphrey (1989) for more detailed descriptions of the process, and the means by which its effectiveness can be measured. The use of software inspections is not a "silver bullet" that will solve all problems. It is only one tool of many intended to increase software quality and productivity, but it is one of the most effective.

4. Experience in the STSDAS Group

Design reviews and source code inspections were formalized and implemented in the Science Software Group at ST ScI during the past year. While objective, quantitative data on quality control processes are generally difficult to come by, we have endeavored to keep detailed records of the review results, as well as of the process itself, in order to assess their efficacy. To date, six applications and one system-level utility have been reviewed. Reviews of modest-sized applications have taken approximately one half to one week of staff time to complete.

This quality management process has been very successful at detecting and eliminating software defects early in the development process, and has unquestionably resulted in higher quality software products. Information on defect rates, defect categories, etc. is still a bit sketchy. The process itself has also been generally well received by the programming staff. As a result, a number of sociological benefits are being realized, including a transfer of coding expertise (usually) from the more experienced to the less experienced programmers, a broadening of knowledge within the group about the software system as a whole, and a greater sense of teamwork and shared responsibility for the software products. It is fair to say that the group is still gaining experience with this process, and its value to the organization is still being evaluated. But as of this writing, our interest and commitment to this process is quite strong.

References

Freedman, D. P. & Weinberg, G. M. 1990, Handbook of Walkthroughs, Inspections, and Technical Reviews, 3rd Ed. (New York, NY: Dorset House Publishing)

Gibbs, W. W. 1994, Scientific American, 271, 86

Humphrey, W. S. 1989, Managing the Software Process, Chapter 10 (Reading, MA: Addison-Wesley)

McConnell, S. 1993, Code Complete (Redmond, WA: Microsoft Press)

Jones, C. 1986, Programming Productivity (New York: McGraw-Hill)

Astronomical Data Analysis Software and Systems VI
ASP Conference Series, Vol. 125, 1997
Gareth Hunt and H. E. Payne, eds.

Overview of the Ftools Software Development Philosophy

W. Pence

NASA/GSFC, Code 662, Greenbelt, MD 20771

Abstract. This talk will describe some of the features that have lead to the success of the ftools project and have enabled it to produce and manage a large package of software with low development costs. The ftools software package is a modular, platform-independent set of programs for analyzing FITS format data files in general, with a strong emphasis on data from high energy astrophysics missions. Ftools development began in 1991 and has produced the main set of data analysis software for the current *ASCA* and *XTE* space missions and for other archival sets of X-ray and gamma-ray data. One of the original requirements of ftools was to support both the IRAF and non-IRAF user communities, therefore the ftools software can be built either as an IRAF package or as a set of stand-alone executables that can be run directly from the host operating system. Platform-independence has been achieved in part by adopting FITS as the run-time data analysis format. Other external projects are now developing their own ftools analysis tools that can be layered on top of the existing ftools packages. We encourage further such collaborations as a cost effective way to develop new software.

1. Background

Development of the ftools software package started in 1991 at the HEASARC (High Energy Astrophysics Science Archive Research Center) to support analysis of data from the *ASCA* X-ray Satellite. During the past five years, ftools has grown into a large system containing both generic utilities for examining any data file in FITS format, as well as mission specific analysis tools for current and past X-ray and gamma-ray astrophysics missions. Because the ftools project could not afford to spend much effort on new systems-level development, it has adopted existing infrastructures when possible.

One of the original requirements for ftools was to be able to support both the PROS/IRAF and XANADU user communities, therefore we adopted the IRAF-style parameter file mechanism as the main interface between the application program and the user. This allows the software to be built as a package under IRAF, but in addition, the availability of a standalone parameter interface library, initially from SAO, also enables the ftools source code to be built as standalone executable programs. In this way we were able to satisfy both the IRAF and non-IRAF user communities.

The other major initial decision was the choice of analysis data format. We adopted FITS for several reasons, including the fact that most of our data files are already in FITS format, and FITS is a versatile format well suited for the type of "event list" data common in high energy astrophysics. FITS is also a good format choice because it is platform independent and files can simply be copied from machine to machine without any translation. This choice was only practical because the FITSIO (and later CFITSIO) interface library was available for application programs to easily read and write FITS files. Some initial concerns about the efficiency of using FITS formats (since FITS was originally developed as an archive and interchange data format, not an on-line analysis format) have largely been dispelled. Recent throughput tests of CFITSIO show that it can read or write FITS images or tables with rates in the range of 2–5 MB s^{-1} on current generation workstations which is very close to the limit imposed by the underlying magnetic disk.

The latest ftools v3.6 release now contains more than 220 tasks subdivided into a dozen packages (e.g., FITS utilities, images, timing, *ASCA*, *ROSAT*, *XTE*). The software is supported on a wide variety of operating systems (e.g., SunOS, Solaris, Alpha/OSF, Ultrix, SGI/IRIX, Linux, and VMS) and more than 200 registered sites worldwide have installed it.

2. Development Philosophy

The quality of the programming staff is of course one of the primary factors in the success of any software development project. All the ftools programmers are required to have a strong science background and most in fact have a M.S or Ph.D. in the physical sciences. Having a science background enables the programmers to understand the requirements of a new application program faster without the need for a minutely detailed set of requirements.

In a project like ftools, which lasts for many years, there will inevitably be some staff turnover, thus we have found it important to require cross training and overlapping duties to ensure continuity in case a key programmer leaves the project. We have also learned that it is getting more difficult to find and hire experienced FORTRAN programmers, presumably because fewer students in universities are learning FORTRAN. We have always supported both FORTRAN and C application programs in ftools, but we have shifted to more emphasis on writing software in C rather than FORTRAN as the skill mix of the programmers has evolved.

The ftools project has never had any formal requirements, design, or code reviews. Instead, we usually rely on the close interaction of an individual project scientist and a ftools programmer to develop the final product. In a few more complex or critical tasks, we may have a somewhat larger subgroup all working on a project, but the vast majority of ftools tasks have been single person projects. In effect, the ftools development is a distributed system with minimal central oversight or management bureaucracy. This approach does have some drawbacks in that the resulting ftools package is somewhat less homogeneous, and the quality of the code can be somewhat uneven. However, the increased productivity resulting from of this type of management more than compensates for this. The only formal management structure imposed on the project is a bi-weekly group meeting to maintain communication among programmers and scientists.

The ftools coding standards are mainly driven by the requirement to support the software on a wide variety of platforms and compilers. The primary coding requirement is to adhere to very strict ANSI C or FORTRAN-77. Even so, the differences between compilers continues to cause headaches and is a major expense in terms of having to build, test, and debug the software on all the different supported platforms. Because of this, we do not guarantee support for all possible compilers, but will at least support the GNU gcc and g77 compilers, which are freely available for most platforms. Because of these portability issues, we are currently reluctant to use FORTRAN 90 or C++ in any of the ftools software. If we do move to these languages in the future we will probably stick to GNU g++ (and g90 if it ever becomes available).

3. Code Management and Distribution Procedures

For most of the history of the ftools project there has been no formal configuration control of the software. Instead, we assigned one programmer to manage the "official" version of the software, and all the other programmers would deliver any changes to this person for installation. This worked satisfactorily, but in the past year we have switched to using the GNU version control software package called CVS. This turned out to have relatively low start up cost and now allows each developer to safely check-in new versions of their software whenever necessary. CVS also enables better tracking of different release versions of ftools and allows new enhancements to be checked in without affecting the frozen release version which may be undergoing final beta testing.

We rely on the Web and anonymous ftp to distribute the ftools software to users. From the ftools web page[1] users can download selected components of the ftools or the entire package. Currently we provide the source code and an intelligent build script that the user runs to compile and link the software. This works well in the majority of cases, but there are always a few users who experience difficulties, mainly due to configuration problems on the user's machine that are beyond our control. To better address these users, in the future we plan to distribute executable versions of the software, at least for the major platforms. This can introduce its own set of installation problems, but at least the user will not require a functioning C and FORTRAN compiler on their system in order to install ftools. In order to make the size of the executable files more manageable, a technique for combining several related ftools tasks into one executable program has been developed. This is transparent to the user but can reduce the disk space required for ftools by a factor of ten.

4. Support for External Ftools Packages

The success of the ftools project has lead other groups to consider adopting the same approach for their software development. Currently, several external X-ray groups have developed their own ftools-compatible packages, or are strongly considering doing so. We encourage further collaborative efforts like this, and view it as a very cost effective way for other projects to rapidly implement a new package of application software.

---

[1] http://legacy.gsfc.nasa.gov/ftools

The ftools group itself is continuing to make ftools more modular, so that it will become even easier for externally developed packages to be layered on top of the core ftools. The main requirements for developing a new ftools package are simply to learn to use the parameter file interface for communicating with the user, and the FITSIO or CFITSIO libraries for accessing FITS files. Most of the other necessary pieces can easily be implemented by copying the directory structures and the make procedure files that are in the current ftools package. The HEASARC is more than willing to work with any interested group in setting up their own ftools project.

5. The Future of ftools

The HEASARC is strongly committed to supporting and enhancing the ftools for the foreseeable future. We are currently starting development of a new analysis package for the "*Astro E*" X-ray mission (the successor to *ASCA*).

As well as developing the traditional C and FORTRAN based tasks which make up the core of ftools, we are also continuing to explore newer GUI applications, based on Tcl/Tk. Current examples of such applications are the Xselect and fv tasks in the ftools release. We are also considering ways to provide ftools-type analysis capabilities over the Web using Java.

The compatibility with IRAF is an important feature of ftools, so we are watching with interest the ongoing development of OpenIRAF. We fully expect to remain compatible with IRAF as it evolves, especially since the goals of OpenIRAF to provide better direct support for FITS files, and to decouple the IRAF tasks and software libraries from the CL, closely parallel the ftools philosophy.

Astronomical Data Analysis Software and Systems VI
ASP Conference Series, Vol. 125, 1997
Gareth Hunt and H. E. Payne, eds.

Design and Implementation of CIA, the ISOCAM Interactive Analysis System

S. Ott,[1] A. Abergel,[2] B. Altieri,[1] J-L. Augueres,[3] H. Aussel,[3] J-P. Bernard,[2] A. Biviano,[1,4] J. Blommaert,[1] O. Boulade,[3] F. Boulanger,[2] C. Cesarsky,[5] D. A. Cesarsky,[2] A. Claret,[3] C. Delattre,[3] M. Delaney,[1,6] T. Deschamps,[3] F-X. Desert,[2] P. Didelon,[3] D. Elbaz,[3] P. Gallais,[1,3] R. Gastaud,[3] S. Guest,[1,7] G. Helou,[8] M. Kong,[8] F. Lacombe,[9] J. Li,[8] D. Landriu,[3] L. Metcalfe,[1] K. Okumura,[1] M. Perault,[2] A. M. T. Pollock,[1] D. Rouan,[9] J. Sam-Lone,[3] M. Sauvage,[3] R. Siebenmorgen,[1] J-L. Starck,[3] D. Tran,[3] D. Van Buren,[7] L. Vigroux,[3] and F. Vivares[2]

[1]*ISO Science Operations Centre, Astrophysics Division of ESA, Villafranca del Castillo, Spain*

[2]*IAS, CNRS, University of Paris Sud, Orsay, France*

[3]*CEA, DSM/DAPNIA, CE-Saclay, Gif-sur-Yvette, France*

[4]*Istituto TESRE, CNR, Bologna, Italy*

[5]*CEA, DSM, CE-Saclay, Gif-sur-Yvette, France*

[6]*UCD, Belfield, Dublin, Ireland*

[7]*RAL, Chilton, Didcot, Oxon, England*

[8]*IPAC, JPL and Caltech, Pasadena, CA, USA*

[9]*DESPA, Observatoire de Paris, Meudon, France*

Abstract. This paper presents an overview of the Interactive Analysis System for ISOCAM (CIA).[1] With this system ISOCAM data can be analysed for calibration and engineering purposes, the ISOCAM pipeline software validated and refined, and astronomical data processing can be performed. The system is mainly IDL-based but contains FORTRAN, C, and C++ parts for special tasks. It represents an effort of 15 man-years and is comprised of over 1000 IDL and 200 FORTRAN, C, and C++ modules.

[1]CIA is a joint development by the ESA Astrophysics Division and the ISOCAM Consortium led by the ISOCAM PI, C. Cesarsky, Direction des Sciences de la Matiere, C.E.A., France.

1. Introduction

ESA's Infrared Space Observatory (*ISO*) was successfully launched on November 17th, 1995.[2] *ISO* is a three-axis-stabilised satellite with a 60-cm diameter primary mirror (Kessler et al. 1996; Maldari et al. 1996). Its four instruments (a camera, ISOCAM, an imaging photo-polarimeter, and two spectrometers) operate at wavelengths 2.5–240 μm at temperatures of 2–8 K.

ISOCAM takes images of the sky in the wavelength range 2.5–18 μm (Cesarsky et al. 1996). It features two independent 32×32 pixel detectors: the short-wavelength channel (2.5–5.5 μm), and the long-wavelength channel (4–18 μm). A multitude of filters and lenses enable the observer to perform measurements at different wavelengths, with different fields of view or polarizers.

2. Requirements and Constraints

The requirements on CIA were, in decreasing order of importance:

1. to calibrate ISOCAM,
2. to perform fast data reduction to assess the ISOCAM performance during the Performance Verification phase of *ISO*,
3. to monitor the health of ISOCAM,
4. to provide the means to perform any sort of investigation requested for problem diagnostics,
5. to assess the quality of ISOCAM pipeline data products,
6. to debug, validate and refine the ISOCAM pipeline, and
7. to perform astronomical data-processing of ISOCAM data.

External constraints were the extremely tight schedule (see §4) and the operating system. For historical reasons, VAX/VMS was chosen as operating system for the pipeline. It was also decided that data files within the processing environment be in a VAX-specific variant of FITS. Therefore, in order to stay as close as possible to the ISOCAM pipeline, the operational CIA version had to run under VMS. To achieve sufficient performance, it was decided to opt for VMS/Alpha instead of a classical VAX operating system, and accept the minor porting efforts required.

3. Design and Evolution

Given the time constraints and the team-structure, it was decided to re-use the existing IDL experience and code as much as possible. Therefore IDL (V3.6)

---

[2] *ISO* is an ESA project with instruments funded by ESA member states (especially the PI countries: France, Germany, the Netherlands and the United Kingdom) and with the participation of ISAS and NASA.

and high level languages (C, C++) for special IA applications and FORTRAN (programming language for the pipeline) were chosen.

In order to be able to meet the schedule, it was decided to break the development down into three steps: a mimimum system, covering requirements 1–4; an operational system, covering requirements 1–6; and the final astronomical data-processing system, covering all requirements listed in the previous section. The core system was intended for expert users, having only a very limited use of graphical user interfaces for display purposes.

Standardized module headers were designed to provide help to the users. Their contents feed into

- a document generator to create a user's reference and a programmer's reference manual,
- a comfortable, widget-based help-browser, using also the hierarchical organisation of the modules, to guide users, and
- the standard IDL help widgets.

For the full astronomical data-processing system, a UNIX version tuned to the needs of astronomers was introduced, together with more user-friendly, widget-based applications.

4. Schedule and Cost

The development of CIA began in June 1994 and followed standard software engineering practice (ESA PSS-05). User requirements were consolidated in September 1994 and the first version of the architectural design document agreed in November 1994. The minimum system was completed to meet the target date for IA readiness in April 1995, and the operational system in November 1995 prior to the launch of *ISO*. Since then, work has continued for the astronomical data-processing system, the next version is expected to be released in December 1996.

Currently, the system is comprised of over 1000 IDL modules totalling over 210,000 lines, around 200 FORTRAN files taken from the pipeline processing system, and around 10 FORTRAN, C, and C++ files for specialized (mainly cpu-intensive) applications.

Around 15 man-years have been spent in the development of the IA core system, excluding the time for algorithmic research. For comparison, 30 man-years went into the calibration of ISOCAM and 5 man-years into ISOCAM related pipeline processing. This amounts to around 3% of the overall cost of *ISO* Science operations or 0.4% of the total *ISO* cost to completion.

5. Architecture and Design

A quasi-object-oriented approach was taken, with functions communicating via standardized data structures. The same data structures are used by the pipeline, calibration procedures, and astronomical data analysis. This commonality improves considerably the speed of algorithmic development within CIA.

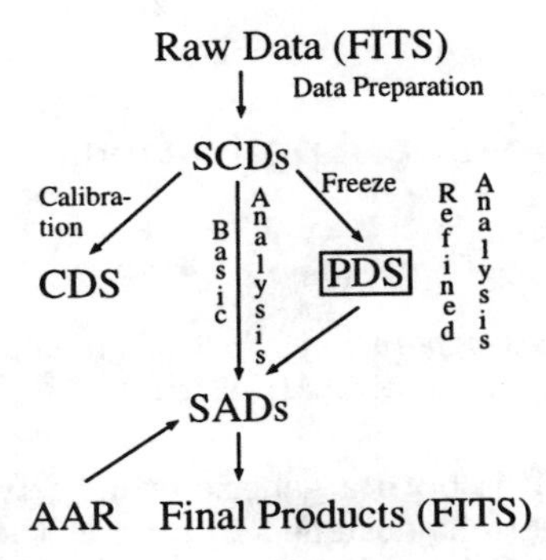

Figure 1. Data Flow and Architecture.

SCDs (Science CAM Data) are self-contained entities, holding *all* data of an ISOCAM state.[3] They are generic, i.e., independent from the observation mode and can be re-arranged to suit all needs.

CDSs (Calibration Data Set) are self-contained entities, holding the calibration results. They are used directly within CIA, or transformed into calibration files, accessed by the pipeline.

SADs (Science Analysed Data) are similar to SCDs, but hold the results either from Auto Analysis (AAR), or following an analysis by Interactive Analysis.

SCDs, CDSs and SADs are implemented as complex IDL data structures using pointers and therefore need dedicated access functions.

PDSs (Prepared Data Structures) are self-contained entities, holding the data currently recognized as relevant for data reduction and their result for an ISOCAM observation. Their flavour depends on their purpose, determined by the ISOCAM observation mode. PDSs are implemented as IDL structures.

References

Cesarsky, C., et al. 1996, A&A, 315, 32

Delaney, M. ed., ISOCAM Interactive Analysis User's Manual, ESA Document

ESA Software Engineering Standards, ESA Document, Reference PSS-05

Kessler, M., et al. 1996, A&A, 315, 27

Maldari, P., Riedinger, J., & Estaria, P. 1996, ESA Bulletin, number 86

Siebenmorgen, R., et al. ISOCAM Data User's Manual, ESA Document, Reference SAI/95-221/DC

[3]A state is the atomic unit of CAM activities. An *ISO* pointing and all CAM parameters are fixed within a state.

Astronomical Data Analysis Software and Systems VI
ASP Conference Series, Vol. 125, 1997
Gareth Hunt and H. E. Payne, eds.

The OPUS Pipeline Applications

James F. Rose[1]

Computer Sciences Corporation, Space Telescope Science Institute, 3700 San Martin Drive, Baltimore, MD 21218

Abstract. OPUS is both a generic event-driven pipeline environment and a set of applications designed to process the spacecraft telemetry at the Space Telescope Science Institute in Baltimore, Maryland. This paper describes those OPUS applications which process the telemetry, validate the integrity of the information, and produce standard FITS (Flexible Image Transport System) data files for further analysis. The applications are to a great extent table-driven in an effort to reduce code changes, improve maintainability, and reduce the difficulty of porting the system. The tables which drive the applications are explained in some detail to illustrate how future missions can take advantage of the OPUS systems.

A science pipeline can be roughly described as a sequence of six basic steps which range from telemetry unpacking to database dredging to data formatting. Under normal conditions turning a raw telemetry stream into a usable science dataset is not even an interesting challenge: you know the input formats, and you know the output formats, and it is often a simple task to understand how to move the bytes around to accomplish that task.

But that's not the whole task. If a pipeline is expected to handle a large amount of data, continuously and robustly, or when that pipeline is designed to run consistently in an unattended environment, it must anticipate problems and take some reasonable action automatically: developers must assume "normal conditions" are the exceptions. Most of the work, and most of the code in such a system is designed to handle the abnormal case.

The OPUS applications which process a telemetry stream into standard FITS files fall into the following six components.

Data Partitioning: This is the front-end workhorse of the telemetry processing. Its function is to scan the incoming telemetry for known patterns and to segment the telemetry stream into its basic constituents.

The OPUS software assumes the telemetry is "packetized"—that is, the stream of data are made up of discrete chunks, and each chunk has a few bytes of information which identifies the beginning of the chunk and some information about the chunk. Chunks can be of different sizes and types. Packets of different types, which can arrive in any order, are segmented and put into their own files.

Here is where most of our "bad data" checks are done. The software asks a number of questions: Can we identify the observation? Can we identify the header? Can we identify the beginning of a packet? Can we tell how long the packet is? Can we tell if the data are in order? Can we determine if any of the

[1]rose@stsci.edu

data are missing? Do we have enough packets to continue? Do we have a mixed set of packets? etc. In addition to the raw telemetry, two drivers help guide the partitioning process: Telemetry_Patterns, and Telemetry_Locations. Each of these is described briefly in Appendix A.

Data Quality Editing: If telemetry gaps are indicated, this pipeline process constructs a data quality image to ensure the subsequent science processing does not interpret fill data as valid science data.

OPUS must anticipate that telemetry data may sometimes have problems: dropped packets, missing segments, corrupt data. Rather than attempt to "correct" the data, OPUS must simply flag bad or suspect telemetry. This is a much more important issue when dealing with images and spectra than with time-tagged photon events. Data which has an implicit order rather than an explicit time-tag order is susceptible to misinterpretation if whole segments are out of order. In addition to the science packet file and the error indications file, this process also relies on the Telemetry_Patterns table to guide processing.

Support Schedule Keywords: This step produces observation-specific datasets for a single exposure or for any time interval (e.g., an operational shift's worth). Information from proposals/experiments which have been previously logged into a relational database is extracted for the given observations.

By "keywords" we mean the standard FITS[2] (Flexible Image Transport System) definition which allows an eight character keyword name, a value, and a comment. For example:

```
RA_TARG =       215.59415 / right ascension of the target (deg) (J2000)
DEC_TARG=      -12.725682 / declination of the target (deg) (J2000)
```

Keywords and their values, which describe the science, are at the heart of OPUS processing. This information describes the configuration of the instrument, the timing of the exposure, and the existence of associated exposures. It also controls the downstream calibration—that is, all calibration steps beyond the OPUS Generic Conversion determine the parameters of the science exposure from these keywords. Two tables drive this package: Keyword_Names and Keyword_Source.

Data Validation: In addition to the database, another source of keyword information is the telemetry itself. The Data Validation stage actually has two purposes: first to dredge the required words from the telemetry, converting them to meaningful values on the fly, and then to ensure that the actual values and the "planned" values from the database are consistent.

In addition to the telemetry packet files and the support values produced by the Support Schedule Keywords process, four tables drive the data dredging and validation: Keyword_Names, Keyword_Source, Telemetry_Location, and Telemetry_Conversions. See Appendix A below for a brief description.

World Coordinate System: Pointing an orbiting observatory is a well understood problem, but the accurate pointing of a particular aperture for a particular instrument requires some further processing. This task takes information from the database concerning the pointing of the vehicle and, using a table which relates the location of the apertures to the orientation of the observatory, converts the pointing to the standard FITS coordinate system parameters.

---

[2] http://www.gsfc.nasa.gov/astro/fits/fits_home.html

All information required by this process comes from processes upstream: the support values from the Support Schedule Keywords, and the telemetry values from the Data Validation step. In addition the table describing the aperture locations specified by the instrument engineers is accessible in the Aperture_Location table.

Generic Conversion: This process puts it all together, unscrambling the input data, potentially Doppler-shifting the photon events, and constructing FITS files with the appropriate keywords describing the data.

Generic Conversion is the process which converts the unformatted data into FITS files. This involves writing the raw science FITS files as well as the "support" FITS files from the data contained in the dataset: the science packet file, the data quality packet file (if any), the support values, the telemetry values, and the World Coordinate values.

In addition to needing the science packet files and the data quality packet files, this process requires the support values from the Support Schedule Keywords, the telemetry values from the Data Validation step, and the World Coordinate System values. Additionally the following tables drive the system: Keyword_Names, Keyword_Source, Keyword_Order, and Keyword_Rules.

Data Collector: Besides the six basic pipeline steps, a seventh package has been developed to control the flow of exposures through the pipeline. Some calibrations require that a series of exposures be associated with a separate calibration exposure, such as a wavelength calibration. The Data Collector pauses the processing of individual members of an "association" until all members are present in the pipeline.

Appendix A. Drivers

A system which claims to be table-driven obviously requires a number of tables to drive the system. These tables contain the detail which is specific to a particular mission/experiment. Providing the specifications in these tables limits the amount of mission-specific modifications that would otherwise be required.

Telemetry_Patterns: This is a table which describes a hierarchy of telemetry headers, and their relative position and size. Currently the hierarchy is limited to File/Image/Packet/Segment, however extensions are certainly possible where warranted. This table also describes how to distinguish each telemetry stream by looking at the first "few" bytes of the stream.

Telemetry_Location: Within a telemetry stream—especially within the segment, packet, image, and file headers—are words of special interest when interpreting the stream, and interpreting the science. This table describes the location of each mnemonic, or telemetry item, giving its position in terms of byte offsets and bit offsets. A restriction is that each mnemonic must consist of contiguous bits.

Telemetry_Conversions: This table contains parameters for the three types of telemetry conversions that are accommodated within the OPUS system. Discrete conversion assumes the telemetry mnemonic is an index to a table of strings; it is useful for converting a status monitor to "On" or "Off," for example. Piecewise linear conversion is done for simple temperature or voltages. And polynomial conversion is used for more complicated parameters.

Keyword_Names: This table contains the set of keywords for each instrument and for each mode, and information regarding their datatypes, order in the header, default values, and one-line descriptions. This is the essential information required to build the science headers. More complete descriptions of the keywords should reside in an independently maintained keyword database.

Keyword_Source: This table specifies the source of the keyword value. This may be a database relation and field, or a telemetry mnemonic and conversion mnemonic. This table eliminates the need to hard-code database queries, allowing the OPUS system to be more dynamic and more customizable.

Keyword_Order: Keyword Order provides information to enable Generic Conversion to write the header keywords in the correct order in the FITS files. A keyword can only occur once for a particular instrument, but the order of keywords may be different depending on the mode of the exposure. A spectrographic exposure will require a different set of keywords than will a simple image or a more complex time-tagged event product. This table completes the design of the FITS file headers.

Keyword_Rules: Consistent with the OPUS goal of reducing hard-coded algorithms and making the system more table driven, many of the keyword values are derived from others which have already been determined. The keyword rules table employs a simple rule-based parser which allows the value of these keywords to be modified without writing any code. However, complex algorithms to populate a small percentage of keyword values must still be hard-coded.

References

Rose, J., Choo, T. H., & Rose, M. A. 1996, in ASP Conf. Ser., Vol. 101, Astronomical Data Analysis Software and Systems V, ed. G. H. Jacoby & J. Barnes (San Francisco: ASP), 311

Rose, J., et al. 1995, in ASP Conf. Ser., Vol. 77, Astronomical Data Analysis Software and Systems IV, ed. R. A. Shaw, H. E. Payne, & J. J. E. Hayes (San Francisco: ASP), 429

Nii, H. P. 1989, in Blackboard Architectures and Applications, ed. V. Jagannathan, R. Dodhiawala, & L. Baum (San Diego, CA: Academic Press), xix

Astronomical Data Analysis Software and Systems VI
ASP Conference Series, Vol. 125, 1997
Gareth Hunt and H. E. Payne, eds.

The OPUS Pipeline Toolkits

C. Boyer, T. H. Choo

Space Telescope Science Institute, 3700 San Martin Dr., Baltimore, MD 21218

Abstract. The OPUS pipeline, which employs a blackboard architecture, has been processing Hubble Space Telescope (HST) data for nearly a year. OPUS was designed for both reusability and extensibility, as well as portability to different platforms and projects. OPUS contains a toolkit of resource files and programs which provide the users with the ability to customize their own pipeline. ASCII resource files can be used to define the configuration of the system, and to add processes to the pipeline dynamically. The OPUS callable routines provide applications with even more flexible methods for interfacing with the OPUS blackboard. This paper will discuss how the OPUS toolkit—both the resource files and the software libraries—is used to configure an OPUS data processing pipeline.

1. Introduction

The purpose of the OPUS toolkit is to provide a mechanism for processes to be easily incorporated in the OPUS data processing pipeline (Rose et al. 1994). The OPUS toolkit contains software built upon the OPUS resource file concept, which allows the user to tailor the pipeline without any software changes. The toolkit contains C callable routines that provide access to the pipeline's blackboards and the OPUS resource files, GUI applications for monitoring and managing the blackboards, and an OPUS shell that can be used to dynamically add a third party software application to the OPUS pipeline.

Currently three types of resource files—the pipeline stage file, the process resource file, and the path—file are supported. The pipeline stage file defines the processing steps, or "stages," available in the pipeline. Each processing stage is described in a process resource file. This file contains the specific behavior, and input/outputs, of a processing step. Finally the path file describes the data flow of the pipeline: directories where data are stored and obtained, and global pipeline information.

Resource files allow the user to customize the scope of the pipeline, process specific attributes, and disk management, using only a text editor. It is possible to create pipelines that feed data to other pipelines, pipelines that merge or separate based on the data being processed, and pipelines that distribute the finished data product based on some characteristic of the data. The OPUS toolkit is built on both the existence of these resource files and access to the blackboard.

The OPUS system employs two blackboards to support the communication between pipeline and operators. One blackboard is used for processes activities, and the other is used to track the progress of observations in the pipeline. The

C callable toolkit contains two packages (the Process Status Flag (PSF) package and Observation Status Flag (OSF) package) that allow an application to interface directly with these blackboards. The C callable toolkit also contains a resource package that is used to manage the OPUS resource files, needed to support the OPUS distributed pipeline.

The GUI applications that come with the toolkit are the Process Status Manager and Observation Status Manager. The Process Manager (PMG) is used to monitor the PSF blackboard and to send command messages to processes. The OMG (OSF Manager), on the other hand, is used to monitor and manage the OSF blackboards. Both the PMG and OMG are discussed in detail in Rose et al. (1995).

2. Process Status Flag (PSTAT) Package

Each process in the OPUS pipeline is assigned a PSF message on process initialization. The PSTAT package provides an interface to the PSF message on the blackboard.

The PSF message consists of the components, PID, Process name, Activity status, Start Time, Reinitialized Time, Path name, CPU, and Command. The PID is set at process initialization with the process ID assigned by the operating system. The process name is the name of the process, and it is the same as the process resource file name. The activity status is updated by the process to reflect the current activity, and it could be set to, for example, INITIALIZING, SUSPENDED, WORKING, or Observation name. The start time is the process start time. The reinitialized time is time when the process last reinitialized itself. The path name is the name of the path that the process runs on. The CPU is the name of the CPU which executes the process. The command is the current user directive. This field is set by user, and it may be set to HALT, SUSPEND, RESUME, or REINITIALIZE. The process then reacts to this command, and performs the appropriate action.

Accessing the PSF message package must be done through the PSTAT package. The PSTAT package contains functions that allow creation, modification, deletion, and searches of PSF messages on the blackboard. As multiple PSF messages are updated, created, and removed by processes and users, message contention can become an issue. The PSTAT package internally resolves the message contentions on the blackboard.

3. Observation Status Flag (OSF) Package

The Observation Status Flag (OSF) package provides an interface to the OSF blackboard. Each exposure that enters the pipeline is assigned a unique OSF which is used to track the progress of observation processing.

An OSF message consist of the components, Start time, Status, Dataset name, Data ID, DCF, and Command. The Start time is the OSF creation time. The Status shows the progress of an observation through the pipeline stages. The field can be retrieved as a whole or broken down to derived stages in the pipeline.stage. As an observation proceeds down a pipeline, different stages of the status field can be updated by various processes. The Dataset name is the name of the observation. The Data Class assigns an OSF to a class of data. The

DCF field is a HST specific number and refers to the number assigned by DCF to the exposure. The Command is the current user directive. For example, the user may set this field to "HALT," halting an exposure from further processing. Processing is restored via the "RESUME" directive.

The OSF package also contains functions that allow creation, modification, deletion, and searches of OSF message on the blackboard. Since OPUS is a distributed and parallel processing pipeline, multiple processes may update the same OSF message at the same time. The OSF package resolves the simultaneous accessing of the same OSF message, and encapsulates the message contention from the application.

4. Process Resource File (PRSC) Package

The PRSC package provides an interface to the OPUS resource files. The OPUS process resource file defines the specific behavior of a process without regard to the rest of the processes in the pipeline. All processes are run in a path defined by a path resource file. Resources in the process resource file can be superseded by the resources in the path file. This allows the user to enforce specific behaviors for all processes selected to run in the same path.

The PRSC package contains routines that perform the symbol substitution, and resource hierarchy superseding, and maintain the resources in a dynamic structure that can be easily retrieved by the application at run time. An example of symbol substitution, and resource superseding performed by the PRSC package is shown below. These are the values in path and resource files:

`RED.PATH` File	`INGEST.RESOURCE` File
`*.RETRY = 5`	`RETRY = 1`
`LEVEL = LEVEL1`	`UPDATE = LEVEL`
`INGEST.LEVEL = LEVEL5`	
	`DATA_DIR = disk$scratch:[in]`
`HRS_DIR_1 = disk$sratch:[red]`	`IN_DIR = HRS_DIR_1`

Keyword values from the PRSC package for the INGEST process are:

`RETRY = 5`	The `RETRY` value specified in the `INGEST.RESOURCE` file, is superseded by the value in `RED.PATH`
`LEVEL = LEVEL5`	The symbol `LEVEL` is substituted with value of `ST.LEVEL` specified in the `RED.PATH`
`IN_DIR = disk$scratch:[red]`	The symbol, `HRS_DIR_1` is substituted `with disk$scratch:[red]` from `RED.PATH` file.
`DATA_DIR = disk$scratch:[in]`	The `DATA_DIR` value in `INGEST.RESOURCE` is not substituted or overriden by any path file values.

5. OPUS Shell

OPUS Shell is an executable that allows third party applications that have no knowledge about the OPUS environment to run in the pipeline. The third party software can be a VMS DCL script, IRAF script, or an executable. The OPUS shell manages both the blackboards and the resource files on behalf of the third party software. A process is added to the pipeline by making an addition to the pipeline.stage file and creating a process resource file. The pipeline.stage file assigns a character of the OSF's status column to track process activities. The process resource file defines the type of OPUS event that triggers the OPUS shell into action. Currently OPUS supports an OSF event, a time event, file event, and a command event. A single process can be made to respond to multiple types and instances of OPUS events.

An OSF event is the presence of an OSF message of a certain pattern on the blackboard, and the message pattern is described in the process resource file. When the OPUS shell encounters the OSF event, it will update the PSF message Status field with the observation dataset name, and then execute the third party software. On completion of the third party software, the OPUS shell will reset the PSF's Status field to IDLE and update the OSF message to reflect the processing result.

A File Object event is defined as the presence of a file whose file name pattern and directory are defined in the process resource file. Whenever there is a File Object event, OPUS shell will first update the PSF's Status field with the name of the file object before executing the third party software. On completion, it will reset the PSF's Status field to IDLE.

OPUS shell can also be configured to execute the third party software based on a time interval, and this is called a time event. The interval between execution can be set in the process resource file.

A command event is defined as the presence of a non-blank string in the PSF's Command field. The OPUS shell has been configured to recognize the following predefined commands: SUSPEND, HALT, and RESUME. When a HALT command is detected, the OPUS Shell terminates the process. If a SUSPEND command is detected, OPUS shell will update the PSF's Status field to SUSPENDED, and the OPUS shell remains in a sleep mode until a RESUME or HALT command is detected.

References

Rose, J., Choo, T. H., Rose, M. A. 1996, in ASP Conf. Ser., Vol. 101, Astronomical Data Analysis Software and Systems V, ed. G. H. Jacoby & J. Barnes (San Francisco: ASP), 311

Rose, J., et al. 1995, in ASP Conf. Ser., Vol. 77, Astronomical Data Analysis Software and Systems IV, ed. R. A. Shaw, H. E. Payne, & J. J. E. Hayes (San Francisco: ASP), 429

Astronomical Data Analysis Software and Systems VI
ASP Conference Series, Vol. 125, 1997
Gareth Hunt and H. E. Payne, eds.

Multiwave Continuum Data Reduction at RATAN-600

O. V. Verkhodanov

Special Astrophysical Observatory, Nizhnij Arkhyz, Karachaj-Cherkessia, Russia, 357147

Abstract. The RATAN-600 radio telescope allows us to carry out simultaneous multi-frequency observations using wide-band radiometers. Thus, we can account for the influence of the atmosphere using high frequency observations and the Galactic background using low frequencies, or observe instantaneous spectra of radio sources. Using these features a flexible astronomical data processing system (FADPS), operating on different UNIX platforms, was created. Methods of the FADPS construction are described. The system supports Gauss-fit analysis (to estimate parameters of sources and to decompose them into individual components), non-linear smoothing and averaging, background calculation and subtraction, interpolation, convolution, visualization using X-Windows and PostScript, and reading and writing of data in FITS format. FADPS commands may be executed directly in UNIX, in scripts, or using a special shell for our data processing system in the X-Window system. FADPS also operates on spectra of radio sources, recorded as FITS TABLE data. The system allows us to fit a spectrum by a set of curves with weighted spectral points, and to calculate spectral indices and fluxes. This system is connected to the CATS database, operated at the RATAN-600. Different stages of data reduction and results are shown in figures.

1. Introduction

The RATAN-600 (Parijskij & Korolkov 1986) is a transit radio telescope with a variable profile allowing simultaneous multi-frequency observations using different feed horns. At present, the wide-band radiometers at 0.968, 2.31, 3.95, 7.69, 11.1, and 22 GHz are mounted on feeder cabin No. 1. The large number of frequencies allows observations of instantaneous spectra of radio sources. A high sensitivity to extended sources permits deep sky surveys, detection of new radio sources, and studies of the fluctuations of the microwave background radiation. At the higher frequencies (11.1, 22 GHz) we may account for the influence of the atmosphere, and at lower frequencies (0.968 GHz) we may obtain the Galactic background level, and clean the data at intermediate frequencies (Parijskij & Korolkov 1986).

The result of a single observation at the RATAN-600 from one receiver is a one-dimensional scan of the sky. This data vector contains points of measured antenna temperatures of the sky with a constant registration interval in sidereal time.

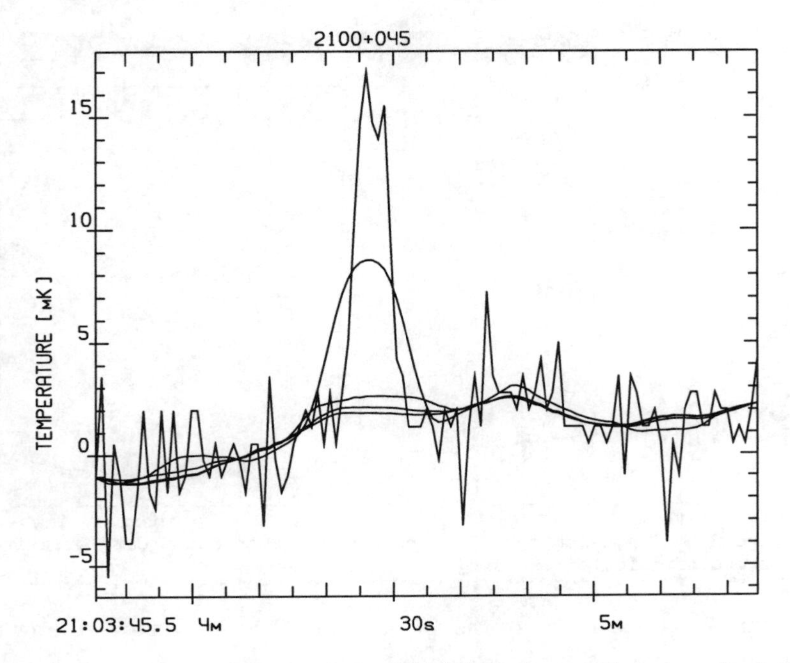

Figure 1. Four iterations of the FADPS background curve calculation.

Based on these features we have created a "Flexible Astronomical Data Processing System" (FADPS, Verkhodanov et al. 1993), operating under different flavors of UNIX (Linux, Solaris, etc.).

2. Data Reduction

The usual stages of the RATAN-600 data reduction are the following. After observation, the registration program automatically removes interference spikes by resampling the raw data using robust methods (Shergin et al. 1997). The user may then further smooth the data using the same algorithms to estimate the low noise component or the background (Figure 1). After subtraction of the background from the primary data, Gauss-fit analysis may be used to find components of the source and estimate their parameters.

3. Realization

Our approach in the construction of FADPS follows the main principles of operation of programs in a data processing system and the rules of operation in UNIX.

The command languages of UNIX (Bourne shell and C shell) are used as the command languages of our system. These languages have loop, condition, and local transition operators. The process of data reduction is a consecutive application of operators (commands) to the user's data recorded as equidistant

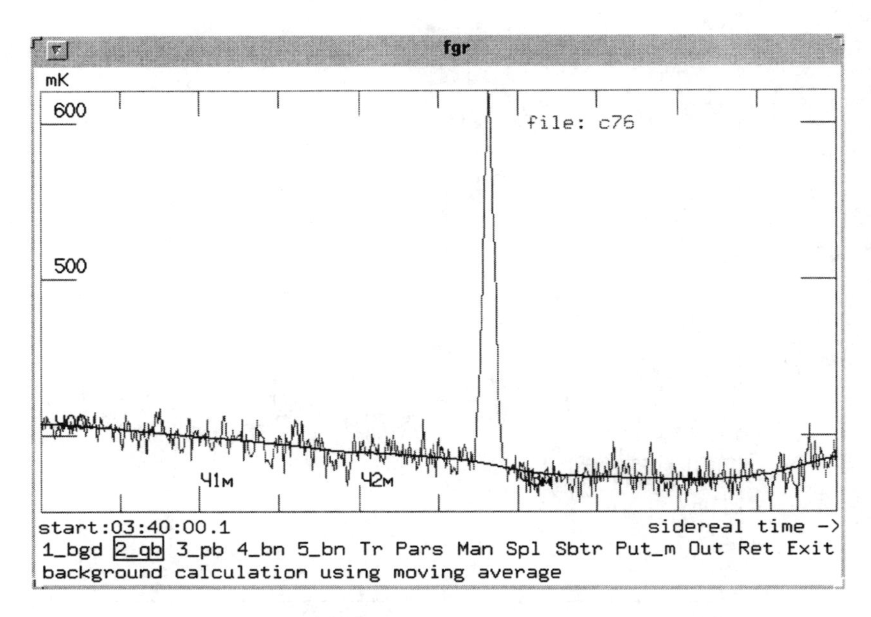

Figure 2. The screen of the *fgr* program with the menu and drawn data.

measurements and stored in separate files. Each operator corresponds to a specific command called a "module." Using a set of such modules and a basic knowledge of UNIX the user may construct his/her own data processing system or simply type a sequence of modules to obtain the desired result. The command modules can read data from standard input and write data to standard output and may be used as filters in UNIX pipes. Let us consider some vector containing observational data of a radio source. To obtain the parameters of a source we have to fit a Gaussian to our record. The standard analysis consists of the consecutive application of two operators: the background subtraction operator F_{bgd} and the Gauss-fit operator F_{gauss}:

$$\mathbf{R} = F_{gauss}(F_{bgd}(\mathbf{D})),$$

where **D** is the initial vector and **R** is the resulting vector. Under UNIX the application of both operators corresponds to the following command string:

```
bgd  -w  10  <  D  |  gauss  -n  0.020  >  R,
```

where *bgd* is the program for background subtraction, *gauss* is the program for Gaussian fitting, the key "`-w`" defines that the subsequent parameter is the size of the window for smoothing, the key "`-n`" defines that the subsequent parameter is the lower amplitude level of the Gauss-fit.

4. Basic Functions

The system includes the commands to perform the following: Gauss-fit analysis (to estimate the parameters of sources and to decompose them into separate components), fitting of tabular beam patterns, non-linear smoothing and averaging on the basis of robust algorithms (Shergin et al. 1997), background calculation and subtraction, interpolation, convolution, Fourier cleaning, visualization in X-Windows or in PostScript, and reading and writing files in FITS and FITS BINARY TABLE format.

These commands may be executed as UNIX scripts or using our own graphics shell for our data processing system in the X-Window system. The shell scripts may be used in ASCII terminal mode. The interactive mode of the FADPS is available in the graphics program *fgr* (Files GRaphics), where a user may select required operations from a menu and see the result of the execution in the X-Windows screen of *fgr* (Figure 2). *fgr* also permits the execution of external users' filters.

FADPS also operates with spectra of radio sources recorded as FITS TABLEs. The system supports fitting a spectrum with a set of curves using different weights for different spectral points. It calculates spectral indices and fluxes at any desired frequency. The graphics program *spg* (SPectra Graphics) operates with these functions (see Figure 2 in Verkhodanov et al. 1997). The *spg* menu provides a least-squares fit of a spectrum with several curves. The choice among these curves is either automatic or via mouse clicks. The mouse also allows a manual fitting (when the cursor delineates the curve) and the deletion of unreliable or unwanted data points (or attaching a zero weight to them). Data points may be weighted in different ways: using equal weights, weights by flux density errors, or on a point-by-point basis. This system is connected with the CATS database of radio catalogs (Verkhodanov et al. 1997) operated at the RATAN-600.

5. Epilogue

At present the FADPS is used at the RATAN-600 and in some others Russian institutes as the system to operate on one-dimensional vectors of data recorded in FITS format. An additional Tcl/Tk interface over the FADPS modular structure is being developed. The software is written in the algorithmic language "C" and may be freely distributed.

Acknowledgments. This work was supported partly by grant No 96-07-89075 of the Russian Foundation for Basic Research. The author thanks ISF-LOGOVAZ Foundation for the travel grant, the SOC for the financial aid in the living expenses and the LOC for the hospitality. The author is thankful for H. Andernach for critical notes on this paper.

References

Parijskij, Y. N., & Korolkov, D. V. 1986, Ap.Sp.Sci.Rev., Ser. D 5, 40

Shergin, V. S., Verkhodanov, O. V., Chernenkov, V. N., Erukhimov, B. L., & Gorokhov, V. L. 1997, this volume, 46

Verkhodanov, O. V., Erukhimov, B. L., Monosov, M. L., Chernenkov, V. N., & Shergin, V. S. 1993, Bull. of SAO, 36, 132.

Verkhodanov, O. V., Trushkin, S. A., Andernach, H., & Chernenkov, V. N. 1997, this volume, 322

Astronomical Data Analysis Software and Systems VI
ASP Conference Series, Vol. 125, 1997
Gareth Hunt and H. E. Payne, eds.

NOVIDAS and UVPROC II—Data Archive and Reduction System for Nobeyama Millimeter Array

T. Tsutsumi, K.-I. Morita

Nobeyama Radio Observatory, National Astronomical Observatory of Japan, Minamimaki, Minamisaku, Nagano 384-13, Japan

S. Umeyama

Surigiken Corporation, 1901-1 Ryuo-shinmachi, Ryuo-cho, Nakakoma, Yamanashi 400-01, Japan

Abstract. A workstation-based data archive and reduction software system is developed for the Nobeyama Millimeter Array (NMA). NOVIDAS is used for archiving the output data from the interferometer onto 8 mm tapes. It also provides a search utility for the archived database. The calibration and reduction of the data are handled by UVPROC II. It uses the *AIPS* file system via a set of interface modules called *vif library* developed by us in order to make an easier transition from the calibration processing to further imaging and analysis using *AIPS*. Basic functions of the UVPROC II are: correcting bandpass characteristics and time variations of complex gains, correction of variations in system temperature due to atmosphere, subtraction of continuum from spectral data, data flagging, and flux scaling.

A key concept for designing the software is that the system should be user-friendly. We adopted a GUI based on X-windows system with a menu-button feature for execution of commands and PGPLOT for graphical display. This allows users to examine the data quality visually and edit the data interactively. The *vif library* consists of simple FORTRAN programs to open/close and read/write the *AIPS* files. Therefore it can also be used by users to develop their own codes for manipulating the *AIPS* data without calling the *AIPS* routines directly.

The system currently runs on SUN-Solaris and will be ported to SGI-Irix with a disk array.

1. Introduction

Nobeyama Millimeter Array (NMA) is an interferometer consisting of six transportable 10 m antennas operated in the frequency bands of 100 GHz, 150 GHz, and 230 GHz by Nobeyama Radio Observatory (NRO). We recently upgraded the correlator system which consists of two types of digital spectroscopic correlators, FX (with 1024 channels, bandwidth of 32 MHz) and XF (128 channels, bandwidth of 1 GHz) correlators. The cross correlated signals from each pair of the array elements need to be calibrated to correct for instrumental and atmo-

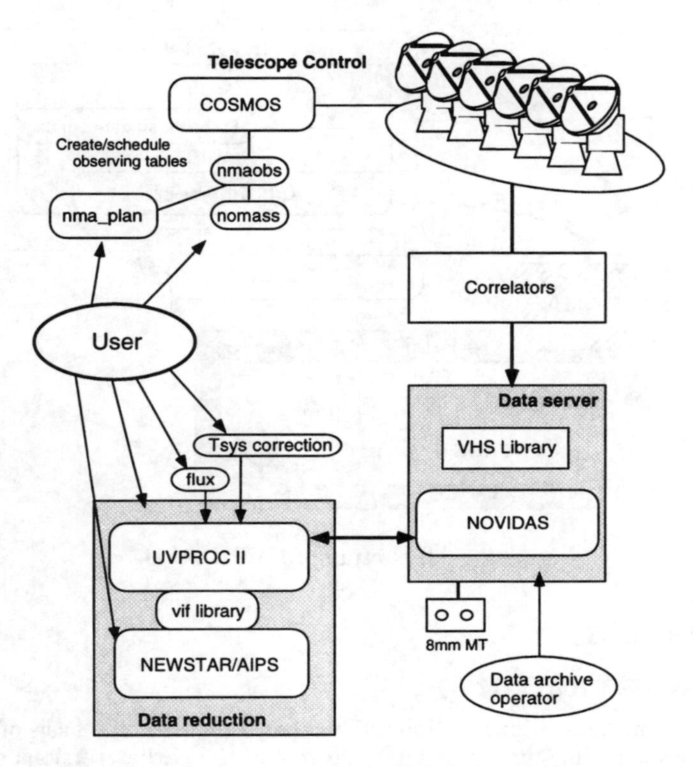

Figure 1. Software systems for NMA.

spheric effects and to be edited for bad data before applying a Fourier transform to produce images.

UVPROC II and NOVIDAS in relation to the other software systems for NMA are schematically described in Figure 1. Since we have the 45 m Telescope and the array at the same site, some of the software is shared between the two instruments.

For observation scheduling, users use *nma_plan* and *nomass* interactively. COSMOS is the telescope controlling system for both the 45 m and NMA. The output data from the correlators are written in the standard FITS format (random group FITS). A single FITS file is created for each source for each observing run. The data archived by NOVIDAS are calibrated using UVPROC II. The calibrated data are processed using NEWSTAR[1] for imaging and analysis. The core of NEWSTAR is the $\mathcal{AIPS}$ package. But it has been customized for the reduction of the 45 m single dish and NMA data by implementing X-windows-based GUI and by adding new tasks to be used for the single dish data.

[1]It stands for Nobeyama Engineering Workstation Systematic Tool for Astronomical Reduction. It should not be confused with NFRA's software known by the same acronym.

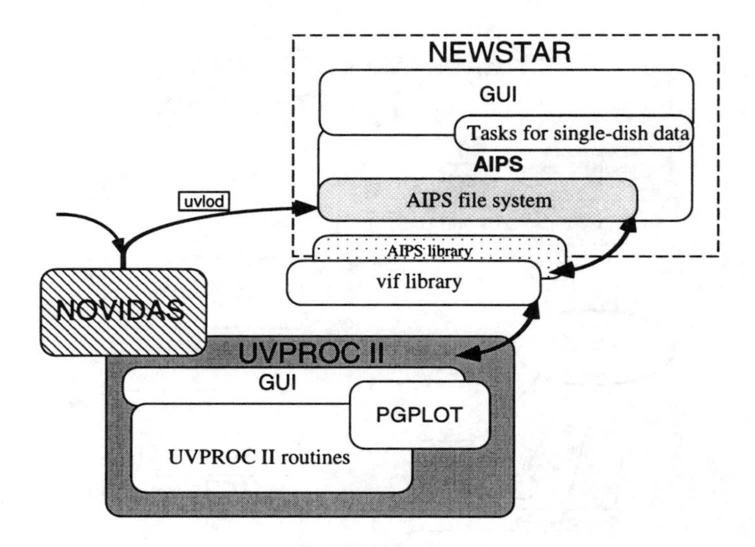

Figure 2. Structure of UVPROC II.

2. UVPROC II

2.1. Why UVPROC II?

As described in the previous section, *AIPS* has been used as a basis of the data reduction tool for imaging at NRAO. However, the earlier versions of *AIPS* did not have utilities for calibration and editing of visibility data. Also, writing new programs and implementing in *AIPS* was difficult for those not familiar with the system.

UVPROC, which was the predecessor of the UVPROC II, was developed at NRO as the calibration software for the NMA data. It was designed to run on IBM compatible main-frame computers. Lack of a GUI and use of its own file system made it somewhat difficult to use. UVPROC II was developed to replace its predecessor with a full GUI support based on X-windows system and use of *AIPS* file system.

The basic structure of UVPROC II is shown in Figure 2. *AIPS* does all the file management for the data used in UVPROC II. Thus for users, UVPROC II appears as a subsystem of *AIPS*.

2.2. Key Features

UVPROC II runs on SUN-Solaris platform. An example of UVPROC II windows is shown in Figure 3. For graphics, PGPLOT was implemented with functions such as interactive flagging of data points and zooming in/out by defining a region by a mouse. To use the *AIPS* file system, the interface FORTRAN library (*vif library*) was developed. Some utilities included are standard calibration processes such as bandpass calibration and time-dependent gain calibration.

Because of the *vif library*, the addition of new tasks to UVPROC II becomes relatively easy. The library can also be used to access the *AIPS* data files outside UVPROC II.

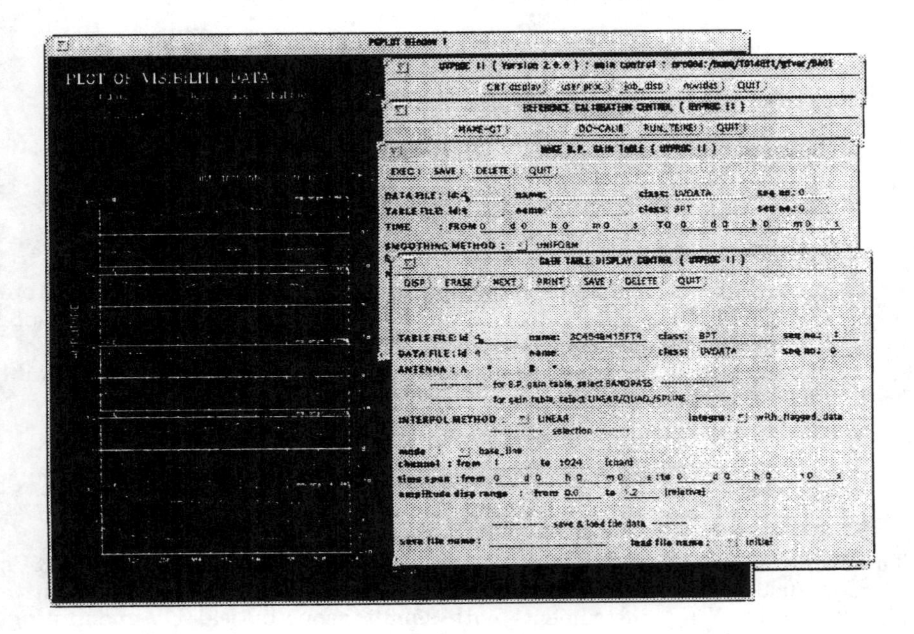

Figure 3. Example of UVPROC II windows.

3. NOVIDAS

NOVIDAS is an X-windows-based data archiving system for the NMA. The output data from the correlators transferred via an FDDI network to a data server (Sun Sparc-1000 compatible). The raw data are currently stored on 8 mm tapes but we plan to upgrade the data storage system to 1 TB VHS tape library. When NOVIDAS archives the data, the FITS headers are added to the database. NOVIDAS also provides a search utility for the archived database.

Core programs of NOVIDAS are installed on the data server. However the search utility (data search, downloading from the MT) can be accessed from individual user's workstation by starting up a new panel within UVPROC II.

4. Upgrade Plans

Documentation and on-line help need to be enhanced. Currently there are no history or log files to record what UVPROC II processes have applied to the data, thus this needs to be changed. Also, for better performance we plan to port UVPROC II to an SGI-IRIX disk array system.

Astronomical Data Analysis Software and Systems VI
ASP Conference Series, Vol. 125, 1997
Gareth Hunt and H. E. Payne, eds.

The XMM Survey Science Centre

Clive G. Page

Dept. of Physics & Astronomy, Leicester University, UK

Abstract. The XMM Survey Science Centre will process all data from the XMM X-ray observatory to produce standard data products for observers and catalogues of serendipitous sources, all of which will be included in the XMM science archive. This paper introduces the XMM mission and outlines the plans for software development.

1. The XMM Mission

The X-ray Multi-mirror Mission (XMM) is the second cornerstone in ESA's Horizon 2000 programme. The observatory is equipped with a set of sensitive high-resolution X-ray instruments with simultaneous optical/UV monitoring. XMM is due for launch in late 1999 with consumables designed to last 10 years.

The X-ray telescope consists of three modules each having 58 nested thin mirrors to provide a total collecting area of 4500 cm^2 and a field of view 30′ across. Each module consists of an X-ray CCD camera imaging the 0.2–10 keV band with a resolution of ≈10″ (FWHM); two modules are also fitted with an X-ray reflection grating spectrometer providing concurrent spectroscopy over the 0.35–2.5 keV band.

The Optical Monitor (OM) is a 30-cm Ritchey-Chretien telescope equipped with an image-intensified CCD detector, which should be able to detect stars of 24^m in a 1000 s exposure. More detailed descriptions of the XMM instruments are given in Lumb et. al. (1996) and the XMM Home Page.[1]

2. The XMM Survey Science Centre

XMM, besides providing excellent X-ray, optical, and UV data on the chosen target, will also detect large numbers (50–200) of other X-ray sources in each pointing. These serendipitous sources will constitute a deep sky survey which, over the course of the mission, will become a major resource offering potential for a wide range of astrophysical studies.

ESA has established the XMM Survey Science Centre (SSC) to ensure that all the scientific data are processed uniformly and systematically, and that the wealth of information from the serendipitous detections is properly catalogued and followed up for the benefit of the whole community.

---

[1] http://astro.estec.esa.nl/XMM/xmm.html

The SSC Consortium, led by Mike Watson of Leicester University, consists of eight astronomical institutes in the UK, France, and Germany.[2]

3. Survey Science Centre Tasks

Before launch the SSC is helping to design and build the science data analysis software in conjunction with ESA's XMM Science Operations Centre (SOC) at ESTEC. A common set of software modules will be used in the pipeline processing system to produce the standard products, and in the interactive analysis system which will be made available to guest observers.

After launch, the SSC will process all XMM observations and slew datasets (perhaps 600 MB/day) to produce source lists, images, spectra, and time series. The quality, reliability, and completeness of these pipeline products are of high importance, and the SSC team members will inspect and validate the results. The SOC will distribute the validated products together with raw data and calibration files to observers, and will add them to the XMM science archive.

Another major responsibility of the SSC is that of finding identifications for the serendipitous detections in the X-ray images, and of supporting a programme of follow-up observations on data which have entered the public domain. This is designed to complement, rather than compete with, individual observer's own programmes, and the results will be publicly accessible through the XMM science archive at the SOC.

4. Software Development Issues

The high spatial and spectral resolution of the XMM instruments and the availability of simultaneous optical/UV data, present many challenges to the design of a data analysis system. In the pipeline a predefined sequence of tasks must execute with high reliability, with all operations carefully logged, while the interactive system needs a user-friendly GUI-style front-end. The pipeline will run on a Sparc/Solaris platform, but portability is of great importance since the interactive analysis software must run on a wide range of platforms.

The existing astronomical software environments (IRAF, MIDAS, Starlink/ADAM, AIPS) provide a wealth of applications, but most are designed to handle optical or radio datasets rather than the photon-event lists which X-ray telescopes produce. Each environment has also invented its own file formats, none of them sufficiently flexible or widely-used to be an attractive prospect. Indeed, no single environment has attracted unequivocal support from the X-ray community.

For this reason, XMM software will be not be dependent on any environment (except the operating system) and will handle data stored in FITS files, principally FITS binary tables. Interfacing by FITS simplifies many aspects of the design, and also allows many other astronomical packages to be used easily, including FTOOLS, XSPEC, and SAOimage. We have some concerns over the

---

[2]Department of Physics & Astronomy, Leicester; Mullard Space Science Laboratory, University College London; Observatoire de Strasbourg; Service d'Astrophysique (CEA/DSM/DAPNIA), Saclay; CESR, Toulouse; Max-Planck Institut für Extraterrestrische Physik, Garching; Astrophysikalisches Institüt Potsdam; and the Institute of Astronomy, Cambridge

I/O overheads of the FITS format, but with the use of buffers and disc caches where appropriate, we expect these to be tolerable. XMM software is likely to make use of mature and well-supported libraries such as FITSIO, PGPLOT, and SLALIB. New software will be developed, where appropriate, under the general guidance of the ESA Software Standard, PSS-05. Special attention is being paid to software quality management and the software testing programme.

We foresee the need for a database management system at Leicester to manage the data flow through the pipeline, and there are similar requirements at Strasbourg and MSSL. We are investigating object-oriented and object-relational DBMS to see if these can handle XMM scientific data, as well as the management tasks. Commercial software licences will not, however, be needed by users of the interactive system.

The main processing pipeline is likely to contain at least 30 tasks and operate on up to 1000 raw data files each day. The recovery from failures in such a system can be very labour-intensive when conventional scripts and batch jobs are used. We are therefore investigating the use of an event-driven scheduler, such as the OPUS system used at STScI (Rose et al. 1996).

5. Programming Languages: C, C++, and FORTRAN

The SSC will use FORTRAN (which, all earlier forms being obsolete, now means FORTRAN 90) as the primary language for new scientific software. Since this goes against the current fashion, it may be appropriate to explain our reasons.

Object-oriented (OO) programming, it is often claimed, reduces development and maintenance costs particularly by promoting code re-use. The experiences reported in the 1995 ADASS meeting of those using OO programming in astronomy are, however, not wholly encouraging. See, for example, Glendenning (1996) on the AIPS++ project. Clearly, the up-front costs of switching to the OO paradigm can be substantial, especially in training the software development staff. Although C++ is the most widely used OO language, there are many who consider it to be by no means the best.

In my view, apart from the inheritance of methods, which is essential for OO programming, C++ has only two significant advantages over FORTRAN: a built-in exception-handling mechanism, and templates (which make it easier to create generic functions than in FORTRAN). For scientific programming, however, FORTRAN clearly has many substantial advantages over C++ (and indeed C):

1. FORTRAN arrays are first-class objects: this allows whole-array expressions and assignments, a compact array-section notation, and intrinsic functions which operate element-wise. C and C++ arrays turn into pointers when passed to a procedure, which is especially awkward when multi-dimensional arrays are involved.

2. FORTRAN has a better set of built-in data types, for example `logical` type distinct from `integer`, true character-string type (not just arrays of single `char`), and has identical facilities in single and double precision.

3. FORTRAN has better encapsulation of data and procedures in its modules than C++ has in classes, because there is no artificial split between code and header files. FORTRAN's `use` statement also provides better name-space management.

4. Dynamic memory can be allocated and accessed without using pointers in FORTRAN, where pointers are only needed to handle complex dynamic structures like trees and linked-lists, and are entirely type-safe.

5. FORTRAN allows user-defined operators, whereas in C++ one can only overload existing operators.

6. FORTRAN allows generic names to be used for procedures in a predictable and efficient manner since all names are resolved at compile-time.

7. FORTRAN programs are more portable than those in C++, because an ISO Standard exists and it carefully defines many features left as *implementation dependent* in the C++ standard, which is so far only a draft.

8. FORTRAN 90 provides many more features which allow errors to be detected at compile-time, for example argument `intent`, and explicit interfaces to procedures.

9. FORTRAN has a simpler syntax which makes it easier to write, and far fewer syntactic pitfalls than in C++, which it inherited from C.

10. FORTRAN programs usually execute faster because of built-in array features and powerful intrinsic functions, and because C/C++ pointers inhibit optimisation within loops. FORTRAN 90 also has better support for parallel or multi-processor architectures, a feature of increasing practical importance.

It is true that some of these deficiencies can be remedied by writing (or purchasing) a suitable class library, for example, one to handle arrays. But compilers are unlikely to generate code which is as efficient as if arrays had been built-in to the language. And if one makes use of higher-level class libraries from commercial or free-ware sources, they are unlikely to be based on the same array class, which can easily lead to code duplication, or other problems.

In choosing FORTRAN, we also took into account the need to make use of many existing libraries coded in FORTRAN 77, which are easier to access from FORTRAN 90 than from C++, and on the fact that good FORTRAN 90 compilers are now readily available for all major platforms, including PCs.

References

Glendenning, B. E. 1996, in ASP Conf. Ser., Vol. 101, Astronomical Data Analysis Software and Systems V, ed. G. H. Jacoby & J. Barnes (San Francisco: ASP), 271

Lumb, D., et. al. 1996,[3] X-ray Multi-mirror Mission—an overview, Proc. 1996 SPIE Conference

Rose, J., Choo, T. H., & Rose, M. A. 1996, in ASP Conf. Ser., Vol. 101, Astronomical Data Analysis Software and Systems V, ed. G. H. Jacoby & J. Barnes (San Francisco: ASP), 311

---

[3]ftp://astro.estec.esa.nl/pub/sciproj/xmm_mission.ps.gz

Astronomical Data Analysis Software and Systems VI
ASP Conference Series, Vol. 125, 1997
Gareth Hunt and H. E. Payne, eds.

The DRAO Export Software Package

L. A. Higgs, A. P. Hoffmann and A. G. Willis

Dominion Radio Astrophysical Observatory, National Research Council of Canada, Penticton, B.C., Canada

Abstract. Over the years, the computing staff at the Dominion Radio Astrophysical Observatory (DRAO) have developed an extensive package of software for the analysis of radio-astronomical data. Recently, the Observatory, in conjunction with a number of Canadian and foreign institutions, has embarked upon an extensive survey of radio emission from the Galactic Plane. This has required the conversion of the DRAO software environment for visibility and image processing into an exportable form that is relatively easily installed at foreign sites. This package is now freely available to the general astronomical community by anonymous ftp. Although the visibility-analysis portions of the package are of little use outside of DRAO, other portions are of general interest. The general philosophy behind the software and its scope are outlined.

1. Introduction

The DRAO export software package came into being not by deliberate plan, but as a result of users' needs—precipitated by the funding of the Galactic Plane Survey project. It therefore suffers somewhat from not being engineered from the beginning as exportable code!

Also, it differs somewhat from well-known radio-astronomy packages such as AIPS, GIPSY, MIDAS, or MIRIAD, in that it is not a self-contained, well-defined package. Rather, it is a suite of observatory application programs bound together by an underlying software environment. This environment is defined mainly by the way in which "header information" is ascribed to datasets.

The end result is a collection of application programs that operate on a wide variety of datasets, and that all have a similar "feel" to the user.

2. The DRAO Concept of a "Dataset"

In the DRAO convention, a "dataset" is any one, two, or three-dimensional collection of data that is uniformly sampled, and that is contained in homogeneous binary files. The binary files can be 16-bit integer, 32-bit integer, or 32-bit floating point. By convention, one and two-dimensional data are contained in single files, and three-dimensional data are contained in an ensemble of files where each file corresponds to a given Z coordinate. Therefore, any data that are stored in this format can be accessed directly by the DRAO software without any format conversion.

```
MADR Version 3.7, November 1995
?> s f1

Definition (FILE) no.:   1
  File name:              CYGNUSX.74CM
  Data type:              R
  Sizes: E = 512( 1, 512, 1): R = 512( 1, 512, 1)
  Effective sizes:        E  512   R  512
  No-data value:          .00000E+00
  Units:                  .01 K (Tb)
  File type:              RD                    Sky projection:    S (1950)

  Coordinates:                    R.A.                   DEC
  Reference coord.:       x: 20H 30M 0.00S      y: 40D 30' 0.0"
  Reference pixel:        x: 257.00 (pix)       y: 257.00 (pix)
  Delta coord.:           x: 1' 10.000"         y: 1' 10.000"

  Spatial res.:           5' 24.00"      x 3' 30.00"   at 90.000D
  Velocity res.:          .0000 km/s
  Central freq.:          408.00000 MHz         Stokes parameter:  I
  Map bandwidth:          4.0000 MHz            Observation epoch: 1988.00
  Last changed by: convreals          on 24-OCT-1991

  408-MHz image of Cygnus X
?> exit
```

Figure 1. A typical file definition for a 2-D dataset.

The parameters that are usually associated with a dataset are stored independently of the data, in a master-directory file known as a "file-definition file." Entries in this file can easily be created, modified, etc., by the DRAO software. Linkage between an entry in this file and a dataset is by file name(s). A separate file-definition file exists in each directory used by a user.

An example of a dataset definition (commonly known as a file definition) is shown in Figure 1. Here, astronomical parameters have been provided to describe a dataset with the file name "CYGNUSX.74CM".

3. The "File Header" System—The Binding Glue

The underlying "glue" that binds the DRAO software system together is the "file header" system—a library of routines that define, modify, or delete header information that is contained in "file definitions." These routines also allow applications programmers to specify what types of auxiliary information are required for particular applications, and to set defaults and limiting checks on these items of information. If a user runs an application program without sufficient specification of data parameters, the application will request the missing information from the user.

Accompanying the file-header library is a standard library of common subroutines used by all applications. Their use ensures that all applications have a similar "feel" to the end-user.

Most subroutines and applications programs are written in a FORTRAN-like preprocessor language, developed at DRAO by Geoff Croes many years ago. Known as FORCE, this language produces FORTRAN as output. The basic structure of the DRAO file-header and standard libraries was also developed by Geoff Croes, although there has been evolution over the years.

4. Two Basic Applications: MADR and PLOT

4.1. MADR

Although many application programs exist within the export package, two are of very general use. The **madr** program ("manipulation and data review") is a general-purpose program for quick-look analysis of data and simple data presentation. It references a very minimal set of data descriptive parameters and its use is therefore not restricted to astronomical data. Datasets can be manipulated by using only their file-definition numbers, and this allows a modest amount of array arithmetic, for example. Datasets can be "subsetted" for various operations, and data elements can easily be edited. The parameters that specify data display operations in **madr** are also stored in "definition files," so that operations can be recalled later to be used again.

Some of the data manipulation operations that can be done using **madr** are: data subsetting, data listing and editing, transposition, Fourier transforms, array arithmetic, collapsing along a principal axis, derivation of statistical parameters, contour and shaded displays, and ruled-surface and "strip chart" displays.

The **madr** program is very useful for the creation of datasets for testing purposes. For example, if dataset #1 is defined to be a 128×128 array, the following commands to the dataset **manipulator** in **madr** will create a test Gaussian of width 10 pixels, centered at (65,65):

f1 = (X−65)∗(X−65)+(Y−65)∗(Y−65)
f1m = EXP(−2.7726∗(**f1**/100))

(The two-step procedure is a result of limited internal buffering in **madr**). Similarly, one dataset, say with definition number 5, can be multiplied by another, say with definition number 6, by the command:

f5m = f5∗f6

The datasets can be 1-D, 2-D, or 3-D.

4.2. PLOT

A second program for dataset presentation is **plot**. Somewhat similar to **madr**, this program is designed to produce publication-quality diagrams. Like **madr**, it uses a modified PGPLOT as its graphics backbone. Diagrams are represented as the superposition of objects, each of which can be defined and the definition stored in definition files for future usage. Some of the basic objects that can be defined are: display boxes; astronomical grids; text boxes; contour, shading (both continuous and stepped), ruled-surface, vector and X/Y plot displays; wedge and pie graphs; plots of object positions; histograms; ellipses, arrows and lines; and polygons. Examples of **plot** displays can be found in Higgs et al. (1994), for example, where all the graphical figures were produced using this program.

5. Suite of Applications Programs

The DRAO export package includes some 80 application programs. These include programs for:

1. image generation from DRAO visibility data,
2. coordinate transformations,
3. display and analysis of continuum (2-D) images,
4. display and analysis of spectral data cubes (3-D),
5. display and analysis of 1-D spectra,
6. modeling of data,
7. input and output of data, and
8. several miscellaneous programs.

Most application programs are designed with two stages in their execution: an interactive definition stage in which program parameters are established, and an execution stage which may be spawned as a batch task. Many of the applications programs would be of general utility in most radio observatories, although those few that deal with DRAO visibility data would be of limited interest.

6. Visualization

Included in the DRAO export package are a few visualization tools that have been implemented in IDL. If a remote site does not have an IDL licence, these tools will not be available, but there will be no difficulties in installing the rest of the DRAO package. These tools include **imview** for 2-D images, **newcube** for 3-D datacubes, and **specview** for collections of 1-D spectra.

7. Distribution of Software

The export package is available by anonymous ftp, from **cygnus.drao.nrc.ca**, in the file *pub/nwexpsoft.tar.gz.* Installation instructions are given in the file *pub/drao_install_proc.* Installation is supported for IBM AIX, Sun OS, and SGI IRIX systems, and has been carried out on Sun Solaris and DEC Alpha systems. The current status of bug fixes, etc., is documented under *Status of DRAO Export Software* on the DRAO home page.[1] The exported tar file is about 2 MB, and the full, compiled DRAO system requires about 200 MB.

References

Higgs, L. A., Wendker, H. J., & Landecker, T. L. 1994, A&A, 291, 295

[1] http://www.drao.nrc.ca

Astronomical Data Analysis Software and Systems VI
ASP Conference Series, Vol. 125, 1997
Gareth Hunt and H. E. Payne, eds.

The SRON-HeaD Data Analysis System

C. P. de Vries

Space Research Organization Netherlands (SRON), Sorbonnelaan 2, 3584 CA Utrecht, Netherlands, E-mail: C.deVries@sron.ruu.nl

Abstract. A data analysis system has been developed at SRON, which has been designed to allow rigorous control of the quality of its processed data products. In order to fulfill this requirement, all data processing steps are recorded in a central database. The system will initially be used for analysis of *SAX*-WFC and *XMM*-RGS data at SRON.

1. Introduction

A critical requirement for data analysis systems used for massive routine data processing is that the system is able to deliver processed data of controlled quality in an automatic fashion. In order to control quality of data, it is necessary to have the ability to trace the heritage of all data products. This means recording all parameters of all steps which lead to the establishment of the final data products (Figure 1). To check processing status and initiate subsequent processing steps, one should easily be able to generate overviews of all available data and intermediary products based on data descriptions and processing heritage.

The SRON-HeaD (SRON High Energy Astrophysics department Data analysis) system has been developed to fulfill these requirements, and is based on earlier experiences with the *CGRO*-Comptel data analysis system (de Vries 1995).

2. Requirements

The following basic requirements were defined:

- Full traceability of data processing. Storage of all parameters of all processing steps, including complete software configuration.
- Complete catalogue of all available data products. Proper user-interface for manual processing and access to data descriptions and data heritage.
- Automatic processing based on data catalogue and processing status.
- Automatic archiving/retrieval of bulk data from mass storage devices.
- Use FITS format data files, where possible.
- Allow external analysis packages (e.g., FTOOLS, IDL, etc.) in the system.
- Separate "test environment" for testing of all system aspects and data processing programs. Capable of running on a variety of UNIX systems.

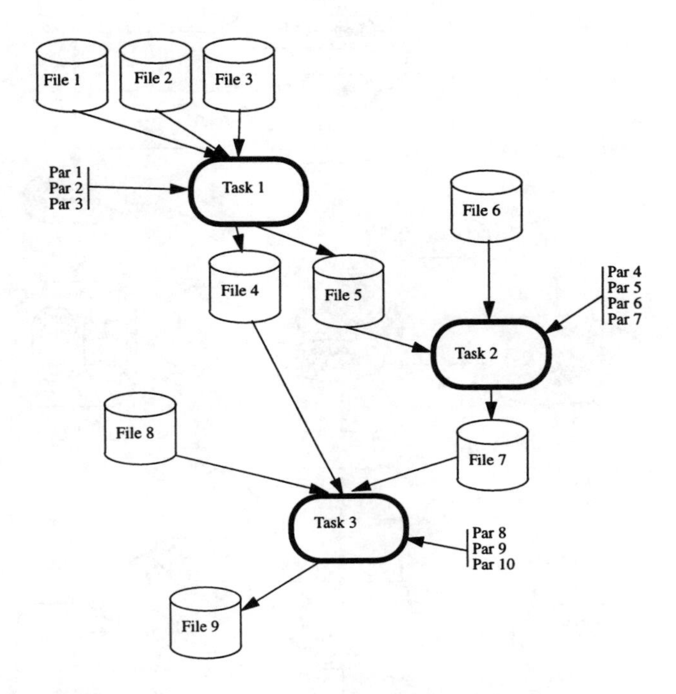

Figure 1. A processing pipeline is a sequence of tasks with several input/output files and parameters, which pass data from one task to the next. Heritage of any file (e.g., file 9) can be established by recording all input/output files and parameters of all steps and by uniquely identifying each task, requiring thorough configuration control.

3. Implementation

The core of the system is the recording of data descriptions and data heritage in the database, where this information may be queried via user interfaces or the routine processing pipeline to start new jobs (Figure 2). The dataset heritage consists of actual processing parameter values and the software configuration used. Since data processing parameters are available either in the FITS headers or from input parameter files used by the generating programs, these parameters can be recorded after actual data files have been created. This means that no connection to a database is required during data processing and that external packages can easily be incorporated into the system. In addition, externally generated (FITS) data files can easily be imported into the system, as well.

The basic processing module is a "task" or program executable, called from a script or "job," which may also call other tasks. The script defines the control flow and communication between tasks within a job. The job script may prompt the user for task parameter input. Actual parameters are passed to tasks via IRAF format parameter files. Automatic recording of processing parameters is done for each output file at the end of job processing.

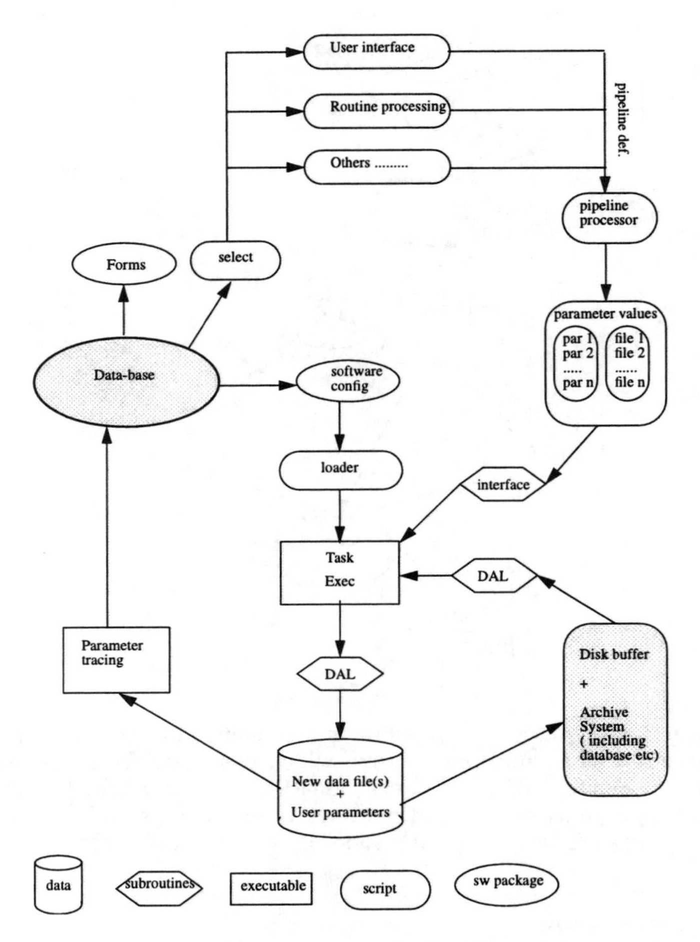

Figure 2. Top level functional breakdown of the HeaD system.

An Oracle client-server database architecture serves as the central database system (RDBMS), which holds the data catalogue and heritage, processing parameters description and software configuration. Software configuration at the system level is maintained through use of the RCS system, and use of this system is enforced by the appropriate user interfaces.

The Tcl/Tk package plus extensions (TclX,OraTcl) is used to create the windows-based user interface (Figure 3). On-line help is available by means of a Tcl/Tk HTML viewer. Direct selection of data files for input to the appropriate user interfaces can be made by selecting from lists resulting from the execution of SQL procedures. These can be taken from a library of procedures which base selections on a variety of data descriptions, heritage, or processing status. In addition, Oracle-Forms can be used for direct database queries.

Data processing "jobs" can be initiated manually by explicitly entering task parameters for specific tasks, or automatically through manual or automatic

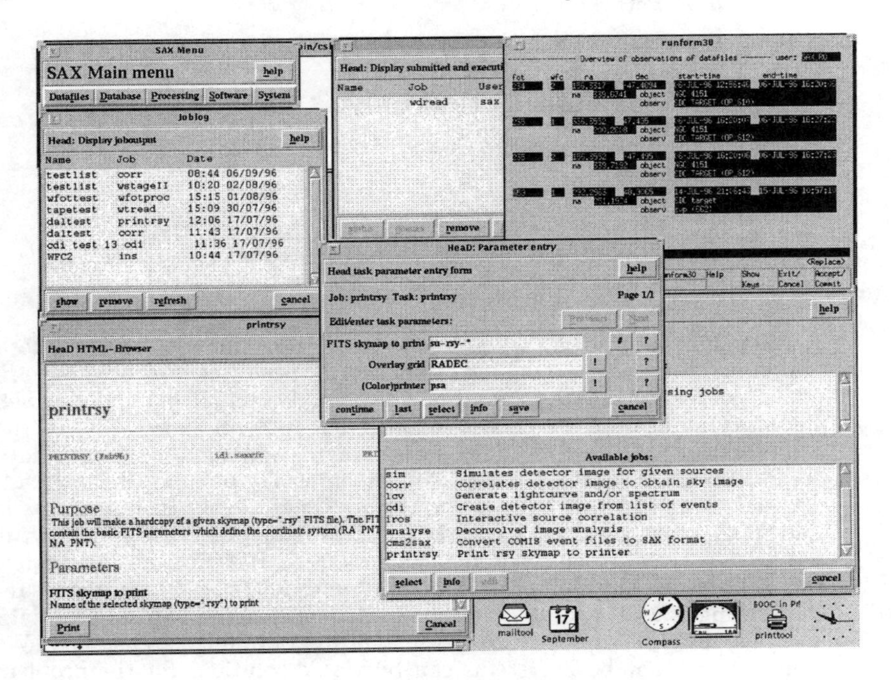

Figure 3. HeaD user interface.

marking of individual datasets or dataset types for further processing. In that case, special database tables define the processing flow.

A data access layer (DAL) is available which separates the actual scientific code from basic data I/O, allowing for greater system portability. This layer is partially composed of the FITSIO library, modified to allow for communication with the archive system, and specially developed routines.

Currently the system contains specially developed data processing programs as well as tasks taken from general packages like FTOOLS, IDL, SAOimage, etc. The system has been initially implemented on Sun (SunOs, Solaris) and HP (HP-UX) systems.

More information can be found on HEAD Home Page.[1]

References

de Vries, C. P. 1994, in ASP Conf. Ser., Vol. 61, Astronomical Data Analysis Software and Systems III, ed. D. R. Crabtree, R. J. Hanisch & J. Barnes (San Francisco: ASP), 399

[1] http://ws13.sron.ruu.nl:8080/head/Welcome.html

Astronomical Data Analysis Software and Systems VI
ASP Conference Series, Vol. 125, 1997
Gareth Hunt and H. E. Payne, eds.

Three-dimensional Data Analysis in IRAF and ZODIAC+

P. L. Shopbell

Astronomy Department, California Institute of Technology, Mail Code 105-24, Pasadena, CA. 91125

Abstract. Despite the widespread use of IRAF for astronomical data reduction, there are a number of special-purpose image processing applications for which the general-purpose IRAF system may not be well suited. One example of this is imaging Fabry-Perot data. Due to the complex three-dimensional data sets, many astronomers with such data employ image processing systems which can easily be extended with additional "home-grown" software routines. Chief among these systems is ZODIAC, a mathematical parser-based data reduction environment written by George Miyashiro of the University of Hawaii. In this paper, we examine the suitability of `IRAF 2.10` and `ZODIAC+ 1.0` (a recent restructuring of `ZODIAC 1.38`) to the problem of imaging Fabry-Perot data analysis, in terms of both ease of use and programming extensibility. We demonstrate that both systems can be viable solutions for the problem of reducing complex three-dimensional data sets.

1. Motivation

In the field of optical Fabry-Perot astronomy, many observers reduce their data with a little-known package called ZODIAC (Shopbell 1996). Originally written by George Miyashiro of the University of Hawaii, it is an excellent interactive environment for the manipulation of 0-, 1-, 2-, and 3-D data structures. It also has a relatively simple programming interface. As such, it is well-suited to the small community of Fabry-Perot astronomers, who have extended ZODIAC with special-purpose routines for reducing and analyzing large 3-D data sets.

The simplistic nature of ZODIAC, while a strength in terms of usability and extensibility, places limits upon one's data analysis, particularly in the areas of line and image graphics, networking, and image information handling. This has motivated the development of a Fabry-Perot package in IRAF (Bland-Hawthorn, Shopbell, & Cecil 1992). IRAF can provide the astronomer with a number of basic features not found in ZODIAC, such as a built-in graphics, as well as a large set of familiar tasks for 1-D and 2-D data analysis.

The purpose of this paper is to illustrate a few important differences between ZODIAC and IRAF, as have been encountered during development of the IRAF Fabry-Perot package. We have found that while IRAF is a superior platform from a user standpoint, there are a number of useful ZODIAC features which could be valuable in the IRAF environment.

2. Features of IRAF and ZODIAC

One must be careful when comparing software systems whose applications are broad in scope. While it may be feasible to evaluate two implementations of an FFT algorithm, for example, large packages such as IRAF are applied to a wide variety of problems in many fields. We must therefore take care to examine only generic issues and basic functionality, such as the user interface or line graphic support, and not implementation choices, such as the organization of tasks. A wide variety of approaches to the complex problems of astronomical image processing are often necessary due to constraints such as hardware, but such variety is also beneficial for the end user, providing alternatives for personal preference. The current proliferation of scripting packages (e.g., `Perl`, `python`, `tcl`) are evidence of this well-known software development concept.

2.1. User Interface

Both IRAF and ZODIAC are command-line based systems. Common user-level features include: (*i*) a command-line interpreter with environment variables, command history, expression parsing, optional command-line editing, (*ii*) images are stored as individual disk files, (*iii*) a simple syntax for accessing image sub-arrays, (*iv*) on-line task-specific help, and (*v*) extensibility through a simple scripting language. Features particular to IRAF include: (*i*) substantial use of environment variables (e.g., `stdimage`, `editor`, `graphcap`), (*ii*) organization of routines into package/task tree structure, (*iii*) image headers that maintain a data description and processing history, and (*iv*) a parameter system that provides a standard interface to all tasks. Features particular to ZODIAC include: (*i*) an interpreter with knowledge of mathematical expressions, for easy manipulation of data, regardless of dimension, (*ii*) a simple binary disk format for image data, and (*iii*) support for 0-D constants.

From a user point of view, IRAF is substantially more powerful and robust than ZODIAC. The presentation of each task in a standard parameter system and the organization of tasks into packages based upon function are extremely useful, given the large numbers of tasks in these systems. (However, search routines, such as the `APROPOS` task, are still necessary, as in `UNIX`.) The lack of image header information in ZODIAC is a major problem, especially as space-based observatories proliferate and ground-based instruments and telescopes become more complex. The ability of ZODIAC to mathematically manipulate multi-dimensional data (e.g., $a = b + c$, where a, b, and c have arbitrary dimensions) is much more intuitive than the `IMEXPR` and `IMARITH` tasks.

2.2. Programming Interface

Both IRAF and ZODIAC include support for the addition of user code through interpreted scripts and compiled programs. Common features include: (*i*) a scripting language with simple flow control (e.g., `for` and `while` loops), (*ii*) an image access library interface, and (*iii*) execution is permitted outside the interactive environment. Features particular to IRAF include: (*i*) a simple, but limited `IMFORT` interface, and (*ii*) a complex `SPP` interface with an extensive set of standard libraries. Features particular to ZODIAC include: (*i*) a simple language-independent interface, and (*ii*) independent integration of each task.

The simplicity of the ZODIAC API has led many Fabry-Perot astronomers to choose it over IRAF. In fact, the ZODIAC interface is only slightly more powerful than the IMFORT interface of IRAF, so external libraries must be used for many purposes, such as plotting (e.g., pgplot) and image display. While this reduces portability and uniformity of the user interface, the steep learning curve of SPP is often more daunting. The OpenIRAF initiative should alleviate this difficulty by eliminating the need to learn another language.

2.3. Fabry-Perot Data Support

Both IRAF and ZODIAC contain basic support for three-dimensional data handling: (*i*) multi-dimensional array support, (*ii*) no data structure size restrictions, and (*iii*) efficient 3-D data access. Features particular to IRAF include: (*i*) a large library of complex fitting and interpolation routines, (*ii*) well-integrated graphics, and (*iii*) multi-axis WCS support. Features particular to ZODIAC include: (*i*) a small number of interface layers to provide simple, fast data access, and (*ii*) extensive support for image masks.

Although the foundation for three-dimensional data analysis exists in IRAF, support is not universal at the application level. For example, IRAF has an excellent pixel masking interface (PMIO), but the standard tasks do not support masks—substantial re-coding would be required to do so. In contrast, ZODIAC contains full support for image masks throughout its tasks, but is lacking in well-integrated graphics and mathematical routines.

3. Summary

Both IRAF 2.10 and ZODIAC+ 1.0 contain features which facilitate the analysis of large three-dimensional data sets. Strengths of each system include:

IRAF	ZODIAC
powerful user environment	intuitive user interface
extensive portable libraries	simple software interface
large user base	fast development and execution

The primary weaknesses of these systems are common among current image processing packages: difficult debugging of external software, steep learning curves for users, etc. However, these weaknesses and others are being addressed, as we have seen at the ADASS conferences and in the FADS discussions. Future improvements to these two systems, such as the OpenIRAF initiative, increasing use of GUIs and image masks in IRAF, and adoption of image headers and "standard" libraries in ZODIAC, should allow both systems better to support the reduction of large three-dimensional datasets. Although the user's choice may still remain one of preference, we expect the more extensive popularity of IRAF to lead to its wider use in the area of Fabry-Perot astronomy.

References

Bland-Hawthorn, J., Shopbell, P. L., & Cecil, G. 1992, in ASP Conf. Ser., Vol. 25, Astronomical Data Analysis Software and Systems I, ed. D. M. Worrall, C. Biemesderfer, & J. Barnes (San Francisco: ASP), 393

Shopbell, P. L. 1996, Zodiac+ User's Manual, 1st ed.

Astronomical Data Analysis Software and Systems VI
ASP Conference Series, Vol. 125, 1997
Gareth Hunt and H. E. Payne, eds.

Interactive Data Analysis Environments BoF Session

Joseph Harrington and Paul E. Barrett

Codes 693 and 660, NASA Goddard Space Flight Center, Greenbelt, MD 20771-0001

Abstract. We conducted a discussion of interactive environments in which scientists handle data. Traditional astronomical packages (AIPS, IDL, IRAF, MIDAS, etc.) lack many modern language and interactive features, whereas modern interactive/scripting/rapid prototyping environments (Perl, Tcl, etc.) lack convenient and efficient numerical capability and a way to access existing astronomical code. Our focus was thus on how best to merge the capabilities of modern environments with astronomical data processing. To follow developments in this evolving field, we have set up a Web page that compares the available environments.

1. Introduction

Even with the data-volume explosion of modern astronomy, desktop computers can easily keep pace with the calculations we require of them. However, our workstations do not themselves know what calculations must be performed. Developing an analysis approach generally takes an individual astronomer much longer than the observations and calculations themselves. Reducing this bottleneck should therefore be a high priority. In recent years the community of programmers on the Internet has created interactive environments that accelerate the development of data handling software. Although these systems are often extremely powerful and flexible, they tend to have better support for textual data manipulation than for the numerical calculation and image display that astronomers need. Most astronomical environments, rich in task-specific packages of routines and capable of the necessary numerics, were written long before the advances in interactive environments and feel cumbersome when compared to the newer systems. The need for individual astronomers to handle larger and more complex data sets continues to grow, so it is time to bring the state of the art in interactive environments to astronomy.

We conducted a Birds-of-a-Feather (BoF) discussion about interactive environments for handling astronomical data. Representatives from environments both within and from outside of astronomy formed a panel that took audience questions. Panel members were Brian Glendenning for Glish/AIPS++, Wayne Landsman for IDL, Michael Fitzpatrick for IRAF, Arnold Rots for Khoros, Klaus Banse for MIDAS, Lee Rottler for Python, and Nicholas Elias for Tcl. We wanted to find out what the community wanted in an interactive environment, and how it felt about certain specific issues. We also wanted to inform the community of the possibilities now available with state-of-the-art systems outside of astronomy. The BoF ran for 90 minutes and maintained an attendance of about 50 participants, with some flux into and out of the room. The range of expertise in the audience was large, so the discussion was punctuated by a number of well-made points rather than an approach to consensus or disagreement. In this article we will focus on the points made during the discussion. Length restrictions preclude making a detailed case for a change in astronomical data environments or for discussing specific systems.

2. Context: Our Investment in Software

The task of bringing modern language and interactive capability to astronomical data analysis is difficult for two reasons. First, the developments in computing range from standardized command-line interfaces to network-transparent inter-process communication. We need to find out what capabilities are available, which are most important to researchers, and what the relative difficulties of implementation may be. Second, we cannot afford to abandon or manually convert the huge quantity of code we currently use. We will not likely agree on a single environment, yet we will want to be able to use each others' code more than we can now. Any solution we choose must not leave us in a position to suffer from the same problem again in the future.

Our discussion focuses on languages and interactive environments, but we must always keep a larger, second problem in mind: We desire a solution that accommodates the simultaneous use of multiple languages and allows for easy transition from one to the next as languages evolve. One possibility is the "Gaming Table" concept of Noordam & Deich (1995), Noordam (1997). Similar strategies are currently under consideration by several of the major astronomical packages, most notably AIPS++. Any system that is not capable of using code written in other languages (particularly compiled languages), or of being run by code written in other systems, is a bad, short-term investment.

3. Ideas from the Discussion

Key features of the interactive environment include command-line editors (standards like GNU Readline are not yet supported by the astronomical packages), real-time numerical efficiency, debuggers, and graphics display. Scripting language issues include friendly syntax, fast learning curve, good array notation, and access to compiled code and code written in other scripting languages.

Although they are different concepts, the issues of interactive environments and scripting languages are tied together. A script may have been prepared in advance and developed with a plan, or it may evolve as the user thinks of things and types them. Whereas a script file is read monotonically from beginning to end, an interactive user may need to abandon a train of thought, backtrack, correct errors, etc. Interactive users are more efficient when they can re-use portions of earlier commands, when they can call up information such as documentation, when they type to standard interfaces, and when their overall typing burden is reduced. The author of a non-interactive script (one contained in a file) can select the editing environment to suit personal taste and is usually much less concerned about extra typing than an interactive user. For example, many non-interactive languages (such as C or Perl) use semicolons to signal the end of a line. Signaling the end of an interactive line with anything more than a return is a nuisance. Likewise, non-interactive function calls often have the form `function(`*arg1*`, `*arg2*`, ...)`. The most common interactive environments (such as the shells) dispense with the parentheses and commas and just use spaces as delimiters. There are many reasons to desire the interactive and scripting languages to be the same, but the best interactive syntax may be very different from the best script syntax. Flexible systems, like IRAF cl, allow both styles.

Language syntax was a contentious area. Individuals preferred different styles and were unwilling to agree on a single language. Some were not willing to learn a new language unless there was overwhelming reason to do so. Some liked object-oriented languages, others thought that a waste of time, at least for scripting. Some were picky about the exact style of a language's syntax, others said they could adapt to anything without much cost.

A strong appeal of analysis languages like IDL and Python is the ability to manipulate arrays simply. If `a` and `b` are arrays, then `c=a*b` multiplies corresponding elements of `a` and `b` and `b=a[20:30,19:29]` extracts a sub-array from `a`. This "syntactic sugar" makes the code much easier to read and write than function calls or explicit loops. Furthermore, the operation runs at machine speed, making it practical to reduce large datasets using an interpreted language. However, complicated expressions like $a = b \times (c + d/2)^e$ are done as a series of simpler whole-array operations. Modern compilers can take advantage of machine instructions that optimize such expressions, passing the data through the CPU only once. We hope to see interpreters that do this soon.

Those who value rapid prototyping (quickly developing scripts that do a job well, then converting them to a compiled language that runs fast) prefer a language that is as close as possible to the ultimate compiled language, but that is interactive, doesn't have declarations, has conveniences like array operations in the syntax, etc. Some thought that researchers should want to write something quickly in a high-level system and have it run reasonably well without further work. They felt rapid prototyping was for production work, not research.

As soon as each new system arrives, people reinvent a lot of code, such as a FITS file processor. This is a foolish waste of time. We should be learning how to work between languages, and spend our valuable time writing new code.

The point was made that if a large project chooses a proprietary system for its basic calibration environment, it makes access to their project much more difficult and/or costly to data customers who do not have licenses for that particular product. Likewise, the project places its investment in coding at risk should the provider of that product cease to support it or cease to exist. ISO was cited as having committed this error in choosing IDL. Basing work on packages distributed under the GNU General Public License assures access for anyone.

No one (not even panelists) claimed to prefer the interactive environments of existing astronomical systems over the modern languages. However, the financial constraints on the programming groups (particularly IRAF) that support these packages are very tight and programming teams are small—usually much fewer than five people. Their home institutions generally hold the development of new application code for that institution as highest priority, regardless of the desires of the programming group and the community at large (even if grant money supports the project). Thus, a replacement of the IRAF cl is unlikely anytime soon, at least from NOAO. The command line interpreters for astronomical environments were designed for a special purpose, and not as general programming languages. One panelist suggested that given the opportunity to redesign the command line shell, they would choose a more modern design or even an existing general purpose language. This is the route chosen by the AIPS++ group, who use Glish.

Some groups are making their systems' core routines available in C or FORTRAN libraries. This would allow compiled code and interactive languages with dynamic linking capability to use those routines. However, it may be some time before all the applications of a given system are available in this way. Some limitations of the system designs, like storing arrays in disk files rather than memory, may make linking against these packages' routines from external programs less attractive to those who demand efficiency.

The major analysis environments strongly influence the work of many people, but rarely do their prospective users have the opportunity to influence the product during the critical design and development stages. These projects are community investments, and there is much expertise to be gained from the community. We encourage developers to hold open design reviews and to take seriously the comments they receive.

4. Conclusions

The list of desirable features is long, and we are told that modernizing old systems is unlikely. However, several of the new languages (including Glish, GUILE, Perl, and Python) have, or are gaining, the numerical features needed for data analysis. What remains to be provided are the means to connect languages to each other and to existing code, and we can expect to see some experimental systems here soon. In the mean time, we should develop our new code with an eye toward the future. Codes requiring top efficiency or guaranteed longevity are best written in compiled languages. Modularity will pay off because tools such as SWIG[1] will read application source code and automatically generate wrapper scripts that make the routines available to a number of interactive languages. This means the programmer need not write a main routine or graphical user interface, since the interactive languages provide them.

Because this is a fast-developing situation, we have created a WWW site[2] that compares many systems. Each has an entry in a feature comparison table, a textual description, and links to relevant sites. Several have programming examples that were prepared by the BoF panelists. We welcome additions and corrections.

References

Noordam, J. E., & Deich, W. T. 1996, in ASP Conf. Ser., Vol. 101, Astronomical Data Analysis Software and Systems V, ed. G. H. Jacoby & J. Barnes (San Francisco: ASP), 229

Noordam, J. 1997, this volume, 73

[1] http://www.cs.utah.edu/~Ebeazley/SWIG/swig.html

[2] http://lheawww.gsfc.nasa.gov/users/barrett/IDAE/table.1.html

Astronomical Data Analysis Software and Systems VI
ASP Conference Series, Vol. 125, 1997
Gareth Hunt and H. E. Payne, eds.

FADS II: The Future of Astronomical Data-analysis Systems BoF Session

J. E. Noordam

Netherlands Foundation for Research in Astronomy (NFRA), P.O.Box 2, 7990 AA, Dwingeloo, The Netherlands, E-mail: jnoordam@nfra.nl

Abstract. The discussion about the future of astronomical data analysis systems is essentially boring, but nevertheless essential. There is a large measure of agreement that the key word is "inter-operability," and that the future will look roughly like the Gaming Table of Figure 1. It is generally accepted that Distributed Objects will play a major role. But for inter-operability, "universal" binding mechanisms will be needed, along with agreed interfaces for things like images and tables. Despite the shining example of FITS, the astronomical community will probably not be able to generate and enforce those on its own. So, while waiting for standards like CORBA to emerge from industry over the next few years, projects like IRAF and AIPS++ are experimenting with the new concepts in various ways, and gathering experience. On the basis of this, they will be able to outline more detailed plans at FADS-III at ADASS VII.

1. Introduction

The series of yearly discussions about the Future of Astronomical Data analysis Software (FADS) was started in at ADASS V. A significant part of the success of FADS-I was due to a very lively e-mail discussion in the weeks before the conference. This caused participants to think about the issues beforehand, and arrive with well-formulated points of view. Unfortunately, the e-mail discussion before FADS-II was less lively, which had a clear impact on the actual BoF session. All e-mail messages can be perused on the FADS Web page.[1]

2. Summary of Discussions

The abstract basically says it all. Doug Tody (IRAF) and Brian Glendenning (AIPS++) gave short introductions, in which they professed to be in total agreement about the general structure of the future, and described how their respective projects were moving towards it. After this, the participants were invited to address the following list of issues: The four main elements of the Gaming Table (see Figure 1) with the emphasis on binding mechanisms, the role of Java (a language, not a binding mechanism), and likely scenarios of development.

---

[1] http://www.astro.umd.edu/~teuben/fads/

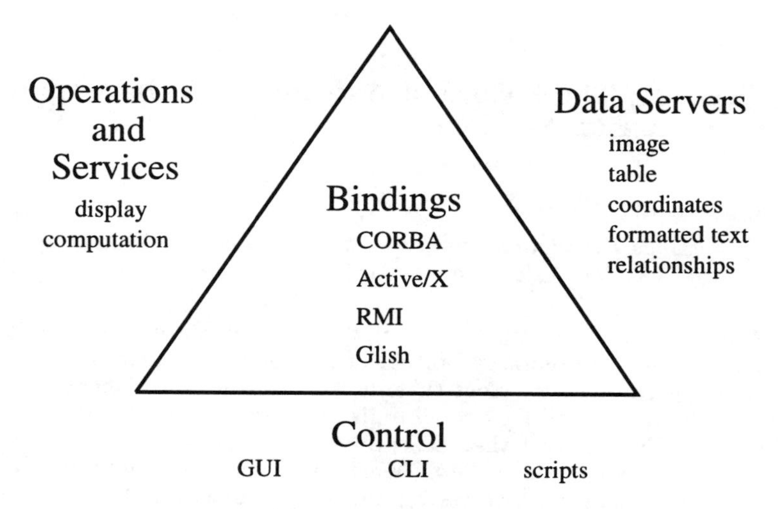

Figure 1. The Gaming Table.

There were frequent references to an earlier talk in the same conference by Don Wells about "Speculations on the future of FITS." This subject is all the more important for FADS since, even if a universal binding mechanism (CORBA?) emerges from industry, our community may still have to decide upon a common data interface (images, tables, etc.).

3. Conclusion

The FADS process will continue with FADS-III at ADASS VII. It is expected that the situation surrounding binding mechanisms and interfaces will have become a little more clear by then. We will try to incite a little more real debate by starting a few weeks earlier with the preliminary e-mail discussion, and by starting it off with a few stimulating propositions.

Acknowledgments. The author is grateful to Doug Tody and Brian Glendenning for giving short introductions, and for generally helping the FADS process with advice and comments. The Gaming Table concept was originally proposed at FADS-I by Jim Coggins. Will Deich set up the FADS e-mail exploder, and Peter Teuben manages the FADS Web page, and also made notes during the discussion itself.

References

Noordam, J. E., & Deich, W. T. 1996, in ASP Conf. Ser., Vol. 101, Astronomical Data Analysis Software and Systems V, ed. G. H. Jacoby & J. Barnes (San Francisco: ASP), 229

Wells, D. C. 1997, this volume, 257

Part 2. Science Software Applications

Astronomical Data Analysis Software and Systems VI
ASP Conference Series, Vol. 125, 1997
Gareth Hunt and H. E. Payne, eds.

Difmap: An Interactive Program for Synthesis Imaging

M. C. Shepherd

Owens Valley Radio Observatory, California Institute of Technology, Pasadena, CA 91125

Abstract. Difmap is a stand-alone program used by the radio-astronomy community to produce images from radio interferometers. It reads and writes the standard UV FITS file format produced by packages such as AIPS, and provides convenient ways to inspect, edit, and self-calibrate visibility data while incrementally building up a model of the sky. It has proven to be a popular alternative to larger packages that do much more—so part of this paper will focus on why this is so, and on whether the lessons learned could benefit future projects. Some time will also be spent describing the general structure of Difmap, particularly how being contained within a single-process is exploited to achieve speed, portability, and a degree of interactivity that is hard to match in larger systems.

1. Introduction

In June of 1992 I was asked to write a new program for the Caltech VLBI package. The new program would implement an iterative mapping technique called "difference mapping," but would rely on existing programs for data inspection, data editing, and image display. I didn't want users to have to continually switch between programs, so I decided instead to write an integrated difference mapping environment in which all of the functionality of the Caltech VLBI package would be incorporated within a single program. Three months later I had written the minimum quorum of commands needed to implement difference mapping, packaged under a scripting interface that I had written previously. The resulting program was called Difmap.

Difmap quickly gained users from the Caltech VLBI group, where it was initially used to augment the Caltech VLBI package. Meanwhile Difmap continued to subsume the functionality of the Caltech VLBI package, and in June of 1993—one year after the project started—I made the first external release of Difmap. Difmap quickly spread throughout the VLBI community, and later, after Difmap had been upgraded to support multi-dimensional AIPS UV-FITS files, it became a popular alternative to more general packages, such as AIPS.

No significant development has taken place since the release of the FITS version of Difmap in April 1995, owing to a reduction in NSF funding. However, Difmap has retained a strong following and has required very little maintenance.

2. Interferometry in a Nutshell

What follows is a brief introduction to radio interferometry. Many of the terms that are introduced here will be used throughout the rest of the text.

The essential steps needed to construct an image from signals received by a radio interferometer are:

1. Point an array of telescopes at a radio source.
2. Measure the complex wavefronts that arrive at each telescope with radio receivers.
3. Insert artificial delays in the signal paths of each telescope so that the telescopes all appear to be at the same effective distance from the source.
4. Interfere signals from pairs of telescopes and record the resulting complex fringe visibilities as a function of time.
5. Note that the visibilities from a given pair of telescopes sample spatial frequencies on the sky in proportion to the projected separation of the telescopes. Also note that this separation changes as the Earth turns, so each pair of telescopes samples a locus of spatial frequencies as a function of time. This is called *Earth rotation synthesis.*
6. Use the observations of a bright source to calibrate the visibilities.
7. Interpolate the visibilities onto a 2-D spatial-frequency plane called the *UV plane.*
8. Calculate the FFT of the UV plane. The resulting image is called a *dirty map*, and constitutes an image of the source corrupted by the point-spread function of the interferometer.
9. Form a model of the source through CLEAN deconvolution of the dirty map, or by fitting a model to the visibilities.
10. Convolve the model by an elliptical Gaussian approximation of the point spread function of the interferometer, and add the result to the un-modeled residual noise in the dirty map. This is called a *clean map.*
11. Publish the map.

3. What Does Difmap Do?

3.1. Data Inspection and Editing

A significant fraction of Difmap is devoted to providing users with convenient tools for interactive data inspection and editing. A family of graphical commands displays observed and model visibilities from a variety of perspectives, and allows the user to flag or unflag visibilities directly with the mouse. Simple key-bindings are combined with mouse positioning to quickly navigate through telescopes, baselines, bands, and sub-arrays, to edit visibilities collectively or individually, to zoom in on parts of the data, and to perform many other command-specific operations. These facilities are important because:

- They allow users quickly to identify and excise corrupt data.
- They provide a direct comparison between the model and the data.
- They help to familiarize users with the peculiarities of their datasets.
- They provide beginners with visualization tools to learn how given visibility profiles produce given maps.
- They can be used to inspect dynamically changes made to the visibilities and the model during processing.

3.2. Difference Mapping

Difmap was named after a mapping technique called *Difference Mapping.* This technique was originally pioneered in the Olaf package written at the Nuffield Radio Astronomy Laboratories.

When a model of a source is subtracted from a dirty map, what remains is known as a *difference* or *residual* map. This is a key component in Difference Mapping.

The source model that is subtracted from the dirty map is built up in an iterative fashion, usually through CLEAN deconvolution of successive versions of the residual map. Alternatively a model can be fitted to the observed visibilities.

When CLEAN is used to build up the source model, each iteration of CLEAN not only adds a delta-function to the model, but also subtracts that delta-function and its PSF from the residual map. Thus CLEAN automatically keeps the residual map up to date with the changes that it makes to the model.

Conversely, when model fitting is used to build up the model, or when a user edits either the visibilities or the model between successive iterations of CLEAN, then the residual map becomes out of date. When this happens, Difmap automatically re-calculates the residual map by taking the 2-D FFT of the *difference* between the observed and the model visibilities.

At this point, if visibilities were edited by the user, then the current model may retain minor features that were deconvolved from those visibilities. With the offending visibilities removed from the data, these artifacts only remain in the model, so when the model is subtracted from the data, the revised residual map contains an inverted version of the artifacts. Subsequent iterations of CLEAN thus erase them from the model. This typically involves both positive and negative CLEAN components.

The ability to transparently continue deconvolution after modifying the model or the observed visibilities highlights a fundamental difference between difference mapping and traditional techniques. The traditional approach would require the user to restart deconvolution from scratch, whereas the difference mapping approach encourages the user to edit or re-calibrate the data on-the-fly as ever weaker artifacts appear in successive versions of the residual map. It also enables users to incrementally add clean windows. This is especially useful when features that were initially obscured by the point spread functions of brighter features, are revealed in the residual map.

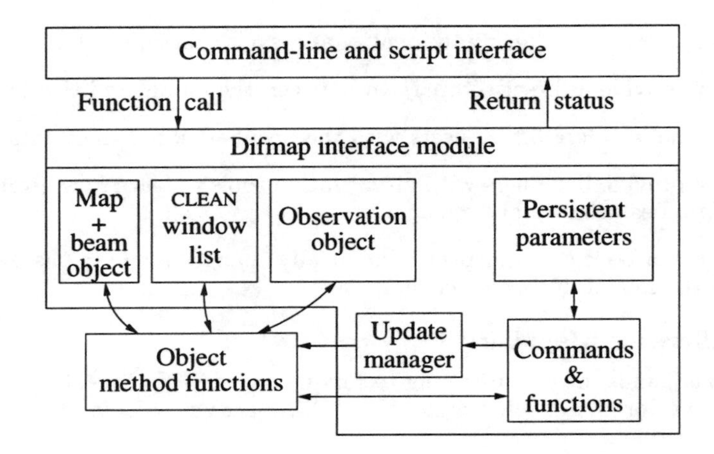

Figure 1. An overview of Difmap's architecture.

4. Difmap Architecture

An overview of Difmap's internal architecture is depicted in Figure 1. At the top is the command-line scripting interface with which users interact. This performs expression evaluation, parameter checking, etc., then delegates work to functions in the Difmap interface module. To eliminate dependencies on the user interface, Difmap functions never call back to the scripting interface.

Functions within the Difmap interface module maintain a number of objects that collectively contain the state of Difmap. Operations on those objects are delegated to method functions.

4.1. The Observation Object

The most significant object in Difmap is the Observation object. This encapsulates the observed visibilities (contained mainly in paging files), the observation parameters, model components and their associated visibilities, self-calibration corrections, and the pseudo-continuum channel that is the unit of in-core processing in Difmap.

The constructor for an `Observation` object reads a UV FITS file and records all but its visibilities in memory. The visibilities are recorded in a scratch file. The `select` command is then used to select or change which polarization and range of spectral-line channels to process next. This results in the construction of one pseudo-continuum channel per band. The corresponding visibilities are stored in a paging file, ordered such that Difmap processing commands can sequentially page into memory one band at a time, for processing.

Modifications to the visibilities, such as self-calibration corrections, phase-center shifts, UV coordinate frequency scaling, and weight scales are recorded in memory, then applied on-the-fly whenever a new band is paged into memory. They are also applied on-the-fly if an observation is written to disk as a new UV FITS file. The only modifications that are made to the scratch files are editing operations. Even these are temporarily buffered in memory to reduce file I/O to a minimum. As a result, un-doing most visibility corrections is simply a matter of clearing them from memory.

Model visibilities for each band are also recorded in a paging file. The associated image-plane model components are stored as lists in memory, partitioned into tentative and established models. The tentative model records components that have not yet been established as model visibilities, and is automatically established in the UV plane whenever a method function needs the model visibilities, or when the `keep` command is called. Until then, it can be discarded with the `clrmod` command. Thus it is possible to call `clean`, look at the resulting residual map, and then decide to discard the latest additions to the model. Note that to reduce the number of components that have to be transformed to the UV plane, identically positioned delta-components are combined into single components.

The model may be further split into normal and continuum models. Continuum subtraction in Difmap exploits difference mapping techniques to perform the model subtraction on-the-fly, without changing the recorded observed visibilities. This is important, not only because it gives the user the ability to modify or discard the continuum model, but also because self-calibration can then continue to operate on the original observed visibilities. The only practical difference in the treatment of the continuum and the normal models is that the continuum model is not added back to the residual map when the clean map is constructed.

4.2. The Update Manager

The update manager acts a bit like the UNIX `make` facility. It has the responsibility of keeping the objects of the Difmap interface module up-to-date with each other. It does this in two stages:

1. When a user command modifies an object in such a way as to invalidate the contents of another object, then the latter object is marked as out-of-date.
2. Later, when a user command requires a given object, it asks the update manager for it. If the object has been marked as invalid, then the update manager causes the object to be updated before passing it back to the requesting command.

For example, when the `shift` command is called to move the phase center of the observed and model visibilities, the map object is marked as out of date by the `shift` command. If the `clean` command is then invoked, it asks the update manager for the map object, and the update manager interpolates a call to the `invert` command to re-compute the residual map.

The update manager thus manages relationships between otherwise unrelated objects. This frees users from the tedious bookkeeping that traditional packages require, and this makes it safer for users to experiment.

The update manager enforces consistency among its objects by interpolating calls to user commands (hereafter referred to as *command interpolation*). On the rare occasions when the default running parameters of these commands are not appropriate, users can call them explicitly.

While it is easiest to describe the update manager as a single entity, in reality, it is a collection of facilities operating at different levels to implement the rules described above.

5. What Distinguishes Difmap from its Peers?

5.1. Difmap is a Single Program

Whereas most large packages split tasks into separate programs, Difmap implements them as functions within a single program. The advantages of keeping everything within a single process are hard to ignore. In particular:

- The use of memory objects and function calls instead of intermediate files, interprocess communications, and multiple processes: (*i*) reduces overhead dramatically, (*ii*) removes a major source of unportability, (*iii*) eliminates a significant amount of code, and (*iv*) simplifies the user interface.
- Since a lot of state information resides permanently in memory, sanity checking can trivially be extended to cross command boundaries. This makes command-interpolation possible.
- The low overhead of invoking command functions means that: (*i*) minor commands are justifiable on their own, rather than having to be redundantly tacked on to other commands, and (*ii*) redundancy is reduced because commands can trivially delegate work to other commands via function calls.
- Bugs in a single process package often manifest themselves quickly through cumulative effects. This means that most bugs are caught during prerelease tests.

5.2. Dynamic Resource Allocation

Difmap doesn't employ hard-wired limits. All resources are dynamically allocated to take advantage of whatever resources a given computer has available. This includes memory for objects and paging files for visibilities.

5.3. Scripting Facilities

Most packages provide some kind of scripting language, and Difmap is no exception in this respect. However, Difmap exploits its scripting interface in a couple of novel ways.

Log files: When Difmap is started, a log file is created, and everything that the user types, along with the resulting output messages, are recorded in this file. By recording command-lines verbatim and output messages as comments, a log file doubles as a script. This script can then be used to re-play a Difmap session.

Recording log files as scripts is not only useful to users, but really comes into its own when a user encounters a bug. When this happens, the script can be used directly to re-produce the bug.

Save files: The Difmap `save` command saves the state of Difmap by writing a script that contains the user-commands needed to restore the running parameters. In addition, the `save` command saves the visibility data, model components, etc., in files, and adds to the script the commands needed to re-load them.

Saving parameters via scripts eliminates the need for special parameter-file formats.

5.4. Difmap Doesn't Obfuscate the Environment

Traditional packages redundantly implement their own file-systems, internal file formats, printer spooling facilities, and tape I/O facilities. Difmap doesn't. It lets users use the facilities that they are familiar with, and, in the process, simplifies package management and the amount of work that users have to do to get data into the program.

In particular, Difmap directly reads UV FITS files, so that users can start inspecting and processing new observations within seconds of starting the program. In traditional packages, these files would first have to be converted to an internal file format and then written to a package-specific file-system. Difmap doesn't need an efficient internal format because its single process design removes the need to re-read files each time that a new command is invoked.

6. Conclusions

Difmap is a popular alternative to more capable packages such as AIPS because:

- It is tightly focused on a particular processing scheme.
- It is highly interactive.
- It is fast for small to moderate sized data sets.
- UV FITS files can be processed directly without the overhead of conversion to an internal format and without requiring users to learn new file management utilities.
- It is easy to install and manage.
- It encourages learning through experimentation via its use of command-interpolation to correct for missed steps, and the ability to undo certain operations.
- Its small set of simple commands is easier to use and understand than the complex tasks that traditional packages provide.

From a programming standpoint, experience with Difmap has shown that there is no need for packages to partition tasks into separate programs. Object based techniques provide sufficient partitioning on their own. It is even more feasible to do this now than it was when Difmap was written, because:

- The concurrency gained by spawning tasks as separate programs can now be achieved much more efficiently in a single program by using POSIX threads.
- To allow multiple programmers to work on a single program, and to keep the size of the resulting executable small, commands or modules can now be written as dynamically loadable entities, which can then be loaded on-the-fly into the running kernel of the program. At this time, there is no single standard for dynamic loading that works on all machines, but the dynamic loading facilities of popular environments such as Tcl and Perl have shown that this is not a big obstacle.

Acknowledgments. As a graduate student at Jodrell Bank, in England, I used both Jodrell Bank's Olaf package and NRAO's AIPS package for data reduction. The interactive and difference mapping aspects of Olaf, together with the painful lack of the same features in AIPS, inspired me to design Difmap in the manner that I did.

Algorithm development for some of the core commands in Difmap would have been much more difficult if the code of the Caltech VLBI package hadn't been at hand for comparison. Similarly, the algorithm underlying the `selfcal` command was derived from SDE code that Tim Cornwell supplied.

Difmap owes its graphical capabilities to the PGPLOT library written by Timothy Pearson, and exploits the algorithms in Patrick Wallace's SLALIB library for precessing source coordinates.

Most importantly, Difmap would not have been as useful as it came to be were it not for the encouragement and suggestions of Difmap users, both at Caltech and externally. In particular I would like to thank Greg Taylor both for writing the Difmap cookbook, and for some invaluable discussions. I would also like to thank Anthony Readhead and Timothy Pearson for giving me the freedom to write Difmap.

Astronomical Data Analysis Software and Systems VI
ASP Conference Series, Vol. 125, 1997
Gareth Hunt and H. E. Payne, eds.

CENTERFIT: A Centering Algorithm Library for IRAF

L. E. Davis

IRAF Group, NOAO,[1] P.O. Box 26732, Tucson, AZ 85726

Abstract. This paper presents an overview of the CENTERFIT image centering algorithm library currently under development in the IRAF system.

1. Introduction

Centering algorithms are important to many astronomical applications. One dimensional (1-D) centering algorithms are used in spectral line detection, wavelength calibration, and radial velocity applications. Two dimensional (2-D) centering algorithms are used in object detection, image registration, photometry, and astrometry applications. The centering algorithms employed by these applications must satisfy a wide range of efficiency and accuracy requirements, and function robustly on data sets spanning a broad wavelength, resolution, and signal-to-noise regime. This paper presents an overview of CENTERFIT, a library of 1-D and 2-D centering algorithms currently under development in the IRAF system.

2. The Science Requirements

The CENTERFIT algorithms satisfy a broad range of scientific requirements. The algorithm choice was guided by the literature (Stone 1989; Lasker et al. 1990, and references therein) and by experience with existing IRAF centering routines. Algorithms optimized for efficiency (e.g., moment analysis), precision (e.g., functional fits to a Gaussian), and flexibility (e.g., functional fits to a SuperGaussian), are included. Wherever possible, 1-D and 2-D versions of the same algorithm are included, permitting efficiency versus precision tradeoffs on the part of the calling application (e.g., moment analysis of the marginals versus elliptical Gaussian fits to the original data). All of the CENTERFIT algorithms compute error estimates, and size, shape, and orientation information, where appropriate.

The CENTERFIT fitting routines are flexible. The calling applications can include or reject data from the fit, fix or fit the background value, impose minimum signal-to-noise criteria, fix or fit the values of selected function parameters

[1] National Optical Astronomy Observatories, operated by the Association of Universities for Research in Astronomy, Inc. (AURA) under cooperative agreement with the National Science Foundation

(e.g., the Gaussian FWHM or the Moffat beta values), select one of several builtin weighting schemes, and set the fit convergence criteria.

The CENTERFIT library is self-contained. All of the required supporting algorithms, including a coarse centering algorithm, a selection of sky fitting algorithms, and a signal-to-noise estimation algorithm, are part of the CENTERFIT library. The required mathematics libraries are part of the IRAF core system MATH package.

3. The Software Requirements

The CENTERFIT library is easy to use. The number of interface routines is small, the calling sequences are simple, the algorithm parameters are initialized to reasonable default values and may be examined or reset at any time, image I/O and memory are managed internally, and all the routines return error information.

The CENTERFIT library also provides access to a set of numerical fitting routines, permitting applications to manage their own image I/O and memory requirements, and perform their own parameter initialization.

The CENTERFIT library is extensible. New centering and sky fitting algorithms can be added by writing the required low level routines, and making the appropriate entries in the algorithms.

4. The Algorithms

The 1-D CENTERFIT algorithms operate directly on 1-D image data or on the marginal distributions derived from 2-D image data. The 1-D algorithms list includes: moment analysis, optimal centering with a sawtooth function, and fits to Gaussian, Moffat, and SuperGaussian functions (integrated and non-integrated). The 2-D CENTERFIT algorithms operate directly on 2-D image data. The 2-D algorithms list includes: moment analysis, and fits to elliptical Gaussian, Moffat, and SuperGaussian functions (integrated and non-integrated).

Function fitting is performed by minimizing the chi-squared statistic using the Levenberg-Marquardt non-linear least squares fitting algorithm. Three builtin weighting schemes are supported: uniform, noise, and 8-fold symmetry weighting (Lasker et al. 1990). Integrated function evaluation is performed using a four point Gauss-Legendre integration scheme (Abramowitz & Stegun 1970).

Several sky fitting algorithms are supported including: setting a constant sky value, direct computation of the mean and median of the sky pixel distribution, moment analysis and optimal centering of the sky pixel histogram, and fitting a planar sky.

5. The Interface Routines

The principal CENTERFIT interface routines are listed below. These routines are used to allocate and deallocate the CENTERFIT data structures, set and get the CENTERFIT algorithm parameters, locate an object, extract the image data, estimate the background value, and compute the center.

```
ct   =  ct_cinit (calgorithm, cbox, salgorithm, sibox, sobox, sky)
        ct_free (ct)
        ct_set[i|r|s] (ct, param, val)
val  =  ct_stat[i|r] (ct, param)
        ct_stats (ct, param, strval, maxch)
ier  =  ct_getsky (ct, im, xin, yin, sky, sigma, skew)
ier  =  ct_sbuf (ct, im, xin, yin, sbuf, ibuf, nsky, snx, sny)
ier  =  ct_fitsky (ct, sdata, idata, nsky, snx, sny, sky, sigma, skew)
ier  =  ct_getcenter (ct, im, xin, yin, xout, yout, xerr, yerr)
ier  =  ct_locate (ct, im, xin, yin, xout, yout, cbuf, cnx, cny,
            xcbuf, ycbuf)
ier  =  ct_cbuf (ct, im, xin, yin, cbuf, cnx, cny, xcbuf, ycbuf)
ier  =  ct_fitcen (ct, cdata, cnx, cny, xin, yin, xcbuf, ycbuf, sky,
            sigma, xout, yout, xerr, yerr)
```

The following routines are examples of the 2-D numerical interface routines. These routines compute the weights, initialize the function parameters using 2-D moment analysis, and fit the selected function to the data using the Levenberg-Marquardt non-linear least squares fitting code.

```
ct_2dsetwgts (cdata, weights, cnx, cny, emission, weighting,
    datamin, datamax, threshold, gain, rdnoise)
ct_2dmoments (cdata, cnx, cny, xc, yc, sigma, ratio, theta,
    datamin, datamax, threshold, emission)
ct_2dsetpars (cdata, cnx, cny, function, pars, amplitude,
    xc, yc, sigma, ratio, theta, exponent, sky)
ct_2dnlfit (x, y, cdata, weights, npts, function, pars, perrs
    plist, npars, maxiter, tol, nreject, reject, niter, ier)
```

6. Results

A comparisons of four of the most popular CENTERFIT algorithms: moment analysis of the marginals, Gaussian function fits to the marginals, radially symmetric Gaussian function fits to the data, and elliptical Moffat function fits to the data, was made using a 1024 pixel square artificial star field created with the IRAF ARTDATA package. The test field contained 500 stars uniformly distributed in position and in brightness over a 10 magnitude range. A radially symmetric Moffat function with FWHM = 2.0 pixels and beta = 2.5 was used to model the stars, and realistic Poisson and Gaussian read noise were added to the image.

For all four algorithms the bright object measurement precision was limited by the psf model template spacing (0.01 pixels) and sub-pixel gridding (10×10) to ~0.005 pixels. The three function fitting algorithms exceeded this expected accuracy with mean bright object measured (fitted − model) centering errors of $\pm\sim0.004$ pixels. The 1-D moments algorithm performed significantly worse, with a mean bright object measured centering error of $\pm\sim0.009$. The bright object predicted centering errors were limited by how well the assumed model fit the data. For the 1-D and radially symmetric Gaussian function fits, the predicted centering errors were much larger than the measured errors. By contrast, the bright object predicted centering errors for the model independent 1-D moments and elliptical Moffat function fitting algorithms were in reasonable agreement with the measured errors. Faint object measurement precision was limited by signal-to-noise, with the 2-D algorithms performing significantly, but not overwhelmingly, better than the equivalent 1-D algorithms, in agreement with Chiu (1977). The faint object predicted centering errors agree well with the faint object measured errors.

The four algorithms vary widely in efficiency, with the 1-D Gaussian, radially symmetric Gaussian, and elliptical Moffat function taking $\sim$3.0, 6.4, and 13.1 times as long, respectively, as the 1-D moments algorithm, which required $\sim$2.2 cpu seconds to fit 500 stars on a SPARCstation2.

For this particular data set, no significant improvement in the centering errors was obtained by using the integrated forms of the 1-D and 2-D Gaussian and 2-D Moffat functions. Better formal centering error values were obtained, but only at the expense of significantly increased execution times.

7. Current Status and Future Plans

Coding and preliminary testing of the current CENTERFIT algorithm set is complete, although detailed numerical testing is still in progress. Future plans for the package include adding function fitting criteria other than chi-squared minimization, and fitting techniques other than the Levenberg-Marquardt algorithm.

References

Abramowitz, M., & Stegun, I. A. 1970, Handbook of Mathematical Functions (New York: Dover), 916

Chiu, L.-T. G. 1977, AJ, 82, 842

Lasker, B. M., Sturch, C. R., McLean, B. J., Russell, J. L., Jenkner, H., & Shara, M. M. 1990, AJ, 99, 2019

Stone, R. C. 1989, AJ, 97, 1227

Astronomical Data Analysis Software and Systems VI
ASP Conference Series, Vol. 125, 1997
Gareth Hunt and H. E. Payne, eds.

A Method for Obtaining Reliable IRAS-LRS Data via the Groningen IRAS Server

S. J. Chan[1] and Th. Henning

Max Planck Society, Research Unit "Dust in Star-forming Regions," Schillergäßchen 2-3, D-07745 Jena, Germany

R. Assendorp

Astrophysical Institute Potsdam, An der Sternwarte 16, D-14482, Potsdam–Babelsberg, Germany

Abstract. A method for processing IRAS-LRS data via the Groningen IRAS server is presented. This is part of an effort to search for objects with an emission feature at 21 μm from the IRAS data base. Using the GIPSY software, we are able to obtain and to select reliable LRS spectra from the IRAS database. The scientific results have been recently reported elsewhere (Henning, Chan, & Assendorp 1996).

1. Introduction

The Infrared Astronomical Satellite (IRAS) surveyed about 95% of the sky in four broad bands at 12, 25, 60, and 100 μm during a 10-month period in 1983. A Low-Resolution Spectrometer (LRS) was in operation during the mission, observing in parallel with the survey detectors. The LRS is a slitless prism spectrometer which contains five detectors: three short-wavelength (SW) detectors (7.7–13.4 μm) and two long-wavelength (LW) detectors (11–22.6 μm). The LRS aperture has a size of $15' \times 6'$. The LRS spectral elements were produced when the source crossed the $6'$ wide aperture. Since the dispersion direction is the same as the IRAS scanning direction, the output is a convolution of the source structure and the spectra. The projected detector size is $\sim$ 15–18$''$. Spectra for sources smaller than 15$''$ are not affected by the convolution (cf. Assendorp et al. 1995).

To date, we have searched the LRS spectra of all possible IRAS point sources with a 21 μm feature in the IRAS LRS data base. Our search procedure is described in detail by Henning et al. (1996). New spectra with a 21 μm feature were extracted from the IRAS data base using the Groningen IRAS server.

---

[1]Present address: Instituto de Astrofísica de Canarias, C/Via Láctea s/n, E-38200 La Laguna, Tenerife, Spain

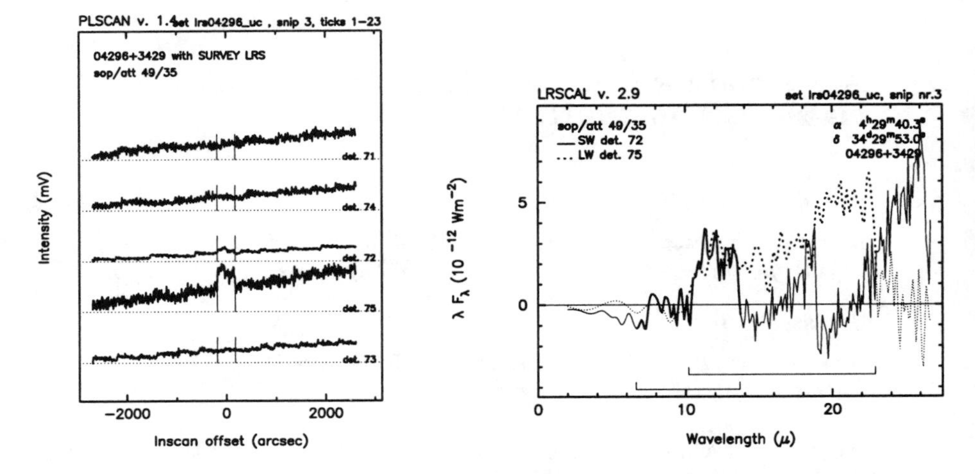

Figure 1. The PLSCAN snip-spectra (left) and the calibrated snip-spectra (right) of IRAS0496+3429 respectively.

2. How to Obtain Reliable IRAS-LRS Data

2.1. GIPSY and IRAS Data Structure

GIPSY (Groningen Image Processing System, van der Hulst et al. 1992) is the main tool for the analysis of the extracted LRS spectra in this study. Raw IRAS data in GIPSY are stored as 4-D data sets (cf. Figure 2 in Assendorp et al. 1995) in IRDS (IRAS Data Structure). The data are ordered in the following way: sample axis; ticks axis; sequence detector number axis (SDET): the detector number in the LRS are 71, 72, 73 (SW) and 74, 75 (LW); snip axis: a snip is part of a scan that has an overlap with a user-selected area of the sky.

Four programs in GIPSY (TRACKS, PLSCAN, LRSCAL, and GDS2TEXT) have to be heavily used for processing the extracted LRS data. TRACKS is for inspecting the trajectory of snips. PLSCAN is used to find out which detectors actually crossed a source, or to make a high-resolution in-scan view of the spatial structure around a source of interest. The outputs are uncalibrated snip-spectra (see Figure 1). LRSCAL is used to calibrate the data and to match the LW and SW regions. The calibrated spectra have only three dimensions, which are wavelength, SDNET, and snip. The outputs are calibrated snip-spectra (see Figure 1) and an average spectrum. GDS2TEXT is used for converting the LRS data from an IRDS format to an one-dimensional ASCII format.

2.2. How to Produce a Good Average Spectrum

Usually, a source is observed in several LRS scans, ranging in number from 1 to 10. Plots of all the individual calibrated snip-spectra, individual PLSCAN snip-spectra, and of the default averaged spectrum of each target source are examined. If the shape of any snip-spectrum is different from the other calibrated snip-spectra, and/or PLSCAN snip-spectra, or if the signal is too weak for the purpose of examination, it is rejected. If all snip-spectra and also all PLSCAN snip-spectra of a target source were consistent with each other, a final averaged spectrum is produced.

2.3. Selection of Candidates with a 21 micron Feature

After all the averaged spectra of objects are made, the final averaged spectrum, the selected snip-spectra and/or the selected PLSCAN snip-spectra are examined again. If a feature appears around 19–21 μm and if the width (FWHM) of the feature is ~3 μm, the LRS spectrum of this object is accepted as a 21 μm feature candidate. Altogether, 16 objects out of 498 sources are selected as sources which may have a 21 μm feature. In order to compare our preliminarily selected candidates with the spectra of objects which are reported to show a 21 μm feature, we also extracted the LRS data of all these sources. Typically, the shapes of all or some of the calibrated snip-spectra and/or PLSCAN snip-spectra of the newly selected sources are similar to those PLSCAN snip-spectra of already reported 21 μm feature objects.

2.4. Final Selection

IRAS survey data images. For all selected objects including those reported in the literature, we have processed the raw IRAS survey data at 12 and 25 μm in order to check whether the objects are extended or if there is evidence for other objects within 6′. Since IRAS is a scanning instrument, an image should be made by adding adjacent scans into a two-dimensional plane. The baselines of the individual scans are usually not equal, and may have a drift over the extent of the image. In order to have a good image, the IRAS scans must be "destripped." We have done this using the program IMAGE. It turned out that several objects are located in, or at the edge of, extended regions of emission. This means that one should be very careful in interpreting the LRS spectra. Only if the source is much brighter than the background can one reliably determine the global profile and individual spectral features.

Offset LRS spectra processing. Since the LRS is a slitless spectrometer, the wavelength of the spectrum is established by the in-scan positional offset between the source and the spectrometer on the sky. For sources extended in the in-scan direction, the spectrometer projects flux from the extended wings across the wavelength scale, thus smearing positional and wavelength information. Further, the LRS samples the sky every 7″. In fact, an error in the IRAS position of only 7″ would result in a spectrum that is shifted by half a spectral element, about 0.2 μm, and at the wavelength extremes of the detectors the quoted flux would be incorrect. In order to investigate whether a different central position may affect the resulting LRS spectrum, we have extracted spectra for all of our samples and the reported 21 μm objects for different central positions in steps of 7″ in declination up to the major radius of its IRAS error ellipse. After all plots of the individual offset PLSCAN snip-spectra and offset-calibrated snip-spectra were carefully examined, we found that this problem also appears in reported 21 μm objects.

After the above examination and analysis, we finally selected twelve objects as our best new candidates which probably show a 21 μm feature (cf. Figure 8 in Henning et al. 1996), to add to the objects which were already known. We summarize offset spectra processing with these rules of thumb:

- If a feature is real, it is still present or not distorted severely after the offset.
- If a feature is spurious, it disappears after the offset, or is severely distorted.
- When we consider only those sources whose signal falls entirely into the detector, the probability of misinterpreting the existence of a feature is low.
- Following-up observations to confirm a feature are necessary, and will hopefully performed with *ISO*.

Acknowledgments. The IRAS data were obtained using the IRAS data base server of the Space Research Organisation of the Netherlands and the Dutch Expertise Centre for Astronomical Data Processing funded by the Netherlands Organisation for Scientific Research. This work was partially supported by the German Bundesministerium für Bildung, Wissenschaft, Forschung and Technologie (Förderkennzeichen 05 2JN13A). S. J. Chan, thanks the Program Organizing Committee for the Sixth Annual Conference on Astronomical Data Analysis Software and System for offering her full financial support to attend the Conference.

References

Assendorp, R., Bontekoe, T. R., de Jonge, A. R. W., Kester, D. J. M., Roelfsema, P. R., & Wesselius, R. R. 1995, A&AS, 110, 395

Henning, Th., Chan, S. J., & Assendorp, R. 1996, A&A, 312, 511

van der Hulst, Th., Begeman, K. G., Zwitser, W., & Roelfsema, P. R. 1992, in ASP Conf. Ser., Vol. 25, Astronomical Data Analysis Software and Systems I, ed. D. M. Worrall, C. Biemesderfer, & J. Barnes (San Francisco: ASP), 131

Astronomical Data Analysis Software and Systems VI
ASP Conference Series, Vol. 125, 1997
Gareth Hunt and H. E. Payne, eds.

Near-IR Imaging of Star-forming Regions with IRAF

S. J. Chan and A. Mampaso

Instituto de Astrofísica de Canarias, C/Via Láctea s/n, E-38200 La Laguna, Tenerife, Spain

Abstract. We report on our ongoing project "Infrared Study of H II Regions Associated with Small Clouds." Several tasks are being developed under the IRAF environment for efficiently reducing and analyzing the near-infrared data obtained from the Teide observatory with the IAC infrared camera (IAC-IRCAM).

1. Introduction

Several galactic H II regions associated with small molecular clouds were observed with the IR-Camera on the 1.5 m CST (Carlos Sánchez Telescope) of the Teide Observatory in October 1996. These IR data, combined with optical imaging (which will be obtained from the IAC-80 telescope), IRAS survey maps, and existing CO observations of the associated clouds, will allow us to undertake a detailed investigation of the gas-to-dust mass ratio, initial mass function and the star-formation efficiency in these complexes. Studies of the embedded clusters also allow us to derive the luminosity function and thus, using an appropriate mass-luminosity relation, to determine the stellar mass spectrum. The near-IR imaging study will allow us to search systematically for exciting sources and the low-mass stars associated with the nebulæ.

Some data were obtained during the test run of the IAC-IRCAM on February 1996. We are using these data to test our new IRCAM package, developed under the IRAF environment.

2. IRCAM Package

2.1. IAC-IRCAM

The IAC-IRCAM is based on a 256×256 NICMOS 3 array and has a plate scale of 0.4″/pixel. Currently, the camera contains seven filters covering the wavelength range from 1.2 μm to 2.32 μm. The 3σ 60 s limiting magnitudes for point sources were 18.8 mag at J, 18.5 mag at H, and 16 mag at K.

2.2. Preliminary Tasks in the IRCAM package

At the time of writing this paper, eighteen preliminary tasks have been written. A common characteristic of these tasks is ease in handling the bookkeeping of data reduction/analysis history. The tasks are as follows:

- APCOR: estimate the aperture correction,

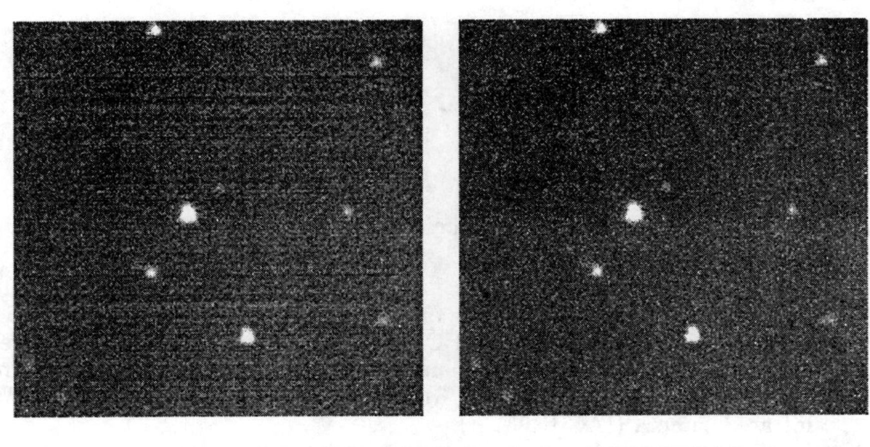

Figure 1. Part of the S269 field: left—before MDRIPNEW processing; right—after MDRIPNEW processing.

- CHECKPSF: check the PSF (Point Spread Function) model,
- DUMPMARK: mark and label objects on the displayed frame,
- EXAMPSFGP: examine the PSF model,
- FINMISS: find missing stars not found with DAOFIND,
- FINSHIFT: find the image shifts with respect to the reference image,
- FINFIT: fit all the stars, including the missing stars, with the final PSF model,
- FIRSTPSF: obtain the first iteration of a PSF model,
- MARK3FR: display all the frames and mark the stars,
- MATCHMARK: find the stars which match all the frames,
- MATCHPHOT: obtain the photometry from the match frame and produce a final list,
- MDRIPNEW: reduce the "drip" noise produced by the IAC-IRCAM (see Figure 1),
- NITERPSF: obtain the nth iteration PSF model,
- PARFIND: find the critical parameters from the frames,
- SELBRIGHT: select well isolated bright stars,
- SELPST1: select first iteration PSF candidates, and
- SUBMARK: subtract the neighbours and their friends from the PSF stars and display the subtracted image.

2.3. Future Development of IRCAM

Future tasks include:

- writing help pages for the tasks,
- using new IRCAM data to test every detail of the package,
- write more tasks based on IMAGES, IMMATCH, and DIMSUM packages (brand-new tasks will be written if necessary),
- combine several small tasks into a single task, and
- if satisfied with the performance of this package, release it to the public.

Acknowledgments. S. J. Chan thanks the Program Organizing Committee for the Sixth Annual Astronomical Data Analysis Software and System Conference for offering her full financial support to attend the Conference.

Astronomical Data Analysis Software and Systems VI
ASP Conference Series, Vol. 125, 1997
Gareth Hunt and H. E. Payne, eds.

ETOOLS: Tools for Photon Event Data

M. Abbott, T. Kilsdonk, C. Christian,[1] and E. Olson

Center for EUV Astrophysics, 2150 Kittredge Street, University of California, Berkeley, CA 94720-5030

M. Conroy, J. Herrero, and R. Brissenden

Smithsonian Astrophysical Observatory, 60 Garden Street, Cambridge, MA 02138

Abstract. We present an overview of the ETOOLS project and describe some of the design features of the ETOOLS system.

1. Goals

Our primary goal is to provide a set of generic tools for the basic manipulation of event data. We take the word generic to have several implications. The tools should be generic in function, meaning they are useful with any kind of event data and do not depend on details that may be specific to the telescope or instrument that produced the data. However, the tool set should be extensible to include new, possibly instrument-specific, tools.

The tools should also be generic in environment. By that, we mean that the tool set should function both stand-alone and within common data reduction environments, such as the IRAF CL or IDL.

Finally, the tools should not impose any particular format on the event data files, but, instead, be able to adapt to any format.

2. What is Event Data?

Event data refers to the very rich data sets produced by photon counting detectors such as those commonly used in high-energy telescopes. Some recent examples of missions that have produced, or will produce, event data are *Einstein*, *ROSAT*, *EUVE*, *ASCA*, *XTE*, and *AXAF*.

In event data, each event (such as a photon) on the detector is tagged with its arrival time, detector position, and other information. These tags are retained through the later processing of the events, so that a scientist analyzing the data can use filtering techniques to examine subsections of the complete data set, or form secondary data products such as images or light curves by binning the data in any desired manner.

Another typical feature of event data sets is the presence of auxiliary information from the telescope or the spacecraft on which it may be operating, in ad-

[1] also Space Telescope Science Institute, 3700 San Martin Dr., Baltimore, MD 21218

dition to the events themselves. The auxiliary data are most often time-ordered and tabular in nature, and are frequently used to determine the "quality" of the event data they accompany.

Because of this structure, event data lend themselves both to a very detailed and flexible analysis, in which examining small numbers of (or even individual) events can be useful, and to the more usual kinds of image analysis. Event data sets will typically consist of several event lists and tables of auxiliary information, along with various filters that may have been established, and images and light curves that have been computed.

3. The Scope of ETOOLS

Levels of abstraction are imposed on event data in order to analyze and understand it:

1. The data are stored in basic objects such as images and tables, which provide appropriate manipulations (e.g., filtering of tables, sectioning of images).

2. Relationships and interactions are supported between basic objects, which implement additional functionality needed to fully represent event data (see §4). For example, a "filter" object contains the time intervals to be used to filter a table of events, and is persistently associated with the event table.

3. Scientific meaning is attached to event data objects. For example, a particular filter object is identified as the "SAA removal filter," or a particular table is identified as containing a light curve of the source (a light-curve object).

ETOOLS addresses the intermediate, second level. We build upon existing libraries, which support basic objects, and provide applications that manipulate event data. But we leave the assignation of scientific meaning and manipulation of higher-level objects to tools (and possibly libraries), which are layered on top of ETOOLS.

The ETOOLS I/O library implements the Event Data Model in an application programming interface, which is then called from the ETOOLS applications. The I/O library calls on existing, basic object libraries.

4. The Event Data Model

We have developed a model for event data that describes the objects and operations needed for their analysis. An overview is given here (not all details are described).

Dataset. The dataset is an object that contains all of the component parts of an event data observation in a directory-like structure. It supports operations to add, remove, and list contained objects.

Property List. Every object in the dataset has an associated property list, which is a set of zero or more properties. A property is a string keyword and an associated value in one of several data types.

Image. Image objects are largely identical to well-known existing image objects such as FITS or IRAF images. Each is composed of pixel data and an associated WCS object. Operations will exist to read and write from the image or sections of it.

WCS. WCS objects encode the transformation between two coordinate systems. They are used in the data model in conjunction with image objects and the binning of tables.

Table. Table objects start with the capabilities of familiar table objects such as ST Tables or FITS Bintables. Additional supported features include a property list associated with each table column, binning of tables into images, and filtering.

Binning is supported by allowing table objects to be opened as image objects. Each table has a list of binning keys; each key indicates the columns defining the bins, the column to sum, and the WCS object to attach to the resulting image.

Filtering is supported as an integral part of reading tables. A list of filtersets is associated with a table; the active set is applied at the time of a read. Each filterset specifies the type of filter object to be applied and the table columns it applies to.

Interval Filter. An interval filter is a one-dimensional filter that specifies a set of valid ranges in a quantity. Interval filters are associated with tables through filtersets.

Region Filter. A region filter is a two-dimensional filter that defines a valid area in the space defined by two quantities. Region filters are associated with tables through filtersets.

5. The ETOOLS I/O Library

In order to achieve our goals of environment and file format independence, the ETOOLS I/O library is implemented in two layers: generic code and kernels. These layers provide a simple way to isolate the code, which depends on the details of a particular file format.

A kernel must be written for each underlying set of file formats and/or basic object libraries that will be used. Each kernel is responsible for implementing a set of functions—the kernel programming interface or KPI—which are called from the generic code. Kernels must implement missing functionality in the underlying libraries to provide the basic object capabilities assumed by the generic code.

To extend the ETOOLS library and applications to work with a new file format, only a new kernel is needed. See the paper on the implementation of the ASC Data Model (Herrero et al. 1997) in this volume for some examples of kernels under development.

The ETOOLS library and applications are written in ANSI C for maximal portability. Kernels may be implemented in any desired manner as long as the KPI is supported. Kernels may have dependencies on certain run-time environments. ETOOLS sites can select any desired kernels for installation. For example, a site that does not have IRAF installed can build ETOOLS by not including the IRAF-based kernels (perhaps working entirely with FITS).

6. ETOOLS Applications

A suite of non-interactive tools will perform all of the operations supported by the event data model. These are aimed primarily at scripting uses, but will also be suitable for users who prefer a command line interface.

A small number of interactive tools with graphical user interfaces will support the display and editing of datasets, tables, and interval filters. We have adopted SAOtng as the image display for ETOOLS.

Applications are written in ANSI C for portability. In order to function properly inside certain common data reduction environments, applications may contain a very thin "front end" of environment-specific code.

The ETOOLS applications will function in at least two environments: as a package of IRAF tasks and as a set of UNIX programs. Further environments can be supported with proper additions to the front end.

7. Project Status

The design of the library is completed and coding is ongoing. We are using a prototype kernel based on ST Tables and IRAF images to validate the software. Kernels based on FITS and on the IRAF QPOE file format are in planning and development stages at the ASC, separate from the ETOOLS effort.

Non-interactive tools that use the completed sections of the library have been developed. More will be added as the library progresses. Work is now concentrating on the interactive applications.

The ETOOLS project is planned for completion in early 1997. All products will be available from the Center for EUV Astrophysics and Smithsonian Astrophysical Observatory WWW and ftp sites, as well as other locations to be determined. Look for future announcements of ETOOLS beta and final releases on the ADASS newsgroups.

Acknowledgments. ETOOLS is supported by NASA contract NAS5-32698 through the Astrophysics Data Program. The Center for EUV Astrophysics is a division of the UC Berkeley Space Sciences Laboratory.

References

Herrero, J., Oberdorf, O., & Conroy, M. 1997, this volume, 469

Astronomical Data Analysis Software and Systems VI
ASP Conference Series, Vol. 125, 1997
Gareth Hunt and H. E. Payne, eds.

Data Processing for the Siberian Solar Radio Telescope

S. K. Konovalov, A. T. Altyntsev, V. V. Grechnev, and E. G. Lisysian

Institute of Solar-Terrestrial Physics, Lermontov St. 126, Irkutsk, Russia 664033, E-mail: root@sitmis.irkutsk.su

A. Magun

Institute of Applied Physics, University of Bern, Switzerland, 3012 Bern Sidlerstrasse 5, E-mail: magun@sun.iap.unibe.ch

Abstract. The SSRT is a cross-shaped interferometer. The way of forming its response is rather specific, so various programs to process its data were required. Furthermore, fast observations provide us with a large amount of data; therefore, implementation of special techniques for previewing and processing of data is required. We believe IDL to be the most convenient for these purposes. A set of IDL programs was developed which allows us to look through the records and view data in various representations; to measure a radio source size, flux, and its location on the Sun; to build high-sensitivity 1-D images, etc.

1. Introduction

The Siberian Solar Radio Telescope (SSRT) (Smolkov et al. 1986) is a large astronomical instrument. It is a cross-shaped interferometer consisting of 128×128 parabolic 2.5-meter antennas, equally spaced with a separation of 4.9 m. The SSRT is located 220 km from Irkutsk in a beautiful forest valley lying between the two chains of Sayan Mountains. This radio telescope is devoted to the study of solar activity in the microwave range (5.7 GHz), where processes in the solar corona are visible across the entire solar disk. In 1992, fast observations were started for studying of the fine time structure of flare manifestations, such as spikes (Altyntsev et al. 1994). After introduction of a fast data acquisition system developed together with the Institute of Applied Physics (Bern) (Altyntsev et al. 1996), the SSRT can record processes as short as 14 ms with a spatial resolution of down to $15''$.

The response of the SSRT is formed in a specific way, unlike that of most other aperture synthesis telescopes. The orientation of the interferometer's beam depends on the receiving frequency, making it possible to scan the solar disk by tuning the receiver in a series of steps. By making this stepwise tuning quick enough, we obtain a series of one-dimensional images (scans) taken across the Sun. This method—frequency scanning—provides fast sampling in one direction. It is used in the fast observations at the SSRT. The Sun passing through the fan beam provides sampling in the other direction. This sampling, together with frequency scanning, is used for 2-dimensional mapping of the Sun. Two-D

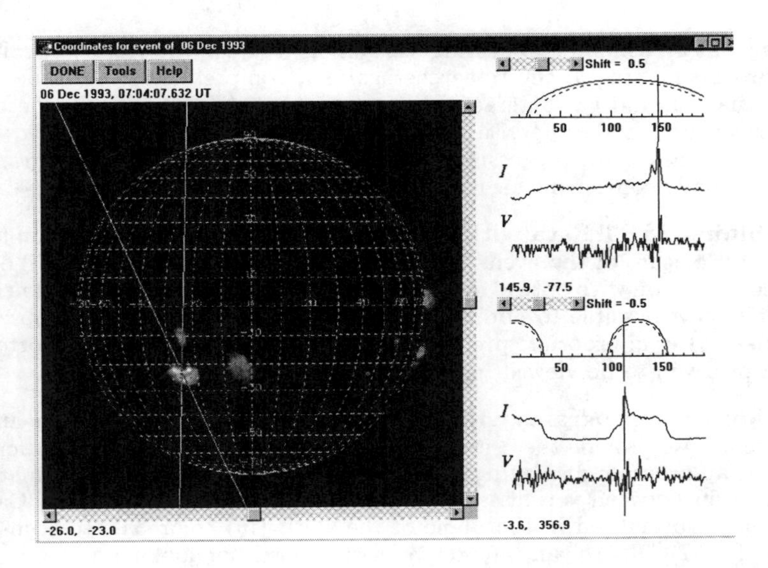

Figure 1. The display window of a program which allows correlating SSRT 1-D images with YOHKOH SXT maps and other images.

mode observations started in autumn, 1995, and have been providing images every day since spring, 1996.

Observations produce a large amount of data, which require special techniques for processing and previewing. To solve this problem, we had to develop our own special programs. Considering the advantages lent by IDL, such as efficiency, easy coding, powerful graphics, and wide acceptance among astronomers, we have chosen this language. As a result, a set of IDL programs was developed to provide these capabilities:

- looking through records and viewing data in various representations;
- measurement of a radio source size, flux, and location on the solar disk;
- building high-sensitivity 1-D images from a number of single scans;
- transformations of solar maps, such as rotation within the plain of the sky and around the polar axis of the Sun; and
- viewers and conversion programs for special and standard files, etc.

An example display from one of the programs is shown in Figure 1.

2. Program Features

To program more effectively, we have been trying to follow some guidelines:

Widget-based interface, a few graphics windows. Most programs are provided with a widget-based graphical interface. By presenting a few graphical windows, reflecting various aspects of the data and different processing stages, the researcher gains a comprehensive overview of the situation, and can interactively and simply select an object, control the procedure, and supervise intermediate and final results. It allows him to expedite data processing and to monitor progress. The programs are especially convenient for rapid data analyses.

Combining SSRT data with optical or X-ray images. To obtain a thorough picture of the event, it is useful to view and process SSRT data in combination with optical or X-ray images. By means of the global network, it is now possible to obtain the required images from various observatories. After appropriate preprocessing, these images could be imported into a program to be viewed together with the SSRT data.

Providing the programs with supporting tools. For convenience and efficiency, we provide the user with various supporting tools: (*i*) a calculator; (*ii*) an program displaying current SSRT parameters; (*iii*) coordinate conversion between various solar coordinate systems (heliographic, Carrington's, normalized in the plane of the sky); (*iv*) color palette adjustment (XLOADCT); (*v*) an interactive array viewer for looking at data in various representations (wire-mesh surface, shaded surface, halftone image, contour, and their combinations); and (*vi*) temporary exit into the shell.

Use of the library worked out by our group. The unique features of the SSRT influence the programming needs of all of our researchers. For this reason we have developed, supplemented, and corrected some routines from the IDL library, resulting in our own library. This library contains special routines for our use, as well as routines which could be of common interest and applicability (Konovalov et al. 1997). The widget-based data processing programs use this library extensively.

3. Implementation

Our widget-based programs are complex: a typical size ranges from 10 to 50 kB. Each program consists of three to six routines and uses some dozens of local variables having different types and dimensions. Because it is unlikely that a user would run two or more copies of such a program simultaneously, most of them exchange variables by means of common blocks. These variables are collected into structures. When the program is terminated, and the respective widget dies, the memory is freed by setting all these variables into scalar values.

IDL does not provide for retaining the scaling in graphics windows; therefore, when working with a few windows, one has to save the state of the system variables !X, !Y, !Z, !MAP for all the graphics windows being in use, and restore them when the cursor is placed onto a particular window.

Besides completely widget-based programs, we have also developed a hybrid program using both widgets and conventional graphics windows. This approach makes it possible to use the IDL library routines (such as PROFILES, ZOOM, etc.) in a straightforward way, and expedites program development.

All the programs can be run under IDL 3.0.1 or later versions. We welcome any interest and cooperation.

Acknowledgments. We are grateful to Drs. B. Kliem (AIP, Germany), V. P. Maximov, T. A. Treskov, S. V. Lesovoi, V. G. Miller, A. V. Bulatov, and Yu. M. Rosenraukh (ISTP, Russia) for the helpful discussion and assistance in providing us with the data.

This work was supported by the grants International Science Foundation (ISF) RLD000 and RLD300, ESO C&EE Programme A-03-049 and A-05-013, Swiss National Science Foundation 20 29871.90, and INTAS (INTAS-94-4625). Our special thanks to ISF for supporting our participation at this conference.

References

Altyntsev, A. T., et al. 1994, A&A, 287, 256

Altyntsev, A. T., et al. 1996, Solar Physics, 168, 145

Konovalov, S. K., Altyntsev, A. T., Grechnev, V. V., Lisysian, E. G., Rudenko, G. V., & Magun, A. 1997, this volume, 447

Smolkov, G. Ya., Pistolkors, A. A., Treskov, T. A., Krissinel, B. B., & Putilov, V. A. 1986, Ap&SS, 119, 1

Astronomical Data Analysis Software and Systems VI
ASP Conference Series, Vol. 125, 1997
Gareth Hunt and H. E. Payne, eds.

Java: The Application and Data Distribution Vector for Astronomy

A. Micol, R. Albrecht

ST-ECF, ESA/SSD/SA, ESO

B. Pirenne

ST-ECF, ESO

Abstract. Among the tasks of the Space Telescope European Coordinating Facility (ST-ECF) is providing to the astronomical community software for the retrieval, calibration, and analysis of HST data. This has been plagued by notorious problems of programming language incompatibility, platform incompatibility, installation problems, on-line help/documentation and user interface inconsistencies, etc.

The combination of Java with state-of-the-art network browsers has the potential to solve many of these problems. In addition, it eliminates potentially paralyzing throughput problems by pushing the processing load to the client machine. This is important for post-retrieval re-calibration of multiple data sets from science data archives.

First experiments were conducted in the area of NICMOS support tools, and for the preview of data from the HST archive. Examples are shown, and experiences are discussed.

1. Introduction

One of the tasks of the Space Telescope European Coordinating Facility (ST-ECF) is to provide to the astronomical community software for the calibration and analysis of HST data. From long experience with such attempts, we find that the following problems exist:

- Programming language compatibility (different languages/compilers, etc.).
- Installation problems.
- On-line help/documentation.
- User interface (incompatible window managers or operating systems).
- Platform incompatibility.
- Remote maintenance (the need to notify/distribute a new version).

There have been many attempts to solve the above problems, some more successful than others. A successful approach has been to use a *data analysis host system* of the MIDAS or IRAF type which, if installed properly on the target site, will overcome most, if not all, of the above mentioned problems, except for the remote maintenance. The main disadvantage of this approach, however, is that it cannot benefit at all from software development going on in other disciplines, or in the commercial sector: with the exception of IDL, none of the major astronomical data analysis systems is being used outside of astronomy.

2. Java on the Web

Astronomers were quick to recognize the advantages of the World Wide Web for the distribution of information and documentation, and the Web has had a tremendous impact on astronomy.

While the distribution of information (text, images, video, sound) was thus solved, there still remained the problem of distributing what might be called computational functionality. Workarounds were introduced: a CGI (Common Gateway Interface) compliant software program was invoked on the server via a fill-in form, and the result was sent back as a Web page.

We note that this approach already solves many of the problems listed in the previous section: all users will be able to invoke this service as long as they have, for their particular platform, a Web browser which supports HTML forms. The problem of software system versions is also tackled, since there is now only one version: the one on the server. However, the most serious obstacles—that invoking the service puts a computational load on the server, and the stateless nature of the HTTP protocol—are still present.

In 1995, Sun Microsystems released Java. Its main advantage is that it produces architecture-neutral compiled code (applet) which can be transferred across the Web and can be statefully executed within the Web browser. If a link to such a page is followed, the page (encoded in ASCII) and the applet (encoded in ASCII-like pseudo-binary, called bytecode) are transferred to the client. The page contains the information required to operate the software, and the object code executes within the Web browser of the client, using it as a virtual machine.

The advantages of this approach cannot be overstated:

- No programming language incompatibilities (no re-compilation).
- Installation problems are eliminated: the software is ready to execute as soon as it arrives across the net.
- On-line help and documentation can seamlessly be provided through the hypertext capabilities of the Web.
- HTML allows easy user interface creation. In addition, the Web browsers provide a good UIF paradigm.
- Platform incompatibility is eliminated. Any platform which supports a Java capable browser becomes usable. The penetration of the Internet by the Web will make sure that the number of supported platforms will grow, and the capabilities of the browsers will be increased (since the commercial sector is very interested in the public having easy and cheap access to the Web).

- Remote maintenance. Any first link traversal and any reload forces a transfer of the current, i.e., most recent version of the page, plus the applet. This means that any change made to the software by the original authors is propagated to the users in a rapid and painless manner.

3. First Experiments

When the above considerations became obvious in late 1995, we started a process of familiarization with Java. This was made difficult by the fact that Java was a moving target at that time: even books on the subject printed as late as October 1995 were already partially obsolete. And, we only had a beta release of the Java capable Netscape browser.

After overcoming these initial difficulties, however, it was possible to implement a few test applications in a relatively painless manner. In fact, we found that for many applications which do not require the software to run continuously, it was advantageous to use Javascript rather than Java applets: the code is more readable than the code for an applet, and it is visible on the HTML level, so that users can convince themselves of the correctness of the computation. They also can save the code locally and enhance or modify it.

This approach is particularly appealing for incorporating applets into technical documents, which normally only provide formulæ and, at best, examples. The Instrument Handbooks for the science instruments of the Hubble Space Telescope are a good example for this: they contain numerous formulæ and algorithms to calculate or estimate integration times, signal-to-noise, background, unit conversion, etc. We intend to produce applets for many of these functions, and to provide links to them from the electronic version of the handbooks.

3.1. Magnitude/Jansky Converter

As a working example, we developed the Magnitude/Flux/Jansky Converter.[1] The tool is meant to help the community during proposal preparation for the NICMOS (Near Infrared Camera and Multi-Object Spectrometer) instrument, installed in the Hubble Space Telescope (HST) during the 2nd Servicing Mission in February 1997.

This tool is based on a fill-in form handled by a Javascript. The Javascript can recognize and respond to user events to behave like a spread-sheet: as soon as an input field is changed, all the others are automatically re-computed according to the chosen units. An error message is displayed if the input value is not appropriate. This feature, in particular, is not implementable with the CGI approach, where the form has to be submitted before the actual verification can be performed.

3.2. Archive Data Preview

For several years, the HST archive has allowed previewing the data before requesting the retrieval of the entire data set. For images, this has been little more than a static display, but for spectra, the preview facility included a set of quick-look analysis tools like zooming and examining actual pixel values. As

---

[1] http://ecf.hq.eso.org/nicmos/tools/nicmos_units_tool.html

these relied on the availability of certain utility software at the remote site, those tools needed to be distributed together with the user interface.

With a Java applet adapted from a template found on Web, it was possible to prepare a very useful—if primitive—tool to visualize our spectra and provide some data visualisation aid. With a simple mouse click, the software is loaded and starts to run. The first thing it does is to load the data from the ST-ECF database. The data are then displayed in a graphical form within the Web page. The software then sets itself into a mode allowing user interaction: using the mouse and keyboard keys, basic functions such as zooming and pointing are available. All the interaction and display is obviously done on the local client machine, with no more references to the http server.[2]

4. Long Term Aspects

At this point in time there are limitations to what can be done with this technique, mainly having to do with security considerations: it is impossible for an applet to access a local file. Another limitation is that Java is an interpreted language, and, as such, is 20 times slower than "C." However, these limitations apply not only to astronomy, but to other fields, as well. We can, therefore, expect that the market will solve this problem for us.

This approach opens up a wide range of possibilities. For instance, the traditional approach of data archiving has been to archive raw data, plus data calibrated soon after the observations were carried out. Increasingly, we see a trend towards on-the-fly calibration, which means that the data are re-calibrated during the retrieval process using the most appropriate calibration reference files and the most recent version of the calibration software. The fact that even modest computers today are capable of performing the calibration of large data sets in a reasonable time makes this approach feasible. It is quite conceivable that future observational facilities will soon eliminate the storing of calibrated data.

The on-the-fly calibration is being performed at the archive machine, and, so far, we have been able to support this service. With the increased appreciation of data archives, and with more sophisticated and larger data sets, it is quite evident that the archive machine will become hopelessly overloaded. Applets provide the elegant solution of pushing the computational load out to the client site.

Acknowledgments. Thanks are due to Richard Hook and Michael Naumann for solving a number of Java related installation problems.

References

Albrecht, R., et al. 1997, this volume, 443

Pirenne, B., Benvenuti, P., & Albrecht, R. 1997, this volume, 290

---

[2]http://archive.eso.org/wdb/wdb/preview/preview_hst/query/Z0E35308T

Astronomical Data Analysis Software and Systems VI
ASP Conference Series, Vol. 125, 1997
Gareth Hunt and H. E. Payne, eds.

The ISOPHOT Interactive Analysis PIA, a Calibration and Scientific Analysis Tool

C. Gabriel[1]
European Space Agency - Astrophysics Division

J. Acosta-Pulido[1]
Max-Planck Institut für Astronomie, Heidelberg, Germany

I. Heinrichsen,[1] D. Skaley[2]
Max-Planck Institut für Kernphysik, Heidelberg, Germany

H. Morris
Rutherford-Appleton Laboratory, Chilton, Didcot, UK

W.-M. Tai
Dublin Institute for Advanced Studies, Dublin, Ireland

Abstract. ISOPHOT is one of the instruments on board the Infrared Space Observatory (*ISO*), launched in November 1995 by the European Space Agency. ISOPHOT **I**nteractive **A**nalysis, **PIA**, is a scientific and calibration data analysis tool for ISOPHOT data reduction. The PIA software was designed both as a tool for use by the instrument team for calibration of the PHT instrument during and after the *ISO* mission, and as an interactive tool for ISOPHOT data analysis by general observers. It has been jointly developed by the ESA Astrophysics Division (responsible for planning, direction, and execution) and scientific institutes forming the ISOPHOT consortium.

PIA is entirely based on the Interactive Data Language IDL (IDL92). All the capabilities of input/output, processing, and visualization can be reached from window menus for all the different ISOPHOT sub-systems, in all levels of data reduction from digital raw data, coming from the *ISO* telemetry, to the final level of calibrated images, spectra, and multi-filter, multi-aperture photometry. In this article, we describe the structure of the package, putting special emphasis on the internal data organization and management.

---

[1]Currently at the *ISO* Science Operations Centre, Villafranca, Apdo 50727, E-28080 Madrid, Spain

[2]Now at the Max-Planck Institut für Radioastronomie, Bonn, Germany

1. Introduction

ISOPHOT (Lemke et al. 1996) is one of the four instruments on board *ISO*, the Infrared Space Observatory (Kessler et al. 1996), launched in November 1995 by the European Space Agency. As the first true space observatory for infrared astronomy, *ISO* is making the majority of its observing time available to the astronomical community.

ISOPHOT is the *ISO* instrument with the widest wavelength coverage—it is also the most complex. It contains several sub-instruments for performing photometry, polarimetry, and imaging in the infrared range of 2.5–240 μm, and low resolution spectro-photometry in the range 2.5–11.6 μm. Spacecraft raster observations and beam chopping are included in different modes.

The different ISOPHOT capabilities lead to a large number of observing modes and corresponding data products, as well as a large number of calibration files needed both for commanding the instrument (uplink), and for performing data analysis. As a consequence, the instrument calibration and general data analysis are rather complex tasks.

The general requirements, main design concepts, and development of an ISOPHOT interactive analysis system for calibration and scientific analysis are described elsewhere (Gabriel et al. 1996). In this paper, we concentrate on the specific requirements for interactive data reduction by general ISOPHOT observers, and how this is achieved within PIA.

2. ISOPHOT Data Reduction

ISOPHOT data reduction consists basically of three parts:

1. **Removal of instrumental effects.** Instrumental features are corrected using either calibration files or parametric algorithms. Calibration files were established by the ground calibration, and are partly updated in flight using PIA (Gabriel et al. 1997b). The different parameters used by the data correction algorithms are changeable and testable by dedicated menus, showing graphically and textually the results of applying a correction algorithm.

2. **Flux calibration** can be performed by using either default detector responses or calibration measurements, which run in association with almost all astronomical observations (Laureijs et al. 1996). Both possibilities are included in PIA, the second performed by analysing calibration measurements with dedicated tools for reviewing the results.

3. **Combination of data into images/spectra/photometric and polarimetric tables.** The final results of an observation can be obtained by combining data from:

   a) different raster positions, view angles, and detector pixels within a measurement to obtain images (Gabriel et al. 1997a);

   b) different pixels of the ISOPHOT spectrometer within a measurement to obtain spectra and parameterizing functions describing them; or

   c) measurements from different filters, apertures, and polarimeters to obtain photopolarimetric tables.

3. Internal Data Handling and Reduction

The scientific data corresponding to an ISOPHOT measurement is incorporated by PIA into a single entity, an IDL "structure" contained in a corresponding internal buffer. A structure is a collection of several variables of different types. The main variables are pointers to one or two-dimensional arrays.

Data reduction is organized by the various levels of processing. Every structure can be reduced to the next level, also contained in its corresponding buffer. PIA recognizes five main levels of data reduction. Any number of structures can be contained in the internal buffers, allowing access to data from different measurements/levels at any time for comparisons, correlation analysis, etc.

On every reduction level, several data corrections can be performed. PIA is capable of testing and configuring all the parameters used for those corrections, giving information on the outcome in graphical and statistical form. Several selection criteria for discarding data can be applied, which, in combination with the graphical capabilities, represent the optimal data reduction. Information about instrument configuration, observation characteristics, and reduction steps already applied to the data can be recalled from every level.

Input and output are possible in FITS format (FITS 1993), as well as in a PIA internal format, for all levels. This allows for data re-processing from any level in a later session, exporting data to other astronomical software packages, accessing and processing data from the ISOPHOT Standard Product Generation (Laureijs et al. 1996), at every level of reduction.

The graphical user interface (GUI), covering all package features, is user-friendly, easy to understand, and usable by every *ISO* observer without any knowledge of IDL, the PHT instrument, or the data structures. However, it is possible to leave the GUI, and still access all package routines and buffers interactively, at the plain IDL level. Re-entering PIA after performing data reduction in IDL is also possible, allowing PIA users to expand the package.

4. Documentation, On-line Help and Distribution

The PIA User Manual (Gabriel et al. 1996), written in hypertext format, is available via a Web browser of the user's choice from the PIA main menu. Individual chapters are linked to the corresponding graphical interfaces, to be recalled context sensitive. All the interfaces are fully described in the manual, which includes a tutorial, processing algorithm descriptions, data structure definitions, etc. A special chapter on the use of PIA structures and programs outside the GUI is also provided.

In addition, documentation for every PIA routine is given, available via the IDL on-line help system. This facility is intended to allow an "advanced" PIA user to run PIA single programs at the IDL interactive level, or to replace routines in the PIA graphical environment.

The package is freely distributed,[3] including executables, calibration files, and all documentation. It is available for platforms running IDL (from version 3.6 on), and its installation is extremely simple.

[3]During the operational phase of *ISO* PIA is distributed by the ISOPHOT Data Center at MPIA, Heidelberg, Germany. E-mail: phthelp@orion.mpia-hd.mpg.de

References

FITS 1993, Definition of the Flexible Image Transport System (FITS), Standard, NOST 100-1.0, NASA / Science Office of Standards and Technology, Code 633.2, NASA Goddard Space Flight Center, Greenbelt, Maryland

Gabriel, C., Acosta-Pulido, J., Heinrichsen, I., Morris, H., Skaley, D., & Tai W.-M. 1997, in Proceedings of the 5th International Workshop on Data Analysis in Astronomy, Erice, Italy, in press

Gabriel, C., Haas, M., Heinrichsen, I., & Tai, W.-M. 1996, ISOPHOT Interactive Analysis User Manual, available from ESA/*ISO* Science Operations Centre, VILSPA, or MPI Astronomie Heidelberg

Gabriel, C., Heinrichsen, I., Skaley, D., & Tai, W.-M. 1997a, this volume, 116

Gabriel, C., Schulz, B., Acosta-Pulido, J., Kinkel, U., & Klaas, U. 1997b, this volume, 112

IDL 1992, IDL is a registered trademark of Research Systems, Inc., copyright 1992

Kessler, M. F., et al. 1996, A&A, 315, 27

Laureijs, R., Richards, P. J., & Krüger, H. 1996, ISOPHOT Data Users Manual, V2.0, available from ESA/*ISO* Science Operations Centre, VILSPA

Lemke, D., et al. 1996, A&A, 315, 64

Astronomical Data Analysis Software and Systems VI
ASP Conference Series, Vol. 125, 1997
Gareth Hunt and H. E. Payne, eds.

Calibration with the ISOPHOT Interactive Analysis (PIA)

C. Gabriel,[1] B. Schulz[1]

European Space Agency - Astrophysics Division

J. Acosta-Pulido,[1] U. Kinkel,[1] U. Klaas[1]

Max-Planck Institut für Astronomie, Heidelberg, Germany

Abstract. The ISO**P**HOT **I**nteractive **A**nalysis (**PIA**) software was conceived primarily as a tool to be used by the instrument team for the calibration of ISOPHOT. This is one of the four instruments on board the European Space Agency's Infrared Space Observatory (*ISO*), launched in November 1995. Actually, PIA has been extended to form an interactive tool for ISOPHOT data analysis in general, which can be distributed to all interested ISOPHOT observers. In this article, we describe the philosophy behind the implementation of the processing sequences dealing with calibration observations, and how this leads to a general calibration assessment and finally the generation of calibration files, both for commanding the ISOPHOT instrument and reduction of scientific data.

1. Introduction

The ISOPHOT Interactive Analysis (PIA), the analysis tool for ISOPHOT scientific data reduction, is described elsewhere in this volume (Gabriel et al. 1997).

ISOPHOT (Lemke et al. 1996) is the photometric instrument on board *ISO*, the Infrared Space Observatory (Kessler et al. 1996), launched in November 1995 by the European Space Agency.

ISOPHOT is a complex instrument. It contains three subinstruments with various observational modes. This is reflected in a large number of calibration files and procedures, needed both for commanding the instrument ("uplink") and for performing data analysis ("downlink"). In order to facilitate the access, visualization, and analysis of the calibration data, PIA was conceived from the very beginning as a user friendly tool, requiring from the calibration scientists neither a deep knowledge of ISOPHOT data structures and files, nor of the software itself.

The ISOPHOT calibration requirements (IOCRD 1995) were implemented as Calibration Implementation Procedures, describing the detailed observation strategy and measurement sequences, and as Calibration Files Derivation Procedures (CFDP 1995) describing the data reduction path and the derivation of

---

[1]Currently at the *ISO* Science Operations Centre, Villafranca, Apdo 50727, E-28080 Madrid, Spain

the calibration files. On the basis of the latter, calibration processing sequences were set up inside the PIA to automate data reduction steps and combine data to produce the final data products. These are:

- Cumulative information in a form suitable for performing consistency checks for calibration assessment and trend analysis.
- Files used to define the instrument settings controlling an observation, called calibration uplink files (Cal U).
- Files used for calibration of the data along the various processing steps from raw data to final data products (Cal G files). These files are used by both ISOPHOT data reduction packages, the ISOPHOT Standard Product Generation ("pipeline") and the PIA itself. They are also distributed to the observers in FITS format (FITS 1993). A description of standard data processing, data products and Cal G files can be found in Laureijs et al. (1996).

2. Processing Data to Derive Calibration Files

All the (variable) parameters necessary for commanding ISOPHOT are contained in the uplink calibration files. Their determination is crucial for the most optimal set-up of the instrument. Even more stringent requirements apply in the case of the assessment of instrumental effects for the downlink calibration. Here, the best achievable accuracy is required for accurately calibrating the detectors.

Calibration measurements are performed routinely, in addition to those of the extensive Performance Verification Phase (the first two months of *ISO* operations). They are used to increase the accuracy of the calibration, monitor the photometric stability, and look for new, unexpected effects. A database of dedicated tables containing the results of individual measurements has been developed within PIA—these are referred to as "cumulated tables." They allow the user to study the long term behaviour of the various parameters, correlations, interdependencies, and trends.

Data reduction of calibration measurements can be done in several ways. At one extreme is the fully interactive method, which allows detailed inspection at each stage of data reduction. At the other extreme is the fully automatic method, which speeds up routine data reduction. For all cases, a graphical interface is available, which allows access to all variables at all levels of data reduction. It also allows the user to change and test various data reduction parameters. The data reduction follows the steps outlined in the calibration file derivation procedures, written by the calibration scientists. However, it has a high level of flexibility, and allows the user to try alternative ways of processing the data.

There are also certain cases of data correction algorithms where the parameters used are not contained in calibration files. These parameters are testable and changeable via dedicated menus within PIA. The same menus allow investigation of effects and efficiency, and were used by the calibration scientists to obtain the default values, used by the pipeline data reduction.

3. The Graphical Interface

The graphical interfaces for the different calibration items (currently a total of eleven, associated with more than 20 calibration files), were designed individually, but following a general common scheme, including:

- Graphical access to:

  a) the values currently used by the system,

  b) the values contained in the actual calibration files, and

  c) new values which are generated.

- Editing of individual calibration values, allowing individual entries to be manually changed, if necessary.

- The production of both uplink and downlink calibration files. In this case the user may add some information into the file headers, specifying author, date, version, and modifications done, for configuration control. Special routines are then used to produce ASCII or FITS files with the correct format required by the different systems.

For data processing defined in a CFDP, automatic processing sequences were implemented. There are basically two different steps: a) reducing data and adding the results to the "cumulated tables," and b) using all the data recorded in those tables to obtain the final calibration numbers.

A dedicated system for handling the "cumulated tables" has been implemented. The tables can be inspected, variables can be plotted against each other, flags to deselect individual measurements can be edited, etc. Processing the data within the tables to obtain final calibration numbers is done via dedicated menus, which allow selection of all the variables present in the tables.

4. A Simple Example: Derivation of "Dark Current" Values

The residual signal measured without illumination is called the "dark current." This current is always present, and has to be subtracted as a constant. Accurate values for this instrumental effect are especially important for the analysis of very faint sources.

There are several effects which could modify the dark current values, like orbital position of *ISO*,[2] detector temperatures, detector and electronics aging, etc. Therefore, it is necessary to monitor them regularly under different conditions.

The derivation of dark currents is rather simple. However, there are certain effects which have to be corrected for in the data. One is the almost continuous disturbances produced by cosmic particle hits into the detectors during an observation, which are particularly relevant at these low signal levels. Another is given by the signal transient effects due to changing illumination history. These

---

[2] *ISO* has an extreme elliptical orbit with a distance to the Earth of 1000 km at perigee and 70,000 km at apogee

effects are present in all ISOPHOT data. The detailed correction steps in PIA are described in the PIA User Manual (Gabriel et al. 1996).

At the end of the data reduction, the average dark current signal and noise are recorded in the "dark current cumulated table" along with several other parameters: *ISO*'s orbital position, temperature of the detector assembly, observation date, processing date, etc. The calibration scientist can determine the average value from all recorded measurements, look for trends with orbital position and temperature, and deselect measurements whose quality is affected by memory effects of a prior strong illumination or data taken at times of high space weather activity.

5. Experience with PIA and Outlook

The availability of a well suited interactive data reduction package from the very beginning of the *ISO* mission has been a big advantage for a complex instrument like ISOPHOT. From the calibration point of view, it has helped most in two ways: a) discovering and solving problems rapidly, both on the instrumental side and in the data reduction, and b) deriving new algorithms for data correction and calibration strategies. Its user friendliness, provided by the graphical user interface, has been especially important for the first point. The second capability is mainly provided by the flexible use of internal structures and single programs outside the graphical interface. This last point represents one of the main directions that further PIA development is taking.

References

CFDP 1995, Calibration Files Derivation Procedures, PHT-Instrument Dedicated Team internal document

FITS 1993, Definition of the Flexible Image Transport System (FITS), Standard, NOST 100-1.0, NASA / Science Office of Standards and Technology, Code 633.2, NASA Goddard Space Flight Center, Greenbelt, Maryland

IOCRD 1995, ISO In Orbit Calibration Requirements Document, ISO-SSD-9003

Gabriel, C., Acosta-Pulido, J., Heinrichsen, I., Morris, H., & Tai, W.-M. 1997, this volume, 108

Gabriel, C., Haas, M., Heinrichsen, I., & Tai, W.-M. 1996, ISOPHOT Interactive Analysis User Manual, available from ESA/*ISO* Science Operations Centre, VILSPA, or MPI Astronomie Heidelberg.

Kessler, M. F., et al. 1996, A&A, 315, 27

Laureijs, R., Richards, P. J., & Krüger, H. 1996, ISOPHOT Data Users Manual, V2.0, available from ESA/*ISO* Science Operations Centre, Vilspa

Lemke, D., et al. 1996, A&A, 315, 64

Astronomical Data Analysis Software and Systems VI
ASP Conference Series, Vol. 125, 1997
Gareth Hunt and H. E. Payne, eds.

Mapping Using the ISOPHOT Interactive Analysis (PIA)

C. Gabriel[1]

European Space Agency - Astrophysics Division

I. Heinrichsen,[1] D. Skaley[2]

Max-Planck Institut für Kernphysik, Heidelberg, Germany

W.-M. Tai

Dublin Institute for Advanced Studies, Dublin, Ireland

Abstract. The ISOPHOT Interactive Analysis PIA is a full astronomical data analysis tool for data reduction and calibration of ISOPHOT, one of the instruments on board the Infrared Space Observatory *ISO*, launched in November 1995 by the European Space Agency.

This article is devoted to a description of the image processing capabilities of PIA, on the basis of the different mapping strategies with ISOPHOT. PIA offers a full graphical interface, giving the user all the informations related to the observation and data selection possibilities. Special flat fielding techniques, extraction of profiles, map rotation and convolution, point source extraction, three dimensional display, etc., are implemented in an interactive way.

1. Introduction

The ISO**PHOT** **I**nteractive **A**nalysis **PIA** is described elsewhere in this volume (Gabriel et al. 1997). It was conceived primarily as a calibration tool for ISOPHOT (Lemke et al. 1996), one of the four instruments on board *ISO*, the Infrared Space Observatory (Kessler et al. 1996). However, the software package has been developed into a full interactive astronomical analysis system for ISOPHOT data.

ISOPHOT performs infrared mapping in different modes using different detector subsystems. A detailed description of these modes can be found in the ISOPHOT Observer's Manual (Klaas et al. 1994). Mapping can be performed using one of the three single pixel photometers (P1, P2, P3) in the range 2.5–100 μm, or one of the two far infrared cameras (C100, a 3×3, and C200, a

[1] Currently at the ISO Science Operations Centre, Villafranca, Apdo 50727, E-28080 Madrid, Spain

[2] Now at the Max-Planck Institut für Radioastronomie, Bonn, Germany

2×2 detector array) in the range 50–240 μm. Several filters in those ranges and apertures (in the case of the single photometers) are available.

The raster capability of *ISO*, pointing sequentially to several positions on a two-dimensional grid, makes it possible to have the combination of individual fluxes and an image of a sky region. ISOPHOT measures continuously with a fixed instrument configuration during the raster performance. The PIA mapping software basically combines the sky brightnesses measured at the different positions to an image in sky coordinates.

2. Mapping with ISOPHOT

A raster observation can have a maximum of 32×32 raster points with a maximum sampling area of 1.6×1.6°. The *ISO* Observer's Manual (ISOOBS 1994) gives full information on different aspects of *ISO*'s raster observations.

There are two mapping modes possible using ISOPHOT:

- **Staring raster mode** consists of several staring observations on a regular grid. It is possible to oversample. The single ISOPHOT photometers can be used with different raster points separated by a minimum of 8″ (during the first year of the *ISO* mission this was 13″). For the long wavelength array detectors, the minimum separation is 15″ (1/3 of a C100, 1/6 of a C200 pixel).
- **Chopped raster mode** can be used with the long wavelength detector arrays, in order to achieve a high spatial resolution, while optimizing the observation time. In this case, data are taken in chopper steps of 15″ in the spacecraft Y-direction. This also allows observing the same celestial position during several raster pointings, therefore making possible the elimination of transients in detector response.

3. Mapping using PIA

As input for mapping processing, PIA uses ISOPHOT data which has been reduced to the level of sky brightness (in MJy/sr) for each raster (and chopper) step and per detector pixel, together with the associated pointing information per raster point. All these data, corresponding to one measurement, are contained in an element of the so called AAP (astrophysical application) data structure. A description of the data reduction from the raw telemetry to this level as done by PIA can be found in Gabriel et al. (1996).

On this level, we must deal with data derived from a measurement: an array of measured brightnesses, their uncertainties, associated sky positions and corresponding observation times of these positions. PIA calculates the positions of the individual detector pixels at different raster/chopper positions during the raster measurement. After this calculation, the values of the detector signals are binned into map pixels. A simple gridding function is applied, which is a trapezoidal function, i.e., the geometric overlap of detector pixel and map pixel is used as the contributing part of the measured detector brightness to the map pixel. For the final image computation, PIA uses the coverage and the time of coverage of all the contributing signals for normalization.

PIA produces three kinds of maps for each measurement:

- A **brightness map**, computed from either mean values, median values, or first or third quartile values, depending on the user's selection. The unit is [MJy/sr].
- An **uncertainty map**, processed from the uncertainty array in the AAP data. The unit is [MJy/sr].
- An **exposure map**, which is the map of exposure times for each map point in [s]. The exposure time is the total of all exposure times from the contributing detector signals, corrected for the map pixel size.

3.1. Options for Producing a Map

PIA allows a free choice for the image binning, although there is a "natural" choice, given by the level of oversampling reached in the observation. The pointing taken for the image computation is given by the measured positions, which can differ slightly from the planned ones.

For data obtained with one of the detector arrays, there is also the possibility of selecting/deselecting detector pixels to use only part of the data, or obtaining maps from individual detector pixels. This may help judge the quality of parts of the map, reveal flat fielding problems, etc.

Flat fielding of the detector arrays is in principle given, since measurements of the internal fine calibration sources are performed before and after the raster source measurement. Nevertheless, the possibility of using an additional flat fielding technique is given. If this option is chosen, individual maps are produced for every detector pixel and the central, common region of the maps taken for obtaining flat fielding factors.

3.2. Displaying a Map

Once the map is produced, the PIA graphical interface offers several possibilities for enhancing the quality of the image display. Starting from a Map Display Window, several context sensitive menus allow changing the color tables, interpolating image pixels, zooming every map region with different zoom factors, setting cut values, overplotting contours to the map, obtaining profiles, extracting flux values and positions from the map, and extracting possible point sources.

It is also possible to obtain a three-dimensional surface from the map, using an interface allowing rotation about each axis and super-position of contours.

3.3. Map Transformations

PIA includes the option of convolving a map to a given spatial resolution. To facilitate comparisons between maps obtained in different wavelengths, conversion to the resolution of every PHT-filter is available. A two-dimensional Gaussian approximation to the point spread function is used for the convolution.

Maps can be also rotated by PIA to every angle with respect to the RA-DEC plane.

3.4. Input/Output

The main input for mapping with PIA is the AAP data, resulting from the reduction of a raster measurement. All of the mapping capabilities just described may be applied to these data.

Several additional formats can be used both as input to and output from the PIA imaging software:

- Internal format data, for use within PIA in a future session.
- FITS format (FITS 1993), providing an interface to basically every other package for instrument independent image processing, and to PIA itself.
- PostScript output to a file or printer, and GIF output to a file.

4. Further PIA Development on Mapping

The very simple trapezoidal gridding function used for obtaining an image by PIA will be complemented with enhanced imaging methods. Using the redundant information from a single detector pixel should lead to better spatial resolution. The same redundant information will be used for modeling detector response transients during an observation. Flat-fielding may also be enhanced by taking a user defined area (preferably a background area) for computing the flat-fielding factors. Measured point spread profiles will be used for image convolution and for point source extraction.

References

FITS 1993, Definition of the Flexible Image Transport System (FITS), Standard, NOST 100-1.0, NASA / Science Office of Standards and Technology, Code 633.2, NASA Goddard Space Flight Center, Greenbelt, Maryland

ISOOBS 1994, *ISO* Observer's Manual,[3] *ISO* SOC Team.

Gabriel, C., Acosta-Pulido, J., Heinrichsen, I., Morris, H., & Tai, W.-M. 1997, this volume, 108

Gabriel, C., Haas, M., Heinrichsen, I., & Tai, W.-M. 1996, ISOPHOT Interactive Analysis User Manual[3]

Kessler, M. F., et al. 1996, A&A, 315, 27

Klaas, U., Krüger, H., Heinrichsen, I., Heske, A., & Laureijs, R. (eds), 1994, ISOPHOT Observer's Manual[3]

Laureijs, R., Richards, P. J., & Krüger, H. 1996, ISOPHOT Data Users Manual,[3] V2.0

Lemke, D., et al. 1996, A&A, 315, 64

---

[3]available from ESA/*ISO* Science Operations Centre, VILSPA, or MPI Astronomie Heidelberg.

Astronomical Data Analysis Software and Systems VI
ASP Conference Series, Vol. 125, 1997
Gareth Hunt and H. E. Payne, eds.

IMAGER: A Parallel Interface to Spectral Line Processing

Doug Roberts
University of Illinois/NCSA

Dick Crutcher
University of Illinois/NCSA

Abstract. We report on the development and use of the IMAGER system at NCSA. IMAGER is an interface to parallel implementation of imaging and deconvolution tasks of the Software Development Environment (SDE) of the NRAO. The interface is based on the MIRIAD interface of the BIMA array and it allows for interactive and batch operation. The parallelization is carried out by distributing independent spectral line channels across multiple processors. The use of this system on the SGI Power Challenge array at NCSA is presented. For most problems, the speed-up with number of processors is nearly linear. We present results from our use of IMAGER on a data reduction of multiple-pointing BIMA observations of molecular gas around the "Sickle" near the Galactic center. We also report on the general use and timing results of this system at NCSA.

1. Introduction

There are many problems in observational astronomy that can be solved only with high performance computers. Because computers are the optical element of a synthesis telescope, radio synthesis data reduction has been one of the most computer intensive operations in observational astronomy. The sizes of data sets can frequently be large. In the some cases, such as pulsar observations, short time samples lead to large data sets. In the common case of radio spectral line observations, large numbers of frequency channels lead to large amounts of data. It is not uncommon with such instruments such as the VLA and the BIMA telescopes to have spectral line data sets in excess of a gigabyte. Astronomers need access to fast processing to allow the analysis of such large data sets. It is especially important to have analysis capabilities that allow astronomers to use different methods of non-linear deconvolutions in a timely basis to properly interpret their observations.

The analysis of spectral line data, in which each channel is independent from every other channel, is an embarrassingly parallel problem. Such problems have little or no data dependence and require very little communication between individual processors. Thus, an additional motivation for this project is the need to demonstrate that parallel execution of embarrassingly parallel problems can be accomplished in a straightforward manner. In order to prototype a solution to spectral line data reduction and to provide a substantial performance increase from conventional packages such as AIPS and MIRIAD, the IMAGER package was developed.

2. IMAGER Implementation

We decided to implement a three dimensional data reduction package starting with existing software to carry out 2-D data reduction. This involved writing a user interface and control package to send the data and data reduction software for separate channels to individual processors and collect the results into a data cube.

We intended to use an efficient 2-D data reduction package and decided on the SDE package written by Tim Cornwell of NRAO. SDE code was written assuming that the visibility and image data could fit into the virtual memory. We developed IMAGER to run optimally on the SGI Power Challenge Array of NCSA. The total physical memory is 4 GB; thus almost all problems will fit into physical memory and will not require the program swapping to disk. SDE requires that visibility data be in UVFITS format with each channel in a separate file. Additionally, multiple pointing data sets have all pointings in a special "mosaic" database. Thus, functionality for data conversion is included in the IMAGER package.

In addition to the underlying 2-D programs, we needed to create an interface to the package that handled user inputs, documentation, and execution. We decided to use the MIRIAD shell for the interface to IMAGER. The MIRIAD interface allows inputs to be saved and recalled, help documents to be browsed, and IMAGER programs to be executed. The documentation and user input system includes normal and expert modes. The normal mode hides less frequently used inputs (and the associated documentation), and the expert mode provides access to all inputs and help.

The final part of the package carries out parsing the inputs, distributing the channels to the individual processors and accumulating the results into a data cube. History and logging information are also managed. Two large Perl scripts written to carry out these actions constitute the core of the IMAGER system.

These data reduction processes were implemented in IMAGER:

- **uvmap**—Inverting the visibility data into the image plane, without subsequent deconvolution (i.e., creating "dirty" images from visibility data).
- **clean**—Deconvolution of image plane data created with the **uvmap** program using the CLEAN algorithm.
- **mosaic**—Iterative imaging and MEM deconvolution; includes mosaics of multiple fields.

The data reduction steps can be carried out in a pipelined manner or they can be done one step at a time. For instance, a dirty cube can be created in one step (using **uvmap**) and saved; then **clean** can be run several times to different output files using different parameters, (e.g., clean iterations, flux limits). Alternately the **uvmap** and **clean** steps can be pipelined together and executed on one step. Both interactive and batch execution are supported.

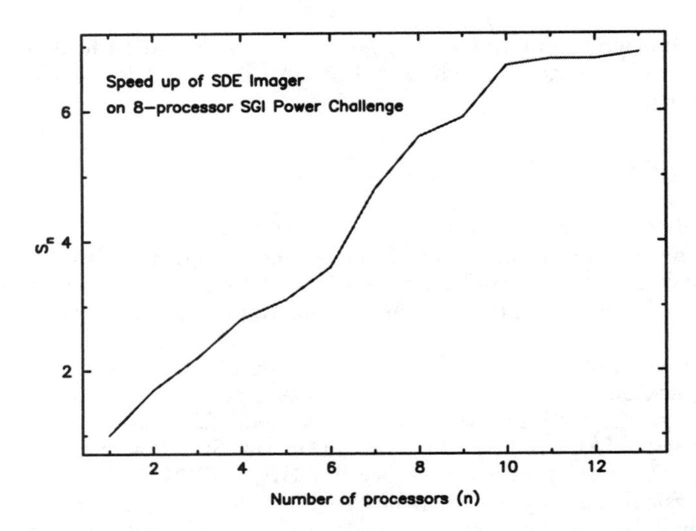

Figure 1. Speedup of IMAGER program with number of processors.

3. Results

The IMAGER system has been used by astronomers at the University of Illinois to carry out analysis of data from the VLA and BIMA telescopes. In all projects, there is an initial overhead for the data conversion into the proper format. The worst case is that of multiple-field spectral line data, in which initially data for all channels of a single pointing are in separate files (one file for each pointing). The data conversion stage separates each pointing and channel and recombines the pointings together, resulting in a separate file for each channel. In the worst case, the overhead may be 50% of the total execution time. However, the IMAGER package is intended to give astronomers the power for iterative data reduction; thus in all real cases this step is done only once and subsequent steps (those optimized for parallel execution) are carried out repeatedly.

The projects have used the SGI Power Challenge at NCSA to carry out computer processing. In addition to production scientific processing, we carried out tests in order to determine the speedup on the 8-processor SGI Power Challenge. Our test problem was the imaging (**uvmap**) and CLEAN deconvolution (**clean**) of a single pointing observation of the molecular gas (the CS line) associated with the "sickle" HII region near the Galactic center. The data were acquired with the BIMA interferometer. The visibility data set was about 300 MB and the output images were 256 × 256 pixels × 100 channels. All timing tests were carried out after the initial conversion of data formats. The number of threads were varied from 1 to 13. The speedup ($S_n = t_1/t_n$, t_1 is time for a single processor run, t_n is time using n processors) as a function of the number of threads is shown in Figure 1. The speedup is nearly linear and saturates at 7; this is one less than the number of processors because the control program is running on one processor. The slope is a bit less than one because when the control script detects that a channel has been completed, a new channel is sent, but only after a bit of delay.

Also compared in this test was the performance between the same operations carried out with the AIPS, MIRIAD, and IMAGER packages. The single processor timing between the three packages is shown in Table 1. The AIPS package was by far the slowest. However, the computational problem was carried out in AIPS in a manner that required the mapping program (HORUS) to be run once for the cube and the CLEAN deconvolution program (APCLN) run once for each *channel* (i.e., for this test 100 times). The AIPS code was written assuming a very small memory model machine; AIPS writes many scratch disk files for each program execution. The I/O caused by these scratch files dominates the execution time of the programs and is responsible for the poor performance relative to MIRIAD and IMAGER. The I/O for the various programs is given in Table 1 as well.

Table 1. Timing results between packages.

Parameter	AIPS	MIRIAD	IMAGER
Execution time (sec)	1041	247	261
I/O overhead	43%	2%	2%

4. Discussion and Future Plans

The IMAGER package is currently supported on the SGI Power Challenge array at NCSA. Astronomers are using the IMAGER system to analyze radio synthesis data. Anyone wishing further information on this can view the on-line documentation[1] or contact Doug Roberts directly.

One of the goals of this project was to demonstrate the ease of programming and use of a embarrassingly parallel data reduction package. We feel that this has been accomplished. The next step was to use this experience and design a plan for the coding of embarrassingly parallel algorithms and compute intensive algorithms of the AIPS++ project. The plan for parallelization of AIPS++ has been created. Work will begin after the beta release of AIPS++ in Jan 1997. Parallel functionality at some level should be available in AIPS++ by the time of the first full release of AIPS++ in mid-1997.

[1] http://monet.ncsa.uiuc.edu/hpcc/imager/imager.html

Astronomical Data Analysis Software and Systems VI
ASP Conference Series, Vol. 125, 1997
Gareth Hunt and H. E. Payne, eds.

Tcl- and [Incr tcl]- Based Applications for Astronomy and the Sciences

Nicholas M. Elias II
United States Naval Observatory, Navy Prototype Optical Interferometer, 3450 Massachusetts Avenue NW, Washington, DC 20392-5420, e-mail: nme@fornax.usno.navy.mil

Abstract. Tcl is a shell/script language used to create both instrumental control systems and interactive data reduction programs. It has extensions for creating GUI (Tk) and object-oriented ([incr tcl]) applications. Several Tcl-based tools that may be used for astronomical and other scientific applications have been created and are discussed. One example is *ptcl*, which registers PGPLOT functions as Tcl commands, creating a powerful interactive plotting package. The Tcl-astronomy WWW homepage and majordomo mailing list server are also described.

1. Introduction

Tcl is a general-purpose shell and script language. It is maintained by J. K. Ousterhout,[1] and is available free of charge over the World Wide Web.[2]

Tcl consists of a powerful set of "core" commands that look like a cross between C functions and (t)csh commands. This core includes math functions, variables, associative arrays, string and list commands, flexible parsing/substitution, program control (for, while, if, etc.), regular expressions, I/O, and error handling. New Tcl commands can be added either by "registering" C/FORTRAN functions or using the "proc" Tcl core command. Tcl can also access UNIX shell commands without awkward escape sequences. Therefore, Tcl may be used as "glue" to create exciting new software packages.

Tcl has some advantages over compiled languages such as FORTRAN, C, or C++. For example, programming ideas can be tested "on the fly" (while the application is running), decreasing development time. Also, Tcl installs on any UNIX workstation that has an ANSI-C compiler (versions for MS-Windows and Macintosh exist as well), making Tcl-based applications very portable. In addition, Tcl can be bundled with other software with no restrictions or charges.

2. The Most Common Tcl Extensions, Tk and [Incr tcl]

Most scientists use X-Windows based window managers to interact with the UNIX operating systems on their workstations. Creating graphical user inter-

[1] http://www.sunlabs.com:80/people/john.ousterhout

[2] http://www.sunlabs.com/research/tcl

faces (GUIs) in C is straightforward but very time consuming, especially in the debugging stage. J. K. Ousterhout's answer to this problem is the Tcl extension called Tk,[3] an X-Windows toolkit. This toolkit consists mainly of window widgets registered as Tcl commands. With Tk, all the annoying details of X-Windows are hidden from the programmer, allowing him/her to create GUIs in minutes to hours instead of hours to days. More complicated "mega-widgets" (typically combinations of the standard widgets) can be created by writing C code, if desired. Like Tcl, Tk installs on any UNIX workstation (it must have X-Windows, of course) that has an ANSI-C compiler, and Tcl/Tk scripts are totally portable (no machine dependent C code is necessary).

Object-oriented programming systems (OOPS) are becoming more and more prevalent because of their many advantages, such as configurable class templates, inheritance, and "members-only" manipulation of data (encapsulation). [Incr tcl][4] is an object-oriented extension of Tcl, created by M. J. McLennan.[5] In addition to the above advantages, [incr tcl] is useful for grouping a large number of Tcl commands into classes, making a command-line application much more manageable.

3. Tcl-Based Tools and Software

Many software libraries have been created over the years by scientists for scientists, and have become standards in the community. These packages can be made interactive using Tcl, thus extending their usefulness. For example, PGPLOT[6] (the Caltech plotting package written by T. J. Pearson) library functions have been registered as Tcl commands in the *ptcl*[7] package. Also, HDS[8] (the Hierarchical Database System of the Starlink Project) functions have been registered as Tcl commands in the *htcl*[9] package.

A sample *ptcl* script and its corresponding PGPLOT FORTRAN subroutine are shown in Figure 1. The Tcl script and FORTRAN code are very similar, which means that someone who is already familiar with PGPLOT can learn *ptcl* quickly. Also, recall that Tcl scripts can be modified and tested while an application is still running, which means that the Tcl procedure quad_plot can be modified more quickly than the FORTRAN subroutine QUAD_PLOT.

Software for the Navy Prototype Optical Interferometer (NPOI) laser metrology system[10] has been created using [incr tcl], *ptcl* and *htcl*. The general structure of these programs (bottom to top) is: low-level C code, mid-level C code, C code that registers mid-level functions as Tcl commands, and [incr

---

[3]http://www.sunlabs.com/research/tcl

[4]http://www.tcltk.com/itcl/index.html#moreInfo

[5]http://www.tcltk.com/itcl/mmc.html

[6]http://astro.caltech.edu/~tjp/pgplot

[7]ftp://fornax.usno.navy.mil/dist/ptcl/ptcl.html

[8]http://star-www.rl.ac.uk

[9]ftp://fornax.usno.navy.mil/dist/htcl/htcl.html

[10]http://aries.usno.navy.mil/ad/npoi/npoi.html

```
proc quad_plot {x} {

  set n [llength $x]

  set y ""
  foreach x2 $x {
    lappend y [expr $x2*$x2]
  }

  pgbeg 0 /xs 1 1

  pgsci 1
  pgsch 1.3

  set xmin [lindex $x 0]
  set xmax [lindex $x [expr $n-1]]
  set ymin [lindex $y 0]
  set ymax [lindex $y [expr $n-1]]
  pgenv $xmin $xmax $ymin $ymax 0 0

  pglab x y "Quad Plot"

  pgsci 2
  pgpt $n $x $y 17

  pgsci 4
  pgline $n $x $y

  pgend

}
```

```
      SUBROUTINE QUAD_PLOT( N, X )

      INTEGER*4 N
      REAL*4 X(N),Y(N)

      DO 10 I=1,N
        Y(I)=X(I)*X(I)
   10 CONTINUE

      CALL PGBEG( 0, "/xs", 1, 1 )

      CALL PGSCI( 1 )
      CALL PGSCH( 1.3 )

      CALL PGENV( X(1), X(N), Y(1),
     :     Y(N), 0, 0 )

      CALL PGLAB( "x", "y",
     :     "Quad Plot" )

      CALL PGSCI( 2 )
      CALL PGPT( N, X, Y, 17 )

      CALL PGSCI( 4 )
      CALL PGLINE( N, X, Y )

      CALL PGEND

      RETURN

      END
```

Figure 1. A simple *ptcl* procedure to plot the square of an array. For the sake of comparison, a FORTRAN subroutine using PGPLOT subroutines is also shown.

tcl] classes corresponding to each type of data (logs, raw data, averaged data, configuration information, etc.). The [incr tcl] classes can be accessed directly from the Tcl command line or from GUIs. There are two main programs in this package, FAKE (creates simulated data) and INCHWORM (laser interferometer and environmental sensor analysis). Both programs use the same C code, Tcl commands, and [incr tcl] classes; only the GUIs are different. A library of Tcl scripts are also included.

4. Tcl and the World Wide Web

New "bytecode" languages, such as Java, are now being used to create exciting Web-based applets. Tcl can now be considered a "Web language", since Tcl/Tk bytecode "plug-ins"[11] have been created for Netscape, i.e., Netscape can be used to run Tcl/Tk applet scripts.

A Tcl-astronomy homepage on the World Wide Web is now being maintained by the author. *ptcl* and *htcl* can be obtained here, and links to other Tcl pages are provided as well. In addition, De Clark (de@ucolick.org) maintains a majordomo mailing list manager called "tclastro"; check the Tcl-astronomy home page for directions on how to subscribe.

[11] http://www.sunlabs.com/research/tcl/plugin

Astronomical Data Analysis Software and Systems VI
ASP Conference Series, Vol. 125, 1997
Gareth Hunt and H. E. Payne, eds.

POW: A Tcl/Tk Plotting and Image Display Interface Tool for GUIs

L. E. Brown

Hughes STX for NASA/GSFC, Code 664, Greenbelt, MD 20771, E-mail: elwin@redshift.gsfc.nasa.gov

L. Angelini

USRA for NASA/GSFC, Code 664, Greenbelt, MD 20771, E-mail: angelini@lheavx.gsfc.nasa.gov

Abstract. We present a new Tcl/Tk based GUI interface tool which features plotting of curve and image data and allows for user input via return of regions or specific cursor positions. The package is accessible from C, Tcl, or FORTRAN. POW operates on data arrays, passed to it as pointers. Each data array sent to POW is treated as either an Image object or a Vector object. Vectors are combined to form Curves. Curves and Images may then be combined to form a displayed Graph. Several Graphs can be displayed in a single Tk toplevel window. The Graphs can be rearranged, magnified, and zoomed to regions of interest by the user. Individual graph axes can be "linked" to implement a "multiple y-axis" (or x-axis) plot. The POW display can be written out in PostScript, for printing.

1. Introduction

With the development of several GUI based tools for High Energy Astrophysics at the HEASARC, we found a need for a interface tool that supports combined curve plotting and image display in a native Tcl/Tk environment. POW is our answer.

2. Design Goals

The principal design goals for POW were:

1. must coexist with (almost) any language (FORTRAN, C, Tcl/Tk),
2. must allow plotting of both images and curves on a single graph,
3. must allow display of multiple graphs at once,
4. must allow return of arbitrary information to calling program,
5. should have an intuitive GUI interface, and
6. should be built with widely supported free software.

3. Implementation

We used the existing Tk extension VISU for our image display engine, and cannibalized existing plotting code written for the *fv* project for the curve plotting and axis drawing. All "drawing" is done as "items" on a Tk "canvas." The result is a simple and powerful tool which fits smoothly into the highly customizable Tcl/Tk environment.

The power of Tk then allows for a very general interface to receive input from the user. Tk allows the developer to query raw cursor positions on the canvas. POW provides the developer with several utilities for translating cursor positions into physical coordinates and in which graph a given cursor position falls. This allows the developer to write whatever sort of interface makes the most sense for his application, using simple Tk code.

4. Structure

All elements in POW may be thought of as "Objects" (see Table 1)

Table 1. The object types in POW.

Object	Created from
Data	a (void) pointer to some data
Vector	Data plus units
Curve	2 to 6 Vectors
Image	Data plus origin, size of pixels, and units
Graph	List of Images and Curves (this is the Displayed object)

5. GUI

The end user can perform many activities on displayed Graphs via the POW GUI:

- graph objects can be moved around, zoomed, magnified, and panned,
- individual items (lines, text, images) can be deleted or changed from the GUI,
- applications programmers or end users (via a dialog box) can add features by sending arbitrary scripts to the Tcl interpreter, and
- the POW display can be dumped to a file in PostScript or PPM formats.

6. POW "Methods"

These methods are implemented as C functions, TCL procs, or FORTRAN subroutines (via the black magic of cfortran.h). (The C/FORTRAN versions carry a status variable to fit FTOOLS conventions.)

6.1. Object Constructors

- PowInit(): Creates a POW window.
- PowCreateData(data name, data array, data type, length, copy): Passes a pointer to some data to POW and gives it a name. Pow can make its own copy of the data or use yours, whichever best suits your memory management needs.
- PowCreateVector(vector name, data name, offset, length, units): Gives 1-D physical meaning to a chunk of Data. There is also a function to create a vector and its associated data given a start value and an increment.
- PowCreateCurve(curve name, x vector, x error, y vector, y error, z vector, z error): combine vectors into a Curve.
- PowCreateImage(image name, data name, xoffset, yoffset, width, height, xorigin, xinc, yorigin, yinc, xunits, yunits, zunits): Give 3-D physical meaning to a chunk of Data.
- PowCreateGraph(graph name, curves, images, xunits, yunits, xlabel, ylabel, xdim display, ydim display, xmin, ymin, xmax, ymax): Display a list of curves and images as a Graph on the POW canvas. Most of these parameters are optional and will default to "sensible" values if left NULL.

6.2. General Utilities

This is only a partial list.

- powPlotCurves(graph, curves): Adds the list of Curves to an existing Graph.
- powPlotImages(graph, images): Adds the list of Images to an existing Graph
- powMagGraph(graph, newmagstep): Resizes a Graph to a given magstep (magstep = 1 is the "natural" size of the graph. Magsteps must be integers or 1/integers).
- powStretchGraph graph factor: Shrinks or expands a Graph to the allowed magstep nearest to (current magstep × factor).
- powFindCurvesMinMax(curves, axis): Takes a list of curves and an axis and returns the minimum and maximum values of the curve along that axis.

6.3. Cursor Positions

The developer can bind mouse clicks (or other X events) to send him cursor positions. The following routines can then make sense of the positions.

- powGraphToCanvas(graph, axis, coordinate): Takes a physical coordinate and returns the corresponding position on the POW canvas.
- powCanvasToGraph(graph, axis, coordinate): Takes a canvas coordinate and returns the corresponding physical coordinate. The second argument specifies x or y axis.
- powWhereAmI(x, y): takes an (x, y) pair of canvas coordinates and returns which graph (if any) in which they fall.

6.4. Linked Axes

POW allows you to "link" together any number of axes on different graphs. The resulting set of linked axes is called a "chain." Each axis can be a member of only one chain. Linking an axis from one chain to an axis in another chain has the effect of merging the two chains. Zooming on a region of interest on one graph will affect the linked axis on any other graph. There are several utility routines. Also, the GUI allows the user to view links as lines connecting linked axes.

- powLinkAxes(graph1, axis1, graph2, axis2): Links two axes.
- powBreakLink(graph, axis): Removes the specified axis from its chain.
- powAlignChain(graph, axis, orientation): Moves all graphs belonging to the same chain as the specified graph so that they are aligned on the canvas (i.e., it "stacks" the graphs into a column or lines them up in a row on the user's screen).

7. Availability

An early version of POW is integrated into *fv* in the FTOOLS3.6 distribution and the standalone *fv* distribution. A much improved version of the POW library should be available by the time these proceedings are published. All of this software is available from the HEASARC Web site[1] under the link labeled "Software".

"The road of excess leads to the Palace Of Wisdom"
The Marriage of Heaven and Hell
William Blake

[1] http://heasarc.gsfc.nasa.gov/

Astronomical Data Analysis Software and Systems VI
ASP Conference Series, Vol. 125, 1997
Gareth Hunt and H. E. Payne, eds.

An ExpectTk/Perl Graphical User Interface to the Revision Control System (RCS)

R. L. Williamson II

Space Telescope Science Institute, Baltimore, MD 21218,
E-mail: ramon@stsci.edu

Abstract. In order to track changes in the TABLES/STSDAS code, as well as for Quality Assurance and to simplify the code release process, the GNU Revision Control System (RCS) was chosen as the method of revision control. Due to special needs of our group and the desire to have an easy-to-use interface to the RCS system that could be used not only for TABLES/STSDAS but for personal code as well, a suite of Perl scripts and an Expect/Tk graphical user interface was written. This user interface will be described as well as the functionality of the Perl scripts.

1. Introduction

Due to the size and complexity of the TABLES/STSDAS system, it was decided to place the packages under Revision Control. A survey of the Revision Control software available led to the choice of the GNU Revision Control System (RCS) as the method of revision control. This choice has many advantages:

- Free,
- Ease of Use,
- Secure, but not rigidly so,
- Tracks versions of code by version number or any number of logical version names, and
- Version numbers can be placed in comments in the code and executables, either manually or automatically by RCS so that the pedigree of executables can be determined.

Of course, as with many choices there are disadvantages:

- Only works at most with one directory at a time—doesn't do trees,
- Wild cards will do all files—including files like libraries and executables that are not under revision control,
- No mechanism for removing files from a distribution, yet leave the file in the system for checking out older versions of the packages, and
- Can be difficult to use if the source directory and the working directory are different.

Solution: A set of Perl scripts that enable directory tree check in and out, excluding some files and placing and keeping track of links, as well as retiring files. For ease of use, a User Interface is also needed.

2. Perl Scripts

Three Perl scripts control most of the actual checking in, checking out, and retiring of files.

2.1. checkin

This script is used to initiate the revision control and update changes to the code in the revision control system. Object files, libraries, executables, and directories are excluded from the checkin, and unless told otherwise will descend trees, checking in all files found there.

2.2. checkout

This script is used to check files out of the RCS and places them in a working directory to be examined and edited. Unless told otherwise, checkout will also descend trees, checking out all files found there.

2.3. doretire

This script takes the path to a file to be retired, and removes it from the distribution, while retaining it in a Crypt directory for access by checkin and checkout. Checkin and checkout automatically look for this Crypt directory and check the versions of the files there to determine whether they should be checked out as well.

3. Expect/Tk Graphical User Interface

The part of the system that the user sees is the Graphical User Interface. This code was written in Expect/Tk in order to take advantage of the ability to have an interaction window where the output from the RCS commands are shown, but also the user can reply to queries by the code.

This GUI connects to the checkin, checkout, and doretire commands through the action buttons in the middle of the GUI. Here each widget is described with respect to the number labels in Figure 1.

1. Package Radiobuttons

   These radiobuttons are used by Check Out and Do a mkpkg to determine the name of the package to act on. Only one option can be chosen at a time. These buttons are turned off when Check In is chosen, since they are not used by this task.

2. System Directory Radiobuttons

   These radiobuttons toggle which major branch of the chosen package is checked out or mkpkg done on. For STSDAS or TABLES for example, you can choose top, lib, or pkg. "Top" refers to the top-level directory.

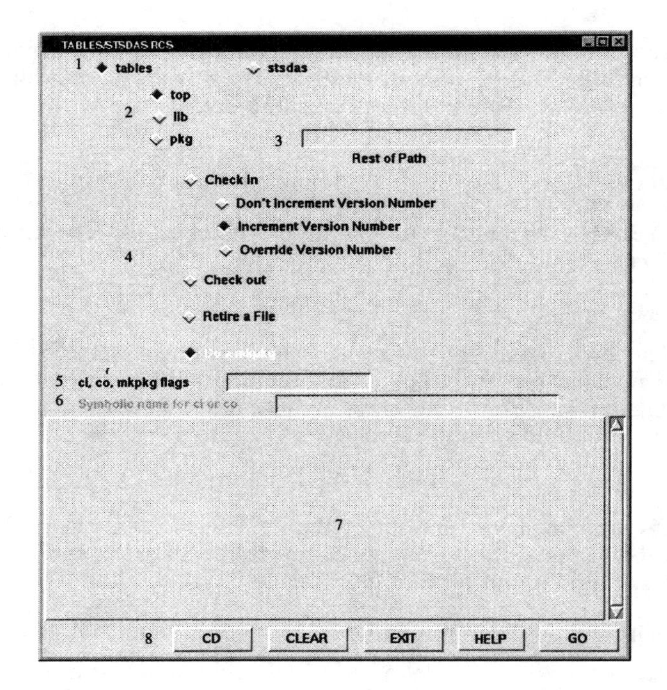

Figure 1. RCS Graphical User Interface.

Only one can be chosen. These buttons are turned off when Check In is chosen, since they are not used by the task.

3. Rest of Path Entry Widget

   This entry widget is for typing in the path to the package desired past the lib, pkg, etc. For example, if you would like to checkout the Synphot package, found in stsdas$pkg/hst_calib/synphot, you would type "hst_calib/synphot" in this widget after setting the Package and System Directory radiobuttons for stsdas and pkg. This widget is deactivated when Check In is chosen, since it is not used by this task.

4. Action Radiobuttons

   These radiobuttons toggle which action to take on the files. You can check filesinto the RCS system, Check files out of the RCS system, Retire a file or do a mkpkg within the configured directories for the packages. Only one can be chosen.

5. ci, co, mkpkg flags Entry Widget

   This entry widget is for typing any flags or arguments that might be recognized by RCS's ci or co, or IRAF's mkpkg. This widget parses the entry. Say, for example, you want to use mkpkg and the flag you want to use is -p tables and an argument is update. Then in the widget type `-p tables update`. The widget parses this and sends the parameters to mkpkg correctly. One flag added to this GUI is the only flag. Used in coordination with checkin or checkout, this flag tells the task to only check in or out the directory chosen, no subdirectories. This is especially useful when dealing with files in the top-level directory.

6. Symbolic name for ci or co Entry Widget

   This entry widget is for typing in the symbolic name for the code to be checked in or out. Multiple values can be placed in the widget space delimited for assigning multiple symbolic names to checked in code. This widget parses the Symbolic names and adds "-n" to the name for use by ci or co. Information in this widget are not used by mkpkg, so this widget is deactivated when the "Do a mkpkg" radiobutton is chosen.

7. Interaction Widget

   This widget is similar to a console for the tasks spawned from this GUI. All output from the tasks are sent to this widget, and any prompts for input will go here as well. To reply to the task, simple place the mouse cursor in this widget and type your reply.

8. Control Buttons

   These buttons control many of the aspects of the GUI.
   CD—Changes the working directory.
   CLEAR—Clears all entry widgets, and resets all buttons to the default values.
   EXIT—Kills all processes spawned from the GUI and exits the GUI.
   HELP—On-line help for the widgets.
   GO—Initiates the chosen action.

4. Personal Packages

The STSDAS RCS system can be customized using two configuration files which control the colors of the GUI and the packages under revision control. Almost any number of packages can be under configuration by adding the packages to these configuration files.

5. Availability

The STSDAS RCS system may be obtained by contacting the author at ramon@stsci.edu. Included will be instructions for installation as well as a user's guide.

Astronomical Data Analysis Software and Systems VI
ASP Conference Series, Vol. 125, 1997
Gareth Hunt and H. E. Payne, eds.

The Portable-CGS4DR Graphical User Interface

P. N. Daly

Joint Astronomy Centre, 660 N. A'ohōkū Place, Hilo, HI 96720, USA

Abstract. This paper describes the elements and style of the Portable-CGS4DR graphical user interface built around some extensions to Tcl and the Tk toolkit.

1. Introduction

CGS4DR (Daly 1995) is a suite of tasks designed to reduce and analyze data, automatically, from the UKIRT long-slit spectrometer. These co-operative tasks use a message system (AMS) and a shared memory noticeboard system (NBS). There is also a command line interface (ICL), although it is cumbersome to use for this application.

Tcl (Ousterhout 1994) is an embeddable scripting language and *Tk* is a graphical user interface toolkit. Both have become accepted in a range of disciplines. They are freely available, run on a variety of architectures, and are extensible. These features make them attractive to the astronomical community.

Since the port of CGS4DR to UNIX, the question of a user interface has been reviewed. Several extensions to Tcl (Terrett 1995a,b) have been introduced that provide access to AMS and NBS. It is these extensions that are the basis for the CGS4DR graphical user interface.

2. Extensions to Tcl

The extensions to Tcl are known collectively as STARTCL. They comprise the message system extensions, the noticeboard system extensions, the graphics window manager widget, and the *awish* and *atclsh* utilities (with the extensions embedded therein) for prototyping interfaces to tasks.

2.1. The Message System Extensions

The message system extensions provide access to inter-task communications via the AMS library. Rather than require this level of knowledge, a suite of procedures has been added that provides higher level access to the extensions. These procedures, used in CGS4DR, are best described by analogy with their ICL counterparts as shown, in brief, in Table 1. Note that the *%V* tag is one of a number of possible returns to the user and that there are also ways to communicate with tasks synchronously.

The tasks do not communicate with the user interface directly but go through a relay script, written in Tcl, called *adamMessageRelay*. The reason

Table 1. Icl and the Tcl Extensions for the Message System.

Action	**Commands**
Load task	ICL> loadw /star/bin/cgs4dr/cred4 (cred4_alias) TCL> adamtask $taskname /star/bin/cgs4dr/cred4
Obey action	ICL> send (cred4_alias) obey status TCL> $taskname obey status "" –inform "puts %V"
Cancel action	ICL> send (cred4_alias) cancel reduce TCL> $taskname cancel reduce "" –inform "puts %V"
Set parameter	ICL> send (cred4_alias) set sky_wt 1.0 TCL> $taskname set sky_wt 1.0 –setresponse "puts %V"
Get parameter	ICL> send (cred4_alias) get verbose (value) TCL> $taskname get verbose –getresponse "puts %V"

for this is that AMS hangs if it expects to receive a message when none is in the queue. This would hang the interface, so the relay script handles the communications, enabling the interface to remain responsive.

2.2. The Noticeboard System Extensions

The noticeboard system stores arrays of bytes in shared memory. Each item within a noticeboard has a datatype and dimension, although it is up to the application to handle this information correctly. The NBS extensions to Tcl are shown, again by analogy with their ICL counterparts, in Table 2. In STARTCL—although not in ICL—it is possible to monitor a NBS item, and couple that to a Tcl *trace variable* command to achieve a variety of effects.

Table 2. Icl and the Tcl Extensions for the Noticeboard System.

Action	**Commands**
Put value	ICL> putnbs ((p4_nb_alias)&'.port_0.autoscale') TRUE TCL> nbs put ${P4NoticeBoard}.port_0.autoscale TRUE
Get value	ICL> getnbs ((p4_nb_alias)&'.port_0.title') (txtvar) TCL> set txtvar [nbs get ${P4NoticeBoard}.port_0.title]
Monitoring	ICL> TCL> nbs monitor ${P4NoticeBoard}.port_0.plot_whole tclvar

3. The Graphical User Interface

Using Tcl/Tk and STARTCL, a graphical user interface has been added to CGS4DR in as little as 12,000 lines of code. Various aspects of the interface are documented elsewhere.[1] In essence, there is a separate window for each of the four tasks (one-to-one mapping), plus the graphics widget. Although it is possible to control more than one task from a single window (one-to-many mapping) this has not been implemented in CGS4DR for reasons of clarity and aesthetics. The default style of the CGS4DR interface is shown in Table 3 although these can be re-defined by the user via the file ${HOME}/cgs4dr_configs/cgs4dr.xopts.

Table 3. The Portable-CGS4DR Graphical User Interface Style.

Class	**Item**	**Default Style**
Backgrounds	Generic	Wheat
	Textpane	Snow
	Scrollbar	Lightyellow (inactive)
		Palegreen (active)
	Radiobutton	Wheat (inactive)
		Palegreen (active)
	Checkbutton	Wheat (inactive)
		Grey (active)
	Button	Pink (inactive)
		Palegreen (active)
	Menus	Wheat (inactive)
		Palegreen (active)
Functionality	Keyboard traversal	Enabled (pull-down menus)
	OK button	Do something
	Cancel button	Abort something
	Dismiss button	Abort help
Cursors	Green arrow	Normal
	Orange pirate	Dialogue box is open
	Red watch	Task is busy
	Yellow pencil	Help box is open
Mouse Buttons	MB1:	
	Single click	Do something (generic)
	Double click	Do something (listbox)
	MB2:	
	Single click	Set dynamic default
	Double click	Clears entry widget
	MB3:	
	Single click	Invokes help from WWW page

[1] http://www.jach.hawaii.edu/ukirt_sw/cgs4/cgs4dr/

4. Special Bindings

The control task is responsible for starting, pausing, and stopping the software in response to a user request. The special bindings here are that the START button is replaced by a STOP button when the software is started (and vice-versa). These buttons and the associated PAUSE button provide a number of toggles in the software. So that the user understands the present state, a status label is updated regularly. This label displays either STOPPED on a red background, RUNNING on a green background, or PAUSED on an amber background, utilising the graphical nature of the interface.

The main action for the plotting task is to plot a dataset specified in the DATA entry widget. This widget has been bound to the PLOT button using a carriage return to avoid unnecessary mouse movements. Available plotting surfaces are represented using bitmaps rather than digits. The graphics window manager widget comes supplied with a variable colour palette, cross hair and a PRINT button offering several hardcopy output formats.

The queue manager task is a generic task for storing any time-stamped information and can be used for other jobs not related to CGS4DR. This means, however, that entering a range of reduction commands into the queue invokes a number of OBEY actions in the task (with different strings to write each time). It does this by tumbling through a pair of Tcl scripts while incrementing a counter. The problem is that if the user enters a large number of observations in error, the interface does not return until the sequence is complete which may take some time. To avoid this problem—a pseudo-hang—an INTERRUPT button has been added to grab the widget immediately, thus stopping the sequence of events.

Mouse button 3 is bound to invoke help text. The text invoked is extracted from a WWW page, dynamically, thus ensuring consistency with the Web document and the need to create only one level of help text.

Acknowledgments. D. L. Terrett (RAL) and B. D. Kelly (ROE) are responsible for the STARTCL extensions to Tcl.

References

Daly, P. N. 1995, in ASP Conf. Ser., Vol. 77, Astronomical Data Analysis Software and Systems IV, ed. R. A. Shaw, H. E. Payne, & J. J. E. Hayes (San Francisco: ASP), 375

Ousterhout, J. K. 1994, Tcl and the Tk Toolkit (Reading, MA: Addison-Wesley)

Terrett, D. L. 1995a, SUN/186, Starlink Project, CLRC, UK

Terrett, D. L. 1995b, in ASP Conf. Ser., Vol. 77, Astronomical Data Analysis Software and Systems IV, ed. R. A. Shaw, H. E. Payne, & J. J. E. Hayes (San Francisco: ASP), 395

Astronomical Data Analysis Software and Systems VI
ASP Conference Series, Vol. 125, 1997
Gareth Hunt and H. E. Payne, eds.

RoadRunner: An Automated Reduction System for Long Slit Spectroscopic Data

Susan P. Tokarz and John Roll

Smithsonian Astrophysical Observatory, Cambridge, MA 02138

Abstract. Driven by a dramatic, almost fivefold, increase in data flow, we developed a system, RoadRunner,[1] for the automatic and rapid reduction of CCD two-dimensional long slit data taken at the 1.5 meter telescope at the Fred L. Whipple Observatory on Mt. Hopkins. The typical 70 to 100 observations made in an average night are now routinely reduced in a few hours the following day. The reduction process takes the data from raw two-dimensional images to one-dimensional wavelength calibrated spectra with measured redshifts which have been checked for accuracy. The reduced spectra are stored in an on-line archive and are entered into a relational database. Standardizing observing protocols proved to be a requirement for designing an automated reduction system. RoadRunner, which is based on IRAF data reduction routines, has many internal checks for errors and anomalies and produces a data set of high quality.

1. Introduction

In January 1994 a new spectrograph, FAST (Fabricant et al. 1997), was mounted on the 1.5 meter telescope on Mt. Hopkins. This two-dimensional longslit CCD spectrograph replaced the Z-machine, a one-dimensional Reticon detector which had been in operation for almost 16 years and was used primarily to measure galaxy redshifts (e.g., the First CFA Redshift Survey). In 2.75 years of operation, 22,889 FAST spectra have been reduced. By comparison, over the nearly 16 year lifetime of the Z-machine, 27,169 spectra were reduced.

The huge increase in the amount of data obtained from the 1.5 meter, combined with the more complex reduction procedure required by two-dimensional data made it necessary to automate our reduction. A further complication for the reduction is that the higher efficiency and greater flexibility of FAST allows a vastly wider range of observing programs; there are now 55 programs using FAST and they require different combinations of gratings, CCD pixel binnings, slits, and grating tilts. In addition to the redshift surveys, these programs include observations of symbiotic stars, ISO quasars, AGNs for monitoring, dwarf novae, etc.

---

[1]RoadRunner was named after a bird of that species who for a time lived near the 1.5 meter and was always a source of inspiration before a night's observing to one of the authors

2. Observing Protocols

As a first and essential step in automating reduction we *set observing protocols.* The FAST Observing Protocols are a list of instructions about how and when to take biases, flats, darks, comps, skys and standards; how to determine object integration times; and how to handle setup changes. These protocols ensure uniformity and completeness of the data and make it possible to reduce data automatically from queued observations and for a variety of observers. Although our remote observers do at least 80% of the observing, FAST is also used by graduate students, postdoctoral fellows, and staff scientists. Without protocols, automated reduction would not be possible.

3. Automated Reduction

The data are transferred automatically at the end of a night's observing from Mt. Hopkins to SAO in Cambridge, Mass. Data from a night can be reduced and entered into a database in just a few hours of processing. A recent typical night had 287 raw data files: 40 biases, 40 flats, four darks, five skys, 83 comps, and 115 object exposures. There were two binning modes: in some cases, the CCD pixels were binned by two in the spatial direction, in others by four.

The automated reduction system, RoadRunner, consists of a series of IRAF routines combined in processing scripts. Many of the routines used, such as CCDPROC, RESPONSE, REIDENTIFY, APALL, etc., are widely available. Additional routines were written as necessary for automated processing. Some of the new routines are fairly straightforward: for example, there is code to pass the calculated shift between two comparison lamps from FXCOR, where the calculation is made, to REIDENTIFY, where it is used. Thus we are able to use an old wavelength solution to effortlessly compute good wavelength solutions for our current comparison frames. Other programs are more complex: FINDALL, written by Doug Mink, adjusts size and background parameters used in APALL enabling us to extract a variety of galaxy and stellar spectra automatically. FINDALL is able to extract the correct object and a satisfactory background even when there is more than one object on the slit.

3.1. RoadRunner

RoadRunner, the original script, is completely automated and performs all the preprocessing necessary for reduction. It accepts raw data from one night of observing and produces two-dimensional, wavelength calibrated spectra ready for extraction. Because the data may be taken using a variety of setups at the telescope, the script first sorts the data into sets (e.g., 3″ slit, binned by four, 300 l/mm grating, grating tilt = 607 is a set) and each set is reduced separately.

Processing steps included in RoadRunner are:

- overscan and bias subtraction
- trimming
- normalization
- illumination correction
- wavelength calibration
- transformation

There are many checks in RoadRunner to guarantee the accuracy and integrity of the data. For example, there is a plot of the means of the bias frames that can show at a glance when there is something awry in the biases for a night. Most of the other checks are arithmetic rather than visual. Limits are set for computed values and a log is maintained and checked to make certain that all is going as expected. A high residual in a comparison can mean lower signal or some other problem in the lamp. An unusual value in the chip gain/read noise calculation can mean an undocumented change in the grating tilt. The program will also notify the user when something is wrong as when, for example, there is no comparison file for an object.

3.2. BeepBeep

A second major script, BeepBeep, combines automated and interactive utilities which are necessary to complete the reduction. This script accepts the two-dimensional, wavelength calibrated images from RoadRunner and extracts one-dimensional, wavelength calibrated spectra with measured redshifts ready to be archived and entered in our database.

The steps in this script include:

• FINDALL	extract spectra	*automatic*
• QSPEC	check and correct FINDALL	*interactive*
• PFAST	remove cosmic rays	*interactive*
• XCSAO	compute redshifts	*automatic*
• QPLOT	check redshifts	*interactive*

Although it may seem somewhat strange to switch back and forth between automated and interactive programs in one script, doing so has allowed us to efficiently use programs which must be interactive while automating where possible.

4. Output

The reduced one-dimensional, wavelength calibrated spectra with measured redshifts are transferred to a permanent archive on disk. The data are maintained in IRAF format in the archive. From the archive, the data are entered into a relational database which consists of IRAF keywords pulled from the headers. The database consists of flat ASCII files in the starbase database format (Roll 1995). This format permits easy and fast retrieval of information for data analysis.

5. Summary

Automated reduction has become necessary due to the large increase in the amount of data we now collect as well as the increasingly complex data that new instruments provide. The system we have built allows us to handle this data in an efficient way. Automation increases the reliability of the reduced data by ensuring greater consistency in data reduction. An added bonus of automation has been rapid feedback to the observer allowing problems to be addressed at an early stage.

Acknowledgments. We wish to thank Doug Mink, who over the years has written routines which made automating our reduction easier and who is always generous with his time and energy. Mike Kurtz has participated in many discussions about reduction technique and his sage counsel has been much appreciated. We thank Dan Fabricant for his thoughtful and valuable critique of this paper. Margaret Geller was the guiding force behind automating reduction and we are grateful for her generative spirit which stimulates, encourages and fosters growth.

References

Fabricant, D., Cheimets, P., Caldwell, N., & Geary, J. 1997, PASP, in press

Roll, J. 1996, in ASP Conf. Ser., Vol. 101, Astronomical Data Analysis Software and Systems V, ed. G. H. Jacoby & J. Barnes (San Francisco: ASP), 536

Part 3. Algorithms

Astronomical Data Analysis Software and Systems VI
ASP Conference Series, Vol. 125, 1997
Gareth Hunt and H. E. Payne, eds.

Variable-Pixel Linear Combination

Richard N. Hook
Space Telescope - European Coordinating Facility, European Southern Observatory, Karl Schwarzschild Str.-2, D-85748, Garching, Germany, E-mail: rhook@eso.org

Andrew S. Fruchter
Space Telescope Science Institute, 3700 San Martin Drive, Baltimore, MD 21218, USA, E-mail: fruchter@stsci.edu

Abstract. We have developed a method for the linear reconstruction of an image from undersampled, dithered data. The algorithm, known as Variable-Pixel Linear Reconstruction (or informally as "drizzling"), preserves photometry and resolution, can weight input images according to the statistical significance of each pixel, and removes the effects of geometric distortion both on image shape and photometry. In this paper, the algorithm and its implementation are described, and measurements of the photometric accuracy and image fidelity are presented. We also describe experiments in which the method is extended to dynamically detect and suppress the effects of cosmic-ray events on individual frames.

1. Introduction

Many imaging systems in astronomy have detectors which undersample the image that falls on them. An important example, which we consider here, is the Wide Field Planetary Camera 2 (WFPC2) on the Hubble Space Telescope (HST). Although the optics of WFPC2 provide a superb Point Spread Function (PSF), the detectors at the focal plane severely undersample the image. This problem is most acute for the three WF chips, where the width of a pixel equals the FWHM of the optics in the near-infrared, and greatly exceeds it in the blue.

Much of the information lost in sampling can be recovered by combining images which have been shifted by fractions of a pixel between successive exposures. Such images are referred to as "dithered," and this observing strategy is commonly used with WFPC2. Several methods have been proposed for such a combination, including methods based on iterative, maximum likelihood reconstruction methods. Such methods can be effective but tend to be slow and unable to handle effectively geometrically distorted images such as those produced by WFPC2. They also produce images in which the noise properties are correlated from pixel to pixel and the statistical errors are difficult to estimate. To avoid these problems, and to handle the major image combination problem posed by the "Hubble Deep Field" (HDF, Williams et al. 1996), we have considered the family of techniques we refer to as "linear reconstruction." The most commonly used of these techniques are shift-and-add and interlacing. However,

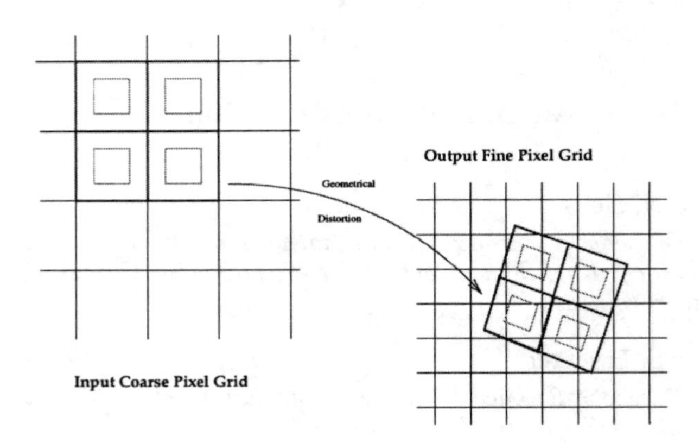

Figure 1. A Schematic representation of drizzling.

due to poor placement of the sampling grid or the effects of geometric distortion, true interlacing of images is often not feasible. On the other hand, the other standard technique, shift-and-add, convolves the image yet again with the original pixel, further adding to the blurring of the image. Here we present a method which has the versatility of shift-and-add and yet aims to maintain the resolution of interlacing.

2. The Method

"Drizzling" maps pixels in the original input images onto the pixel grid of a subsampled output image, taking into account shifts and rotations between images and the optical distortion of the camera. However, in order to avoid convolving the image with the large pixel "footprint" of the camera, we allow the user to shrink the pixel before it is averaged into the output image. The new shrunken pixels, or "drops," rain down upon the subsampled output image. In the case of HST/WFPC2 images with multiple dither positions (such as the Hubble Deep Field), the drops typically have linear dimensions one-half that of the input pixel—slightly larger than the size chosen for the output subsampled pixels. The flux in each drop is divided up among the overlapping output pixels with weights proportional to the areas of overlap. This procedure is shown schematically in Figure 1. Note that if the drop size is sufficiently small not all output pixels have any data added to them from an input image. A drop size must therefore be chosen that is small enough to minimize degradation of spatial resolution but large enough that the coverage is fairly uniform after all the images have been drizzled.

When a drop with value i_{xy} and user defined weight w_{xy} is added to an image with pixel value I_{xy}, weight W_{xy}, and fractional pixel overlap $0 \leq a_{xy} \leq 1$, the resulting value of the image I'_{xy} and weight W'_{xy} is

$$W'_{xy} = a_{xy} w_{xy} + W_{xy} \quad (1)$$

$$I'_{xy} = \frac{a_{xy} i_{xy} w_{xy} + I_{xy} W_{xy}}{W'_{xy}} \quad (2)$$

This algorithm preserves both surface and absolute photometry so that flux density can be measured using an aperture whose size is independent of position on the chip. The weighting arrays also allow missing data, due to cosmic ray hits and hot pixels, to be handled in a totally flexible way. The linear weighting scheme employed is statistically optimum when inverse variance maps are used as weights. These weights may vary spatially to accommodate changing signal-to-noise ratios across input frames (e.g., due to variable scattered light). The final output weighting image (an inverse variance map) is saved as well as the combined image frame and can be used in further analysis.

The method also minimizes resolution loss, and largely eliminates the distortion of absolute photometry produced by the flat-fielding of the geometrically distorted images (see §4).

To obtain more information about the drizzling method, as well as a well-tested version of the software which will run under IRAF, the drizzling Webpage[1] should be consulted.

3. Image Fidelity

The drizzling algorithm was designed to obtain optimal signal-to-noise on faint objects while preserving image resolution. These goals are, unfortunately, not fully compatible. For example, non-linear image restoration procedures which attempt to remove the blurring of the PSF and the pixel by enhancing the high frequencies in the image (such as such as the Richardson-Lucy and MEM methods) directly exchange signal-to-noise for resolution. In the drizzling algorithm, no compromises on signal-to-noise have been made, and the weight of an input pixel in the final output image is entirely independent of its position on the chip. Therefore, if the dithered images do not uniformly sample the field, the "center of light" in an output pixel may be offset from the center of the pixel, and that offset may vary between adjacent pixels. This effect is seen in the HDF images, where some pointings were not at the requested position or orientation. Furthermore, large dithering offsets which may be used for WFPC2 imaging, combined with geometric distortion, produce a sampling pattern that varies across the field. The output PSFs produced by the combination of such irregularly dithered datasets using drizzling may show substantial variations about the best fit Gaussian due to the effects of non-uniform sampling. Fortunately, these variations do not noticeably affect aperture photometry performed with typical aperture sizes.

4. Photometry

The WFPC2 optics geometrically distort the images: pixels at the corner of each CCD subtend less area on the sky than those near the center. However, after application of the flat field, a source of uniform surface brightness on the

[1] http://www.stsci.edu/~fruchter/dither/dither.html

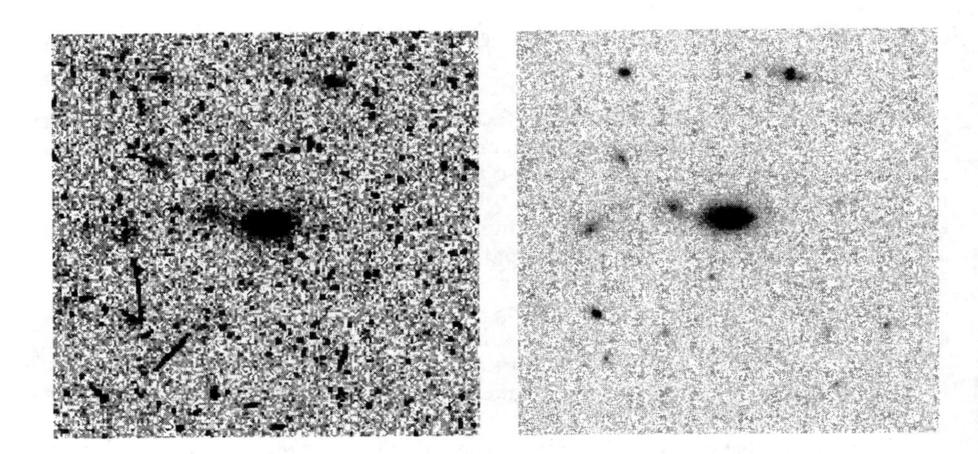

Figure 2. Cosmic ray removal on dithered images using drizzling

sky produces uniform counts across the CCD. Therefore point sources near the corners of the chip are artificially brightened compared to those in the center.

This effect has been studied by performing photometry on a grid of 19×19 artificial stellar PSFs which had their counts adjusted to reflect the effect of geometric distortion—the stars in the corners are up to ~4% brighter than those in the center of the chip. This image was then shifted and sampled on a 2×2 grid and the results combined using drizzling and typical parameters. Aperture photometry on the 19×19 grid after drizzling reveals that the effect of geometric distortion on the photometry has been dramatically reduced: the RMS photometric variation in the drizzled image is 0.004 magnitudes.

5. Cosmic Ray Detection

Few HST observing proposals have sufficient time to take multiple exposures at each of several dither positions. So if dithering is to be of widespread use, one must be able to remove cosmic rays from data where few, if any, images are taken at the same position on the sky. We have therefore been examining the question of whether we can adapt the drizzling procedure to the removal of cosmic rays. Although the removal of cosmic rays using drizzling is still very much work in progress, we have developed a procedure which appears quite promising. Figure 2 shows the result of such processing on a set of twelve dithered deep WFPC2 images taken from the archive.

References

Williams, R. E., et al. 1996, AJ, 112, 1335

Astronomical Data Analysis Software and Systems VI
ASP Conference Series, Vol. 125, 1997
Gareth Hunt and H. E. Payne, eds.

Heuristic Estimates of Weighted Binomial Statistics for Use in Detecting Rare Point Source Transients

James Theiler and Jeff Bloch
Astrophysics and Radiation Measurements Group, MS-D436 Los Alamos National Laboraotry, Los Alamos, NM 87545 e-mail: jt@lanl.gov, jbloch@lanl.gov

Abstract. The ALEXIS[1] (Array of Low Energy X-ray Imaging Sensors) (Priedhorsky et al. 1989) satellite scans nearly half the sky every fifty seconds, and downlinks time-tagged photon data twice a day. The standard science quicklook processing produces over a dozen sky maps at each downlink, and these maps are automatically searched for potential transient point sources. We are interested only in *highly significant* point source detections, and, based on earlier Monte-Carlo studies (Roussel-Dupré et al. 1996), only consider $p < 10^{-7}$, which is about 5.2 "sigmas." Our algorithms are therefore required to operate on the far tail of the distribution, where many traditional approximations break down. Although an exact solution is available for the case of unweighted counts (Lampton 1994), the problem is more difficult in the case of weighted counts. We have found that a heuristic modification of a formula derived by Li & Ma (1983) provides reasonably accurate estimates of p-values for point source detections even for very low p-value detections.

1. Introduction

We test the null hypothesis of no point source (assuming a spatially uniform background) at a given location by enclosing that location with a source kernel (whose area A_{src} is generally matched to the point-spread-function of the telescope) and then enclosing the source kernel with a relatively large background annulus (area A_{bak}). Given N_{src} photons in the source kernel, and N_{bak} photons in the background annulus, the problem is to determine whether the number of source photons is *significantly* larger than expected under the null.

More sensitive point source detection is obtained by weighting the photons to match the point-spread function of the telescope more precisely. Further enhancements are obtained for ALEXIS data by weighting also according to instantaneous scalar background rate, pulse height, and position on the detector. In this case, we ask whether the weighted sum of photons in the source region is significantly larger than expected under the null.

---

[1] http://nis-www.lanl.gov/nis-projects/alexis/

2. Unweighted Counts

If counts are unweighted (i.e., all weights are equal), then it is possible to write down an exact, explicit expression for the probability of seeing $N_{\rm src}$ or more photons in the source kernel, assuming $N_{\rm total} = N_{\rm src} + N_{\rm bak}$ is fixed. This is a binomial distribution, and Lampton (1994) showed that the p-value associated with this observation can be expressed in terms of the incomplete beta function: $p = I_f(N_{\rm src}, N_{\rm bak}+1)$, where $f = A_{\rm src}/(A_{\rm src}+A_{\rm bak})$. See also Alexandreas et al. (1994), for an alternative derivation of an equivalent expression (the assumption that $N_{\rm total}$ is fixed is replaced by a Bayesian argument).

If the count rate is high (or the exposure long), so that $N_{\rm src}$ and $N_{\rm bak}$ are large, then an appropriate Gaussian approximation can be used. In general, this involves finding a "signal" and dividing it by the square root of its variance.

Case 1u. The most straightforward approach uses the signal $N_{\rm src} - \alpha N_{\rm bak}$, where $\alpha = A_{\rm src}/A_{\rm bak}$. Under the null hypothesis, this signal has an expected value of zero, and a variance—if $N_{\rm src}$ and $N_{\rm bak}$ are treated as independent Poisson sources—of $N_{\rm src} + \alpha^2 N_{\rm bak}$. To get a p-value, use

$$p = \mathcal{S}\left(\frac{N_{\rm src} - \alpha N_{\rm bak}}{\sqrt{N_{\rm src} + \alpha^2 N_{\rm bak}}}\right), \tag{1}$$

where $\mathcal{S}(s) = \frac{1}{2}(1 - \mathrm{erfc}(s/\sqrt{2}))$ converts "sigmas" of significance into a one-tailed p-value.

Case 2u. An alternative approach, suggested by Li & Ma (1983), treats the sum $N_{\rm total} = N_{\rm src} + N_{\rm bak}$, as fixed, so that $N_{\rm src}$ and $N_{\rm bak}$ are binomially distributed. In particular, choose the signal $N_{\rm src} - fN_{\rm total}$, and note that the variance of $N_{\rm src}$ is given by $f(1-f)N_{\rm total}$, while the variance of $N_{\rm total}$ is by definition zero. In that case

$$p = \mathcal{S}\left(\frac{N_{\rm src} - fN_{\rm total}}{\sqrt{f(1-f)N_{\rm total}}}\right) = \mathcal{S}\left(\frac{N_{\rm src} - \alpha N_{\rm bak}}{\sqrt{\alpha N_{\rm src} + \alpha N_{\rm bak}}}\right). \tag{2}$$

Case 3u. By looking at a ratio of Poisson likelihoods, Li & Ma (1983) also derived a more complicated equation

$$p = \mathcal{S}\left(\sqrt{2\left\{N_{\rm src}\ln(N_{\rm src}/\hat{N}_{\rm src}) + N_{\rm bak}\ln(N_{\rm bak}/\hat{N}_{\rm bak})\right\}}\right), \tag{3}$$

where $\hat{N}_{\rm src} = fN_{\rm total}$ and $\hat{N}_{\rm bak} = (1-f)N_{\rm total}$. This is considerably more accurate than Eqs. (1,2) when $N_{\rm src}$ and $N_{\rm bak}$ are not large, but is still just an approximation to Lampton's exact formula. Abramowitz & Stegun (1972) provide several approximations to the incomplete beta function, one of which (25.5.19) is an asymptotic series whose first term looks very much like the Li & Ma formula. The left panel of Figure 1 compares these cases, along with the Lampton (1994) formula, using a Monte-Carlo simulation.

3. Weighted Counts

Define $W_{\rm src} = \sum_{i\in{\rm src}} w_i$ and $Q_{\rm src} = \sum_{i\in{\rm src}} w_i^2$, where w_i is the weight of the i-th photon. Notice that when all the weights are equal to one, we have $Q_{\rm src} =$

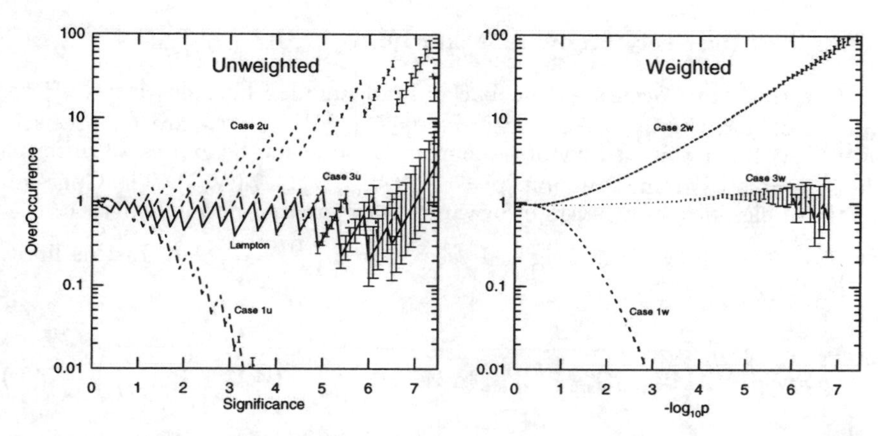

Figure 1. Results of Monte-Carlo experiments with $N = 100$ photons, with $f = 0.1$, and with $T = 10^7$ trials. For the weighted experiment, N weights were uniformly chosen from zero to one, and assigned to the N photons. The photons were randomly assigned to the source kernel or background annulus with probabilities f and $1 - f$ respectively. Values of $W_{\rm src}$, $W_{\rm bak}$, $Q_{\rm src}$, and $Q_{\rm bak}$ were computed, and a p-value was computed using the formulas for the three cases. As the p-values were computed, a cumulative histogram $H(p)$ was built indicating the number of times a p-value less than p was observed. Since we expect $H(p) = pT$, we plotted $H(p)/pT$ as the frequency of "overoccurrence" of that p-value. The plot is this overoccurrence as a function of "significance," defined by $-\log_{10} p$.

$W_{\rm src} = N_{\rm src}$ and $Q_{\rm bak} = W_{\rm bak} = N_{\rm bak}$. Note also that $W_{\rm src}/N_{\rm src} = \langle w_i \rangle_{i \in {\rm src}}$, and that $Q_{\rm src}/W_{\rm src} = \langle w_i^2 \rangle / \langle w_i \rangle$. We do not make any assumptons about weights averaging or summing to unity. (We define $W_{\rm bak}$ and $Q_{\rm bak}$ similarly.)

Generalizing **Case 1u**, we define the signal as $W_{\rm src} - \alpha W_{\rm bak}$ and then treating source and background as independent, we can write the variance as $Q_{\rm src} + \alpha^2 Q_{\rm bak}$. We can similarly generalize **Case 2u** and obtain:

$$\textbf{Case 1w:} \qquad p = \mathcal{S}\left(\frac{W_{\rm src} - \alpha W_{\rm bak}}{\sqrt{Q_{\rm src} + \alpha^2 Q_{\rm bak}}} \right). \tag{4}$$

$$\textbf{Case 2w:} \qquad p = \mathcal{S}\left(\frac{W_{\rm src} - \alpha W_{\rm bak}}{\sqrt{\alpha Q_{\rm src} + \alpha Q_{\rm bak}}} \right). \tag{5}$$

Case 3w: It is not as straightforward to generalize Eq. (3), but we have tried the following heuristic:

$$p = \mathcal{S}\left(\sqrt{\left(\frac{2W_{\rm total}}{Q_{\rm total}} \right) \left\{ W_{\rm src} \ln(W_{\rm src}/\hat{W}_{\rm src}) + W_{\rm bak} \ln(W_{\rm bak}/\hat{W}_{\rm bak}) \right\}} \right), \tag{6}$$

where $\hat{W}_{\rm src} = fW_{\rm total}$ and $\hat{W}_{\rm bak} = (1 - f)W_{\rm total}$. The Monte-Carlo results shown in Figure 1 indicate that this heuristic provides reasonably accurate p-values even for very small values of p.

4. Limit of Precisely Known Background

An interesting limit occurs as the background annulus becomes large. Here, $A_{\rm bak} \to \infty$, and the expected backgrounds $\hat{N}_{\rm src}$, $\hat{W}_{\rm src}$, etc. are all precisely known. For the unweighted counts, the exact p-value can be expressed in terms of the incomplete gamma function: $p = 1 - \Gamma(N_{\rm src}, \hat{N}_{\rm src})/\Gamma(N_{\rm src})$. The Gaussian estimate of significance is straightforward[2] both for the unweighted case, $p = \mathcal{S}\left(\frac{N_{\rm src}-\hat{N}_{\rm src}}{\sqrt{\hat{N}_{\rm src}}}\right)$, and for the weighted case: $p = \mathcal{S}\left(\frac{W_{\rm src}-\hat{W}_{\rm src}}{\sqrt{\hat{Q}_{\rm src}}}\right)$. In this limit, Eq. (6) becomes

$$p = \mathcal{S}\left(\sqrt{2\left(\hat{W}_{\rm src}/\hat{Q}_{\rm src}\right)\left(W_{\rm src}\ln(W_{\rm src}/\hat{W}_{\rm src}) - (W_{\rm src} - \hat{W}_{\rm src})\right)}\right). \quad (7)$$

Marshall (1994) has suggested an empirical formula $p = \mathcal{S}\left(\frac{W_{\rm src}-\hat{W}_{\rm src}+\Delta}{\sqrt{\hat{Q}_{\rm src}+\Delta}}\right)$, where $\Delta = 0.7\hat{Q}_{\rm src}/\hat{W}_{\rm src}$, which produced reasonable results in his simulations, but does not appear well suited for p-values at the far tail of the distribution.

Acknowledgments. This work was supported by the United States Department of Energy.

References

Abramowitz, M., & Stegun, I. A. 1972, Handbook of Mathematical Functions (Dover, New York), 945

Alexandreas, D. E., et al. 1993, Nucl. Instr. Meth. Phys. Res. A328, 570

Babu, G. J., & Feigelson, E. D. 1996, Astrostatistics (Chapman & Hall, London), 113

Lampton, M. 1994, ApJ, 436, 784

Li, T.-P., & Ma, Y.-Q. 1983, ApJ, 272, 317

Marshall, H. L. 1994, in ASP Conf. Ser., Vol. 61, Astronomical Data Analysis Software and Systems III, ed. D. R. Crabtree, R. J. Hanisch & J. Barnes (San Francisco: ASP), 403

Priedhorsky, W. C., Bloch, J. J., Cordova, F., Smith, B. W., Ulibarri, M., Chavez, J., Evans, E., Seigmund, O., H. W., Marshall, H., & Vallerga, J. 1989, in Berkeley Colloquium on Extreme Ultraviolet Astronomy, Berkeley, CA, vol 2873, 464

Roussel-Dupré, D., Bloch, J. J., Theiler, J., Pfafman, T., & Beauchesne, B. 1996, in ASP Conf. Ser., Vol. 101, Astronomical Data Analysis Software and Systems V, ed. G. H. Jacoby & J. Barnes (San Francisco: ASP), 112

[2]Babu & Feigelson (1996) incorrectly suggest $p = \mathcal{S}\left((N_{\rm src} - \hat{N}_{\rm src})/\sqrt{N_{\rm src} + \hat{N}_{\rm src}}\right)$.

Astronomical Data Analysis Software and Systems VI
ASP Conference Series, Vol. 125, 1997
Gareth Hunt and H. E. Payne, eds.

A Computer-Based Technique for Automatic Description and Classification of Newly-Observed Data

S. Vasilyev

SOLERC, P.O. Box 59, Kharkiv, 310052, Ukraine

Abstract. A technique allowing automatic representation by a relatively small number of independent parameters based on the principal component analysis of data sequences is presented. In some instances the parameters can serve as an independent description, classification, and compression of the observational results.

1. Introduction

In recent years the spectrum of observed astronomical data can be characterized as greatly varied. In particular, this can be explained by the rapid growth in the number of objects studied and the appearance of new data types due to the progress in space-based observations. In this instance, the problem of initial description and classification of newly-observed data becomes most urgent, especially if there is a lack of preliminary observational material and theoretical expectations.

The literature developing methods to treat statistical data sequences is ample but many studies are based on the interpretation of data images and have difficulties determining the minimum set of independent parameters appropriate for further analysis. The well-known principal component method of the multivariate statistical data treatment can be extended in order to obtain a tool for determination of the independent parameters applicable for reliable data representation and further comprehensive analysis.

2. Proposed Approach

The distinction of the approach consists in the direct use of the observed data records as input parameters in composing the initial and covariance matrices involved in further analysis. Each of n observational dependencies forms a row in the initial matrix and is represented by a vector in m-dimensional space in accordance with the number of observed dependencies and the number of points on each curve, respectively. These n vectors determine the dimension of the covariance matrix and thus the number of its eigenvectors and eigenvalues. The eigenvectors are orthogonal and, consequently, each row of the covariance matrix, as well as that of the initial one can be only expressed by their linear combination. It is advisable to normalize the eigenvectors by dividing them by their lengths, which are equal to the square roots of the corresponding eigenvalues. This puts the eigenvectors on the same scale.

Finding the eigenvalues often makes it possible to represent the initial matrix with needed accuracy by taking a linear combination of a relatively small number of principal components corresponding to the largest eigenvalues and so bearing most of the information on the data (Genderon & Goddard 1966). The other principal components are usually responsible for the random noise in observations and can be neglected from consideration (Lorge & Morrison 1938). The quality of data representation can be controlled with a test matrix composed of linear combinations of the principal components multiplied by the corresponding eigenvalues.

Generally, analytical functions for the selected principal components can be found by fitting, and we obtain the data being analytically represented in addition. In this case, all of the initial dependencies can be easily calculated as linear combinations of the functions, which are presumably described by a minimum parameter set. We have developed an interactive computer package which can automatically describe some kinds of data by principal components. The software also performs the preliminary data fitting on the measurements, which are not uniformly tabulated observational records. The procedure of finding the largest eigenvalues and corresponding eigenvectors, or principal components, is based on the algorithm presented in Simonds (1963). The approach has been successfully applied to describe all of the variety of the asteroid polarization phase dependencies (Vasilyev 1994) and tested for some other types of data. We have found two principal components which are adequate to represent any polarization curve belonging to the analyzed assemblage and even for data not involved in the initial analysis. The corresponding eigenvalues λ_1 and λ_2 can be considered as new parameters instead of the widely used system of four interdependent parameters ($P_{\min}$, $\alpha_{\min}$, α_0, and h), and are more suitable for further analysis (Vasilyev 1996).

Expressions obtained for the principal components can be used to describe new data and gappy observational dependencies. In the case of asteroids, the method allows the synthesis of the polarization curves using only three observations and shows a better fit to the data compared to other fitting techniques tried. The power of this method is its intrinsic ability to find the principal peculiarities appropriate to all the data under study. It can be efficiently used for restoring truncated data records and rationally planning further observations.

Although the principal component method itself does not imply knowledge of the physical nature of the analyzed data, it often allows connection of the principal components with the physical parameters of the objects. In particular, we have found that both of the principal components of the family of asteroid polarimetric phase curves have physical meanings and the fit correlations were determined.

Additional useful possibilities give the correlative diagrams of the largest eigenvalues corresponding to the first principal components. These diagrams do not exhibit any mutual correlation between the eigenvalues, of course, as it is required by the technique. However, they may reveal the differences in properties among the analyzed data and can be successfully used for independent classification of the studied objects (Tholen 1984; Vasilyev 1996).

As the number of the principal components and corresponding eigenvalues used for the data representation is usually much smaller than that of the observed dependencies, we obtain an alternative tool for compact data storage. The problem of finding the balance between the required accuracy of the data restoration and the needed compression ratio is the subject of a separate study. Our preliminary results show that the use of the principal component technique can reduce the data volume by up to several times (Vasilyev 1995). In the instance of the asteroid polarimetric data the compression ratio was increased to a factor of five, while the differences between the initial and restored data did not exceed the errors in observations. Furthermore, as the principal components keep the data structure, this ratio can be increased by the subsequent application of any other archiving software.

3. Conclusion

The technique based on the principal component analysis of the data records may serve as a powerful tool for the initial statistical data treatment allowing data inter/extrapolation, analytical representation, classification, and compact storage. Among the advantages of the technique are the stability of the obtained eigenvectors when adding new data and the minimizing the *rms* errors in data representation. It is important that the application of this approach does not require any *a priori* assumption, either about the objects or about the physical mechanisms under study. In order to make possible such a multipurpose application of the technique for some types of the newly observed data in the automatic mode we are currently developing an integrated program package PCMAD (Principal Component Method for Astronomical Data) including the most of the described features. It should be noted that the method has no special requirements of computer performance except during the first stage of its application when the matrix operations are performed.

References

Genderon, R. G., & Goddard, M. G. 1966, Photogr. Sci. Eng, 10, 77
Lorge, J., & Morrison N. 1938, Science, 87, 491
Simonds, J. L. 1963, J. Opt. Soc. America, 53, 968
Tholen, D. J. 1984, Ph.D. Thesis, Univ. of Arizona
Vasilyev, S. V. 1994, BAAS, 26, 1173
Vasilyev, S. V. 1995, Vistas in Astronomy, 39, 275
Vasilyev, S. V. 1996, Ph.D. Thesis, Kharkiv St. Univ., Ukraine

Astronomical Data Analysis Software and Systems VI
ASP Conference Series, Vol. 125, 1997
Gareth Hunt and H. E. Payne, eds.

Unified Survey of Fourier Synthesis Methodologies

P. Maréchal, E. Anterrieu, and A. Lannes

Laboratoire d'Astrophysique and CERFACS, Observatoire Midi-Pyrénées, 14, Avenue Édouard Belin F-31400 Toulouse (France)

Abstract. This paper deals with theoretical and practical results on aperture synthesis. Full attention is devoted to the Fourier synthesis operation, which proves to be the central issue with regard to the global problem. Different regularization techniques are shown to derive from a unique principle: the Principle of Maximum Entropy on the Mean (PMEM). This is the case for WIPE, as well as for the traditional Maximum Entropy Method (and for some others). In order to compare the performances of different regularizers, some numerical simulations are presented.

1. Introduction to Aperture Synthesis

Let $u \equiv u(j,k)$ denote the spatial frequency corresponding to the baseline (j,k) of the interferometric device, and let $\psi(j)$ be the aberration phase term of pupil element j. The central problem is to reconstruct both the brightness distribution ϕ (of some object) and the pupil phase function ψ, by using as data complex-visibility measurements of the form $\mathcal{V}(u(j,k))$. The latter are related to ϕ and ψ by the equation $\mathcal{V}(u(j,k)) = \exp i\beta(u(j,k))\ \widehat{\phi}(u(j,k))$, in which $\beta(u(j,k)) := \psi(j) - \psi(k)$ and $\widehat{\phi}$ is the Fourier transform of ϕ. Solving for ϕ (assuming that β is known) is a Fourier synthesis operation, while solving for β corresponds to a phase calibration operation. The whole problem must be solved in such a way that the solution is not too sensitive to unavoidable measurement errors. In other words, the inverse problem under consideration must be regularized, and it is essential that the chosen methodology provides an estimation of the reconstruction stability. In fact, the Fourier synthesis operation proves to be the core of the problem. It has the form of a linear inverse problem. The corresponding "measurement equation" may be written as $y = Ax$, in which $y \in R^m$ is the real-valued data vector associated with $\mathcal{V}$, A is the Fourier sampling operator, and x is the vector formed with the components of ϕ in an adequate basis (Maréchal & Lannes 1996).

2. Survey of Regularization Techniques

In practice, equation $y = Ax$ fails to have a unique and stable solution, and is therefore replaced by an optimization problem of the form

$$\min_{x \in R^n} \left\{ g(x) := \frac{1}{2}\|y - Ax\|^2 + \alpha f(x) \right\} \tag{1}$$

in which f is a measure of roughness of x, and α is the so-called regularization parameter. The quadratic term in (1) forces the object to fit the data, while the regularization term stabilizes the solution with respect to small variations in y (and of course ensures uniqueness). Denoting by δx the variation of the solution $\bar{x}$ induced by a variation δy, the stability is governed by an inequality such as

$$\|\delta x\| \leq \rho \, \|\delta y\|. \tag{2}$$

The trade-off between fitting the data and stability depends in a crucial manner on the nature of f and on the value of α. We emphasize that the regularization term should also be designed so that g has a physical meaning, so that the solution $\bar{x}$ can be easily interpreted.

Let us now review some classical examples. The well-known Tikhonov regularization corresponds to the case $f(x) = \|x\|^2/2$. It can be generalized by choosing $f(x) = \langle x, Qx \rangle /2$, in which Q is a symmetric non-negative $n \times n$ matrix. When x can be interpreted as a (discrete) probability density, one often takes the Shannon entropy $f(x) = \sum x_j \ln x_j$ or the Kullback measure $f(x) = \sum x_j \ln(x_j/x_{0j})$, in which x_0 represents a prior knowledge of the object. If x is only assumed to be positive, one can use the Generalized Cross-Entropy (GCE) $f(x) = \sum \left(x_j \ln(x_j/x_{0j}) + (x_{0j} - x_j)\right)$. Let us also mention the Itakura-Saito criterion $f(x) = \sum \left(x_j/x_{0j} - \ln(x_j/x_{0j}) - 1\right)$, frequently used in spectral analysis.

Recently, a new Fourier synthesis method has been developed for radio imaging and optical interferometry (Lannes, Anterrieu, & Bouyoucef 1994, 1996; Lannes, Anterrieu, & Maréchal 1997): WIPE. The name of WIPE is associated with that of CLEAN. To some extent (Lannes, Anterrieu, & Maréchal 1997), WIPE can be regarded as an updated version of CLEAN. In particular, the robustness of the reconstruction process is well controlled. The main aspects of WIPE are its regularization principle (for controlling the image resolution) and its matching pursuit strategy (for constructing the image support). The regularizer of WIPE is defined by the relation

$$f(x) = \|Bx\|^2/2 = \langle x, B^T Bx \rangle /2, \tag{3}$$

in which $\|Bx\|^2$ represents the energy of x in the high frequency band (Lannes, Anterrieu, & Bouyoucef 1994). Clearly, this regularization principle belongs to the Tikhonov family, since $B^T B$ is a symmetric non-negative $n \times n$ matrix. It is very closely related to the notion of resolution. The function to be minimized in (1) takes the form $\|y_0 - A_0 x\|^2/2$ with $A_0 = [A; B]$ and $y_0 = (y; 0)$. This kind of function is efficiently minimized by a conjugate-gradients algorithm, which has the advantage of providing (with negligible additional computing time) an estimate of the condition number and, thereby, control of the stability of the reconstruction process. As for the matching pursuit strategy, we simply mention that the main difference from CLEAN is that it can be conducted at the level of the scaling functions of the object workspace (Lannes, Anterrieu, & Bouyoucef 1994)

3. Unification Results

We now turn to the unification of WIPE with the methodologies mentioned in §2 by the Principle of Maximum Entropy on the Mean. Note first that the optimization problem (1) is equivalent to

$$\min_{(x;b)\in R^{n+m}} \left\{ \frac{1}{2}\|b\|^2 + \alpha f(x) \;\middle|\; y - Ax = b \right\}$$

in which we have explicitly introduced the error term b. In the PMEM, $(x;b)$ is a random vector to which a prior probability measure $\mu \otimes \nu$ is assigned. The Maximum Entropy Principle is then used to infer a posterior probability on $(x;b)$, and finally, we choose the expectancy of x under the inferred density $\bar{p}$ as the solution to our inverse problem. The core of the PMEM is an infinite dimensional linearly constrained optimization problem (Maréchal & Lannes 1996) in which the functional to be minimized is (the continuous version of) the Kullback information measure. As explicitly shown in Maréchal & Lannes (1996), it can be solved by means of a dual strategy. It is then possible to demonstrate that, for particular choices of the priors μ and ν, the PMEM gives rise to some of the most classical regularization techniques. For example, a Gaussian ν with αI as the covariance matrix, associated with a Gaussian μ with covariance Q, gives rise to a Tikhonov regularization technique. In particular, if $Q = (B^T B)^{-1}$ (provided that $B^T B$ is positive definite), we retrieve WIPE. Now, taking the multidimensional Poissonian distribution with vector parameter x_0 as μ yields the GCE regularizer (Lannes, Anterrieu, & Bouyoucef 1994), and the Gamma law with vector parameter x_0 gives rise to the Itakura-Saito criterion. Note that in this description, the quadratic fit term derives from the standard Gaussian prior measure on b. If another kind of noise were to corrupt the data, this prior may, of course, be replaced by the appropriate one.

Many other criteria could be derived from the above scheme, taking into account the probabilistic description of the problem. However, it is important to keep in mind that the design of f (i.e., the choice of the corresponding μ) must be governed by the nature of the imaging operator A. The next paragraph illustrates the importance of this point.

4. Simulations and Conclusion

In the simulations presented here, the frequency coverage consists of 211 points. The object to be reconstructed, shown in Figure 1(a), is the original object convolved by a point-spread function corresponding to the selected resolution limit (i.e., to the corresponding frequency coverage to be synthesized; see Lannes, Anterrieu, & Bouyoucef 1994). Three reconstructions were performed: with WIPE (Figure 1(b)), the Shannon entropy on the support determined by the matching pursuit strategy of WIPE (Figure 2(a)), and the GCE in which the prior model x_0 is a smoothed (and normalized) version of the characteristic function of the previous support (Figure 2(b)). For the Shannon entropy and the GCE, the regularization parameter α was adjusted in such a way that the final fit term is equal to that reached by WIPE. The best reconstructed images, shown in Figures 1(b) and 2(b), are quite similar, the values of the stability parameter ρ being reasonably small. In both cases, the inverse problem is well regularized. However, for the selected resolution, the best stability is obtained with the WIPE regularizer.

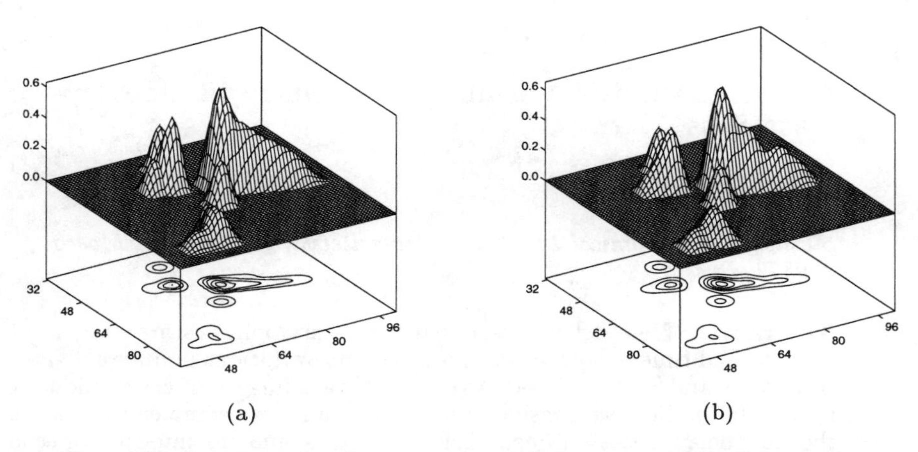

Figure 1. a: image to be reconstructed; b: reconstructed image by WIPE ($\rho = 2.96$).

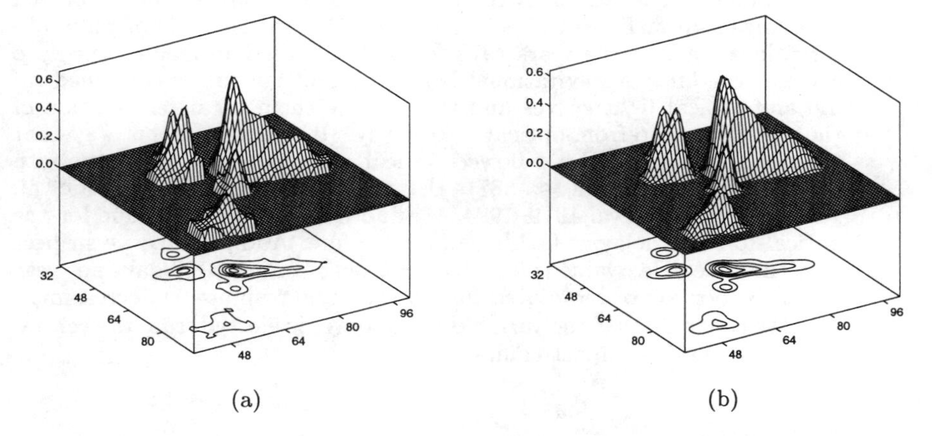

Figure 2. Reconstruction via the GCE; a: with a uniform prior x_{0j} on the support provided by WIPE ($\rho = 4.41$); b: with a smoothed version of the previous prior ($\rho = 3.28$).

References

Lannes, A., Anterrieu, E., & Bouyoucef, K., 1994, J. Mod. Opt., 41, 1537

Lannes, A., Anterrieu, E., & Bouyoucef, K., 1996, J. Mod. Opt., 43, 105

Lannes, A., Anterrieu, E., & Maréchal, P., 1997, A&AS, in press

Maréchal, P., & Lannes, A. 1996, Inverse Problems, 12, 1

Astronomical Data Analysis Software and Systems VI
ASP Conference Series, Vol. 125, 1997
Gareth Hunt and H. E. Payne, eds.

Determination of Variable Time Delay in Uneven Data Sets

V. L. Oknyanskij

Sternberg Astronomical Institute, Universitetskij Prospekt 13, Moscow, 119899, Russia

Abstract. Time delay determinations in astrophysics are used most often to find time delays between flux density variations in different spectral bands and/or lines in AGNs, and different images of gravitationally lensed QSOs. Here we consider a new algorithm for a complex case, when the time delay is itself a linear function of time and the intensity of echo response is exponential function of the delay. We apply this method to the optical-to-radio delay in the lensed double quasar Q0957+561.

Radio-optical variability correlation in Q0957+561 was first reported by Oknyanskij & Beskin (1993, hereafter OB) on the basis of radio observations made in the years 1979 to 1990. OB used an idea to take into account the known gravitational lensing time delay to get combined radio and optical light curves and then to use them for determination of the possible radio-from-optical time delay. It was found this way that radio variations (5 MHz) followed optical ones by about 6.4 years with high level of correlation (≈ 0.87). Using new radio data (Haarsma et al. 1997), for the interval 1979–1994, we find nearly the same value for the optical-to-radio delay as had been found before. Additionally, we suspect that the time delay value is linearly increasing at about 110 days per year while the portion of reradiated flux in the radio response is decreasing.

We conclude that the variable radio source is ejected from the central part of the QSO compact component.

1. Introduction

Time delay determinations in astrophysics are used most often to find time shifts between variations in different spectral bands and/or lines in AGNs, as well as time delays between different images of gravitationally lensed QSOs. In most cases, the task is complicated by uneven spacing of data, so that standard cross-correlation methods become useless. Two different methods are most often used: CCF (Gaskell & Spark 1986) and DCF (Edelson & Krolik 1988), which are based on line interpolation of data sets or binning of correlation coefficients, respectively. We have introduced several simple improvements to CCF (Oknyanskij 1994) and this modernized MCCF combines the best properties of CCF and DCF methods. With MMCF we calculated regression coefficients as functions of time shift. Here, this calculation is generalized for the more complex case where the time delay is a linear function of time, and a portion of the flux density is itself a power-law function of the delay. We apply this method to the optical-to-radio time delay in the gravitationally lensed double quasar Q0957+561. The data sets used here were obtained to determine the gravitational lensing time delay τ_o. Our results are nearly identical for values of τ_o in the interval of 410–550 days. In the discussion below, we take $\tau_o = 425$ days.

2. Method and Results

Our method includes several steps, which are briefly explained below:

Combined light curves. We take the radio (Haarsma et al. 1997) and optical (Vanderriest et al. 1979; Schild & Thompson 1995) data sets for A and B images and determine (using MCCF) the line regression coefficients k(τ) and m(τ). Then we transform $A(t_i)$ values into the B image scale system for the known value of τ_o:

$$B'(t_i - \tau_o) = k(\tau_o) \cdot A(t_i) + m(\tau_o). \tag{1}$$

We combine these values B' with the usual B ones, sorting by time. The resulting optical light curve was then smoothed by averaging in 200 day intervals with steps of 30 days. This accounts for the physical argument that radio sources should be bigger than optical ones. The value of 200 days for smoothing was taken as about optimal from the autocorrelation analysis of light curves.

Correction for change of time delay and radio flux. Taking the optical-to-radio time delay τ_{or} to be a linear function of time, let V be the change of optical-to-radio time delay τ_{or} per year. We fix some moment of time as t_0. It is attractive to choose t_0 so that it falls near a strong maximum in the optical light curve (here J.D. 2445350), which obviously correlates with the high maximum in the radio light curve if take $\tau_{or}(t_0) = 2370$ days. So we can calculate the needed correction:

$$S(t) = \frac{V \cdot (t - t_0)}{365^{\mathrm{d}}} \tag{2}$$

to be added to dates in the optical light curves:

$$t'_i = t_i + S(t_i) \tag{3}$$

Assuming that a portion of radio flux decreases as a power-law function of time with exponent α. We should also correct the optical flux for that fading before computing the cross-correlation function:

$$I'_{\mathrm{op}}(t'_i) = I_{\mathrm{op}}(t_i) \cdot (1 + S(t_i)/2370^{\mathrm{d}})^{-\alpha} \tag{4}$$

Computing MCCFs. We compute an array of MCCFs for combined radio and optical light curves, varying V and α.

Map cross-correlation as a function of V and α. For points (V, α) we map the MCCF values (see Figure 1). The best correlation occurs for $V \approx 110$ days/year, and $\alpha \approx 0.7$.

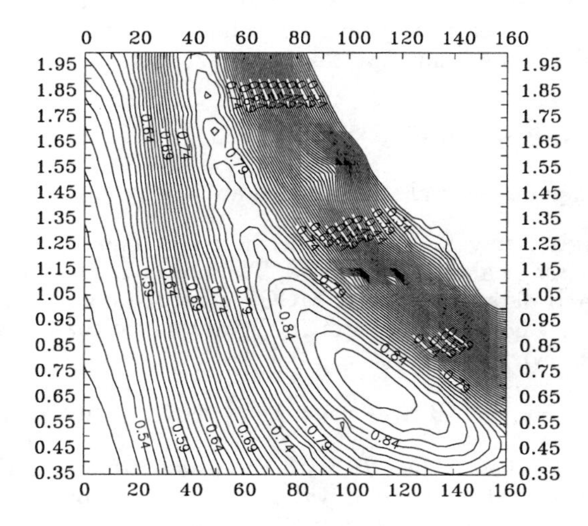

Comparison of optical and radio light curves. We correct the optical combined light curve using (3) and (4) with the parameters $V = 110^{\mathrm{d}}$ and $\alpha = 0.70$, shift ahead by $\tau_{\mathrm{or}}(t_0) = 2370^{\mathrm{d}}$, and then fit to the radio data by analogy with (1). The corrected optical light curve is shown with with the radio light curves in Figure 2. Most features in both light curves coincide quite well. So the investigation supports our assumption on the lengthening of the optical-to-radio time delay. As a result we can give an expression for the optical-to-radio time delay as a linear function of time:

$$\tau_{\mathrm{or}} = 2370^{\mathrm{d}} + 110^{\mathrm{d}} \cdot \frac{(t - t_0)}{365^{\mathrm{d}}} \tag{5}$$

3. Conclusion

We have calculated the time delay between radio and optical flux variations using a new method. In addition, we have investigated the possibilities that (1) there is a change of the time delay that is a linear function of time, and (2) the radio response has power-law dependence on the time delay value.

Finally, let us stress some additional consequences from our results:

1. For some objects, optical-to-radio time delays were probably not found because they were too long compared to the duration of monitoring program.

2. Optical-radio correlations may not have been recognized in some objects since the time delays as well as response functions probably were variable. This possibility has never been entertained before.

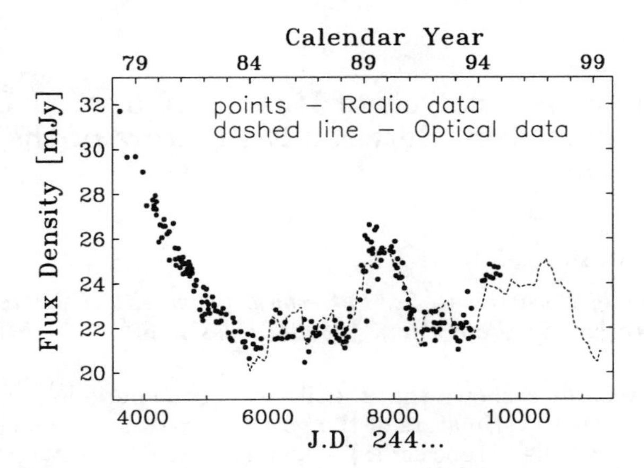

Figure 2. Radio and optical combined light curves. (The optical light curve is corrected as described in the text.)

3. The variable radio flux in Q0957+561 may originate in a very compact jet component moving away from the optical source. Only after another jet component appears (whose time delay value will of course be different) will the QSO again show some optical-radio correlation.

4. If several compact jet components exist simultaneously in a QSO then we may have no chance to find any radio-optical correlation. Only if these jet components move toward the observer closely to the line of sight (as it is probable for Blazars), will radio-optical correlation have a chance to be found, since the time delays for all these variable radio components would be very close to zero.

Acknowledgments. In conclusion, we thank Debborah Haarsma for sending us the preprint with the new radio data for Q0957+561 before publication.

References

Edelson, R. A., & Krolik, J. H. 1988, ApJ, 333, 646

Gaskell, C. M., & Spark, L. S. 1986, ApJ, 305, 175

Haarsma, D. B., Hewitt, J. N., Lehar, J., & Burke, B. F. 1997, ApJ, 479, 102

Oknyanskij, V. L., & Beskin, G. M. 1993, in Gravitational Lenses in the Universe: Proceedings of the 31st Liege International Astrophysical Colloquium, eds. J. Surdej et al. (Liege, Belgium: Universite de Liege, Institut d'Astrophysique), 65

Oknyanskij, V. L. 1994, Ap&SS, 222, 157

Schild, R. E., & Thomson, D. J. 1995, AJ, 109, 1970

Vanderriest, C., et al. 1989, A&A, 215, 1

Astronomical Data Analysis Software and Systems VI
ASP Conference Series, Vol. 125, 1997
Gareth Hunt and H. E. Payne, eds.

Time Series Analysis of Unequally Spaced Data: Intercomparison Between Estimators of the Power Spectrum

V. V. Vityazev

Astronomy Department, St. Petersburg University, Bibliotechnaya pl.2, Petrodvorets, St. Petersburg, 198904, Russia.

Abstract. It is shown that the likeness of the periodogram and the LS-spectrum (both estimators of the power spectrum are widely used in the spectral analysis of time series), depends on the properties of the spectral window $W(\omega)$ corresponding to the distribution of time points. The main results are: a) all the estimators evaluated at frequency ω are identical if $W(2\omega) = 0$; b) the Schuster periodogram differs from the LS-spectra at the frequencies $\omega = \hat{\omega}_k/2$, where $\hat{\omega}_k$ are the frequencies at which the spectral window has large side peaks due to irregular distribution of time points. Two examples for situations typical in astronomy illustrate these conclusions.

1. Introduction

In various branches of astronomy, we face the problem of finding periodicities hidden in observations. If the data are regularly spaced in time, the Schuster periodogram is the basic tool for evaluating the power spectra (Marple 1987; Terebizh 1992). Unfortunately, astronomical observations are irregular for various reasons: day-time changes, weather conditions, positions of the object under observations, etc. Present day theory and practice of the spectral analysis of unequally spaced time series are based on two approaches. The first one employs the Schuster periodogram (Deeming 1975; Roberts et al. 1987). The second one uses the procedure of the least squares fitting of a sinusoid to the data (Barning 1962; Lomb 1976; Scargle 1982) with resulting estimators known as the LS-spectra. The most valuable feature of the LS-spectra is well defined statistical behavior. At the same time, the LS-spectra lose very important properties: description in terms of the spectral window, connection with the correlation function, etc. On the other hand, the Schuster periodogram of a gapped time series satisfies all the fundamental relations of the classical spectral analysis, but its statistical properties are complicated as compared to the case of regular data. It is worth mentioning that despite different theoretical foundations, the Schuster periodogram and the LS-spectra frequently turn out to be almost identical. This similarity requires an explanation, and we are trying to find situations when the Schuster periodogram and the LS-spectra are very close to each other or differ greatly. The ultimate goal of this study is to clarify the properties of various techniques which are used to derive the periodicities from the unequally distributed data.

2. Two Estimators of the Power Spectrum

For a set of N observations $x_k = x(t_k)$, $k = 0, 1, \dots, N-1$ with zero mean obtained at arbitrary times t_k, we can set up the model

$$f(t) = \sum_{i=1}^{2} a_i \phi_i(t), \tag{1}$$

where

$$\phi_1(t) = \cos \omega t, \qquad \phi_2(t) = \sin \omega t. \tag{2}$$

Using the following notation

$$(p, q) = \frac{1}{N} \sum_{k=0}^{N-1} p(t_k) q(t_k), \qquad \|p\|^2 = (p, p), \tag{3}$$

the "classical" estimator of the power spectrum (the Schuster periodogram) can be written in the form

$$S(\omega) = (x, \phi_1)^2 + (x, \phi_2)^2 = \frac{1}{N^2} \mid \sum_{k=0}^{N-1} x_k e^{-i\omega t_k} \mid^2 . \tag{4}$$

If the signal contains a sine function of frequency ω_0, then the product $x_k e^{-i\omega t_k}$ makes a large contribution to S provided that $\omega = \omega_0$. In other words, the Schuster periodogram, to the limit of normalizing factor, is a square of the correlation coefficient between the data and a harmonic function.

The alternative estimator of the power spectrum based on the least squares fitting of the sine function to the data was proposed by Lomb (1976) and Scargle (1982). Their approach is based on the introduction of the new time points

$$\hat{t_k} = t_k - \frac{1}{2\omega} \arctan \frac{\sum_k \sin 2\omega t_k}{\sum_k \cos 2\omega t_k}, \tag{5}$$

where the time shift provides the orthogonality of the functions

$$\hat{\phi}_1(t) = \cos \omega \hat{t_k}, \qquad \hat{\phi}_2(t) = \sin \omega \hat{t_k}. \tag{6}$$

Under this assumption the LS-spectrum looks as follows:

$$L(\omega) = \frac{1}{2} \left[\frac{(x, \hat{\phi}_1)^2}{\|\hat{\phi}_1\|^2} + \frac{(x, \hat{\phi}_2)^2}{\|\hat{\phi}_2\|^2} \right]. \tag{7}$$

Thus we see that the Schuster periodogram differs from the LS-spectra by definition.

The intercomparison between the Schuster periodogram and the LS-spectrum is given by the following

Theorem. *At the set of frequencies that satisfy equation*

$$W(2\omega) = 0, \tag{8}$$

where the spectral window $W(\omega)$ *is*

$$W(\omega) = \frac{1}{N^2} \mid \sum_{k=0}^{N-1} e^{-i\omega t_k} \mid^2 \tag{9}$$

the Schuster periodogram and the LS-spectra are identical.

3. The Spectral Windows for Typical Distribution of Time Points

In this section we consider two typical distributions of points for which the frequencies that satisfy Eq. (8) do exist.

3.1. Time Series with Periodic Gaps

Astronomical observations are often performed with periodic gaps. Ground-based observations are interrupted by day-night alternation; the observations from a space vehicle are usually stopped when the satellite enters the radiation belts. To model the situations we suppose that, in the set of observations with a constant interval Δt, one has n successive observations and p successive missing points, and the group of $n+p$ points is repeated m times, so the period of gaps is $\Delta T = (n+p)\Delta t$. In the previous paper (Vityazev 1994) it was shown that in this case the spectral window looks as follows

$$W(\omega) = \frac{\sin^2(n\omega\Delta t/2)}{n^2\sin^2(\omega\Delta t/2)} \frac{\sin^2(m\omega\Delta T/2)}{m^2\sin^2(\omega\Delta T/2)}. \tag{10}$$

It is easy to find that the frequencies

$$\omega_j = \frac{2\pi}{m\Delta T}j, \quad j = 1, 2, ... \tag{11}$$

satisfy Eq. (8), provided that $j \neq m/2, m, ...$ if m is even and $j \neq m, 2m, ...$, if m is odd.

3.2. Observations with a Long Gap

Considered here is a situation where two sets of observations (each one consisting of n successive points) are separated by p missing points forming the gap. As earlier, all the points are supposed to be regularly spaced over the time interval $\Delta t = const$. Now, for the spectral window we have (Vityazev 1994)

$$W(\omega) = \frac{\sin^2(n\omega\Delta t/2)}{n^2\sin^2(\omega\Delta t/2)} [\cos((n+p)\omega\Delta t)]/2. \tag{12}$$

It is not difficult to show that the frequencies

$$\omega_j = \frac{\pi}{(n+p)\Delta t} (j + \frac{1}{2}), \ j = 0, 1, \ ... \ , n+p-1, \tag{13}$$

satisfy the condition of Eq. (8).

4. Conclusions

The LS-spectra gained popularity due to the fact that they retain the exponential distribution of the classical periodogram when the time series is assumed to be white noise. Now we see that at frequencies that satisfy Eq. (8) the Schuster periodogram retains the exponential distribution too.

The Schuster periodogram differs from the LS-spectra only at the frequencies that satisfy the condition $1 - W(2\omega) \ll 1$. It means that the discrepancies between the Schuster periodogram and the LS-spectra are large when the time series contain a harmonic of the frequency, the double value of which coincides with the frequency at which the spectral window has a large side peak. In the case of periodical gaps it happens when the period of a signal hidden in the data is one half the period of the gaps. In this situation (Vityazev 1997a), the spectral estimation faces unrealistic intensities of the spectral peaks and the strong dependence of the heights of peaks on the phase of the signal. It is very important to emphasize that these problems come not from the choice of the tool to evaluate the power spectrum; they originate from mixing two sources of the periodicities: one is the physical process that we observe and another one is a periodical interruption of observations. In astronomy, the rotation and revolution of the Earth impose diurnal and annual cycles on the Earth-based observations. The periods hidden in observations of the Sun, stars, quasars, etc., are hardly connected physically with the periods specific to the Earth. For these observations, the probability of mixing periodicities is negligible. On the contrary, if we study the Earth from the Earth (such is the case with astrometric observations of the Earth's rotation parameters), then the semi-annual period in the Earth's rotation interferes with the annual gaps in observations.

For further details the reader is referred to Vityazev (1997a, 1997b).

References

Barning, F. J. M. 1962, B.A.N., 17, N1, 22

Deeming, T. J. 1975, Ap&SS, 36, 137

Lomb, N. R. 1976, Ap&SS, 39, 447

Marple, Jr., S. L. 1987, Digital Spectral Analysis with Applications (Englewood Cliffs, NJ: Prentice-Hall)

Roberts, D. H., Lehar, J., & Dreher, J. W. 1987, AJ, 93, 968

Scargle, J. D. 1982, ApJ, 263, 835

Terebish, V. Yu. 1992. Time Series Analysis in Astrophysics (Moscow: Nauka)

Vityazev, V. V. 1994, Astron. and Astrophys. Tr., 5, 177

Vityazev, V. V. 1997a, A&A, in press

Vityazev, V. V. 1997b, A&A, in press

Astronomical Data Analysis Software and Systems VI
ASP Conference Series, Vol. 125, 1997
Gareth Hunt and H. E. Payne, eds.

The Time Interferometer: Synthesis of the Correlation Function

V. V. Vityazev

Astronomy Department, St. Petersburg University, Bibliotechnaya pl.2, Petrodvorets, St. Petersburg, 198904, Russia.

Abstract. A comparative study of the fundamentals, problems and techniques common to the spectral analysis of time series and interferometry is presented. The aperture synthesis technique well known in radio astronomy is adapted to the spectral analysis of the time series.

1. Review of the Fundamentals

Suppose that a time series $X(t)$ is a stationary stochastic process with zero mean defined by a set of realizations

$$X(t) = \{x_p\}_{p=1}^{N}, \quad 0 \leq t \leq T < \infty. \tag{1}$$

To describe gaps in observations we introduce the *time window function*

$$h(t) = \begin{cases} 1 & \text{if at time } t \text{ the data exists} \\ 0 & \text{if at time } t \text{ the data is absent.} \end{cases} \tag{2}$$

With this notation the observed time series can be represented as

$$Y(t) = h(t)X(t). \tag{3}$$

Calculate the *periodogram*:

$$D(\omega) = \frac{1}{2\pi T} < \left| \int_0^T y_p(t) \exp(-i\omega t) dt \right|^2 >, \tag{4}$$

where $<>$ denotes averaging over the set of realizations. Under the conditions stated above, the relationship between the periodogram $D(\omega)$ and the power spectrum $G(\omega)$ is given by the convolution

$$D(\omega) = \int_{-\infty}^{+\infty} W(\omega')G(\omega-\omega')d\omega', \tag{5}$$

where the *spectral window function* $W(\omega)$ is the periodogram of the time window function

$$W(\omega) = \frac{1}{2\pi T} \left| \int_0^T h(t) \exp(-i\omega t) dt \right|^2. \tag{6}$$

Now, from Eq. (5), for the correlogram $k_D(\tau)$ and the correlation window $H(\tau)$, introduced as inverse Fourier transforms of $D(\omega)$ and $W(\omega)$ respectively, one has:

$$k_D(\tau) = H(\tau)k(\tau), \tag{7}$$

where $k(\tau)$ is the auto-correlation function of $X(t)$. For further details the reader is referred to Jenkins & Watts (1968), Deeming (1975), Otnes & Enocson (1978), Marple (1978), and Terebish (1992).

In interferometry (Esepkina et al. 1973; Thompson et al. 1986), the *intensity of radiation* from any source on the sky can be described in terms of position (α, δ), wavelength (λ), and time (t). Each measurement is an average over a band of wavelengths and over some span of time. Here, the one-dimensional, monochromatic, and instantaneous approximation is used to simplify the discussion. The resulting specific intensity $T_b(\vartheta)$ that describes the distribution of the source brightness along the arc of a circle (ϑ is the angular coordinate) has a Fourier transform $\hat{T}_b(u)$, which is called *the spectrum of spatial frequencies.* Correspondingly, when an interferometer measures *the visibility data* $\hat{T}_a(u)$, the image, or *the map*, $T_a(\vartheta)$ can be calculated by the Fourier transform of $\hat{T}_a(u)$. Two fundamental relations are valid:

$$T_a(\vartheta) = A(\vartheta) \otimes T(\vartheta), \tag{8}$$

$$\hat{T}_a(u) = \hat{A}(u) \times \hat{T}(u), \tag{9}$$

where $A(\vartheta)$ is *the beam* of interferometer and $\hat{A}(u)$ is *the transfer function.* Eq. (9) determines an interferometer as a *filter of spatial frequencies*, whereas Eq. (8) explains why, due to convolution of $T_b(\vartheta)$ with the beam $A(\vartheta)$, the resulting image is called a *dirty map.*

Upon close examination one can see that in both sciences the rigorous (theoretical) quantities are introduced at the first level. In the spectral-analysis case they are the *power spectrum* and the *correlation function*; in the interferometry their counterparts are the *distribution of brightness* and the *spatial spectrum.* At the second level we have estimators of the strict quantities. In spectral analysis, they are the *periodogram* and the *correlogram*, whereas in the interferometry these are the *map* and the *visibility data*, respectively. Finally, equations which connect the quantities of the two levels are identical (convolution and multiplication) and include the characteristics of observations: the *beam* and the *transfer function* and their analogs, i.e., the *spectral window* and the *correlation window.*

In reality, due to the finite dimensions of mirrors and the finite time spans of observations we cannot get the true quantities, and all we can do is to find their optimal approximations. In optics or in radio astronomy, when filled apertures are used, the maps are produced directly in the focal plane of a telescope. Analogously, when the time series is given at all points of some interval or at time points regularly spaced within the interval, the evaluation of the periodogram can be made quite easily. When an interferometer is used, the aperture is not solid, and what we can measure is the visibility data, i.e., the estimator of the spectrum of spatial frequencies. The longer the baseline, the smaller the area of the $(u - v)$-plane (u-domain in the one-dimensional case) filled, and the more dirty the resulting map becomes.

To overcome this, various techniques of *aperture synthesis* are used, and this leads to complete solution of the problem since the $(u - v)$-plane is completely filled. If the aperture synthesis provides partial filling of the $(u - v)$-plane, the *cleaning* procedures can be used with the aim of eliminating the artifacts of the "holes" in the $(u - v)$-plane from the map. We have the same problems in the spectral-analysis case, when the time points are distributed irregularly or have long gaps. In this case the correlograms cannot be determined for all values of time lag τ, and this would give false features in the resulting periodograms. The main aim of the present paper is to answer the question: is it possible to apply the aperture synthesis method to spectral analysis of time series?

2. Synthesis of the Correlation Function

It is known that to do aperture synthesis one must have an interferometer with the changeable baseline. Of all the schemes of the aperture synthesis the one proposed by Ryle (1960) is the most suitable for us. This consists of two antennas A and B fixed at separation L. The third antenna C is moving inside the interval $[L/2, L]$. At each position of the moving antenna one obtains two interferometers (CB) and (AC) with the baselines l and $L/2 + l$, and they yield the values of the visibility data $\hat{T}_a$ at the points $u = l/\lambda$ and $u = U/2 + l/\lambda$, where $U = L/\lambda$. Obviously, while antenna C sweeps all the interval $[L/2, L]$, the visibility data $\hat{T}_a(u)$ become available at all points of the interval $[0, U]$.

Now we apply this idea to time series analysis. Every two points spaced at the distance τ may be called *the Time Interferometer* with variable baseline τ, since each such pair yields the estimation of the correlation function $k_s(\tau)$ by averaging the products $X(t)X(t + \tau)$ over the set of realizations. The value $k_s(\tau)$ can be obtained no matter where the points t and $t + \tau$ are located inside the interval $[0, T]$. This follows from our assumption that $X(t)$ is a stationary stochastic process, and that is crucial for our study.

Assume that the time series is given at two points t_1 and $t_1 + T/2$. This Time Interferometer allows us to get the correlation function only at the point $\tau = T/2$. Let us make new observations at the points $t_1 + T/2 < t < t_1 + \mathrm{T}$. It is clear that each new point $t = t_1 + T/2 + \tau$ yields two additional values of the correlation function, namely, at the points τ and $T/2 + \tau$. Obviously, when the observations cover the interval $[t_1 + T/2, t_1 + T]$, the values of $k_s(\tau)$ become available at all points of the interval $[0, T]$.

Now we see that the fixed antennas A and B in the Ryle interferometer are the counterparts of the boundary points t_1 and $t_1 + T$, while each new position of the moving antenna C is nothing else but the new point of observations. Since Ryle's interferometer makes the synthesis of the visibility data, it is a good reason to call the described procedure the *synthesis of the correlation function.*

Obviously, complete synthesis and the evaluation of $k_s(\tau)$ from an even time series are the same. Thus, with $h(t) \equiv 1$ elsewhere for the synthesized periodogram we get

$$D_s(\omega) = \frac{1}{\pi} \int_0^T (1 - \frac{\tau}{T}) k_s(\tau) \cos(\omega\tau) d\tau. \tag{10}$$

This estimator yields the clean spectrum, i.e., free of the false peaks that come from the "holes" in the τ-domain.

To proceed further, assume that we have a gap in the observations. Let the length of the gap be l and the longest distance between the borders of the gap and the boundary points of the interval $[0, T]$ be a. If $l \leq a$ then the correlation function can be evaluated at all points $\tau \in [0, T]$, otherwise only on the subintervals $[0, a]$ and $[l, T]$. In this case the correlation function turns out to be synthesized at all points except the "hole" of the length $l - a$, and, consequently, the synthesized periodogram $D_s(\omega)$ calculated from Eq. (10) will not be clean. Nevertheless, it is less contaminated than the periodogram calculated directly from Eqs. (4). Thus we see that to clean the spectrum completely, one needs to perform more observations until the condition $l \leq a$ becomes true.

When only one realization is available, averaging over the ensemble should be replaced with averaging over time. For further development of the method the reader is referred to Vityazev (1996).

References

Deeming, T. J. 1975, Ap&SS, 36, 137

Esepkina, N. A., Korolkov, D. N., & Pariysky, Yu. N. 1973, Radio telescopes and Radiometers (Moscow: Nauka)

Jenkins, G. M., & Watts, D. G. 1968, Spectral Analysis and its Applications (San Francisco: Holden-Day)

Marple, Jr., S. L. 1987, Digital Spectral Analysis with Applications (Englewood Cliffs, NJ: Prentice-Hall)

Otnes, R. K., & Enocson, L. 1978, Applied Time Series Analysis (New York: Wiley-Interscience)

Ryle, M., & Hewish, A. 1960, MNRAS, 120, 220

Terebish, V. Yu. 1992, Time Series Analysis in Astrophysics (Moscow: Nauka)

Thompson, A. R., Moran, J. M., & Swenson, G. W. Jr. 1986, Interferometry and Synthesis in Radio Astronomy (New York: Wiley)

Vityazev, V. V. 1996, Astron. and Astrophys. Tr., in press

Astronomical Data Analysis Software and Systems VI
ASP Conference Series, Vol. 125, 1997
Gareth Hunt and H. E. Payne, eds.

A New Stable Method for Long-Time Integration in an N-Body Problem

Tanya Taidakova

Crimean Astrophysical Observatory, Simeiz, 334242, Ukraine

Abstract. The most serious error in numerical simulations is the accumulation of discretization error due to the finite stepsize. Traditional integrators such as Runge-Kutta methods cause linear secular errors to the energy, the semi-major axis, and the eccentricity of orbiting objects. Potter (1973) described an implicit second-order integrator for particles in a plasma with a magnetic field. We have used this integrator for an investigation of the dynamics of particles around a planet (or star) in a co-rotating coordinate system. A big advantage of this numerical integrator is its stability: the error in the semi-major axis and the eccentricity depends only on the step size and does not grow with an increasing number of time steps. The argument of pericenter changes linearly with time and more slowly than in the case of the Runge-Kutta integrator. In addition, this implicit integrator takes much less computing time than the second-order Runge-Kutta method. We tested this method for several astronomical systems and for motion of an asteroid in a 1:1 Jupiter resonance during 200 million time steps (about 5 million years or 800 thousand periods of asteroid resonance motion).

1. Introduction

In the last few years there has been great interest in the numerical study of long term evolution of bodies of the Solar system. As the integration time increases, the numerical results become more contaminated by various errors. The most serious error is the accumulation of the discretization (truncation) error due to a finite stepsize (or the replacement of continuous differential equations by finite difference equations). The conventional integrators such as Runge-Kutta, multistep and Taylor methods, generate linear secular errors in orbital energy and angular momentum. This means that the semi-major axis and the eccentricity change linearly with time and the linear secular error in the semi-major axis produces a quadratic secular error in the planetary longitude. A new symplectic integrator produces no secular truncation errors in the actions of a Hamiltonian system.

2. Implicit Integrator

In this paper we briefly discuss an implicit numerical integrator. The discretization errors in the energy, the semi-major axis, and the eccentricity by the implicit second-order integrator show only periodic changes. The truncation error in the argument of pericenter grows linearly in time. The equations of motion of a particle in the gravitational field of the Sun and the planet with mass m_{pl} in the corotating coordinate system take the form:

$$\begin{aligned} \ddot{x} &= 2\dot{y} + x + F_x \\ \ddot{y} &= -2\dot{x} + y + F_y \\ \ddot{z} &= F_z \quad , \end{aligned} \tag{1}$$

where:

$$\begin{aligned} F_x &= -\frac{x}{(x^2+y^2+z^2)^{3/2}} - \frac{m_{pl}(x-x_{pl})}{\left[(x-x_{pl})^2+(y-y_{pl})^2+(z-z_{pl})^2)\right]^{3/2}} \\ F_y &= -\frac{y}{(x^2+y^2+z^2)^{3/2}} - \frac{m_{pl}(y-y_{pl})}{\left[(x-x_{pl})^2+(y-y_{pl})^2+(z-z_{pl})^2)\right]^{3/2}} \\ F_z &= -\frac{z}{(x^2+y^2+z^2)^{3/2}} - \frac{m_{pl}\,z}{\left[(x-x_{pl})^2+(y-y_{pl})^2+(z-z_{pl})^2)\right]^{3/2}} \quad . \end{aligned}$$

Here the total mass of the Sun and the planet is taken as the unit of mass, and the distance between the planet and the Sun is taken as the unit of length. The unit of time is chosen in such a way that the angular velocity of orbital motion of the planet is equal to unity, and, hence, its orbital period is 2π. Let $v^* = v_x + i\,v_y$, $x^* = x + i\,y$, and $F^* = F_x + i\,F_y$. We obtain rather than (1):

$$\begin{aligned} \frac{dz}{dt} &= v_z \\ \frac{dv_z}{dt} &= F_z \\ \frac{dx^*}{dt} &= v^* \\ \frac{dv^*}{dt} &= -2\,i\,v^* + x^* + F^* \quad . \end{aligned} \tag{2}$$

We may solve the equation $\frac{dU(t)}{dt} + R(U(t),t) = 0$ with initial conditions $U(t_0) = U_0$ by the implicit second-order integrator described in Potter (1973):

$$v^{[n+1]} = v^{[n]} - \frac{1}{2}\left(R^{[n]} + R^{[n+1]}\right)\Delta t \quad . \tag{3}$$

In our equations (2), R is a function of x, y, z. We will calculate this function R in space-time points $n + 1/2$. From (2) by the use of Eq. (3) we derive the equations for new integrator (Taidakova 1990, 1995):

$$v_x^{[n+1]} = \frac{v_x^{[n]}(1-\Delta^2 t) + (2v_y^{[n]} + x^{[n+\frac{1}{2}]} + F_x^{[n+\frac{1}{2}]})\Delta t + (y^{[n+\frac{1}{2}]} + F_y^{[n+\frac{1}{2}]})\Delta^2 t}{1+\Delta^2 t}$$

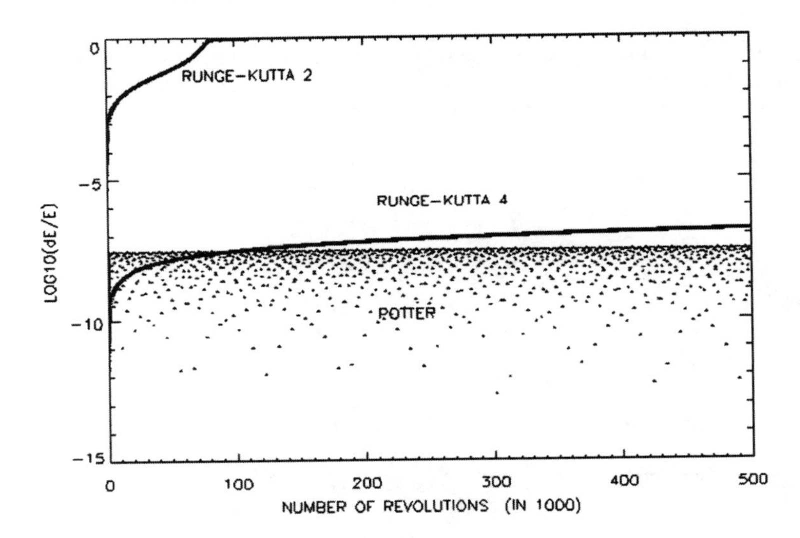

Figure 1. Numerical errors in the energy with second order Runge-Kutta, fourth order Runge-Kutta and second order Potter integrator.

$$\begin{aligned}
v_y^{[n+1]} &= \frac{v_y^{[n]}(1-\Delta^2 t) - (2v_x^{[n]} - y^{[n+\frac{1}{2}]} - F_y^{[n+\frac{1}{2}]})\Delta t + (x^{[n+\frac{1}{2}]} + F_x^{[n+\frac{1}{2}]})\Delta^2 t}{1+\Delta^2 t} \\
v_z^{[n+1]} &= v_z^{[n]} + F_z^{[n+\frac{1}{2}]}\Delta t \\
x^{[n+1]} &= x^{[n]} + (v_x^{[n+1]} + v_x^{[n]})\Delta t/2 \\
y^{[n+1]} &= y^{[n]} + (v_y^{[n+1]} + v_y^{[n]})\Delta t/2 \\
z^{[n+1]} &= z^{[n]} + (v_z^{[n+1]} + v_z^{[n]})\Delta t/2 \quad , \qquad (4)
\end{aligned}$$

where: $x^{[n+\frac{1}{2}]} = x^{[n]} + v_x^{[n]}\Delta t/2$; $y^{[n+\frac{1}{2}]} = y^{[n]} + v_y^{[n]}\Delta t/2$;
$z^{[n+\frac{1}{2}]} = z^{[n]} + v_z^{[n]}\Delta t/2$; $F_{x,\,y,\,z}^{[n+\frac{1}{2}]} = F\left(x^{[n+\frac{1}{2}]}, y^{[n+\frac{1}{2}]}, z^{[n+\frac{1}{2}]}, t^{[n+\frac{1}{2}]}\right)$.

3. Numerical Examples

We tested this method for several astronomical systems (Taidakova 1995). In order to see the properties of the implicit integrator, we first choose the 2-body problem. Figure 1 shows the numerical errors in the energy $\log(\Delta E/E)$ in the numerical integrations of the motion of the particle in circular orbit around the Sun with different integrators. Figure 2 shows the errors in the parameter $\Delta_i = \sqrt{(x_0 - x_i)^2 + (v_{x_0} - v_{x_i})^2 + (v_{y_0} - v_{y_i})^2}$, where i is the number of revolutions, x is the coordinate ($y = 0$), and v_x, v_y are the velocities in the numerical integration of the orbital motion of asteroid in the 1:1 Jupiter resonance during 200 million steps. The computer time with the implicit second-order integrator is about 1.52–1.94 times faster than that with second-order Runge-Kutta integrator.

The calculation were carried out in a PC with DX4/100.

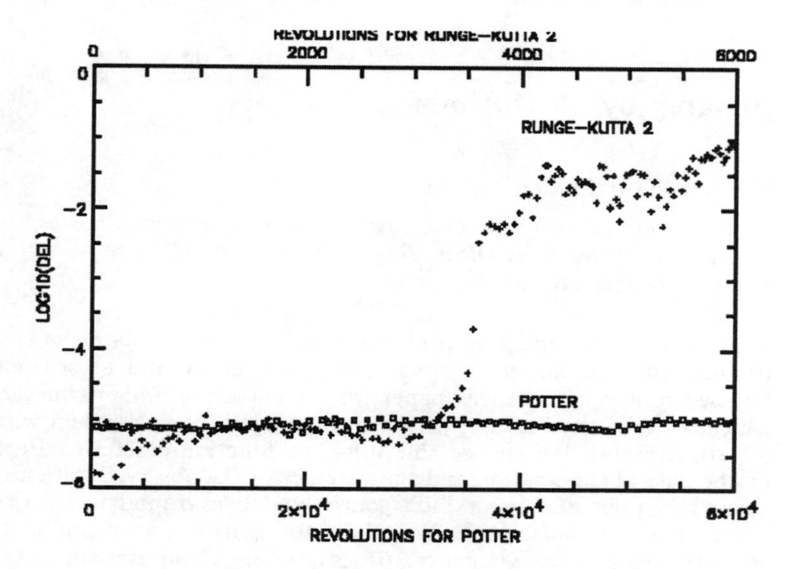

Figure 2. Numerical errors in the parameter Δ_i with second order Runge-Kutta method and second order Potter integrator.

Acknowledgments. The author thanks the Local Organizing Committee and U.S. Civilian Research & Development Foundation (Grant # 96023) for support.

References

Potter, D. 1973, Computational Physics (New York: Wiley)

Taidakova, T. 1990, Nauch.Inform.Astrosoveta Akademii Nauk SSSR (Riga: Zinatne), 68, 72

Taidakova, T. 1995, Ph.D. Thesis, Moscow State University

Astronomical Data Analysis Software and Systems VI
ASP Conference Series, Vol. 125, 1997
Gareth Hunt and H. E. Payne, eds.

Imaging by an Optimizing Method

Y. Chen, T. P. Li, and M. Wu

High Energy Astrophysics Lab, Institute of High Energy Physics, Chinese Academy of Sciences, Beijing 100039, PRC,
E-mail: cheny@astrosv1.ihep.ac.cn

Abstract. The imaging problem can be described as an optimizing problem in mathematics. Thus optimizing theory and algorithms can be used to solve it. In this paper we present an optimizing method, in which we take the imaging problem as an optimizing problem with linear constraints. We choose the objective function carefully. Both the mathematical expectation and the variance of the observed data are considered. Upper and lower limit source and background intensities can be conveniently considered. We adopt an algorithm very similar to the affine scaling approach in convex programming. Computer simulations of rotating modulation collimator imaging show that the quality of images from this method is better than that from the traditional cross-correlation method. Both point and extended sources can be imaged in the same field of view. We also apply the algorithm to ROSAT PSPC pointed observation data of the Crab nebula. The image quality is improved significantly. The extended structure of the Crab nebula can clearly be seen.

1. Introduction

The imaging problem may be described as inferring the sky brightness distribution from observations and prior knowledge (Cornwell 1992). In this paper, we will introduce an optimizing method and adopt it to the imaging problem. Then we will apply it to simulations of a rotating modulation collimator (RMC), and ROSAT PSPC (the Position Sensitive Proportional Counter) observations.

2. Imaging Problem

The imaging problem is:

$$d = Pf + n \tag{1}$$

where d is the observational data, P is the point spread function, f is the unknown sky, and n is the noise. Usually, there are some constraints for f and n in this linear system of equations. The optimizing problem is:

$$min.\ \ F(x) \tag{2}$$

$$subject\ to\ \ \ Ax = b \tag{3}$$

$$x \geq 0 \tag{4}$$

where $F(x)$ is an objective function.

In astronomy, the noise n_k usually follows a Poisson or Gaussian distribution. Thus it has a certain expected value and variance. The sky intensity f may have a upper limit *up* and a lower limit *low*. Then we can make an objective function

$$F(f,n) = (\sum_i^k n_i^2/d_i - k)^2 + a(\sum_i^k n_i)^2 - b\sum_i^m (\ln(f_i - low_i) + \ln(up_i - f_i)) \quad (5)$$

where a and b are coefficients, k is the number of bins of observational data, and m is the number of sky bins. The constraint condition is

$$\sum_i^m p_{ji} f_i + n_j = d_j \quad (j = 1, ..., k) \quad (6)$$

Both f_i and n_j are unknown. This problem is similar to the convex programming problem.

3. Affine Scaling Algorithm

The affine scaling (AS) algorithm is one of the simplest and most efficient of interior point method algorithms (Dikin 1967). For the optimizing problem (2) $\sim$ (4) the AS algorithm in detail is:

1. Try to find an initial solution.
2. Calculate
$$H_k = [\nabla^2 f(x^k) + X_k^{-2}]^{-1} \quad (7)$$
$$[AH_k A^T]\omega^k = AH_k \nabla f(x^k) \quad (8)$$
$$s^k = \nabla f(x^k) - A^T \omega^k \quad (9)$$
3. Check whether the stopping criteria is satisfied.
4. Find a transition direction.
$$d_x^k = -H_k s^k \quad (10)$$
5. Calculate the step length α_k. Search for that α_k which minimizes the objective function.
6. Move to a new solution.
$$x^{k+1} \leftarrow x^k + \alpha_k d_x^k \quad (11)$$
7. Let $k \leftarrow k + 1$ and go to Step 2.

We developed an algorithm based on the affine scaling algorithm (Goldfarb 1991; Fang 1993) for problem (5) $\sim$ (6).

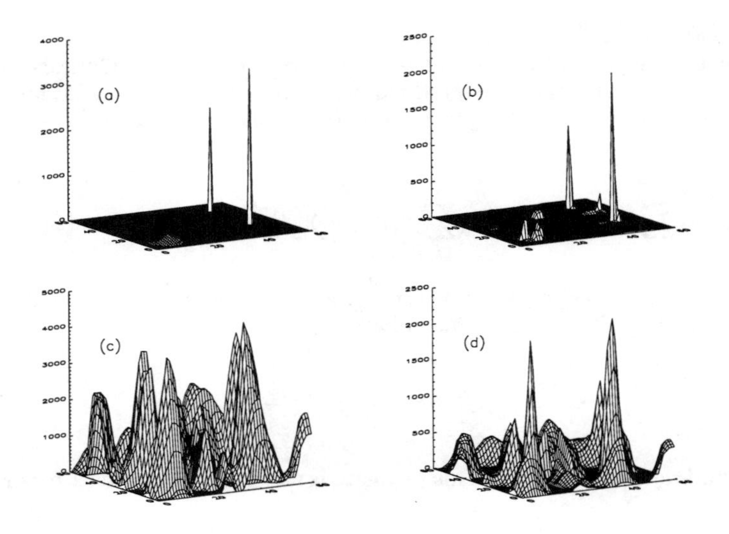

4. Application

4.1. Rotating Modulation Collimator

We have simulated a rotating modulation collimator. The configuration of the RMC is shown in Table 1. The background is assumed to be 0.09 ph cm^{-2} s^{-1}. The fluxes of the two point sources are assumed to be 1.0×10^{-2} and 6.7×10^{-3} ph cm^{-2} s^{-1}, and the total flux of the extended source 3.5×10^{-2} ph cm^{-2} s^{-1} (Figure 1a). The observing time is assumed to be one day. Figure 1b shows the

Table 1. The Configuration of the RMC

distance between strips (cm)	distance between grid planes (cm)	FOV (^{o})	total active area (cm^2)
1.0	34	6×6	1000

result of the AS algorithm, while Figure 1c and d result from the cross-correlation method and the CLEAN cross-correlation, respectively. Both the point sources and the extended source can be seen in Figure 1b. The angular resolution and image quality in Figure 1b are much better than in the cross-correlation images.

4.2. ROSAT PSPC Image of Crab

We used this algorithm to reconstruct ROSAT PSPC data. The results are shown in Figure 2. Figure 2a is the original image observed by PSPC. The

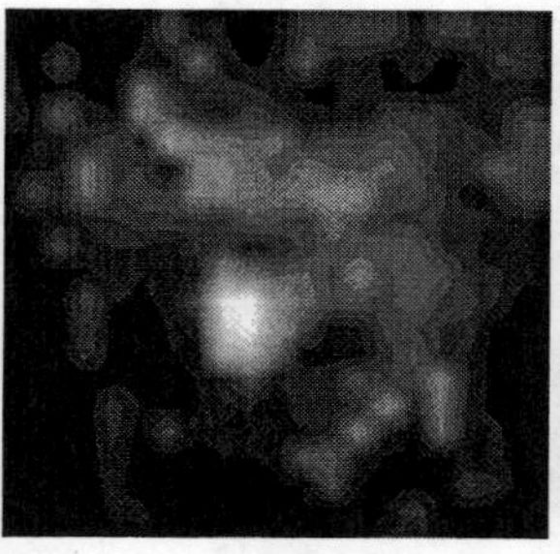
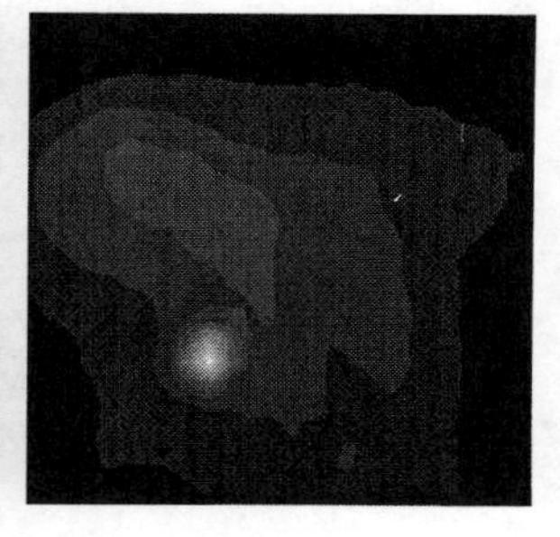

Figure 2. a) The original ROSAT PSPC Crab nebula image. b) The image obtained from the AS algorithm. The FOV is 100×100 arcsec. c) A ROSAT HRI image of the Crab nebula. The FOV is 100×96 arcsec.

result of the AS algorithm is shown in Figure 2b, where the extended structure is clearly seen. This extended structure can also be seen by ROSAT HRI (the High Resolution Imager) (Figure 2c).

5. Discussions and Conclusions

The calculation time for the AS algorithm depends on the initial solution. We can use the solution, from some other algorithm such as Richardson-Lucy iteration or cross-correlation, as the initial solution in order to reduce the calculation time.

In this paper we developed an AS algorithm and applied it to the imaging problem. The results show that this algorithm can be used in the imaging problem and usually results in a better image than that obtained from a traditional method such as cross-correlation. This algorithm can be also used in the data reconstruction of an imaging instrument (e.g., ROSAT PSPC).

Acknowledgments. We thank X. J. Sun for the help on the ROSAT data analysis.

References

Cornwell, T. J. 1992, in ASP Conf. Ser., Vol. 25, Astronomical Data Analysis Software and Systems I, ed. D. M. Worrall, C. Biemesderfer, & J. Barnes (San Francisco: ASP), 163

Goldfarb, D., & Liu, S. 1991, Mathematical Programming, 49, 325

Fang, S.-C., & Puthenpura, S. 1993, in Linear Optimization and Extensions (Englewood Cliffs, NJ: Prentice Hall)

Astronomical Data Analysis Software and Systems VI
ASP Conference Series, Vol. 125, 1997
Gareth Hunt and H. E. Payne, eds.

Non-parametric Algorithms in Data Reduction at RATAN-600

V. S. Shergin, O. V. Verkhodanov, V. N. Chernenkov, B. L. Erukhimov, and V. L. Gorokhov

Special Astrophysical observatory, Nizhnij Arkhyz, Karachaj-Cherkessia, Russia, 357147

Abstract. Non-linear and non-parametric algorithms for data averaging, smoothing and clipping in the RATAN-600 flexible astronomical data processing system are proposed. Algorithms are based on robust methods and non-linear filters using an iterative approach to smoothing and clipping. Using robust procedures to detect faint sources is proposed also. This detector is based on the ratio of two statistics, characterizing the noise and signal, in the given interval. These methods allow us to accelerate the process of the data reduction and to improve the signal/noise ratio. Examples of operation of these algorithms are shown.

1. Introduction

Obtaining a reliable result on the background from different types of interferences is one of the main problems for the observational astronomy. The question "what is useful signal and what is noise?" is especially essential when we begin "dumb" data processing.

The ordinary way to obtain a good signal/noise ratio is to apply the standard average for vectors of data observed on the same sky strip. This way is the most optimal and realizes the maximum likelihood estimation with $\sqrt{n}$ improvement. But the real observational data have a distribution far from normal. This is caused by the presence of power spikes, "jumps," and slow trends. The source of this interference is human activity, atmosphere, and possible instability of the receiver. Therefore, observers prefer a manual method of data reduction, because a single bad record spoils the resulting sum when doing "dumb" standard averaging. To automate the users' procedure of data quality checking, special algorithms have been worked out.

The background problems are absent for ordinary average. But when we use the robust (non-parametric) average, the correct background subtraction (or smoothing) is the main problem. Therefore, the first problem to be solved is to find the correct (from the viewpoint of an observer) smoothing. Moreover, the knowledge about the background around the sources is very important for some problems in radio astronomy. The use of standard procedures (fitting with splines) very often can not help us in this situation. Thus, special non-linear algorithms for smoothing were developed.

Another type of processing where similar algorithms can be used is data compression (when we have a surplus of points per beam and can compress the data keeping useful information) and searching for sources.

2. Smoothing

The history of algorithms, based on the robust methods and non-linear filters using an iterative approach to smoothing and clipping, began twelve years ago in the data reduction at the RATAN-600. Since then they have been further developed and are used in modern computer systems in the RATAN-600 data reduction system FADPS (Verkhodanov et al. 1993; Verkhodanov 1997) and in the MIDAS (Shergin et al. 1995).

The algorithm uses several start parameters: noise dispersion, smoothing interval, iteration count, and type of smoothing curve. Practically there is no effect on a real signal in a given interval, and the background is accurately calculated.

The detailed description of the smoothing and clipping (SAC) algorithm is given in Erukhimov et al. (1990) and in Shergin et al. (1996). Briefly, the SAC algorithm consists of the following steps: (*i*) calculation of an input vector $\overline{C}_i$ for $\subset$-th smoothing iteration as $\overline{C}_i = \mathcal{F}(\overline{S}, \overline{W}_{i-1})$ where $\overline{W}_{i-1}$ is a vector of weights calculated in the previous iteration. In the simplest case the function $\mathcal{F}$ is calculated as a product of each vector component $\mathcal{F}_k = S_k * W_k$, where k is k-th component; (*ii*) smoothing of $\overline{C}_i$: $\overline{B}_i = \Lambda(\lambda)\,\overline{C}_i$, where $\Lambda_i(\lambda)$ is the smoothing operator, and λ is the input parameter; (*iii*) subtraction $\overline{D}_i = \overline{S} - \overline{B}_i$, where $\overline{S}$ is the input data vector; (*iv*) calculation of a new function of weights as a non-linear transformation $\Phi_i(\overline{\sigma})$ which is a function of input noise $\overline{\sigma}$ and number of iteration in a general case: $\overline{W}_i = \Phi_i(\overline{\sigma})\,\overline{D}_i$; and (*v*) transition to the next iteration. There are several possible smoothing and weighting methods. In practice we use the following methods for smoothing: simple boxcar average; convolution with Gaussian profile (see Figure 1 in Verkhodanov 1997), median average (Erukhimov et al. 1990) (see Figure 2 in Verkhodanov 1997); weighted least squares polynomial approximation of 1/15°. For calculation of weights some empirical transformations are used. They may be specialized to cut emission, absorption or both.

3. Averaging

After background subtraction the procedure of robust averaging is applied for records. Usually we do it by the Hodges-Lehmann (HL) method (Huber 1981, Erukhimov et al. 1990) with estimation of middle value:

$$X = med\left[\frac{X_i + X_j}{2}\right]$$

Robust averaging with the HL method is illustrated in Figure 1.

4. Compression

Another type of data reduction procedure is data compression. The result of data compression on the base of Hodges-Lehmann (Huber 1981) estimates is illustrated in Figure 2. A similar robust algorithm was proposed and applied in the RATAN-600 data registration system by Chernenkov (1996).

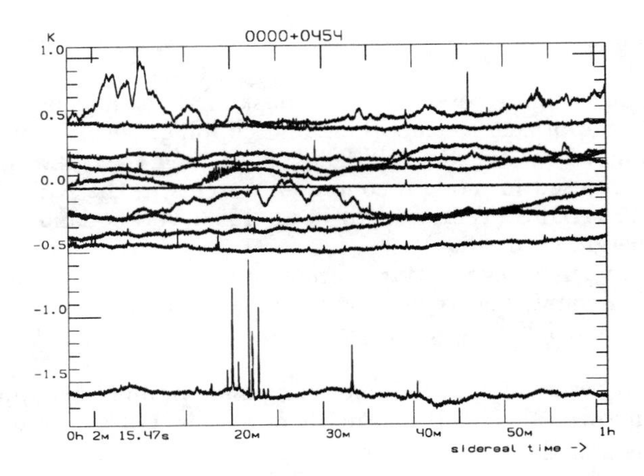

Figure 1. Average of 20 records. The record at zero level is a result.

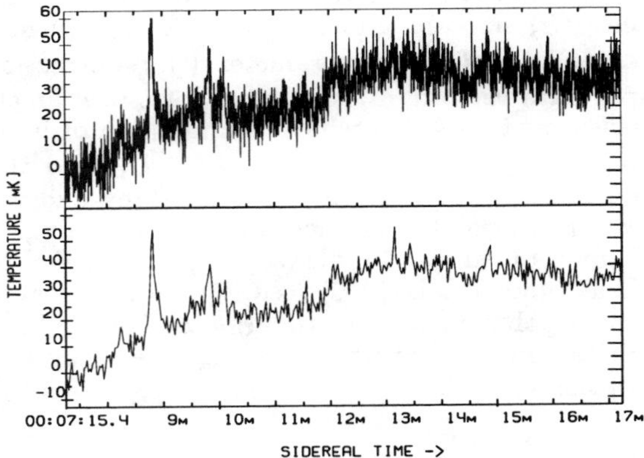

Figure 2. Compression with factor 5.

5. Detection

Based on robust procedures, a detector for faint sources (extremal-median signal detector: EMSD) in a series of synchronous scans (Verkhodanov & Gorokhov 1995). is proposed also. This detector is based on the ratio of two statistics, characterizing the noise and signal, in the given interval.

For this algorithm several statistics are calculated: $z_{ij} = \min_t\{\mathbf{D}_{i,j+t}\}$, $v_{ij} = \max_t\{\mathbf{D}_{i,j+t}\}$, where $i = 1, m; j = 1, n; t = 1, r$; $\mathbf{D}$ is the matrix of m vectors, m is the number of scans, n is the number of points in a scan, r is the number of points in a signal search interval. Using these two extremal statistics the medians and the following statistics are computed in each interval: $\theta_{v_j} = \text{med}_i\, v_{ij} - \text{med}_i\, z_{ij}$, $\theta_{z_j} = \text{med}_i\, z_{ij}$, where $i = 1, m; j = 1, n$; then their ratio is used as a test statistic:

$$q_j = \theta_{v_j} / \theta_{z_j}$$

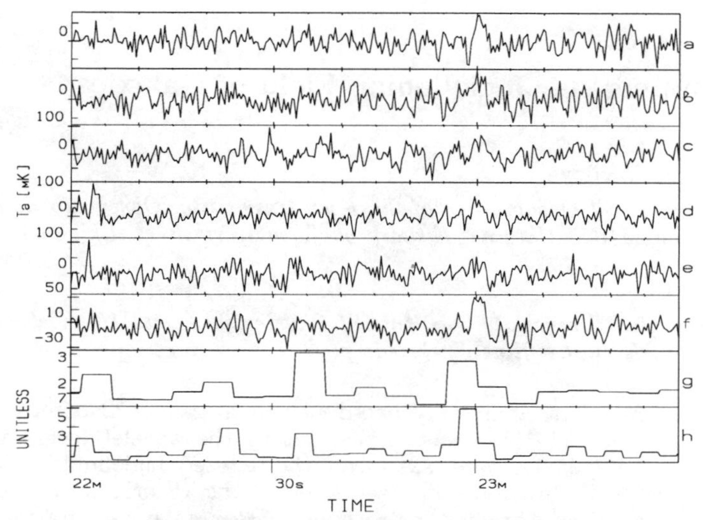

Figure 3. Results of EMSD operation for the different input intervals: g—interval of two beam patterns, h—one beam pattern, for real data of five synchronous scans (a–e). To compare results the scan with robust average of these data is shown (f). The antenna temperature is for the first six scans, and ratio (5) is for the two last scans plotted along the Y-axis.

where $j = 1, n$. In this case the hypothesis of object detection is adapted when the quantity q_j exceeds a certain quantity **c**. The threshold **c** can be either set by user or computed in the program. The example of EMSD operation is illustrated in Figure 3.

Acknowledgments. O. Verkhodanov thanks ISF-LOGOVAZ Foundation for the travel grant, the SOC for the financial aid in the living expenses and the LOC for the hospitality.

References

Chernenkov, V. N. 1996, Bull. of SAO, 41, 150

Erukhimov, B. L., Vitkovskij, V. V., & Shergin, V. S. 1990, Preprint SAO RAS, 50

Huber, P. J. 1981, Robust Statistics (New York: Wiley)

Shergin, V. S., Kniazev, A. Yu., & Lipovetsky, V. A. 1996, Astron. Nach., 317, 95

Verkhodanov, O. V., Erukhimov, B. L., Monosov, M. L., Chernenkov, V. N., & Shergin, V. S. 1993, Bull. of SAO, 36, 132

Verkhodanov, O. V., & Gorokhov, V. L. 1995, Bull. of SAO, 39, 155

Verkhodanov, O. V. 1997, this volume, 46

Astronomical Data Analysis Software and Systems VI
ASP Conference Series, Vol. 125, 1997
Gareth Hunt and H. E. Payne, eds.

Mapping the Jagiellonian Field of Galaxies

Irina B. Vavilova

Astronomical Observatory of the Kiev University, Observatornaya str., 3, Kiev 254053 Ukraine, E-mail: vavilova@rcrm.freenet.kiev.ua

Piotr Flin

Pedagogical University, Institute of Physics, ul.Lesna 16, Kielce 25-509 Poland, E-mail: sfflin@cyf-kr.edu.pl

Abstract. The analysis of two-dimensional galaxy distribution in the Jagiellonian Field (JF) was carried out by the wavelet technique. The positions of galaxies were taken from the Revised Jagiellonian Field Catalogue (RJFC) based on digitized scans of the JF original photographic plates. We discuss briefly the procedure of star/galaxy separation applied by us to the RJFC, the algorithm and the first results of mapping of the selected parts of the RJFC.

1. What is the Revised Jagiellonian Field Catalogue of Galaxies?

The region of the original Jagiellonian Field (JF) sky survey (Rudnicki, Dworak, & Flin 1973) was selected by Zwicky (Zwicky 1962) as a region which is enriched by clusters of galaxies and contains a small number of the brighter galaxies. The JF sky survey (photographic plates in three colors with coordinates of center $\alpha = 11^h20^m$, $\delta = +35°26'$ (2000.0) taken with the 1.25 m Palomar Schmidt telescope) coincides with the CGCG field No. 185. The central part of the JF ($4° \times 4°$) was scanned and automatically processed by the COSMOS machine. This procedure allows one to digitize the images on the plate and to receive a set of parameters fitting the inertia ellipse to each image (e.g., Stobie 1980; MacGillivray & Stobie 1984).

Constructing a modern complete galaxy catalogue in the JF region, we decided to carry out an automated procedure of star/galaxy separation on the digitized images. We experimented with the possible combinations of parameters usually applied for such purposes on similar photographic data (e.g., Hewett 1981; Heydon-Dumbleton, Collins, & MacGillivray 1989; Odewahn et al. 1992; Weir, Fayyad, & Djorgovski 1995 with a brief review of discrimination philosophy).

Before applying the procedure, all small objects with the *COSMOS area* parameter smaller than 60 (in increments) were removed from our sample and about 35,000 objects remained to be analyzed.

Among the possible discriminators only three gave a clear separation of the loci of stars and galaxies. There were plots presenting the dependence on the *logarithm of image area* denoted as log(*area*), the *width of Gaussian fit* (S) and the *intensity weighted second moment* (K_w) versus *COSMOS image magnitude* parameter *COSMAG*: $COSMAG = -2.5 \log(\sum(I_i - I_{sky})/I_{sky})$, where I_i and I_{sky} are the intensity of i-pixel of the image and the intensity of the sky at the image centroid respectively.

For bright and intermediate magnitudes we used the *log(area) vs. COSMAG* discriminator. Due to lower surface brightness for this range of magnitudes, the galaxies lie above the stellar sequence on this plane.

For the intermediate magnitudes we applied the *intensity weighted second moment* discriminator K_w giving an estimate of the filling of the image by an ellipse fitted to its semiaxes: $K_w = \pi a_i b_i/(area)$, where a_i and b_i are the intensity weighted semi-major and semi-minor axes of the inertia ellipse respectively. The more centrally concentrated images (stars) have the smallest values of this parameter and occupy a tight locus under the galaxy sequence.

For the fainter and intermediate objects we used the *width of a best Gaussian fit* (S) to the threshold intensity and maximum image intensity: $I_{th} = I_{max} \exp(-(area)/2\pi S)$, where I_{th}, I_{max} are the threshold and maximum image intensity respectively. On this diagram the Gaussian width for stars is determined by the point spread function; galaxies lie above this sequence.

We checked all the brightest objects by visual inspection.

The principal point of the automated star/galaxy classification is: the statistical weight of stars exceeds that of galaxies for the whole range of considered magnitudes, and stars occupy the tighter sequence for each discriminator mentioned above. Therefore, for each magnitude in this range, it is possible to construct the histogram of distribution of the number of objects corresponding to the value of chosen discriminator. This histogram gives the main points for a spline to separate the star and galaxy sequences on the plot of discriminator. Notice that, for intermediate magnitudes, the loci of stars and galaxies are more separated, so we did a spline interactively.

So, the Revised Jagiellonian Field Catalogue (RJFC) is based on the digitized scans of the original JF photographic plates. The RJFC contains information on about 20,000 galaxies (Flin & Vavilova 1997, in preparation).

2. Mapping the Galaxy Distribution in Region of the RJFC

Modern astronomical data such as, e.g., DPS and digitized POSS II surveys, constitute an excellent base for finding large-scale structures. This structure search must be objective, which requires appropriate mathematical methods. The wavelet technique, being well-suited for approximation of data with sharp discontinuities, seems to be one such promising tool. For this reason we decided to verify the applicability of the wavelet analysis, having at our disposal data similar to DPS data, by carrying out the mapping of a small region of the sky.

Unlike Fourier analysis, the wavelet algorithms process data at different scales (resolutions). In the case of applying the wavelets for detachment of the structure of galaxy cluster, the task is to convolve the two dimensional galaxy distribution (*signal function*, $s(r)$) on a grid of N×N pixels by the *analyzing wavelet* $F(r, a)$, where a denotes the scale of the wavelet. This parameter determines an effective radius of wavelet or, in other words, the extent of spreading of the density distribution of galaxies in space of *wavelet coefficients* (WC). So, such convolution leads to the "spread" galaxy distribution in terms of a wavelet, and the further analysis of this distribution can be performed using just the corresponding WC. Adapting the best *analyzing function* $F(r,a)$ to data or cutting the WCs below a threshold we sparsely represent our data.

From the whole RJFC we chose small regions containing Abell clusters: A1226, A1228, and A1257, two of them never studied before. For the galaxy distribution in these three regions the *analyzing wavelet* $F(r,a)$ known as the "Mexican Hat" was adopted. Avoiding the edge effects where discontinuities occur, we analyzed a region greater than the cluster itself. The "Mexican Hat" function has a radial shape permitting however the detection of non-circular structure: $F(r,a) = (2 - r^2/a^2)\exp(-r^2/2a^2)$, where r is the distance between center and point (x,y) in which the "Mexican Hat" is calculated. This approach has already been described and performed for identification of structure of clusters of galaxies (e.g., Slezak, Bijaoui, & Mars 1990; Escalera et al. 1994).

Our procedure had several principal steps:

(*i*) normalizing the galaxy distribution data in the working zone of the wavelets through a linear transformation into the range $[-1, 1]$, so the radius of the analyzed zone is $R_f = 1$;

(*ii*) applying the "Mexican Hat" formula to the normalized data;

(*iii*) using a full set of scale wavelet parameters: the analysis starts at the largest scale $a = 0.25R_f$, where the structure of the whole cluster is detected, and ends at the smallest one, where only one galaxy is located inside the region;

(*iv*) performing a Monte-Carlo simulation using the same number of galaxies, distributed in the same zone, to verify the reality of detected structures and substructures;

(*v*) picking up the galaxies inside the detected structure of the cluster to study the properties of galaxy cluster.

3. Conclusion

The first results of mapping of the RJFC gave additional support for the correctness and efficiency of the wavelet analysis both for detachment of the structure of galaxy clusters itself and for the further investigation of their morphological properties (Flin & Vavilova 1995, 1996).

Notice that, in our case, the existence of the structures had been known *a priori* and the task here was to adapt the analyzing function. The problem of finding the structure of the galaxy cluster and the galaxy group (e.g., with low richness, having different geometrical scales, and lying close to each other in two- or three-dimensional space) in a general distribution is not easy. For mapping the general galaxy distribution in the whole RJFC, some other families of wavelet systems and approaches should be applied. It looks as if, in this case, it would be correct to use a hierarchical algorithm, sometimes called a *pyramidal algorithm* (e.g., Bijaoui 1996, private, communication), providing the multiscale vision of a sky survey like the RJFC.

Acknowledgments. IBV thanks the ADASS VI Organizing Committee for the financial support enabling her to participate in so exciting and fruitful conference, as well as Pedagogical University in Kielce for hospitality during her stay there.

References

Escalera, E., Biviano, A., Girardi, M., Giuricin, G., Mardirossian, F., Mazure, A., & Mezzetti, M. 1994, ApJ, 423, 539

Flin, P., & Vavilova, I. B. 1995, SISSA Ref. 65/95/A, ed. G. Giuricin, F. Mardirossian, M. Mezzetti

Flin, P., & Vavilova, I. B. 1996, Astrophys. Letters & Communications, to be published

Hewett, P. 1981, report for internal distribution ROE

Heydon-Dumbleton, N. H., Collins, C. A., & MacGillivray, H. T. 1989, MNRAS, 238, 379

MacGillivray, H. T., Stobie, R. S. 1984, Vistas in Astron., 27, 4, 433

Odewahn, S. C., Stockwell, R. L., Pennington, R. M., Humphreys, R. M., & Zumach, W. A. 1992, AJ, 103, 318

Rudnicki, K., Dworak, T. Z., & Flin, P. 1973, Acta Cosmologica, 1, 7

Slezak, E., Bijaoui, A., & Mars, G. 1990, A&A, 227, 301

Stobie, R. S. 1980, JBIS, 33, 323

Weir, N., Fayyad, U., & Djorgovski, S. 1995, AJ, 109, 2401

Zwicky, F. 1962, in Problems of Extra-galactic Research, ed. G. C. McVittie (New York: Macmillan), 347

Astronomical Data Analysis Software and Systems VI
ASP Conference Series, Vol. 125, 1997
Gareth Hunt and H. E. Payne, eds.

Asteroseismology—Observing for a SONG

Rob Seaman

IRAF Group,[1] NOAO,[2] PO Box 26732, Tucson, AZ 85726

Caty Pilachowski, Sam Barden

Kitt Peak National Observatory

Abstract. The Stellar Oscillations Network Group (SONG) seeks to study p-mode (acoustic) oscillations in solar type stars. These are difficult phenomena to detect due to the limited amplitude of the oscillations in integrated light. Success will require continuous observing sessions over many pulsation cycles, preferably with multiple telescopes staggered in longitude, similar to the GONG project.

Such oscillations are best detected as a beat frequency relative to some very regular inertial observing cadence. The phase of the cadence must be maintained, both between widely separated telescopes and between observing sessions that may be separated by months.

We discuss techniques to improve the observing efficiency and the likelihood of detection. Precisely identical data handling and reduction steps for tens or hundreds of thousands of spectra are critical to success.

1. Introduction

The detection and study of acoustic oscillations in solar-like stars offer a new constraint on stellar models, beyond the global properties of mass, radius, age, and chemical composition. For stars with well-measured global properties, knowledge of frequencies and frequency splittings of p-mode oscillations will permit detailed comparison with stellar models at a level unprecedented outside the solar system. For stars whose global properties are less well known, the addition of asteroseismological data may allow the determination of mass and age for individual field stars. The observation of many solar-like stars will also help us to understand how solar p-modes are excited and damped.

The Stellar Oscillations Network Group (SONG) at NOAO has undertaken to develop methodology for precise equivalent width measurements of Balmer lines in solar type stars. Typical oscillation frequencies in solar type stars are a few mHz, or a period of a few minutes. The amplitude of spectral line variations in integrated sunlight is about 5 mÅ, but larger amplitudes (20–30 mÅ) are expected in warmer stars. Time series of spectra are obtained over many

[1] Image Reduction and Analysis Facility, distributed by NOAO

[2] National Optical Astronomy Observatories, operated by the Association of Universities for Research in Astronomy, Inc. (AURA) under cooperative agreement with the NSF.

nights at precisely timed intervals. Exposure times are limited to a few minutes or less by the need to preserve time resolution. Very high S/N ratios (~1000) are needed per exposure to detect the weak signal. With telescopes of modest aperture (4 meters or less), only the brightest stars can be observed.

Such measurements can be made with commonly available spectrographs. With the expectation of developing a world-wide network of telescopes to make asteroseismic observations, we are developing techniques of data acquisition, data reduction, and analysis that can be easily propagated to observatories around the world. Of greatest concern is high observing efficiency so that oscillations can be detected during observing runs of reasonable length.

2. Observations and Reductions

While the potential of asteroseismology is great, so are the difficulties in detecting oscillations in solar type stars. In integrated light, the amplitudes of oscillations are small: a few meters per second in radial velocity or a few to a few dozen parts per million in the strength of spectral lines. New techniques in spectroscopy, data reduction, and analysis are needed to obtain this precision.

2.1. Exposure Scheduling

Complications are encountered when attempting to implement even the simplest requirements for conducting a valid sequence of asteroseismological observations. SONG hopes to detect stellar oscillations with periods of a few minutes or tens of minutes. Naively, the two or three orders of magnitude improvement over this provided by the normal 1 second time resolution of the KPNO CCD cameras should be quite sufficient for such a detection.

For this to be true, however, the *scheduling* of the individual CCD exposures also has to match at least this level of accuracy. Our first proof-of-concept observing night at the telescope relied on a long free-running sequence of several hundred exposures to provide this scheduling. Figure 1a is a plot of the running average of the cycle time (from shutter-open to shutter-open) versus sequence number through the night. The repeatability of the cycle time is respectable—the mean deviation is well within the 1 second precision of the camera system.

The only remarkable thing about Figure 1a is a slight increasing trend of the average cycle time from about 68.3 s at the beginning of the observing session to about 68.7 s at the end of the session several hours later. We attribute this 0.6% effect to increasing hard disk seek times as the data partition filled up.

This small slope has a disproportionately dramatic effect, however, on the quality of the science that can be performed. Figure 1b is a plot of the residuals of the start time of each exposure compared to an evenly partitioned grid of precisely identical cadence from the start to the end of the observing sequence. These residuals sweep out more than 120° of phase in the nominally regular observing cadence. Without a regular cadence, the straightforward (in principle) detection of stellar oscillations as simple beat frequencies becomes more difficult. The much more complicated mathematical treatment required by unevenly gridded observations jeopardizes the reliability of any detection whatsoever, and at best comes at the expense of observing efficiency.

Clearly, some way is required to schedule the observations on regular clock ticks. We refer to this as "cadencing" the observations. A short delay is added to

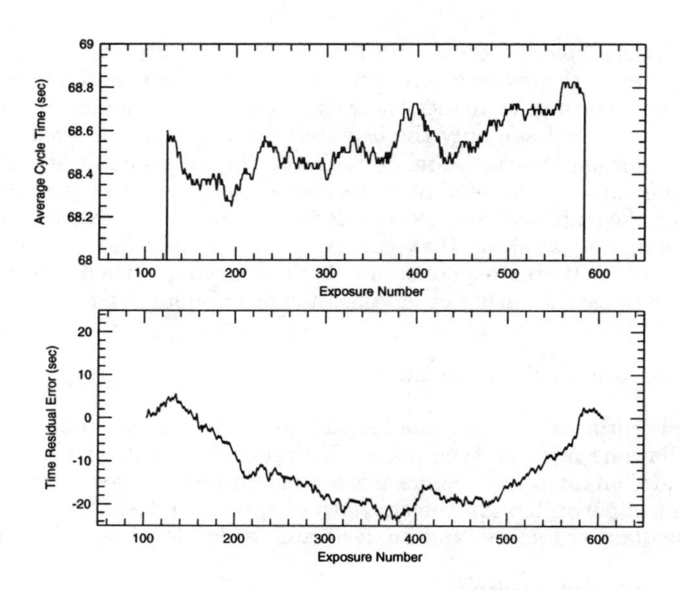

Figure 1. Non-cadenced data—a free-running observing sequence.

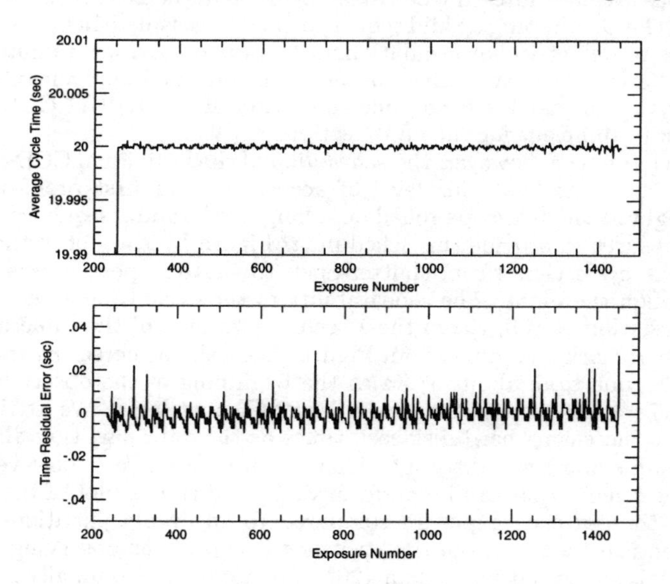

Figure 2. Cadenced data—shutter timed to better than 0.01 s.

each exposure such that the shutter is opened on the precise clock tick desired. Figures 2a and 2b show the improvement realized after cadencing.

Having addressed the question of enforcing a regular observing cadence for a single observing sequence, the next complication arises. Our telescopes do not provide an inertial observing platform. Unless we account for the 1000 s light travel time amplitude of the Earth's motion around the Sun, we cannot hope to easily combine data from observing runs separated by any significant fraction of a year. The problem is even more fundamental than this, since, except when a particular target is very near opposition to the Sun, the heliocentric correction can vary by several seconds over the course of a single day.

Note that the amplitude of the barycentric correction for the Solar system is about 10 light seconds—about 1% of the heliocentric correction. Jupiter contributes about half of the barycentric leverage and Saturn most of the rest. At any given epoch the absolute value of the barycentric effect may be significantly less than the full 10 s, depending on Saturn's position relative to Jupiter.

It is the non-zero second derivative of these motions that will smear out the phase of the observations, since otherwise they could be treated like the typical target star's radial velocity term of a few $\mathrm{km\,s^{-1}}$, which can just be compensated for as a small Doppler correction. For instance, Procyon's radial velocity is $-3\,\mathrm{km\,s^{-1}}$; this will just require a 1 part in 10^5 Doppler correction.

2.2. CCD Readout Mode Variations

As another way to increase efficiency, we investigated the use of a continuous readout mode (CRM) for spectra of bright stars to eliminate the time spent with the shutter closed. A major concern is the variation of the light across the spectrum during the observations. We simulated CRM observations using the Artdata package in IRAF. We created a CRM spectrum using the known pixel-to-pixel sensitivity variations of the CCD, and allowed the artificial spectrum to vary in intensity, position on the CCD, and width due to seeing during "readout." The simulated spectra contain errors of up to 4% in intensity. The S/N in the final spectrum is ~500, no better than a normal observation in the same time.

A second approach for improving the observing efficiency is a stepped CCD readout mode producing "miniframe" images. Using a narrow region of interest covering the spectrum, multiple integrations can be obtained within a single image frame. The spectrograph shutter is closed between integrations, and the region of interest is read out into a buffer; repeated integrations are appended to the same buffer. With this stepped readout mode, a total integration time of 25–30 seconds can be obtained each minute without saturating the CCD. This compares to a total integration time of only 15–20 seconds using a conventional readout, nearly doubling our observing efficiency.

References

Stellar Oscillations Network Group[3]

Massey, P., et al. 1996, An Observer's Guide to Taking CCD Data with ICE[4]

---

[3]http://www.noao.edu/noao/song/

[4]http://www.noao.edu/kpno/manuals/ice/ice.html

Astronomical Data Analysis Software and Systems VI
ASP Conference Series, Vol. 125, 1997
Gareth Hunt and H. E. Payne, eds.

Titan Image Processing

Nailong Wu and John Caldwell

Dept. of Physics and Astronomy, York University, 4700 Keele St., North York, Ontario, M3J 1P3, Canada

Abstract. Images of Titan by the Hubble Space Telescope (HST) are very small in size and have very low spatial resolution. Special methods of image processing are required to extract information from these images. In this presentation, these methods are described and results are reported.

1. Introduction

Titan, one of the satellites of Saturn, has a planet-size solid body, and a thick atmosphere with molecular nitrogen as its principal constituent and with a considerable amount of methane (CH_4) (Beatty et al. 1990).

Currently, in searching for features on Titan's disk imaged by the HST PC1 detector, we are faced with two problems: (1) Low spatial resolution (Figure 1). The diameter of Titan's disk is only approximately 20 pixels. The pixel size corresponds to about 290 km at the center of Titan's disk. (2) Poisson noise on Titan's disk, which makes it difficult to detect features. For example, it is impossible to detect a point source having an amplitude of 12 DN (digital number) and sitting on Titan's disk with a uniform pixel value of 2000 DN, because the SNR (signal-to-noise ratio) is only 1.0. In contrast, it would be easy to detect the same source if it sat on an empty background, because the SNR is 13.1.

The two aforementioned problems necessitate processing Titan images by special methods to extract information for analysis. Each of the methods used by us is described in one of the following four sections, including the problem addressed, the method, the key computer programs (mainly in IRAF/STSDAS and IDL), and the results. In the conclusion we summarize our experience, and give suggestions for observing Titan in the future.

2. Image Restoration Using MEM and MLM

Problem: *Limb-brightening and -darkening.* The variation of brightness radially from the center of Titan to the limb is called limb-brightening or limb-darkening, depending on whether this variation is increasing or decreasing. Its determination is important for modeling Titan's atmosphere.

Method: *Remove the smoothing effect of the PSF (point spread function) of the HST.* Because of this smoothing effect, limb-darkening will be enhanced, while limb-brightening will be weakened so that a radial profile in an observed image may appear to be falsely limb-darkened or neutral. Therefore, in the case of (apparent) limb-darkening, it is required to restore the image to find out

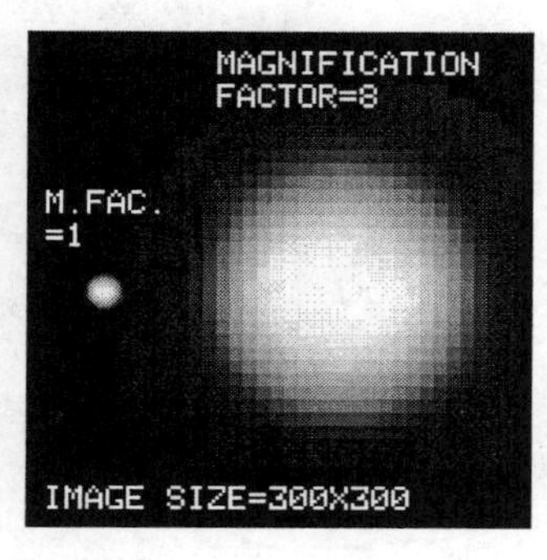

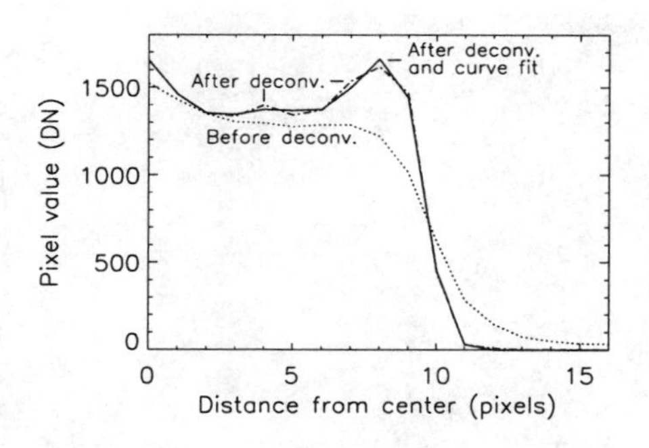

Figure 2. Profiles at position angle 0°.

the truth. This is accomplished by deconvolution using MEM (the Maximum Entropy Method) or MLM (the Maximum Likelihood Method). Furthermore, for accurate modeling, deconvolution is advisable in any case.

Programs: Two tasks in IRAF/STSDAS, **mem** and **lucy**, implement deconvolution by MEM and MLM, respectively, and give similar results in our case. The PSF associated with the observed image, necessary for executing these tasks, is generated by running the stand-alone program **TinyTim**.

Results: For a CH_4 image at 889 nm wavelength, out of twelve radial profiles at position angles 0° (Titan North), 30°, ..., 330°, seven appear to be limb-brightened, while five are limb-darkened in the observed image before deconvolution. In contrast, after deconvolution, four (at 0°, etc.) of the latter five profiles become limb-brightened, and the other is nearly flat to the limb.

Three radial profiles at 0° are plotted in Figure 2 to show the effect of deconvolution. Curve fitting is used for further clarification of limb-brightening after deconvolution.

3. Image Enhancement by Subtraction and Filtering

Problem: *Temporal brightness changes in images.* By detecting brightness changes in Titan images with time, we can discover transient features. The difficulty is that the changes may well be small compared with the noise, and the spatial resolution of the images is very low.

Method: *Taking differences between sequential images.* Before doing this, we must interpolate the images to reduce the pixel size, and register (align) them at the subpixel level. After the subtraction operation, we lowpass-filter (smooth) the difference images to improve SNR and eliminate small-scale fluctuations.

Programs: The key tasks used in IRAF/STSDAS are: **magnify** for interpolation; **crosscor** and **minmax** to find the relative shifts between images; **imshift** to shift images for registration; **imarith** for subtraction; and **gauss** for lowpass-filtering (convolution).

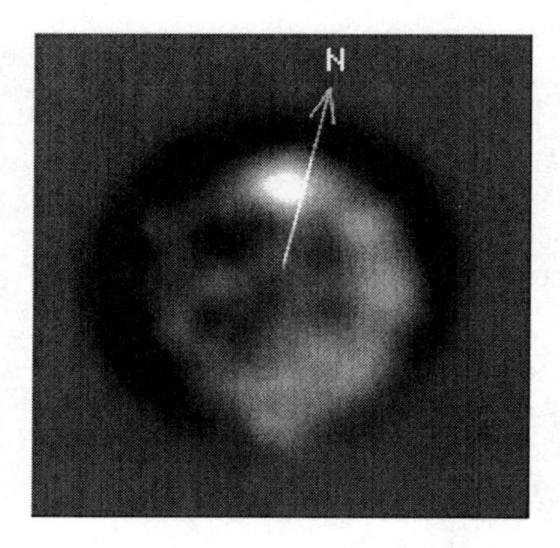

Figure 3. One of the six filtered difference images.

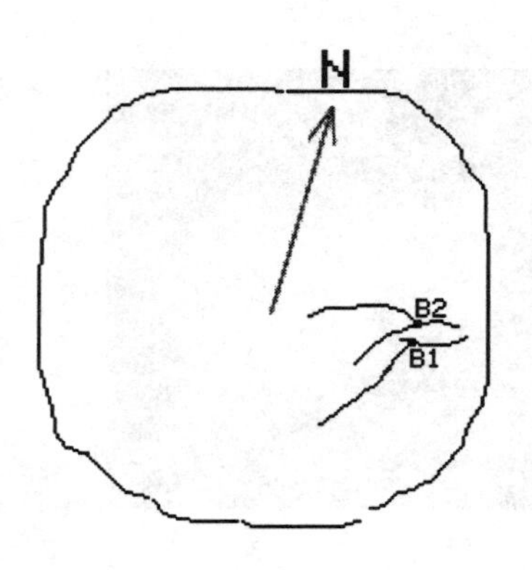

Figure 4. Bifurcation points in two groups of images.

Results: Two groups of images at 673 nm wavelength, with a time difference of $2^{d}9^{h}20^{m}$ are processed. The groups have two and three images, respectively. The magnification factor is 9, i.e., the pixel size is reduced to 1/9. The FWHM (full width at half maximum) of the Gaussian filter is 18 (reduced size) pixels.

A bright area is detected in all six (2×3) difference images, being centered at position angle $+8\overset{\circ}{.}3$ with respect to Titan North, latitude $+45\overset{\circ}{.}4$. One of the six filtered difference images is shown in Figure 3. The pixel values in the bright area increased noticeably during this period of time. Similar results have been reported and discussed by Lorenz et al. (1995).

4. Image Enhancement by Edge Detection and Morphological Processing

Problem: *Wind velocity.* Measuring the wind velocity in Titan's atmosphere is difficult. Theoretical wind speeds ranging up to 360 km/hour are possible, but the image pixel size is 290 km. Results may be questionable.

Method: *Detect cloud edges and their motions in a time interval.* To increase accuracy, the pixel size must be reduced by interpolation. The time interval should be reasonable. If it is too small, then no significant motion will be detected. If it is too large, on the other hand, clouds may disperse so much that no corresponding edges can be identified. Furthermore, to determine the velocity (including speed and direction) at least two crossed edges in each image must be detected. This is accomplished by image edge detection (enhancement) and morphological processing.

Programs: The key programs are: **magnify** in IRAF for interpolation; and **sobel** and **thin** in IDL for detecting edges and then thinning (morphological processing) them, respectively.

Results: Two groups of CH_4 images at 889 nm wavelength, with a time difference of 1.017 hours are processed. The two images within each group are combined to improve SNR. The magnification factor is 9, resulting in a (small) pixel size of 31.8 km.

Numerous edges were detected, but discounted because of apparent alignment with rows, columns or diagonals. Only one bifurcation point detected in each of two groups is considered to be significant (Figure 4). There is a displacement from B_1 to B_2 of 7.0 small pixels. The apparent speed is $31.8 \times 7.0/1.017$ $\sim$220 km/hour. The direction is $-5^{\circ}\!.3$ with respect to Titan North.

5. Dithering

Problem: *Low spatial resolution.* The pixel size on the CCD chip (PC1 in our case) is large compared with the width of PSF, resulting in "undersampling."

Method: *Combine images shifted by subpixel amounts.* Increasing spatial resolution means reducing the pixel size. We can change the pointing of the telescope so that successive images are shifted along each axis by subpixel amounts, say a multiple of 1/2 or 1/3 pixel size, then combine these images to obtain a single image having a smaller pixel size on a finer grid. This technique is called dithering (HST WFPC 2 Handbook 1995).

Programs: The program **TinyTim** is used to generate PSFs associated with dithered images. These images and PSFs are input to the task **acoadd** in IRAF/STSDAS to get a single image with improved resolution.

Results: Simulation gives good results, but our HST data do not. The reasons include: insufficient pointing accuracy of the HST, and insufficient SNR in the images. Also, better software for dithered image processing may be needed.

6. Conclusion

Image processing techniques do help to extract information from HST images. MEM and MLM can be used for deconvolution. The method *subtraction and filtering* can be used to detect changes between images. More experiments using the method *edge detection and morphological processing* are necessary. For the sophisticated method *dithering* to succeed, every effort should be made to achieve the nominal pointing accuracy of the HST, 3 milliarcseconds. At least two observations must be made at each position to increase SNR and facilitate cosmic ray removal.

References

Beatty, J. K., & Chaikin, A., eds. 1990, The New Solar System (Cambridge, MA: Sky Publishing Corp.), Chap. 14

Hubble Space Telescope WFPC2 Instrument Handbook, v. 3.0, June 1995

Lorenz, R. D., Smith, P. H., & Lemmon, M. T. 1995, BAAS, 27, 1104

Astronomical Data Analysis Software and Systems VI
ASP Conference Series, Vol. 125, 1997
Gareth Hunt and H. E. Payne, eds.

Generalized Spherical Harmonics for All-Sky Polarization Studies

P. B. Keegstra[1], C. L. Bennett[2], G. F. Smoot[3], K. M. Gorski[4], G. Hinshaw[2], L. Tenorio[5]

[1]*Hughes STX Corporation*

[2]*Laboratory for Astronomy and Solar Physics, NASA/Goddard Space Flight Center*

[3]*Lawrence Berkeley Laboratory, University of California, Berkeley*

[4]*Theoretical Astrophysics Center, Denmark and Warsaw University Observatory, Poland*

[5]*Universidad Carlos III de Madrid, Spain*

Abstract. When whole-sky linear polarization is expressed in terms of Stokes parameters T_Q and T_U, as in analyzing polarization results from the Differential Microwave Radiometers (DMR) on NASA's Cosmic Background Explorer (*COBE*), coordinate transformations produce a mixing of T_Q and T_U. Consequently, it is inappropriate to expand T_Q and T_U in ordinary spherical harmonics. The proper expansion expresses both T_Q and T_U simultaneously in terms of a particular order of generalized spherical harmonics. The approach described here has been implemented, and is being used to analyze the polarization signals from the DMR data.

1. Definition and Motivation

Generalized spherical harmonics are an extension of ordinary spherical harmonics, intended for expansion of functions whose transformation properties at each point on the sphere are more complex than just scalars. The general form

$$T^{\ell}_{n,m}(\theta,\phi) = e^{im\phi} P^{\ell}_{n,m}(\theta)$$

has three indices ℓ, m, and n where $-\ell \le m \le \ell$ and $-\ell \le n \le \ell$ (Gel'fand et al. 1963). The forms appropriate for expanding complex Stokes parameters T_Q and T_U are (Sazhin & Korolëv 1985)

$$T_Q + iT_U = \sum_{\ell=2}^{\infty} \sum_{m=-\ell}^{\ell} D_{\ell,m} T^{\ell}_{2,m}(\theta,\phi)$$

$$T_Q - iT_U = \sum_{\ell=2}^{\infty} \sum_{m=-\ell}^{\ell} E_{\ell,m} T^{\ell}_{-2,m}(\theta,\phi)$$

Since T_Q and T_U are real and $T^{\ell}_{-2,m} = T^{\ell}_{2,-m}$, the two expansions are degenerate, and we may restrict our consideration to the first form. Thus, for the DMR case we need only consider generalized spherical harmonics with $n = 2$, which we will henceforth refer to as T^{ℓ}_{m}.

The $D_{\ell,m}$ are complex expansion coefficients, analogues of the $a_{\ell,m}$ of ordinary spherical harmonic expansions of scalar quantities. Like them, the $D_{\ell,m}$ for a given ℓ transform among themselves in a coordinate transformation, but the absolute sum $\sum_m D_{\ell,m}\overline{D_{\ell,m}}$ is invariant.

Following recent work by Zaldarriaga & Seljak (1997) and Kamionkowski et al. (1997), we can partition the $4\ell+2$ independent real parameters per value of ℓ into those associated with even-parity solutions and odd-parity solutions, called E-like and B-like respectively by Zaldarriaga & Seljak. The formula appropriate for the phase convention used here is

$$\begin{aligned} D^E_{\ell,m} &= -(D_{\ell,m} + (-1)^{\ell+m}\overline{D_{\ell,-m}})/2 \\ D^B_{\ell,m} &= i(D_{\ell,m} - (-1)^{\ell+m}\overline{D_{\ell,-m}})/2 \end{aligned} \tag{1}$$

2. Properties and Computation

- Generalized spherical harmonics start at $\ell = 2$, and for each ℓ, $-\ell \leq m \leq \ell$.
- $P^{\ell}_{2,-m}(\theta) = P^{\ell}_{2,m}(180° - \theta)$.
- $P^{\ell}_{2,m}$ is real for m even, and pure imaginary for m odd.
- All functions are zero at the poles except $P^{\ell}_{2,2}$, which is nonzero at the North Pole ($\theta = 0°$), and $P^{\ell}_{2,-2}$, nonzero at the South Pole ($\theta = 180°$).
- Functions are normalized such that for any value of ℓ, the integral over the sphere of the sum of squares for all m gives unity. Thus the "strength" of an individual function decreases as ℓ increases when contrasted with the usual normalization for ordinary spherical harmonics, where each m individually integrates to unity.
- Function evaluation is by recursion. Recurrences on ℓ and then on m are used to reach each particular function. (Note that θ here refers to the colatitude, not the latitude.)

$$\begin{aligned} &\frac{\sqrt{(\ell+m+1)(\ell-m+1)(\ell+j+1)(\ell-j+1)}}{(2\ell+1)(\ell+1)}P^{\ell+1}_{j,m}(\theta) + \frac{mj}{\ell(\ell+1)}P^{\ell}_{j,m}(\theta) + \\ &\qquad \frac{\sqrt{(\ell+m)(\ell-m)(\ell+j)(\ell-j)}}{\ell(2\ell+1)}P^{\ell-1}_{j,m}(\theta) = \cos\theta P^{\ell}_{j,m}(\theta) \end{aligned} \tag{2}$$

$$\begin{aligned} &\sqrt{(\ell+m+1)(\ell-m)}P^{\ell}_{j,m+1}(\theta) - \sqrt{(\ell+m)(\ell-m+1)}P^{\ell}_{j,m-1}(\theta) = \\ &\qquad 2i\frac{m\cos\theta - j}{\sin\theta}P^{\ell}_{j,m}(\theta) \end{aligned} \tag{3}$$

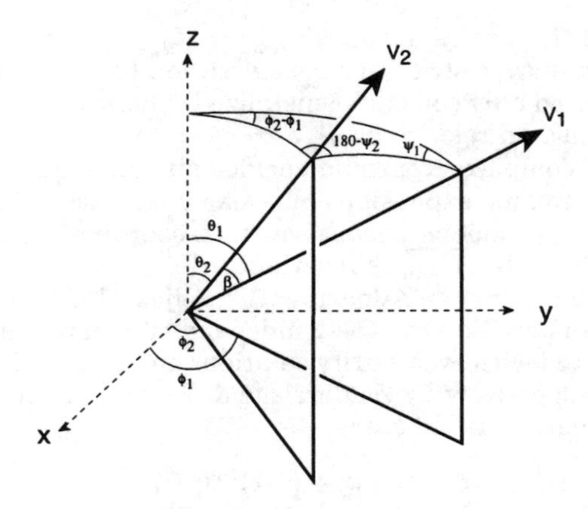

Figure 1. Geometry for definitions of ψ_1 and ψ_2 (Kosowsky 1996).

- The recursion is anchored by explicit formulas for the generalized quadrupole $P^2_{2,m}(\theta)$. (The recursion in ℓ at $\ell = 2$ defines $\ell = 3$, since the coefficient for $\ell = 1$ vanishes.)

$$\begin{aligned}
P^2_{2,-2}(\theta) &= \frac{1}{4}(\cos\theta - 1)^2 \\
P^2_{2,-1}(\theta) &= \frac{i}{2}\sin\theta(\cos\theta - 1) \\
P^2_{2,0}(\theta) &= \sqrt{\frac{3}{8}}(\cos^2\theta - 1) \\
P^2_{2,1}(\theta) &= \frac{i}{2}\sin\theta(\cos\theta + 1) \\
P^2_{2,2}(\theta) &= \frac{1}{4}(\cos\theta + 1)^2
\end{aligned} \tag{4}$$

3. Sum Rules and Correlation Functions

Generalized spherical harmonics obey a sum rule analogous to a familiar one for ordinary spherical harmonics, but it includes an explicit phase factor which depends on the orientation of the two lines of sight. That phase factor depends on the angle ψ which carries the reference direction for line of sight $\vec{v}_1$ into the reference direction for line of sight $\vec{v}_2$. The geometry of $\vec{v}_1$ and $\vec{v}_2$ is illustrated in Figure 1. $\psi = \psi_1 + \psi_2$, which has the following geometric interpretation. The reference direction is rotated by ψ_1 into the great circle from $\vec{v}_1$ to $\vec{v}_2$, translated to $\vec{v}_2$, and then rotated through ψ_2 to bring it into alignment with the reference direction at $\vec{v}_2$. (β is the angle between $\vec{v}_1$ and $\vec{v}_2$).

$$\begin{aligned}
\cos\beta &= \cos\theta_1\cos\theta_2 + \sin\theta_1\sin\theta_2\cos(\phi_2 - \phi_1) \\
\sin\psi_1 &= \sin\theta_2\sin(\phi_2 - \phi_1)/\sin\beta
\end{aligned}$$

$$\begin{aligned}\cos\psi_1 &= (\sin\theta_2\cos\theta_1\sin(\phi_2-\phi_1)-\sin\theta_1\cos\theta_2)/\sin\beta\\ \sin\psi_2 &= \sin\theta_1\sin(\phi_1-\phi_2)/\sin\beta\\ \cos\psi_2 &= (\sin\theta_1\cos\theta_2\sin(\phi_2-\phi_1)-\sin\theta_2\cos\theta_1)/\sin\beta \qquad (5)\end{aligned}$$

With that definition of ψ, the sum rule relating generalized spherical harmonics along two lines of sight to the angle β between those lines of sight is

$$P^{\ell}_{2,2}(\beta) = e^{-2i\psi}\sum_{m=-\ell}^{\ell} T^{\ell}_{2,m}(\theta_1,\phi_1)\overline{T^{\ell}_{2,m}(\theta_2,\phi_2)}.$$

If this phase factor is included in the definition of the spherical average over all directions $\vec{v}_i$ and $\vec{v}_j$ separated by an angle β

$$C(\beta) =< T_Q(\vec{v}_i)T_Q(\vec{v}_j) + T_U(\vec{v}_i)T_U(\vec{v}_j) >= \sum_{\beta_{ij}=\beta} e^{-2i\psi} Z(\vec{v}_i)\overline{Z(\vec{v}_j)}$$

where $Z = T_Q + iT_U$, then this allows us to define rotationally invariant analogues C^P_ℓ to the power spectra C_ℓ:

$$C(\beta) = \sum_\ell C^P_\ell P^\ell_{2,2}(\beta) = \sum_\ell P^\ell_{2,2}(\beta)\sum_m D_{\ell,m}\overline{D_{\ell,m}}.$$

Additionally, we can construct analogous sums of $D^E_{\ell,m}$ and $D^B_{\ell,m}$, which we denote as C^E_ℓ and C^B_ℓ respectively. These are the appropriate quantities to use for comparison to theoretical treatments of polarization. The partitioning into C^E_ℓ and C^B_ℓ is pertinent since Zaldarriaga & Seljak (1996) and Kamionkowski et al. (1996) both show that scalar perturbations cannot produce a nonzero C^B_ℓ.

It is interesting to note that $P^l_{2,2}(\cos(180°)) = 0$, which implies that correlations between physical polarization signals vanish at the antipodes.

Acknowledgments. The National Aeronautics and Space Administration (NASA)/Goddard Space Flight Center (GSFC) is responsible for the design, development, and operation of the Cosmic Background Explorer (*COBE*). Scientific guidance is provided by the *COBE* Science Working Group. GSFC is also responsible for the development of analysis software and for the production of the mission data sets.

Fruitful discussions with M. Jacobsen, University of Maryland Department of Mathematics, and B. Summey, Hughes STX, are gratefully acknowledged.

References

Gel'fand, I. M., Minlos, R. A., & Shapiro, Z. Y. 1963, Representations of the Rotation and Lorentz Groups and their Applications (New York: Pergamon Press)

Kamionkowski, M., Kosowsky, A., & Stebbins, A. 1997, Phys. Rev. Lett., 78, 2038

Kosowsky, A. 1996, Ann. Phys., 246, 49

Sazhin, M. V., & Korolëv, V. A. 1985, Sov. Astron. Lett., 11, 204

Zaldarriaga, M., & Seljak, U. 1997, Phys. Rev. D, in press

Astronomical Data Analysis Software and Systems VI
ASP Conference Series, Vol. 125, 1997
Gareth Hunt and H. E. Payne, eds.

Image Reconstruction with Few Strip-Integrated Projections: Enhancements by Application of Versions of the CLEAN Algorithm

M. I. Agafonov

Radiophysical Research Institute (NIRFI), 25 B. Pecherskaya st., Nizhny Novgorod, 603600, Russia, E-mail: agfn@nirfi.nnov.su

Abstract. Iterative algorithms with non-linear constraints are very attractive in image reconstruction with only a few strip-integrated projections. We present research into various versions of the iterative CLEAN algorithm for the solution of this problem. We suggest a method to determine the area of permissible solutions in complicated cases for two CLEAN algorithms. This procedure was named 2-CLEAN Determination of Solution Area (2-CLEAN DSA).

1. Introduction

Two dimensional image reconstruction from 1-D projections is often hampered by the small number of available projections, by an irregular distribution of position angles, and by position angles that span a range smaller than about 100°. These limitations are typical of both lunar occultations of celestial sources and observations with the fan beam of a radio interferometer, and also apply to greatly foreshortened reconstructive tomography.

2. Deconvolution Problem

The problem requires the solution of the equation

$$G = H * F \ (+noise) \ , \qquad (1)$$

where $F(x,y)$ is the object brightness distribution, $H(x,y)$ is the fan (dirty) beam, and $G(x,y)$ is the dirty (summary) image. The classical case (Bracewell & Riddle 1967) needs a number of projections $N \geq \pi D/\varphi$, where φ is the desired angular resolution, and D is the diameter of the object. The incomplete sampling of $H(u,v)$ requires the extrapolation of the solution of $F'(x,y)$ using non-linear processing methods.

2.1. The Iterative Algorithms

The general scheme of an iterative algorithm is

$$F^{k+1} = r_\alpha CF^k + \lambda \left(G - HCF^k\right) , \qquad (2)$$

where λ is the loop gain ($0 < \lambda < 2/max\ H(u,v)$), $C = C_1C_2...C_n$ are the limitations, and r_α is the stabilizer.

The simple standard CLEAN (Högbom 1974) is the best known realization of iteration schemes in radio astronomy. But the algorithm has defects (stripes and ridges) in areas of extended emission. CLEAN was used for the image reconstruction of the Crab Nebula from four lunar occultation profiles (Maloney & Gottesman 1979; Agafonov et al. 1986). However, more complete information is needed for an extended object. Trim Contour CLEAN (TC-CLEAN) (Steer et al. 1984) gave hope for the improvement of image quality with extended features, but it needed a study in different practical cases (Cornwell 1988).

2.2. Numerical Modeling

The process of solution convergence by σ (ERROR of initial and control 1-D profiles) minimization with variation of parameters λ and TC (Trim Contour level) was investigated (Agafonov & Podvojskaya 1989; Agafonov & Podvojskaya 1990) for both algorithms using of the following procedure:

- **2-D object model → 1-D profiles → Dirty image**
- **CLEAN (λ)** or **TC-CLEAN (λ, TC)** using the **Dirty beam**
- **Control test** from clean maps: **Calculation of σ** (ERROR of control and initial 1-D profiles)
- **Correction of λ or λ, TC to** $min\ \sigma$

The process of parameter (λ or λ, TC) optimization to $min\ \sigma$ was shown very well graphically (Agafonov & Podvojskaya 1989).

CLEAN The dirty map peak is the target of each iteration. The choice of loop gain λ was not clear in the original scheme, but it significantly influences the solution. For example, by modeling the test object (Crab Nebula map at 1.4 GHz) the optimum range of λ was found to be about 0.05–0.10 (from the dependence of $\sigma(\lambda)$ in this map). But the optimum value depends on the object structure. The algorithm has high instability for distributed objects. Changes in the resulting maps as a function of σ with small changes of λ testifies to this instability. We attempted to increase stability by: (*i*) choosing solutions with $min\ \sigma$ from the optimum interval and then averaging; and (*ii*) by special processing—such as a complex method like speckle-masking (but increasing the computational efforts).

TC-CLEAN A smooth function is subtracted at each iteration. Trim contour (TC) is used for the choice of components per iteration. TC must be low, but above the level corresponding to the true object dimensions. The algorithm has high stability, a simple choice of λ and TC, converges in few iterations, and is computationally efficient.

3. Discussion and Conclusions

A simple object (consisting of the peaks) may be successfully restored by the standard CLEAN. The results obtained by both methods are practically identical for a simple object consisting of individual components, but TC-CLEAN is more computationally efficient. A difference between the solutions is observed for complicated objects with small components in areas of extended emission. The standard CLEAN reconstruction has a "grooved" structure for such areas.

For smoothed 1-D profiles with small "hillocks," the solution can be obtained from the isolated individual components (CLEAN), and also from the more smoothed components (TC-CLEAN) by using the same $min\ \sigma$ for the initial and control profiles. CLEAN increases the contrast of small components, but the extended background decreases because of the "grooves." If $min\ \sigma$ (CLEAN) $\cong min\ \sigma$ (TC-CLEAN), the solutions will be formally equivalent for both algorithms, and so we have two choices: (*i*) to prefer the result corresponding to the physical peculiarities of the object in accordance with *a priori* information; or (*ii*) to assume the existence of a probable class of solutions between the "obtuse" (smooth) one from TC-CLEAN and the "sharp" one from CLEAN.

CLEAN forms the solution from the sum of peaks, and the result is the sharpest variant permissible within the established constraints. On the other hand, TC-CLEAN accumulates its result from the most extended components that satisfy the constraints, producing the smoothest solution (Agafonov & Podvojskaya 1990).

We carried out the study of TC-CLEAN applied to image reconstruction with few projections. TC-CLEAN proved to be an effective and stable solution. CLEAN emphasizes only individual features and needs a special treatment to obtain the solution stability (except the object from the peaks). We suggest determining the area of permissible solutions of complicated objects with the help of both algorithms. This procedure was named 2-CLEAN Determination of Solution Area (2-CLEAN DSA). It is also useful to study the possibility of reconstruction (the reality of the components on the map) for any new case with poor UV-filling by using a similar method used for our observational test object. The 2-CLEAN DSA procedure can show, in complicated cases, a range of possible images from "obtuse" to "sharp" variants satisfying imposed constraints and poor *a priori* information.

4. Maps from Real Lunar Occultation Observations

In certain cases, usually at lower frequencies, the angular resolution of synthesis radio telescopes is insufficient. However, observations of an object during lunar occultations provide 1-D brightness profiles with high angular resolution. This is also useful in observations with optical instruments. The lunar limb is approximated by a plane screen, moving through different position angles. The first images of the Crab Nebula from four projections of lunar occultations was presented by using the standard CLEAN (Maloney & Gottesman 1979; Agafonov et al. 1986). But the maps had defects in extended areas due to the reconstruction algorithm. The application of TC-CLEAN was used in the reconstruction of the Crab Nebula map at 750 MHz with angular resolution $20\times35''$ (Agafonov et al. 1990). The 1-D profiles were obtained by observations using the 70 meter dish (RT-70) in West Crimea. The method of 2-CLEAN DSA allowed us to determine that the area of permissible solutions lies formally between the "sharp" (CLEAN) and "smooth" (TC-CLEAN) variants. The two maps were generally similar. The standard CLEAN increased the contrast of small components, while the TC-CLEAN map gave a better agreement with known *a priori* information about the Nebula, and was closer to the true brightness distribution of the Crab. The CLEAN variant of the map can be used only for information about the location of the small components.

Acknowledgments. I am grateful to the National Radio Astronomy Observatory for the support which made possible this presentation and my special gratitude to R. Simon and C. White for their endurance in the organization of this Conference.

References

Agafonov, M. I., Aslanyan, A. M., Gulyan, A. G., Ivanov, V. P., Martirosyan, R. M., Podvojskaya, O. A., & Stankevich, K. S. 1986, Pis'ma v AZh, 12, 275

Agafonov, M. I., & Podvojskaya, O. A. 1989, Izvestiya VUZ. Radiofizika, 32, 742

Agafonov, M. I., & Podvojskaya, O. A. 1990, Izvestiya VUZ. Radiofizika, 33, 1185

Agafonov, M. I., Ivanov, V. P., & Podvojskaya, O. A. 1990, AZh, 67, 549

Bracewell, R. N., & Riddle, A. C. 1967, ApJ, 150, 427

Cornwell, T. J. 1988, A&A, 202, 316

Högbom, J. A. 1974, A&AS, 15, 417

Maloney, F. P., & Gottesman, S. F. 1979, ApJ, 234, 485

Steer, D. G., Dewdney, P. E., & Ito, M. R. 1984, A&A, 137, 159

Astronomical Data Analysis Software and Systems VI
ASP Conference Series, Vol. 125, 1997
Gareth Hunt and H. E. Payne, eds.

Refined Simplex Method for Data Fitting

Y.-S. Kim

Optical Science Laboratory, Department of Physics and Astronomy, University College London, London, WC1E 6BT, U.K.

Abstract. The simplex method, a data fitting method to any type of function, is refined by eliminating a redundant process. The refined method is applied to Zernike polynomials in Cartesian coordinates, which describe an optical surface or wavefront in terms of aberrations. The advantages and disadvantages of the simplex method are discussed.

1. Introduction

The simplex method was first introduced by Nelder & Mead (1965) for finding the minimum value of functions. O'Neill (1971) produced a FORTRAN implementation of the simplex method, and Caceci & Cacheris (1984) applied the simplex method to curve fitting and produced a PASCAL program. When the algorithm of the simplex method was re-examined, a redundant mechanism in the process was found. This fact has been checked by using Zernike polynomials in Cartesian coordinates.

Zernike polynomials are well-known and broadly used functions for describing the wavefront of optical systems in terms of aberrations. Data fitting to Zernike polynomials has mainly been done by the least-squares method (Malacara et al. 1990; Rayces 1992). However, in order to use the least-squares method, the function to be fitted should have a derivative function, and the function terms should be orthogonal or transformed to be orthogonal (Rayces 1992), to make the calculation easy. The Simplex method needs neither the derivative function nor the orthogonality condition. The simplex method is used for curve fitting to Zernike polynomials in Cartesian coordinates.

2. Variance

The variance (σ^2) of function $y = f(x_i)$, $i = 1$–n over k data sets is,

$$\sigma^2 = \sum_{j=1}^{k} \{y_j - f(x_{1j}, x_{2j}, x_{3j}, ..., x_{nj})\}^2. \tag{1}$$

The coefficients that minimize the variance give the best-fitted function. If we take a simple 2-dimensional function $y = ax + b$, then the variance will be,

$$\sigma^2 = \Sigma(y_j - ax_j - b)^2 = \text{minimum}. \tag{2}$$

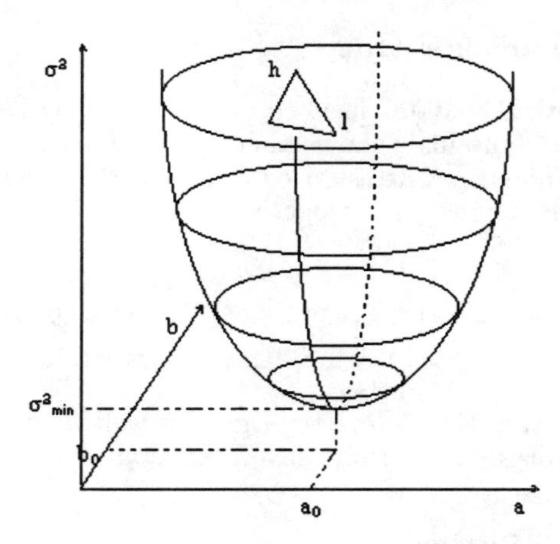

Figure 1. A response surface of σ^2 versus a and b.

A graph of the variance versus the coefficients a and b is shown in Figure 1, which is called a response surface. Each ellipse is a contour of equal variance. The lowest point on the figure will give the appropriate value of the coefficients.

3. The Simplex Method

The algorithm and codes for the simplex method are described in other papers (Nelder & Mead 1965; O'Neill 1971; Caceci & Cacheris 1984), so I will not explain them in full detail.

The only thing I want to mention is the four mechanisms of vertex movements. Vertices are moved toward the minimum point by the four mechanisms, reflection, expansion, contraction, and shrinkage. Reflection moves the highest (worst) vertex to the opposite side of the center of the other vertices. Expansion moves the highest vertex to twice the distance from the center, which is used when a reflected vertex is lower than the lowest vertex and the expanded vertex becomes the lowest one. Contraction moves to the middle between the highest vertex and the center, which is used when a reflected vertex is higher than the highest vertex. When the contracted vertex is still higher than the highest vertex, all the vertices except the lowest vertex move toward the lowest one by half, which is called shrinkage.

4. The Refined Simplex Method

When the simplex algorithm was examined, it was found that shrinkage does not occur when curve fitting. The variance is the sum of squares, so the response surface is always an $(n + 1)$ dimensional paraboloid. So, either the reflected vertex or contracted one should be lower than the highest one.

5. Zernike Polynomials Fitting

The wavefront of optical systems used to be described by Zernike polynomials, as each term of the polynomials shows aberrations individually.

Zernike polynomials in Cartesian coordinates (Cubalchini 1979; Genberg 1983) of four and nine terms are, respectively,

$$z = c_0 + c_1 x + c_2 y + c_3(2r^2 - 1), \tag{3}$$

$$\begin{aligned} z = \; & c_0 + c_1 x + c_2 y + c_3(2r^2 - 1) + c_4(x^2 - y^2) + 2c_5 xy + c_6 x(3r^2 - 2) \\ & + c_7 y(3r^2 - 2) + c_8(6r^4 - 6r^2 + 1), \; \textit{where } r^2 = x^2 + y^2. \end{aligned} \tag{4}$$

In order to check whether the simplex method fits well with Zernike polynomials in Cartesian coordinates, simulations have been done for mirror data.

6. Simplex Curve Fitting

The simulated mirror is spherical, with diameter of 760 mm and radius of curvature 5400 mm. The mirror surface is divided into 51×51 grid lines, whose intersections mark 1959 points on the mirror surface.

The simulation program is written in C, converted from Caceci & Cacheris's (1984) PASCAL program. Zernike polynomials in Cartesian coordinates are adapted.

The first step of the simulation is to give initial values to the coefficients and calculate the height error on each point on the mirror surface from the mirror data. Then, from the height error data, new coefficients are derived by the simplex method. The result of the simulation can be judged by comparing the derived coefficients to the true coefficients.

The initial values of the coefficients are all set zero, since the ideal mirror would be aberration-free. The calculation is terminated when the difference between the maximum and the minimum coefficient value on each term is less than 10^{-4}.

Table 1 shows the result of the simulations. The first two simulations are done for four term Zernike polynomials and the other two for nine term Zernike polynomials. The table shows that the derived coefficients are quite accurate.

During the simulations described above, the number of iterations for each mechanism is counted, in order to check whether shrinkage is used in the simplex calculation. Table 1 also gives the number of iterations on each mechanism. The "Iteration" row shows the number of total iterations of the simplex process, and the last four rows give the number of iterations of each of the four mechanisms. It clearly shows that shrinkage never occurs.

7. Conclusions

The simplex method is one of the best curve fitting methods, having several advantages. First, it is simple to use. Compared to the well-known least-squares method, the simplex method does not need the derivative function, nor does the orthogonality condition apply. Second, it is easy to adopt any function in the computer program—only the function evaluation needs to be changed. And since the response surface is a paraboloid, there are no local minima giving false

Table 1. Simulation of Zernike polynomial fitting by the simplex method.

	Simulation 1		Simulation 2		Simulation 3		Simulation 4	
	Given	Derived	Given	Derived	Given	Derived	Given	Derived
c_0	−0.2137	−0.2137	−0.0739	−0.0739	0.0115	0.0115	0.2169	0.2169
c_1	−0.0977	−0.0977	0.1999	0.1999	0.2340	0.2340	−0.0617	−0.0617
c_2	0.1357	0.1357	−0.1600	−0.1600	0.0262	0.0262	0.0146	0.0146
c_3	−0.1117	−0.1117	−0.1229	−0.1229	−0.0186	−0.0186	−0.1362	−0.1362
c_4					0.0276	0.0276	−0.0187	−0.0187
c_5					0.1930	0.1930	0.0152	0.0151
c_6					0.0036	0.0036	−0.0151	−0.0150
c_7					−0.0378	−0.0378	0.0145	0.0145
c_8					0.0049	0.0049	−0.0822	−0.0822
Iteration		127		108		652		1584
Reflection		55		46		444		1077
Expansion		21		16		70		219
Contraction		51		46		138		288
Shrinkage		0		0		0		0

solutions. However, the simplex calculation is a little slower than the analytic least-squares calculation. And when the algorithm is examined, it is found that one of the four mechanisms, shrinkage, is not necessary. This fact is confirmed by testing Zernike polynomials in Cartesian coordinates. Eliminating the shrinkage mechanism would make the computing time a little bit shorter.

References

Caceci, M. S., & Cacheris, W. P. 1984, Byte, May, 340

Cubalchini, R. 1979, J. Opt. Soc. Am., 69, 972

Genberg, V. L. 1983, SPIE, 450, 81

Malacara, D., Carpio-Valadéz, J. M., & Sánchez-Mondragón, J. J. 1990, Optical Engineering, 29, 672

Nelder, J. A., & Mead, R. 1965, Comp. J., 7, 308

O'Neill, R. 1971, Applied Statistics, 20, 338

Rayces, J. L. 1992, Appl.Optics, 31, 2223

Part 4. Modeling

Astronomical Data Analysis Software and Systems VI
ASP Conference Series, Vol. 125, 1997
Gareth Hunt and H. E. Payne, eds.

Numerical Simulations of Plasmas and Their Spectra

G. J. Ferland, K. T. Korista, and D. A. Verner

Physics & Astronomy, University of Kentucky, Lexington, KY 40506

Abstract. This review centers on the development and application of Cloudy, a large-scale code designed to compute the spectrum of gas in photoionization or collisional balance. Such plasma is far from equilibrium, and its conditions are set by the balance of a host of microphysical processes. The development of Cloudy is a three-pronged effort requiring advances in the underlying atomic data base, the numerical and computational methods used in the simulation, culminating in the application to astronomical problems. These three steps are strongly interwoven.

A complete simulation involves many hundreds of stages of ionization, many thousands of levels, with populations determined by a vast sea of atomic/molecular processes, many with accurate cross sections and rate coefficients only now becoming available. The scope of the calculations and the numerical techniques they use, can be improved as computers grow ever faster, since previous calculations were naturally limited by the available hardware. The final part is the application to observations, driven in part by the revolution in quality of spectra made possible by advances in instrumentation. The galactic nebulae represent laboratories for checking whether the physics of nebulae is complete and for testing galactic chemical evolution theories, and can validate the analysis methods to be used on the quasars. Models of ablating molecular clouds, the likely origin of some gas in quasars, are tested by studies of the Orion complex. Finally, the quasars themselves present the ultimate challenge: to deduce the composition of their emitting gas from the spectrum, determine the dependencies of this spectrum on the shape of the ionizing continuum, correlating this with observed changes in the spectral energy distribution, and finally to understand its dependencies on luminosity (the Baldwin effect). Once the emission line regions of quasars are understood, we will have a direct probe of the $z \leq 5$ universe.

Cloudy is widely used in the astronomical community, with roughly 50 papers per year employing it. The code is a general spectroscopic tool whose development has an impact well beyond out specific studies.

1. Introduction

The quasars are the most luminous objects in the universe and the highest redshift objects we can observe directly. Understanding their emission lines has a cosmological imperative, since their spectra depend on luminosity (Baldwin 1977; Boroson & Green 1992; Osmer et al. 1994). Once we can directly measure their luminosity, the quasars will gauge the expansion of the universe at redshifts

$z \leq 5$. At the same time, the origin of the chemical elements remains a central theme across much of stellar, galactic, and extragalactic astrophysics. Quasars probe early epochs in the formation of massive galaxies and their emission lines can reveal the composition of the interstellar medium (ISM) when the universe had an age well under a billion years. Within our galaxy, HII regions and planetary nebulae define the end points of stellar evolution. Detailed analysis of their emission line spectra can reveal both the nebula's composition and the luminosity and temperature of the central star(s), and can validate the analysis methods to be used for the quasars.

Deducing reliable abundances and luminosities of galactic and extragalactic emission line objects is the central theme of the development of Cloudy. Emission lines are produced by warm ($\sim 10^4$ K) gas with moderate to low density ($n \leq 10^{12}$ cm^{-3}). Such gas is far from equilibrium, and its physical conditions cannot be known from analytical theory. Rather, the observed spectrum is the result of a host of microphysical processes which must be simulated numerically.

The ionization, level populations, and electron temperature are determined self-consistently by solving the equations of statistical and thermal equilibrium. Lines and continua are optically thick and their transport must be treated in detail. Predictions of the intensities of thousands of lines and the column densities of all constituents result from the specification of only the incident continuum, gas density, and its composition. By their nature, such calculations involve enormous quantities of atomic/molecular data describing a host of microphysical processes, and the codes involved are at the forefront of modern computational astrophysics. Although the task is difficult, the rewards are great, since numerical simulations make it possible to interpret the spectrum of non-equilibrium gas on a physical basis. Cloudy has been developed as an aid to this interpretation, much as an observer might build a spectrometer.

The next few years will witness first light of a large number of new optical to infrared observational facilities, and we will be in a position to obtain spectra of faint objects with unprecedented precision. The basic atomic data base is growing in both precision and size, and high-end workstations have the power of yesterday's supercomputers—large scale numerical simulations and spectral synthesis can now be done with unprecedented precision and facility. Cloudy is openly available, and other astronomers use it to publish roughly 50 refereed papers per year. It is a general spectroscopic tool whose development has an impact well beyond the specific studies undertaken here.

2. Development of Cloudy

2.1. Web Access

Cloudy's source (about 110,000 lines of FORTRAN) and its documentation *Hazy, a Brief Introduction to Cloudy* (UK Physics Internal Report, 461 pages), are freely available on the Cloudy home page.[1] This Web page also has group preprints, and Dima Verner's Atomic Data for Astrophysics[2] (ADfA) Web page provides public access to the numerical forms of the atomic data used by Cloudy.

---

[1] http://www.pa.uky.edu/~gary/cloudy

[2] http://www.pa.uky.edu/~verner/atom.html

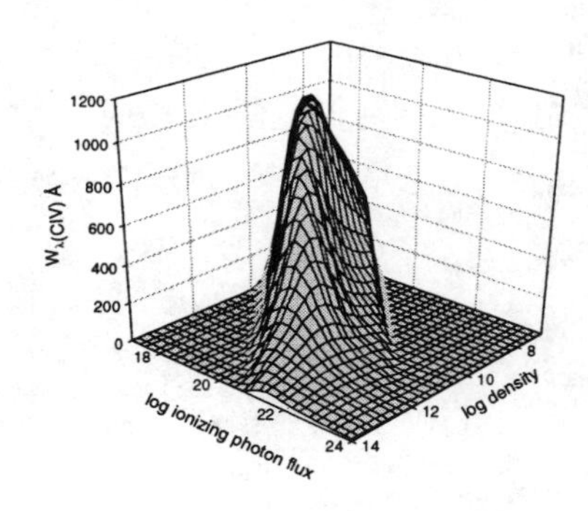

Figure 1. CIV 1549 equivalent width for a wide range of densities and flux of photons.

2.2. Cloudy's Stability

A major use of spectral synthesis calculations is to deduce the conditions and abundances in the matter producing a spectrum. There is always a question of uniqueness since there may be more than one way to get any particular result, and we must work backwards to deduce the question (properties of the object) from the answer (the spectrum). Examining predictions over the very broadest possible range of physical parameters is vital to really understand what a spectrum is telling us. Today we usually use the code to generate very large grids, involving thousands of complete simulations, to generate contour or 3-D plots like that shown in Figure 1. This is an example, based on Baldwin et al. (1995), showing that one of the strongest quasar emission lines is most efficiently produced over a very narrow range of conditions. The peak visibility occurs for the parameters long ago deduced as "standard" quasar conditions. We argued that this is just a selection effect.

A *major* effort has gone into making it possible to generate such large grids on a routine basis. Issues include the following:

The code must have enough intelligence to autonomously converge for models with very different conditions, without user intervention. This goal has largely been realized. All of the predictions shown here are the result of fully autonomous calculations with no outside intervention.

The code must respond appropriately, no matter what conditions it is asked to compute. It goes to the Compton, molecular, ISM, and LTE limits. The temperature limits are $2.8\,\mathrm{K}<T<10^9\,\mathrm{K}$, and density limit $n \leq 10^{12}\,\mathrm{cm}^{-3}$ because of the approximate treatment of line transfer, and since most ions are treated as two-level systems. The code will identify conditions which are inappropriate for its use.

The code must examine its results and predictions to confirm that the simulation is valid. It simply is not possible to go over the many tens of megabytes of results generated in large grids, after the fact, to confirm that all went well. As the code is developed, potential problems are identified and logic written to check for this in future calculations.

The code *will probably* converge a simulation with no problems, and it *will certainly* identify any problems at the end of the calculation. These are essen-

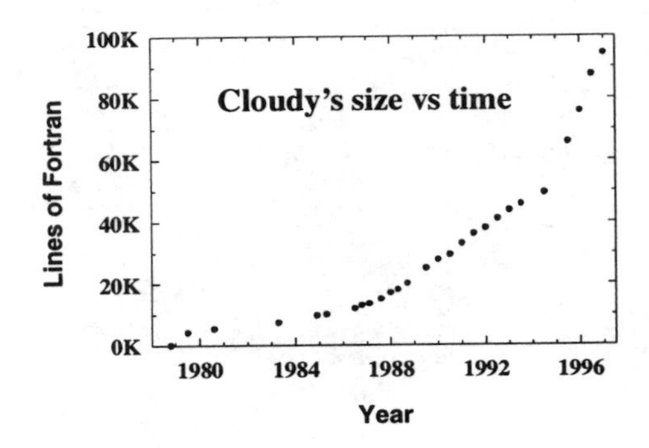

Figure 2. Cloudy's size as a function of time. This shows only the number of executable lines of FORTRAN. The total distributed source now constitutes ~110,000 lines of code.

tial core features—vital if very large grids are to be computed reliably. Each simulation ends with a summary of all remarkable or surprising features, an analysis of any convergence problems, and checks that the range of validity was not exceeded. One specific example of these internal checks is that the code now tracks timescales for all heating-cooling and ionization-recombination reactions. At the end of the calculation the code will identify the longest timescale. A warning is produced if the time-steady assumption is not valid.

2.3. Completeness of the Simulations

The code includes $\sim 10^4$ resonance lines from the 495 possible stages of ionization of the lightest 30 elements—an extension that required several steps. The charge transfer data base was expanded to complete the needed reactions between hydrogen and the first 4 ions and fit all reactions with a common approximation (Kingdon & Ferland 1996). Radiative recombination rate coefficients were derived for recombination from all closed shells, where this process should dominate (Verner & Ferland 1996). Analytical fits to Opacity Project (OP) and other recent photoionization cross sections were produced (Verner et al. 1996). Finally, rescaled OP oscillator strengths were used to compile a complete set of data for 5971 resonance lines (Verner, Verner, & Ferland 1996).

Many other improvements are summarized in Ferland et al. (1997). Figure 2 shows a partial indicator of the scope of this activity, the number of lines of executable FORTRAN, as a function of time. The growth of the code in the past few years has been explosive, thanks to modern workstations.

2.4. Reliability in the Face of Complexity

This is the central difficulty in any large simulation since analytical answers are seldom known. Predictions are affected by the numerical approximations and atomic data used, as well as the presence of bugs which are certainly present in any large program. The best way to verify results is to compare completely independent calculations. We organized a workshop in 1994 to bring together researchers who have developed nebular codes. The result was the set of benchmarks published in the STScI Osterbrock-Seaton commemoration (Ferland et

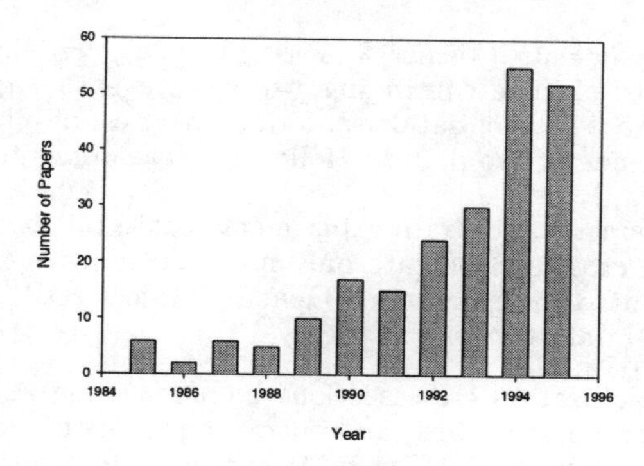

Figure 3. Refereed papers acknowledging Cloudy.

al. 1995). These benchmarks provide reference points to aid validation of any plasma code. We also make a continuous comparison of Cloudy with another independently written photoionization code, LineSpec (Verner & Yakovlev 1990).

2.5. Community Use

Cloudy is widely used by others in their analysis and theory of spectroscopic observations. Figure 3 shows the number of refereed papers acknowledging the use of Cloudy through 1995. At least 138 papers were published in 1993–1995. Although I was not a co-author on these projects, the code did play some role in their execution.

3. Future Development of Cloudy

The long-term goal is for the ionization and thermal equilibria of all species and the radiative transfer of continua and lines to be exact for all conditions. By this standard, the code is well over half-way complete.

3.1. The Underlying Atomic Data Base

A major effort has gone into Verner's *Atomic Data for Astrophysics* (ADfA) database. The results of any fully non-equilibrium calculation are no better than the underlying atomic data. New data appear throughout the physics and chemistry literature; maintaining the data base is a major ongoing effort, but one that does not directly result in publications. The ADfA contains basic atomic data required for calculation of the ionization state of astrophysical plasmas and for quantitative spectroscopy. The home page had more than 3000 visits and more than 4000 data files were retrieved during October 1995–October 1996. The contents of ADfA are driven by Cloudy's needs, which are defined by the spectroscopic problems. Because of this, and the public access we provide to our data, it has broad application to anyone working in quantitative spectroscopy.

3.2. Photoionization Cross sections, Recombination Coefficients

We have placed a great deal of effort into deriving accurate fits to the best photoionization cross sections (Verner et al. 1996), the corresponding radiative recombination coefficients (Verner & Ferland 1996), and incorporating them into Cloudy. This effort is continuing, with the eventual goal of a complete photoionization cross section database, for all energies, of all atoms and ions of the first thirty elements. We plan the following developments:

- For some elements, accurate experimental data on the ground-state photoionization cross sections are only now becoming available. We are replacing the fits to the smoothed Opacity Project (OP) data by the fits to experimental data for such species.
- Some photoionization cross sections include strong and wide resonances which cannot be smoothed, and energy positions of these resonances are accurately known by experiment. We are extending the treatment of such resonances in our fits.
- We are improving our current photoionization fits for the outer shells of non-OP elements up to zinc ($Z = 30$) by use of isoelectronic interpolation of the fits for the OP elements.
- Our current fits reproduce theoretical non-relativistic high-energy asymptotes which cannot be applied for energies above 50 keV. We plan extensions to relativistic cross sections based on available X-ray experimental data (see, e.g., Veigele 1973).
- We are creating fits for excited shell photoionization cross sections of atoms and ions based on the OP data near thresholds, and on new Hartree-Dirac-Fock calculations far from threshold. The new data are being calculated by Band and Trzhaskovskaya (St. Petersburg). This is needed for dense environments, and for accurate radiative recombination rate coefficients.

We have calculated radiative recombination rates for all H-like, He-like, Li-like and Na-like ions of elements from H through Zn, and fitted them with convenient analytical formulae (Verner & Ferland 1996). Using our new fits to the excited shell photoionization cross section data described above, we shall complete the calculations of radiative recombination rates for ions of all isoelectronic sequences of all elements up to zinc ($Z = 30$), and fit them by analytical formulae. This requires partial cross sections, and cannot be done with total cross sections, such as those available from the Opacity Project.

3.3. The Line Data Base

The majority of the $\sim 10^4$ heavy element lines now use collision strengths from $\bar{g}$-approximations (Gaetz & Salpeter 1983; Mewe 1972). These can be computed on the fly for any line, but should only be used for transitions with no quantal calculations. The Iron Project (IP; Hummer et al. 1993) is producing large numbers of collision strengths which must be fitted and incorporated, if this remarkable data set is to be exploited. We will fit the IP data with a procedure drawing on our previous work on photoionization cross sections (Verner et al. 1996) and the asymptotic limits of Burgess & Tully (1992). We will develop a generalized scheme to represent the full temperature dependence in terms of analytical fits, with coefficients that can be stored along with other line information.

We have compiled accurate wavelengths, energy levels, and atomic transition probabilities for the permitted resonance lines of all ionization states of astrophysically important elements (Verner, Verner, & Ferland 1996), and these data are in Cloudy. Thus, the resonance line list is fairly complete. However, the set of forbidden and intercombination lines is not complete, especially for high ionization transitions of 3rd row and higher elements. Using our compilation of experimental energy levels, we will calculate accurate wavelengths for a complete list of forbidden and intercombination transitions between the lowest levels of all ions and elements included in Cloudy. We will compile atomic transition probabilities and collision strengths for them from literature and from available atomic databases, including the very recent database CHIANTI (Dere et al. 1996). All the collision strengths will be averaged over Maxwellian electron distribution, checked for asymptotic behavior by use of the Burgess & Tully (1992) method, and fitted over temperature with analytical formulae.

4. Applications to Quasars

4.1. The LOC Approach

The "Locally Optimally-emitting Clouds" (LOC) model is outlined in Baldwin et al. (1995). The homogeneity of AGN spectra was long a mystery (Baldwin & Netzer 1978). Baldwin et al. (1995) showed that the average quasar spectrum can be produced by simply integrating over all possible clouds. Selection effects due to the line visibility function (Figure 1) ensure that most quasars will have very similar spectra even if their distributions of cloud properties are different.

The new aspect of this approach is allowing a range of cloud properties at a given radius, including ranges in column density and gas density. This seems much more realistic than just single-valued functions. Clouds of different gas densities could exist side-by-side for several viable confinement mechanisms (magnetically or shock confined, or dissipating clouds).

These integrations have been carried out over the full range of possible parameters. The line visibility functions are so strongly peaked that the spectrum of a typical quasar can be easily matched. This is not an entirely positive conclusion—we want to be able to deduce conditions in the quasar from the observations. The advantage is that there is no longer any hidden hand needed to adjust cloud parameters—we are dealing with a known, calculable, set of line visibility functions, which introduce powerful selection effects.

Two approaches are being taken. First we set up specific "straw man" distribution functions, compute the resulting spectrum, and compare this with observations. As an example, Broad Line Region (BLR) clouds could be radiatively driven outflows; we find that this can drive mass from main sequence stars near quasars. Such winds are Rayleigh-Taylor unstable, and they should cut off at an optical depth of unity in the Lyman continuum. We have carried out such an LOC integration, and find that the predicted lines can match observations. Radiatively driven winds are consistent with the observations.

The second approach is to use observations to try placing limits on possible distribution functions. We are working with Keith Horne (St. Andrews) to adopt our LOC grids, Keith's maximum entropy techniques (Horne 1994), and Jack Baldwin's line profile observations, to deduce the cloud distribution function. An integration over the full LOC plane produces results which are in good agreement with "typical" spectra of quasars (Baldwin et al. 1995). There is a large dispersion from object to object about this mean, which, together with detailed line profile studies, suggests that this plane is not necessarily fully populated in a given object. A goal is to determine observational limits to the cloud distribution functions. The combination of Keith's methods, Jack's observations, and our cloud simulations makes it possible to quantify which portions of the plane are populated, and whether this depends on luminosity.

For both approaches, we will first generate large data cubes containing predicted line equivalent widths as a function of gas density, column density, and flux at the illuminated face of the cloud. There will be a large number of these data cubes, with various continuum shapes and metallicities. Each will be incorporated into an interpolating routine, so that arbitrary parameters can be specified and the resulting spectrum obtained. Calculations of this core set of predictions is now underway and an initial atlas has been produced (Korista et al. 1997).

Our goal in these data cubes is to explore the implications of the LOC. However, this work has application well beyond our specific LOC models. Any other model of the BLR is only a subset of the LOC. Our grids will be made publicly available, so it will be possible for anyone to predict the spectrum resulting from a favored kinematic-spatial-magneto-hydrodynamic model, without becoming involved in the details of photoionization modeling.

4.2. Quasar Luminosity, Continuum Shape, and Metallicity Correlations

The long-term emphasis is to use luminous quasars to probe the high redshift universe. The basic problem is to understand luminosity correlations such as those in Baldwin, Wampler, & Gaskell (1989), and Osmer et al. (1994), on a physical basis. Luminosity-line correlations are complicated by other correlations, such as luminosity with continuum shape, most obviously $\alpha_{\rm ox}$ (Avni & Tananbaum 1986; Worrall et al. 1987; however LaFranca et al. 1995 find no dependence, and Avni, Worrall, & Morgan 1995 find a complicated dependence). The NV/CIV relation discussed by Hamann & Ferland (1992, 1993) suggests a metallicity-luminosity correlation (Ferland et al. 1996; Korista et al. 1997).

The LOC data cubes will be generated for many different continuum shapes, which we characterize by a Big Bump temperature, $\alpha_{\rm ox}$, and slopes for the bump and X-ray continuum. From this set, we can search for line ratios more sensitive to the continuum shape than to the LOC integration or metallicity. The large data sets available from CTIO will be used, together with our theoretical data sets, archival X-ray and IR observations, to establish constraints on the change in the continuum shape. The CIV/Lyα ratio is an example of lines that are sensitive to changes in the $\sim$30 eV continuum (Clavel & Santos-Lleo 1990). For the high redshift quasars, the high energy end of the Big Bump is a more important ingredient than the X-rays, and we can make explicit predictions for how line equivalent widths change with continuum shape. This work will be extended using cloud distribution functions, and the absence/presence of Baldwin effects in various lines, to quantify changes in the Big Bump with luminosity.

Acknowledgments. The development of Cloudy would not have been possible without the continued support of NASA and the NSF.

References

Avni, Y., & Tananbaum, H. 1986, ApJ, 305, 83

Avni, Y., Worrall, D. M., & Morgan, W. A. 1995, ApJ, 454, 673

Baldwin, J. A. 1977, MNRAS, 178, 67p

Baldwin, J., Ferland, G., Korista, K., & Verner, D. 1995, ApJL, 455, L119

Baldwin, J. A., & Netzer, H. 1978, ApJ, 226, 1

Baldwin, J., Wampler, J., & Gaskell, C. M. 1989, ApJ, 338, 630

Boroson, T. A., & Green, R. F. 1992, ApJS, 80, 109

Burgess, A., & Tully, J. A. 1992, A&A, 254, 436

Clavel, J., & Santos-Lleo, M. 1990, A&A, 230, 3

Dere, K. P., Monsignori-Fossi, B. C., Landi, E., Mason, H. E., & Young, P. R. 1996, BAAS, 28, 961

Ferland, G. J., Baldwin, J. A., Korista, K. T., Hamann, F., Carswell, R. F., Phillips, M., Wilkes, B., & Williams, R. E. 1996, ApJ, 461, 683

Ferland, G. J., et al. 1995, in The Analysis of Emission Lines, ed. R. Williams & M. Livio (Baltimore, MD: STScI), 83

Ferland, G. J., Korista, K. T., Verner, D. A., Ferguson, J. W., Kingdon, J. B., & Verner, E. M. 1997, PASP, in press

Gaetz, T. J., & Salpeter, E. E. 1983, ApJS, 52, 155

Hamann, F., & Ferland G. J. 1992, ApJ, 391, L53

Hamann, F., & Ferland G. J. 1993, ApJ, 418, 11

Horne, K. 1994, in ASP Conf. Ser., Vol. 60, Reverberation Mapping of the Broad Line Region in Active Galactic Nuclei, ed. P. M. Gondhalekar, K. Horne, & B. M. Peterson (San Francisco: ASP), 23

Hummer, D. G., Berrington, K. A., Eissner, W., Pradhan, A. K., Saraph, H. E., & Tully, J. A. 1993, A&A, 279, 298

Kingdon, J. B., & Ferland G. J. 1996, ApJS, 106, 205

Korista, K., Baldwin, J., Ferland, G., & Verner, D. 1997, ApJS, 108, 401

La Franca, F., Franceshini, A., Cristiani, S., & Vio, R. 1995, A&A, 299, 19

Mewe, R. 1972, A&A, 20, 215

Osmer, P. A., Porter, A. C., & Green, R. F. 1994, ApJ, 436, 678

Veigele, W. J. 1973, Atomic Data Tables, 5, 51

Verner, D. A., & Ferland, G. J. 1996, ApJS, 103, 467

Verner, D. A., Ferland, G. J., Korista, K. T., & Yakovlev, D. G. 1996, ApJ, 465, 487

Verner, D. A., Verner, E. M., & Ferland, G. J. 1996, Atomic Data Nucl. Data Tables, 64, 1

Verner, D. A., & Yakovlev, D. G. 1990, Ap&SS, 165, 27

Worrall, D. M., Giommi, P., Tananbaum, H., & Zamorani, G. 1987, ApJ, 313, 596

Astronomical Data Analysis Software and Systems VI
ASP Conference Series, Vol. 125, 1997
Gareth Hunt and H. E. Payne, eds.

On Fractal Modeling in Astrophysics: The Effect of Lacunarity on the Convergence of Algorithms for Scaling Exponents

I. Stern

Smithsonian Astrophysical Observatory, Cambridge, MA 02138

Abstract. Fractals and multifractals are used to model hierarchical, inhomogeneous structures in several areas of astrophysics, notably the distribution of matter at various scales in the universe. Current analysis techniques used to assert fractality or multifractality and extract scaling exponents from astrophysical data, however, have significant limitations and caveats. It is pointed out that some of the difficulties regarding the convergence rates of algorithms used to determine scaling exponents for a fractal or multifractal, are intrinsically related to its texture, in particular, to its *lacunarity*. A novel approach to characterize prefactors of cover functions, in particular, lacunarity, based on the formalism of *regular variation (in the sense of Karamata)* is proposed. This approach allows deriving bounds on convergence rates for scaling exponent algorithms and may provide more precise characterizations for fractal-like objects of interest for astrophysics. An application of regular variation based fractal modeling to apollonian packing, which can be useful in cosmic voids morphology modeling, is also suggested.

1. Introduction

Structures with fractal characteristics are important for astrophysics. Fractal-like structures may be created whenever nonlinear dynamical mechanisms are at work. Such mechanisms may generate inhomogeneity on a large range of physical scales and "incomplete space filling." For example, certain aspects of the distribution of matter in the universe are well described by fractals at small and medium scales (distribution of galaxy number counts, microwave background fluctuations, etc.) (Borgani 1995). However, the problem of a crossover from "fractality" to "homogeneity" at large scales, as required by the cosmological principle (Peebles 1993), raises some questions related to formal aspects of the fractal modeling involved and the ways in which the data are analyzed (Labini et al. 1996). At this point in time, it is recognized that fractal models with pure scale invariance are not sufficient to describe the morphology of large scale structure. In addition, the details of the mentioned crossover, if it exists, are open. In fact, it was pointed out recently (Mandelbrot 1995) that the research regarding the large scale morphology can benefit from a better understanding of the finer morphological features of fractals, namely, of their texture, and in particular, their *lacunarity*.

2. Lacunarity of a Fractal Structure

For a fractal, the fractal dimension alone is an incomplete description of the degree and nature of space filling. One can generate fractals with exactly the same fractal dimension and very different space filling structure or lacunarity. In a family of fractals with the same fractal dimension, the lower the "lacunarity" of a fractal the closer its appearance is to "homogeneity." Thus, as pointed out by Mandelbrot, large scale homogeneity may be mimicked by a fractal structure with low lacunarity (Mandelbrot 1983, 1995). Moreover, texture and even the fractal dimension can change with scale (a structure can be fractal without being scale invariant). The existence of "transient fractals" (for which the "dimension" changes with scale) is known in the fractal literature.

In principle, the concept of lacunarity should capture quantitative aspects related to the distribution of magnitudes of *holes* ("empty" spaces) in a fractal structure. Mandelbrot pointed out that there is no unique all-encompassing way to formally define a measure of lacunarity, and proposed to characterize texture properties of fractals using the behavior of their cover functions. In particular, the *prefactor oscillations of a Minkowski-Bouligand cover of a fractal*, can be used to describe *shell lacunarity*. However, save for some particular classes of mathematical fractals (e.g., Theiler 1988; Mandelbrot 1995), the properties of cover functions and their prefactors are poorly understood and a systematic approach to their study is lacking. Moreover, some cover functions may even be nondifferentiable and the presence of high lacunarity, has the effect of degrading the convergence properties of data analysis algorithms used to assert fractality and extract scaling exponents.

3. Using the Regular Variation Formalism for Fractal Texture

The aim of this note is to point out that the use of the mathematical Theory of Regular Variation (in the sense of Karamata) (RVT) and its extensions (de Haan theory) (Bingham et al. 1987), provides a formal framework to study more systematically the properties of cover function prefactors for fractal-like structures and thus, of fractal texture, including lacunarity. Some of the benefits of the RVT based approach are a) more precise characterizations of a variety of morphological features of fractal-like structures, not captured by fractal dimension, in particular, a taxonomy of fractal-like structures based on the "roughness" of their cover functions, b) means to quantify differences among fractals with the same fractal dimensions, in particular, a rigorous approach to shell lacunarity, c) bounds on rates of convergence for some basic algorithms used to evaluate fractal dimensions and other scaling exponents, and a base to develop novel algorithms for this purpose, and d) means to detect and characterize scale dependence using "fine" properties of cover function. The texture characterization methodology based on regular variation can be used for both, *mono-* and *multi*-fractals, as well as for deterministic and random fractals.

Below, a very brief overview of the main point of the RVT based approach is stated. The formal details and examples, including some astrophysical applications regarding the above mentioned points, will be presented in an extended version of this note (Stern 1997, in preparation).

Let $N_S(\epsilon)$ be the smallest number of balls of the same diameter ϵ, needed to cover a fractal set S. Then $N_S(\epsilon)$ is a *cover function* which provides the *box dimension* of S as $D_B = lim_{\epsilon \to 0} \ln N_S(\epsilon)/\ln(1/\epsilon)$, if the limit exists. The box dimension is the most common form of "fractal dimension" encountered in applications. Denote, $x = 1/\epsilon$; x is a convenient scale variable for the RVT approach. Asymptotically, the power-law form of this cover function, is, $N(x) \approx \wp(x)x^{\alpha}$, where $\alpha \geq 0$. In general, it is assumed in literature that the various cover functions used to extract scaling exponents are *power-law*-like functions. The main reason why RVT provides a natural framework to study cover functions is that it is a general mathematical theory focusing on asymptotic properties of functions under rescaling transformations of the argument. In particular, RVT provides a general theory of *power-law*-like behavior, including "correction factors" and other fine properties of such functions.

The function $\wp(x)$ has a slower growth than x and is called a *prefactor function.* For simple, self-similar fractals the prefactors, $\wp(x)$, are periodic functions in $\ln(x)$. However, various other forms of prefactor behavior reflecting textural properties of the fractal structure, are possible. This wide spectrum of behaviors can be understood and classified using RVT.

Let $C(x), x \in [x_0, \infty)$ be a generic cover function. The most basic type of behavior of $C(x)$ in RVT is known as *regular variation.* $C(x)$ is *regularly varying at infinity (RV)* if $\lim_{x\to\infty} f(\lambda x)/f(x) = \phi(\lambda)$, where $\phi(\lambda)$ depends only on λ, and has the form $\phi(\lambda) = \lambda^{\rho}$, $\rho \geq 0$, for any $\lambda > 0$. $C(x)$ is *slowly varying (SV)* if $\phi(\lambda) = 1$, i.e., $\rho = 0$. A slowly varying function cannot grow or decrease "as fast" as a power of x. The exponent ρ is called in RVT, the *regular variation index* of $C(x)$.

The "simplest" fractals, from RVT point of view, have a cover, $C(x)$, which is RV. Such fractals are called here *soft.* For soft fractals ρ coincides with the fractal dimension. Soft fractals are important for applications, since they have "low lacunarity" in the sense that the prefactors are SV functions, and many random fractals of practical interest are likely to be soft, because "randomness" may wash out to some extent strong prefactor oscillations. For soft fractals, a detailed formalism for convergence rates of algorithms for scaling exponents can be provided. The study of fractals with stronger prefactor oscillations than allowed by RV, needs the use of the more involved concepts of *extended regular variation* and *O-regular variation* (see Bingham et al. 1987; Stern 1997, in preparation).

4. A Toy Model for Cosmic Voids Based on Apollonian-like Packing.

The above sketched RVT approach can help to better understand the properties of solid packings of finite regions of 3-D space. The possibility of such a modeling approach to galaxy distributions was suggested by Mandelbrot in his celebrated book (Mandelbrot 1983). Consider, for example, the cover of such a region with nonoverlapping, tangent spheres of different radii ("cosmic voids"). The residual set remaining after removing these spheres is a geometrical construct which is known to be a geometrical multifractal, if the packing is "solid," that is, if the Lebesque measure of the residual set is zero. If the covering spheres are generated deterministically and recursively, the packing is "apollonian" (suggested by Apollon of Perga, 200 BC). Stochastic processes leading to solid random packings of this kind are also possible. The infinite sets of tangency points of the spheres in these constructions form multifractal point sets, providing nontrivial examples of fractal "hierarchical clustering" geometries, potentially relevant to distributions of galaxy counts. These sets can also serve as support for multifractal mass distribution models as well as, a class of generic toy models alternative to more familiar "beta"-type multifractal models common in the galaxy count distributions literature. A visual examination of such tangency point sets suggests that some clustering patterns in such point distributions can mimic, crudely, structures like "walls," "filaments," and "voids" on "all scales," etc. Few generic results exist in the mathematical literature regarding such point sets. For a 3-D apollonian covering with spheres, a crude estimate of the fractal dimension exists (Boyd 1973 gives the value of 2.4). An RVT based approach was undertaken to refine this estimate and investigate the lacunarity of such models, using the properties of a ranked series of radii of the 3-D covering spheres. The formal details and results will be reported elsewhere.

5. Conclusions

The specific texture of a fractal cannot be ignored in the study of fractal-like structures of interest for astrophysics. Textural features carry information about the generating dynamics of a fractal. Information about texture, contained in prefactors of cover functions can be analyzed if the sampling of the structure in "resolution space" is not too sparse. RVT provides a useful formal framework to understand the variety of possible prefactor behaviors and their impact on convergence rates of algorithms for scaling exponents.

A class of multifractal structures based on apollonian-like solid packings of space with spheres may provide an interesting modeling approach to some scaling properties of the large scale distribution of matter. The RVT based methodology, provides some novel ways to study such structures in more detail.

References

Bingham, N. H., Goldie, C. M., & Teugels, J. L. 1987, Regular Variation (Cambridge: Cambridge Univ. Press)

Borgani, S. 1995, Phys. Reports, 251, 1

Boyd, D. 1973, Mathematics of Computation, 27, 369

Labini, F. S., Gabrielli, A., Montuori, M., & Pietronero, L. 1996, Physica A, 226, 195

Mandelbrot, B. B. 1983, The Fractal Geometry of Nature (San Francisco: W. H. Freeman)

Mandelbrot, B. B. 1995, in Progress in Probability, Vol. 37, Fractal Geometry and Stochastics, ed. C. Bandt, S. Graf, & M. Zahle (Basel: Birkhauser)

Peebles, P. J. E. 1993, Principles of Physical Cosmology (Princeton: Princeton Univ. Press)

Theiler, J. 1988, Thesis, California Institute of Technology

Astronomical Data Analysis Software and Systems VI
ASP Conference Series, Vol. 125, 1997
Gareth Hunt and H. E. Payne, eds.

Using Massively Parallel Processing of a NLTE Spectrum Synthetic Code and an Automated Comparison with Observations to Determine the Properties of Type Ia Supernovae from their Late Time Spectra

R. Smareglia and P. A. Mazzali

Astronomical Observatory of Trieste, Via Tiepolo,11, 34131 Trieste, Italy

Abstract. The nebular spectra of SNeIa can be used to determine the mass, density, and distribution of the SN ejecta, and the distance and reddening to the SN. Synthetic spectra can very rapidly be computed in NLTE and compared to the observed ones. Given the relatively large number of parameters, it is convenient to explore the parameter space in an automated form. The nebular spectrum synthesis code can be run in parallel on a workstation cluster. Each synthetic spectrum is immediately compared to the observed one. Exploration of the parameter space can be guided by the information available on some quantities (e.g., line width). The results for a few cases of "normal" and possibly peculiar SNeIa are presented and discussed.

1. Introduction

In the nebular phase, the (emission) spectrum of a SN Ia is formed in the densest, innermost part of the ejecta, and is dominated by forbidden lines of Fe II and Fe III. Modeling this phase thus offers a unique opportunity to investigate the properties of the central region of the ejecta, where explosive nuclear burning is thought to be most efficient. Moreover, since nebular emission depends directly on the density, a successful model of this emission should yield values of the masses of the elements visible in the spectrum and, through the cooling effect of other elements, also of the total mass in the nebula. The parameter necessary to convert from density to mass is the velocity (the epoch is known), which can be inferred from the FWHM of the emission lines under the assumption of optically thin gas. Our model is based on a NLTE treatment of the rate equations for a nebula of uniform density. This simplifying assumption is reasonably well justified since all explosion models predict that the density near the centre of the ejecta is a much weaker function of radius than in the outer layers. The necessary input is then limited to the mass and composition of the ejecta, the sphere's outer velocity, and the time since explosion. Heating of the nebula comes from the deposition of energy generated in the radioactive β-decay of ^{56}Co into ^{56}Fe. This energy is produced in the form of γ-rays and positrons, which lose their energy in the ionization of the gas and in heating the plasma through collisions with electrons, in a cascade process. Eventually, a fraction of the produced energy is thermalized, and the heated gas cools by net emission of radiation. The electron density and temperature, as well as the population of the atomic levels and the emission line fluxes, are determined simultaneously from the conditions of statistical and thermal equilibrium.

The synthetic spectra are then corrected for the distance and reddening to the observed SN, and compared to the observed spectrum.

The basic treatment of nebular emission in a SN Ia was given by Axelrod (1980). A similar version of the code used here was employed by Ruiz-Lapuente & Lucy (1992), to which the reader is referred for further details. Our atomic data for the nebular spectrum model are from the sources quoted in Ruiz-Lapuente & Lucy (1992), with the exception of the [Fe II] collision strengths, which were obtained from Pradhan & Zhang (1993).

A further refinement included in this work is the introduction of the possibility that the gas is not distributed uniformly within the sphere. This is expressed in the form of a filling factor, which represents the fraction of the volume occupied by the gas. The effect of using a filling factor smaller than unity is to increase the local gas density, and thus to change the ionization regime.

In practice, the epoch and the velocity give the size of the sphere, while the mass and the filling factor give the density. The Ni mass gives the energy input. Since the epoch is known, and since we adopt a constant composition, apart from the decay of ^{56}Ni into Co and Fe, there are three parameters that can be changed in the building of the physical model: velocity, mass, and filling factor. The remaining two parameters, distance and reddening, can be changed when comparing a synthetic model to an observed one. In this paper, we report on the modeling of a normal SN Ia: SN 1992A.

2. Implementation on a Parallel Computer Network

There are five unknown parameters: velocity V, mass M, filling factor ff, distance μ, and reddening $E(B-V)$. We need to determine three of these to establish the SN proprietes. The grid can be set with a knowledge of the most likely range of properties for a normal SN Ia, while the distance and reddening grids are determined on the basis of observational estimates for the object under study, as show in Table 1.

Table 1. Range of values

Parameters	Range	Step
V	6000–11000	500
M	0.3–1.0	0.05
ff	0.1–1.0	0.1
μ	30.0–33.0	0.25
$E(B-V)$	0.0–0.2	0.02

We have a total 235,950 possible models. A single model on an high performance (57 MFLOPS) workstation such as an HP 735/125, takes 6 seconds, plus the time for the comparison with the observed spectrum. The simulation program can be split into two parts:

- **Physical model:** the spectrum at the source is created using the parameters (M, V_b, ff). This requires about 5 cpu seconds.
- **Synthetic spectrum:** the physical model is corrected for the observational parameters (μ, $E(B-V)$). This requires about 1 cpu second.

Finally, a χ^2 determination is used to check the goodness of the fit.

We used a very simple "master/slave" parallel architecture based on the LAM/MPI software:

Master: it first checks the available guest hosts, then it organizes the sharing of the tasks among the slaves. Every single task has a defined value for the parameter triplet (M, V_b, ff). When the entire parameter space has been scanned, the Master organizes all results into a single ASCII file containing the model parameters and the fitting test results.

Slave: it acquires the values of (M, V_b, ff), and creates the "physical spectra." It then uses an IDL procedure to convert from the "physical spectra" to the "synthetic spectrum," for values of μ and $E(B-V)$, which is compared to the observed spectrum. The results of the fit, which can be applied to the entire spectrum or to a particular region, are passed back to the Master.

Trieste Astronomical Observatory has a heterogeneous workstation environment, with 6 Sun and 14 HP computers available. For this work we used 9 homogeneous HP computers with a total power of 273 MFLOPS. The full computional work (i.e., simulation plus best fit test) was run at night at low priority and required about 16 hours, against the 493 hours which would have been required to run it sequentially.

3. Conclusions

In the case of SN 1992A, the velocity was first determined as $v = 8500\,\mathrm{km\,s^{-1}}$ from fitting the line width. Since the (small) reddening is known, ($E(B-V) = 0.02$), good fits can be obtained for different values of the mass and the distance, if different values of the filling factor are used.

The interesting result is that the best fits for the various values of ff are practically indistinguishable, as shown in Figure 1. Some physical insight must therefore be applied. If we assume that we know from explosion models the mass ejected with velocities below the observed value, then an estimate of the most likely values of distance and filling factor can be derived (Figure 2).

In the standard SNIa explosion model (W7, Nomoto et al. 1984), $1.4\,M_\odot$ of material is ejected, $0.7\,M_\odot$ of which has velocities below $8{,}500\,\mathrm{km\,s^{-1}}$. The best fitting model for this value of enclosed mass has μ ~31.5 mag and ff ~0.55.

Using the tools available at the Observatory (IDL and a distributed password system), and a public domain software package (LAM/MPI Parallel Computing), it is possible to create, optimize, and automate the comparison of synthetic spectra with observed Supernova spectra. The scientific user must then decide (as pointed out above) the correct physical properties if best fits having similar appearance, but different physical characteristics, are present.

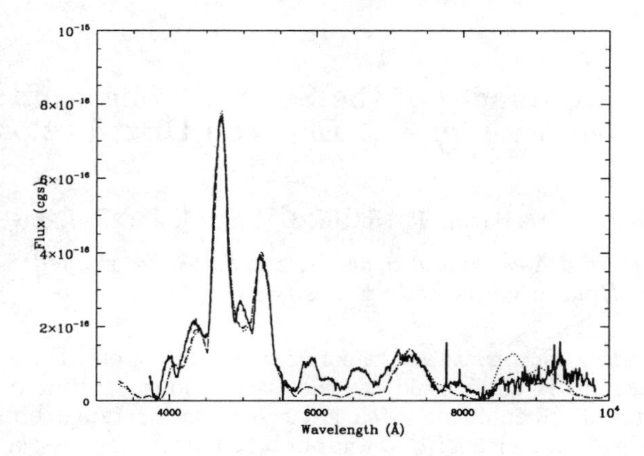

Figure 1. The spectrum of SN 1992A (continous line) compared to the best fitting models for $ff = 0.5$ (dotted line) and $ff = 1.0$ (dash-dotted line). The corresponding values of mass and distance can be read from Figure 2.

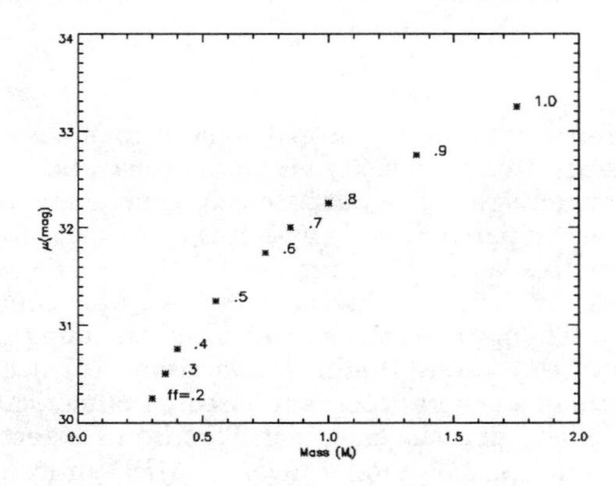

Figure 2. Best fit for various filling factors.

Acknowledgments. The technology group of the OAT is gratefully acknowledged for the support given during this work. The project for parallel implementation of scientific codes has been financed by a CRA grant.

References

Gregory, D. B., Raja, B. D., & James, R. V. 1994, LAM: An Open Cluster Environment for MPI, Supercomputing Symposium '94, Toronto

Nomoto, K., Thielemann, F. K., & Yokoi, K. 1984, ApJ, 286, 644

Ruiz-Lapuente, P., & Lucy, L. B. 1992, ApJ, 400, 127

Astronomical Data Analysis Software and Systems VI
ASP Conference Series, Vol. 125, 1997
Gareth Hunt and H. E. Payne, eds.

Synthetic Images of the Solar Corona from Octree Representation of 3-D Electron Distributions

D. Vibert, A. Llebaria, T. Netter, L. Balard, and P. Lamy

Laboratoire d'Astronomie Spatiale, BP 8, 13376 Marseille CEDEX 12, France, E-mail: vibert@astrsp-mrs.fr

Abstract. Empirical and theoretical modeling of 3-D structures in the solar corona is confronted by the tremendous amount of data needed to represent phenomena with a large dynamic range both in size and magnitude, and with a rapid temporal evolution. Octree representation of the 3-D coronal electron distribution offers the right compromise between resolution and size, allowing computation of synthetic images of the solar corona.

1. Introduction

The photometric representation of coronal structures is based on a geometrical model of the spatial electron density in the corona and computation of the corresponding scattered light. The purpose is to generate synthetic photometric images which can be compared directly with images of the corona obtained either during eclipses or with space coronagraphs.

Density models for the coronal structures have traditionally been obtained by inversion of the Thomson scattering integral, assuming *a priori* shapes of the modeled structures. Later, Bohlin & Garrison (1974) adopted the direct approach in the form of a general program based on numerical integration along specified lines of sight to find the emergent Thomson scattered white light.

Today, the coronal plasma model involves MHD processes. However, numerical physical models of the corona cannot be implemented for high resolution in size and magnitude with classical finite element methods, due to the amount of data one has to cope with. In order to be able to simulate images of the corona at high resolutions we concentrate on the simulation program, improving the methods of Bohlin & Garrison, and incorporating an octree representation of the electron density.

In this paper, we first explain the coronal spatial electron density model and the equations to compute the scattered white light. We then describe the octree representation and coding used, and the algorithms used to integrate along a line of sight through the octree data structure. Finally, we present a sequence of computed images displaying the corona over a solar rotation.

2. The Model

2.1. Principle

Given the electron density in the coronal plasma, we compute the photospheric light scattered by the coronal electrons:

$$B(x,y) = C \int_{-\infty}^{+\infty} N_e[2A - (A - B)\sin^2\theta]dz \tag{1}$$

where N_e is the electron density, C is a constant depending upon electron scattering cross-section and mean solar brightness, A and B are two coefficients derived from the limb-darkening of the Sun and the distance from the electron to the Sun center, dz is a path element along the line-of-sight, and θ is the scattering angle—that is, the angle between the line-of-sight and the line from Sun center to scattering point.

2.2. The Models

The models attempt to explain the observed coronal streamers as structures associated with the neutral magnetic sheet of the Sun, seen edge-on (Saito et al. 1993).

The corresponding electron density is written as a product of two independent functions:

$$N_e = N_{radial}(r) \times N_{shape}(d). \tag{2}$$

N_{radial} is the radial decrease along the axis of the streamer and depends only upon r, the distance from the considered point to the Sun center. This function has been derived from eclipse observations of streamers. N_{shape} represents the shape of the neutral sheet and is restricted to a variation with the distance d from the considered point to the neutral sheet. We have chosen for N_{shape}:

$$N_{shape}(d) = e^{-(d/d_0)^4}, \tag{3}$$

a smooth function sharper than a simple Gaussian. The parameter d_0 represents half the "thickness" of the streamer.

The shape of the neutral sheet is then defined as the radial extension of the neutral line observed on the photosphere, for one Carrington rotation.

3. The Octree Representation

3.1. Octree Structure Description

An octree is a tree-structured representation that can be used to describe a set of volumetric data enclosed by a bounding cube. The octree is constructed by recursively subdividing each cube into eight sub-cubes, starting at a single large cube represented by the root-node in the tree. A cube will have descendants only if its associated volume of object space is not homogeneous. The recursion continues until either all sub-cubes are homogeneous or until the required resolution is achieved.

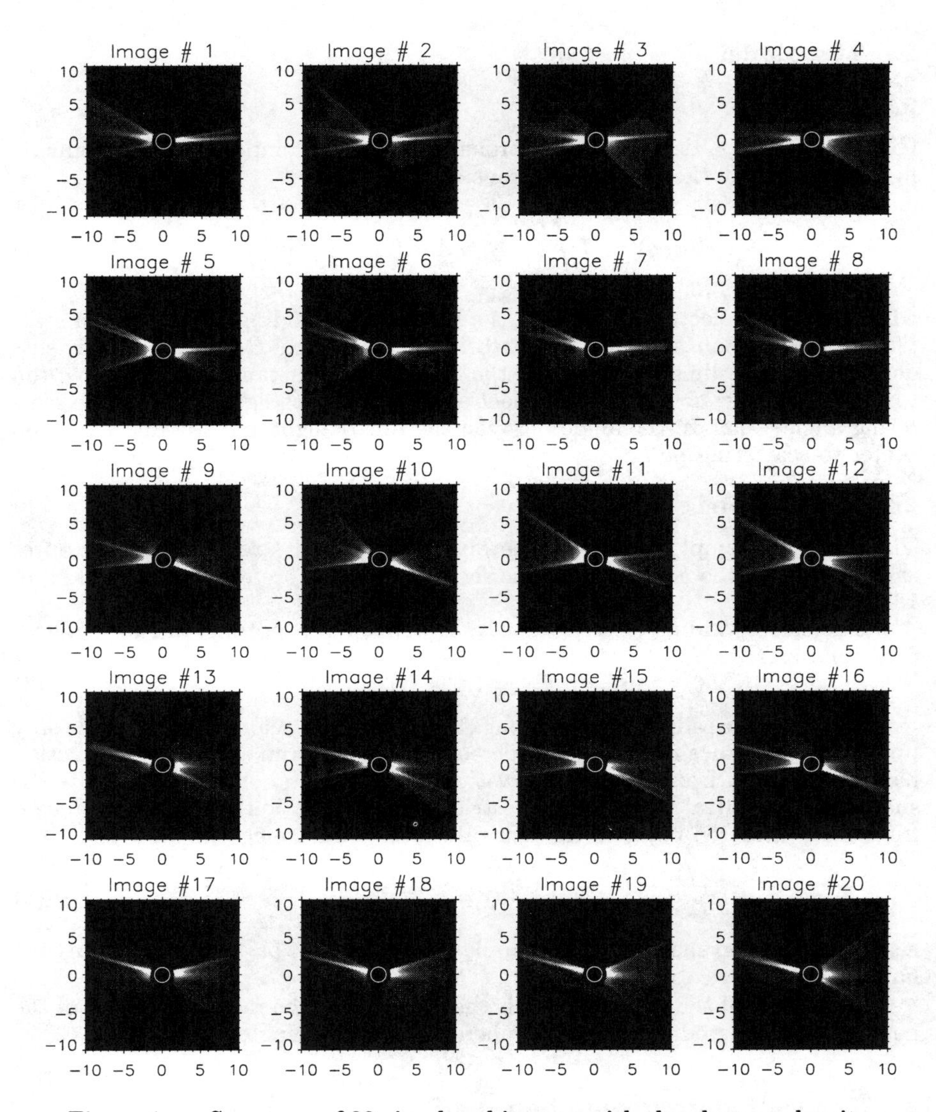

Figure 1. Sequence of 20 simulated images with the electron density based upon the neutral sheet for Carrington rotation 1901 (20 Sep–27 Oct 1995). Frame increases first from left to right and second top to bottom. The time interval between images is one day. The rapid radial decrease in brightness has been compensated. Source for the neutral line: Solar Geophysical Data, no. 617 (January 1996), synoptic chart of the source surface field on p. 39.

3.2. The Octree Encoding

A number of efficient schemes have been developed to represent and encode octrees. The S-tree type (Jonge, Scheuermann, & Schijf 1994) linear encoding was implemented in this work. The tree is built directly from the continuous expression of the coronal electron density given by the mathematical expressions (Eq. 2). The subdivision criterion for a cube is the variation of the electron density within the cube.

4. The Octree Ray-tracing Algorithm

The traversal of the octree data structure to determine the leafs hit by a line-of-sight relies on the HERO algorithm (Agate, Grimsdale, & Lister 1991). At each level of the octree, the algorithm generates the addresses of child voxels (volume elements) in the order they are penetrated by the ray, avoiding the time consuming test of determining the intersection between the ray and the eight children for every node in the tree.

5. Results

Such octree coding and ray-tracing methods lead to significant improvements in terms of resolution, compression, and speed. The generation of an octree of size $1024 \times 1024 \times 1024$, based on smallest voxel size, with 4 byte floating point values, takes approximately two hours (using a standard low cost workstation with 64 MB of memory) and requires 60 MB of disk space. The same data uniformly sampled at the same resolution would have required 4096 MB.

The ray-trace reconstruction of the former octree-based data produces white light images with a resolution of 512×512 pixels, in less than half an hour. A uniform sampling (e.g., by 1024) of every line-of-sight (every pixel) would have led to a far longer computing time for the same output.

This relatively short time necessary to compute one image allows to compute a temporal sequence of images showing corona aspect variations with Sun rotation (Figure 1).

References

Agate, M., Grimsdale, R. L., & Lister P. F. 1991, in Advances in Computer Graphics Hardware IV, ed. R. L. Grimsdale & W. Strasser (Dordrecht: Springer-Verlag), 61

Bohlin, J. D., & Garrison, L. M. 1974, Solar Physics, 38, 271

de Jonge, W. P., Scheuermann, P., & Schijf, A. 1994, Comput. Vision Graphics and Image Processing, 24, 1

Saito, T., Akasofu, S. I., Takahashi, Y., & Numazawa, S. 1993, Journal of Geophysical Research, 98, 5639

Astronomical Data Analysis Software and Systems VI
ASP Conference Series, Vol. 125, 1997
Gareth Hunt and H. E. Payne, eds.

Error and Bias in the STSDAS fitting Package

I. C. Busko
Space Telescope Science Institute, Baltimore, MD 21218,
E-mail: busko@stsci.edu

Abstract. The *fitting* package in STSDAS (Space Telescope Science Data Analysis System) relies on two basic techniques (one linear and one non-linear) for fitting functions to data. In this work, the statistical properties of both the fitted function coefficients and their errors are examined, using a Monte Carlo approach. Results show that both methods may generate biased coefficients (at the ~few percent level), and over or under-estimate error bars (at the ~10 percent level), in particular from low signal-to-noise data.

1. Introduction

The two basic techniques used in the STSDAS *fitting* package are: (*i*) linear functions (Legendre and Chebyshev polynomials, cubic splines) are fitted by minimizing χ^2, solving the normal equations by the Cholesky method. Function coefficient errors are computed directly from the covariance matrix. This technique is provided by the IRAF (Image Reduction and Analysis Facility) library *curfit*. (*ii*) *any* function, linear or non-linear in its coefficients, can be fitted by minimizing χ^2 using the *downhill simplex* method, also known as *amoeba*. This method is unable by design to compute coefficient errors. The package relies on an independent technique (*bootstrap resampling*) to estimate these errors.

This work aims at assessing the reliability of coefficient estimates and error bars generated by both methods.

2. Method

Artificial data sets were generated from two functions: a 3rd degree power-series polynomial and a sum of two Planck functions. Each function was replicated 30 times and in each replication an independent noise realization was added at a fixed signal-to-noise level. Noise types included pure Gaussian, pure Poisson and a mixture of Gaussian plus Poisson. A separate, 30-element data set was created for each signal-to-noise level studied. Signal-to-noise spanned the range from ~1000 down to ~1 in logarithmic steps.

Each individual data set was fitted by the appropriate *fitting* task and results were output to tables. The power series polynomial was fitted by both linear and non-linear methods.

From the 30 measurements of each function coefficient c_i, and its estimated error e_i, $i = 1, 2 \ldots 30$, three statistics were computed:

- The average of coefficient error estimates $\overline{e} = 1/30 \sum_i e_i$

- The standard deviation of coefficient measurements $\sigma_c = \sqrt{\sum_i (c_i - \overline{c})^2/29}$
- The average "residual" coefficient (measured minus true) $c_{bias} = 1/30 \sum_i (c_i - c_{true})$

If the coefficient computation is unbiased, c_{bias} should distribute itself around zero. If any bias is present, the average of the c_{bias} distribution will depart from zero. On the same grounds, if the coefficient error estimates are unbiased, that is, if they reflect the population true standard deviation, the difference $\overline{e} - \sigma_c$ should also distribute itself around zero.

3. Results

Results are summarized in the Figures.

When applied to a power-series polynomial with Gaussian noise, both linear and non-linear techniques generated unbiased coefficients at the level of $< 1\%$ for any tested S/N ratio. Poisson noise introduced underestimation bias at a level of a few percent for lower (~2–5) S/N data. The largest bias was seen on polynomial's zero-order term, induced perhaps by the non-symmetric nature of the Poisson distribution. A different behavior was seen when fitting the strongly non-linear sum of two Planck functions. The dominant (in intensity) black-body had its temperature determined with almost no bias down to S/N $\sim$ 2. The weaker component, however, showed significant overestimation bias at low S/N. The dominant black-body's amplitude showed large (~8–10%) underestimation bias, and the weaker amplitude did not give significant results. This can be interpreted as the result of both amplitudes being confused into a single one by the fitting algorithm, thus resulting in a biased estimate for one of them. The noise model seems to play no role in these results.

When fitted by the non-linear algorithm, the power-series polynomial errors showed a systematic underestimation of ~10–20 percent, seemingly independent of noise type. The linear algorithm, on the other hand, delivers errors which are off from the "true" ones by amounts that depend on noise type, and might also depend on the coefficient values themselves. The double black-body function fit showed errors that lie close to the true ones in the Gaussian noise case, and with systematic overestimation of ~20 percent in the Poisson noise case.

As a general rule one might say that bias at a few percent level should be expected when fitting either non-linear functions, or linear functions with Poisson noise, in particular with low S/N data. Also, error bars generated by both linear and non-linear methods are prone to under or overestimate the "true" errors by as much as 10–20 percent, even with high S/N data. The details, though, seem to be dependent on the functional form and noise type.

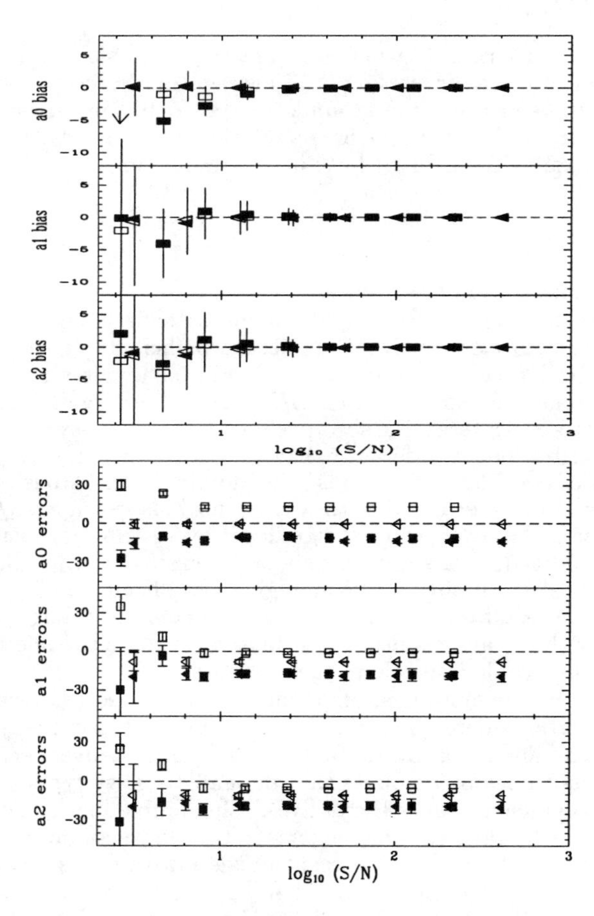

Figure 1. Power-series polynomial bias and errors. Ordinate is "estimated – true" coefficient residual (upper panel) and error bar (lower panel), in a relative (percent) scale, for the three first lower-order polynomial coefficients. Abscissa is signal-to-noise. Each point depicts the average of 30 measures; error bar depicts the standard deviation of the same 30 measures. Solid symbols: non-linear algorithm. Open symbols: linear algorithm. Squares: Poisson noise. Triangles: Gaussian noise. Points below the zero line mean that the coefficient (or its error) is systematically underestimated; above the line it is overestimated.

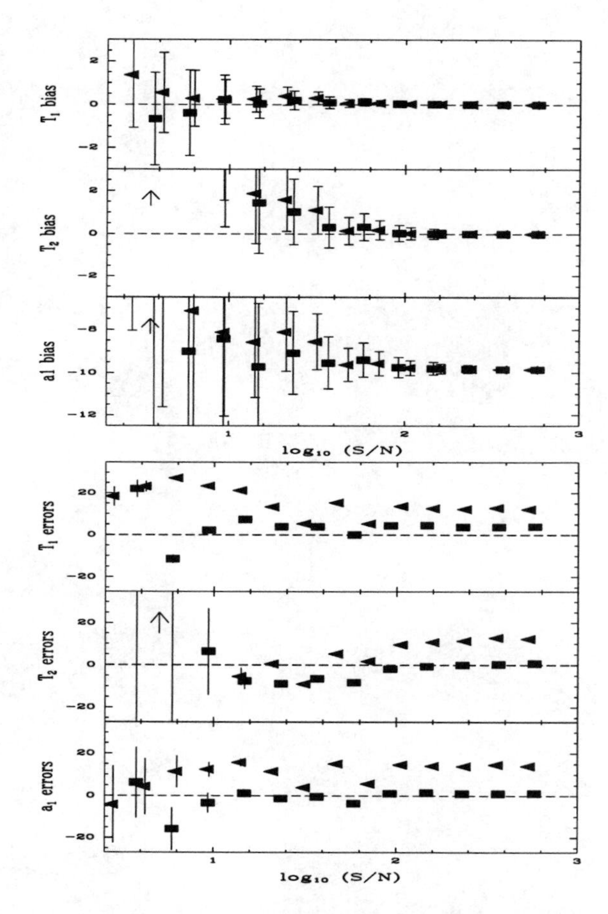

Figure 2. Double Planck function bias and errors. The two temperatures and the largest black-body amplitude are depicted. See caption for Figure 1.

Part 5. FITS—Flexible Image Transport System

Astronomical Data Analysis Software and Systems VI
ASP Conference Series, Vol. 125, 1997
Gareth Hunt and H. E. Payne, eds.

Practical Applications of a Relational Database of FITS Keywords

De Clarke and S. L. Allen

UCO/Lick Observatory, Kerr Hall, UCSC, 1156 High Street, Santa Cruz, CA 95064

Abstract. In the course of designing a major mosaicked instrument (DEIMOS) we need tools to manage the large set of FITS keywords required for instrument control and data reduction/archiving. We have therefore constructed a relational database schema which describes FITS keywords, and have implemented it using Sybase. In this schema FITS keywords and structures are entities, having attributes such as format, units, datatype, minimum and maximum value, semantics, etc. The schema was expanded and generalized to document information flow and database schema (becoming self-referential). Many useful output products (documentation, diagnostics, configuration, and source code) can be generated from one authoritative source of keyword information.

1. Introduction

Using a centralized database or information service to document FITS keywords is not a new idea. However, in previous incarnations of this concept, the object has either been purely documentary or single-purpose (e.g., the SaveTheBits database schema). Our goal was to capture enough information, and to generalize the information sufficiently, that multiple related applications could all use one authoritative, online reference database for information about FITS keywords. For example, live observing and engineering interfaces to an instrument could rely on the same database that was used to generate printed documents, drawings, and even sections of source code.

A more complete documentation of our work, with examples, and live demos, can be found online at Memes (Keywords) Database Homepage.[1]

2. Keywords and Memes

The first challenge was to determine the list of attributes necessary to document a FITS keyword. Some, such as name, FORTRAN datatype, semantics, and min/max values, were obvious. Others were more subtle, and the list of attributes was revised several times as we explored the syntax and semantics of existing FITS headers and our proposed DEIMOS instrument keywords.

[1] http://www.ucolick.org/~de/deimos/Memes

FITS keywords have *provenance* or *context* (i.e., an institution, instrument, author, or standards document which defines their valid use). Knowing the context is essential, given the lack of constraint on duplication in FITS namespace.

FITS keywords have relationships to other keywords in the same dataset (for example, the value of NAXIS controls the occurrence of NAXISn). We developed attributes for a generalized description of these relationships.

FITS keywords occur in groups or "bundles"; a standard set of required keywords must appear in a valid table extension header, for example. A FITS header itself is a special case of such a bundle, as is a FITS table. We established ancillary tables to describe the grouping or bundling of keywords.

We found it necessary to establish special types of keywords called "tuple keywords," whose values are actually a list or other parsable structure of subvalues. We established a general-purpose way of documenting such tuple values, which handles any such keyword by defining subsemantics and subformats, separators, delimiters, etc. We accommodated the case where identical semantics are repeated N times, and the case where N sub-elements have distinct semantics.

Given that our instrument and telescope control systems use a keyword/value architecture, we needed to establish access control for keywords: some are writable, some readable, some are both. Access control attributes are needed for the configuration of dynamically-generated graphical user interfaces.

We found it necessary to establish a special relationship between certain keywords and an "archetypal" keyword of the same semantics. For example, one instrument system may format a value as F8.4 where another will format it as F6.4 or as a string. It is evident that these are really "the same" keyword.

As we struggled with subsemantics and archetypes, it became clear that we were documenting something more abstract than "a FITS keyword." We borrowed the term "meme" to describe "a unit of meaning"—a more general-purpose name for the entities we were manipulating.

This generalization led swiftly to the use of the "Memes" database to document itself, since the fields (columns) of a database table are memes with a large subset of the attributes of a FITS keyword (such as name, datatype, semantics, etc.). Exploration of the relationship between database tables and Meme bundles led equally swiftly to automated translation between these formats.

In other words, we are able to extract (outload) a Sybase table as a FITS table extension, and to import (inload) a defined FITS table extension to a Sybase table. We are able to generate Sybase table definitions for any meme bundle, including standard FITS headers and FITS table extensions. There are many practical applications for these abilities (see Table 1).

Although we cannot adequately discuss the Memes schema within the present page limit, we invite the FITS community to investigate the Memes table definition and related tables, using the demo reports available at the Memes (Keywords) Database Demos[2] page. If you choose "Documentation of Memes by Name" and enter the name Memes, the resulting report will be the current Memes table definition. We propose these attributes as the first draft of a standard set of attributes of FITS keywords, and invite comments from the FITS community.

---

[2]http://www.ucolick.org/~de/deimos/Memes/demos.html

Table 1. Table of Output Products

	Product	Practical Application
1	HTML documentation	We can generate (dynamic) documents describing individual keywords, database tables, FITS tables, and FITS headers. We plan to use such documents for design review requirements. These documents can be regenerated quickly whenever design decisions are changed. The same code generates documents of the database schema itself.
2	Sample FITS headers	We can generate sample (fake) FITS headers which can be used to test FITSIO code.
3	C source	We can generate C source for data structures (such as would be needed to store any given FITS header or table), and for FITSIO writer code (reader code is in progress).
4	Graphic FITS documentation	We can generate digraph-type graphical representations of FITS headers, showing individual keywords and pre-defined bundles.
5	Sybase DDL	We can generate Sybase table definitions (any other database engine could also be easily supported) for the dynamic creation of tables to store data from FITS headers.
6	FITS tables	We can generate conformant FITS table extensions containing data from Sybase tables; we can load data into Sybase tables from FITS table extensions.
7	Information Flow Graphs	We can generate flow charts documenting the movement of information between agents in a system. This generalized tool can document instrument control and data reduction; it also documents itself, or any other system which can be described by the passage of information elements between agents.
8	Online Access (code)	The online Meme database is being used as the live reference for DEIMOS interface prototypes. When the interface user wishes to monitor one or more keywords, the interface software generates a control panel with graphical meters and other widgets according to the attributes of the keywords being monitored. Keyword attributes need not be hardcoded into the interface, and the interface will always adapt itself to engineering changes if those changes are documented via the Memes database.
9	Online Access (user)	We expect to use the online database as a reference and documentary tool for users of the instrument, during actual observing. We expect that the dynamic generation of flow control graphs will assist in diagnosis of problems and improve user understanding of the instrument and its output.

3. Information Flow

The schema was expanded to include a set of ancillary tables describing protocols, formats, event timings, and agents (software, hardware, and human) so that we could document the flow of information through any large system.The Agents schema permits us to document the source language, revision, authorship, and other attributes special to software agents. (This model corresponds well with the "gaming table" model of Noordam & Deich 1996.) As with Memes, a URI field permits the attachment of detailed documentation to any Agent.

A table of Paths documents the passage of Memes between Agents. The attributes of a Path are the sending and receiving agent, a controlling agent, the Meme ID, and a cluster of attributes describing the transaction: format, protocol, event timing, and elapsed time. Since human beings can be agents and Key Entry is an acceptable protocol, it is possible to document the Memes/Agents toolset itself using this schema.

We use a digraph generation package (see Acknowledgments) to generate graphical representations of information flow. A hierarchy of agents and paths was required in order to generate both simplified and detailed drawings. Accordingly, the schema was adjusted to permit the definition of "superAgents" which consist of multiple Paths.

The Agents/Paths database can be used not only to generate functional diagrams, but to assist in diagnostic procedures. It can be used to trace any FITS keyword value back to the originating or authoritative source; conversely, it can reveal which agents handle a keyword and thus may be affected by changes to the syntax or semantics of that keyword; it can reveal which keywords are handled by an agent, and which "downstream" agents may be affected if an agent is disabled or altered.

Currently we are addressing the distinction between syntactic and functional relationships among Memes. We have addressed, for example, the fact that the value of NAXIS controls the appearance of keywords NAXISn (a syntactic relationship). We have now to model those relationships in which a set of FITS keywords acts (in use, rather than in syntactic specification) as a state engine. This is particularly relevant for instrument control systems, in which (for example) a HATCHCLO keyword and a HATCHOPN keyword cannot both be true, but may both be false. The solution to this problem will almost certainly require one additional ancillary table in the schema.

Acknowledgments. We would like to acknowledge the invaluable contributions of the free software community, with particular gratitude to Tom Poindexter (Talus Technologies) for the SybTcl package, Stephen North and Eleftherios Koutsoufios (Lucent) for their digraph toolkit, Per Cederqvist and friends for CVS, John Ousterhout (Sun) for Tcl/Tk, and Mark Diekhans (Information Refinery) and Karl Lehenbauer (Neosoft) for TclX.

References

Noordam, J. E., & Deich, W. T. 1996, in ASP Conf. Ser., Vol. 101, Astronomical Data Analysis Software and Systems V, ed. G. H. Jacoby & J. Barnes (San Francisco: ASP), 229

Astronomical Data Analysis Software and Systems VI
ASP Conference Series, Vol. 125, 1997
Gareth Hunt and H. E. Payne, eds.

Multiple World Coordinate Systems for DEIMOS Mosaic Images

S. L. Allen, De Clarke
UCO/Lick Observatory

Abstract. The Deep Extra-galactic Imaging Multi-Object Spectrograph (DEIMOS) under construction for Keck II will employ a detector comprised of a mosaic of eight CCDs. World coordinate system (WCS) information is needed for mapping between FITS pixels and many other systems. To permit an arbitrary number of WCS mappings, we extend the FITS metaphor by placing the WCS information in a FITS table. For the purposes of simulation and planning, the scheme can be generalized to accommodate mappings between non-FITS coordinates.

1. WCS Needs of the DEIMOS Instrument

DEIMOS[1] will be installed on the Nasmyth platform of the Keck II telescope in 1998. The instrument will sample a 16×5 arc-minute off-axis region of the telescope field for imaging and multi-slit spectroscopy. The slitmasks will be manufactured from flat metal sheets; these are curved cylindrically while in use to approximate the shape of the Nasmyth focal surface. The detector will be a mosaic of eight 2048×4096 pixel two-amplifier CCDs.

When examining data from the instrument, the observer will need a real-time conversion from FITS pixels to numerous other coordinate systems. Data reduction pipelines and archival data users will also need this WCS information. Coordinates of various CCD defects will be constant per amplifier and/or per chip. Vignetting effects will be constant in the overall dewar focal plane. Effects of the hexagonal Keck pupil depend on the telescope elevation and DEIMOS position angle. Identification of slitlets requires coordinates on the slitmask. Identification of objects requires coordinates on the sky. The guide camera FOV will sample subregions of the slitmask and CCD mosaic. Identification of spectral features requires calculation of dispersion. For the sake of simplicity, DEIMOS will use a single scheme for manipulating, transmitting, and archiving all of the WCS mappings.

2. Adaptation of Existing WCS Schemes

A single WCS mapping is easily documented using the formalisms of the original FITS paper (Wells et al. 1981), and these formalisms are widely recognized by FITS image viewers. The eight-character keyword name space of FITS headers

[1] http://www.ucolick.org/~deimos/

limits the number of WCSs which can be stored within an image HDU. The conventions in the draft WCS proposal (Greisen & Calabretta 1996) permit up to nine different mappings from FITS pixels to sky, and they store the information in repeating groups of FITS indexed keywords. These draft conventions are not intended to provide the generality necessary for conversions from FITS pixel to instrumental coordinate systems, nor do they address the issues of conversion between coordinates where neither system is FITS pixels.

Some of the data products from the DEIMOS mosaic will need more than nine complete WCS mappings. Our initial attempts to describe these within the image HDU involved keywords with unilluminating names consisting largely of indexing digits. The keywords were also incapable of handling the amount of WCS data which could be generated by mosaics more complex than DEIMOS.

As an alternative, we store the bulk of the WCS information in a separate FITS table associated with the image HDUs. This is similar to the scheme used in the HST Data Archive of WF/PC images,[2] but for DEIMOS we use standard FITS tables rather than adaptations of FITS random group parameters. We thus avoid the problem of consuming large amounts of the precious name space of indexed keywords in FITS headers.

The DEIMOS design includes a relational database (RDB) as an active component of its operation. The documents defining FITS tables (Harten et al. 1988; Cotton et al. 1995) permit a very close mapping between astronomical data sets and RDB tables. We treat the separate WCS FITS table as an instance of a RDB table. We treat the keywords in the image header as if they were fields from a single row of another RDB table (one example of such a table would be the header catalog of the NOAO Save the Bits[3] archive project). We follow the principles of good RDB design (Date 1990) by avoiding the use of repeating groups of WCS information in the image HDU.

3. Operational Principles

Consider the simple case of operation with only a single FITS array. The principal WCS, typically the sky, is stored in the image header. The WCS table contains the same principal WCS as the fields of one row, plus other rows with mappings from FITS pixels to various instrumental systems. In addition to the fields which contain the WCS information there is another field, `WCS_DEST`, which serves as part of the primary key of the WCS table. Values for `WCS_DEST` can indicate a reference frame for the sky (e.g., `FK5_2000`) or for the instrument (e.g., `SLITMASK`). The user interface will permit selection of any desired WCS from a menu of available values of this key.

Actual DEIMOS FITS files will consist of numerous IMAGE and TABLE extensions. Each sub-raster readout region of each CCD will be stored with a unique value of the keyword `SRRID`. A single WCS table will contain mappings for all the IMAGE extensions; it will include the column `SRRID` to serve as another component of the primary key used to select the WCS coefficients.

---

[2]http://archive.stsci.edu/keyword/

[3]http://iraf.noao.edu/projects/stb/

The WCS table contains at least the following fields: SRRID, WCS_DEST (as composite primary key); CRPIX*n*, CRVAL*n*, CDELT*n*, CTYPE*n*, CUNIT*n*, CROTA*n*, CD*i_j*, CD*iiijjj*, PC*iiijjj*, RADECSYS, EQUINOX, and MJD-OBS (parameters for the WCS). Note that we can supply information as may be needed by both the traditional and draft WCS specifications. We can also include additional fields, e.g., a field which documents the provenance of the rest of the WCS information. Thus the WCS table scheme permits archival documentation of arbitrary numbers of WCS mappings for the FITS pixels of a given image.

4. Extension to Non-FITS Coordinates

The observation planning and simulation software for DEIMOS shares many WCS requirements with the archival FITS software. In particular, slitmask fabrication requires the mapping between CCDs, sky, and slitmask. Neither of the latter two of these systems is represented as FITS pixels. Because there are eight separate CCDs, it is impractical to implement the mappings from CCD to sky and CCD to slitmask as a two-step process. Instead, we intend to use the existing WCS formalisms to encode mappings such as the one between slitmask and sky.

The origin for these mappings is not an actual FITS image, and we must specify the information which permits proper selection and interpretation of the simulation WCS tables. The field WCS_ORIG is added as another component of the primary key. WCS_ORIG specifies the original reference frame whenever it differs from FITS pixels. We always place the sky on the WCS_DEST side of the mapping, and pixels on the WCS_ORIG side. Thus we preserve the general sense of FITS WCS mappings. In the simulation WCS tables we must also provide bounding information to serve the purpose filled by NAXIS*n* in actual images. And we include CCDLOC and AMPLOC to indicate individual elements of the CCD mosaic.

5. Discussion

Work for the DEIMOS project (Clarke & Allen 1997) has led us to explore various other possibilities of employing FITS files as portable instances of RDB tables. We require only a few restrictions on the structure of FITS tables and image headers, and the attributes of RDBs provide interesting possibilities for user interfaces. We find that this provides sufficient versatility to outweigh the drawbacks of separating WCS information from the image HDU, and these same tables continue to perform their traditional archival role. We are experimenting with a generic scheme for classifying FITS keywords and table columns as primary keys and foreign keys. We expect that further examination of the relationships between FITS tables and RDBs could lead to general formalisms for grouping FITS HDUs.

Acknowledgments. The construction of DEIMOS is supported by the National Science Foundation, the Center for Particle Astrophysics, and the California Association for Research in Astronomy.

References

Clarke, De, & Allen, S. L. 1997, this volume, 241

Cotton, W. D., Tody, D. B., & Pence, W. D. 1995, A&AS, 113, 159

Date, C. J. 1990, An Introduction to Database Systems Vol. I (Reading, MA: Addison-Wesley)

Greisen, E. W. & Calabretta, M. 1996, Representations of celestial coordinates in FITS[4]

Harten, R. H., Grosbøl, P., Greisen, E. W., & Wells, D. C. 1988, A&AS, 73, 365

Wells, D. C., Greisen, E. W., & Harten, R. H. 1981, A&AS, 44, 363

[4]http://fits.cv.nrao.edu/documents/wcs/wcs.all.ps

Astronomical Data Analysis Software and Systems VI
ASP Conference Series, Vol. 125, 1997
Gareth Hunt and H. E. Payne, eds.

WCSTools: Image World Coordinate System Utilities

Douglas J. Mink

Harvard-Smithsonian Center for Astrophysics, Cambridge, MA 02138,
E-mail: dmink@cfa.harvard.edu

Abstract. The WCSTools are portable C utility programs and subroutines for setting and using the world coordinate system (WCS) of a FITS or IRAF image. The WCS describes the relationship between sky coordinates, such as right ascension and declination, and image pixels, and can be described using standard keywords in an image header. The subroutine library is currently used by the image display and browsing programs SAOimage, SAOtng, and Skycat to translate between image pixels and sky coordinates. Some subroutines have been improved and new ones have been added to read and write all legal FITS image data types and to deal with IRAF .imh (OIF) files as easily as FITS files. The programs in this package set the WCS of an image, convert between pixel and sky coordinates in an image, and find catalog stars in the region of the sky which the image covers.

1. Introduction

After a means of translating image coordinates to sky coordinates was added to the SAOimage display program (Mink 1995), the same subroutine library was added to the SAOtng (Mandel 1994) and Skycat (Albrecht et al. 1997) image display programs. Users of these programs were then faced with the fact that world coordinate system (WCS) descriptions exist for only a few sets of data which optical astronomers routinely use: images from the Digitized Sky Survey, and spacecraft such as the Hubble Space Telescope and the Infrared Astronomy Satellite. Virtually no ground-based optical observatories add information to their image headers to translate pixel coordinates to sky coordinates.

Individual astronomers have used a variety of methods to match image stars with reference catalogs to set world coordinate systems for small sets of images. All widely-known methods required extensive user interaction until Elwood Downey reported (Downey & Mutel 1996) that the University of Iowa Automated Telescope Facility was automatically adding world coordinate system information to the header of each image it took. Immediately, steps were taken to apply his SETWCS program to data from telescopes used by astronomers at the Center for Astrophysics. As the Iowa software was adapted to deal with more types of images, a whole suite of portable utilities and subroutines was developed to set and use image world coordinate systems.

2. Setting World Coordinate Systems

The sequence of processes used in SETWCS was retained in the revised program, IMWCS:

1. Read a FITS or IRAF image and its header;
2. Find all stars in the image and sort them by brightness;
3. Find all stars in a reference catalog in the region of the sky where the image header says the telescope is pointing and sort them by brightness;
4. Match the reference stars to the image stars;
5. Fit a standard WCS function to the matches;
6. Add the WCS information to the image header;
7. Write the revised image and header

While its overall structure was maintained, the program was modified extensively to handle all FITS image data types, access IRAF .imh (OIF) files as well as FITS files, use a standard WCS library, and handle a variety of reference catalogs. In addition to the HST Guide Star Catalog, the USNO A1.0 Catalog (Monet 1996), and user-generated reference catalogs in Starbase format (Roll 1996) may be used. Image stars are now sorted by integrated rather than peak flux. Instead of matching stars using a grid of possible offsets, each possible set of offsets between reference and image stars is checked, and the most common offset is used as an initial value for the fit. The WCS fit has been expanded to include plate scale and rotation angle, and many fixed parameters, such as tolerance for reference/image position matches can now be changed on the command line.

In one-third to one-half of the $10' \times 10'$ images we used for testing, there were too few Guide Stars to fit a world coordinate system. Figure 1 shows such an image with only two HST Guide Stars, positioned according to a nominal WCS based on the telescope pointing direction and an assumed plate scale. To deal with this situation, the star-finding code in IMWCS was broken out into a new program, IMSTAR, which catalogs stars with positions from images with a known WCS, such as those created from the Digitized Sky Survey (DSS). The DSS thus serves as a giant catalog, the depth of which can be varied as needed by altering the parameters used by IMSTAR. To make it easier to obtain DSS images covering the same region as a telescope image, a new program, IMSIZE, was written. It translates pointing information from the telescope image to the format needed by the DSS GETIMAGE program to extract an image covering the correct region. Figure 2 shows the same image as Figure 1, after it was processed by IMWCS. The positions of the stars extracted from a DSS image of the same region of the sky are plotted as circles. Where it is available, the USNO A1.0 catalog, with 491,848,883 sources, can be used directly, instead of the extracted DSS catalog, to fit a WCS with similar accuracy. Figure 2 shows the DSS stars found and fit to those in the image by IMWCS.

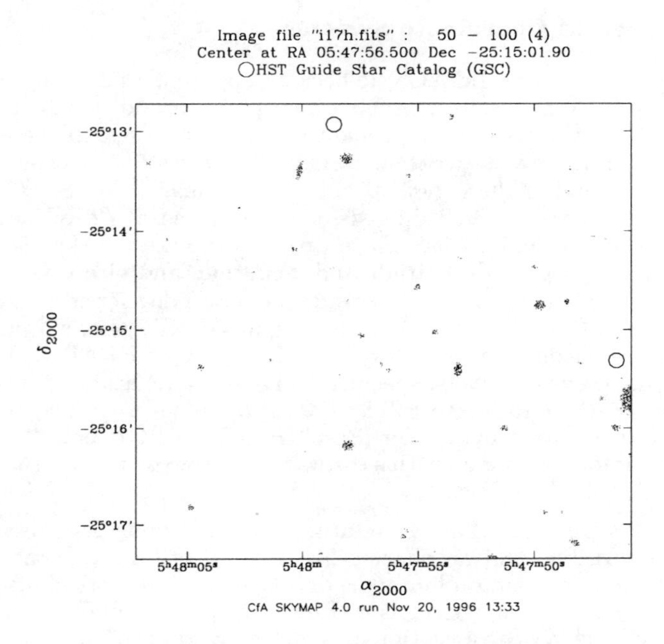

Figure 1. An image with only two HST Guide Stars.

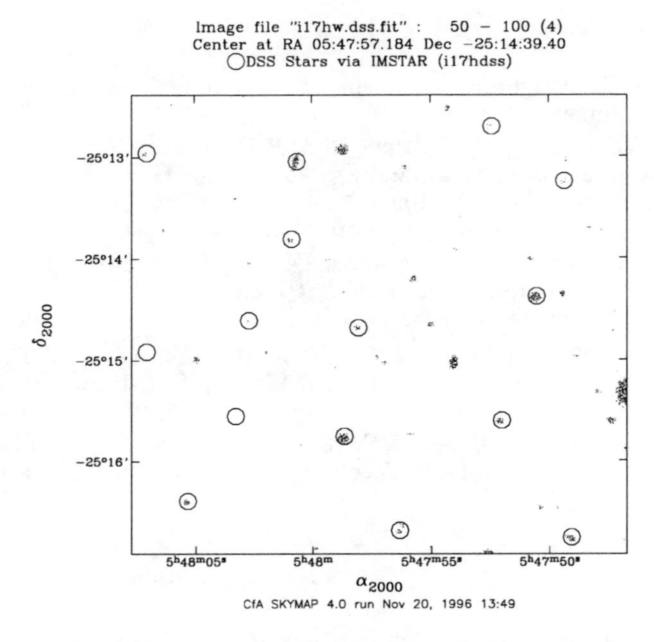

Figure 2. The same image with stars from the Digitized Sky Survey.

3. Using World Coordinate Systems

In the process of testing IMWCS, it became apparent that many of the pieces of that program were useful in their own right. In addition to the IMSTAR star-finding program used above, separate programs to search the GSC, USNO A1.0, or user catalogs were written. IMCAT lists the stars in a specified catalog which can be found in the region specified by an image's WCS. SCAT searches a catalog over a region specified on the command line. XY2SKY and SKY2XY translate coordinate pairs from image pixel coordinates to sky right ascension and declination (or galactic latitude and longitude) and vice versa.

Several image manipulation programs were also developed. EDHEAD edits the header of a FITS or IRAF image using any UNIX text editor. IMHEAD lists the entire header of a FITS or IRAF image. GETHEAD and SETHEAD print and change, respectively, specified FITS or IRAF header parameters. I2F converts an IRAF .imh file to FITS. For all of these programs, the IRAF file may be of either binary byte order (Sun/HP or IBM PC/Dec).

Further information about this software and how to obtain it may be found on the WCSTools Web Page.[1]

Acknowledgments. I am grateful to Elwood Downey for proving that this project was possible, and for sharing his code. Scott Kenyon patiently tested the WCS fitting program on hundreds of images until it worked all of the time. He also provided the image used as an illustration in this paper. Allan Brighton of ESO tested the WCS translation code on many kinds of images, enabling me to make it as general as possible.

References

Albrecht, M. A., Brighton, A., Herlin, T., Durand, D., & Biereichel, P. 1997, this volume, 333

Downey, E. C., & Mutel, R. L. 1996, in ASP Conf. Ser., Vol. 101, Astronomical Data Analysis Software and Systems V, ed. G. H. Jacoby & J. Barnes (San Francisco: ASP), 109

Mandel, E., & Tody, D. 1995, in ASP Conf. Ser., Vol. 77, Astronomical Data Analysis Software and Systems IV, ed. R. A. Shaw, H. E. Payne, & J. J. E. Hayes (San Francisco: ASP), 125

Monet, D. 1996, BAAS, 28, 905

Mink, D. J. 1996, in ASP Conf. Ser., Vol. 101, Astronomical Data Analysis Software and Systems V, ed. G. H. Jacoby & J. Barnes (San Francisco: ASP), 96

Roll, J. 1996, in ASP Conf. Ser., Vol. 101, Astronomical Data Analysis Software and Systems V, ed. G. H. Jacoby & J. Barnes (San Francisco: ASP), 536

---

[1] http://tdc-www.harvard.edu/software/wcstools

Astronomical Data Analysis Software and Systems VI
ASP Conference Series, Vol. 125, 1997
Gareth Hunt and H. E. Payne, eds.

The SAOtng Programming Interface

E. Mandel

Smithsonian Astrophysical Observatory, Cambridge, MA 02138

Abstract. The architecture of the *SAOtng* image display program is centered on supporting communication with external programs and processes. Easy to use C, FORTRAN, and UNIX interfaces provide direct access to *SAOtng* information, data, and control functions. Through these interfaces, developers can utilize such advanced *SAOtng* features as direct *FITS* transfer, shared memory *FITS*, WCS-enabled positioning, and region of interest demarcation. In this paper, we describe the *SAOtng* programming interface and show, by example, how easy it is to integrate *SAOtng* into astronomical analysis systems.

1. Introduction

SAOtng (SAOimage: The Next Generation) is an updated and improved version of the popular *SAOimage* display program for X Windows. It utilizes the Xt Toolkit and Xt-based widgets, the Gterm-image widget, and the NOAO widget server as the basis for its graphical and imaging functionality (Mandel & Tody 1995). *SAOtng* supports direct display of *IRAF* images and *FITS* images (and easily can support other file formats), multiple frame buffers, region/cursor manipulation, add-on scaling algorithms, many colormaps, and communication with external analysis tasks. It is highly configurable and extensible to meet the evolving needs of the astronomical community.

2. The SAOtng Public Access Interface

SAOtng uses the X Public Access mechanism (*XPA*) to give external processes access to its data and algorithms (Mandel, Swick, & Tody 1995). *XPA* allows an Xt program such as *SAOtng* to define named "public access points" through which data and commands can be exchanged with other programs. Each public access point in *SAOtng* supports the exchange of arbitrary amounts of information through the call-back paradigm already familiar to Xt application programmers.

XPA extends the *SAOtng* graphical user interface beyond the customary boundaries of the program. Usually, actions in a graphical program are initiated by mouse and keyboard events input by the user. But in *SAOtng*, these actions can be initiated by external programs using the public access programming interface. In addition, new actions that are not part of the GUI can be added to this interface.

The public access interface to *SAOtng* supports UNIX, C, and FORTRAN calls. For example, the UNIX public access commands are:

```
    # does SAOtng exist? (returns ''yes'' or ''no'')
    csh) xpaaccess SAOtng

    # send data or commands to SAOtng
    # access points can accept modifying parameters
    csh) <data> | xpaset SAOtng <access_point> [parameters]
or  csh) imset <access_point> [parameters]

    # retrieve data or info from SAOtng
    csh) xpaget SAOtng <access_point> [parameters]
or  csh) imget <access_point> [parameters]
```

These programs can be used interactively on the command-line or in batch scripts. They also can be used directly in application programs by means of the *system()* routine. The latter mechanism has the drawback that the *system()* routine must start a new shell process each time it is called. In cases where efficiency is at a premium, external programs can use the C or FORTRAN public access interface to communicate with *SAOtng*.

The C and FORTRAN interfaces define a set of calls similar to the UNIX commands described above. For example, the C interface contains the following routines:

```
/* open a connection to the xpa server */
void *xpafd = OpenXPA(''SAOtng'');

/* send data or commands to SAOtng */
SetXPAValue(void *xpafd, char *paramlist, char *data, int len);

/* retrieve data or info from SAOtng */
GetXPAValue(void *xpafd, char *paramlist, char *data, int *len);

/* close connection */
CloseXPA(void *xpafd);
```

When *OpenXPA()* is called, an *XPA* server program is started to mediate communication between the calling program and *SAOtng*. This server program remains active until a call is made to *CloseXPA()*. It significantly improves the communication speed between *SAOtng* and external programs over the repeated use of *system()* to call to xpaset and xpaget. Note also that the use of this intermediary server program obviates the need for applications to link the X11 libraries directly.

More than thirty-five public access points in *SAOtng* (a selection of which is shown in Table 1) support communication with external processes. Several of these are designed specially to provide commonly-used information and data to astronomical analysis programs. It is this combination of broad flexibility and specific useful functionality that makes the public access interface a valuable part of *SAOtng*'s image display services.

3. Examples of the SAOtng Public Access Interface

New image files can be loaded using the *file* public access point:

```
csh) imset file ../data/coma.fits
```

Table 1. A Selection of Public Access Points in SAOtng

Name	Description
analysis	run analysis command on current image
blocking	set/get blocking info for extracting image data
colormap	set/get colormap for current frame
coords	get coordinate values
file	display image file in current frame
frame	create or change display frame
gif	create gif from current frame
print	set execute print command in current frame
redisplay	redisplay selected image in current frame
regions	get/set regions markers in current frame
rescale	rescale selected image in current frame
shm	map FITS image in shared memory
tcl	execute Tcl code
zoom	set/get zoom factor for current frame

Sending an image file name to the *file* access point causes *SAOtng* to load and display that image. If the image file is in *FITS* format, the data can be loaded directly. Otherwise, *SAOtng* calls the appropriate external image access program to extract a section of the image and send it back to *SAOtng* for display.

The *file* access point can contacted from the command line or from programs such as archive servers and file browsers to display local files. For example, the *XDir* file browser issues a *file* command to *SAOtng* when the user double-clicks on an image file. UNIX scripts also can use *file* to process a list of images.

For processes requiring the fastest possible display of image data, the *shm* (shared memory) access point can be used to share in-memory *FITS* with *SAOtng*. For example, a data acquisition program can collect *FITS* data into a shared memory segment. The program sends *shm* information to *SAOtng* so that the latter can access the same shared memory:

```
sprintf(paramlist, ''shm %s %d %d'', name, shm_id, shm_segsz);
SetXPAValue(xpafd, paramlist, NULL, 0);
```

The *shm* message instructs *SAOtng* to display the contents of the shared memory as a *FITS* file. Then, as data are collected over time, the *redisplay* public access point can be used to update the evolving image.

SAOtng supports the ability to mark regions of interest on images. Geometric markers (circles, rectangles, ellipses, polygons, points, and lines) and text can be placed interactively on an image and then moved, resized, reshaped, etc. Markers can have attributes of "include" or "exclude" and "source" or "background." They are used in systems such as *IRAF/PROS* and *HEASARC* xselect to specify spatial filtering of image data (Mandel et al. 1993).

The *regions* public access point supports external control of these markers. Using this access point, markers can be loaded from a file:

```
csh) cat regions.file | xpaset SAOtng regions
```

Once the shape, size, position, color, etc., of these markers are finalized, they can be saved for later use:

```
csh) xpaget SAOtng regions degrees > newregions.file
```

As indicated above, the coordinates used in the region descriptors can be loaded or saved in hms, degrees, or pixel format.

Finally, the *coords* public access point allows a user to retrieve the spatial position (and associated data value) pointed to by the mouse:

```
GetXPAValue(xpafd, ''coords'', cbuf, 132);
```

When a request is made to the *coords* access point, *SAOtng* waits for the user to position the mouse on the desired pixel and then press the *Return* key. The coordinates and raw data value at that point are returned to the calling program.

4. Summary

SAOtng and *XPA* are available as part of the SAO R&D software suite. Other programs contained in this package include *ASSIST* (a uniform interface to heterogeneous analysis systems), *XDir* (an X directory and file browser which supports user-defined actions for different file types), and *ncl* (*IRAF* cl with line-editing). This software suite has been ported to Sun, SGI, HP, Dec Alpha, PC/Linux, and PC/FreeBSD systems, and is available via anonymous ftp.[1]

The SAO R&D software suite is an embodiment of an evolving software cooperation philosophy that we hope to bring to astronomy and other disciplines. It reflects our understanding about how software systems (and researchers and developers) can act in concert without sacrificing their independence.

Acknowledgments. This work was performed in large part under a grant from NASA's Applied Information System Research Program (NAGW-3913), with support from the AXAF Science Center (NAS8-39073).

References

Mandel, E., Roll, J., Schmidt, D., VanHilst, M., & Burg, R. 1993, in ASP Conf. Ser., Vol. 52, Astronomical Data Analysis Software and Systems II, ed. R. J. Hanisch, R. J. V. Brissenden, & J. Barnes (San Francisco: ASP), 430

Mandel, E., & Tody, D. 1995, in ASP Conf. Ser., Vol. 77, Astronomical Data Analysis Software and Systems IV, ed. R. A. Shaw, H. E. Payne, & J. J. E. Hayes (San Francisco: ASP), 125

Mandel E., Swick R., & Tody D. 1995, The X Resource, Vol. 13, (Boulder: O'Reilly), 235

[1] ftp://sao-ftp.harvard.edu/pub/rd/saord.tar.Z

Astronomical Data Analysis Software and Systems VI
ASP Conference Series, Vol. 125, 1997
Gareth Hunt and H. E. Payne, eds.

Speculations on the Future of FITS

Donald C. Wells
National Radio Astronomy Observatory,[1] *Charlottesville, VA 22903-2475, E-mail: dwells@nrao.edu*

Abstract. The history and philosophy of FITS are reviewed, with emphasis on the lessons-learned and on the archival requirements. Opinions are offered on the likely outcome of current FITS negotiations, such as the year-2000 problem and the WCS proposal, and on possible subjects of future data interchange format negotiations in astronomy. `BINTABLE` schemas in third-normal-form are advocated. The long-term importance of the `BINTABLE` format as a platform for future layered-convention agreements is stressed.

1. On the Philosophy of FITS (Lessons-learned)

FITS [Flexible Image Transport System] provides a common canonical *language* for talking about astronomical data structures and, as such, it has a profound positive influence on software design practice in astronomy. By negotiating FITS as a *family* of similar data formats, Basic-FITS (1979), random-groups (1980), generalized extensions (1983), `TABLE` (1984), `BINTABLE` (1991) and `IMAGE` (1992), we have minimized the negotiation, documentation and training costs for our community. Our most effective negotiating strategy has been to try to achieve bi-lateral agreements. We include "escape hatches" in our agreements, in places where we expect to negotiate future agreements. The history of FITS shows that it is not possible to fully transmit *meanings*. Instead, the purpose of FITS is agree on the *syntax* of a language for talking about astronomical data, and to agree on the *semantics* in only a limited range of cases. Agreement on syntax permits basic portability and interchange of data, and the users are able to bridge the semantic gaps.

Newcomers to FITS often ask: "Why doesn't FITS have a `VERSION` keyword?" Our answer is: "It does, but the value is always 1.0 by default." The point is that the introduction of a `VERSION` code would be incompatible with the use of FITS as an archival format, because designers of new software would be tempted to support only recent versions. The FITS committees will never knowingly obsolete existing conforming FITS files. This policy is often summarized as "once FITS, always FITS."

Seventeen years of production experience with FITS have demonstrated that only a few actual mistakes were made in the design of Basic-FITS, and that they have not hurt us (yet). A minor mistake is that we specified keyword

[1]The National Radio Astronomy Observatory is a facility of the National Science Foundation, operated under cooperative agreement by Associated Universities, Inc.

EPOCH instead of EQUINOX. A more serious mistake is that DATExxxx='31/12/99' was specified, which exposes the FITS community to the infamous "year-2000" problem; we must correct this within the next three years.[2] The author (one of the original designers) wishes he could change two "mistakes" of style: (1) we should have specified use of SI units more clearly and, in particular, we should have specified radians, the SI auxiliary unit for angles, instead of degrees, and (2) we should have explicitly advocated use of a hierarchical keyword notation, such as the "HISTORY VLACV MAP METHOD='FFT'" notation which appears on line 2/4 of Fig. 1 of Wells, Greisen, & Harten (1981).

2. FITS as an Archival Format

"A data set that is not used by its creator in its archived form is notoriously unreliable."

FITS is not only a way to talk to remote astronomers in the here and now, it is also a way to talk to future astronomers. The FITS standards have been published, and copies will be available in libraries around the world *forever.* The human-readable (self-documenting) headers of FITS, with 60% of the characters *reserved* for comments, complement the published rules of FITS. One alternative interchange format, HDF [Hierarchical Data Format], uses an API [Application Programming Interface] with registered binary tags instead of human-readable self-descriptions in the bitstream. This type of architecture is not as safe as FITS for archival applications because we cannot predict the future in computer languages and operating systems over periods of *decades.* Therefore, *our archival format must always be defined at bitstream level, as FITS is, not by an API.*

3. Repeating Groups Considered Harmful

Our BINTABLE extension is a superb exchange format for normalized databases (sets of related tables). Consider a telescope with multiple detectors operating in parallel, each producing a matrix, each with different dimensionality and WCS parameters. If these detectors are dumped at the same timestamp, should all of the matrices be recorded in the same row of a BINTABLE or should they be recorded as multiple rows with one matrix per row? Only the latter schema is capable of becoming a *normalized* relational database, i.e., of being cast into Third Normal Form, the simplest and most compact schema concept. The first schema (multiple matrices in the same row) is an example of a *repeating group.* Repeating group schemas are harder to design and program, more costly to maintain, and do not support flexible query techniques; the database industry has deprecated them for the past twenty years (Martin 1977, p. 245). Repeating group schemas require that we invent complex conventions to form subscripted column labels for matrix dimensionality, WCS parameters, etc. This complex keyword notation should not be necessary in BINTABLE, because any repeating group schema can be re-designed as a normalized relational schema. Let's apply Occam's razor!

---

[2] A recent poll of our community has produced the consensus that we will support DATExxxx='1999-12-31', with optional ISO-8601 time notation.

4. FITS Evolution—Work-in-progress

"The purpose of standardization is to aid the creative craftsman, not to enforce the common mediocrity."

Clever pieces of craftsmanship like the CHECKSUM proposal (Seaman 1995) can greatly enhance FITS without actually changing it. The author recommends that CHECKSUM be implemented in astronomy data systems. The FITS community expects to define and implement a new syntax for DATExxxx value strings before 1999-12-31, while agreeing to continue to support the old syntax. We expect to also agree that optional time values can be appended to the date strings. We continue to work toward a celestial coordinates WCS [World Coordinate Systems] agreement. The 25 projections of the sphere onto a 2-D FITS image as specified by Greisen & Calabretta (1996) have been implemented in four different languages (FORTRAN, C/C++, IDL, Java) already. It is likely that we will eventually also agree on spectroscopic and time-series coordinate conventions. Probably we will agree to allow non-printing codes like CR/LF to be used in undefined fields of TABLE extensions, in order to make it easier to upload TABLE bodies into commercial database software. It appears likely that the BINTABLE variable-length vector convention[3] (Cotton et al. 1995) will be widely implemented and used in the future.

5. FITS Evolution—Some Future Possibilities

It is easy to speculate about future FITS agreements—it is much harder to actually negotiate them! The following items are some ideas that the author considers to be possibilities, but which he may or may not support in future negotiations. First, there are a number of ways in which we could agree to "loosen" FITS header syntax, e.g., move the "=" around, support lowercase keywords, longer keywords, hierarchical keywords, longer string values, header line continuation convention, etc. We should be *very* cautious about most such header syntax changes, but it is a fact that we could make many of them in such a way as to preserve backward compatibility. We could agree to allow extended character sets (probably the UTF-7/RFC1642 version of ISO-10646/Unicode) in string values of keywords like OBSERVER and in TABLE extensions. We could agree to support BITPIX=1. We could adopt a wide variety of conventions layered on top of BINTABLE, such as codings for high performance image compression algorithms, or the Jennings et al. (1995) hierarchical grouping proposal; the author expects that almost all future FITS object types will be layered on BINTABLE. We could agree to support XTENSION='MPEG' or other MIME-coded types in order to associate such objects with our datasets (a FITS generalized extension is capable of encapsulating *any* other bitstream format). In particular, XTENSION='JAVA' might enable us to transmit *portable methods* along with our data objects.

---

[3]This feature of BINTABLE is an especially clever piece of craftsmanship.

6. Has FITS Outlived Its Usefulness?

We need an interchange and archival format more than ever, so the short answer to the question must be "No!" Therefore, the real question is whether we should decide to adopt some other existing format or should negotiate a new format agreement. The author's opinion is that the potential alternative formats are only slightly stronger than FITS in their areas of strength, and are significantly weaker than FITS in its areas of strength.The costs of re-designing FITS (negotiation, R&D, documentation, retraining, coding support in hundreds of applications) would be enormous. It is very unlikely that the possible gains of re-design could ever be worth all of these costs. Indeed, we would incur most of these costs even if we adopted an existing design from another discipline. Furthermore, it may no longer be possible to negotiate a general interchange and archival format for a community as large, diverse and sophisticated as astronomy now is. Perhaps we were lucky: 1979 was a moment when there were very few vested interests and when several of the largest software projects were still in their startup phases, and were able to adopt FITS as their external canonical form at about the same time. The author expects that these conclusions about the role of FITS will remain true for several more decades.

"We must indeed all hang together, or, most assuredly,
we shall all hang separately."

References

Cotton, W. D., Tody, D., & Pence, W. D. 1995, A&AS, 113, 159

Greisen, E. W., & Calabretta, M. 1996, Representations of celestial coordinates in FITS[4]

Jennings, D. G., Pence, W. D., & Folk, M. 1995, in ASP Conf. Ser., Vol. 77, Astronomical Data Analysis Software and Systems IV, ed. R. A. Shaw, H. E. Payne, & J. J. E. Hayes (San Francisco: ASP), 229

Martin, J. 1977, Computer Data-Base Organization (Englewood Cliffs, NJ: Prentice-Hall)

Seaman, R. 1995, in ASP Conf. Ser., Vol. 77, Astronomical Data Analysis Software and Systems IV, ed. R. A. Shaw, H. E. Payne, & J. J. E. Hayes (San Francisco: ASP), 247

Wells, D. C., Greisen, E. W., & Harten, R. H. 1981, A&AS, 44, 363.

---

[4] http://fits.cv.nrao.edu/documents/wcs/wcs.all.ps

Astronomical Data Analysis Software and Systems VI
ASP Conference Series, Vol. 125, 1997
Gareth Hunt and H. E. Payne, eds.

FV: A New FITS File Visualization Tool

W. Pence

NASA/GSFC, Code 662, Greenbelt, MD 20771

Jianjun Xu, Lawrence Brown

HEASARC/HSTX, Greenbelt, MD 20771

Abstract. A new software tool, called fv, for visualizing the contents of any FITS format file is now available.

Fv is a general software tool that can be used for viewing, plotting, and editing FITS format data files. The first release of fv has a graphical user interface (GUI) that provides spreadsheet-like widgets to display data in any FITS table or image, a text widget to display FITS header keywords, and a graph widget, POW, to display images. In addition, users can plot values in two or more columns of a FITS table/image, and export the plot to a PostScript file. Users can also save listings of table/image data to an ASCII file. Future releases of fv will allow users to edit any keyword or data value in the FITS file and save the changes. The current image and graph display will also be greatly enhanced. Fv is built as a standard part of the FTOOLS distribution for release 3.6 and higher, or is available standalone from the ftools home page.[1]

[1] http://legacy.gsfc.nasa.gov/ftools

Astronomical Data Analysis Software and Systems VI
ASP Conference Series, Vol. 125, 1997
Gareth Hunt and H. E. Payne, eds.

FITS++: An Object-Oriented Set of C++ Classes to Support FITS

A. Farris[1]

Space Telescope Science Institute, Baltimore, MD 21218

Abstract. FITS++ is a object-oriented set of C++ classes, supporting both a sequential and random processing model, designed to be the basis of applications that manipulate FITS files. All aspects of the FITS standard are supported, allowing users to access any portions of FITS headers or data. Extensive error handling is also available.

1. Introduction

The purpose of this set of C++ classes is to give programmers the tools necessary to manipulate FITS files for a broad range of applications. Its primary purpose is to import and export data while adhering strictly to the FITS standards. All FITS data structures are supported, including random groups, image extensions, and binary tables. On input, errors are detected and reported but corrections are attempted. On output, all errors that might result in writing an invalid FITS file are detected.

Unlike the previous version (Farris 1995), these classes support two processing models, sequential and random, integrated into one coherent framework. The particular processing model, open mode (ReadOnly, WriteOnly, ReadWrite) and other file attributes, are selected when the FITS file is opened. The sequential processing model is useful for tape, pipes, and standard I/O. FITS header-data-units are read or written in sequential order.

The random processing model is restricted to devices that support lseek. Any FITS header-data-unit can be accessed randomly. Any keyword in the header or any portion of its associated data can be read or written. The random model allows the user considerable flexibility in manipulating random access FITS files. An append mode also allows users to add data to the end of a FITS file and update size information in the FITS header when the writing of the data is complete, all while maintaining the internal consistency of the FITS file. The random mode maintains a directory of header-data-units, much like an operating system maintains a directory of files. This directory is created when the file is opened.

A subset of processing commands may be used in both the sequential and random modes. Programs that only use this subset can be written to work with either mode. The only difference is the statement that opens the file. Commands that are extended beyond this subset apply only to the random

---

[1]E-mail: farris@stsci.edu

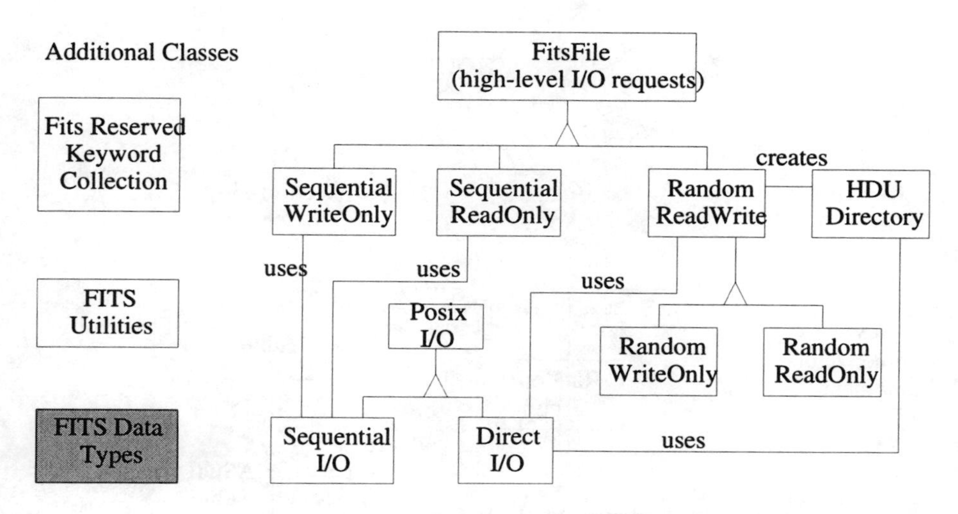

Figure 1. FitsFile Family of Classes.

processing model. Attempts to use these extended commands on a sequential file are flagged as errors.

All aspects of the FITS standard are supported, including image extensions, random groups structure, general extensions, ASCII and binary tables. The variable length array facility and the multidimensional array convention, used in conjunction with binary tables, are also supported.

FITS++ is implemented in C++ and will compile on most available C++ compilers. Language features recently added by the ANSI standardization committee and only available on newer compilers are avoided. GNU's C++ compiler, g++, running on SUN workstations is the default development environment. FITS++ runs on UNIX, VMS, and Windows-95/NT operating systems. Under VMS, DEC's C++ compiler is used. Both DEC VAX and Alpha platforms are supported and all VMS floating-point options (IEEE_float, G_float, and D_float) are supported.

2. Basic Design Considerations

Within FITS++, a FITS header-data-unit (HDU) can be formed in two ways: from an existing file, or from a FITS keyword list. In the first way, the HDU is initially associated with a file. In the second way, it is not. Regardless of how the HDU is created, once it is created it is an independent object in its own right. It may be modified or written to other FITS files. The header and data are read (or written) by separate operations.

On input (or output), all of the HDU's data does not have to be read (or written) at once. For output purposes, an HDU may be associated with any file. The HDU classes contain the knowledge of the internal structure of the data, as described by the FITS keywords. Their job is to provide access to the descriptive keywords and to the data themselves in the form of operations that

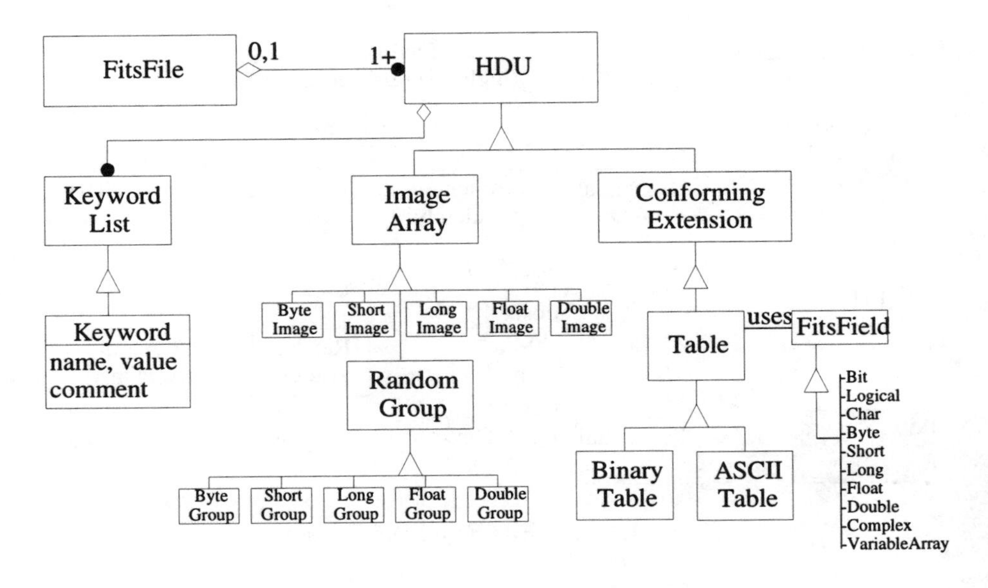

Figure 2. FITS HDU Family of Classes.

are meaningful to a data structure of that particular type. They know nothing about how this kind of HDU is formatted in a FITS file.

The FitsFile classes, on the other hand, know nothing about the internal structure of HDU data; they do know everything about how those HDUs are structured within FITS files. Furthermore, they know how to translate FITS headers into keyword lists and vice versa. System level I/O commands are encapsulated in a class that implements Posix I/O functions. This class is designed to be operating system independent, and runs under both UNIX and VMS.

Figure 1 shows the FitsFile family of classes. These classes perform basic I/O functions. The "open" function creates a FitsFile object. Destroying the object flushes buffers and closes the file. On input, high-level I/O commands allow one to select a particular HDU, determine its type, and create an HDU object, if desired. Generic or specific HDUs may be created. An ImageHDU may be created if one only wishes to access physical data and disregard the raw data type. Likewise, if one wishes to access a table, and ignore whether it is an ASCII table or binary table, one may do so. Low-level I/O commands provide mechanisms for accessing and modifying any portion of a FITS file with no restrictions. These can be used to cope with unusual situations.

Figure 2 shows the HDU family of classes. These classes implement the basic FITS data structures. All FITS data structures are supported, including full support for random groups and all standard extensions. An ImageHDU is automatically converted to either a primary HDU or an image extension, as appropriate. Image arrays of arbitrary dimensionality are supported, but are optimized for 1, 2, and 3 dimensional arrays. One can access either physical data (after applying BSCALE and BZERO) or raw data.

Tables are read by rows. Access to fields within a row is similar to programming interfaces to relational DBMS tables. The multi-dimensional and variable length array conventions, used in conjunction with binary tables, are fully supported. The append mode is particularly useful in creating tables. Rows of a table can be added in the append mode. When this mode is closed, the number of rows in the FITS header is automatically updated. Because the append mode maintains an externally consistent view of the FITS file, a separate process can read such a file as it is being updated. To such a process, the appended records appear to be FITS "special" records.

3. FITS Keyword Handling and Error Handling

Within these classes, FITS keywords are implemented using an encapsulated linked list. Uses of a keyword depend only on the keyword name, value, and comment. Detailed FITS formatting rules are invisible to the user. A FITS keyword-to-card translator is built into the FitsKeyword classes. They also use a table of properties of reserved FITS keywords. The list of keywords can be manipulated in a wide variety of ways. A keyword's name, index (if any), value, or comment may be changed. On input, an exact copy of the FITS keyword string is retained, and the order of keywords is maintained. On output, the user has the option of formatting the keyword string rather than relying on built-in formatting functions.

Errors are arranged in order of increasing severity and divided into: warnings (technically not errors, but are undesirable), minor errors (correctable, or not resulting in data loss), serious errors (conditions that will result in loss of data), severe errors (usually halting processing, such as I/O errors), and fatal errors (halting processing). These classes use a flexible method of error handling to give the user control over coping with error conditions. Global error conditions and error messages can always be accessed via global functions. In addition, the user can install an error handling function that is called if an error condition is raised. This function can be applied to global conditions or can be installed to catch errors pertaining to a particular FITS file.

4. Current Status

FITS++ is being used internally at STScI in the pipeline calibration and archive processing software for STIS and NICMOS, both of which are currently under development.

References

Farris, A. 1995, BAAS, 27, 909

FITS 1993, "Definition of the Flexible Image Transport System (FITS)," Standard, NOST 100-1.0, NASA / Science Office of Standards and Technology, Code 633.2, NASA Goddard Space Flight Center, Greenbelt, Maryland

Wells, D. C., Greisen, E. W., & Harten, R. H. 1981, A&AS, 44, 363

Astronomical Data Analysis Software and Systems VI
ASP Conference Series, Vol. 125, 1997
Gareth Hunt and H. E. Payne, eds.

The FITS List Calculator and Bulk Data Processor

Elizabeth B. Stobie and Dyer M. Lytle

NICMOS Project, Steward Observatory, University of Arizona, Tucson, AZ 85721, E-mail: bstobie@as.arizona.edu, dlytle@as.arizona.edu

Abstract. The FITS List Calculator and Bulk Data Processor (FLC) is an IDL program with a graphical user interface developed for processing large groups of Near Infrared Camera and Multi-Object Spectrometer (NICMOS) test data efficiently. The program is built around the simple concept of defining a new IDL variable type, the file list. Lists of files may be assembled with assigned variable names and displayed with subwidgets. These variables may then be arguments to predefined functions and procedures in the main widget as well as IDL functions and procedures. They may also be used in arithmetic expressions. Images and arrays may be defined by direct loading or as the result of calculation. When processing is repetitive, scripts may be written to process many groups of data in exactly the same way.

1. Introduction

A total of 126,370 observations were produced during the NICMOS instrument characterization tests in the Brutus thermal vacuum chamber at Ball Aerospace during the summer of 1996. These observations provided performance information as well as preliminary calibration reference data for the instrument. Software was needed to process large datasets in routine ways with minimal effort. FLC was developed for this purpose by the authors in a time period of approximately two months.

2. Data Types

There are four basic data types used in FLC:

1. lists
2. images
3. arrays
4. scalars

The user may customize the number of lists, elements in a list, images, and arrays by editing the user_defs file. Structures are allocated at run time for the maximum number of lists, images, arrays, and scalars. One must be careful of the number of images allocated as they are the largest memory hog. Figure 1 shows the data structures used to hold lists and images.

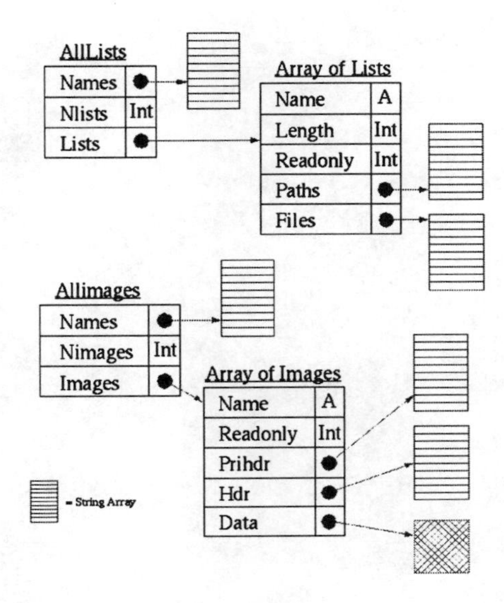

Figure 1. List and Image structures in FLC.

2.1. Lists

Lists contain lists of FITS filenames (no other filetype is supported), which are stored in memory. The files themselves are stored on disk and read in one at a time when needed. List names may be used in FLC functions and procedures or used in arithmetic expressions. The default limits are 100 file lists with 100 filenames in each. They are stored as IDL structures and these structures are, in turn, stored in a parent IDL structure containing all lists.

2.2. Images

Images may be produced as the result of arithmetic operations, functions, or defined procedures or may be directly loaded from FITS files residing on disk. The default size of the image data is 256×256, optimized for NICMOS data. The image header is also loaded into memory. For files with image extensions the primary header and the first image extension header are stored separately. Any subsequent image extensions are ignored. Newly created images may be saved to disk in FITS format. The default number of images is 50. The IDL structures for images are similar to the structures for lists except that the image headers and image data arrays are kept in the structure.

3. Display

There are two display functions in FLC. The first, shown in Figure 2, is for detailed examination of individual images. The window includes a scrolling text region for examining the image headers, a histogram/plot display that can show a histogram of the entire image, a histogram of an individual quadrant, or plots

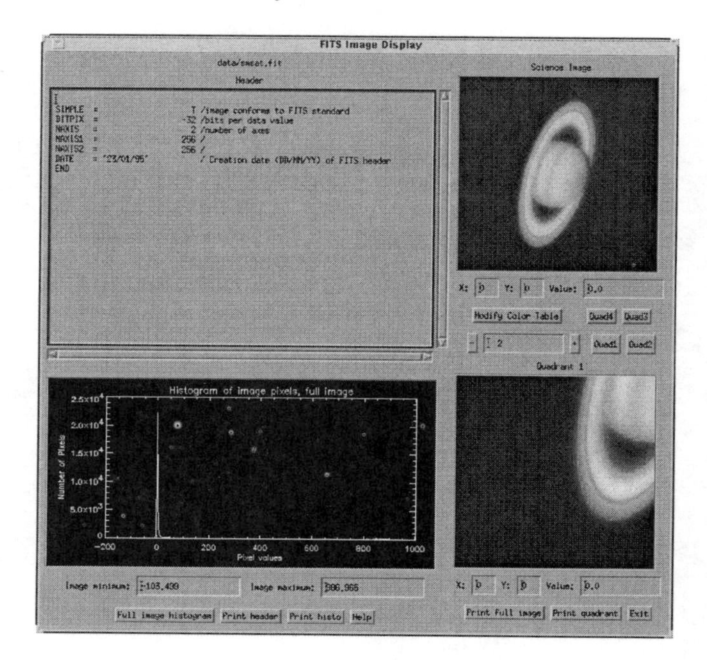

of cross sections of the image, vertical or horizontal. Finally there are two image displays, one showing the entire image and one showing only a quadrant of the image. The image displays also have (x,y) value readouts below them.

The other display function is called the "scanner" and its primary purpose is for quickly scanning through a list of images to look for irregularities in the data sets. It has one image display and forward and reverse scan buttons. Along the bottom of the window, the median and standard deviation of each image is shown.

4. Arithmetic

Standard arithmetic operators can be used in FLC expressions. Operators like plus, minus, multiply, divide, and exponentiation are generally overloaded where overloading makes sense. For example, an image can be added to a list resulting in a new list where each member of the new list is the sum of an image from the original list and the added image. In other situations, where overloading would be ambiguous, those constructs cause errors; for example, trying to add a 1-D array to a list of images.

5. Functions

Several functions for operating on lists, images, and arrays are predefined for the convenience of the user. They include makelist, getlist, savlist, and dellist for creating, saving, and deleting lists. Similarly, getimage, getmask, savimage, and delimage are predefined for images and extract, getrow, getcol, savarray, and delarray for arrays. Other functions defined specifically for lists include averaging images in a given list, computing a median image from the images in a given list, summing the images in a given list, computing image statistics (mean, standard deviation, median, skew, and kurtosis) images across the images in a given list, and computing the same statistics for each image in a given list.

6. Procedures

Several routines do not require a target for storing a result. The user may plot a specified array, display a specified image or element in a list, scan all images in a specified list, fit a linear expression by linear regression to each pixel of the images in a specified list, or execute a script.

7. Scripts

Often many datasets are processed in an identical manner which can require several repetitions of a set of commands. Scripts were implemented to avoid repetitious typing and minimizing the opportunity for mistakes. Any command allowed at the command line may also be used in scripts and comments are supported when prefaced by a semicolon (;). Scripts may call other scripts.

8. Parser

Most commands, excluding scripts and IDL procedures, may be invoked by selecting them from the appropriate widget menu or by typing the commands in the command window. The FLC parser processes commands as they are entered in the command window. All variables, functions, and procedures are identified as well as arithmetic operators. Variables are then replaced with internal variables that are associated with the variable names and the IDL EXECUTE command is used to execute the command. No command may extend beyond one line (80 characters in length) and no conditionals or do loops are allowed.

9. Future Plans

The current version of FLC has only been tested on computers running IDL version 4.0 or later under various flavors of the UNIX operating system. With a little work, it could be made to run on the Apple Macintosh and Intel processor based systems running Windows.

FLC will continue to be used within the NICMOS project during the coming years. If there is a demand, the program could be generalized somewhat to handle a wider variety of data.

Acknowledgments. We are grateful to Tony Ferro for helping test the program.

Astronomical Data Analysis Software and Systems VI
ASP Conference Series, Vol. 125, 1997
Gareth Hunt and H. E. Payne, eds.

FITS BoF Session

Peter Teuben

Astronomy Department, University of Maryland, College Park

Don Wells

National Radio Astronomy Observatory, Charlottesville

Abstract. We summarize the discussions that took place during the FITS Birds of a Feather (BoF) session, which was held Monday, September 23, 1996 at 16:00. For completeness we also include a summary of the 1995 FITS BoF, which was never published.

1. Introduction

This year marked the second FITS BoF at the ADASS. In previous years the WGAS meetings at the AAS Summer meetings provided a regular discussion ground, but with the introduction of the BoF sessions at the ADASS and a much more targeted audience, it was decided to hold these meetings at the annual ADASS. Most FITS-related communications take place on the newsgroup `sci.astro.fits` (and the associated "`fitsbits`"[1] e-mail exploder).

2. FITS BoF session at ADASS VI

As always, a number of oral and poster presentations specifically discussed FITS-related items. Oral papers by Clarke, Allen, Mink, Mandel, and Wells were presented in a special FITS afternoon session, and posters were presented by Pence, Farris, Tripicco, and Stobie; see the papers by these authors elsewhere in this volume.

The NOST/FITS Support Office is responsible for two major documents. First, the NOST standard is intended to define FITS in more precise terms. The technical panel met on September 11 and its chairman (Bob Hanisch) reported on some of the major changes which were decided. The new version of the standard is expected to be out for comments early in 1997. Second, the User's Guide provides background information, especially examples, to aid in producing and interpreting legal FITS datasets. Version 4 is now being prepared (also available early in 1997); the IMAGE, BINTABLE and blocking agreements are now part of the regular text. It now also includes discussion of the World Coordinates draft (Greisen & Calabretta 1995).

---

[1] To subscribe, send an E-mail message to "`majordomo@majordomo.cv.nrao.edu`" containing the command "`subscribe fitsbits`" in the body of the message.

The FITS Support Office also maintains a Web site,[2] which is one of the major information clearinghouses for FITS. It includes the registered FITS extension types. The home page is accessed about ten times a day, and the NOST standard and User's Guide are retrieved about once a day.

Grosbøl reported on a recent `sci.astro.fits/fitsbits` discussion of the "year-2000" problem, which resulted in a proposal by the European FITS committee which will be discussed by other regional FITS committees and the IAU-FWG. One possible starting date for a new format for the DATE-OBS (and related) FITS keywords will be January 1, 1998.

Wells discussed the proposed MIME codes for FITS files. Efforts are still in progress to register several FITS data types with the Internet Assigned Numbers Authority [IANA].

This past year has also seen some major and minor improvements in existing software, and the introduction of some new and interesting tools:

- FITSIO, FTOOLS, fv (Pence et al.)
- CFITSIO (Pence)
- MRDFITS in the IDL astronomy library updated (McGlynn)
- WCSLIB updated (Calabretta)
- FITSview (Cotton)
- JAVA browser (Lu, NCSA) and FITSWCS Java Package (NCSA)

A new version of the World Coordinate System draft is available, and all FITS users are urged to study it. Hanisch and Tody discussed some of their concerns with the current WCS proposal during the BoF session. It became clear that some issues, particularly with respect to large scale astrometric projections and mosaicking, will need to be looked at in more detail.

3. FITS BoF Session at ADASS V (Tucson)

[A report on the 1995 BoF was not included in the ADASS V volume.] From the FITS Support office, Schlesinger reported on the status of the NOST Standard and the Users Guide. Version 1.1 of the NOST Standard has now been approved by the Accreditation Panel (after some units revisions). The Technical Panel will start to work on Version 2, which will now include BINTABLE, IMAGE, and blocking. The FITS User's Guide is at version 3.1 with Version 4.1 under way. Schlesinger also presented an overview of the new Web site as a primary source of information for FITS and registration of FITS extension types.

3.1. News items

- NOST standard approved by NASA, FITS committees to consider it for formal approval, superseding the four A&A papers.
- BINTABLE paper just came out! (Cotton & Tody 1995)

[2]http://www.gsfc.nasa.gov/astro/fits/basics_info.html

- IDL: MRDFITS (Tom McGlynn, Goddard/CSC), FX* (Bill Thompson, Goddard/ARC), READFITS and WRITEFITS
- FITSIO (V4.08 October 1995) has been updated, and FTOOLS is running at version 3.3
- The WCS paper has been revised again, and now has an Appendix-A. Wide-field survey people in particular should review this.
- Pence introduced a (HEASARC based) proposal to adopt the celestial coordinate keywords as defined in Greisen & Calabretta (1995) WCS paper.
- Wells discussed the planned application to IANA for FITS MIME code(s), probably to be `image/fits`, `application/fits-table` and `application/fits-image`.

3.2. Individual Contributions

Short contributions were presented by the following authors:

- **Don Jennings**
  - FITS hierarchical grouping proposal.
  - FITS and HDF communications (w/ Bill Pence - see also their poster)
- **Immanuel Freedman** FITS Data Compression: developments in Generic Data Compression, These developments relate to work performed on the COBE project. (10 minutes)
  - Field-level compression extensions.
  - ROW or COLUMN pixel order.
- **Rob Seaman** Checksum proposal.

Acknowledgments. We would like to thank Barry Schlesinger (GSFC/ADF FITS Support Office) who could unfortunately not be present, for preparing his news items.

References

Greisen, E. W., & Calabretta, M. 1996, Representations of celestial coordinates in FITS[3]

Cotton, W. D., Tody, D., & Pence, W. D. 1995, A&AS, 113, 159

Wells, D. 1997, this volume, 257

[3] http://fits.cv.nrao.edu/documents/wcs/wcs.all.ps

Part 6. Data Archives

Astronomical Data Analysis Software and Systems VI
ASP Conference Series, Vol. 125, 1997
Gareth Hunt and H. E. Payne, eds.

The XTE Data Finder (XDF)

A. H. Rots

Universities Space Research Association, XTE Guest Observer Facility, Code 660.2, Goddard Space Flight Center, Greenbelt, MD 20771

K. C. Hilldrup

Hughes STX, XTE Guest Observer Facility, Code 660.2, Goddard Space Flight Center, Greenbelt, MD 20771

Abstract. All *RXTE* telemetry will be archived in a hierarchical database consisting entirely of FITS Binary Tables. We describe an interactive tool, the XTE Data Finder (XDF). Though initially designed purely as a navigational tool, XDF has proven to be equally successful as a means to provide users full access to the complete *RXTE* mission database through the Internet.

1. Introduction

The Rossi X-ray Timing Explorer (*RXTE*) is a High Energy Astrophysics mission, launched in December 1995. It carries two pointed instruments (PCA and HEXTE) that together cover the range 2–250 keV with μs time resolution and moderate spectral resolution. A third instrument (ASM) will monitor most of the X-ray sky every 90 minutes in the 2–10 keV range.

RXTE's on-board science data systems provide considerable processing power and unprecedented flexibility in telemetry data modes. Events are processed on-board in several simultaneous data modes, chosen from a large repertoire. New data modes may be added during the mission. Consequently, it is a challenge to keep track of the database contents, and to provide a mechanism to select data that satisfy selection criteria expressed in physical terms.

That challenge is met by the design of a database that exclusively uses FITS binary tables (for storage of observational data as well as metadata) and of a tool for navigating that database.

2. Design of the FITS Database

The *RXTE* FITS Database (XFDB) is described in more detail bye Rots (1996). Here we will only provide a brief description of design features that are essential for understanding the data finder.

Our basic design contains a three level hierarchy: Master Index, Subsystem Index, and Data Table. Table 1 outlines this structure. In all tables, the "vertical" (or "row") axis is time (or observation—which amounts to the same), while the "horizontal" (or "column") axis distinguishes spacecraft subsystem units, or data sources.

Table 1. Elements of the FITS Database

Table	Rows	Columns
Master Index	Observations	Subsystem Indices[a]
Subsystem Index	Observation segments	Data Tables[a]
Data Table	Time stamps	Data items

[a]References (in the form of file names) to these tables

Starting at the top of the hierarchy, each row in the Master Index contains, for a single observation, references to all the Subsystem Indices that, in turn, contain references to Data Tables belonging, or pertaining, to that observation. There are Subsystem Index tables for the science instruments, for the spacecraft attitude control system, for the clock corrections, for the orbit ephemerides, for the system of calibration files, etc. In addition, the Master Index contains columns with observational parameters, such as Observation ID, start and stop times, and source information.

The Subsystem Index tables contain rows that correspond to segments of the observation during which all telemetry data for that subsystem were deposited in the same set of Data Tables. Each row contains references to those Data Tables, as well as data mode and configuration information when relevant.

3. Database Navigation

Given a set of selection constraints, such as selected source(s) and time range(s), it becomes fairly simple to navigate through the system and find the data items one is looking for. The Master Index acts as an observation catalog with references to Subsystem Indices, while the latter contain configuration information and references to the Data Tables.

We have written an XTE Data Finder (XDF) as an itcl script with some C code. It allows the user to navigate the database interactively, using the Master Index and Subsystem Indices, producing a list of data files that satisfy the user's selection criteria. As such, it is the data selection interface to subsequent data analysis tools. A sample of XDF's user interface is shown in Figure 1.

4. Data Retrieval

Even though XDF was designed purely as a database navigation tool, it quickly became clear to us that its power would be greatly enhanced if we would incorporate data retrieval capabilities—something that was very easy to do, as a matter of fact. For its navigation functions XDF only needs the information from the index tables; its final product is a list of data tables. It is trivial to incorporate the actual retrieval of these tables (by ftp) to this function. The next step is that not even all index tables need to be resident locally; XDF can

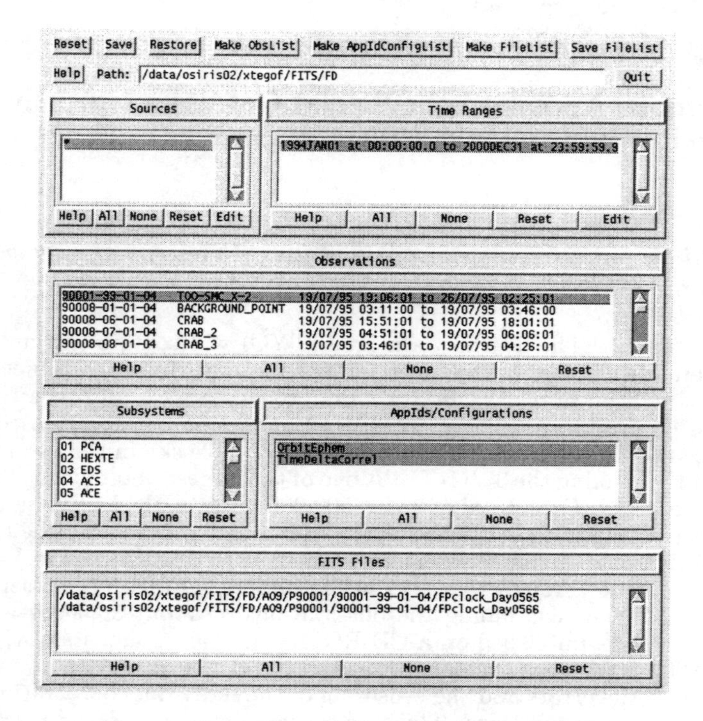

Figure 1. XDF graphical user interface.

decide which ones are required and, again, download them as needed. One might even imagine that, if it cannot find the Master Index Table, it will retrieve that as well.

The result is that XDF allows the user unlimited access to the entire *RXTE* mission archive (at least, the part that is public), while requiring mass storage space only for those data that are actually selected. We would like to emphasize that XDF is primarily the data selection interface to analysis software, and hence the data retrieval capabilities are made transparent to the user, without the need for a separate Web browsing or Internet searching tool. The issue of needing only to store data that are actually going to be used is an important one since the *RXTE* (compressed) archive data volume is of the order of 1 GB/day.

Finally, one more implementation detail. To save space, and to increase data transfer speeds, all data files in the archive are compressed by the utility gzip. XDF will automatically take care of the uncompressing of the data files, too.

References

Rots, A. H. 1996, in ASP Conf. Ser., Vol. 101, Astronomical Data Analysis Software and Systems V, ed. G. H. Jacoby & J. Barnes (San Francisco: ASP), 540

Astronomical Data Analysis Software and Systems VI
ASP Conference Series, Vol. 125, 1997
Gareth Hunt and H. E. Payne, eds.

Remote Access to the Tycho Catalogue and the Tycho Photometric Annex

A. J. Wicenec

Institute for Astronomy and Astrophysics, University of Tübingen, FRG, e-mail: wicenec@astro.uni-tuebingen.de

Abstract. The TYcho Catalogue (TYC) will contain astrometric and mean (de-censored) photometric data of some 1,058,000 objects. The TYC will be published in June 1997 together with all other Hipparcos mission products.[1] The number of observations per object used to produce the TYC is 130 in the mean over the whole sky in the two bands B_T and V_T, making the TYC/TEPA one of the largest photometric databases in the world. For a total of about 500,000 objects, the brightest and some selected fainter objects, the single observations will be published in the so called Tycho Epoch Photometry Annex B (TEPA/B). This Annex will be available through the CDS in Strasbourg, France. A subset thereof, the TEPA/A, containing the observations of about 36,000 selected objects will be published on a CD-ROM. The measurements of the fainter 500,000 TYC stars are not considered to be of a quality to be published, because the errors and the censoring of the individual observations of the faint stars are too large. This paper presents an access and research tool for the TYC and the TEPA to be used locally with widget based GUIs or remotely by HTML-form based WWW-pages.

1. Motivation

Especially the TEPA/B data base needs an easy to use, powerful access tool providing data for selected objects. The Tycho Data Analysis Consortium (TDAC) is using an access tool, where the core is a collection of IDL routines controlled by an IDL-widget based GUI. The size of the TEPA/C (all observations for all TYC objects) sums up to about 50 GB binary data and it thus resides on a single host within TDAC. To provide access for all the TDAC groups there is a WWW-layer above the IDL access tool which copies some of the functionalities to a small number of HTML-form pages. The interface is currently being updated with client side image maps, JavaScript verification and Java tools to limit the network traffic and the server load. If support by the host institute is approved this interface will be opened to the WWW[2] without restrictions at

---

[1] A complete description of the Hipparcos and Tycho Mission and data reduction may be found in (Perryman 1989), some more detailed papers on Tycho and Hipparcos in two series of papers from various authors in (A&A258, 1992 and A&A304, 1995)

[2] http://astro.uni-tuebingen.de/

the time when the Hipparcos and Tycho catalogues are published, i.e., in June 1997.

Since the Tycho instrument on-board the Hipparcos satellite was operated at a fixed sampling rate (integration time), observations for faint stars are censored, i.e., the magnitude distribution of the observations has a clear cut-off caused by the used SNR limit. For the de-censoring and variability studies the number, time, and background of all unsuccessful observations for a given star contains very important information. The two Annexes, TEPA/A and TEPA/B will contain all successful and unsuccessful observations for about 36,442 selected stars (TEPA/A) and the 481,553 brightest stars (TEPA/B) of the TYC. The mean number of observation per star in these catalogues is 191, giving a total of more than 6 million and 90 million observations for TEPA/A and TEPA/B respectively. Access to this photometric data base should be easy to use, yet powerful and flexible.

2. Data Structure

The TYC is a quite normal star catalogue very similar to the GSC in numbering and sorting. Stars common to TYC and GSC version 1.2 do have the same primary (region) and secondary (running) numbers in both catalogues. Yet the TYC contains a third identification number in order to keep double stars resolved by Tycho under the same GSC region and running numbers. The identification numbers in TYC are called TYC1, TYC2 and TYC3, where TYC3 is just 1 for most of the stars. Because of the similar numbering and sorting scheme any catalogue browser capable of browsing through GSC should be able to access the TYC without major changes. On the other hand the TEPA catalogues are much different and they need a special tool to provide access and working capabilities such as time series analysis (variables) or selection of single observations. The TEPA/A will be published on a CD-ROM together with basic access software, i.e., it will be possible to retrieve all observations for one of the TYC-stars contained on the CD-ROM. This software might also be used to retrieve data from the TEPA/B which will only be published through astronomical data centres (CDS, Strasbourg).

The TEPA catalogues contain two different kinds of records: star header records and observation records. Each star header record is followed by a number of observation records. The star header records contain some fundamental data extracted from the TYC, such as the star numbers, and the magnitudes, but no coordinates. The contents of the star headers is merely a compromise between necessary and useful contents and the need to keep the record length at the same length as the observation records are. The most important field of a star header is the number of observation records following the star header. The length of the observation records is adjusted to contain all relevant data for the single observations. Since the TEPA catalogues are intended to be used together with the TYC[3] the user will have access to all published data of a particular star. The TYC and TEPA/A CD-ROMs and the TEPA/B when ordered from CDS will be accompanied by some index files giving a two level indexing of the complete

[3]The TYC contains a field telling the user whether there are single observations available in TEPA/A or TEPA/B

catalogue and its Annex. These index files might be used by a quite simple program to get very fast access to the data of a particular star in TYC and TEPA if one knows the TYC-number (where TYC1 and TYC2 are identical to the GSC number for that star). A more user friendly access should also give the possibility to retrieve the catalogues via coordinates and to produce something like maps from TYC and light curves from TEPA data. Such possibilities are provided by CAT_{MAP}^{TRANS}

3. Core Routines and Catalogue Interfaces

The core routines of CAT_{MAP}^{TRANS} are written in IDL; they use common structures for data exchange and keywords for customising their behaviour. There are catalogue mapping routines which are capable of producing maps, catalogue overlay maps, and image overlay maps. There are a number of selection and coordinate conversion routines as well as an interface to DSS-fits files as produced by the ESO on-line DSS. The routines are virtually independent of the underlying catalogue due to the usage of interface routines between the catalogue input and the internal data structures of CAT_{MAP}^{TRANS}. All the routines are also usable as stand-alone routines and any new routine might use a common initiallisation routine yielding access to the common structures. The access to catalogues is not part of the core routines, but belongs to the catalogue interface. Thus CAT_{MAP}^{TRANS} is able to use every catalogue if the interface routine fills up the common data structure. Due to the flexibility of IDL, interface routines may be written in C, Perl and/or FORTRAN. The catalogue interfaces are responsible for the correct contents of the common structures. Since every catalogue has its special contents some of the core routines are also able to provide access to the complete catalogue information for every object on a map produced from the selected catalogue. The main advantage of using IDL as the programming language of CAT_{MAP}^{TRANS} is the very good portability of IDL-code. This is even enhanced by some special environment files used by CAT_{MAP}^{TRANS} describing the path- and file-names of the available catalogues. The core routines only use string variables for accessing files on the local file-system and are thus independent from file name conventions of the operating system. New catalogues may be easily added to CAT_{MAP}^{TRANS} through some preparation tools providing the interface routines.[4]

4. User Interface

The main user interfaces to the core routines are IDL-widget-based GUIs. They build an easy-to-use layer above the core routines and a offer most of their capabilities by mouse interaction. There are two main GUIs: one for the TYC and other star catalogues and another for the TEPA catalogues. Both are usable as stand-alone interfaces and as an integrated tool. Thus it is possible, for instance, to create a Bright Star Catalogue (BSC) map with an overlay from TYC. By clicking on a star on the map, the TEPA observations for this star will

[4] one of these tools is using a description file for the columns of a catalogue in order to produce an IDL-interface routine

be loaded and the TEPA window will gain control. A click on a button on the latter will pop up an additional window which allows some interactive period investigation by means of a periodogram routine, a minimum entropy routine, and, of course, some plotting routines to visualize the results.

5. WWW Interfaces

The other user-interface to the core routines is WWW-based and runs on every browser that has form and client-side image map capability. There are some simple HTML-forms which ask for coordinate or TYC number input plus some auxiliary input, such as the diameter of the map to be produced. The `<SUBMIT>` button will send this query to the HTTP-server. The underlying CGI mechanism produces an IDL routine from the query string and runs IDL with this routine as a startup file. The output to the HTTP-server and thus the user is of NPH-HTML type importing a GIF-image of the requested part of the sky. The HTML code contains data to describe the image as a client-side image map, merely the TYC-numbers of the stars on this image. A click onto the map will result in a query to another CGI-interface producing a list of all transits of the star closest to the click position. This interface is currently being updated to a functionality comparable to the IDL widget GUI.

Acknowledgments. This work was supported by the DARA under grants No. 01-OO-85029. I am grateful to the Tycho Data Analysis Consortium and the Tycho Data Analysis Working Group in Tübingen.

References

Perryman, M. A. C. P., et al. 1989, ESA-SP 1111 Vol. I-III

Astronomical Data Analysis Software and Systems VI
ASP Conference Series, Vol. 125, 1997
Gareth Hunt and H. E. Payne, eds.

The LASCO Data Archive

D. Wang,[1] R. A. Howard, S. E. Paswaters,[1] A. E. Esfandiari,[1] and N. Rich[1]

Solar Physics Branch, Naval Research Lab (NRL), 4555 Overlook Ave SW, Washington DC 20375

Abstract. The data archive for the Large Angle Spectrometric Coronagraph (LASCO) and the Extreme Ultraviolet Imaging Telescope (EIT) is designed to contain 1 B of image data in an easy to use CD-R based archive. This paper discusses the planning, implementation and cost considerations of designing the archive. The problem of getting data into the archive and distributing data to Co-I institutes is also discussed.

1. Introduction

The Large Angle Spectrometric Coronagraph Observatory (LASCO) and the Extreme Ultraviolet Imaging Telescope (EIT) are instruments aboard the Solar Heliospheric Observatory (SOHO). SOHO was launched by ESA/NASA on December 2, 1995. The three coronagraphs of LASCO and the EIT telescope produce the equivalent of 100 1024×1024 16-bit images of the Sun and the solar corona out to 30 solar radii each day. Since first light on Jan 2, 1996 approximately 40,000 images of the corona of the sun have been taken. The LASCO data archive is responsible for storing all of the data and distributing all of LASCO data.

2. Planning and Implementation

Planning for the LASCO Data Archive started in 1994 with the purchase of a Sybase database, development of database tables and understanding what the database could and could not do. By 1995 we were ready to make hardware purchases. Since LASCO is a joint effort between the Naval Research Lab and three major European co-investigator institutions, the problem of sharing the data was an important consideration. Each institute wanted a copy of the data in a timely fashion. Magnetic tape was clearly the cheapest solution for distribution but difficult to work with since the data would have to be staged on and off of magnetic disk to be useful. The amount of data predicted was clearly beyond what could be budgeted for magnetic disk at each institute (in early 1995 magnetic disk was approximately $1/MB). Optical disk was a possibility but had uncertain support for all the various computers and operating systems used by our co-investigators.

---

[1] Interferometrics Inc., 14120 Parke Long Court—Suite 103, Chantilly VA 20151

The recordable CD-ROM (CD-R) was the ideal solution. The format was usable by almost any computer. The disks are archival with an estimated 100 year lifetime with proper storage. The amount of data was convenient with each CD-R containing several days of data and easy to browse. The media was inexpensive with blank media costing only $0.01/MB which is important when five copies of each CD-R must be made (one for each of four institutes and one for archival storage). The cost of 1 TB is only about $10,000; an amount that is affordable. The only disadvantages are that CD-Rs are much slower to access than magnetic disks and require several hours to assemble a CD-ROM image and write the CD-Rs. CD-Rs have become very popular. The price of CD-R recorders has dropped from $2,500 in 1995 to less than $1,000 at present. CD-Rs are so popular that the media manufacturers were overwhelmed with demand earlier this year and there were long back order times.

Large CD-ROM jukeboxes meant that all of our data could be on-line. Solar scientists often wish to conduct studies that span images spread over a solar rotation (27 days) or even several rotations. Having some data off-line makes assembling datasets more difficult and harder to manage. The 500 disk jukebox we chose provides 325 GB of storage. Using smaller jukeboxes would offer faster access times under heavy usage because the ratio of CD-ROMs to drives would be better, but it also would have cost significantly more money for the same storage capacity. The jukebox format was favored over a carousel both because of its capacity and because a carousel rotates all of its CD-ROMs each time a CD-ROM is loaded or unloaded whereas a jukebox handles only a single CD-ROM with each load or unload operation. Although good statistics for the lifetime of CD-ROMs in carousels versus jukeboxes do not exist, it is intuitive that less motion is better for the life of the CD-R. The total cost of the CD-ROM jukebox, software and 500 CD-Rs to fill it was about $0.07/MB in 1995. Even though the cost of magnetic disk has dropped to $0.20/MB in 1996, CD-Rs other advantages still make it a good choice for use as an on-line storage medium.

3. Archive Operations

The LASCO ground support system at Goddard Space Flight Center receives data in three ways. During times when the NASA Deep Space Network (DSN) is in contact, the data packets are handled in realtime. Since the contact period is typically 8 hours, the remaining 16 hours of non-contact time are covered by an on-board tape recorder which is dumped into packet files when contact is re-established. The final method used to input data is a CD-ROM from the Deep Space Network some weeks after the date of observation. This CD-ROM is produced after DSN has applied all of its error correction and packet recovery techniques. The raw packets are processed by the data reduction pipeline which takes the raw telemetry and produces raw image files. The raw image files are decompressed, rotated so that solar north is up and made into FITS files. The FITS files are then made available to observers at Goddard. The entire process is automated and the FITS files are usable and displayed at the LASCO Experiment Operations Facility in about a minute. The raw image files and packet files are also processed at NRL and stored on magnetic disk on the NRL data server. This provides redundant data reduction capability in case some malfunction occurs at NRL or GSFC. The DSN CD-ROMs containing corrected

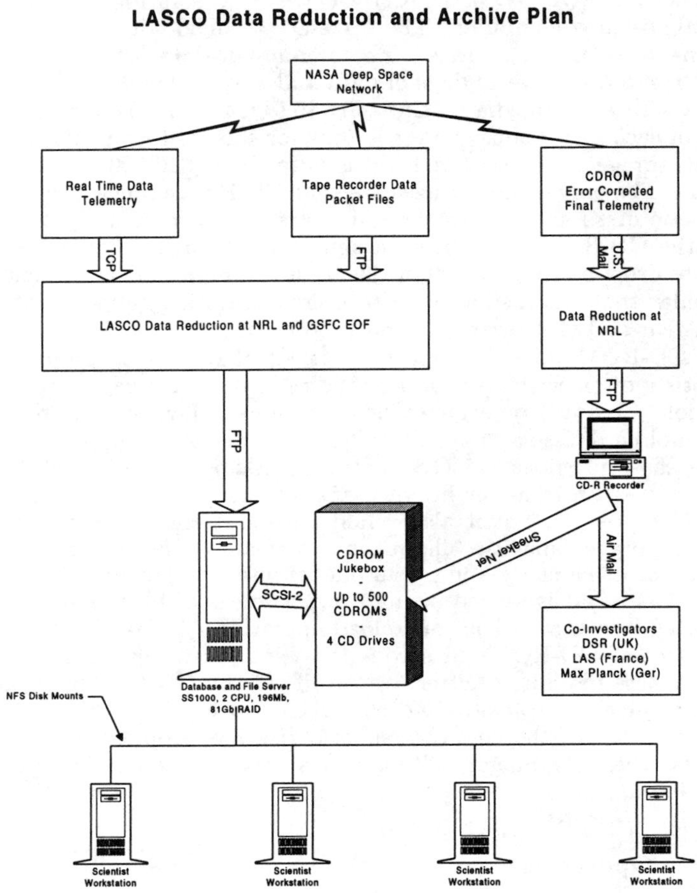

LASCO Data Reduction and Archive Plan

Figure 1. Schematic of the LASCO data archive and data flow.

data are generally processed at NRL as they arrive. This final data is put onto CD-Rs and then into the CD-ROM jukebox. The data flow is shown in Figure 1.

The data server is arranged so that as data is moved from magnetic disk to the CD-ROM jukebox the directory path does not change, so a scientist accessing data via NFS doesn't need to know whether the data is on CD-R or on magnetic disk. The jukebox software maintains a disk cache which stores the directory of each CD-ROM and the first kilobytes of each file making it possible to browse the directories and for file manager software to identify the type of each file (document, image, JPEG image etc.) without mounting the CD-ROM.

The data reduction process also generates a text file list of images taken for our WWW server and a set of SQL commands for updating our database. The text file is commonly used to monitor the data taking in realtime. The

SQL commands update the image information in the database and also store a JPEG browse image. The database and browse images can be accessed through our WWW page[2] using software written by the European Southern Observatory and generally updated within a day or two of the date of observation. Although this could be done as a realtime process, batch processing is preferred because of occasional network outages between our Sybase database server at NRL and the LASCO EOF at Goddard. The database can also be accessed during data analysis by using IDL routines and C routines written at NRL (Esfandiari et al. 1997).

4. Future Plans

Immediate plans for the archive are a network upgrade to Fast Ethernet in November 1996. Future plans include the possibility of adding more CPUs to our data server as the amount of data grows and another CD-ROM jukebox when we outgrow our present jukebox. One nice feature of jukeboxes is that the access time does not increase as jukeboxes are added since the ratio of CD-ROM drives to CD-ROMs can be kept constant. If the new Digital Video Disk (DVD) technology becomes successful it may not even be necessary to buy another jukebox since we could just replace the drives in our present jukebox with new DVD drives. DVD promises 4.7 GB of storage on a CD-ROM sized disk with readers which are backward compatible with CD-ROM and faster access times than CD-ROM. Double sided DVD disks offer over 8 GB on a single disk. Recordable DVD disks are perhaps a year or two in the future but are expected to cost little more than the present CD-Rs.

References

Esfandiari, A. E., Paswaters, S. E., Wang, D., & Howard, R. A. 1997, this volume, 353

[2]http://lasco-www.nrl.navy.mil/lasco.html

Astronomical Data Analysis Software and Systems VI
ASP Conference Series, Vol. 125, 1997
Gareth Hunt and H. E. Payne, eds.

The Evolution of the HST Archive

J. J. Travisano,[1] J. G. Richon

Space Telescope Science Institute, Baltimore, MD 21218

Abstract. The Hubble Space Telescope Archive has been in operation since the launch of HST in April 1990. There have been two generations of archive systems (DMF and DADS) and user interfaces (STARCAT and StarView). In this paper, we describe recent projects and future directions in the continued evolution of the HST Archive.

1. Components of the HST Archive at STScI

The Data Archive and Distribution System—DADS—comprises the core of the Archive (Pollizzi 1995). It stores all HST data onto optical disks. All datasets are cataloged into a Sybase database from the science header keywords and dataset/file information. DADS runs on Digital Alpha and VAX platforms.

StarView is the primary user interface to the HST Archive (Williams 1993). It is used to query the DADS databases, preview selected images and spectra, and submit retrieval requests for data. StarView is supported in CRT and X/Motif modes on Sun and Digital platforms.

World Wide Web (WWW) access is now available. Additional information on this and the HST Archive in general is available through the STScI WWW pages.[2]

2. DADS

2.1. Recent Enhancements

Blackboard Ingest and Distribution The Blackboard Ingest and Distribution projects replaced internal memory lists and mailbox messages with a blackboard trigger mechanism. The OpenVMS cluster-wide filesystem acts as a blackboard, upon which files are created with names indicating work to be done. Individual processes look to this global area for work, claim it, and perform it.

The idea of using blackboards was adopted from the STScI OPUS system (Rose 1995). This architecture gives DADS increased flexibility in that processes can run on different hosts in the cluster. Multiple copies of some programs can be instantiated, improving scalability in dealing with peak loading conditions.

---

[1]Computer Sciences Corporation

[2]http://www.stsci.edu/

Aptec Replacement & Port to Alpha The original DADS system delivered to STScI consisted of a cluster of VAX systems with an Aptec I/O Processor. The Aptec was used for performing I/O to the Sony optical disk drives and for doing data format conversion from STSDAS GEIS format to FITS. This specialized device was not well suited for this application, was unreliable, put a high kernel interrupt load on the host VAX, and was difficult to maintain (both hardware and software).

A project was started in 1995 to replace the Aptec. This project included porting most of DADS to OpenVMS Alpha, replacing the optical disk I/O routines to use native OpenVMS and SCSI services, and using an IRAF/STSDAS task (stwfits) for FITS conversion. The Aptec staging disks were replaced with a RAID array that is local to the Alpha host and cluster-accessible. The resulting system is faster and more reliable, with reduced maintenance costs.

2.2. Work In Progress

HST Servicing Mission 2—February 1997 Significant efforts are underway to support the second HST Servicing Mission. Two new instruments will be installed in the telescope: the Space Telescope Imaging Spectrograph (STIS) and the Near Infrared Camera and Multi-Object Spectrometer (NICMOS).

DADS is being modified to ingest the new FITS keywords in the STIS and NICMOS data files. FITS extensions are used, combining images and tables into single files. There is the new concept of *associations* of datasets, where individual exposures, products from processing these exposures, and ancillary files are grouped together into a collected set.

STIS and NICMOS will produce significantly more data than the current instruments. This and the fact that more observations can be done in parallel will result in three to six times as much data going into the Archive.

Compression One of the biggest problems facing DADS is that the optical disk jukeboxes will soon be filled to capacity, sometime in the spring of 1997. Part of the solution is to compress the files on optical disk. This increases the overall near-line capacity of the Archive, as discussed in the Hubble Archive Re-Engineering Project (HARP) (Hanisch 1997).

2.3. Planned/Future

Archive Engine Redesign The DADS Archive Engine controls the I/O to and from the optical disks and manages the jukeboxes. We are planning to redesign this subsystem to improve overall efficiency, give more control to the operations staff for scheduling/directing requests, allow better segregation of data for near-line vs. off-line storage, and support a magnetic disk cache (Comeau 1997) to reduce contention as well as wear and tear on the jukebox and drive hardware.

Multiple Media DADS uses Sony 12-inch WORM optical disks (6 GB per platter) with Cygnet jukeboxes (four of them holding 131 platters each). We are investigating other archival media and plan to modify DADS to support multiple media. In particular, DVD-ROM looks promising for its storage capacity, lower costs, and the availability of jukeboxes of 500 or more disks. For distribution media, we are planning to support 4mm DAT and CD-ROM in 1997.

System Migration There are more applications to port to the Alpha to complete the migration from VAX. We are considering alternate operating systems for all or parts of DADS. The Alpha hardware can run OpenVMS, Digital UNIX, and Windows NT. Linux is also available for some Alpha systems.

On-The-Fly Recalibration The Space Telescope European Coordinating Facility (ST-ECF) and Canadian Astronomy Data Centre (CADC) have copies of HST data. Both sites support on-the-fly recalibration (Crabtree 1996). We discussed this idea in HARP as a way to reduce the amount of archived data. Due to user support needs, it is not feasible to store only raw science data. However, on-the-fly recalibration will be a useful service for the HST Archive at STScI.

Non-HST Data The Archive contains non-HST data, such as the VLA FIRST Survey (Faint Images of the Radio Sky at Twenty-cm). More than 5000 VLA FIRST datasets have been archived, each catalogued by over 20 parameters (RA, Dec, frequency, etc.). We continue to consider the potential needs of non-HST data, positioning the Archive for new data from HST and other observatories.

HST Servicing Mission 3—1999 The third HST Servicing Mission will add the Advanced Camera for Surveys (ACS). There will be an increase in data volume, estimated to be double the volume after Servicing Mission 2.

3. StarView & The World Wide Web

3.1. Recent Enhancements

World Wide Web Interface After a few prototypes, including one presented at ADASS IV (Travisano 1995), a Web interface to the HST Archive is now available. Users can search for observations and proposal information, preview images and spectra of public data, and retrieve datasets to an FTP area on archive.stsci.edu. The Web interface uses HTML and CGI scripts written mostly in Perl, along with some back-end StarView utilities.

HST Field of View Overlay on The Digitized Sky Survey From StarView or the Web interface, a user can ask for any section of the sky from the Digitized Sky Survey (DSS). The user can now request the HST field of view, containing the primary instrument apertures, as a graphics overlay on the DSS image.

Automatic Retrieval of Calibration Reference Files Retrieving the reference files necessary to recalibrate science data formerly required multiple StarView screens. Now the user can select whether the original and/or best reference files are to be retrieved along with the selected science datasets. The StarView retrieve utility (also used by the Web interface) will look up the appropriate files and add them to the retrieve request sent to DADS.

Improved String/Text Searching Until recently, the only string searching capabilities supported were an explicit string, a wildcarded string, and an **or** list of these two. StarView Release 4.5 includes support for the **and** operator and the **not** qualifier. This improves the ability to find what you are looking for, and exclude what you are not, especially when searching on proposal abstracts.

3.2. Work In Progress

HST Servicing Mission 2—February 1997 StarView and the Web interface are being modified to support STIS and NICMOS. This work includes data dictionary updates, new instrument screens, the HST field of view overlays, screens for associations, additional retrieval options, etc.

Improved Coordinate System Support StarView Release 4.5 includes improvements to display the coordinate system and equinox on query results screens. These displays are based on user selections of how to display coordinate values from the catalog, which are stored in Equatorial J2000.

3.3. Planned/Future

Continued Improvements to Web Interface The Web interface was released in September 1996. We are monitoring usage and feedback to plan new features. Supporting Internet delivery of data will require safeguarding the account information sent from Web browsers to the server.

HST Servicing Mission 3—1999 For the third servicing mission, the StarView and Web interfaces will again be updated. Instrument screens for ACS, the field of view overlays, etc., will be added in support of the mission.

3.4. Concluding Remarks

We continue to improve the HST Archive systems—DADS, StarView, and the new Web interface. The HST Servicing Missions in 1997 and 1999 require much work, as do the continued efficiency improvements to DADS necessary to support operations on a daily basis for the 1000 or so registered HST Archive users.

References

Comeau, T., & Park, V. 1997, this volume, 318

Crabtree, D., Durand, D., Gaudet, S., & Hill, N. 1996, in ASP Conf. Ser., Vol. 101, Astronomical Data Analysis Software and Systems V, ed. G. H. Jacoby & J. Barnes (San Francisco: ASP), 505

Hanisch, R., et al. 1997, this volume, 294

Pollizzi, J. 1995, in ASP Conf. Ser., Vol. 77, Astronomical Data Analysis Software and Systems IV, ed. R. A. Shaw, H. E. Payne, & J. J. E. Hayes (San Francisco: ASP), 162

Rose, J., et al. 1995, in ASP Conf. Ser., Vol. 77, Astronomical Data Analysis Software and Systems IV, ed. R. A. Shaw, H. E. Payne, & J. J. E. Hayes (San Francisco: ASP), 429

Travisano, J. 1995, in ASP Conf. Ser., Vol. 77, Astronomical Data Analysis Software and Systems IV, ed. R. A. Shaw, H. E. Payne, & J. J. E. Hayes (San Francisco: ASP), 80

Williams, J. 1993, in ASP Conf. Ser., Vol. 52, Astronomical Data Analysis Software and Systems II, ed. R. J. Hanisch, R. J. V. Brissenden, & J. Barnes (San Francisco: ASP), 100

Astronomical Data Analysis Software and Systems VI
ASP Conference Series, Vol. 125, 1997
Gareth Hunt and H. E. Payne, eds.

Implementing a New Data Archive Paradigm to Face HST Increased Data Flow

B. Pirenne,[1] P. Benvenuti,[2] and R. Albrecht[2]

Space Telescope – European Coordinating Facility

Abstract. The Hubble Space Telescope Archive at the Space Telescope – European Coordinating Facility[3] (ST-ECF) has undergone several refurbishments in order to cope with user requirements and with the advances in storage technology. In preparation for the installation of the Near Infrared Camera and Multi-Object Spectrometer (NICMOS) and Space Telescope Imaging Spectrograph (STIS) instruments during the 1997 Servicing Mission, it is again necessary to upgrade the archive. This paper describes the adopted strategy and its rationale.

1. Introduction

The science instruments to be installed on *HST* during the 1997 Servicing Mission will substantially increase the current data volume. In order to cope with this increase some of the ECF archive concepts need to be adapted. A second reason for modifying the archive is the high cost of the bulk storage media and the associated hardware. As CD-ROMs have firmly established themselves in the computer market in recent years, it is attractive to look for a solution based on CD-ROM technology. A third reason is that, because of the availability of fast world wide networking, the initial requirement of having in Europe (i.e., at the ECF) the exact copy of the STScI archive can be relaxed. The ECF effort in the archive area should therefore focus on a more dynamic archive environment which complements the work of the STScI. This new approach has been endorsed in the "Report of the ST-ECF Independent Review Group" (May 1996).

2. The Current Situation

Following the transition of the STScI archive to the "DADS" system in 1993, the ECF started to archive *HST* data on the bulk data devices used in DADS, i.e., 12″ Sony optical disks (6.5 GB per platter). The disks are generated at STScI and shipped to the ST-ECF. Currently the ST-ECF receives a full copy of the

---

[1] European Southern Observatory, Garching bei München, Germany

[2] European Space Agency, Space Science Department, Astrophysics Division

[3] http://archive.eso.org/

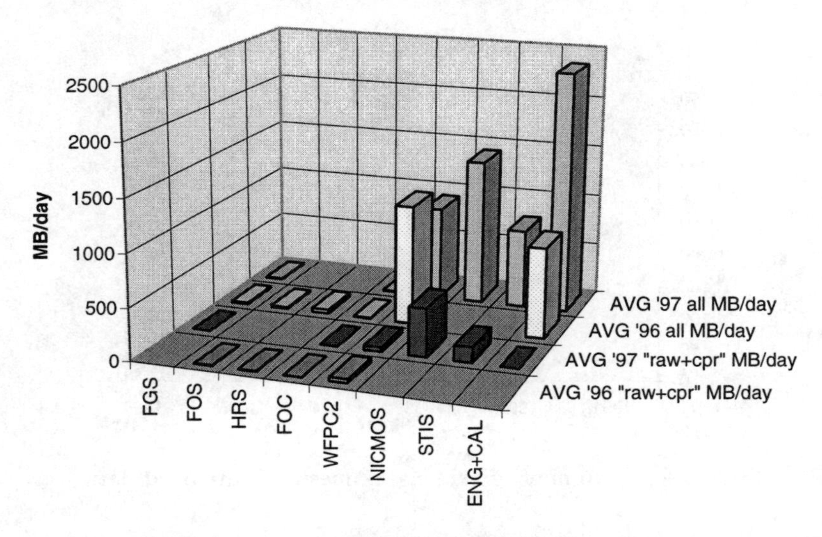

Figure 1. Share of data volume by data type.

archive, including engineering ancillary data. The current average data rate is ~2 GB/day or ~112 Sony disks per year.

Since we receive an exact copy of the archive, software protection mechanisms ensure that proprietary data can only be delivered to authorized users.

3. The Data Rate After the Second Servicing Mission

With the installation of STIS and NICMOS on *HST* in February 97, the data rate is expected to increase up to ~5.4 GB/day or 303 Sony disks per year (270% increase, see Figure 1). It should be noted that a large fraction of the data volume is represented by engineering data (~0.8 GB/day current, ~2 GB/day after 97). The access to engineering data by the (external) European community has been nil.

4. CD-ROM Technology

While CD-ROMs have considerably less capacity than the current Sony optical disks (650 MB/volume vs. 6.4 GB per platter) they are more cost effective owing to their very low unit price (8 USD vs. 300 USD unit cost). In addition, all CD-ROM related hardware (readers and juke-boxes) is cheap, while Sony optical disk hardware—in particular juke boxes—is quite expensive. It is envisaged that the CD-ROMs will eventually be replaced by Digital Versatile Disks (DVD, 3.95 GB per platter; see Scientific American, July 96), which use similar technology and are expected to constitute the next generation CD-ROM standard. DVD readers are expected to be backward compatible with CD-ROMs.

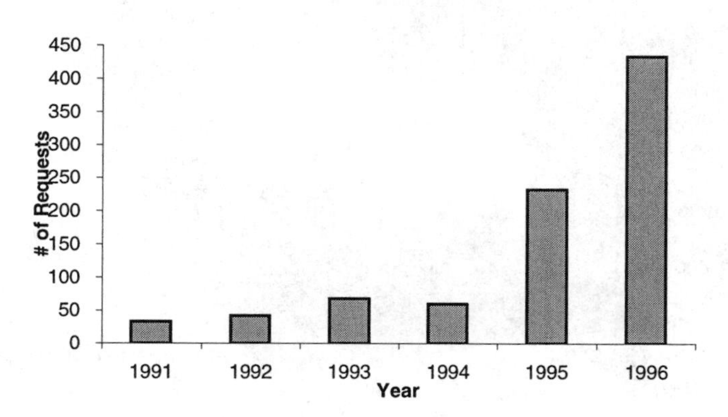

Figure 2. Number of external requests for archived data.

5. On-the-fly Calibration

The Canadian Astronomy Data Centre (CADC) and the ECF developed the capability to calibrate data "on-the-fly," i.e., during the process of retrieving them from the archive (Crabtree et al. 1995). This method has the advantage that the data can be re-calibrated using the best available calibration files, as opposed to the standard calibration done right after taking the observations, when the data have to be calibrated using calibration files taken before the observations. The process has been tested operationally on representative data sets and found to work in a satisfactory manner. Since the capability was announced in May 1996, we noticed an increase of archive requests as users try to improve the quality of the calibration of their data (Figure 2).

Beside offering a valuable user service, on-the-fly re-calibration alleviates the need to transfer and archive calibrated data, considerably reducing the total data volume. However, it also implies that all data distributed out of the archive will have to be calibrated prior to distribution, including multiple re-calibrations for repeated requests of the same data products. On the other hand it makes it possible to have all the raw data on line by copying them to CD-ROMs mounted on jukeboxes (currently, the entire mission *raw science data* is stored on about 60 CD-ROMs). Additional juke-boxes can subsequently be added to the system, making the retrieval and re-calibration process entirely automatic.

Future astronomical projects (e.g., VLT, NGST, ...) are already planning their archives on the on-the-fly calibration concept.

Normal requests and mass retrievals will be handled by spawning off calibration tasks to various archive and ECF computers. This is possible because the ST-ECF Archive Request Handler and the calibration pipeline (OPUS, see Rose et al. 1995) were designed to share the calibration tasks among many machines. The long term solution is to ship compressed raw data and calibration files through the network and perform the decompression and the calibration via client software (e.g., a Java applet) at the user site. But this approach is still beyond our reach.

6. Observation vs. Exposure

So far, the *HST* catalogue only included the notion of individual exposures. The new NICMOS and STIS instrument will introduce the concept of "associations" of exposures (sometimes also called product). Associations group exposures into logical observations, opening up windows of opportunity for further automatic processing of entire observations (e.g., mosaics of neighboring areas of the sky). This concept was not present with previous instruments. The ST-ECF archive is presently retrofitting the existing WFPC2 exposures archive into groups of observations. These groups will enable users to look at logical groups when browsing the catalogue. It will also enable us to provide extra services as part of the on-the-fly recalibration (e.g., cosmic ray rejection).

7. PreView

One of the first major enhancement brought to the Hubble Space Telescope Archive by the CADC and ECF was the concept of PreView: "Imagettes" processed with the current best calibration system are available on-line in the form of PreView (Quick-look) images or spectra that can be viewed instantaneously, thereby helping users assess exposure data quality in a convenient way. PreView will be re-generated on a regular basis so as to use the best calibration available. For the WFPC2, we also plan to use, whenever possible, the cosmic ray-cleaned image.

8. Conclusions

The plan we are describing here involves moving away from expensive 12″ optical storage technology in favor of more economical and stable CD-ROMs. In doing so, we benefit from the availability of the data on-line at a fraction of the current costs. We also open up a window of opportunity to execute large research projects requiring access to a substantial fraction of the archive.

References

Report of the ST-ECF Independent Review Group, ST–ECF Document, 1996

Crabtree, D., Durand, D., Gaudet, S, Hill, N., Pirenne, B., & Rosa, M. 1996, in ASP Conf. Ser., Vol. 101, Astronomical Data Analysis Software and Systems V, ed. G. H. Jacoby & J. Barnes (San Francisco: ASP), 505

Rose, J., Choo, T. H., & Rose, M. A. 1996, in ASP Conf. Ser., Vol. 101, Astronomical Data Analysis Software and Systems V, ed. G. H. Jacoby & J. Barnes (San Francisco: ASP), 311

Astronomical Data Analysis Software and Systems VI
ASP Conference Series, Vol. 125, 1997
Gareth Hunt and H. E. Payne, eds.

HARP—The Hubble Archive Re-Engineering Project

R. J. Hanisch, F. Abney, M. Donahue, L. Gardner, E. Hopkins, H. Kennedy, M. Kyprianou, J. Pollizzi, M. Postman, J. Richon, D. Swade, J. Travisano, and R. White

Space Telescope Science Institute, 3700 San Martin Drive, Baltimore, MD 21218, E-mail: hanisch@stsci.edu

Abstract. The Hubble Data Archive now contains in excess of 2.5 TB of HST data in a system of four optical disk jukeboxes. In addition to providing a WWW-based user interface and removing a custom I/O processor (see Travisano & Richon 1997), STScI has undertaken a high-level effort to improve the operating efficiency, reduce costs, and improve service to archive users. The HARP group is studying data compression, data segregation, large on-line disk caching, on-the-fly calibration, and migration to new storage media. In this paper, we describe the results of our cost-benefit analysis of these and other options for re-engineering the HDA.

1. Introduction

The Hubble Data Archive (HDA) contains over 2.5 TB of near-on-line data. The data are stored on 12-inch WORM optical disks (the OD drives and platters are manufactured by Sony). These disks are mounted in four optical disk jukeboxes (Cygnet). Each jukebox has 131 OD slots. With a storage capacity of 6 GB per platter, the total near-line capacity of the current HDA configuration is 3.1 TB. Data ingest and retrievals are managed by the Data Archive and Distribution System—DADS. DADS operates on a mixed VAX and DEC Alpha architecture cluster, and work is now underway to migrate the entire system to the Alpha OpenVMS environment. Data enter DADS via an FDDI link from the OPUS data processing pipeline. Data are written simultaneously to two optical disks: one that will be used in daily operations, and another than is set aside in a safe area as a backup. Two additional copies of the data are made for the Space Telescope-European Coordinating Facility and the Canadian Astronomy Data Center.

2. Challenges Facing the Hubble Archive

The 12-inch WORM media and optical disk jukeboxes are nearing obsolescence. The jukebox systems are expensive (approximately $150k each), the blank media are also expensive (approximately $300 each), and the optical disk drives are expensive to maintain. We expect this medium to be obsolete once DVD—Digital Versatile Disk—technology is available. We expect DVD media and robotics costs to be much lower. Of course, we have always known that we would have to migrate to new media at some point, and the architecture of DADS should allow for heterogeneous operations as we transition from 12-inch WORM to DVD. The installation of STIS and NICMOS during the 1997 Servicing Mission, and of ACS in 1999, will lead to a factor of 3–6 increase in the volume of data generated by HST. Without changing our approach to archiving, we will be filling optical disk jukeboxes at the alarming rate of one every four months. The alternative is to incur the expense of providing enough operations staff to manage the off-line disk handling and on-demand mounting.

The CADC and ST-ECF have implemented a new "on-the-fly" calibration (OTFC) facility for Hubble data. Using compressed, raw science data as the basic data source, data are calibrated as they are requested by archive users. By calibrating on-the-fly, the most recent or best calibration reference files can be used. Observers are therefore provided with a higher quality result than they might get from analyzing the data archived immediately following the standard pipeline calibration. OTFC is most effective for data from the WFPC and WFPC2, where flat fields and dark-current corrections are often improved in the weeks following an observation. Since the pipeline calibration is normally performed within 24 hours after data are taken, the calibrated data in the archive for WFPC and WFPC2 are almost all sub-optimal.

3. Archive Efficiency Improvements

The HARP team has identified a number of areas in which the efficiency of the Hubble Archive can be increased and costs can be reduced. Areas that are now under consideration for implementation in the coming year are described below.

Data Compression. Currently, data in the HDA are not compressed. Raw data from the current HST instruments can be compressed losslessly by at least a factor of three, and perhaps as much as a factor of ten (depending on instrument and observing mode). Overall, including calibrated science data and engineering data, we expect to achieve compression ratios of about three. Lossy compression techniques can achieve much higher compression ratios, and for WFPC in particular, we are considering at least a simple rounding of low-order bits in the calibrated (floating point) data. Allowing users to retrieve compressed data also eases network loading. Data compression ratios for STIS and NICMOS have not yet been evaluated thoroughly, though our current expectation is that STIS data will compress quite well, and NICMOS data may not compress at all well (owing to strongly non-uniform backgrounds in the IR). Owing to its good overall efficiency, availability of source code, and ubiquitous use in the community, we will use the gzip compression algorithm.

Data Segregation. All types of HST data—raw telemetry, engineering data, and science data—are written to the currently open optical disks. Infrequently accessed data, such as raw telemetry, are intermixed with frequently accessed science data. By writing different types of data to different storage optical disks, or different storage devices, the infrequently accessed data could be moved to off-line storage with little impact on operations. Data segregation has already been implemented via a semi-automated procedure in archive operations.

Secondary Load. Data compression and data segregation can be combined to rewrite the existing archive contents onto a new set of ODs in which all data are compressed, and all infrequently accessed data are not copied. The existing disks would be moved out of the jukeboxes and replaced with disks contained the compressed, segregated data. If off-line data are requested, they can still be provided via the operator. The lifetime of the existing four OD jukeboxes can be extended to well into 1998 if we begin Secondary Load by spring of 1997.

On-the-Fly Calibration. An OTFC facility provides archival researchers with a "best" calibration (the best that has been determined by the time the data are retrieved from the archive) without requiring the researcher to go through a tedious and complex recalibration process themselves. The CADC and ST-ECF have implemented such a facility, and by storing only compressed, raw science data and generating calibrated data on demand, they have reduced the archive of currently public HST data to a volume that fits on ~60 CD-ROMs (Crabtree et al. 1996). The greatest benefit in OTFC, at least as far as efficiency of storage is concerned, comes from WFPC data. Raw WFPC and WFPC2 data compress quite easily by a factor of ten, while the process of calibration increases the total volume of data by another factor of ten. GHRS and FOS data are small in volume, and storing their calibrated data in the archive has virtually no effect on overall storage requirements. FOC data are somewhat larger, but the instrument is used less frequently, and has much more stable calibrations. The computational load of OTFC for the existing HST instruments is quite manageable on a high-end workstation. The case for OTFC is not so clear for STIS and NICMOS. In both cases the calibrated result, i.e., extracted spectra or combined images, is a more compact product. The calibration process requires having large numbers of raw data files available simultaneously, and the CPU requirements are considerably greater than for WFPC and WFPC2.

The support staff at STScI is often required to answer questions from HST guest observers (GOs) concerning the quality of, or artifacts in, their data. Answering such questions requires that the STScI staff has an identical version of the data sent to the GO, or that an identical version can quickly be generated. OTFC can potentially yield different results every time it is used, depending on how frequently the calibration reference files or calibration algorithms themselves are updated.

The HARP team has not yet reached a conclusion on the advisability of OTFC as a means for increasing archive efficiency. It is clearly valuable as a user service, but aside from WFPC and WFPC2, where we can probably achieve sufficient efficiency simply by compression of both raw and calibrated data already in the archive, OTFC does not obviously reduce the overall HDA data volume.

Large Disk Cache. As currently implemented, DADS does not cache data that are frequently accessed or are likely to be accessed. As a result, the optical disk jukeboxes are exercised at a maximum rate, and users must wait periods from minutes to hours before their data are be retrieved. A large disk cache would provide more immediate access to frequently accessed data.

Unfortunately, data access patterns for the HDA are not very simple, and cache implementation is not straightforward (Comeau 1996, Comeau & Park 1997). Current usage patterns indicate that a preloaded cache with a capacity for at least one month's worth of science data (~20–30 GB) would streamline archive performance. Additional disk space would be required in support of data verification and to provide an intermediate time-scale back-up facility.

4. Summary—HARP Study Recommendations

The primary goal of the HARP study group is to reduce HDA operational costs without compromising user service. The key recommendations for achieving this goal include:

- **Segregate science and engineering data.** Engineering data comprise about 1/3 of the archive data storage but are accessed only rarely.
- **Compress all data.** Use gzip on all data sets, both raw and calibrated. Some data sets will not compress very well, but this will apply a standard utility and allows users to retrieve compressed data and uncompress it on their end, thus minimizing network loading.
- **Implement secondary load.** Segregate and compress all data in the existing archive in order to move infrequently used data off-line and conserve jukebox slots.
- **Migrate to DVD technology.** Begin work now on supporting a heterogeneous archive, and eventually migrate all existing data to DVD. CD-ROM will be used as a prototyping medium. We prefer not to transition first to CD-ROM and then to DVD, to avoid operating an archive with three distinct types of media.
- **Implement a Disk Cache.** Handle data retrieval requests that occur within one month of ingest from a preloaded cache rather than requiring full retrieval from the jukeboxes.
- **Implement On-the-Fly Calibration as a User Service.** Provide OTFC to make it easy for users to recalibrate data using the most current calibration reference files and software.

References

Comeau, T. 1996, in ASP Conf. Ser., Vol. 101, Astronomical Data Analysis Software and Systems V, ed. G. H. Jacoby & J. Barnes (San Francisco: ASP), 497

Comeau, T., & Park, V. 1997, this volume, 318

Crabtree, D., Durand, D., Gaudet, S., & Hill, N. 1996, in ASP Conf. Ser., Vol. 101, Astronomical Data Analysis Software and Systems V, ed. G. H. Jacoby & J. Barnes (San Francisco: ASP), 505

Travisano, J., & Richon, J. 1997, this volume, 286

Astronomical Data Analysis Software and Systems VI
ASP Conference Series, Vol. 125, 1997
Gareth Hunt and H. E. Payne, eds.

Integrating the HST Guide Star Catalog into the NASA/IPAC Extragalactic Database: Initial Results

Oleg Yu. Malkov and Oleg M. Smirnov

Institute of Astronomy, Russian Academy of Sciences, 48 Pyatnitskaya St., Moscow 109017, Russia

Abstract. We report initial results of cross-identification of extragalactic objects from the NASA/IPAC database and Guide Star Catalog (GSC). A distribution of galaxies on the sky is discussed as a tool for estimating the probability for a given GSC object to be of a particular type.

1. Introduction

The GSC (with 20 million objects, making it the biggest and most complete sky survey to date) provides accurate object positions. The NASA/IPAC Extragalactic Database (NED)[1] contains positions, extensive data, and 1,023,000 cross-identifications for over 592,000 extragalactic objects (galaxies, quasars, and radio sources). NED currently represents the unique merger of some 40 major catalogs and many shorter listings. Catalogs and lists are being integrated into NED on a continuing basis, following a detailed cross-identification process.

Benefits of NED-GSC cross-identification are obvious for both sides. For many astronomical studies the GSC is unique; but its full potential is still out of reach. At present, the GSC lacks even rudimentary cross-identifications with other astronomical catalogs, and consequently little is known about either the nature of the objects themselves or their relationship to other data already cataloged and independently available in machine-readable form. NED can make use of the accurate positions found in the GSC.

2. General Cross-identification Strategy

The following general strategy is applicable to all kinds of object lists, not only those from NED. By "source list" we will refer to any external list of objects of a homogeneous nature (e.g., NED galaxies, NED IR sources, etc.)

Stage 1: a small sampling of objects from the source list is cross-identified manually, using our GUIDARES (Malkov & Smirnov 1995) and ZGSC software (Smirnov & Malkov 1997) applications. Cross-identification is attempted with both GSC objects, and in the case of multiple entry GSC objects, with individual entries. **Results:** an initial ruleset for automatic cross-identification; confidence estimates.

[1] http://www.ipac.caltech.edu/ned/ned.html

Stage 2 (iterative): the full source list is cross-identified automatically, using the current ruleset. **Results:** Confidently identified objects can be analyzed for hidden dependencies between GSC and external list data. This can yield additional rules and criteria, at which point Stage 2 is repeated. Questionable cases can be analyzed manually.

Final cross-identification results: (*i*) List of unambiguously cross-identified objects, (*ii*) list of ambiguous (one source object, several GSC objects) cross-identifications, (*iii*) questionable cases—for manual analysis, and (*iv*) unidentified objects from the source list.

These results will be provided to the maintainers of the source list or database in question.

3. Initial Results

Here we present results of **Stage 1** cross-identification process for NED galaxies and NED IR sources. Note that the NED list of IR sources does not contain IR sources with an optical counterpart, e.g., galaxies. The 23% below are, therefore, IR sources *with an identified GSC optical counterpart not listed in NED.*

	confidently identified	poorly identified	not identified
galaxies	84%	4%	12%
IR sources	23%	4%	73%

Confident cross-identifications provide an exciting possibility. If a sufficient number of source objects is confidently identified, we can expect to derive specific rulesets that describe the "mean" GSC representation of objects of the same type. These rulesets, coupled with object probability maps (see below), can be applied to the whole GSC in an automatic scan for unlisted objects of the same type. For example, the excellent cross-identification results for NED galaxies suggest that the GSC can yield a wealth of previously unknown galaxies. This is also suggested by two of our preliminary findings:

1. GSC photometry for diffuse objects is, typically, 3–5 magnitudes brighter than actual values. The GSC's formal limiting magnitude of 15^m–16^m is, as far as galaxies are concerned, closer to 18^m–19^m.

2. At least 5% of GSC "stars" are in reality galaxies, nebulae, multiples, etc.

We also give preliminary results on cross-identification of some other types of NED objects with the GSC:

Quasi-stellar objects: too faint to be included in the GSC.

Gamma sources: NED positional errors are too high. It is only possible to indicate the brightest GSC star in the vicinity of the source.

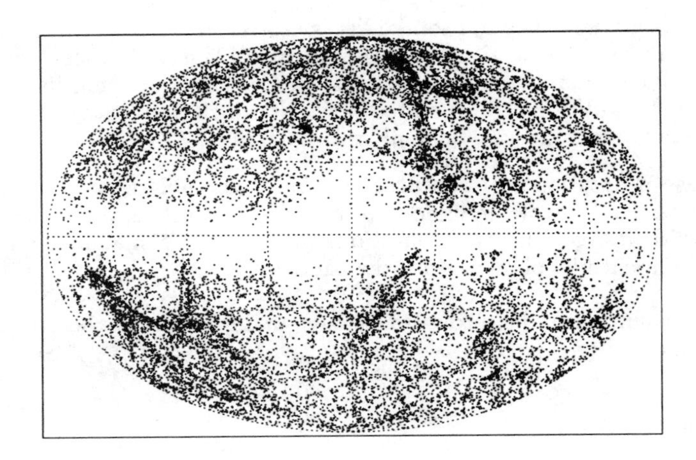

Figure 1. Third Reference Catalogue of Galaxies (galactic coordinates, Aitoff projection).

Clusters of galaxies: Results are very poor.

Planetary nebulae: confidently identified, but no "automatic" discovery is possible, as their GSC counterparts are undistinguishable from stars.

Radio sources: not identifiable at all.

Note again that the NED lists of gamma or radio sources does not contain those sources with a known optical counterpart.

4. Object Probability Maps

Object probability maps (OPMs) are based on the mean density of objects of a particular type at given coordinates and magnitude. OPMs can be used to estimate the probability for a given object to be of a particular type. This provides additional information on the nature of GSC objects (consider, e.g., galaxies' zone of avoidance, or the lack of asteroids at high elliptic latitudes).

To estimate the probability of a given object being a galaxy, we have to know both the stellar distribution and distribution of galaxies. The density of objects is approximated by the function $N = N(m, b)$, where N is the mean number of objects per square degree brighter than m at galactic latitude b (dependence on longitude is assumed to be negligible).

For the stellar distribution, an old but still unsurpassed Seares & Joyner (1928) paper was selected. In this paper, m is given in the international photographic scale. The paper employs the old galactic coordinate system; however, the difference is not significant for our purposes.

To create a distribution of galaxies we had to evaluate available catalogs according to the following criteria: completeness in coordinates, magnitudes, sizes, etc.; large number of objects. After detailed analysis, the Third Reference Catalogue of Bright Galaxies, RC3 (de Vaucouleurs et al. 1991) was selected. It

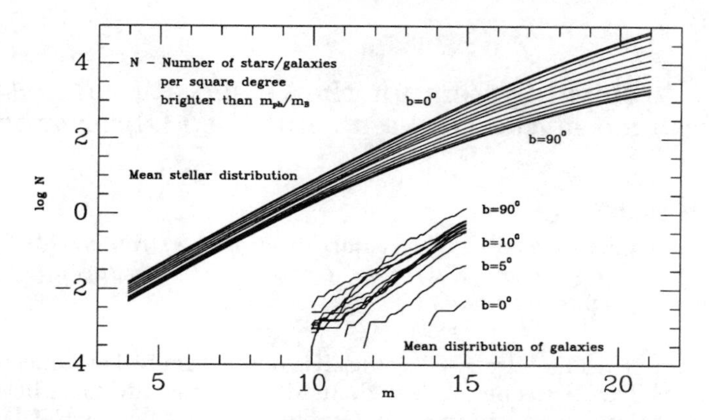

Figure 2. Stellar distribution and distribution of galaxies.

contains 23,011 objects (see Figure 1) and is complete for galaxies having apparent diameters larger than 1′ at the D_{25} isophotal level and total B magnitudes B_T brighter than about 15.5^m, with a redshift not in excess of 15,000 km s^{-1}. The RC3 employs the new galactic system. The catalog has shortcomings: it is inhomogeneous, and not every object is supplied with a magnitude.

The RC3 contains 7 different magnitude bands. We selected m_B, because: (*i*) it yields good correlation with GSC magnitudes, (*ii*) it is more representative in RC3 than other magnitudes, and (*iii*) it is close to the system used in the stellar distribution.

Only 75.6% of RC3 objects have m_B magnitudes. We used six other parameters that were well-correlated (better than 0.59) with m_B, to estimate m_B where it was absent. This produced a total of 22,907 (99.5% of RC3) usable objects. Distributions of galaxies and stars are shown on Figure 2. This should allow us to create the required galaxy distribution by fitting the density of RC3 objects with an analytical formula.

Acknowledgments. Drs. Marion Schmitz (Caltech) and Conrad Sturch (STScI) are gratefully acknowledged for valuable advice and constant help in our work. This presentation was made possible by financial support from the Logovaz Conference Travel Program. OM is grateful to the Russian Foundation for Fundamental Researches for grant No. 16304.

References

Malkov, O. Yu., & Smirnov, O. M. 1995, in ASP Conf. Ser., Vol. 77, Astronomical Data Analysis Software and Systems IV, ed. R. A. Shaw, H. E. Payne, & J. J. E. Hayes (San Francisco: ASP), 182

Seares, F. H., & Joyner, M. C. 1928, ApJ, 67, 24

Smirnov, O. M., & Malkov, O. Yu. 1997, this volume, 429

de Vaucouleurs, G., de Vaucouleurs, A., Corwin, H. G., Buta, R. J., Paturel, G., & Fouque, P. 1991, Third reference catalogue of bright galaxies (New York: Springer-Verlag)

Astronomical Data Analysis Software and Systems VI
ASP Conference Series, Vol. 125, 1997
Gareth Hunt and H. E. Payne, eds.

An Archival System for the Observational Data Obtained at the Okayama and Kiso Observatories II

M. Yoshida

Okayama Astrophysical Observatory, National Astronomical Observatory of Japan, Kamogata, Okayama 719-02, Japan,
E-mail: yoshida@oao.nao.ac.jp

Abstract. The Mitaka-Okayama-Kiso data Archival system second version (MOKA2) has been released. In MOKA2 the interfaces between user and the database engine contain Graphical User Interface (GUI), the SQL generator and the Client/Server system are World Wide Web (WWW) based ones: HTML, cgi, and httpd. This substantially improves its availability on the Internet and makes it easy to maintain the system. All the features developed for the first version of MOKA—e.g., the quick-look system, the name resolving system, and coordinate converter—are also available in MOKA2. MOKA2 offers the material of astronomical research as a prototype of a large-scale data archive of the next generation.

1. Introduction

MOKA was the first full-scale data archive system for the optical astronomy community in Japan (Ichikawa et al. 1995). Is was developed for the observational data taken with the Spectro-Nebulagraph (SNG; Kosugi et al. 1995) attached to the 188 cm telescope at Okayama Astrophysical Observatory and with the prime focus CCD camera of the 105 cm Schmidt telescope at Kiso Observatory.

The development of MOKA was begun in February 1994, and a public operational test started in three places (National Astronomical Observatory of Japan at Mitaka, Okayama Astrophysical Observatory, and Kiso Observatory) in June 1995. Various problems were discovered as a result of the operational test, and the development of MOKA2[1] was begun in February 1996. MOKA2 was made available to the public as a test operation in September 1996. See Horaguchi et al. (1994) and Takata et al. (1995) for detailed technology, development details, and basic idea of MOKA.

2. System Overview

The basic structure of MOKA2 is shown in Figure 1. MOKA2 contains the following three types of data/database:

---

[1] http://www.moka.nao.ac.jp

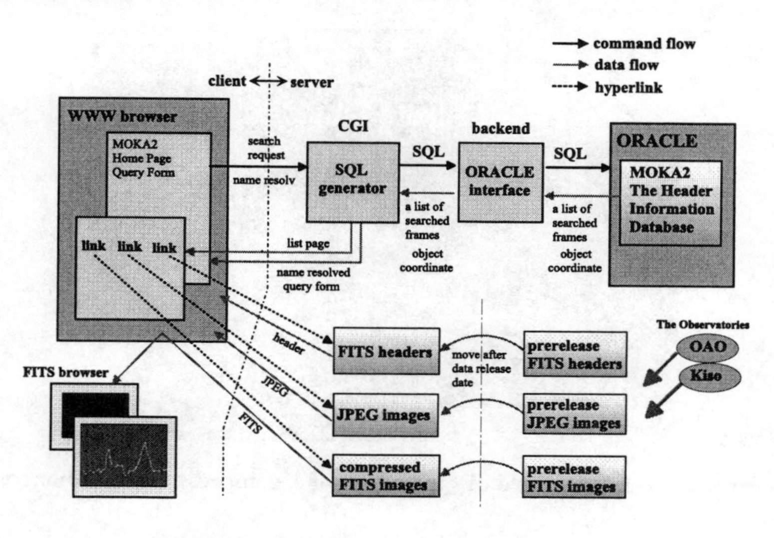

Figure 1. MOKA2 system overview.

- The header information database, which contains the principal information about each frame, is used for quick searches. This database is managed with the database management system ORACLE.
- The FITS header files which are extracted from the raw observational data.
- The quick-look image files, the size of which is reduced from the original data by binning, sampling, and rescaling. Two types of image formats are supported, FITS and JPEG. FITS format images are compressed by gzip.

The user interface is written in Hyper Text Makeup Language (HTML). The Structured Query Language (SQL) generator and the ORACLE interface shown in Figure 1 are newly developed. The SQL generator is called by a CGI program which is written in Perl. It handles the parameters of the data search constraints entered through the home page of MOKA2 (a query form) and generates a SQL command for the data search query. The ORACLE interface transports the SQL command to ORACLE, and receives the search results. The SQL generator receives those results and constructs a Web page according to the query.

3. User Interface

The home page of MOKA2 is a query form for searching the archival data. The basic layout of this page is similar to the GUI of the previous version of MOKA (Takata et al. 1995).

When the "search" button is clicked after setting the search constraints, the data which are matched to the constraints are searched in the header information database. A Web page in which the searched results are listed is then returned. Each line of the list in this Web page corresponds to one CCD frame and contains the basic information of the frame. Three hyperlinks (named

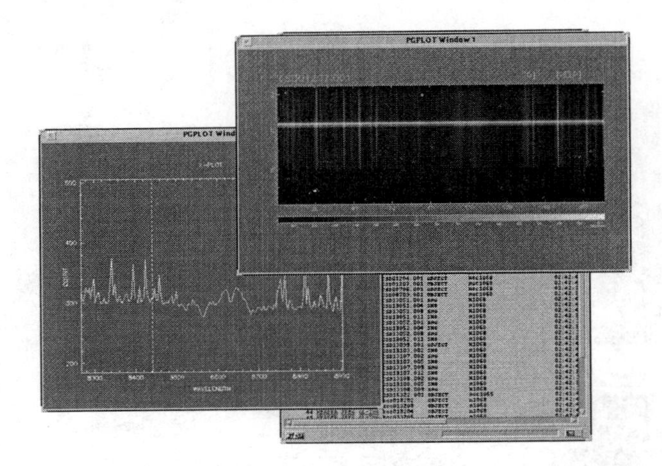

Figure 2. An example of quick-look of a compressed FITS image.

"HEADER," "JPEG," and "QL-IMG") at the head of the each line are linked to a FITS header file, a quick-look JPEG file, and a quick-look FITS file, respectively. Clicking "HEADER" allows the user to browse the original FITS header. Figure 2 shows an example of the quick-look of a compressed FITS image (QL-IMG) using the newly developed image browser (which is provided for SunOS and Solaris). This image browser can treat huge-sized (over 8000 $\times$ 8000 pixels) images.

After clicking the "Request data" button in the page, a data request form is returned as a Web page listing the results of the search. When the "Send request" button is clicked after filling the data request form, request e-mails are sent to the MOKA administrators of Okayama and Kiso observatories.

4. Name Resolver

The function of a name resolver is implemented in MOKA2. It is performed in the following two steps:

1. Convert the name of the object, which the user gives in an arbitrary format, to a standard expression.

2. Get the coordinates of the object from the database for the name resolver.

The database for the name resolver is also managed by ORACLE and currently includes eleven catalogs which were supported in the previous version of MOKA (Takata et al. 1995).

5. Future Plans

MOKA2 was developed under the HTML-CGI scheme to achieve Internet connectivity and it increased the system availability substantially. The efficiency of database access under the HTML-CGI architecture is, however, not so high, because CGI is only an interface between a WWW server and a program on the server. Further, interactive processing through the user interface is not efficient, since the UI is fundamentally a Web page generated by the server. Java can resolve these deficiencies peculiar to HTML-CGI architecture. We therefore plan to develop the next version of MOKA in Java in order to realize more efficient database access and flexible UI.

Future plans for development of MOKA system are follows:

- Development of an advanced user interface and realization of efficient database access using Java.
- Development of an automatic data distribution system.
- Inclusion of the environmental data from the observation sites.
- Dealing with calibrated data.

Acknowledgments. The author is indebted to the development members of MOKA2, E. Nishihara, T. Horaguchi, T. Ito, T. Takata, K. Aoki, S. Yoshida, S. Ichikawa and M. Hamabe. The study on the data archival system is one of the projects promoted by the Japan Association for Information Processing in Astronomy (JAIPA). This work was supported in part under the Scientific Research Fund of Ministry of Education, Science, Sports and Culture (06554001, 07304024 and 08228223), and JAIPA Nishimura foundation. The author thanks S. Nishimura for his continuous encouragement.

References

Horaguchi, T., et al. 1994, Publ. Natl. Astron. Obs. Japan, 4, 1

Ichikawa, S., et al. 1995, in ASP Conf. Ser., Vol. 77, Astronomical Data Analysis Software and Systems IV, ed. R. A. Shaw, H. E. Payne, & J. J. E. Hayes (San Francisco: ASP), 173

Kosugi, G., et al. 1995, PASP, 107, 474

Takata, T., et al. 1995, Publ. Natl. Astron. Obs. Japan, 4, 9

Astronomical Data Analysis Software and Systems VI
ASP Conference Series, Vol. 125, 1997
Gareth Hunt and H. E. Payne, eds.

WIYN Data Distribution and Archiving

Rob Seaman

IRAF Group,[1] *NOAO,*[2] *PO Box 26732, Tucson, AZ 85726*

Ted von Hippel

U. Wisconsin/WIYN

Abstract. The NOAO/IRAF Save the Bits archive has been operating for over three years at Kitt Peak National Observatory and at the National Solar Observatory's nighttime program. Since that time, the W. M. Keck Observatory and the Cerro Tololo Inter-American Observatory have also adopted the software. These first generation Save the Bits installations rely on Exabyte tapes as the archival medium, typically using pairs of drives to produce duplicate copies of the data for heightened protection against data loss.

The upgrade of Save the Bits that is currently in progress to support writable CD-R drives is discussed. In addition to another media option, this expands the role of the package to include data distribution as well as data archiving. Dual CD-R copies are produced as with tapes. One copy is retained for archival purposes, but the second copy of each nightly CD is released to the appropriate institution as the principle means of data distribution from the telescope. The four individual institutions are free to handle their copy of the data in any appropriate way, such as by mounting the disks into a jukebox as they are received. Both raw and mountain-reduced data are included in random access FITS files on the ISO 9660 CD-ROMs. Planned future improvements include support for DVD format disks.

Save the Bits is freely available to outside institutions and is straightforward to install and manage. Hardware requirements are minimal and other storage media should be straightforward to support.

1. Introduction

The WIYN Consortium consists of the University of Wisconsin, Indiana University, Yale University and NOAO—the National Optical Astronomy Observatories. The Consortium manages the 3.5 m optical alt-azimuth WIYN telescope on

[1] Image Reduction and Analysis Facility, distributed by the National Optical Astronomy Observatories

[2] National Optical Astronomy Observatories, operated by the Association of Universities for Research in Astronomy, Inc. (AURA) under cooperative agreement with the National Science Foundation.

Kitt Peak near Tucson, Arizona. The WIYN telescope supports two primary facility instruments, a wide field CCD imager and the Hydra multi-object spectrograph. These instruments are mounted at the two Nasmyth foci of the telescope allowing both instruments to be used throughout a given night. The primary mirror system takes advantage of modern active optics technology through 66 separate actuators that push or pull on the back face of the mirror to maintain the best possible optical figure. A thermal control system maintains the surface of the mirror within 0.2 C of the ambient air temperature, eliminating mirror seeing, which is caused by turbulence in cool air over a warmer mirror surface.

The motivations for a WIYN Archive and Data Distribution System are to enhance the total science output of the telescope over its lifetime and to ease the data distribution and handling process at all four WIYN institutions.

An archive increases the total science output of the telescope as investigators attempt new problems with old data, and as statistically large samples of different classes of objects are accumulated, often from projects initially undertaken for a wide variety of purposes. Indeed, as a large digital imagery and spectroscopy library is accumulated, many new uses will be made of it. One can expect not only statistical studies of commonly observed objects, but also test reductions and analyses by observers contemplating new projects.

Currently WIYN data are recorded by the observers on Exabyte or DAT tape using IRAF tasks. While this procedure does store data, it does not allow for easy recovery or dissemination of the data. Since tapes do not allow random access, since they may not last more than a few years even if carefully stored, and since every observer writes their data to tape differently and has different styles of logging their observations, the result is a cumbersome and heterogeneous data depository. Even the original investigator must load their data to disks, often repeatedly, and often they must spread their data across many disks in order to have access to an entire run at one time.

The system described here is based on an automatic and homogeneous data storage process using random-access media (CD-ROM). The long-term goal is a robust and easy to use WIYN data archive. The near-term goal is a robust and easy to use data distribution system. The raw data CD-ROMs are mounted in jukeboxes at the institution of data ownership and are immediately and continuously readable without transferring the data to hard disk. Additionally, CD-ROMs are stable media and the data are in a standard format.

Observers relying on this mechanism exclusively will be responsible for verifying the content of the archival media containing their data. Tools will be provided to check the image headers and display the images directly from the archival media. Additionally, NOAO backs up all data taken at WIYN and with the KPNO telescopes using *Save the Bits.* This software automatically stages CCD data taken at all Kitt Peak telescopes to a single Sun workstation and its disks, then writes these data as FITS files to Exabyte tape.

2. System Overview

The physical approach proposed is to write WIYN data to two CD-ROMs (for redundancy) simultaneously using two CD-ROM writers, mounted on a dedicated Sun workstation using a version of *Save the Bits.* A separate directory will be written for each instrument (currently Hydra and the Imager) on each CD-ROM as well as directories for mountain-reduced data separate from the raw data. The data files will be copied into these directories as individual FITS files. The data will be staged to a hard disk before writing the CD-ROMs, and the resulting CD-ROMs verified against the hard disk data set and against each other. The mountain-wide *Save the Bits* archive will also still back up WIYN data.

At some point during each day the (now archival) CD-ROMs will be ejected from the drives, such that individual CD-ROMs belong to individual nights, and thus to individual institutions. One copy of each night's data will be kept at NOAO, and the other copy returned to the institution which owns the data. For the CD-ROMs which return to the Universities, the media can be mounted in a jukebox and the observer will effectively have new disk space with their observations. For the CD-ROMs which are stored at NOAO, the media will be available as a backup and for repeated duplication and distribution to the requesting observers.

The advantages of CD-ROMs as the archive media are multiple: they are dependable, easy to store and use, they are random access media, and they are forward compatible with the next generation of DVD "CD-ROMs." Random access is important for an archive as well as for data distribution purposes, as it greatly decreases access time and allows someone to recover, copy and distribute data from widely disparate media locations. Forward compatibility also seems assured with the next generation of higher density DVDs as the principal manufacturers have agreed to this.

The major disadvantage of CD-ROM as an archival media is that a single disk only holds 650 MB. Based on current rates of data taking, this data quantity is large enough, however, such that one CD-ROM will be sufficient to hold the data from a single night on 60% of all nights, while three CD-ROMs will only be needed approximately ten nights a year.

The software to perform the archiving will be based on *Save the Bits*, which will be modified to archive only WIYN data, and to write to CD-ROM writers. *Save the Bits* will also perform the function of creating a growing index file from the headers of all archived data. This will allow for simple searching and post-processing programs to be written. *Save the Bits* is proven software which has preserved more than one terabyte of data from seven telescopes at Kitt Peak over the last two and a half years, and it is also being used at CTIO and at Keck. This software needs only minor adaptations, its use requires minimal interaction from mountain staff, and it already can handle a number of contingencies, including system crashes.

3. Daily Maintenance Chores

The archival system requires the following tasks, generally once, at some point between dawn and mid-afternoon:

1. check the monitor program, *bitmon*, to see that the archive system wrote a good pair of CD-ROMs,

2. eject, date-stamp, and log the just written CD-ROM pair, and

3. load new blank CD-ROMs into the CD-R writers.

In the event that one or both of the CD-ROM disks of the pair were bad, as indicated by the verification passes of *Save the Bits*, new blank CD-ROMs are inserted into the drives, and that dataset is re-archived from the staging disk. Occasionally two (and less often three) CD-ROM pairs per night of data will be required, and the above process will have to be repeated a second (or even a third) time during a given 24 hour period. The system can be allowed to fall behind by several CD-ROMs, if necessary, though this is not desirable.

At the end of each run, the CD-ROM disks will be transferred downtown for a final read verification using a separate CD-ROM drive. This will also provide the opportunity to read detailed observing run information from each disk which will be used to generate an informative label to be printed onto each disk using a thermal printer designed to print directly to normal CD media.

References

NOAO/IRAF *Save the Bits* archive[3]

WIYN Consortium[4]

Exabyte Corporation[5]

CD Information Center[6]

Stinson, D., Ameli, R. & Zaino, N. 1995, Lifetime of KODAK Writable CD and Photo CD Media[7]

DVD: Inside Story[8]

---

[3] http://iraf.noao.edu/projects/stb

[4] http://www.noao.edu/wiyn/wiyn.html

[5] http://www.exabyte.com

[6] http://www.cd-info.com

[7] http://www.cd-info.com/CDIC/Technology/CD-R/Media/Kodak.html

[8] http://www.sel.sony.com/SEL/consumer/dvd/index.html

Astronomical Data Analysis Software and Systems VI
ASP Conference Series, Vol. 125, 1997
Gareth Hunt and H. E. Payne, eds.

Automatic Mirroring of the IRAF FTP and WWW Archives

Mike Fitzpatrick and Doug Tody

IRAF Group,[1] NOAO,[2] PO Box 26732, Tucson, AZ 85726

David L. Terrett

Rutherford Appleton Laboratory, Chilton, Didcot, Oxfordshire OX11 0QX, United Kingdom

Abstract. Large FTP archives have long used mirrors (copies of the network archive maintained on remote hosts) to decrease the load on a particular server or shorten the network path to provide faster download times. Little has been done however to simplify mirroring of WWW (World Wide Web) pages, although many projects and users now rely on Web pages at least as heavily as anonymous FTP services. With the dramatically increasing use of the global Internet in the past year, the network has become overloaded, and network access, especially overseas, is often very slow during peak hours. We present a strategy based on host-independent URLs which allows Web pages to be automatically mirrored to both remote Web hosts and CD-ROMs. Issues affecting a site wishing to mirror a remote archive are discussed.

1. Introduction

The subject of automatic mirroring can be approached in one of two ways: from the standpoint of those wishing to export their archive for mirroring, and of those wishing to be a mirror for an existing archive. Although this paper deals with the specific issues we faced in setting up a mirror of the IRAF archives, the techniques presented are general, and can easily be applied to any similar archive.

On both ends there were some expected setup glitches in trying to verify the thousands of links involved, in bringing both systems to a common understanding about requirements in HTTP server and local software, and in establishing a routine procedure for maintaining the mirror. The initial experiment between the NOAO IRAF Group and the UK Mirror at Rutherford Appleton Labs has worked out many of these problems, and has provided us with the ability to establish other mirrors much more easily. In the first five months of operation, the

[1]Image Reduction and Analysis Facility, distributed by the National Optical Astronomy Observatories

[2]National Optical Astronomy Observatories, operated by the Association of Universities for Research in Astronomy, Inc. (AURA) under cooperative agreement with the National Science Foundation

UK IRAF Mirror Site has distributed more than 4300 files to 120 different nodes in more that ten countries, providing a faster, more reliable link for UK and European sites. Negotiations are underway to establish mirrors in other parts of the world where FTP access to the NOAO Tucson archives or UK mirror sites is prohibitively slow.

The host-independent manner in which the WWW pages are written means that they can also be used from a CD-ROM running on a local machine, in effect duplicating the IRAF archive on any machine. We discuss the limitations and special setup required in this case.

2. Preparing Your Archive for Mirroring

There are only a few steps involved in preparing your archive so it can be easily mirrored elsewhere:

2.1. Host-Independent URLs

The mirroring site will have a Web address different from the original site. If Web pages contain explicit HTTP URLs, then these pages will still refer to the *original* archive when the pages are mirrored, negating the point of the Web mirror. The simplest solution is to substitute file relative URLs in all cases except where one really does want a URL to point to a specific network host. For the exporting site this means each link will need to be examined and changed in the following ways:

- Use "file.html" or "subdir/file.html" link references. Keep it simple, no complex relative paths.
- Since the Web root directory may be different on the mirror node (which is likely serving its own documents), root-relative links such as "/iraf-homepage.html" should be avoided.
- We don't want to require that the mirroring site put documents in a particular directory, so the best compromise is to establish a set of common links for both systems so root-relative paths can be used on either host correctly. In our case we established /iraf/ftp and /iraf/web links pointing to the root of the FTP and WWW areas respectively (this also fits in well with the suggested directory structure for an IRAF installation). It so happens that our Web pages are under the FTP directory tree, but this is not a requirement.

2.2. CGI Scripting

There are several things to be done to make most CGI scripts portable:

- One cannot assume that a mirror node will have all of the custom local software that may be used by CGI scripts, or indeed that it is even the same type of machine. For our Web archive we've created a *bin*, *lib*, and *src* subdirectory containing binaries and source for all programs (mail filters, search engines, etc.) used by the various scripts. All binaries are built from these sources, meaning that versions are current for all platforms and are automatically updated in the mirror site when a new version is installed by the originating site.

- In HTML forms or links, references will be made to a particular script or application. Since a mirror may be running on a type of computer different from the original server, these task names are actually csh scripts which call the binary (in the *bin* subdirectory) appropriate for that platform. In our case, the scripts reference a program called *mget* as a mail filter, the *mget* script figures out what type of machine it's running on and passes the arguments to a *mget.sparc* binary to do the actual work.

- Path names in scripts CGI scripts are often written as scripts of some type (csh, Perl, etc.) which are invoked using an, e.g., `#!/bin/csh` path as the first line. Such paths may not be universal, however. The mirroring site is responsible for creating the system links needed to resolve these paths.

2.3. The Final Steps

You may wish to arrange for mirror site usage logs to be propagated back to the original site. This can be done as a weekly cron job that greps for entries containing a certain keyword in the logs ("iraf" in our case) and automatically mails them to a specific maildrop. If the archive is large, it is best to make a snapshot tape of the full directory tree tree to be mirrored and mail that to the mirroring institution to populate the initial directory tree. Once the initial system is installed and working, updates should be small and will be handled automatically by the mirror software.

3. Setting up a Mirror Archive

Now that the initial IRAF mirror site has been established, we should have worked out most of the bugs in the scripts and documents on our end, but there are still concerns for new sites wishing to establish a mirror:

3.1. Disk Space Required

The complete IRAF archive now requires approximately 3 GB of storage—this will probably increase another 1 GB in the next year as more software is released. Potential mirrors should consider the purchase of a new dedicated disk.

3.2. Mirroring Package Used

The RAL mirror site is maintained using a package called *MIRROR* from Lee McLoughlin of the University of London; other packages are also available. This particular package required Perl 4, which had to be installed specifically to support the package. A cron script is run nightly to update the archive, and a separate script is run once weekly to mail access logs back to Tucson. The archive scripts directory is mirrored separately to a different directory, in part because execute permissions are stripped in the mirroring process and in part so new code may be hand checked, as a security measure.

3.3. HTTP Server Requirements

The UK mirror was already serving Web documents and had a configured HTTP server. New sites, or those using the CD-ROM, may need to configure a server. The only changes required to support the mirror were alias definitions for the IRAF CGI scripts directory. This means editing the httpd/conf/srm.conf file with an *Alias* and *ScriptAlias* definition for the scripts directory which points to the iraf Web scripts directory on the mirror, and aliases for the root-relative links. For example,

Alias	/iraf/web	/iraf/web
Alias	/iraf/ftp	/iraf/ftp
Alias	/scripts	/mirror/iraf/web/scripts/
ScriptAlias	/scripts/	/iraf/web/scripts/

One other problem is that most HTTP servers define a default MIME type as plain text for documents for which the server cannot determine the type from the file name extension. This means that tar files, compressed PostScript files, etc., show up as jumbled text in the browser rather than being identified as binary or starting an external viewer. To work around this, we suggest the following definition in the server's srm.conf file

```
Redirect     /iraf/ftp/     ftp://iraf.noao.edu/
```

This causes the most browsers to create a save pop-up window rather than trying to display the file, which is what is most often desired.

Aside from the initial setup and verification of new scripts, the process is now largely automatic requiring an estimated one hour/month to maintain the mirror. Only rarely has the nightly update not completed successfully; in each case it has succeeded the following night.

4. CD-ROM Issues

While the host-independent nature of the WWW pages means the archive can be distributed on and browsed from a CD-ROM, there are a few issues of concern for viewing the CD-ROM Web pages as though they were a *live* Web site:

- An HTTP server must be running in order for CGI scripting to work, otherwise all documents are accessed with a file URL and scripts will not execute.
- Sites using the CD-ROM as a local archive must also remember to set the /iraf/ftp and /iraf/web links discussed, so links are resolved.
- At present, the archive only contains binaries used by CGI scripts for those mirror platforms we know about. More binaries are needed.

5. Project Status

We welcome inquiries from any sites wishing to set up additional IRAF mirrors, or from sites interested in using the techniques outlined in this paper to mirror their own archives. Contact *iraf@noao.edu* for further information.

Astronomical Data Analysis Software and Systems VI
ASP Conference Series, Vol. 125, 1997
Gareth Hunt and H. E. Payne, eds.

The ROSAT RESULTS ARCHIVE: Tools and Methods

M. F. Corcoran

USRA/LHEA, code 660.2, GSFC, Greenbelt, MD 20771
E-mail: corcoran@barnegat.gsfc.nasa.gov

D. E. Harris

HEA-CFA, MS 3, 60 Garden St., Cambridge MA 02138

H. E. Brunner, J. K. Englhauser

AIP, An der Sternwarte 16, D-14482, Potsdam, Germany

W. H. Voges, T. H. Boller

MPE, Postfach 1603, D-85740 Garching bei München, Germany

M. G. Watson, J. P. Pye

Dept. of Physics and Astronomy, Leicester University, Leicester LE1 7RH, England

Abstract. The *ROSAT* Results Archive (RRA) is a publicly accessible collection of source lists, images, spectra, lightcurves and counting rates derived from pointed-phase observations of the *ROSAT* X-ray satellite observatory. The RRA contains X-ray source data from both the *ROSAT* Position Sensitive Proportional Counter (PSPC) and the *ROSAT* High Resolution Imager (HRI) instruments, using only data processed with the current (REV2) version of the processing system to ensure data uniformity and accuracy. Each detected source is visually inspected and possible problems are flagged. In this paper we describe the methods used to screen the data products and the GUI-based tools used to screen and access the data.

1. Introduction

ROSAT (Röntgensatellit) is a joint German-US-UK satellite observatory for astronomical observations in the soft X-ray through extreme UV band. *ROSAT* was launched in 1990 and continues to obtain wide-field images of X-ray and UV sources.

ROSAT has two X-ray instruments: the Position-Sensitive Proportional Counter (PSPC) and the High-Resolution Imager (HRI). Both offer wide field imaging and sensitive detection of astronomical X-ray sources. The PSPC provides energy-sensitive observations with a 2° field of view; the HRI has a field of view of 1° and provides high spatial resolution imaging but no real energy sensitivity.

The PSPC and HRI on *ROSAT* are the most sensitive X-ray imagers yet flown. This sensitivity means that, in general, a large number of sources are detected in each X-ray image. The *ROSAT* Data Processing System uses an automated detection routine to determine properties of each X-ray source (position, counting rate, variability, etc.) for each observed field. However, due to the complexity of the X-ray sky seen by *ROSAT*, no automated analysis method can be 100% accurate.

In the 6 years since launch, *ROSAT* has scheduled more than 7000 observations. These data have a 1-year proprietary period, after which they enter the public domain and are made available through archive sites in the US,[1] Germany,[2] and the UK.[3]

The *ROSAT* RESULTS ARCHIVE (RRA) is being developed to make available all information ("results") derived from detected *ROSAT* sources identified by the Standard Analysis System Software (SASS) processing system. Before these results are made available to the public they are screened by the data centers, and obvious problems are flagged as a guide for the archive user.

This paper discusses the methods and tools used to screen *ROSAT* source results, and discusses ways in which the data will eventually be made accessible to the general astronomical community.

2. Screening and Archiving

Screening and archiving of the RRA data occurs in the following steps:

1. *ROSAT* pointed data are delivered via an automated pipeline from the *ROSAT* Data Centers to the *ROSAT* screening centers (in the US at the Goddard Space Flight Center and the Smithsonian Astrophysical Observatory; in Germany at the Max-Planck Institute for Extraterrestrial Physics and the Astronomical Institute of Potsdam; and at Leicester University in the UK).

2. After delivery, the data are "pre-screened" by automated scripts which look for obvious problems (sources near edges or other detector structures, sources in regions of high background, or sources below a S/N threshold). Each field and source gets assigned a set of quality flags by these scripts:

    - FIELD FLAGS which depend on characteristics of the entire field-of-view of the observation, and
    - SOURCE FLAGS which depend on the characteristics of individual sources in the field-of-view.

    For each observation, the source characteristics derived by the processing system detection routine and the set of flags are written to an output file.

---

[1] http://heasarc.gsfc.nasa.gov/

[2] http://www.rosat.mpe-garching.mpg.de/

[3] http://ledas-www.star.le.ac.uk/

3. The pre-screened data are then visually inspected by personnel at the screening centers as a final check on source validity. Visual inspection consists of comparing an overlay of flagged X-ray sources to the X-ray images.

4. Screeners can accept the quality flag settings provided by the automated screening software, override these flags, or set additional flags. Screeners may also mark sources missed by the detection software.

5. Consistency checks are done periodically in order to determine the amount of variation from screener to screener.

6. After screening, the data plus quality flags are saved to a file. This file is released to the public, along with additional products (images, spectra, lightcurves, etc.).

3. Screening Tools

Visual inspection of each dataset can be time-consuming, since sequences can have dozens of significant sources. For example, many sequences contain large extended X-ray sources (like supernovae remnants or galactic cluster emission) which can confuse the source detection algorithm resulting in large numbers of spurious sources, all of which must be flagged.

In order to minimize the amount of time spent in visual inspection, software tools have been written which allow easy access to the derived source characteristics and the X-ray images. These GUI-based tools allow the screener to select data sets and sources, overplot sources on X-ray images of the field, and view and set quality flags. There are two software tools currently in use, one for PSPC data and one for HRI data:

- the PSPC screening tool operates in the Munich Interactive Data Analysis System (MIDAS) environment, using the Tcl/Tk toolkit, and

- the HRI screening tool operates in the Interactive Data Language (IDL) environment using the IDL widget toolkit.

4. User Access

The most basic product in the RRA is the source list (source properties + quality flags) produced for each dataset. However, users also want to access the combined list of all sources from all *ROSAT* observations, and to access appropriate data products for each significant detected source. Users may also want an easy way to identify an X-ray source with an optical, radio, or UV source.

Thus, access software must be fairly sophisticated. One method of access which provides most of the required functionality is the BROWSE interface, currently supported by the High Energy Astrophysics Science Archive Research Center (*HEASARC*). Using BROWSE, a list of sources can be identified and selected, and data products can be extracted from an on-line archive for further analysis. In addition, a version (W3BROWSE) which uses the WWW as a convenient graphical interface, is currently available. Source lists from the RRA will be made available to the community from BROWSE and W3BROWSE.

In addition, the *ROSAT* project is currently developing other tools which can provide more flexible and extended access to the RRA. One tool under development lets the user select sources interactively using user-specified quality flag criteria, and, optionally, by position. The user can display selected sources in an X-ray image, and may display an optical image (from the Digitized Sky Survey, created with SkyView) with an overlay of detected X-ray sources. In addition, users can retrieve a list of SIMBAD catalogued sources, along with the separation between the catalogued source and the currently selected X-ray source, as an aid to source identification.

5. For More Information

For more detailed information see the following Web sites:

ftp://heasarc.gsfc.nasa.gov/rosat/data/qsrc/www/RRA.html
http://www.aip.de:8080/~rra/rra.html
http://www.rosat.mpe-garching.mpg.de/~jer/rra/rra.html

Astronomical Data Analysis Software and Systems VI
ASP Conference Series, Vol. 125, 1997
Gareth Hunt and H. E. Payne, eds.

A Database-driven Cache Model for the DADS Optical Disk Archive

T. Comeau and V. Park

Space Telescope Science Institute, 3700 San Martin Drive, Baltimore, MD 21218, USA

Abstract. The Data Archive and Distribution System (DADS) manages the Hubble Data Archive (HDA), a WORM Optical Disk Archive which contains over two terabytes of Hubble Space Telescope data. One fortunate side effect of retrievals from the HDA is that all retrieval requests are permanently logged in database tables. Queries against these tables provide a complete history of requests serviced by DADS. We describe a system which uses the database request logs as input to a flexible cache model. The model permits changes to the size of the cache, replacement strategy, and preloading the cache. We also discuss possible cache sizes and replacement strategies, and their effect on DADS performance in the post-SM97 system.

1. Introduction

Throughput of the Hubble Data Archive (HDA) is an important bottleneck for overall Data Archive and Distribution System (DADS) performance. If a significant fraction of retrieve requests could be serviced from a cache, this bottleneck would be bypassed.

Additionally, retrieve requests currently interfere with the process of ingesting new datasets. This is a potentially serious problem in the post Servicing Mission 97 (SM97) era, when ingest rates are close to the limit of this same archive bottleneck.

All retrieval requests are permanently logged in two database tables. The *requests* table contains information about each DADS retrieval request; the *request_files* table contains information about each file retrieved to satisfy that request.

Queries which select from the *requests* table by *req_date*, and from the *request_files* table by request number (where *req_reqnum* = *rqf_reqnum*) and sequence number (where *rqf_reqnum* specifies a dataset for that request) provide a complete history of requests serviced by DADS.

Likewise, the table *archive_data_set_all* includes all the datasets ingested into DADS. Selecting from this table in *ads_data_receipt_time* order provides a complete history of DADS ingests. Additional information about ingests, including the program for which science data were taken, is kept in the *proposal* table.

2. Model Components

The Cache Model consists of three major components. First, a small set of C routines parse input parameters and call database stored procedures. Second, stored procedures implement the cache management scheme. Finally, two tables implement the cache: a single cache control record with global information, and a cache index table which describes the current contents of the cache.

Keeping the cache index and the routines which manage the index in the database makes modifying the cache behavior simpler. Since a stored procedure is simply a collection of SQL statements with flow-of-control language, writing such a procedure is at least as simple as equivalent C code. Additionally, stored procedure queries are precompiled to improve performance. Stored procedures can easily be modified and reloaded. Keeping the cache index in a database allows queries to monitor and evaluate the effectiveness of the cache, both during runs and postmortem. It also makes the cache index persistent: later runs can use all or part of the cache left over from previous runs.

The model provides the basis for an operational cache index since it uses the same data as DADS uses to obtain datasets from optical disks.

3. Requests Caching

When examining datasets for possible inclusion in a cache, we quickly identified three broad types of requests:

- **DADS Operations requests**, submitted by DADS Operations staff, principally for data verification. Since one primary goal of these requests is to verify that the data are correct on optical disk, the request must be serviced directly from optical disk, and the use of a cache is inappropriate. DADS Operations retrievals are excluded from the model.
- **Auto PI requests**, generated by DADS itself, using database information about recently ingested data, to automatically send recently completed observations to the Principal Investigator of an HST Science Program. Since these requests by definition use recently ingested data, they should be effectively serviced by a cache of recently ingested science datasets.
- **Everybody else**, including internal users retrieving recently ingested data to monitor instrument or observatory health, or to respond to user questions, internal users retrieving non-science data (including science classes for non-science proposals) or older data for a variety of reasons, and internal and external users retrieving older data.

The datasets can also be divided into three broad categories:

- **Calibration data**, used to calibrate science data.
- **Science and science-related data**, the "interesting" datasets to most users. Interesting datasets can be further divided into science programs and non-science programs. The latter include Calibration, Engineering, Orbital Verification, Science Verification, and Early Release Observation programs.

- **Nonscience data**, including everything from intermediate products of the OPUS pipeline, to copies of previous releases of the DADS software.

Calibration data are by far the most popular data in the archive, used by internal, external, and Auto PI requests. Nonscience data is rarely retrieved.

4. Cache Performance Scenarios

Eight cache scenarios were executed against the data for August, 1996:

			Hit		Datasets		Total	
Case	Req's	Hits	Rate	Unhit	1 Hit	Multi-hit	Datasets	MB
1	19052	5338	.28	0	10711	2997	19052	95287
2	19052	8598	.45	7095	10717	2997	20809	137466
3	19052	8578	.45	1339	10616	2993	14948	99350
4	4807	4066	.85	1928	3095	126	5149	18994
5	4807	4066	.85	569	3095	126	3790	12868
6	14245	2055	.14	0	7867	1534	9401	73311
7	14245	3163	.22	2065	7867	1534	11466	79413
8	19052	6976	.37	509	3072	695	4276	29996

Case One simulates an unpreloaded cache with all retrieved datasets inserted into a cache of unlimited size. This is the unfiltered retrieval case, with most datasets retrieved only once. Calibration data are, as expected, the most hit datasets, with the interesting datasets accounting for all but a tiny fraction of the remaining requests.

Case Two duplicates Case One with the addition of preloading new datasets as they arrive, adding an additional unretrieved 42 GB to the cache in order to save one disk access for the datasets which will be retrieved by Auto PI.

Case Three preloads only interesting and calibration datasets. This eliminates from the cache the non-science classes, leaving only the very popular calibration data, and the interesting classes. Only 20 fewer hits are obtained by eliminating uninteresting data from the cache, and the 42 GB of uninteresting data are never inserted.

Case Four examines the option of an Auto PI only cache, preloading the same interesting datasets, but including only Auto PI retrievals. This is the best scenario on a hit rate basis, but there are still datasets being preloaded that are uninteresting, despite being in science classes.

Case Five is also an Auto PI only cache, but only science *proposals* are preloaded. This is the "pure" Auto PI cache. This is both the smallest cache, and the best hit rate cache. Of the 15 percent of the datasets unretrieved, none is more than four days old.

Case Six examines the same cache insertion rule with non Auto PI retrievals. Not suprisingly, 84 percent of these datasets were retrieved exactly once. Also, data retrieved more than once were usually re-retrieved within 14 days of being loaded into the cache. Since this scenario did not preload, this means that if a dataset was not re-retrieved within two weeks, it probably never would be.

Case Seven duplicates Case Six with preloading of new, interesting, science proposal generated datasets. There are additional hits against preloaded data, and these are the same data that would generally be hit by Auto PI: Newly ingested science datasets.

Based on the age of likely-to-be-hit datasets, we selected a cache size of 30 GB, of which roughly half would be Auto PI data. Some of the Auto PI data will also be retrieved by other (Institute internal) users.

Case Eight suggests that preloading datasets for Auto PI also preloads a few datasets for other users. (These are presumably internal users monitoring instrument health or assisting PIs with interpreting their data.) There are still a significant number of one hit datasets older than four weeks, which suggests that a 30 GB cache is more than adequate for a month of retrievals, but that a larger cache would not be much more effective in servicing archive requests.

5. Conclusions

Of the 20,809 datasets distributed in the model scenarios, 436 were calibration datasets used by both Auto PI and other archive users. These datasets total just 1663 megabytes, but account for 2466, or 13 percent of all requests. A cache preloaded with substantially all of the current calibration datasets would give a significant boost to retrieval performance while consuming about ~8–10 GB. Auto PI always uses these datasets; Starview users may now get the appropriate calibration datasets automatically.

While the 37 percent hit rate obtained in the final scenario is disappointing, it is largely a function of how users make requests of DADS. Internal users, including Auto PI, retrieve the data while they are new. Other users request the same kinds of data, but the ages of those data vary, which makes them difficult to cache.

The two calibration classes and the five science related classes account for all but a tiny fraction of retrieval requests. Additionally, science *proposals* account for the vast majority of retrieval requests. Thus, a retrieval cache can be limited to these seven classes for Science Proposals without significantly reducing the hit rate.

A modest cache (~30 GB) preloaded with new (~2–4 weeks old) datasets, and calibration datasets, would service essentially all Auto PI requests. Such a cache would also serve the many internal users that request roughly the same data. A caching scheme that serves older data, however, is probably not possible.

Acknowledgments. I wish to thank Lisa Gardner (STScI), who was extremely helpful in setting up the cache database and assisting in query design.

Astronomical Data Analysis Software and Systems VI
ASP Conference Series, Vol. 125, 1997
Gareth Hunt and H. E. Payne, eds.

The CATS Database to Operate with Astrophysical Catalogs

O. V. Verkhodanov and S. A. Trushkin
Special Astrophysical Observatory, Nizhnij Arkhyz, Karachaj-Cherkessia, Russia, 357147

H. Andernach
INSA; ESA IUE Observatory, Apdo. 50727, E-28080 Madrid, Spain

V. N. Chernenkov
Special Astrophysical Observatory, Nizhnij Arkhyz, Russia, 357147

Abstract. A public database of astrophysical (radio and other) catalogs (CATS), has been created at the Special Astrophysical Observatory (SAO). It allows a user to execute a number of operations in batch or interactive mode, e.g., to obtain a list and parameters of catalogs, to extract objects from one or several catalogs by various selection criteria, to perform cross-identification of different catalogs, or to construct radio spectra of selected sources. Access to CATS is provided in both dialog mode (non-graphical), and graphics mode (hypertext, via Tcl/Tk or possibly Java in future). The result of CATS operation can be sent to the user in tabular and graphical formats.

1. Introduction

Different attempts have been made to combine many astronomical catalogs in unified databases—NED, SIMBAD, ESIS, ADS, etc. (see reviews by Andernach et al. 1994; Andernach 1995). Important shortcomings of these databases are (a) the incompleteness of the information stored compared to that offered in the original publication and (b) the necessity of copying the whole catalog (if available at all) for dedicated work with it. We propose a new solution to this problem with a "CATalogs supporting System" (CATS) at Special Astrophysical Observatory (SAO). It allows a user to operate with catalogs coded in plain ASCII, to cross-identify different radio catalogs, to calculate spectral indices, to construct and fit spectra. This database is now running on the server *ratan.sao.ru* of SAO of Russian Academy of Sciences (RAS), and constitutes part of the data bank of SAO RAS (Kononov 1995).

2. Realization of the database

The present active database of catalogs, version 1.0 (Figure 1), is a unification of catalogs, descriptions of catalogs, and programs operating on these executing

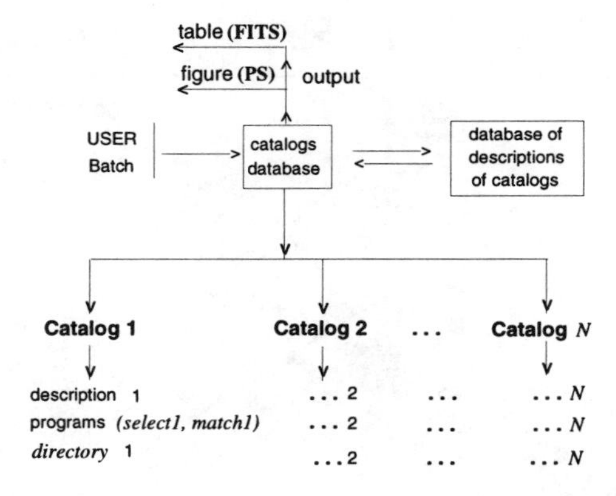

Figure 1. The scheme of the CATS database.

on the freely distributed version of UNIX (Linux) running on a Pentium server (Verkhodanov & Trushkin 1995) at the RATAN-600. The programs are coded in C and are freely sharable except for commercial use.

New catalogs may be added to CATS in conformity with the following rules: 1) every new catalog of objects has to be contained in the UNIX directory having the same name as the catalog of objects; 2) the description of the catalog must also be in that directory; 3) the programs for local operation on the catalog of objects must also be in the same directory; 4) brief characteristics and names of the programs and file with the description of the catalog must be stored in the file `cats_descr`. The described manner of catalog storage eases the database development, its expansion with new data, and the fine-tuning of the supporting programs.

Virtually all catalogs have a different format and list different observables. It will be a major challenge to provide uniform access to such a heterogeneous collection of data sets based on different methods, using different notations and units (in the absence of a "standard" to create catalogs). Except for parameter-dependent derived quantities, we intend to use all different fields (i.e., the columns of data tables) as they were published.

3. Capabilities and Access

CATS can accomplish the following tasks: 1) Provide a short description and characteristics of each catalog; print the full list of catalogs relevant for a given sky area. 2) Select objects from one or several catalogs matching user-specified criteria, such as equatorial and galactic coordinates, flux densities and spectral indices, observing frequency, names of catalogs unified in mixed catalogs as Dixon's Master Source List, and object type (if provided by the catalog). 3) Cross-identification of different catalogs and selection after calculation of spectral indices. 4) Drawing radio spectra of selected sources in PostScript.

The result of the object selection can be sent to the standard output or saved in the following formats: 1) The original format of the input catalog. 2) A standard format, common for all catalogs and used further for unification

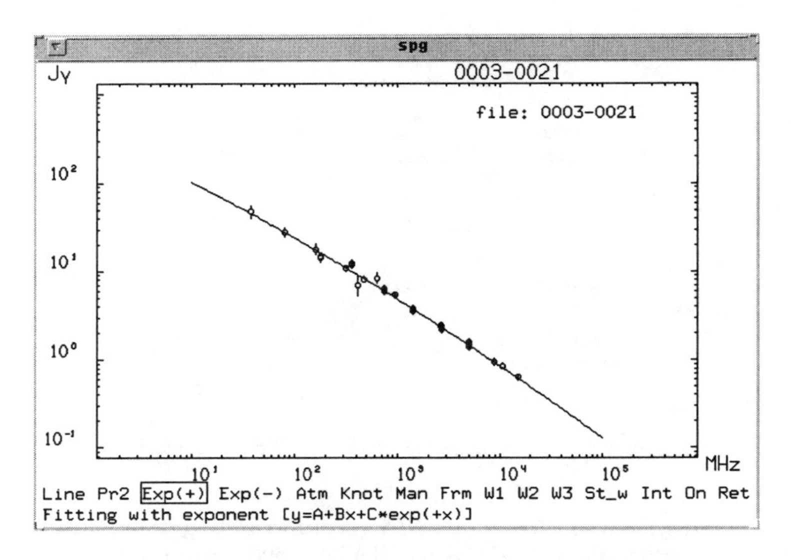

Figure 2. The *spg* screen with the menu and the radio spectrum.

and operation with spectra or other parameters. The standard FITS TABLE header describing the various data fields of the table may be recorded with the resulting file. 3) X-window codes or tape archives (TAR format) of compressed PostScript files for graphical spectra of radio sources. The result of the CATS operation is an ASCII file sorted according to object characteristics. It can be used for subsequent investigation of the radio source spectra or statistical source properties in the RATAN-600 data processing system (Verkhodanov 1997).

On-line access to CATS will be provided in several modes: 1) Dialog mode (non-graphics) is the only mode established at present. Several UNIX *shell* scripts (Verkhodanov & Trushkin 1995) permit the execution of the database-supporting programs via TCP/IP and NFS protocols in the local computer net of SAO. 2) Access via TCL/Tk scripts on the basis of *shell*. This mode also allows operation on the figures or profiles of radio spectra or statistical distributions by different parameters of the selected samples. 3) Hypertext access will eventually allow remote Internet users to operate with CATS via hypertext transfer protocol (HTML). It will allow execution of all described operations in graphic mode and will probably take advantage of the Java language.

Presently we are working on a further type of access to CATS by e-mail request, permitting the user to send a message with his/her requests. The latter will be read automatically and sent for execution to the CATS scripts. The result will be sent automatically to the user via e-mail. Very bulky results will be placed in a public FTP area and the user will be informed about the FTP address of the file(s). The e-mail messages may have several formats describing the search window, the coordinates of the center, the epoch, the type of output format, and the type of catalogs to search. CATS also permits the copying of entire catalogs (via FTP) to the user's local computer and to operate with them "at home."

CATS operates with continuum spectra of radio sources recorded as FITS tables. The graphics program *spg* (SPectra Graphics) allows a user to fit a spectrum with a standard set of curves: 1) $y = A + Bx$, 2) $y = A + Bx + Cx^2$, 3) $y = A + Bx + C \times exp(x)$, 4) $y = A + Bx + C \times exp(-x)$, where $x = log\ \nu$, $y = log\ S$, ν is the frequency (MHz), S is the flux density (Jy).

The menu of the *spg* program (Figure 2) allows a user to choose among these curves either automatically (by a least-squares fit), or by manual selection of the fitting function, or by manual fitting using the mouse (where the curve follows a cursor). Data points may be weighted in different ways: setting equal weight, setting weights by flux density errors, or filling a form with a table of frequencies, flux densities, and weights.

4. Conclusions

The development of CATS will provide a simple and convenient access to astrophysical information and accelerate the process of obtaining characteristics of celestial objects. Operation with the database will permit astronomers to search for peculiar objects and study physical processes in sources of cosmic radiation.

CATS allows a user not only to get accurate positions of radio sources to study radio spectra, but also to derive different statistical properties of object samples. Trushkin & Verkhodanov (1995) recently demonstrated such possibilities in a cross-identification of two large catalogs in two different frequency ranges: the IRAS Point Source Catalog in the infrared and the UTRAO survey at 365 MHz. Two of us (H. A. and O. V.) are working on the cross-identification and eventual optical identification of UTR survey sources detected at 12–25 MHz. We plan to solve the problem of very large error boxes by a stepwise cross-identification progressing from low frequencies and angular resolution to higher ones until the error box size permits the optical identification.

CATS is being expanded continuously. Presently it comprises over 50 catalogs including all the RATAN-600 catalogs and occupies ~250 MB. The database system could be an essential part of a bigger project of the first publicly accessible database of radio sources, proposed by Andernach et al. (1997).

Acknowledgments. This work is supported by the Russian Foundation of Basic Research (grant No 96-07-89075). O. Verkhodanov thanks ISF-LOGOVAZ Foundation for the travel grant, the SOC for the financial aid in the living expenses and the LOC for the hospitality.

References

Andernach, H., Hanisch, R. J., & Murtagh, F. 1994, PASP, 106, 1190

Andernach, H. 1995, Astroph. Lett. & Comm. 31, 1

Andernach, H., Trushkin, S. A., Gubanov, A. G., Verkhodanov, O. V., Titov, V. B., & Micol, A. 1997, Baltic Astronomy, 6, 259

Kononov, V. K. 1995, Preprint 111T, SAO RAS, 22

Trushkin, S. A., & Verkhodanov, O. V. 1995, Bull. SAO, 39, 150

Verkhodanov, O. V. 1997, this volume, 46

Verkhodanov, O. V. & Trushkin, S. A. 1995, Preprint 106 SAO RAS, 66

Part 7. Database Applications

Astronomical Data Analysis Software and Systems VI
ASP Conference Series, Vol. 125, 1997
Gareth Hunt and H. E. Payne, eds.

Dynamic Dynamic Queries (DDQ)

Peter Teuben

Astronomy Department, University of Maryland, College Park

Abstract. We describe an implementation of Dynamic Queries (DQ), a recently developed method as an alternative to the perceived complicated SQL-type operations on relational databases by a very intuitive graphical interface.

This prototype implementation, written within the NEMO package, uses ASCII sliders and the PGPLOT graphics interface and uses a novel interactive analysis back-end tool that is dynamically associated with the query. We call this technique Dynamic Dynamic Queries (DDQ); it can be easily implemented in other existing table manipulation applications.

1. Introduction

Dynamic Queries (DQ) is a recently developed querying technique (Shneiderman 1993) which proved to be a good alternative to the often difficult to learn but very general and flexible SQL-type interface to a relational database. The query in DQ is formulated using a number of graphical widgets that represent a column from a table (columns can be numeric as well as textual). Using two sliders a range search can be set for any column, and the resulting view of this multiply **and**-ed selection is dynamically displayed in a canvas. The canvas can display the data as simple points, but additional information can optionally be added to these points (color, size, etc.). Once sufficient selections have been made, some implementations of DQ then allow the displayed data to be queried in more detail by, e.g., clicking on the point. A number of specific examples of DQ have been discussed in recent years, in particular paying attention to user interface issues.

This paper describes a generic (ASCII-widget) implementation where the input dataset can be an arbitrary table-like dataset. A selection of the viewing coordinates X and Y is made (as a virtual column computed from the existing columns) as well as the columns that are allowed to be be queried. In addition to the above described features, we also implemented a generic (user definable) interactive analysis tool that is dynamically associated with the query. We call this technique Dynamic Dynamic Queries (DDQ).

2. Data Structures

The input dataset has M parameters (columns) for given N observations (rows). Sticking to the two-dimensional aspect of a display screen (see Swayne et al. 1991

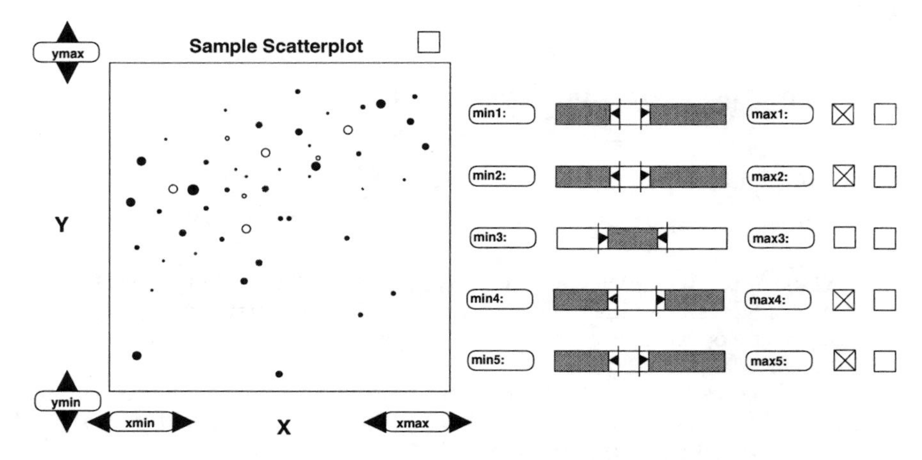

Figure 1. Overall layout of a DDQ GUI tool; sliders on the right (see also Fig. 2), the dynamic view of the selected X and Y coordinates in a scatter plot on the left. The X-Y viewport can be controlled by the xmin/xmax and ymin/ymax buttons.

for an alternative view[1] of these multi-variate data using the XGobi program), we choose two virtual columns (mathematical expressions derived from any of the M parameters) for display and any of the M columns as sliders for DQ interaction.

3. Implementation

The current toy version is implemented in the NEMO[2] package under the name `tabview`. The graphics interface used was that of PGPLOT, although being flexible, a little slow for this kind of application. In addition, rapid querying of large databases can be optimized with more elaborate database structures, such as the multilist, the grid file, k-d tree and the quad tree.

3.1. Startup

At startup a number of sliders (columns) are selected for DQ interaction, and the program overviews simple statistics and all their possible correlations are computed. A sample output with 6 sliders would be:

```
    % tabview in=p100.list xcol=%1 ycol=%2

Reading table p100.list with 6 columns and 100 rows

### Overview statistics:
Slider     : Npts      min       max      mean     sigma skewness kurtosis
1          :  100  -7.1859    5.7947  1.43e-07    1.2811  -1.1083   12.732
```

[1] http://lib.stat.cmu.edu/general/XGobi/

[2] http://www.astro.umd.edu/nemo/

```
2         :  100  -2.3444   7.3121 -5.83e-08   1.1374   3.5281    19.521
3         :  100  -3.4209   5.2131 -1.97e-07   1.1072   1.0584    5.8222
4         :  100 -0.78246   1.1218 -4.76e-08   0.3856   0.3441   -0.1828
5         :  100 -0.95056   1.0854  7.13e-08   0.3906  -0.0894   -0.1902
6         :  100  -1.1176   1.0182  4.87e-08   0.3763  -0.0878    0.3303

### Pearson correlation matrix:
   :     1     2     3     4     5     6
 1 :   1.00
 2 :   0.22  1.00
 3 :   0.54  0.14  1.00
 4 :  -0.07 -0.03 -0.10  1.00
 5 :   0.07  0.02 -0.06  0.06  1.00
 6 :  -0.12 -0.07 -0.08  0.02 -0.06  1.00
Xvar %1: displayed min= -7.51038 max= 6.11921
Yvar %2: displayed min= -2.58581 max= 7.55352
```

The prompt in "command-line mode" is that of the currently activated slider:

```
Slider 1 [-7.18587 5.7947 0 5.7947]:
```

with the following commands implemented:

```
l <num>      modify 'lo' slider
h <num>      modify 'hi' slider
l s <step>   set step in 'lo' slider
h s <step>   set step in 'hi' slider
b <step>     step both 'lo' and 'hi' slider
+/-          change stepping sign/direction
<digit>      change slider to interact with
r            reset lo/hi to min/max for this slider
s            show min/lo/hi/max for all sliders
u            update screen new
f            interactive cursor based flagging of points
o <file>     output table, w/ optional override out= filename
x            swap lo and hi (invert logic)
q            quit
!cmd         execute a shell command 'cmd'
|cmd         pipe visible data as ASCII table to 'cmd'
?            this help plus status
```

3.2. Post-Processing

At each step of the interaction the visible data (or its complement) can be processed. `tabview` implements this by manually piping the resulting data through a tool of the user's choice, but one can imagine this tool to be dynamically associated with the drawing engine, in which case it brings a whole new dimension to the exploration of this kind of data. Here is a sample session:

```
Slider 2 *[1 5.5 7 7]: |tabhist - 2 xmin=0 xmax=24 yapp=1/xs
[1090 data piped]
1090 values read
min and max value in column 2: [74 : 509]
Number of points     : 1090
Mean and dispersion  : 305.88 122.864
Skewness and kurtosis: -0.537889 -0.920467
```

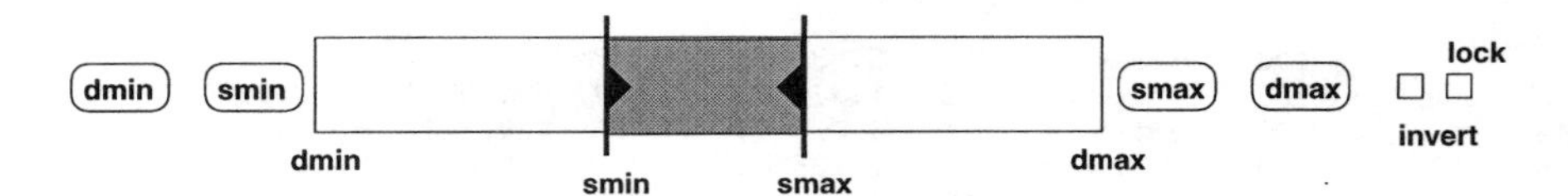

Figure 2. An expanded view of the slider widget that would be needed. The four buttons "dmin," "dmax," "smin," "smax" on either side of the slider can also be edited. The lock and invert check buttons on the right determine the way the slider moves (lock will fix the width (smax−smin) of the slider) and if the inside or outside of the slider is selected (greyed).

4. Conclusions

A definite advantage of DQ is its fast exploration capability of a multi-dimensional database (see also XGobi). Users can quickly discover which regions of this $M \times N$-dimensional database are densely and sparsely populated, find out which variables correlate, and how they depend on other variables in the dataset etc. With added analysis tools DQ will be a very powerful visualization and analysis tool.

An often quoted disadvantage of DQ has been its poor match of current hardware and software, although new implementations by IVEE (see also their excellent Spotfire Java demo[3]) prove that this may be a thing of the past. DQ needs a rapid search algorithm and display strategy (screens must be updated within 100ms in order to satisfy users). To the software side we can add the work required to integrate DQ into existing database systems, although this paper attempts to remedy this somewhat.

A second disadvantage is the limitation of range queries on numeric values. The HCI Lab. at the University of Maryland has experimented with a filter/flow metaphor which then provides full boolean functionality, albeit at a great cost of users to comprehend the system. This is a field in development.

Acknowledgments. We would like to acknowledge the HCIL at the University of Maryland for their inspiring work, and also Erik Wistrand (IVEE Development AB) for the use of their sample graphics during the poster session.

References

Shneiderman, B. 1993, IEEE Software, 11, 70 (DQ)

Swayne, D., Cook, D., & Buja, A. 1991, User's Manual for XGobi, Bellcore Technical Memorandum (XGobi)

Teuben, P. J. 1995, in ASP Conf. Ser., Vol. 77, Astronomical Data Analysis Software and Systems IV, ed. R. A. Shaw, H. E. Payne, & J. J. E. Hayes (San Francisco: ASP), 398 (NEMO)

---

[3]http://www.ivee.com

Astronomical Data Analysis Software and Systems VI
ASP Conference Series, Vol. 125, 1997
Gareth Hunt and H. E. Payne, eds.

Access to Data Sources and the ESO SkyCat Tool

M. A. Albrecht, A. Brighton, T. Herlin, P. Biereichel

European Southern Observatory

D. Durand

Canadian Astronomy Data Center, DAO/NRC

Abstract. SkyCat is a tool that combines image visualization with access to astronomical catalogs and data archives. SkyCat uses a standardized URL syntax to access a large number of data sources on the net. This paper addresses the issues that were solved in defining and implementing such standardized scheme. More information on SkyCat can be found at the SkyCat Web site.[1]

1. SkyCat Features

The ESO SkyCat Tool includes the following features:

- open and visualize a variety of FITS images including support for World Coordinate System (WCS), interactive measurement of offsets, and other standard visualization functions (SAOimage-like),
- overlay and edit color graphic objects on the image, like "tagging" sources with text, arrows, circles, or other graphic elements such as masks,
- PostScript color printing of the display (image + graphics),
- access and load an image from a network server of the Digitized Sky Survey scans,
- access and load catalog information from a number of popular astronomical catalogs, like the HST Guide Star Catalog,
- access local user catalogs,
- save catalog data locally,
- overlay catalog sources on an image,
- interact with Netscape to display more object information when available (support for Mosaic may be added in a future release),

[1] http://archive.eso.org/skycat

- access the observations catalog from the NTT, HST, and CFHT Science Archives,
- access SIMBAD and NED, both as name resolvers and for information on known objects,
- retrieve preview and other compressed images, and decompress them on the fly,
- load compressed images in hcompress, gzip, or UNIX compress format,
- retrieve and plot tabular preview data for a selected object as an (x, y) graph,
- calculate, display, and plot the center position, FWHM, angle and, other information for a selected star or object,
- access SkyCat features from a remote process via socket interface or via Tk "send,"
- access image header and data (FITS format) via SysV shared memory and/or mmap, and
- overlay RA/DEC grid on the image.

We are working on these additional features: (*i*) save canvas graphics, possibly as a FITS extension on the image file, (*ii*) user customizable symbols to "label" objects on the canvas, (*iii*) interact with a WWW browser to access catalog documentation and other documents, (*iv*) load lists of catalogs from sites (ESO, CADC, CDS, local) and allow user to select a preferred default catalog list, (*v*) save user preferences for look-up tables and other settings, and (*vi*) add or edit WCS parameters for a given image.

2. The Software

SkyCat was developed by ESO's Data Management and Very Large Telescope (VLT) Project divisions with contributions from the Canadian Astronomical Data Center (CADC). The tool was originally conceived as a demo of the capabilities of the class library that we are developing for the VLT.

This library consists so far of two major subsystems: (*i*) The Real-time Display (RTD) image classes with the RTD widget (see The Messenger, 81, 1995), and (*ii*) the Catalog Interface classes with the CatSelect widget. These constitute the basic building blocks of SkyCat. Early demonstrations of the tool to users were received so enthusiastically that we decided to polish things up, and produce the current version.

2.1. Java vs. Tcl/Tk

Some users might wonder why we did this development in the Tcl/Tk environment rather than in Java. The main reason is the pragmatic need to have the functionality implemented when the VLT comes on-line (mid-1997). However, great care has been taken to develop as much as possible in the short time with object oriented languages (C++, [incr Tcl]), having in mind that the future lies in tele-scripting rather than in the distribution of binary code.

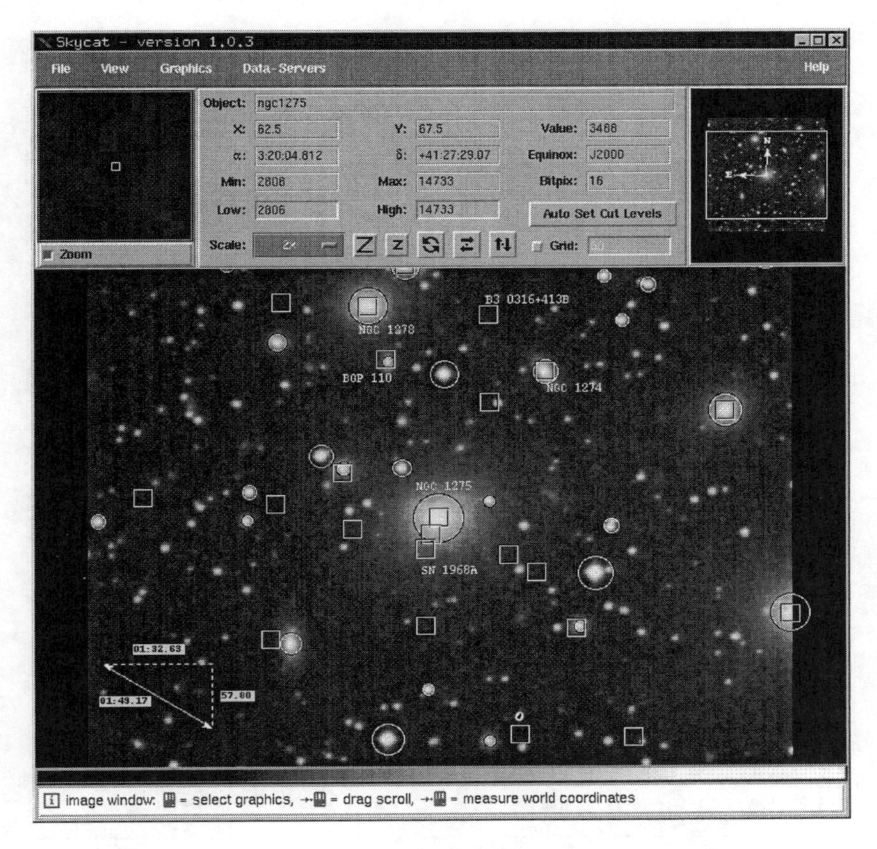

Figure 1. SkyCat showing a DSS image of NGC1275.

2.2. Distribution and Support

SkyCat is freely available to any users who want to download and use the software at their own risk. SkyCat is available as an executable for these platforms: (*i*) SunOS 4.1.3, (*ii*) HP-UX A.09.05, (*iii*) Solaris 2.5 (SunOS 5.5), and (*iv*) HP-UX B.10.10; ESO will maintain the latter two in the longer term—they are the platforms on which VLT software will run. ESO does not have the resources to port and maintain SkyCat on any other platforms. We will be glad to redistribute any port that other people or groups may support, but decline any responsibility for them.

The software is also available as source code for research and other non-profit organizations. If you are interested in obtaining the package, please send us a note. Please report problems or send suggestions to malbrech@eso.org or to abrighto@eso.org.

Acknowledgments. We very much appreciated practicing wishful programming at large, i.e., wishing a utility, a function, or just a code fragment that would just do that bit you badly need, then surfing the net, fetching it, and re-using it in our code. Here is an incomplete list of packages that we either par-

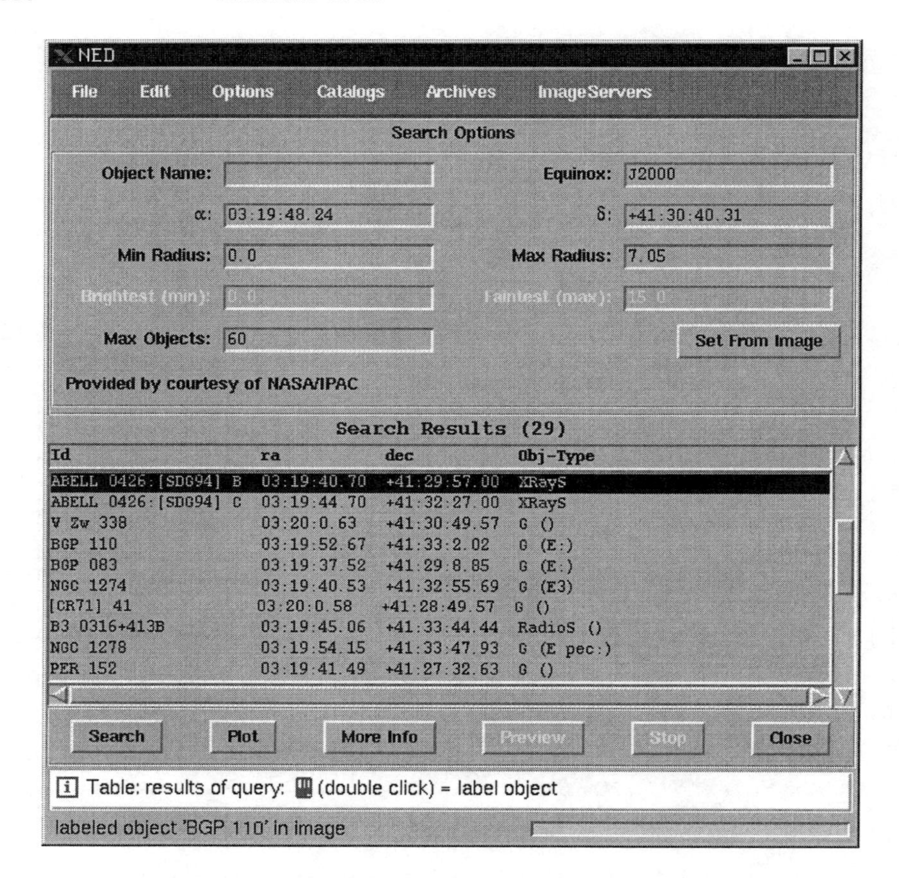

Figure 2. The CatSelect display of the field around NGC1275.

tially re-used or used as a source of inspiration: (*i*) STARCAT contributed many of its internals, (*ii*) Tcl/Tk, TclX, BLT and [incr tcl] give the glue around the C++ classes and widgets and Tk provides the canvas graphics, (*iii*) SAOimage lent the WCS lib, (*iv*) GSC server provided by courtesy of A. Preite-Martinez & F. Ochsenbein, (*v*) SIMBAD and NED client routines allow name resolving, (*vi*) the CADC "press" library is used to inflate compressed (hcompress, gzip, ...) images, (*vii*) Midas routines are used to calculate the centroid position, FWHM, and angle of selected stars/objects, and (*viii*) GNU Emacs routines are used to make a single binary executable from many Tcl/Tk scripts and data files.

References

Chavan, A. M., & Albrecht, M. A. 1997, this volume, 367

Astronomical Data Analysis Software and Systems VI
ASP Conference Series, Vol. 125, 1997
Gareth Hunt and H. E. Payne, eds.

A Java Interface For SkyView

Thomas A. McGlynn,[1] Keith A. Scollick,[2] and Nicholas E. White

NASA/Goddard Space Flight Center, Greenbelt, MD 20771,
E-mail: tam@silk.gsfc.nasa.gov

Abstract. We discuss our experiences in building a Java interface to the *SkyView* virtual observatory. We describe the reasons we choose Java, the problems we encountered, and how we feel this language may be used in the future for Web and stand-alone applications.

1. Introduction

The advent of Java applets has provided a new means by which archive systems can serve users. This paper discusses the reasons for, and experience gained in building a Java interface to the *SkyView* facility developed at the High Energy Astrophysics Science Archive Research Center (HEASARC). We address the capabilities we hoped to achieve in the applet, how the applet paradigm matches with our system, and the strengths and weaknesses of the extent Java implementation. We conclude with a discussion of the success of our preliminary interface, and how we would assess the suitability of Java for other applications.

2. Background

SkyView was designed as a virtual telescope on the Web. The basic concept is that a user specifies a position or object in the Sky and a wavelength regime (or survey), and *SkyView* returns a image with any desired geometry: coordinate system, scale, orientation, projection, ...

Prior to our Java development there were three mechanisms for accessing *SkyView*. By far the most popular is through a set of forms on the World Wide Web (WWW). Three forms of varying complexity allow the user to specify the image desired and return the image as a dynamically generated Web page. There is also a batch interface, where users who have installed some minimal software (a pair of Perl scripts) on their machines may make a request for a *SkyView* image from the command line of their local computer.

These interfaces address that fraction of *SkyView* use where the user understands exactly what image is needed for further analysis. However, a third interface is provided to allow users substantial image manipulation capabilities. *SkyView* can generate images with a large amount of information. *SkyView*'s Interactive Interface allows users to play interactively with the color table, catalog

[1] Universities Space Research Association

[2] Computer Sciences Corporation

overlays, contours, coordinate grids, and other overlays to help understand this information. This interface is implemented as a remote X-windows task started on the *SkyView* server. Because of this implementation, the interface can be extremely slow, does not support Macintosh or PC hosts, and requires users to adjust the security environment of their machines. These problems substantially compromise the usefulness of the interface.

3. Why Java?

Java allows us to re-implement our Interactive Interface in a way which addresses these concerns. Since image manipulation occurs locally, it can be substantially faster than when each image update needs to be transmitted over the Web. Popular Web browsers for all major hardware types support Java. While the security issues are not entirely resolved, they are placed in a context where they can be addressed simultaneously for *SkyView* and other interactive systems.

Beyond these goals, we have hoped that Java will also address concerns we have with our current Web forms. While the idea of filling out a form and then submitting a request is essentially the *modus vivendi* the Web, there are substantial shortcomings in the simple HTML implementations of forms. For example, it is virtually impossible to create a clean-looking form that supports a sophisticated set of options. Since Java supports a full GUI interface for applets, it becomes possible for us to offer a single interface which is simultaneously powerful and easy to use.

Similarly, the HTML form interfaces have less than ideal mechanisms for documentation and error reporting. A user must move back and forth between form and documentation pages. Users must submit a request and get an error document back when an error is made, even when the error is egregious. In a full GUI, we can use pop-up error windows and interface hiding to ensure that users' navigation of *SkyView* is as smooth as possible.

4. Putting SkyView in Java

SkyView's architecture has facilitated a relatively straightforward integration of a Java interface. All of the *SkyView* interfaces can be broken into three elements: the geometry engine which accesses the archive and actually creates images, an image annotater which provides overlays on images and other services such as transforming image formats, and the user interface. The various user interfaces call first the geometry engine, then the image annotater, and finally display the images. Since these elements are already decoupled it is easy to add Java as another user interface.

Our current Java interface incorporates many features that would be difficult or impossible using a purely form driven system including:

- simultaneous retrieval of multiple images,
- blinking of images,
- hiding many complex options within a menu bar,
- color table manipulation,

- displaying the coordinates and value of pixels using the mouse, and
- "greying"-out options that are not currently usable.

Development of additional capabilities is proceeding rapidly. The *SkyView* Java interface is available from the *SkyView* home page.[3]

5. Assessment of Java.

Overall, our experience with Java has been reasonably good. Usable interfaces were built with only a few weeks of work. The language is clear and understandable, and seems to be an excellent *entreé* to the object-oriented world.

The difficulties we encountered were not those we anticipated. The most intractable problems have come from relatively simple tasks, such as laying out the GUI in pleasing fashion, or doing such basic tasks as formatted I/O. These have pointed out annoying, and occasionally alarming, deficiencies in the Java Applications Programmer Interface (API).

Some failings reflect the languages immaturity but, regardless, they make it difficult to write real code. Java does not appear to have made support for scientific programming an important element of its design. There is no support for formatted I/O or array I/O . In general, I/O is handled quite clumsily. The lack of support for operator overloading obfuscates scientific code, where the use of multiplication and addition for non-scalar quantities is well founded. Nor is there any notion of support for simple array operations.

The Abstract Windowing Toolkit (AWT) provides reasonable functionality for individual components. However, it can be extremely difficult to lay out a page in the fashion desired. The `GridBagLayout` class, which is intended to address this issue, is confusing and poorly documented.

Another handicap of programming in Java is the lack of existing libraries. In other languages, this can be addressed, in the interim, by calling existing libraries in other languages. However, Java applet programming does not permit linking to foreign language modules. For *SkyView*, the lack of FITS and geometry libraries is particularly problematic. If Java is to be popular within the astronomical community, a FITS library is essential.

The Java security model is extremely frustrating. While we understand the rationale for the high level of security required for Java applets, the lack of any ability to read from or write to the user file system (in a portable fashion) is extremely annoying. It markedly reduces the functionality we are able to offer in Java and reduces the efficiency of the functions we can offer, e.g., an image must be loaded to the Java interface and then separately downloaded to the user when needed.

6. Future Plans

While we have encountered a number of frustrations in the Java interface, it can clearly meet our original goal of being an effective replacement for our Interactive Interface. Our next step is to continue development, so that it provides

[3]http://skyview.gsfc.nasa.gov

a reasonable alternative to our forms-based interfaces. The ability to hide unneeded capabilities and disable features for which the user has not yet specified prerequisites, the capabilities for immediate feedback that helps users to correct errors in fields, and the possibilities for a really effective context-sensitive help, are all extremely attractive. We do not anticipate ever removing the forms interfaces, but we do feel that Java may become the primary *SkyView* interface as Java becomes more popular and efficient.

Other projects at the HEASARC are also beginning to use Java applets. We have already developed a preliminary applet for plotting the results of database queries, and we are looking at Java to provide an interface which can unify the various elements of our W3Browse database retrieval system.

Our conclusion is that Java can provide the capability to provide users with interactive access to our systems, but it is still very immature. In general, we see most of these applications following the same lines as *SkyView*. The applet provides the user interface to initiating services at our host site. The bulk of the CPU processing and any capabilities requiring large amounts of I/O are done at our site. Results are then returned to the Java interface, which provides capabilities to display and manipulate them locally.

Beyond Java applets, we have also begun exploring the use of Java for full applications. Here a number of the limits to Java (local I/O and linking to non-Java modules) are lifted. Java's strengths in portability, networking, and Web access make it attractive in a number of areas.

Astronomical Data Analysis Software and Systems VI
ASP Conference Series, Vol. 125, 1997
Gareth Hunt and H. E. Payne, eds.

Java, Image Browsing, and the NCSA Astronomy Digital Image Library

R. L. Plante,[1] D. Goscha,[1] R. M. Crutcher,[1] J. Plutchak,[2]
R. E. M. McGrath, X. Lu, and M. Folk

National Center for Supercomputing Applications (NCSA), University of Illinois, Urbana, IL 61820, E-mail: rplante@ncsa.uiuc.edu

Abstract. We review our experience, including some lessons learned, with using Java to create data browsing tools for use with the Astronomy Digital Image Library (ADIL) and related digital library projects at NCSA. We give an overview of our Image Data Browser, a generalized tool under development through a collaboration with the NCSA/NASA Project Horizon in support of access to earth and space science data. Emphasis has been placed on a design that can support a variety of data formats and applications both within and outside of astronomy. ADIL will use this tool in a variety of guises to browse and download images from the Library's collection. We see such a tool filling an important niche as a pipeline from a data repository to specialized, native data analysis software (e.g., IRAF, AIPS).

1. Introduction

Digital library technology for accessing earth and space science data has been the focus of a collaborative effort at NCSA known as *Project Horizon,*[3] a cooperative agreement between NASA and the University of Illinois. Project Horizon includes scientists from the fields of astronomy, atmospheric science, and Web server technology as well as researchers from the NCSA HDF group and the University of Illinois Digital Library Initiative. As a participant in Project Horizon, the NCSA Astronomy Digital Image Library (ADIL)[4] has served as a testbed for solutions to accessing large datasets over the Web.

One of the problems Project Horizon has been studying is how one most effectively browses large image datasets over the Web. One early solution included simple static digests of the data which might include a GIF or JPEG preview of the image. More interactivity was brought to the process with the server-side, on-the-fly generation of images driven by inputs from an HTML form. One shortcoming of this latter method is the load put on the server to

---

[1] Astronomy Department, UIUC

[2] Atmospheric Science Department, UIUC

[3] http://www.atmos.uiuc.edu/horizon/

[4] http://imagelib.ncsa.uiuc.edu/imagelib

generate these custom views. This was demonstrated at UI with early versions of the very popular Weather Visualizer[5] which allowed users to download custom visualizations of weather data. Another problem is that there is limit to the level of interactivity that HTML forms can provide.

The advent of Java provides a way to address some of the problems of data browsing. As a network-smart, object-oriented language, Java allows one to develop highly interactive applications that can interact with a server via a Web browser. Java also brings the advantage of platform-independence to application development. For the specific problems of image data browsing, Java allows a server to pass on some of the CPU chores to the individual clients.

Project Horizon's early explorations with Java produced several examples of how Java could be used to interact with scientific data:

- The ADIL[6] uses a Java applet for browsing large FITS images such as those from the NRAO VLA Sky Survey, providing zooming ability and coordinate tracking.
- The Daily Planet uses a Java version of the Weather Visualizer[7] to visualize a variety of weather data as multiple overlayed images which users may turn on and off.
- The NCSA Hierarchical Data Format (HDF) group has released a FITS Data Browser[8] for browsing local FITS images either as a visual image or as a spreadsheet.
- The HDF Data Browser[9] is also a Java application specially suited for browsing the hierarchical structure of HDF files.

With these early successes, it became clear that each of the Java applications contained features that would be useful for browsing all kinds of scientific images, regardless of data format or field of study with which the data is associated. Project Horizon is now working to combine those features into a single package of reusable Java code called the *Horizon Java Image Data Browser Package.*[10]

2. Horizon: a Java Package for Browsing Images

The Horizon Java package will be a collection of image data browsing solutions in the form of reusable Java classes, along with ready-to-use Java applets and applications. We note that the aim of the Horizon package is *not* meant to duplicate or replace the role of visualizing tools such as SAOimage or Aipsview or of image processing packages such AIPS, IRAF, and others. Instead, Horizon focuses specifically on *data browsing*, that stage of data handling before and

---

[5] http://covis1.atmos.uiuc.edu/covis/visualizer/

[6] http://imagelib.ncsa.uiuc.edu/project/javadoc/96.JC.23.06

[7] http://covis.atmos.uiuc.edu/java/weather0.4/Weather.html

[8] http://hdf.ncsa.uiuc.edu/fits/java

[9] http://hdf.ncsa.uiuc.edu/hdf/java/jhv

[10] http://imagelib.ncsa.uiuc.edu/imagelib/Horizon/

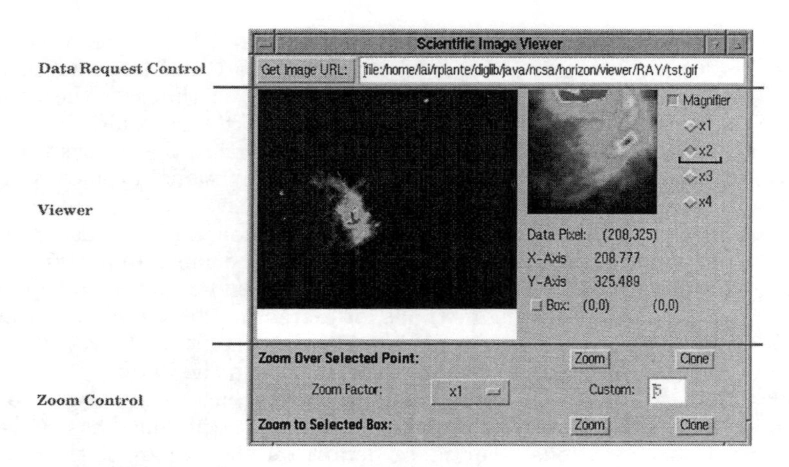

Figure 1. A sample Horizon application for viewing images. Horizontal lines delimit the three Horizon objects that make up the viewer.

after processing and analysis with native software tools. Specifically, the goals of Horizon are: (*i*) to provide a "first-cut" look at image datasets from the network or local disk, (*ii*) to act as a smooth pipeline from data repositories to specialized native software (e.g., IRAF, AIPS, IDL, etc.), (*iii*) to serve as a simple, cheap, and platform-independent image visualizer with basic browsing features for those without access to more sophisticated software, (*iv*) to provide basic image visualization in a collaborative environment, and (*v*) to provide a programming paradigm for adding new features and modifying existing features easily through generalized abstract classes.

Some of the features of the package include: (*i*) basic 2D visualization of multidimensional data, (*ii*) reading/writing data of (in principle) any data format from the network or local disk (initially FITS, HDF, GIF, and JPEG), (*iii*) zooming, sub/super-setting, and animation, (*iv*) pixel value and coordinate position display, (*v*) linking pixels with text-based data, (*v*) spreadsheet browsing, (*vii*) color fiddling, progressive image transmission,[11] (*viii*) overlaying of multiple images, (*ix*) support for collaborative session via the NCSA Habanero Collaboration Tool,[12] and (*x*) easy adaptability and extensibility.

The Horizon package will include applets and application that are ready-to-use or can easily be adapted. For more specific applications, the Horizon classes can be mixed and matched to create new specialized tools.

3. Horizon Design Overview and Implementation Considerations

The Horizon package provides a variety of widgets that can be assembled into a variety of applets and applications. As an example, Figure 1 shows a Viewer ap-

[11] http://www.sal.wisc.edu/~jwp/can.html

[12] http://www.ncsa.uiuc.edu/SDG/Software/Habanero/HabaneroHome.html

plication made up of three Horizon components, a Viewer Panel and two Control Panels. The Control Panels interact with the Viewer Panel (e.g., get selected pixels, request views of subregions, etc.) via public methods of the abstract Viewer class; thus, with very little programming effort, one could remove this particular Viewer and replace it with a specialized version that obeys the same interface. Control Panels will provide specialized functionality such as zoom control, animation, colormap fiddling, and data transfer.

In addition to the GUI components, the Horizon package also provides important classes that work behind the scenes. The most important is the Viewable interface, which serves as the format-independent layer on top of the format-specific reader, providing methods for extracting data and visualizations of the multidimensional data beneath it. Also important is the set of classes that support world coordinate systems associated with the data.

For all its advantages, Java also has limitations which we are trying to keep in mind as we develop the Horizon package, and our experience has suggested some general considerations. For more details on the design of the Horizon package, see the Horizon Design white paper (Plante et al. 1996).

4. Recent Progress

Early fall 1996 saw our first alpha releases. The first was a demo release of classes to illustrate a few of the basic viewers we are currently working on. This was soon followed by the first release of source code in the form of the FITSWCS package. Actually a translation of Mark Calabretta's WCSLIB library, this Java package provides support for the FITS World Coordinate System standard proposed by Greisen & Calabretta (1996). Late fall 1996 will see a major alpha release that will include the basic source code and documentation for building the basic image viewers, including generalized support for coordinates. A full production release is expected by the end of 1997.

In conclusion, we see the Horizon Package filling a variety of niches along the scientific data pipeline through its ability to provide different solutions to different people. Data providers will easily be able to create browsing applets customized for a particular collection of data for a particular audience. Astronomers could use Horizon tools with NCSA Habanero to remotely monitor observations in a collaborative session. Educators and students without access to high-powered graphics workstations could use Horizon applications as a "poor-person's visualizer" of scientific data.

References

Greisen, E., & Calabretta, M. 1995, The draft proposal for WCS syntax in FITS headers[13]

Plante, R., Goscha, D., & Plutchak, J. 1996, Horizon Design White Paper[14]

---

[13] ftp://fits.cv.nrao.edu/fits/documents/wcs/wcs.all.ps.Z

[14] http://imagelib.ncsa.uiuc.edu/imagelib/Horizon/DesignWhitePaper.html

Astronomical Data Analysis Software and Systems VI
ASP Conference Series, Vol. 125, 1997
Gareth Hunt and H. E. Payne, eds.

A Configuration Control and Software Management System for Distributed Multiplatform Software Development

E. Huygen,[1] B. Vandenbussche, and G. Bex

Katholieke Universiteit Leuven, Instituut voor Sterrenkunde, Celestijnenlaan 200B, B-3001 Heverlee.

P. R. Roelfsema, D. R. Boxhoorn, and N. J. M. Sym

Stichting Ruimteonderzoek Nederland, SRON, The Netherlands.

Abstract. A system to efficiently control and manage the software development of a large project is presented. The overall design, implementation and use of the configuration control system (CoCo) and the calibration database (CalDB) used for off-line software development of the Short Wavelength Spectrometer (SWS) in the Infrared Space Observatory (*ISO*) is described.

1. Introduction

CoCo is a software configuration control system especially designed for the development of the *ISO*-SWS interactive analysis data reduction software (IA^3). Since this software is being developed by a team spread out over several institutes, on machines with completely different architectures, we needed a configuration control system that was easily portable and applied a client/server technology.

The CoCo system is capable of efficiently controlling the development of the IA^3 software. The server runs at the satellite tracking station of the European Space Agency (ESA) in VILSPA, Spain, and controls the main repository of what we call CoCo objects, i.e., all source code (FORTRAN, C, IDL, Perl), calibration files, comparison spectra and photometry, documentation, etc. Changes to these objects are saved as new generations and can be requested individually by any CoCo client program.

The calibration database CalDB is constantly and automatically fed with information about all these changes and dependencies of calibration files, algorithms, and data. This information is accessible through a WWW interface.

The software management for IA^3 is greatly simplified by a Software Problem and Modification Report (SPR) system which is built into CoCo. These reports are handled as special objects by CoCo and special commands are implemented to work with them.

[1] E-mail: rik@ster.kuleuven.ac.be

2. Using CoCo

CoCo uses standard configuration control commands to work with its objects, i.e., create/reserve/insert/extract, etc. All changes to objects must be accompanied by a software problem report (SPR) which fully describes the problem, the proposed solution, and the actual change to the object. Changing objects with CoCo requires a mandatory sequence of CoCo commands: a user raises a SPR, software management assigns an action to a developer, the developer reserves the object (which is automatically extracted to his/her local machine), and changes are tested and inserted back into the CoCo system. This will automatically create a new generation for the affected object. The reservation mechanism avoids developers making conflicting versions of the same code, and the SPR system streamlines the reporting of problems from the users to the developers.

All developers and users of the IA^3 data reduction software have individual access and change permission on the CoCo server, which can be granted or revoked by the configuration control manager.

At the consortium institutes, an automatic process starts up a CoCo client every night to update the local system. This means that every site has a consistent version of the software system every morning, including the changes made at any institute the day before.

3. The CoCo Server Implementation

The CoCo server is a dæmon program that accepts commands from any CoCo client by electronic mail. This protocol was chosen because of its simplicity, and because it was, at that time, the only robust way to communicate between the different institutes and the different computer architectures. It is now, for security reasons, the only possible way to communicate with the server, located at VILSPA. Requests from CoCo clients are handled as they appear in the mail queue.

The server is organized in several different command modules (see Figure 1). The command manager parses the requests sent in by a client, and delegates the commands to the proper command modules. Configuration control commands are those which create, extract, reserve, or insert objects in the main repository. They communicate with the Code Management System (CMS) running on a VMS machine and simultaneously feed the calibration database. Info commands provide the user with status information about the system and the individual objects. Administrative commands are used to maintain and debug the CoCo server.

4. The CoCo Client Implementation

The CoCo client is a stand-alone program that accepts user input via a command line interface or a batch file. The user request is converted into a syntactically correct CoCo command and sent to the command manager. The command manager distributes these commands to the proper command modules.

Local commands are handled by the client program itself. These commands do not require interaction with the server, and consist mainly of information commands and commands to check the structure and contents of CoCo objects.

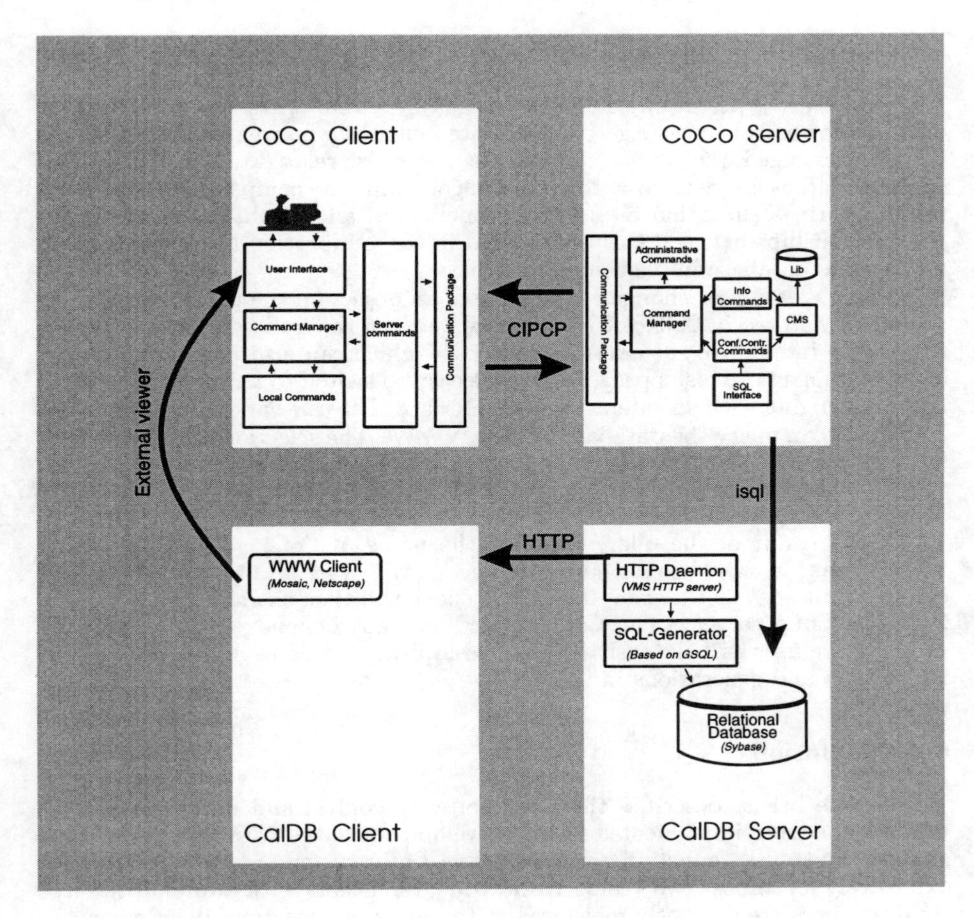

Figure 1. This picture shows the essential design of the complete configuration control and database system.

All commands that require server interaction are sent to the CoCo server. The client program then waits for a response from the server. If this response requires further action from the client, this is done before returning control to the user.

Apart from communication with the server, the client is also responsible for maintaining the local copy of the IA^3 software and calibration data. This is accomplished by an automatic full-update mechanism, whereby clients will send a request to the server to retrieve a list of all the CoCo objects with their last modification time. From this list, the client can decide which objects are out-of-date on the local system, and send the server an update request for each such object.

5. The Calibration Database CalDB

Many different types of objects are managed by CoCo. These objects have not only their own properties (e.g., creation date, scientist who created them, reasons why they changed, purpose, ...), the objects are also related to each other (e.g., calibration files are derived with special algorithms, the comparison spectra are used to derive calibration files). The properties of all the objects in CoCo and the relationships between them are stored in the Calibration Database, CalDB, a relational database implemented in Sybase.

Every creation or change of the state of an object on the CoCo server is automatically stored in CalDB. The interface between the CoCo server and Sybase is isql, the interactive sql interface of Sybase which can also handle batch files of sql statements. This approach makes it easy to switch to any other relational database engine with an interactive sql interface that can handle sql input files.

Users can query the database via the WWW. The selection criteria and the output format are specified by filling out a form and submitting it to the HTTP server on the CalDB host. The script translating the form to sql is based on the NCSA GSQL gateway.

If the result of the query contains the name of CoCo generations, these will be hyperlinks. Following such a hyperlink will start a CoCo client with the external "viewer" mechanism. The CoCo client will then extract the generation from the CoCo server. The CoCo server does not depend on CalDB. CalDB provides the user with extensive query possibilities and extra information about the objects and generations in CoCo.

6. Conclusion

This article briefly describes the CoCo software control and management system designed and implemented for the development of the *ISO*-SWS interactive analysis and data reduction software. The CoCo system has now been used extensively for almost two years. World wide, ten clients with a total of twenty-six developers are currently registered at the server. Since June 1995, when the server was transfered to the ESA satellite tracking station at VILSPA, more than 60,000 client requests were handled by the server, which is more than 100 transactions per day. For more information on CoCo consult the CoCo Homepage.[2]

Acknowledgments. We like to thank all those who supported us during development and maintenance of this system, especially the SWS Instrument Dedicated Team (SIDT) that constantly uses the system and continuously gives feedback on its use and possible improvements.

[2] http://www.ster.kuleuven.ac.be/coco

Astronomical Data Analysis Software and Systems VI
ASP Conference Series, Vol. 125, 1997
Gareth Hunt and H. E. Payne, eds.

Case Study of RDBMS Use on The EUVE Mission

E. C. Olson

Center for EUV Astrophysics, 2150 Kittredge Street, University of California, Berkeley, CA 94720-5030

Abstract. The *Extreme Ultraviolet Explorer* satellite (*EUVE*) mission has employed a Relational Database Management System (RDBMS) in several ways. Shortly after the launch of *EUVE* in 1992, an indexing scheme for the *EUVE* telemetry was implemented to facilitate access to the calibration, in-flight, and supplemental telemetry data. The RDBMS has been used for tracking *EUVE* observations and other administrative information. Over time, it has evolved into the central data repository of the distributed *EUVE* data reduction system. A number of experimental systems also used the RDBMS. In 1992 *EUVE* began participating in the Astrophysics Data System, using the RDBMS as the repository for the *EUVE* bright source list and additional, ancillary information. With the advent of the World Wide Web, *EUVE* used its RDBMS to publish dynamic information for use within the project and across the Internet. More recently, *EUVE* has created prototype client programs to facilitate use of relational databases for scientific applications.

1. Introduction

The *Extreme Ultraviolet Explorer* (*EUVE*) mission (Bowyer et al. 1988; Haisch et al. 1993) employs relational database management systems for several mission-critical activities. Additionally, *EUVE* uses databases for a number of non-scientific, administrative, and other non-mission-critical areas. This paper will focus on the use of database technology for the mission critical aspects.

The Center for EUV Astrophysics (CEA), Space Sciences Laboratory, at the University of California, Berkeley, operates the scientific telescopes and instrumentation on board the *EUVE* satellite. CEA developed ground systems for acquiring, monitoring, archiving, and distributing the *EUVE* science data (Olson & Christian 1992; Christian & Olson 1993; Olson 1995; Olson et al. 1996). The primary mission-critical use of database systems has been in the archiving and distributing of *EUVE* telemetry.

2. Telemetry Archive

A terabyte optical disk jukebox serves as the primary storage facility for the Telemetry Archive (Girouard & Hopkins 1993). The jukebox also supports other hardware devices for various operational reasons beyond the scope of this paper. The jukebox supports the Network File System (NFS) protocol, which greatly simplifies access and manipulation of the archive. The Telemetry

Archive database itself is maintained in the Sybase RDBMS. The Telemetry Archive comprises two components: a general purpose archive database and a telemetry-specific archive database.

The general purpose archive is used for a wide range of purposes. Its primary purpose is to provide a mechanism for the archiving of telemetry. However, a number of supplementary files related to the telemetry and to the general operations of *EUVE* are also archived. This archive behaves like a simple file system that supports several different media. It maintains indexes, based on file names, tracking the location, size, and type of the file, including its media and media-specific information (such as file position and ID on the tape carousel device). This database resolves queries based on a unique ID, returning a specific file. The general-purpose archive also synchronizes processing. Handling telemetry is essentially an event-oriented processing problem. At any time the system must be prepared to receive telemetry. After the initial reception, the telemetry is processed in a variety of ways depending on its type. The features of RDBMS used are transaction locking, consistency checks, reliability, and robustness (particularly in light of machine crashes). Interestingly, the archive does not use the RDBMS relational features, and the system logic is coded in the application software. Essentially, the RDBMS is used as a reliable data store. Routinely, autonomous application software interacts with this database with little direct user interaction. The application software anticipates most routine exceptions and processes them accordingly. Highly unusual exceptions require the manual intervention of the system administrator. Since the system uses unique hardware, it is subject to single point failures. Additionally, the unique hardware and the 24 hour per day operation make testing of the database system difficult.

The telemetry-specific archive is used for maintaining information specifically required for accessing and manipulating the *EUVE* telemetry. The specific information includes time-stamps, exposure durations, and data quality information. This database is fairly straightforward since its organizing feature is time. Typical queries that the telemetry database can answer are related to time-oriented information. For example, the database maintains information about telemetry intervals and gaps in the telemetry stream. As with the archive database, the telemetry database synchronizes asynchronous processes where transaction-locking and reliability are key features. Again, the application software handles the logic, and the database is used as a reliable data store. More significant and less predictable errors can occur with this database. Although routine processing is handled autonomously, deleting badly formed telemetry from the index is problematic.

3. Proposal Database

The Proposal Database of *EUVE* pointed observations is designed to provide not only information about the target, the observation, and the principal investigator for the data, but also information about the observation schedule, the location of the acquired data, and various types of historical data such as the software version used for processing. Over time, the design expanded to include the state of the data reduction processing. In general, information in the Proposal Database handles higher level information than the information found in the Archive Database. The Proposal database has two components: an observation database and a proposal database.

The observation database, closely related to the Archive Database, obtains input data from the telemetry database and from supplementary information about the observations. Using the telemetry interval information in the telemetry database, the observation database constructs higher level data abstractions about observation intervals. For example, these observation intervals tolerate data gaps and track contiguous exposure time intervals on a single target. Additionally, the observation database handles cases in which an observation may be interrupted for, say, calibration pointings. In this case, the database, not the application software, contains most of the logic. The observation database makes significant use of the database relational features as well as triggers for maintaining the integrity of this database. However, the more complex logic required is located in the application software as it was deemed too complex to maintain in SQL. The transaction and synchronization features of the database are used for coordinating access by asynchronous distributed processes. Much of the input for this database comes from operators, and is subject to the usual errors that occur from manual data entry. Therefore, this database also supports a WWW interface for correcting data, and includes a reporting mechanism that facilitates the detection of errors. In addition, the WWW interface allows easy handling of runtime exceptions.

The proposal component of the Proposal Database tracks high-level information about investigators, their proposals, the related observation requests, and completed observations. This database has the greatest amount of operator data entry. Several mechanisms have been used to check the consistency of the data in this database, since errors can directly effect the productivity of the *EUVE* mission. Moreover, the information changes frequently as adjustments are made to observing proposals. Runtime exceptions occur frequently and are handled by operators using a WWW interface. This interface is closely related to the schema structure of the database. Operators must be familiar with the schema structure in order to appropriately correct data in the database. However, this cost is small compared to the cost of identifying and anticipating the multitude of exception conditions that may occur in this database.

4. Discussion

The *EUVE* mission has used RDBMS technology in mission-critical processing. The features used are transactions and synchronization (particularly suited to distributed systems). In some cases the relational features of the RDBMS were also used. However, in many cases complex logical expressions were handled in the application software instead of the RDBMS, and the RDBMS was used as a reliable data store. Overall design, development, and testing of the databases was not difficult.

Providing operators with the capability to access and manipulate data has been an ongoing problem in working with databases on the *EUVE* mission. We experimented with several commercial products, as well as the Astronomical Data System, in an attempt to provide an operator front-end to the RDBMS. None of these systems provided an appropriate solution. Typically such products are designed to provide complete solutions that require extensive development of interface specifications intimately connected to the schema of the database. This type of extensive development was never justified, nor did it appear maintainable since the database schema have continuously evolved. The evolution of the database schema is a natural result of the evolution of the scientific goals of the mission. Therefore, instead of providing a complete solution only applicable at a given moment, the *EUVE* mission has concentrated on providing general-purpose, partial solutions. Specifically, the WWW interface used in the Proposal Database is derived from the database schema. This interface requires operators be trained in the structure of the database, but dramatically reduces development and maintenance costs. More recently, the *EUVE* mission has developed a WWW server prototype (named xdb) that provides structured access to databases based solely on their meta data information.

Acknowledgments. This work was supported by NASA contract NAS5-29298. The Center for EUV Astrophysics is a division of the Space Sciences Laboratory at UC Berkeley.

References

Bowyer, S., Malina, R. F., & Marshall, H. L. 1988, JBIS, 41, 357

Christian, C. A., & Olson, E. C. 1993, in ASP Conf. Ser., Vol. 52, Astronomical Data Analysis Software and Systems II, ed. R. J. Hanisch, R. J. V. Brissenden, & J. Barnes (San Francisco: ASP), 56

Girouard, F. R., & Hopkins, A. 1993, JBIS, 46(9), 342

Haisch, B., Bowyer, S., & Malina, R. F. 1993, JBIS, 46(9), 330

Olson, E. C. 1995, in ASP Conf. Ser., Vol. 77, Astronomical Data Analysis Software and Systems IV, ed. R. A. Shaw, H. E. Payne, & J. J. E. Hayes (San Francisco: ASP), 425

Olson, E. C., & Christian, C. A. 1992, in ASP Conf. Ser., Vol. 25, Astronomical Data Analysis Software and Systems I, ed. D. M. Worrall, C. Biemesderfer, & J. Barnes (San Francisco: ASP), 110

Olson, E. C., Girouard, F., & Hopkins, A. 1996, in ASP Conf. Ser., Vol. 101, Astronomical Data Analysis Software and Systems V, ed. G. H. Jacoby & J. Barnes (San Francisco: ASP), 532

Astronomical Data Analysis Software and Systems VI
ASP Conference Series, Vol. 125, 1997
Gareth Hunt and H. E. Payne, eds.

QDB: An IDL-Based Interface to LASCO Databases

A. E. Esfandiari, S. E. Paswaters, D. Wang
Interferometrics, Inc.

R. A. Howard
Naval Research Laboratory

Abstract. QDB is a collection of IDL and C routines that provides a query interface to the Large Angle Spectrometric Coronograph (*LASCO*) databases maintained under the Sybase database management system. IDL widgets are used extensively to display the databases, tables, columns, and on-line help. This is a fully automated process—no code modification is required to reflect database changes such as adding/dropping databases, tables, or columns. Standard Query Language (SQL) is used to build a query based on the user selection. This query is then passed via Remote SHell (`rsh`) to two C routines that access the Sybase Open Client Database library to execute the query. The result is returned in an IDL structure. Another set of IDL routines optionally displays or manipulates the data in this structure.

1. Introduction

LASCO is one of the first *LASCO* instrument databases to be defined and populated with data. It is relatively change-free now but, while in the test mode, it went through many structural changes. These changes illustrated a need for a general and flexible interface that could be used with any database, regardless of its structure, size, and type of data. We needed a robust interface that did not require editing if, for example, a new table was added, or a column name was changed. This led to the creation of QDB, which interacts with a database without any advance knowledge about its structure or even its existence. QDB is a user friendly software package with much on-line help. It is currently used remotely to build and execute a LASCO SQL query, and then capture the returned data. This paper is devoted to explaining the inner workings of QDB as it performs this task.

2. QDB Details

2.1. Building a Query

QDB may be called with or without arguments. If the user knows SQL and has a query ready, he can simply pass the database name and the query as arguments to QDB, bypassing all the widgets. Otherwise, if no argument is provided, the following events take place. Upon invocation, QDB sends a query to the RDBMS

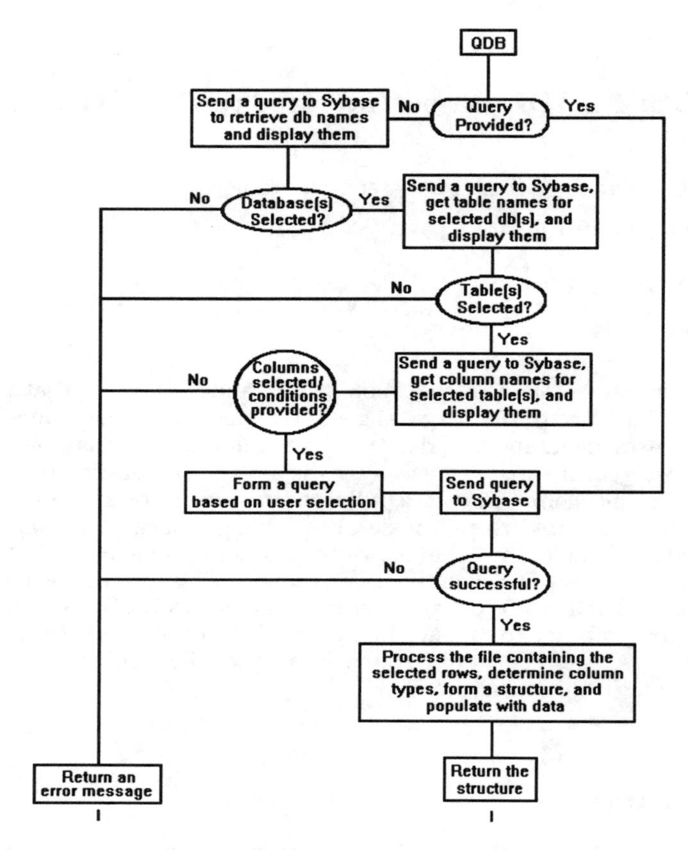

Figure 1. Flow diagram of QDB.

requesting all the existing database names, if any. If successful, it uses the IDL widgets to display them as toggle buttons. The user then clicks on database name(s) to select one or more databases. QDB then sends another query to the RDBMS requesting the table names within the selected database(s). If successful, it uses the IDL widgets again to display the table names using toggle buttons. The user now selects the tables he needs from the displayed list. QDB sends yet another query to the RDBMS, requesting the column names within each selected table, and displaying them using a combination of buttons that allow selection of any column (field) and providing areas for entering optional query conditions. When satisfied with the selections, the user exits the menu. QDB keeps track of all of the user selections, uses them to build a SQL query, and submits it to the RDBMS. At each step of this process, user can seek help using the help buttons, or simply exit the program. Figure 1 shows the flow diagram for QDB.

Because of QDB's dynamic interaction with the RDBMS, it always sees the latest database. Therefore, if a new column is added to a table while QDB is running and is displaying the table names, that new column appears in the columns list when QDB reaches that point. This flexibility makes QDB a low maintenance utility. In fact, only a one-time small code change is required to set conditions for joins between newly added tables, if such joins are desired.

2.2. Interface with RDBMS

QDB itself resides on the client machine but it calls two C routines that access the Sybase Open Client Database library to execute the query and they, along with the RDBMS, reside on the server machine. The query is passed to these routines using UNIX's Remote SHell (`rsh`) service. Besides adding the flexibility of using QDB remotely, `rsh` provides access security since the host machine must know the client login information (user and machine name from which QDB is started) beforehand.

2.3. The Output

The result of the query, along with each column's name, length, and data type, is captured in a text file on the client. QDB parses this file and builds a structure with field names and types corresponding to the selected column names and data types, and populates it with the data. This IDL structure is then returned to the user. Another set of IDL routines is provided to display or manipulate the data in this structure.

2.4. About The Code

This interface consists of about 1200 lines of IDL code and another 550 lines of C code. It is currently setup for use with UNIX and Sybase systems but it can be modified for use with other operating systems and/or relational database management systems. To use another operating system, `rsh` must be replaced with other means of communication with the remote C routines. To use another RDBMS, the C routines must be modified to make calls to the library routines specific to that RDBMS.

3. QDB and LASCO Database

LASCO is one of the *LASCO* instrument databases whose basic structure has been relatively change-free but has been regularly updated with new data. Composed of JPEG browse images and tabular data about these images, it is growing rapidly. Currently, it uses 450 MB of disk space to archive over 80,000 compressed browse images and related information. The database is organized into tables containing compressed images, image names, image header information, image types, image processing steps, image parent information, image history, etc. QDB is currently used to query this database. Figure 2 is a sample display where two tables, *img-browse* and *img-leb-hdr*, have been selected and displayed, and user selection is in progress. In this figure, *filename*, *browse-img*, *data-obs*, *camera*, *filter*, and *polar* fields are toggled on for selection and a time range using the *data-obs* field is entered.

QDB is capable of joining various tables within the LASCO database to produce the desired output, but, more importantly, it is also capable of joining tables from different databases. This is an important feature because we will soon have other databases in production, such as housekeeping, observing,

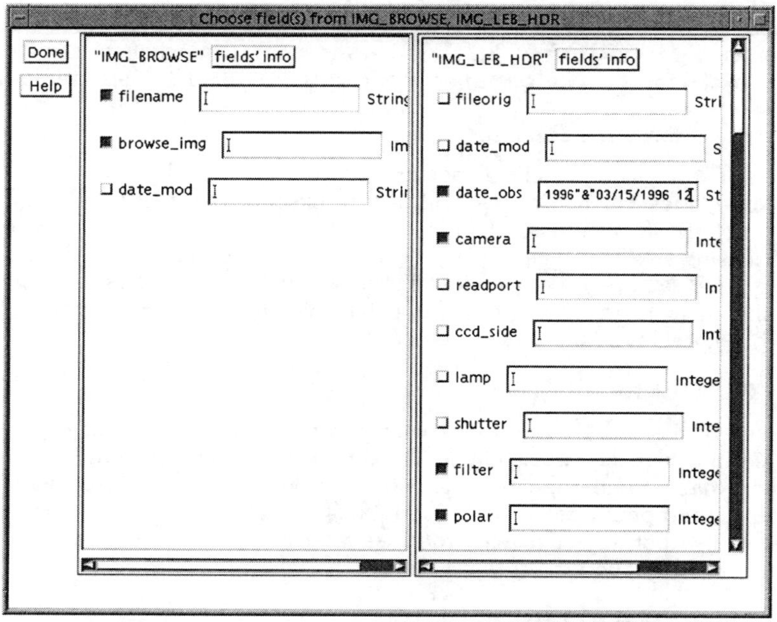

Figure 2. A sample display. Two LASCO tables selected for user interaction.

calibration, user access, and processing tables which may require joins. As mentioned earlier, the dynamic nature of QDB allows for the addition and linkage of these databases with only a minimal amount of additional effort.

4. Conclusion

Because of its dynamic nature and flexibility, QDB can be used remotely to access any database. A user with prior permission, for instance, may run the QDB from his machine over the network, access the LASCO database on our host, and capture the result on his client. Moreover, QDB users need not know SQL in order to use it. All that is needed is the knowledge of which database and tables contain the desired data. In most cases, the database and table names, themselves, convey this information. Furthermore, since result of a QDB query is captured in an IDL structure, it is available and ideal for programmatic use such as displaying, graphing, and data reduction. These features of QDB have allowed us to query the LASCO database frequently and efficiently. We also use QDB to test other databases that are currently under development. In these cases, the only overhead is a one-time minimal code change to handle new joins.

Astronomical Data Analysis Software and Systems VI
ASP Conference Series, Vol. 125, 1997
Gareth Hunt and H. E. Payne, eds.

Astronomical Information Discovery and Access: Design and Implementation of the ADS Bibliographic Services

A. Accomazzi, G. Eichhorn, M. J. Kurtz, C. S. Grant, S. S. Murray

Smithsonian Astrophysical Observatory, 60 Garden Street, Cambridge, MA 02138

Abstract. The NASA Astrophysics Data System integrates a wealth of scientific bibliographic and data resources—originally generated in multiple formats and available from multiple providers—in three discipline-oriented, centralized databases. Search and retrieval of the bibliographies and data sources is possible via a set of World Wide Web forms and interface programs that transparently link the ADS's resources to those of other data providers. The approach followed in designing the ADS system is offered as a paradigm for building flexible networked information and discovery systems. The rationale behind the current technical implementation and the planned enhancements of the system are also discussed.

1. Introduction

The design behind the Astrophysics Data System (ADS) bibliographic databases was mainly dictated by the desire for a powerful and discipline-oriented system featuring sophisticated search capabilities. The main considerations which shaped the final outcome of the system were: the advantages and disadvantages of using a commercial or publicly available RDBMS system versus a custom-build one; the quality and quantity of the data at hand versus the resources available to the project; and the tradeoff between search speed and simplicity on one hand and sophistication on the other.

General-purpose search engines and relational databases were used as part of the abstract service in the first implementation of the search engine, but they were eventually dropped in favour of a home-grown system as the desire for better performance and custom features grew with time (Accomazzi et al. 1995).

The heterogeneous nature of the bibliographic data that had to be entered into our database, and the need to effectively deal with the imprecision in it, lead us to design a system where a large set of discipline-specific interpretations are made. For instance, to cope with the different use of abstract keywords by the publishers, and to correct possible spelling errors and typos in text, sets of words have been grouped together as synonyms for the purpose of searching the databases. Also, many astronomical object names are translated in a uniform fashion when indexing and searching the database.

Because of the large number of features that we have been adding to the abstract service in the last few years, we had to strike a balance between simplicity of the user interface and the creeping featurism syndrome so commonly found in many user interfaces. To avoid overwhelming users with complex search pages, we have devised a design where the main search parameters are always visible within the top part of the screen, with more options to follow. Because of the very nature of the WWW, we have been able to create simpler HTML forms that have much of the additional functionality hidden from the user, and we now even allow users to create and customize their own search form according to their preferences.

In order to provide transparent access to our system from other WWW-based systems, we have provided access interfaces that use bibliographic codes (Schmitz et al. 1995)—or *bibcodes*, as referred to in the rest of this paper—as unique identifiers for references in our databases. Direct HTTP access to our CGI interface programs, and a high-level programming interface implemented as a library of Perl routines, are provided as hooks into our bibliographic search engine.

2. Database Search Interface

The ADS CGI interfaces implement a variety of possibly complex searches of the bibliographic databases, but searches can generally be divided in two classes: reference searches and concept searches.

2.1. Reference Searches

This type of interface allows users to lookup a particular publication or to browse a set of references published in a journal. Access to the program that implements this interface is available by retrieving the URL:

`http://adsabs.harvard.edu/cgi-bin/abs_connect?bibcode=`*bibcode*

where *bibcode* is either a fully qualified, 19-digit bibliographic code, a partial bibcode, or a bibcode pattern possibly containing metacharacters. Consider, for instance, the cases where *bibcode* is one of the following:

- `1996adass...5..558A`: the URL contains a fully-qualified bibcode, and therefore it refers to an individual paper published by the Author at the ADASS V conference in 1996.
- `1996adass...5`: the URL contains a bibcode stem (i.e., truncated bibcode), and will therefore generate the list of publications whose bibcodes begin with the string `1996adass...5`. This list consists of all the papers published at the ADASS V conference.
- `199?adass...?`: the URL consists of a bibcode pattern containing two instances of the "?" metacharacter which matches any single character. The set of references returned by the query will be the list of papers published in all ADASS conferences so far (`1992adass...1`, `1993adass...2`, `1994adass...3`, `1995adass...4`, `1996adass...5`, which currently happen to match the above regular expression.

Other similar programs and HTML forms extend these capabilities by allowing selections based on publication date ranges and journals (see, for instance, the ADS Table of Contents Query Form[1]).

2.2. Concept Searches

Searches based on the identification of a set of references which are relevant to a particular topic or "concept" are implemented in a similar fashion. Because references are structured entities having several attributes (or "fields"), a fielded search is one in which one or more fields are to be searched and one or more terms to be searched for are specified for each field.

Currently the ADS Astronomy database allows users to search by author name, astronomical object name, keywords,[2] words in the title, and words in the abstract text. The general URL syntax for searching for terms in a particular field is `http://adsabs.harvard.edu/cgi-bin/abs_connect?`*field=words* where *field* is the name of the field to be searched and *words* represents the expression to be searched for. For instance, to find the list of all papers published by the Author in the ADS Astronomical Database, one would access the URL `http://adsabs.harvard.edu/cgi-bin/abs_connect?author=accomazzi`.

When specifying more than a single word to be searched in a particular field, the interface allows the user to select whether the resulting list is to include references which contain a subset of the search terms, which search terms must be present, and which should be excluded. When specifying words to be searched in separate fields, the user may choose how the lists of references resulting from the individual field searches should be combined, using a logic similar to the one applied for combining references generated from individual words within a field.

The ADS abstract service search form has many more features and settings that can be set customized, including restricting the search to be performed only on a particular journal or body of literature (e.g., searching on refereed journals only). One immediate application of this is that it provides users with several up-to-date indexes into subsets of the astronomical literature. For instance, to search for all the publications appearing in the ADASS conference series that mention ADS in their abstract, one would simply call the `abs_connect` script with the arguments: `text=ADS&jou_pick=YES&ref_stems=adass`.

3. Links to Bibliographic Resources

One of the most successful features of the design behind the current WWW software agents is that they allow users to transparently browse information available on the Internet via the selection of hyperlinks. In particular, this has created a de-facto standard interface and protocol for accessing network resources available from different institutions, thus becoming the glue between the services provided by different astronomical data centers.

---

[1] http://adsabs.harvard.edu/toc_service.html

[2] Keyword searching is currently not available from the main abstract service search page because of the lack of a uniform and consistent keywording system for the current references.

The ADS databases currently maintain for each bibliography a set of links to both local and network-accessible resources. The following hyperlinks provide interconnectivity between the ADS and other institutions:

- Electronic article links, which point to the full-text electronic version of the current reference, when available from the original publisher or from the ADS article service.
- Data links, which point to the list of electronic datasets published with the article, allowing retrieval of each of them. Currently these resources are available from the following institutions: CDS, NCSA/AIDL, GCIP.
- Object links, which point to the list of objects cited in the article, and available from the SIMBAD database. Available from the CDS.

The relationship between the ADS system and the data centers mentioned above is reciprocal, in the sense that they, in turn, provide hyperlinks from their databases to bibliographic resources available in the ADS, when appropriate.

4. Conclusions

The popularity and usefulness of the ADS bibliographical services is due, in large part, to its discipline-specific features and to the synergy created by several data centers adopting a common language and protocol to link their resources. This cooperation provides astronomers an ever-growing wealth of information and resources that are transforming the way they perform their research.

Because of the size and completeness of its databases, the NASA Astrophysics Data System has become a clearinghouse for astronomical bibliographic resources, and the ADS abstract service has become the bridge between networked resources available from different institutions and societies.

Acknowledgments. This work is funded by the NASA Astrophysics Program under grant NCCW-0024.

References

Accomazzi, A., Grant, C. S., Eichhorn, G., Kurtz, M. J., & Murray, S. S. 1995, in ASP Conf. Ser., Vol. 77, Astronomical Data Analysis Software and Systems IV, ed. R. A. Shaw, H. E. Payne, & J. J. E. Hayes (San Francisco: ASP), 36

Schmitz, M., Helou, G., Dubois, P., LaGue, C., Madore, B., Corwin Jr., H. G., & Lesteven, S. 1995, in Information & On-line Data in Astronomy, ed. D. Egret & M. A. Albrecht (Dordrecht: Kluwer Acad. Publ.), 271

Astronomical Data Analysis Software and Systems VI
ASP Conference Series, Vol. 125, 1997
Gareth Hunt and H. E. Payne, eds.

The Sociology of Astronomical Publication Using ADS and ADAMS

Eric Schulman[1]

National Radio Astronomy Observatory,[2] 520 Edgemont Road, Charlottesville, VA 22903-2475, E-mail: eschulma@nrao.edu

James C. French, Allison L. Powell

Department of Computer Science, School of Engineering and Applied Science, University of Virginia, Charlottesville, VA 22903-2442, E-mail: french@cs.virginia.edu, alp4g@cs.virginia.edu

Stephen S. Murray, Guenther Eichhorn, Michael J. Kurtz

Smithsonian Astrophysical Observatory, 60 Garden Street, Cambridge, MA 02138, E-mail: ssm@cfa.harvard.edu, gei@cfa.harvard.edu, kurtz@cfa.harvard.edu

Abstract. We use the NASA Astrophysics Data System database of astronomical abstracts in seven major astronomy journals to study trends in astronomical publication over the last twenty years. Two of the most interesting trends are the decreasing fractions of papers with one author and the increasing number of authors per paper.

1. Introduction

The sociology of astronomical publication has traditionally been performed by looking for publication trends using every paper published in a few selected journals within a few selected years. For example, Abt (1981) examined the papers published in ApJ, ApJS, AJ, and PASP during the first year of each decade from 1910 to 1980.

By analyzing the NASA Astrophysics Data System[3] (ADS) database of astronomical abstracts we can study a large number of issues in the sociology of astronomical publication while including every paper published in a number of refereed journals during the past twenty years. Here we present preliminary results of a study of astronomical publication trends using papers published in A&A, A&AS, AJ, ApJ, ApJS, MNRAS, and PASP between 1975 and 1995.

---

[1] Jansky Fellow.

[2] The National Radio Astronomy Observatory is a facility of the National Science Foundation operated under cooperative agreement by Associated Universities, Inc.

[3] http://adswww.harvard.edu/

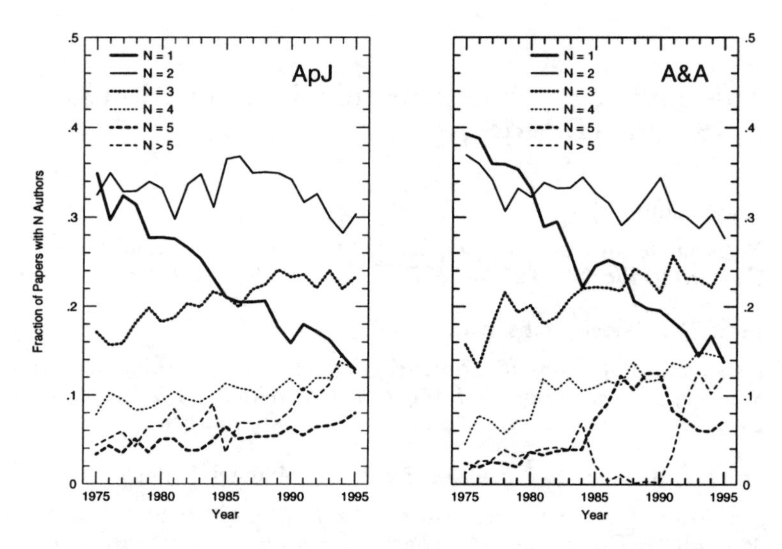

Figure 1. Fraction of ApJ and A&A Papers with N Authors.

2. The Astrophysics Data System Database

The ADS abstract service contains approximately 240,000 abstracts of astronomy and astrophysics papers from more than 1000 journals (Accomazzi et al. 1997). Most ADS abstracts of papers published between 1975 and 1995 were obtained through the NASA Scientific and Technical Information[4] (STI) Program, which compiled papers from the majority of astronomical journals. Although the database is at least 95% complete, there are some systematic errors in the data. For example, the STI author lists were truncated at the tenth author until about 1986 (e.g., Cohen et al. 1975 has fourteen authors, but ADS only lists the first ten). Between 1986 and 1990, ApJ and ApJS author lists were not truncated, but author lists in the other five journals were truncated at five authors, and from 1991 to 1994 the author lists in these five journals were truncated at ten authors. The number of author list truncations can be substantially reduced by comparing the ADS database with the Strasbourg Astronomical Data Center's SIMBAD[5] (Set of Identifications, Measurements, and Bibliography for Astronomical Data) database, a process that is currently underway. The SIMBAD database includes all papers since 1983 that mention at least one astronomical object (excluding Solar System bodies; papers published since 1950 that mention individual stars are also included). SIMBAD currently has information on 85,000 papers from 90 journals and conference proceedings.

[4]http://www.sti.nasa.gov/

[5]http://cdsweb.u-strasbg.fr/Simbad.html

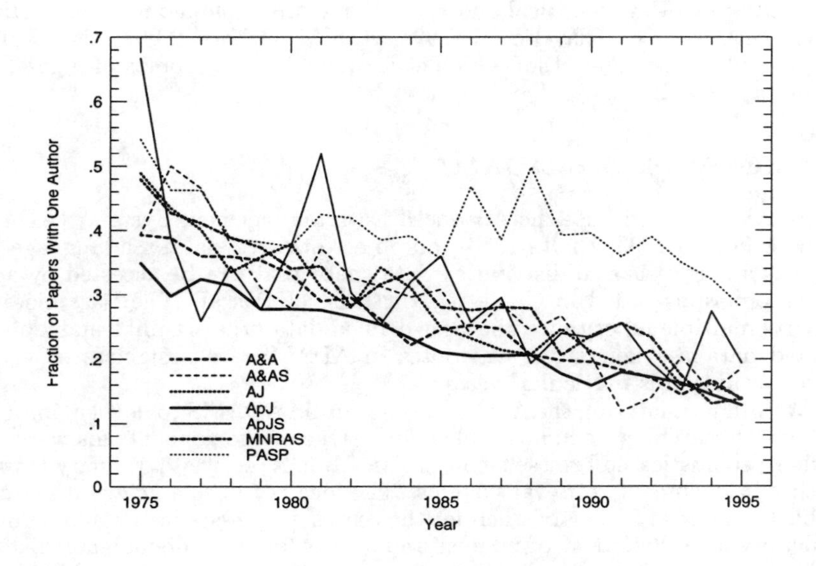

Figure 2. Fraction of Papers with One Author.

3. Number of Authors Per Paper

The fraction of ApJ and A&A papers with one to five authors is shown in Figure 1. The most striking change is the decrease in the fraction of single-author papers from more than $^1/_3$ to about $^1/_8$. The fraction of two-author papers remained fairly constant, while papers with larger numbers of authors became more frequent. The NASA STI truncation of A&A author lists is obvious between 1985 and 1990. The decrease in the fraction of single-author papers has occurred in all seven journals (Figure 2), although the fraction of single-author papers in PASP remains fairly large. Even though the NASA STI truncations make it difficult to analyze the mean number of authors per paper, it is still obvious that there are on average more authors per paper now than there were twenty years ago.

There are a number of possible reasons for decreasing fractions of single-author papers and increasing numbers of authors per paper. One is the growth of multiwavelength astrophysics (Abt 1993), which requires astronomers to be proficient in multiple wavebands or to collaborate with experts in other wavelengths. Another is an increase in the number of papers that present both observations and theoretical interpretations. A third is increasing competition for jobs and grants, which encourages astronomers to write as many papers as possible. Also, in the last five years there has been more research requiring large collaborations, such as using *HST* to determine Cepheid distances to nearby galaxies (e.g., Kelson et al. 1996, with 18 authors). There were nine papers with 50 or more authors in our sample, all of which appeared in the ApJ or ApJS after 1990. Five were a series of papers reporting on intensive *HST*,

IUE, and ground-based optical and near-IR spectroscopic monitoring studies of Seyfert galaxies (e.g., Clavel et al. 1991, with 57 authors). The other four papers, including the 124-author Ahlen et al. (1993), were reports of high-energy cosmic ray experiments.

4. Future Work With ADAMS

In future work we will use the Advanced Data Management System[6] (ADAMS; Pfaltz & French 1993; Pfaltz 1993), an object-oriented database language that supports a single shared, distributed data space that can be accessed by applications programs coded in C, C++, FORTRAN, or Pascal. The class hierarchy supports multiple inheritance and user-defined data types. Unlike many object-oriented database languages, attributes in ADAMS are first class objects so schema evolution is particularly easy.

We will be analyzing the ADS database within ADAMS to develop improved methods of searching document collections. One of the goals of this work is to provide a sophisticated browse facility as an adjunct service to ordinary keyword searches in document retrieval systems. The idea is to use a keyword search to establish an initial focus and then let the searcher access additional documents by identifying a document of interest and asking for more documents "like this one." An underlying topical map will be maintained to support this kind of browse mechanism.

Acknowledgments. We thank Ellen Bouton for her invaluable assistance in finding and describing many different journals, and for very useful conversations about astronomical publication. This research is based on data obtained through NASA's Astrophysics Data System Abstract Service, and is supported by grants from NASA (NCCW-0024), the DOE (DE-FG05-95ER25254), and the NSF (CDA-9529253).

References

Abt, H. A. 1981, PASP, 93, 269

Abt, H. A. 1993, PASP, 105, 437

Accomazzi, A., Eichhorn, G., Kurtz, M. J., Grant, C. S., & Murray, S. S. 1997, this volume, 357

Ahlen, S., et al. 1993, ApJ, 412, 301

Clavel, J., et al. 1991, ApJ, 366, 64

Cohen, M. H., et al. 1975, ApJ, 201, 249

Kelson, D. D., et al. 1996, ApJ, 463, 26

Pfaltz, J. L. 1993, The ADAMS Language: A Tutorial and Reference Manual, Technical Report IPC-93-03, University of Virginia Institute for Parallel Computation

Pfaltz, J. L., & French, J. C. 1993, Data Engineering, 16, 14

---

[6] http://www.cs.virginia.edu/~adams/

Part 8. Proposal Processing

Astronomical Data Analysis Software and Systems VI
ASP Conference Series, Vol. 125, 1997
Gareth Hunt and H. E. Payne, eds.

A Distributed System for "Phase II" Proposal Preparation

A. M. Chavan and M. A. Albrecht

European Southern Observatory, Karl-Schwarzschild-Str. 2, D–85748 Garching, Germany; E-mail: amchavan@eso.org

Abstract. A significant fraction of observing time on ESO's new Very Large Telescope will be spent in service mode; therefore, investigators need to describe in detail what they want to observe, and how. ESO is prototyping a distributed "Phase II" system that will allow astronomers to prepare their observations at their home institutions, while maintaining a central repository for all observation data. Astronomers will be offered a wide array of observation preparation tools, including data entry GUIs, instrument simulators, and catalog interfaces. The system is structured as a set of distributed clients and centralized servers, and it operates as a front-end to the other elements of the VLT Data Flow System.

1. Introduction

In order to maximize the scientific throughput of the system, a complex and expensive observatory like ESO's VLT should let the most demanding observing programmes take advantage of the very best weather conditions. Traditionally, this has not been possible: observers have had to cope with the atmospheric conditions prevailing during the night assigned to them. ESO is experimenting with operating telescopes in *service observing* mode, whereby staff observers execute scientific programmes on behalf of the investigators (this mode of operation is also called *queue observing*).

To make service mode observing possible, scientific programmes must be described in great detail—a process commonly defined as *Phase II* proposal preparation. ESO is developing a Phase II Proposal Preparation system (P2PP) to collect detailed scientific programme descriptions, in the form of *observation blocks* (OB). The system is initially directed to support service mode observing (that is, a significant fraction of the VLT observations), but will eventually be used by "classical mode" observers as well: OBs will be the main input to the VLT's Data Flow System (Grosbøl & Peron 1997).

2. Observation Blocks

VLT observing programmes will be composed of several observation blocks, each coupling a target (*what* should be observed) with a description of *how* the observation itself should be carried on; observation blocks represent a high level view of VLT operations. Several scheduling constraints may be specified for OBs, like

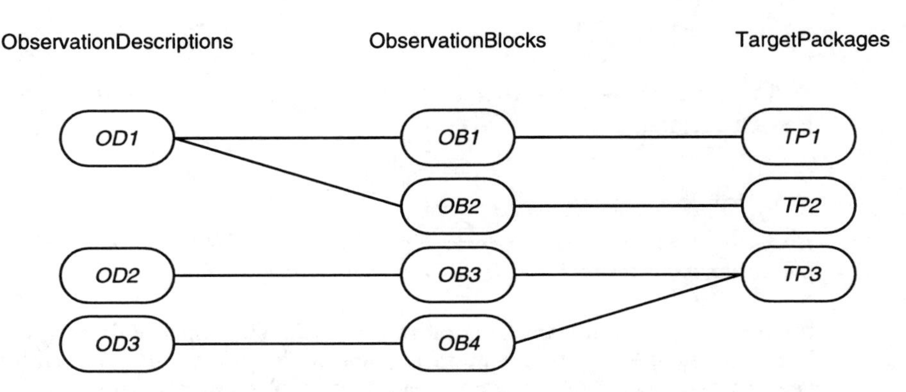

Figure 1. Targets TP1 and TP1 are observed in the same way, since observation blocks OB1 and OB2 share the same observation description, OD1. Conversely, TP3 is shared by two observation blocks, OB3 and OB4. The investigator wishes to observe the same target in two different ways: e.g., with an optical imager and with a spectrograph.

worst acceptable seeing, absolute timing for transient phenomena, chaining of observations, maximum acceptable airmass, etc.

Targets (technically *target packages*) are described in terms of coordinates, proper motion, magnitude, color, and all other information which may be needed to acquire them and track them during exposure. This includes one or more guide stars.

The observation description is composed of series of standard operations, called *templates*: typical templates include target acquisition, science exposures, and calibration frames. For each template to be executed, astronomers need to specify a set of parameters, like filter names, exposure times, CCD readout mode, etc. These parameters specify a full instrumental configuration as well as execution details, like the pattern of images for a mosaic template.

Targets and observation descriptions can be shared among OBs, as shown in Figure 1, thus allowing astronomers to easily group related observations.

3. A Distributed System

For privacy and reliability of information, observation blocks will be stored in a central repository at ESO. Other important services, like instrument simulators, data servers (Albrecht, Brighton, & Herlin 1997), and help pages, will be provided centrally by ESO as well. However, it is imperative that (a) P2PP can take advantage of the investigators' own computing facilities, and (b) astronomers can prepare their observations even when they are not connected to the Internet. These requirements dictate the design of a distributed, client-server system: "lightweight" clients interact with central servers, and can maintain a local cache of information.

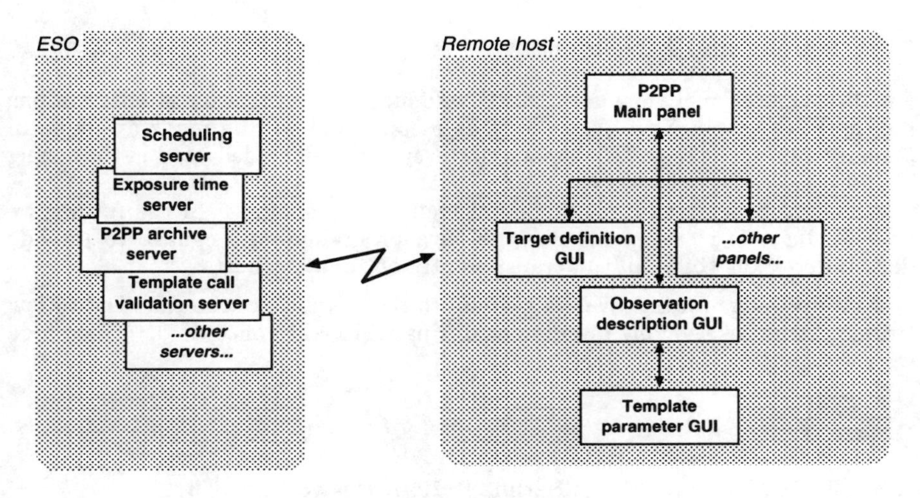

Figure 2. P2PP architecture.

3.1. Architecture

The P2PP system is therefore a collection of independent stand-alone modules, structured in a client/server relationship. The server modules run at ESO and handle centralized services, like database access for OB storage, exposure time calculation, etc., while client modules run on the OB's author's host.

Most client modules generate, display, and process part of an OB's information content: there is a module for describing targets, one for observations descriptions, etc. One special client module is used for global OB operations, like transfer of OBs to and from the central (ESO) database.

All modules offer a graphical user interface. The overall software architecture is shown in Figure 2. P2PP is being implemented in C++ and Tcl/Tk; it uses HTTP as the communication protocol between clients and servers.

4. P2PP and the VLT Data Flow System

ESO's Phase II Proposal Preparation system is an integral part of the VLT Data Flow System (DFS). Observation blocks are built on the basis of information collected by the Phase I Proposal Handling and Reporting System (Chavan 1995), and are fed downstream to the VLT Scheduling Tools (Chavan 1996)—which we are developing in cooperation with the Space Telescope Science Institute—and to the VLT Control Software.

All other subsystems of the DFS—including archiving, pipeline processing, and quality control—deal with observation blocks as well: OBs represent the basic quantum of information flowing through the VLT system.

5. Planning and Development

The first operational version of P2PP will support preparation of observations for the ESO's 3.5m New Technology Telescope, and will be made available in January 1997. The first VLT version of the system will be delivered in the spring of 1998.

The Web is already an important enabling technology. In the future, we hope to be able to use tools like the Java programming language to provide distributed tools that will run transparently on different hosts.

Acknowledgments. We are grateful to the members of the ESO Data Flow Project Team for the long discussions during which the concepts described here were defined.

References

Albrecht, M., Brighton, A., & Herlin, T. 1997, this volume, 333

Chavan, A. M., & Albrecht, M. A. 1995, in ASP Conf. Ser., Vol. 77, Astronomical Data Analysis Software and Systems IV, ed. R. A. Shaw, H. E. Payne, & J. J. E. Hayes (San Francisco: ASP), 58

Chavan, A. M., Johnston, M., & Albrecht, M. A. 1996, in ASP Conf. Ser., Vol. 87, New Observing Modes for the Next Century, ed. T. A. Boroson, J. K. Davies, & E. I. Robson (San Francisco: ASP)

Grosbøl, P., & Peron, M. 1997, this volume, 22

Astronomical Data Analysis Software and Systems VI
ASP Conference Series, Vol. 125, 1997
Gareth Hunt and H. E. Payne, eds.

Filtering KPNO LaTeX Observing Proposals with Perl

David J. Bell

National Optical Astronomy Observatories, Tucson, AZ 85726-6732

Abstract. An automated observing proposal processing system has been in use at KPNO during the past three years. LaTeX proposal templates are filled out by users and sent to KPNO via electronic mail. Observer and proposal information fields in these files are well-marked with LaTeX tags, thus allowing automated extraction and importation to observatory databases.

A significant complication of this process is that although the fields are well-marked, the information they contain often arrives in a variety of formats that must be recognized and standardized. Perl's regular expression and text manipulation capabilities make it an excellent tool for performing these functions. This paper outlines the filtering system in use at KPNO and discusses some of the ways Perl has proven useful for parsing LaTeX documents.

1. Introduction

The Kitt Peak observing proposal handling system (Bell et al. 1996) is an automated system for distributing, receiving, and processing LaTeX observing forms and associated PostScript figures. It has already handled several thousand files for Kitt Peak alone, and has also been in use at two other observatories. Twice a year, as proposals are arriving, information must be quickly imported and updated in observatory databases. In the past this was performed through manual entry while inspecting each paper copy—a tedious and sometimes overwhelming task, particularly when over 100 proposals arrive on the final day.

This paper describes a new and better approach. The proposals are run through a filtering program that locates the desired fields in the proposal form, optionally parses each field into several database subfields, and rearranges the information into a standardized format. When confused about an entry, the program attempts to make a good guess but also flags that item for human inspection. Such an approach can save much time, while still ensuring the accuracy of imported data.

2. Why LaTeX?

LaTeX is a widely and freely available text formatting language that gets significant use throughout the astronomical community. Even inexperienced users can fill in a well-designed template form using any text editor, and submit it with any e-mail program. The completed forms serve a dual role: they can be printed to produce nicely formatted paper documents, and, since needed information is well-tagged by the structural markup, they can also be processed by automated scripts to extract data fields. If needed, they can be edited locally (impractical with PostScript documents, for example), and by modifying a single style file one can quickly reformat hundreds of documents.

However, problems can occur during automated filtering. Some of the fields, such as addresses, consist of just one line on the form, but need to be split into several subfields for the database. The form could be changed, but there are already thousands of existing documents in the present format, and it would be inconsiderate to force users to split entries into many subfields that are all recombined on the printed form. A second problem is that users often embed commands into the fields, and it would be confusing if we mailed out envelopes with raw TeX in the addresses. The biggest drawback, however, is that there are no limits over what users type into the fields, whereas the database needs standardized formats.

3. Why Perl?

Perl is a widely and freely available scripting language that is best known for its many uses for system administration tasks and, more recently, as being *the* CGI-programming language. What makes it so good at these things is that it is a superb text-processing language. It includes some of the most efficient and powerful regular expression and string manipulation operators available anywhere, allowing it to quickly locate and manipulate myriads of tiny parcels of text. Powerful string manipulation code can be written quickly and compactly, making it great for processing LaTeX files (for instance, the popular `latex2html` program is a Perl script). Perl has a familiar C-like syntax and a very forgiving nature. If users leave fields blank, or type in full sentences when the form asked for an integer, the script doesn't bomb out, but can easily be programmed to do something reasonable and move on.

4. Filtering Strategies and Examples

Many of the fields extracted from the forms require very little processing other than reformatting, and this can often be done in just a few lines of code. For example, the lines:

```
$phone = "($1) $2-$3 $4" if ($phone =~
        /^\D*(\d{3})\D*(\d{3})\D*(\d{4})\s?(.*)$/);
```

will recognize a US phone number in practically any likely format, such as "800–5551212ext123," and standardize it to "(800) 555-1212 ext123." Some more interesting examples will be discussed in the following sections.

4.1. Names and Addresses

Name and address entries on the form need to be split into subfields for the database. LaTeX codes are stripped out (e.g., diacritical marks) or replaced (e.g., non-English characters). Names are split into an array based on punctuation and whitespace, and then compared to lists of titles to be thrown away, or

```
\name{Prof.~Dr.~Ant{\^o}nio-Ryan M.\ VAN DE W\O RF, Jr.}
\address{\small Dept.~of Phys.~\& Astr.; Mail Stop 16; rm 101;
     VICTORIA B.C.\ V8 x4 m6~~Canada}

FN: Antonio-Ryan
MI: M
LN: Van De Worf, Jr
A1: Dept. of Phys. & Astr
A2: Mail Stop 16, rm 101
CY: Victoria
ST: BC
ZC: V8X 4M6
CO: CANADA
```

Figure 1. LaTeX Name and Address Fields and Filter Output. Formatting and special-character codes have been stripped and the entries correctly parsed into subfields. The province, postal code, and country name have been standardized.

```
\telescope{ 4.0-meter~~~}
\instrument{Prime focus camera with the new 4$\times$2 CCD mosaic}

TE: 4m
IN: PF
DE: MOSA
```

Figure 2. LaTeX Telescope and Instrument Fields and Filter Output. Regular expression hashes have been used to identify items and return standardized database codes.

```
\optimaldates{2/16-3/2, 4/14-5/1 or 5/13-31}
\acceptabledates{21DEC1996---07JUN1997\hfil}
\optimaldates{\it Feb.~1st--27th or March 2nd--23rd, 1997}
\acceptabledates{Late sept.\ through early-april}

OD: Feb 16 - Mar  2, Apr 14 - May  1, May 13 - May 31
AD: Dec 21 - Jun  7
OD: Feb  1 - Feb 27, Mar  2 - Mar 23
AD? Sep 20 - Apr 10
```

Figure 3. LaTeX Date Fields and Filter Output. Since the last range is somewhat vague, the filter has flagged the field for human inspection.

likely surname components that need to be recombined into a last-name field. Address parsing is more difficult, due to widely varying punctuation and foreign addresses. For this reason, lists of cities and countries from past proposals are first searched, and once a city is found, the parsing is almost always right. When new cities show up, the script makes an attempt at parsing based on punctuation, flags the data, and logs the new city for possible addition to the search list. See Figure 1.

4.2. Telescopes, Instruments, and Detectors

The primary problem in these cases is in recognizing the many ways in which the same item can be requested. This is achieved by stepping through associative arrays in which a standardized code is the key and a regular expression that matches various forms of that item is the value. For instance the hash element

```
RCSP => r[-\.\s]*c[-\.\s]*sp
```

can be used to map user entries such as "r-c spectrograph" or "R. C. Spec" to the standardized code "RCSP." When new instruments become available, the script can be quickly updated simply by adding a new code-regex pair to the hash. See Figure 2.

4.3. Observing Dates

Interpreting date strings requires parsing a string into several date ranges, then each range into dates, and finally each date into a day and month. Commonly used English words like "through" and "or" are first replaced with symbols like "-" and "," respectively, before splitting on these symbols. Unneeded information, such as years and ordinal abbreviations, are removed. Months are then standardized with a series of substitutions, so that the strings "09/", "SEP", "september", etc., will all be turned into "Sep". Once this month string is pulled out, what's left is hopefully a day number—if not, special cases like "mid" are considered. See Figure 3.

5. Conclusion

A Perl script has been developed for filtering KPNO LaTeX observing proposals. Although the desired information arrives in a large variety of formats, Perl's powerful text manipulation capabilities allow the script to accurately identify and reformat entries for database import. Only a few standardization problems remain, and these may be eliminated in the future with pre-configured menus and buttons on an HTML form—such an interface to the LaTeX template is currently being designed.

References

Bell, D., Biemesderfer, C. D., Barnes, J., & Massey, P. 1996, in ASP Conf. Ser., Vol. 101, Astronomical Data Analysis Software and Systems V, ed. G. H. Jacoby & J. Barnes (San Francisco: ASP), 451

Astronomical Data Analysis Software and Systems VI
ASP Conference Series, Vol. 125, 1997
Gareth Hunt and H. E. Payne, eds.

A User Friendly Planning and Scheduling Tool for SOHO/LASCO-EIT

S. E. Paswaters, D. Wang

Interferometrics Inc. 14120 Parke Long Court, Chantilly, VA 22021, E-mail: scott@argus.nrl.navy.mil

R. A. Howard

Naval Research Laboratory, 4555 Overlook Ave. SW, Washington DC, 20375

Abstract. The LASCO/EIT instruments aboard the *SOHO* satellite (launched Dec. 1995) were developed with complicated image processing/acquisition techniques. The Planning and Scheduling Tool was designed allow the user easily to take advantage of all available resources of the four telescopes (LASCO's three coronagraphs and the EIT Telescope) to maximize the use of the telemetry downlink.

The Planning Tool allows users to develop customized observing sequences while monitoring compression factors and on-board processing times to realistically work within the limits of the instruments. The Scheduling Tool then graphically displays scheduled sequences and highlights potential resource conflicts. Sequences are saved in database tables as "as planned observations" and are available to be retrieved at any later date. Outputs include: database updates, inputs to the *SOHO* science activity plan, and the command loads themselves. Statistics are gathered after the images are received so the tool is constantly improving its estimate of processing time and compression factor.

1. Introduction

The LASCO (Large Angle Spectrometric Coronagraph) / EIT (Extreme Ultraviolet Imaging Telescope) instrument packages having been operating continuously since shortly after launch of the *SOHO* (Solar and Heliospheric Observatory) satellite on December 2, 1995. *SOHO* is located at the Lagrangian L1 point providing uninterrupted observations of the sun. Commanding and data acquisition are performed in real time eight hours of every day with the remaining time provided through solid-state recorder dumps (see Wang et al. 1997, for more on the LASCO data archive). Observation schedules are uplinked daily, consisting of a baseline long-range synoptic program as well as specific observations to take advantage of changing targets on the sun, special science programs, or joint observations with other observatories. The LASCO/EIT electronics are extremely flexible, offering various image compression techniques which are traded off with consideration of scientific goals. The Planning Tool (Figure 1) allows users to develop customized observing sequences while monitoring compression factors

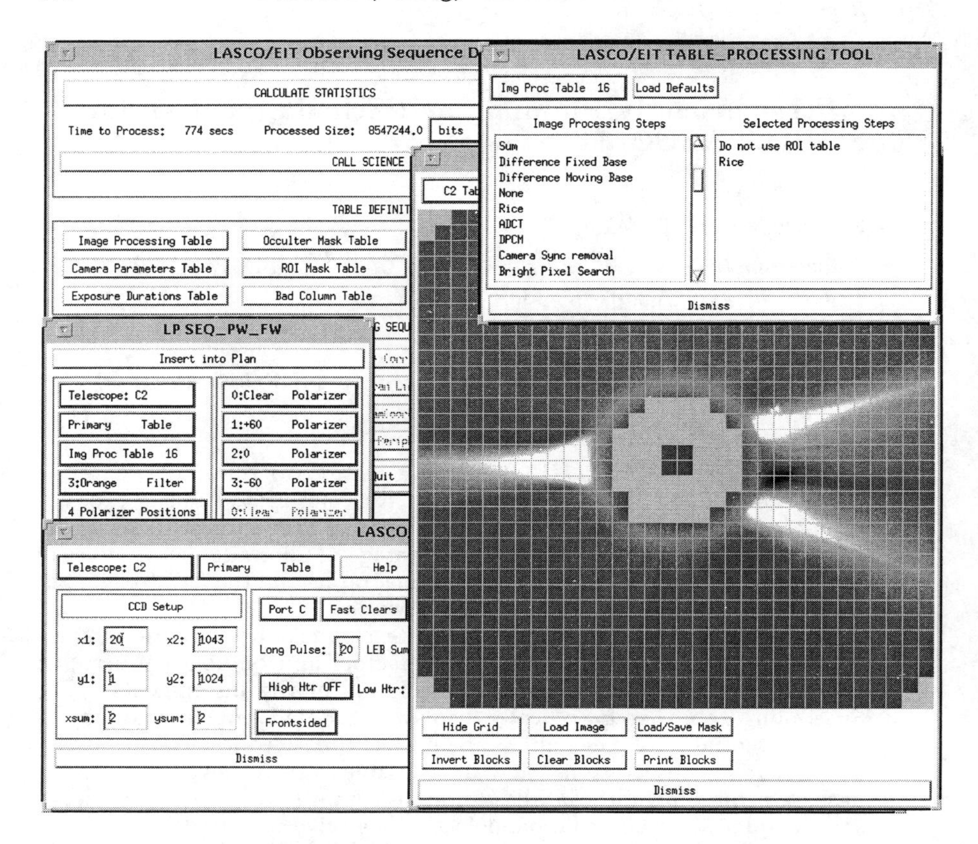

Figure 1. A typical view of the Planning Tool.

and on-board processing times to work realistically within the limits of the instruments. The Scheduling Tool (Figure 2) then graphically displays scheduled sequences and highlights potential resource conflicts. The user can easily add, remove, or modify these sequences.

2. Implementation

The Planning and Scheduling Tool is written in IDL (~12,000 lines of code) with a call to a C routine to generate the binary command strings. The tool is coupled with a relational database (Esfandiari, Paswaters, & Wang 1997) to archive observing sequences and maintain statistics (compression factor, processing time) of science images. Observing sequences are scheduled to maximize the use of LASCO/EIT science telemetry resources ($5\,\mathrm{kB\,s^{-1}}/60\,\mathrm{MB\,day^{-1}}$ compressed science data) while monitoring on-board processing time to avoid over-scheduling.

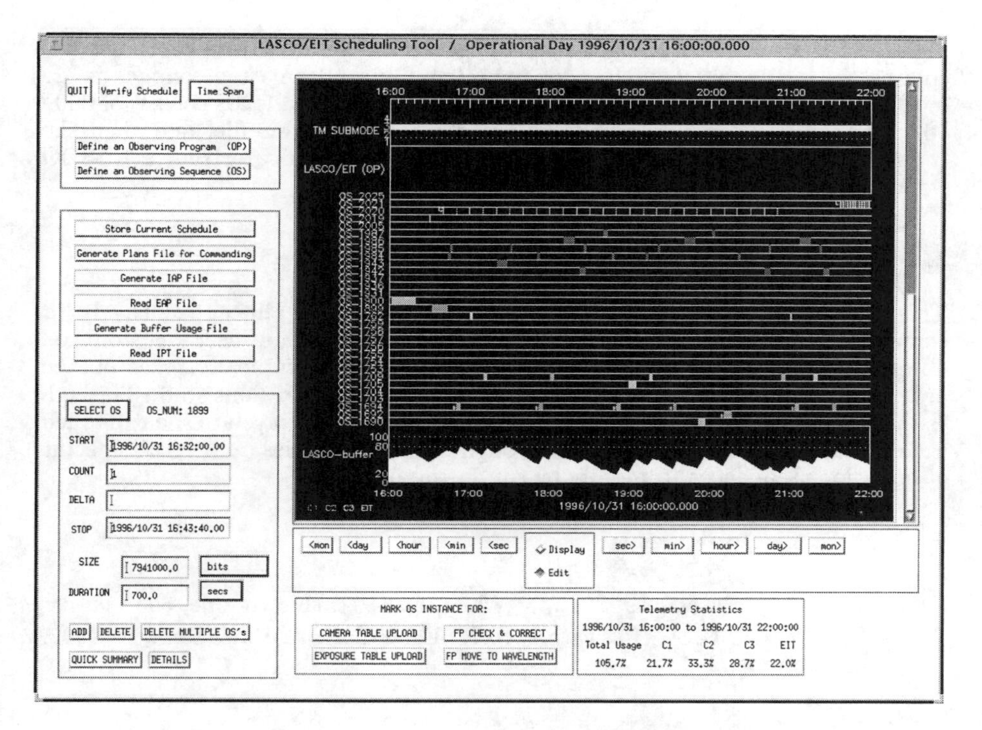

Figure 2. A typical view of the Scheduling Tool.

Table 1. Sample "As Planned Observations" Summary.

Date Obs	Tele	Exp	Nx	Ny	Bin	Filter	Polar	Wavelength	OS
1996/10/31 16:44	C1	100	320	288	1×1	Fe XIV	Clear	5302.4224	1984
1996/10/31 16:46	C1	48	320	288	1×1	Fe X	Clear	6376.4419	1985
1996/10/31 16:50	EIT	9	128	128	1×1	Clear	195A		2020
1996/10/31 16:51	C2	25	1024	576	1×1	Orange	Clear		1694
1996/10/31 17:00	EIT	9	128	128	1×1	Clear	195A		2020
1996/10/31 17:01	EIT	9	1024	1024	2×2	Clear	195A		1762
1996/10/31 17:10	EIT	9	128	128	1×1	Clear	195A		2020
1996/10/31 17:11	C3	19	576	1024	1×1	Clear	Clear		1709
1996/10/31 17:20	EIT	9	128	128	1×1	Clear	195A		2020
1996/10/31 17:21	C1	100	640	640	1×1	Fe XIV	Clear	5302.4224	1943
1996/10/31 17:21	C1	100	640	640	1×1	Fe XIV	Clear	5309.2363	1943

3. Outputs

A summary of "as planned observations" is output as an ASCII readable file providing a top level view of the schedule (Table 1).

A *SOHO* science activity plan file is generated summarizing LASCO's upcoming observations for *SOHO*-wide planning. Binary formatted commands are output into a single command file. Typical daily command loads consist of 300–500 commands. Consistency checks are performed on this file, and then the commands are uplinked through a separate piece of software from a LASCO workstation at GSFC to the instrument.

4. Summary

The Planning and Scheduling Tool is currently being used almost exclusively for commanding the LASCO and EIT instruments. Integrating user's customized observing sequences with the overall synoptic program typically takes on the order of half an hour, thus reducing the burden on the operations staff. Through the use of unique observing sequence numbers, users can correlate their planned observations with the actual data through database queries, and recall specific observations sequences to modify or run in the future.

References

Esfandiari, A. E., Paswaters, S. E., & Wang, D. 1997, this volume, 353

Wang, D., Howard, R. A., Paswaters, S. E., Esfandiari, A. E., & Rich, N. 1997, this volume, 282

Astronomical Data Analysis Software and Systems VI
ASP Conference Series, Vol. 125, 1997
Gareth Hunt and H. E. Payne, eds.

Planning and Scheduling Software for the Hobby•Eberly Telescope

Niall I. Gaffney

Hobby•Eberly Telescope, RLM 15.308, University of Texas at Austin, Austin, TX 78712 E-mail: niall@astro.as.utexas.edu

Mark E. Cornell

McDonald Observatory, RLM 15.308, University of Texas at Austin, Austin, TX 78712 E-mail: cornell@puck.as.utexas.edu

Abstract. We present the initial design of the planning and scheduling software for the 9.2 meter Hobby•Eberly Telescope (HET). These tools are to be used by the PIs to prepare proposals for telescope time and to construct observing plans to be executed by the HET's observing queue, and by the astronomers running the telescope to evaluate and schedule the proposed queued observations. We also outline our model for the operation mode of the HET in queued observing mode.

1. Introduction

The HET[1] (Sebring & Ramsey 1994) is a 9.2 meter telescope located at McDonald Observatory near Ft. Davis, Texas. To reduce construction costs, the telescope is fixed in altitude at 55°. By changing the azimuth of the telescope, different declinations can be reached. Objects can then be tracked with an Arecibo-style tracker. Because of this geometry, the HET can access only a limited range of hour angle at any given declination and can track an object for roughly an hour at a time. Thus, efficient use of the telescope will rely strongly on software tools to know when and how an object can be observed and how best to sequence observations over the course of a night. Once completely operational, 85% of the nights are expected to be used in a queued mode where the night is dynamically scheduled as observing conditions change. In this mode, data are acquired by resident astronomers on behalf of, potentially, many different PIs for many different observational projects each night.

2. Operations

The operations of the HET (Kelton & Cornell 1994; Kelton, Cornell, & Adams 1996) is four phased: a proposal phase, a planning phase, an observing phase,

[1]The HET is operated by McDonald Observatory on behalf of the University of Texas at Austin, the Pennsylvania State University, Stanford University, Ludwig-Maximilians Universitat München, and Georg-August Universitat Göttingen.

and a verification phase. The first two phases take place three times per year, when the database driving the queue for the telescope is constructed, while the last two are executed every night data are taken.

The proposal phase consists of the traditional PI-TAC interaction, where the PI, using our planning tools, creates a telescope proposal which is reviewed and granted time by the TAC. In this phase, the PI needs to be able to predict approximately the ability of the HET to observe a typical object under nominal conditions.

In the planning phase, the TAC informs the HET operations team of the time allocated to the proposal. The HET operations team then gives the PI an account on the operations database server. The PI composes a plan, which is submitted to the database server via e-mail. The plan is checked via a procmail-style e-mail system. In this phase, the PI requires tools to determine how to make observations under both nominal and degraded conditions. Further, the PI must be able to determine what constraints need to be set for the observations, and what effect these may have.

The observing phase begins with the operations team compiling a master database of observing projects. This database is then shipped to the telescope where resident astronomers compile a plan for each night based on the current and forecasted conditions. They then execute this plan, make real time changes to it as conditions change, and gather data. Data are then shipped back to Austin via a KPNO "Save The Bits" style queued ftp transfer protocol (Seaman & Bohannan 1996). For this phase, a manual or automatic scheduling tool is needed to determine what to do tonight, based on the current and forecasted conditions, time allocation shares remaining, TAC rankings, and other constraints.

The verification phase is one in which the PI determines whether the data are as intended. The data are transfered to Austin, where they are bundled for each PI. The PIs are informed by e-mail of new data, which they then retrieve via ftp. Alternately, the PI may use a ftp mirroring package to retrieve new data automatically. The PI then examines the data and informs the operations team of any needed changes in the plan. This phase requires only that PIs be able to retrieve data via ftp and use their own data analysis tools to examine the data.

3. Software Design

Our system contains both planning and scheduling tools. Planning tools are used to determine the feasibility of observations to be made with the HET. Scheduling tools are used to sequence observations during a night and over the course of an observing tri-semester. These tools are being used by both PIs and the operations team to schedule observations on the telescope during all four phases of operations. We have developed tools under SunOS and Solaris operating systems, using freely available packages, to allow simple porting to other UNIX systems.

We have embraced two fundamental concepts in our software development: (*i*) the tools that will be used by the telescope operators/schedulers are the same as those used by the PIs, and (*ii*) that the tools should be easily modifiable as the optimal method of operating this new system becomes apparent. Thus, we have written simple tools which can be linked together to predict and schedule the telescope for any observing plan. Further, we have written Tcl/Tk (Welch

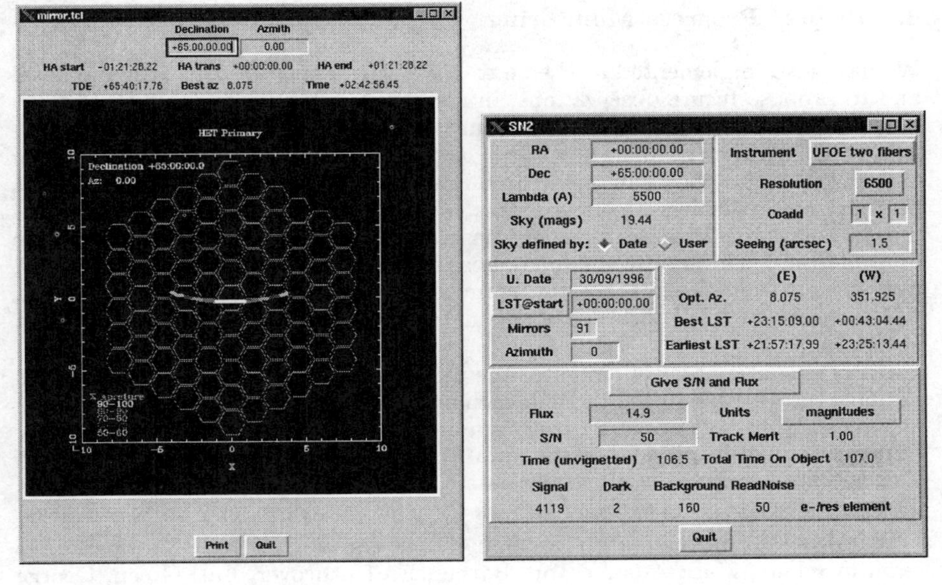

Figure 1. Two HET tools, the mirror tracking tool, which shows a potential track of the tracker across the primary mirror (left) and a Signal-to-Noise estimator (right). The tracking tools shows the TkSteal extension capturing a PGPLOT window and incorporating it into a single window environment while the Signal-to-Noise tool exemplifies the functionality of a TCL/TK GUI running over our base tools.

1994) graphical overlays to assist the user initially in creating plans. The main benefit of this system is that a user, once familiar with the system, can then create scripts in any scripting language which has access to system calls, to make plans for an entire observing program, thus reducing the time required to process a great number of objects in a GUI environment.

We used the TkSteal extension (Delmas 1996) to Tk 4.0 to minimize the number of windows used by each tool. This extension allows the Tcl/Tk script to "steal" any X-window spawned external to the Tcl/Tk script, and embed it in the Tk window as if it were an independent widget. By "stealing" windows spawned by PGPLOT (Pearson 1996), rather than creating a Tcl/Tk widget that interfaces with the PGPLOT or other graphics library, we can create a tool that is not tied to the X-windowing environment, minimize the number of windows, and allow the program that spawned the window to retain interactions with that window without any additional Tcl/Tk code. This allows us to present the user with a complete GUI in a single window, with graphics that are created by a program external to the Tcl/Tk script.

4. Project Progress Monitoring

We have also implemented a scheme to allow PIs to retrieve and examine data, and to propose future observations, in a secure environment. Each PI is given a password-protected WWW/ftp account from which data and status information can be accessed. Thus, PIs can actively monitor their projects via the WWW. Further, they can access public statistics for previous nights, such as partner share, which projects got data, weather/seeing conditions, and instrument availability. However, the PI's password is required to access all sensitive information, such as object names, positions, and setups for observations. A more primitive method of monitoring involves a modified GNUfinger Perl script, which checks to see if the PI has new data, much as the standard finger script checks for new mail.

PIs may also passively monitor the progress of their project, since an e-mail notification will be sent to the PI of each project that acquired data in the previous night. Thus, PIs need not worry about their project until data are acquired. Finally, PIs will be allowed to set up a ftp mirroring (McLoughlin 1994) script that will automatically retrieve any data that had not been previously retrieved. This script may be run as frequently as daily, or as infrequently as once a week.

Acknowledgments. We are thankful to the rest of the HET operations team members, Mark Adams, Tom Barnes, Ed Duthcover, Earl Green, George Grubb, and Phil Kelton. Many thanks go out to Frank Ray for providing us with the initial work on tracking objects with the HET. Further, we would like to thank the Project Scientist, Larry Ramsey, and the myriad of other astronomers who have tested our tools and provided us with useful suggestions.

References

Delmas, S. 1996, TkSteal[2]

Kelton, P. W., & Cornell, M. E. 1994, in A. S. P. Conf. Ser., Vol. 79, Robotic Telescopes, ed. G. W. Henry & M. Drummon (San Francisco: ASP), 136

Kelton, P. W., Cornell, M. E., & Adams, M. T. 1996, in A. S. P. Conf. Ser., Vol. 87, New Observing Modes for the Next Century, ed. T. Boroson, J. Davies, & I. Robson (San Francisco: ASP), 33

McLoughlin, L. 1994, ftp mirroring software (mirror.pl), details and code are available at ftp://src.doc.ic.ac.uk/computing/archiving/mirror/

Pearson, T. J. 1996, PGPLOT User Manual[3]

Seaman, R., & Bohannan, B. 1996, in ASP Conf. Ser., Vol. 101, Astronomical Data Analysis Software and Systems V, ed. G. H. Jacoby & J. Barnes (San Francisco: ASP), 432

Sebring, T. A., & Ramsey, L. W. 1994, in Advanced Technology Optical Telescopes V, SPIE Tech. Conf. 2199

Welch, B. 1994, Practical Programming in Tcl and Tk (New York: Prentice Hall)

---

[2] http://panther.cimetrix.com/sven/tksteal.html

[3] http://astro.caltech.edu/~tjp/pgplot/

Part 9. Real-Time Systems

Astronomical Data Analysis Software and Systems VI
ASP Conference Series, Vol. 125, 1997
Gareth Hunt and H. E. Payne, eds.

CICADA, CCD and Instrument Control Software

Peter J. Young, Mick Brooks, Stephen J. Meatheringham, and William H. Roberts
Mount Stromlo and Siding Spring Observatories, Australian National University, Canberra, ACT, Australia 2611,
E-mail: pjy@mso.anu.edu.au

Abstract. Computerised Instrument Control and Data Acquisition (CICADA) is a software system for control of telescope instruments in a distributed computing environment. It is designed using object-oriented techniques and built with standard computing tools such as RPC, SysV IPC, Posix threads, Tcl, and GUI builders. The system is readily extensible to new instruments and currently supports the Astromed 3200 CCD controller and MSSSO's new tip-tilt system. Work is currently underway to provide support for the SDSU CCD controller and MSSSO's Double Beam Spectrograph. A core set of processes handle common communication and control tasks, while specific instruments are "bolted" on using C++ inheritance techniques.

1. Introduction

Computerised Instrument Control and Data Acquisition (CICADA) is a software system developed to provide access to, control of, and retrieval of data from, instruments mounted on observatory telescopes. User access is provided through a graphical user interface running on a workstation similar to those used for data reduction. Requirement and functional specifications can be found at MSSSO's Web site.[1]

2. Instrument Control Model

As shown in Figure 1, CICADA consists of a process group running on the observer's workstation and potentially more than one process group running on perhaps more than one instrument computer (which may also be the observer's workstation). Each process group consists of a parent control process and a set of subordinate processes, each responsible for the functioning of a subset of the system. The document *CICADA Design*[2] has more detailed information of the CICADA model.

Each instrument is described by a C++ class definition. This class definition is required to provide a set of generic instrument methods as well as specific

[1] http://msowww.mso.edu.au/computing/cicada

[2] http://msowww.mso.edu.au/computing/cicada/cicada_design

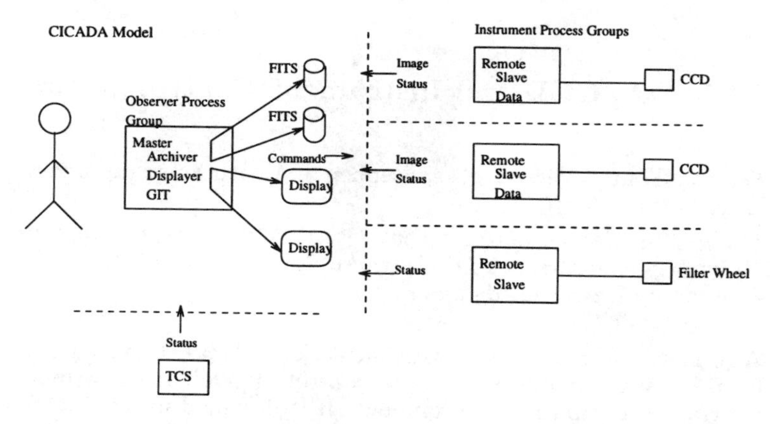

Figure 1. CICADA Process Model.

methods peculiar to the instrument. The slave process for an instrument will instantiate an object for the instrument it is talking to using polymorphism (see §3) and therefore be unconcerned with the actual type of instrument being used.

2.1. Action Requests

CICADA responds to action requests made by the observer. The user interface creates an instance of a control class object that starts the underlying processes and readies the system for action. Requests are then delivered to the `Master` process running on the local workstation which interprets the actions needed before setting the relevant slave process operating. For example, a `readout` request will result in a message being sent to an `Archiver` process to ready a FITS file and another message sent to a `Slave` process to begin the readout. The `Master` process monitors the progress of the request and delivers status information to the observer when required. The user interface process is responsible for initiating all status gathering requests.

Requests are formed using free-form text strings consisting of a verb followed by a number of optional parameter/value pairs. Requests can also be submitted in the form of a Tcl script (see §3), so that sequences of commands can be performed.

2.2. Data Delivery—Data Structures

Each instrument class generates its own request data structures using its own `generate_req` method, which is called by the control class. These data structures are designed using the `rpcgen` programming language, to facilitate the use of `RPC` for inter-machine communication.

An instrument is designed to deliver status information and, for imaging instruments, image data. When an instrument process group is running on the same machine as the observers process group, these data are delivered directly to the control class' mailbox or the waiting FITS file. Otherwise, when the instrument process group is on a remote machine, data are sent via RPC to an RPC server running on the observer's machine.

2.3. Status and Message Passing

The instrument process group maintains a shared memory area (see §3) for each process in the group to store status information. A memory area is also maintained for time-stamped, information and exception messages and is operated as a FIFO queue. Since these messages are only delivered when asked for, the queue can overflow with older messages dropping off.

3. Implementation and Techniques Used

CICADA has been implemented on Sun workstations running the Solaris operating system, Intel x86 machines also running the Solaris OS and partly on VAX/VMS systems. A main objective of the project has been to build the system in a platform independent way, some of the elements of this approach are described below:

- Standards. It is important to adhere to accepted standards so as to achieve platform independence.
- Object Oriented Programming—C++. CICADA makes good use of the object oriented techniques **`encapsulation`**, **`inheritance`**, **`polymorphism`**, and **`exception handling`**.
- RPC for network communications. We have used **`rpcgen`** for painlessly building inter-machine communication links.
- System V Inter Process Communications Facilities. CICADA is a system consisting of groups of cooperating processes. Much use is made of UNIX System V message passing, semaphores, and shared memory.
- Posix Threads. CICADA makes little use of the Posix thread model as yet but this is planned for improved performance and utility. At present multi-threading is used in the **`Archiver`** and **`Slave`** processes to achieve coprocessing during I/O wait situations.
- Tcl for scripting. Tcl is used to provide a command scripting facility. Standard CICADA free-form text requests can be embedded in a `Tcl` script for processing sequences of commands.
- Standard Image Display programs—Ximtool, SAOtng.
- Exception Handling—ANSI C++ standard. Exception handling is an important aspect of CICADA. Throwing an exception at the level it occurs and then catching it at a convenient location is a very elegant solution in a multi-processing, distributed application.
- Real-time scheduling. Sun Solaris provides a real-time scheduling class, use is made of this for time-critical functions in the instrument process group.

4. Current Applications

CICADA has currently been built to support the following instrument types at MSSSO:

- Astromed 3200 CCD Controller.
- MSSSO's New Tip-tilt Secondary.
- San Diego CCD Controller.

More detail on the implementation of the above can be found in the full text of this paper.[3]

MSSSO plan to use CICADA to drive its double-beam spectrograph and a planned 8k $\times$ 8k CCD mosaic, so hardware classes for these instruments are being designed.

5. Adding New Instrument Support

New instrument support from within CICADA requires the programmer to provide a C++ class definition of the hardware, a set of command definitions that are supported and a GUI or CLI module to interface with the defined class. Also important, of course, is the provision of tools for directly testing the operation of the hardware. These tools should be available for running on the instrument computer without any of the CICADA infrastructure, but by using the defined hardware class.

5.1. Creating the User Interface

CICADA has been structured so that the user interface is separated from the underlying machine-specific and hardware-specific code. At the moment, the only interface is a Motif GUI developed using Sun's Sparcworks Visual. All an interface needs to do is generate the command strings expected and understood by the particular hardware class and create an instance of the specific control class to pass these commands to.

6. Graphical Image Tool

A separate stand-alone program designed to provide quick-look image analysis functions has been provided to run with CICADA. This program has an X11 GUI and allows the observer to load images as they are readout from the CCD. It can perform functions such as line plots, statistics and image arithmetic. See *The GIT Users Guide*[4] for a full description.

---

[3] http://msowww.mso.edu.au/computing/cicada/cicada_adass96

[4] http://msowww.mso.edu.au/computing/cicada/git

Astronomical Data Analysis Software and Systems VI
ASP Conference Series, Vol. 125, 1997
Gareth Hunt and H. E. Payne, eds.

Real Time Science Displays for the Proportional Counter Array Experiment on the Rossi X-ray Timing Explorer

A. B. Giles[1]

Code 662, NASA Goddard Space Flight Center, Greenbelt, MD 20771
E-mail: barry@rosserv.gsfc.nasa.gov

Abstract. The Rossi X-ray Timing Explorer (*RXTE*) spacecraft contains a large Proportional Counter Array (PCA) experiment which produces high count rates for many X-ray sources. Telemetry from *RXTE* is returned via the NASA Tracking and Data Relay Satellite System (TDRSS) which provides a stream of nearly continuous real time data packets. This allows opportunity for some serious real time interpretation and decision making by the experiment controller, duty scientist, and Guest Observer (GO), if present. The GOs also have the option of arranging for the remote display of programs at their home institution. This paper briefly describes the available Science Monitoring subsystem display options.

1. Introduction

The PCA is one of three experiments on *RXTE*, and was developed at the Goddard Space Flight Center (GSFC). The *RXTE* experiment software effort was widely distributed, and the Instrument Teams (IT) were contracted to deliver much supporting code for their hardware experiment. During 1992, it was decided to use C++ for all of the software associated with the Science Operations Facility (SOF), but this decision was revised in mid-1993. At this time the Guest Observer Facility (GOF) elected to stay in line with the well established practices of the Office of Guest Investigator Programs (OGIP) and High Energy Astrophysics Science Archive Research Center (HEASARC) activities within NASA at GSFC. The SOF was committed to using C++ and continued with this approach. The SOF build deliveries were driven by the inexorable requirement to keep up with the spacecraft's aggressive schedule, and the need to support various tests and mission simulations. The GOF development was under far less pre-launch pressure. Since almost all GOF code is written in FORTRAN using FITS and FTOOLS, there could be no commonality with the SOF object-oriented C++ environment. This fundamental decision meant that the original requirement for an integrated system with extensive analysis capability in the SOF, was no longer feasible. Resources did not allow the duplication of the functionality of the analysis tools that the GOF was required to produce.

[1]also Universities Space Research Association

The PCA C++ code produced for the SOF was developed using Object Center^{TM} from CenterLine Software Inc. The GUI used is TAE+, which is a commercially available NASA product. Individual graphs were produced using the Athena Tools plotter widget set. Code was delivered to the following SOF subsystems: Command Generation (CG), Mission Monitoring (MM), Health and Safety (HS), and Science Monitoring (SM). In this paper we discuss only the SM subsystem contributions.

2. Design Requirements

The PCA experiment contains five similar detectors, each of which produces identical housekeeping data packets for the HS displays. These packets contain many detector parameters and some X-ray rates, but no spectral information. Most science data packets are generated by the Electronic Data System (EDS), which is provided by the Massachusetts Institute of Technology (MIT). This sophisticated data selection and compression system allows many pre-programmed data modes to be run simultaneously in the six Event Analyzers (EAs) devoted to the PCA. Extensive details on the PCA/EDS combination can be found in the *RXTE* NRA (1995). Further details of the *RXTE* spacecraft and mission can be found in Swank et al. (1994) and Giles et al. (1995). The requirements for the SM displays are summarized in Table 1, in approximate order of increasing scientific interest.

Table 1. PCA Science Monitoring Requirements.

Basic things to examine and look for in real time.
Do spectra look typical? Are they free of spikes, gaps and noise?
Are the spectra from all Xe layers in all detectors similar (total of 30)?
Are all the internal calibration spectra normal?
Has the source been detected?
How does the source intensity vary with time?
What sort of spectral shape does the X-ray source have?
How does the spectrum vary with time?
How do the source intensity and spectrum translate into telemetry loading?
What is the source hardness ratio, and how does it vary with time?
Are "slow" time scale periodic or aperiodic features visible?
Are "faster" time scale features visible in the on-board EDS modes?

Given the practicalities following the SOF/GOF split, the PCA team chose to emphasize the following aspects in SOF displays: real time or near real time graphical displays, visual impact and clarity, multiple options within a predefined set of choices, limited analysis capability but export of data to external tools like IDL, and support for only a few of the many EDS modes, concentrating on Standard Modes 1 and 2. Remember that detailed analysis is not intended with this real time system. If GOs are present at the SOF during their observations and need to do higher level tasks, such as background subtraction or spectral fitting in near real time, the GOF's Fits Formatter (XFF) system can be used. A SOF copy of XFF can be run to produce temporary FITS files on an estimated 15–30 minute time frame. The precise time scale depends on the data flow from the NASA PACOR system into GSFC and "fast" FITS files made in this way may be very incomplete. Normally, a 24 hour period elapses before all late packets are assumed to have arrived. Once the FITS files are created, they can be studied using the GOF FTOOLS.

3. Science Monitoring Displays

The set of display programs developed to address these monitoring requirements is detailed in Table 2. These displays can all be run individually or, more commonly, selected from a main GUI interface.

Table 2. PCA Science Monitoring Displays.

Main group	Sub Group	Description
STD Mode 1	Cal. Spectra	Raw 256 bin spectra every 128 s
	Light Curve (LC)	8 LC's. Each 1024×0.125 s points
Temporal	Power Spectra (PS)	Power spectrum of STD Mode 1 LC
Science	PS History	2-D color coded power, freq. vs. time
	Layer LC's	Select keV range from STD Mode 2
	Energy LC's	Select 4 keV ranges from STD Mode 2
STD Mode 2	Spectra	Xenon and Propane spectra every 16 s.
	Diagnostic Rates	29 rates (× 5 PCU's) every 16 s.
Spectral	Base Summation	Derived from STD Mode 2, Xe signals
Science		added raw or with gain/offsets applied
	Recommend	Base summation mapped to the 6 energy bands of the GOF RECOMMD program
	Colors vs. Time	3 definable keV bands using Base Mode
	Spectral History	2-D color coded intensity, keV vs. time
Others	EDS FFT	On-board EDS FFT spectra mode
	EDS Delta Time	On-board EDS Delta Time Binned Mode
	Event Selection	A tool to examine basic Event Mode
	Slew Detection	A derivative of Event Selection

The main GUI allows multiple instances of programs to be started, e.g., it is common to run four light curve options at the same time—0.125, 1, 8, and 16 s temporal resolution. These displays then span 128 s, 1024 s, 2.28 hours, and 18.2 hours respectively. Standard Modes 1 (mainly temporal) and 2 (mainly spectral) are always present. These displays have too many features and options for a detailed discussion (see User Guide, Rhee 1995; Design Guide, Giles 1995). A document containing sample screens for the PCA real time SM subsystem is referenced on the WWW *RXTE* SOF page. New options are carefully tested before transfer to the SOF configuration controlled environment.

4. Operations

RXTE was a fixed price program that was completed within budget and on time with respect to goals set ~4 years prior to the planned launch. *RXTE* was launched on 30th December 1995 and has already produced a vast amount of high quality timing data on a wide variety of X-ray sources. The various PCA monitoring displays have proved very effective in supporting the mission and have allowed the sort of interactive decision making that was hoped for. The duty scientist and experiment controllers monitor the observation in progress, to try and ensure that it is proceeding as planned, and that modifications to the observing modes are not required. *RXTE* can also be slewed rapidly to point at new Targets Of Opportunity. The PCA instrument team has permanent access to the SOF data flow, to monitor their experiment using the same display programs. Guest observers need not be present at GSFC—they routinely monitor their observation in real time from their home institution using remote displays of the same suite of programs running in the SOF. *RXTE* has already made many public observations and this is expected to continue in the future. The SOF are in the process of providing a mechanism for real time public data to be seen by anyone in the world via a WWW interface to the PCA display programs.

Acknowledgments. The design of these software systems has benefited from much comment and input from my fellow scientists in the PCA Instrument Team. The coding for the PCA SOF subsystems was performed in varying degrees by Vikram Savkoor, Hwa-ja Rhee, Ramesh Ponneganti, David Hon, Aileen Barry, and Arun Simha who were all with the Hughes STX Corporation working under contract for NASA.

References

RXTE 1995, 1st RXTE NASA Research Announcement, January

Giles, A. B. 1995, PCA Science Monitoring, Design Concepts, Screen Functions, SOC User Interface, Version 6.1, December

Giles, A. B., Jahoda, K., Swank, J. H., & Zhang, W. 1995, Publ. Astron. Soc. Aust., 12, 219

Rhee, H. 1995, PCA Science Monitoring User's Guide, Version 4.1.1, August

Swank, J. H., et al. 1994, in NATO ASI Series C, vol. 450, The Lives of the Neutron Stars, eds. M. A. Alpar, U. Kiziloglu, & J. van Paradijs, (Dordrecht: Kluwer), 525

Astronomical Data Analysis Software and Systems VI
ASP Conference Series, Vol. 125, 1997
Gareth Hunt and H. E. Payne, eds.

A Graphical Field Extension for sky

Al Conrad
W. M. Keck Observatory

Abstract. At Keck we use the graphical tool *sky* to plan observations and to control the telescope. Since 1992, *sky* has provided all status via alphanumerics and control via buttons. We recently extended *sky* to provide status via a graphical star field. The user controls the telescope by clicking on stars and an overlayed view of the instrument detectors. Geometrical operations, painful to convey in the alphanumeric *sky*, become trivial using the graphical approach. In this paper we discuss the advantages of the new display.

1. Introduction

Acquisition and guide star selection for the Keck telescope is accomplished via the *sky* program. *Sky* presents two distinct panels, the *main display* for direct control and the *graphical field* for staging an observation.

We developed the main display during early 1992 and have since been using it to control the Keck telescopes. The graphical field is a recent addition motivated by

1. the need for more intuitive pointing and rotation control,
2. the need for a more general method to find suitable guide stars, and
3. AO requirements.

2. The Graphical Field Display

Typically a *sky* user begins by specifying a group of objects, either by reading in a predefined object list or by searching an area of the sky. The group of objects is displayed as a tabular list in *sky*'s main display and as a star chart in *sky*'s graphical field. The *sky* user can overlay an instrument view onto the star chart and use the middle mouse button to drag it and rotate it among the objects in the field.

As the instrument field is dragged with the middle mouse button, its right ascension and declination are displayed next to the cross hair that moves along with it. As the instrument view is rotated with the middle mouse button, its position angle on the sky reads out as ψ in the lower left portion of the display.

These three position values, right ascension (`RA`), declination (`Dec`), and position angle (`PA`), taken together with the pointing origin (`PO`), completely specify an instrument orientation. To point the telescope, the *sky* user selects a pointing origin with the mouse and then drags and rotates the instrument field until the desired orientation appears on the display. At this point, the `RA/Dec/PA/PO` quadruple can be transferred to the main display by clicking on the button labeled "Transfer setup to main display."

The *sky* user can also interact with the graphical display to measure distances, measure angles, and display a particular object's name, visual magnitude, and color. A right mouse button click anywhere in the field positions a red cursor at that point. Rough astrometry data is then displayed for that point, including

1. offset polar coordinates in arcseconds and degrees,
2. offset Cartesian coordinates in arcseconds, and
3. absolute Cartesian coordinates in hours and degrees.

If the right mouse button is clicked on an object, then the object's name, visual magnitude, and color are also displayed. In this case, the astrometry data listed above are taken from the catalog and are therefore more precise.

In summary, the middle mouse button is used to drag the field, the right mouse button is used to interrogate the field, and the left mouse button is reserved for selecting stars, selecting pointing origins, and button clicks.

3. Pointing Origins

To point the telescope, the *sky* user specifies two points: a *sky location* and a *detector location.* After the telescope has been pointed, and an image has been read out from the instrument, the specified sky location will appear at the specified detector location in that image. Moreover, any field rotation that takes place during the exposure will be centered about these two coincident points. The sky location and detector location can be thought of as a pair of points which are pinned together by the act of pointing.

The *sky* program has always provided features, such as catalog search tools, telescope limit displays, and predefined star lists, that help the user specify a sky location. Until recently, however, *sky* had no features to help the user specify a *detector* location. This shortcoming introduced unnecessary confusion into the pointing process. In the new *sky*, the detector location name occupies equal real estate with the target name.

A detector location name, such as "slit" or "center_pixel" serves the same purpose in the instrument coordinate frame as does a target name, such as "Vega" or "NGC1234," in the sky coordinate frame. That is, it provides a named key for referencing a set of precise coordinates. While a target name references the right ascension and declination of a given object, a detector location name references the (x, y) coordinates of a given pixel.

In the previous section we described how the user can drag a cross hair pinned to the instrument's view to adjust the intended right ascension and declination. The point in the instrument view to which the cross hair has been pinned is called the currently selected *detector location*, or, equivalently, the currently selected *pointing origin* (Wallace 1987). The red circles displayed at specific positions in the instrument view indicate the predefined detector locations for the instrument. When the *sky* user clicks on a red circle, the displayed instrument view is shifted to align the specified detector location with the cross hair.

Note that when the user changes pointing origins by clicking on a red circle, the instrument view is shifted to align with the cross hair. Because the cross hair does not move, the intended right ascension and declination do not move either. The notion of a telescope move that does not alter right ascension or declination can be confusing if the display offers only an alphanumeric representation, but is readily apparent with the graphical view.

4. Catalogs (and Other Sources of Objects for sky)

As described in section 2, the objects displayed in the graphical field are either read from a predefined star list or extracted from an online catalog. Many of the features available from *sky*'s graphical field would also be useful with a real time guider display, with images read from `DSS` (Morrison 1994) or with images read from image archives.

Our success in integrating better catalogs, in particular the degree to which we can go faint, will strongly influence our decisions about the alternate sources listed above. In particular, if we can provide catalog coordinates with reasonable coverage down to 18th visual magnitude, the urgency to integrate `DSS` images recedes.

Processed image data from `DSS` is now readily obtained via the Internet. With current Internet bandwidth to Hawaii, these images are barely useful for daytime planning. Even with planned improvements for Hawaii's Internet access, `DSS` images could not be used for night time decision making unless either a complete set is available on a juke box or a subset is available on a conventional disk.

Similarly, the turn around time for accessing past Keck images is sufficiently slow that using these images for planning from within the *sky* graphical field would barely be useful for daytime planning and could only be used for night time work if an appropriate subset of images were loaded onto a conventional magnetic disk before the night's observing.

For both `DSS` and archive images, the level zero method for integrating images into the sky graphical field is straightforward: simply add a button for loading an arbitrary `FITS` image. The question of what do to with that `FITS` image is more subtle. There are three choices:

1. read raw pixel data directly into the sky graphical field,
2. use centroiding and background flattening to generate a stylized version of the `FITS` image, or
3. produce a stylized image as above and, in addition, scale and translate the object locations to match the results of a catalog search from the same area (Mink 1997).

5. Star Lists

While still at their home institutions, visiting observers prepare a list of objects for their run at Keck. In past versions of *sky*, this list contained only the object name, right ascension, declination, and equinox, and, optionally, keyword-value pairs for proper motion or differential tracking rates. In the new version of *sky*, the user can optionally include

1. rotator position angle,
2. detector location name (also known as, pointing origin name), and
3. the positional coordinates of any movable instrument detectors.

Past star lists only described sky objects, and provided no information specific to the telescope or the instrument. With the addition of the above information, each line in a predefined star list can now specify all of the information needed to set up the telescope completely.

6. Conclusion

The graphical field display makes pointing and rotating more intuitive. The display provides point-and-click mechanisms both for setting up an observation and for measuring distances and angles. *Sky* now provides methods for selecting the detector location as well as the sky location; both are needed to set up an observation. The best method for incorporating non-catalog sources requires further study and is influenced by the availability of faint guide star catalogs.

Acknowledgments. We thank Julie Barreto, Tom Bida, Randy Campbell, John Gathright, Tony Gleckler, Bob Goodrich, Wendy Harrison, Hilton Lewis, William Lupton, Jerry Nelson, Gerry Neugebauer, and Pat Wallace.

References

Wright, J. 1993, in ASP Conf. Ser., Vol. 52, Astronomical Data Analysis Software and Systems II, ed. R. J. Hanisch, R. J. V. Brissenden, & J. Barnes (San Francisco: ASP), 495

Christian, C. A., & Olson, E. C. 1993, in ASP Conf. Ser., Vol. 52, Astronomical Data Analysis Software and Systems II, ed. R. J. Hanisch, R. J. V. Brissenden, & J. Barnes (San Francisco: ASP), 56

Wallace, P. 1987, The pointing and tracking of the Anglo-Australian 3.9 metre telescope

Mink, D. 1997, this volume, 249

Morrison, J. E. 1995, in ASP Conf. Ser., Vol. 77, Astronomical Data Analysis Software and Systems IV, ed. R. A. Shaw, H. E. Payne, & J. J. E. Hayes (San Francisco: ASP), 179

Astronomical Data Analysis Software and Systems VI
ASP Conference Series, Vol. 125, 1997
Gareth Hunt and H. E. Payne, eds.

The JCMT Telescope Management System

Remo P. J. Tilanus,[1] Tim Jenness, Frossie Economou, Steve Cockayne[2]

Joint Astronomy Centre (JAC), 660 N. A'ohoku Place, University Park, Hilo, HI 96720, E-mail: rpt@jach.hawaii.edu

Abstract. Established telescopes often face a challenge when trying to incorporate new software standards and utilities into their existing real-time control system. At the JCMT we have successfully added important new features such as a Relational Database (the Telescope Management System—TMS), an online data Archive, and WWW based utilities to an, in part, 10-year old system. The new functionality was added with remarkably few alterations to the existing system. We are still actively expanding and exploring these new capabilities.

1. Introduction

The introduction of GUI-based applications and control systems, Database Management Systems (DBMS), HTML, and the World Wide Web (WWW) have set new standards in the way users expect to interact with computers and data. Living up to those standards with systems which essentially predate these developments presents a noteworthy challenge, in particular if these systems are not UNIX-based.

The real-time control system at the JCMT uses a VMS-based network where control of (and data from) the telescope is handled by a number of VAX computers. As is traditionally the case at many telescopes an observation results in a file on a disk consisting of a "header" part and "astronomical data" part. In addition to astronomically relevant items the header also contains much information about the hardware status of the telescope. This presents an annoying and very time-consuming complication in case problems arise which need tracing or monitoring over significant time-intervals: hundreds of files may need to be accessed in order to extract the relevant parameter.

In order to address these problems we have introduced at the JCMT a Telescope Management System (TMS): a DBMS-based system which keeps permanently available on-line the same information as in the file headers plus any other information deemed of interest. Another way to express the goal of the TMS is that it aims to provide an accurate image of the generalized telescope status for each instance in time. Ideally, the real-time system would interact with the TMS to optimize the real-time performance since from the history con-

[1] Netherlands Foundation for Research in Astronomy (NFRA)

[2] Now at Canadian Astronomy Data Centre (CADC)

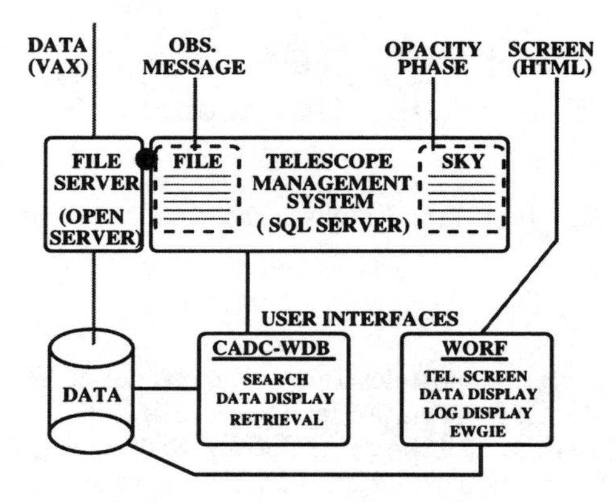

Figure 1. The JCMT Telescope Management System (UNIX).

tained in the TMS it should be possible to make reasonably accurate predictions for settings of sub-systems which (slowly) drift in time.

2. The Telescope Management System

The TMS is based upon Sybase, a relational DBMS, installed on a UNIX host at the JCMT and consists of a set of linked "flat" tables which together contain all the items one wants to keep track of. A key strategy in its design stems from the realization that a telescope in general consists of a number of discrete sub-systems which each change with their own time-constant. Quite often these logical subsystems correspond to physical subsystems of the telescope, but not always: a camera, which otherwise retains its configuration during the night may contain a filter wheel which changes with each observation.

The most efficient and accurate implementation of a TMS is to create a table for each of the (logical) subsystems: when the subsystem changes its configuration only one table needs a new entry, while the others do not change. The telescope can be regarded as providing a continuous, asynchronous stream of information which gets mapped onto the appropriate tables in the TMS. In such a scenario, the astronomical data is just another package delivered, albeit an important one. The complete telescope status can thus be reconstructed from the tables by retrieving records from each table corresponding to the correct time-interval (although a smart use of "indexes" (yuk: "indices" please) rather than the timestamps will in practise be a more efficient approach).

The TMS actually installed at the JCMT represents a compromise between the ideal system described above and speed of implementation. In particular we wanted to add a TMS with minimal changes to the existing real-time system. Most information is read out of the traditional file header (rather than directly from the real-time system) and uploaded to the TMS once the observation is finished. Consequently, the TMS retains the telescope status valid per observation rather than for every instant in time. Upon the completion of the observation and the close of the observation file on the VAX the real-time systems sends a notification to the TMS on the UNIX host which immediately copies the file

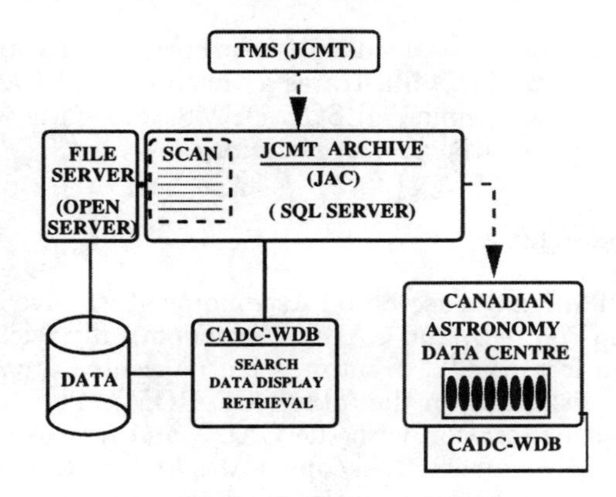

Figure 2. The JCMT Archive.

into a disk archive and uploads the header information to the DB. An exception are the opacity and seeing monitors (and shortly the weather station), which log directly to the DB independent of observations.

In summary, three components were added to the real-time control system: the notification of the TMS via an observation completion message, the direct logging of the atmospheric conditions, and, thirdly, an HTML dump of the telescope status screen at regular intervals for use with the WWW.

Figure 1 schematically illustrates the TMS on the UNIX host. The atmospheric conditions are logged directly into the corresponding DB tables. The insertion of the observation completion message into a DB table causes a "trigger" to be activated which results in the copying of the data file and the upload of the header information to the DB using a Sybase OpenServer. The OpenServer gives the ability to extend the SQL Server with custom processing, e.g., the transparent delivery of data to SQL Clients even while the data resides on disk rather than within the DB. Another application is the sending of a mail notification if cryostat temperatures exceed a certain limit indicating the need to refill cryogens.

3. The User Interface

Any standard SQL-based tool can be used to interact with the TMS and retrieve and display information. However, our general user interface is WDB (Rasmussen 1995) developed by ESO and the Canadian Astronomy Data Centre (CADC) as a Web-based archive query tool (Figure 1). It currently lacks the ability to plot any item against any other item but we hope to add that capability shortly using an existing Java applet. Before WDB we used the Starcat interface which was superb in this respect through the use of Xmgr. In order to extend the use of WDB from astronomical archives only to a general interface to (Sybase-based) DBs we use custom data (pre-)viewing[3] and retrieval tools and

[3]The TMS shares many of the custom modules with WORF2

added user authentication to safeguard the proprietary nature of some of the information. The use of WDB illustrates a very powerful feature of the TMS: since it is based upon a commercial SQL DBMS very little work is needed to provide adequate and versatile user applications.

4. The JCMT Archive

In addition to the TMS, a Sybase-based Astronomical Archive has been installed at the JAC in Hilo (Figure 2) and CADC. Astronomically relevant header information is extracted from the TMS into a science-oriented archive together with the actual data on disk (and, in the future, CD-ROM). The JCMT Archive has been developed in collaboration with the CADC and also uses WDB as a frontend. An automatically synchronised copy of the Archive tables is maintained at the CADC which is the main entry point for searches involving non-proprietary data.

5. Conclusion and Future Developments

A noteworthy aspect of the new features, such as the TMS and WORF2 (Jenness, Economou, & Tilanus 1997), added to the UNIX-based systems is that a state-of-the-art look-and-feel has been built around an existing (VMS-based) real-time system while requiring the latter only to send a single message to the TMS when an observation has finished. Although currently the TMS is not used by the real-time system itself (a possible future development), it provides the observatory staff and observers with the general diagnostic capabilities of a fully integrated RDBMS. Many features, especially the Web based utilities, required little local development and often took a few days rather than months to implement.

In the future we expect Web based utilities to become more integrated with the actual process of observing at telescopes. In particular the on-line availability of remote data archives and catalogs will make it possible to instantaneously correlate new data with existing information. Examples are marking known source positions on a real-time display, solving for plate-solutions, identification of spectral features and overlaying data from different telescopes. With astronomically relevant information increasingly available at one's "finger-tips" at the telescope observing once again should encompass "observation" rather than just data collection followed by a analysis stage much later.

References

Jenness, T., Economou, F., & Tilanus, R. P. J. 1997, this volume, 401

Rasmussen, B. F. 1995, in ASP Conf. Ser., Vol. 77, Astronomical Data Analysis Software and Systems IV, ed. R. A. Shaw, H. E. Payne, & J. J. E. Hayes (San Francisco: ASP), 72

Astronomical Data Analysis Software and Systems VI
ASP Conference Series, Vol. 125, 1997
Gareth Hunt and H. E. Payne, eds.

Remote Eavesdropping at the JCMT via the World Wide Web

Tim Jenness, Frossie Economou, R. P. J. Tilanus[1]

Joint Astronomy Centre, 660 N. A'ohoku Place, University Park, Hilo, HI 96720

Abstract. The James Clerk Maxwell telescope (JCMT), a submillimetre facility on Mauna Kea, has recently adopted flexible scheduling. This is expected to result in fewer astronomers travelling out to Hawaii, and more data being taken by local observatory staff. In order to allow astronomers to monitor their data from their home institutions, JCMT has adopted the WWW Observing Remotely Facility (WORF) already offered by the United Kingdom Infrared Telescope (UKIRT).

WORF allows astronomers to eavesdrop on their data using the Netscape™ browser with minimal impact on observatory staff and computer systems. The UKIRT implementation has been extended to meet the differing expectations of the JCMT community, and has many additions including a telescope status screen, the ability to access the observation log and the use of a conferencing tool between the observer and the eavesdropper(s).

1. Introduction

The James Clerk Maxwell Telescope (JCMT) is a UK-Canada-Netherlands 15-m submillimetre telescope situated on the summit of Mauna Kea, Hawaii. In order to use good observing conditions more efficiently, the JCMT is moving to full flexible-scheduling. This will result in fewer astronomers travelling out to Hawaii, and more observations taken by observatory staff. It also implies that astronomers may only be given a few hours notice before observations in their program are taken. Although JCMT users understand the advantages of flexible scheduling, they are reluctant to relinquish the ability to modify observing strategy on-the-fly as new data come in.

We are anticipating the community's requirement for a way to monitor remotely their observations by developing the World Wide Web Observing Remotely Facility (WORF). WORF had been developed previously for the United Kingdom Infra-Red Telescope (UKIRT), where the HTTP based implementation was shown to have clear advantages (Economou et al. 1995); the JCMT specific development has become known as WORF2.

[1] Netherlands Foundation for Research in Astronomy (NFRA)

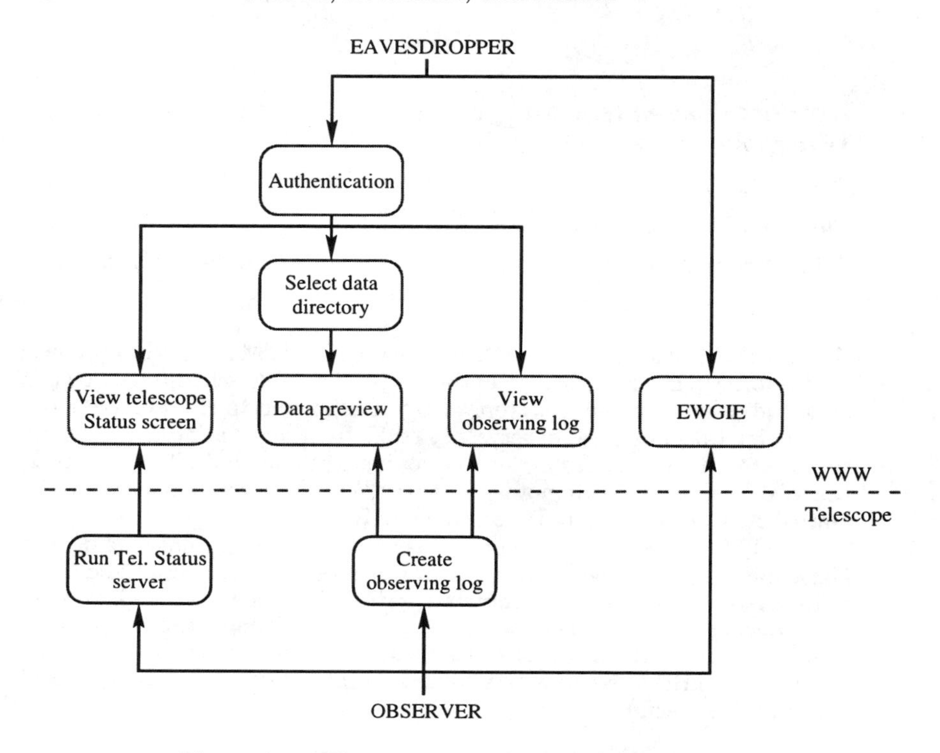

Figure 1. The structure of the WORF2 system.

2. User Interface

Each successful proposal for JCMT time is assigned a user name and password at the beginning of the semester. The first thing the potential eavesdroppers have to do is authenticate themselves, as access to data directories is restricted. They can then enter the WORF2 system, whose user interface consists of four Web pages (the locations of which are shown in Figure 1):

- **Data Preview**. This is the main display window of the WORF2 system. Here the eavesdropper is presented with a list of completed observations from which to choose a particular one and display spectral line data, adjusting the plotting scale as desired.
- **Telescope status**. This is a continuously updating (every 60 s) display of basic telescope parameters. The most important of those are the Right Ascension and Declination, which reassures the eavesdropper that the telescope is indeed pointing at her source. It is also important that the observer can check the current instrumentation configuration.
- **Observation log**. The person actually observing at the telescope can use in-house software to generate a log consisting of basic information about the observations and additional comments on the quality of the data. This window enables the eavesdropper to peek at this log file.

- **Conferencing tool.** A WWW based conferencing tool[2] is also used so that the observer and astronomer(s) can discuss details of the run using both text and diagrams.

3. Implementation

3.1. Data Preview

The data previewing CGI backend is written entirely in Perl5 (Wall, Christiansen, & Schwartz 1996). JCMT observations are stored in a telescope specific GSD format originally developed for single-dish radio telescopes. Using the flexible Perl extension mechanism, we first had to develop a module allowing access to the C GSD library. Data from the JCMT digital autocorrelation spectrometer (DAS) needs to be processed before being displayed in a manner meaningful to an astronomer. The processed data is plotted into a GIF image via pgperl, the Perl PGPLOT module,[3] and is then served via HTTP.

File naming conventions do not provide sufficient information for the eavesdropper to determine the type of observation contained in a file. Rather than reading in every file in the data directory to extract this information, an observing log generated by the scientist at the telescope is used instead. In the absence of such an observing log, only observation numbers are displayed.

3.2. Security

Data privacy is a subject dear to the heart of many an astronomer. The eavesdropper is authenticated by the HTTP server and the login name thus provided is compared against the ownership of each individual file requested thereafter (data ownership is determined by an entry in the file header). This takes advantage of the way observations from different projects are taken and stored at the JCMT.

3.3. EWGIE

We use the public domain **E**asy **W**eb **G**roup **I**nteraction **E**nabler (EWGIE) conferencing tool. This Java based tool provides a "chat area" as well as a "whiteboard" allowing diagrams to be exchanged between multiple participants over HTTP.

3.4. Browsers

The data preview component of WORF2 will run on most graphical WWW browsers. EWGIE requires Java support. The observation log and telescope status screens make use of client pull in order to automatically update, a feature which is not currently supported by many browsers other than Netscape™. These pages can, of course, be updated manually by reloading the page.

---

[2]EWGIE, developed by Kevin Hughes at kevinh@commerce.net. (http://www.eit.com/ewgie/)

[3]http://www.ast.cam.ac.uk/AAO/local/www/kgb/pgperl/

4. Future

JCMT has recently taken delivery of SCUBA, a much anticipated submillimetre bolometer array, which will provide our users with imaging data for the first time. Data from SCUBA is stored in the Starlink N-dimensional Data Format (NDF), which is widely used in the UK community. We have already developed a Perl module allowing access to the FORTRAN Starlink libraries, and intend to take full advantage of the on-line data reduction pipeline written for SCUBA.

WORF2 is already in use, mainly by Dutch astronomers who were the first JCMT partner to move to flexible scheduling. We expect it will be used more widely as the British and Canadian allocated telescope time switches to flexible scheduling later this year.

We are in the process of adding more features to WORF2 but, as with WORF, other aspects of the software group's work are of higher operational priority.

A demo of the data preview component of WORF2[4] is also available.

References

Economou, F., Bridger, A., Daly, P. N., & Wright, G. S. 1996, in ASP Conf. Ser., Vol. 101, Astronomical Data Analysis Software and Systems V, ed. G. H. Jacoby & J. Barnes (San Francisco: ASP), 384

Wall, L., Christiansen, T., & Schwartz, R. L. 1996, Programming Perl, 2nd edn. (Sebastopol, CA: O'Reilly)

[4]http://jach.hawaii.edu/worf2/worf2.csh

Astronomical Data Analysis Software and Systems VI
ASP Conference Series, Vol. 125, 1997
Gareth Hunt and H. E. Payne, eds.

WinTICS-24 Version 2.0 and PFITS—An Integrated Telescope/CCD Control Interface

R. Lee Hawkins

Whitin Observatory, Wellesley College

David Berger and Ian Hoffman

Foggy Bottom Observatory, Colgate University

Abstract. WinTICS-24 Version 2.0 is a telescope control system interface and observing assistant, that provides the ability to control a telescope and guide-acquire module, along with a suite of utilities to assist observers. PFITS is a suite of scripts for taking data in PMIS, with the ability to construct very comprehensive FITS headers by using data supplied to PFITS from WinTICS-24.

1. History

WinTICS is an evolution TICS-24, an integrated telescope control system (TCS) written by Hawkins & Ratcliff (1993) for the Macintosh in HyperCard. While sharing the general design philosophy of TICS-24, WinTICS is much more flexible and extensible.

2. User Interface and Communications

The WinTICS user interface is a windowing interface written in Microsoft Visual Basic for Windows 95/NT. While mimicking somewhat the DFM Engineering TCS screen layout, WinTICS adds to the DFM TCS commands to give a more complete observer interface. Controls are grouped by function, and color is used to indicate normal (green), caution (yellow), or abnormal (red) telescope and instrument conditions. All site-specific settings are configurable by the observer. All observer input is checked for range validity, and whether the target position is closer than an observer-set limit to the Sun and/or Moon. Additionally, WinTICS contains a complete hypertext help system describing every window and control. WinTICS communicates with the remote TCS by sending ASCII commands over a serial port. Although originally developed for DFM TCS, WinTICS could easily be adapted to any other TCS that accepts ASCII commands over a serial port, simply by changing the appropriate commands and rebuilding the package. Each set of controls will be described below. The main WinTICS window is shown in Figure 1.

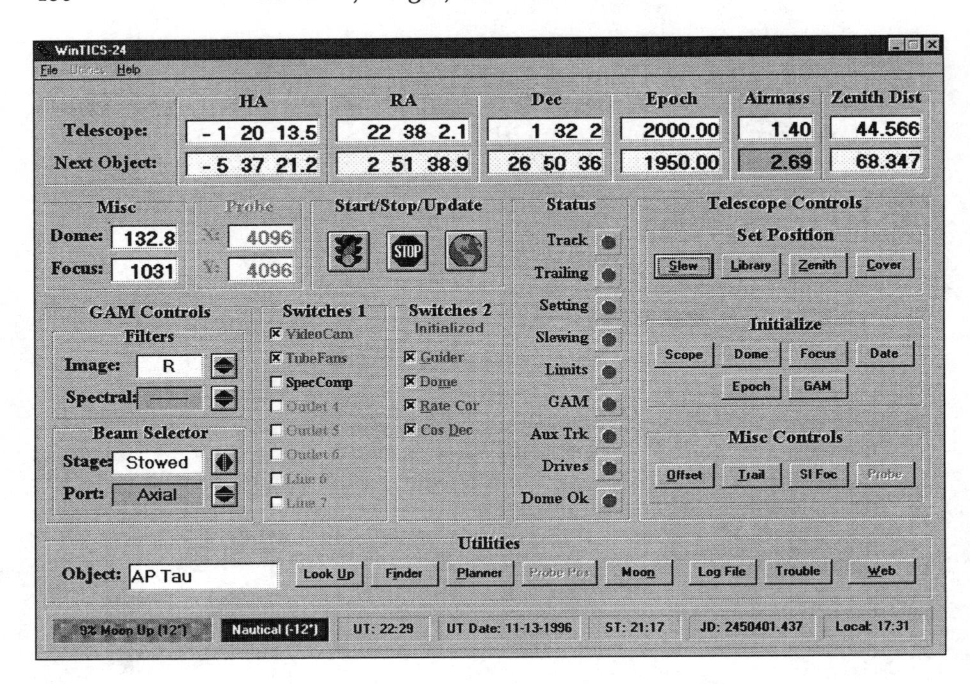

Figure 1. The WinTICS-24 Main Window.

3. Position Readouts

WinTICS communicates with the native telescope TCS to get and display the following position information for both the current telescope position and the next object: Hour Angle, Right Ascension, Declination, Epoch, Airmass, and Zenith Distance. In addition to the coordinates of the telescope, WinTICS also displays the focus position in ADUs, the dome position in degrees, and the X-Y position of the Guide-Acquire Module (GAM) guide probe.

4. Telescope and GAM Control and Status

On the right side of the main WinTICS window are controls for movement and initialization of the telescope. These controls are as follows: Set Position: Controls for slewing the telescope to coordinates, library object by number, the zenith, and the cover position for removal/replacement of the telescope covers. Initialize: Controls for initializing the telescope, dome, and focus positions, setting the date/time and epoch of precession, and moving the GAM filters and mirrors to known positions. Misc. Controls: Controls for offsetting the telescope in RA and Dec, trailing a star along the spectrograph slit, moving to a given focus position, and moving the GAM guide probe to a new X-Y position. The Switches 2 box contains check boxes to turn the Guider flag, dome, rate correction (first derivative of the pointing model), and cosine of the declination for guide rate on/off. Additionally, this box contains messages that tell the observer whether the remote TCS is initialized or not. The Status box contains "virtual LEDs" that display the status of various telescope modes: Tracking, Trailing, Setting, Slewing, Limits, GAM, Aux Track, Drives, and Dome OK. In the Start/Stop/Update box are three buttons, two of which are applicable here. The green traffic light starts a slew, and the red Stop sign cancels an enabled slew. On the middle left of the main WinTICS window are controls for the GAM: Filters: Position of the axial and south port filter wheels. Beam Selector: Stage: The position of the stage that carries the beam selector mirror. In the stowed position, the beam goes through the axial port and imaging filter to a CCD. In the On-Beam position, the beam is directed to either the north, south, east, or west radial ports by a 45° flat mirror. Port: The radial port to which the beam is being directed. Switches 1 gives the status of 8 digital I/O lines on the GAM controller. Six of these are in turn connected to solid state relays which switch outlets on and off.

5. Observing Utilities and Time Status Bar

Perhaps the most useful part of WinTICS is the observing utilities, found in the Utilities box along the bottom of the main WinTICS window. From left to right, these are: Object: The name of an object to look up in the WinTICS object database. If the object is found, then the coordinates are automatically sent to the remote TCS and a slew is enabled. If the object is not found, a window is brought up that allows the observer to add the object to the WinTICS object database, then slew to the object. Look Up: Look up the object given in the Object box. Finder: Bring up a finder chart centered on the current telescope coordinates from the Guide CD ROM, with a box the size of the CCD field of view overlaid on the chart. Planner: Bring up an eclipsing binary star time of minimum prediction for the object given in the Object box (Downey & Hawkins 1995). Probe Pos: Move the guide probe to the predefined position for the the object given in the Object box. Moon: Brings up a graphical view of the phase of the Moon, accurate to a day or so (Craig 1993). Log File: Brings up a window to make an entry into an observing log file to be filled out by the observer, detailing the observers, program, and comments. Trouble: Brings up a window to enter problems into a trouble log file. Web: Brings up a window that communicates via DDE with a Web browser to display frequently accessed Web pages (e.g., weather forecasts and images). In addition to these options, there is a button on the Time Preferences tab which dials the Naval Observatory and resets the PC's internal clock (Craig 1993). (See also NTP support, below.) Below the Utilities section on the WinTICS main screen is a status bar that displays the position of the Sun and Moon, along with the time in various formats. The backgrounds of these fields change color depending on whether the Moon/Sun is up/down and the twilight type (civil, nautical, or astronomical). The fields are updated once per minute, and the times are accurate to ± 1 minute. From left to right, they are: Moon: Percent illumination of the Moon, Moon up/down, and altitude of the Moon in degrees. Sun: Sun up/down or twilight type and the altitude of the Sun in degrees. Times: The current time is displayed in the following formats: UT, UT Date, ST, JD, and Local. All these times except Local come from TCS. Local comes from the WinTICS computer system clock. In turn, the WinTICS computer system clock can be updated via the Network

Time Protocol by supplying the hostname of an NTP server and enabling NTP updates on the Time Preferences tab.

6. Coordinate/Status File

Each time the telescope coordinates are updated or a filter is changed, WinTICS writes a log file entry containing the telescope coordinates, filter positions, and status of the telescope and switches. These are written as FITS format keywords, which can then be read by a CCD control program such as PMIS (Remington 1995) to be merged into the image header for the current observation. This file essentially allows WinTICS to act as the "glue" between the telescope TCS and the data acquisition system. The author and a series of students have written a suite of scripts for PMIS called PFITS (Berger & Hawkins 1995; Hoffman & Hawkins 1996) that greatly simplifies and automates data taking, including incorporating the FITS keywords written by WinTICS into the image headers.

7. Conclusions

When compared to the typical native telescope control system, WinTICS provides an observer interface that is more easily and quickly understood at a glance. In addition, its incorporation of various utilities for observation planning and execution makes WinTICS much more powerful than most vendor-supplied TCSs. Anyone interested in receiving more information on WinTICS should contact the author at lhawkins@wellesley.edu, or have a look at the WinTICS homepage.[1]

Acknowledgments. This project was made possible by NSF-ARI grant AST-9413335 and a grant from the W. M. Keck Foundation to the Keck Northeast Astronomy Consortium.

References

Berger, D., & Hawkins, R. L. 1995, in Proceedings of the Keck Northeast Astronomy Consortium (Poughkeepsie, NY: Vassar)

Craig, J. C. 1993, Visual Basic Workshop (Redmond, WA: Microsoft), 33 and 428

Hawkins, R. L., & Downey, K. F. 1995, BAAS, 27, 1931

Hawkins, R. L., & Ratcliff, S. J. 1993, BAAS, 25, 1364

Hoffman, I., & Hawkins, R. L. 1996, in Proceedings of the Keck Northeast Astronomy Consortium (Wellesley, MA: Wellesley)

Kelly, M. 1995, DFM TCS-24 User's Guide (Longmont, CO: DFM)

Remington, G. K. 1995, PMIS User's Guide (Boulder, CO: GKRCC)

[1] http://www.astro.wellesley.edu/lhawkins/tics/ticsIntro.html

Part 10. Instrument-Specific Software

Astronomical Data Analysis Software and Systems VI
ASP Conference Series, Vol. 125, 1997
Gareth Hunt and H. E. Payne, eds.

Physical Modeling of Scientific Instruments

M. R. Rosa[1]
Space Telescope European Coordinating Facility, European Southern Observatory, D–85748 Garching, Germany, E-mail: mrosa@eso.org

Abstract. This contribution revolves about computer models of astronomical instruments and their application in advanced calibration strategies and observation model based data analysis. The historical connection between calibration and data analysis is reviewed. The success of physical models in the analysis of observational data with strong instrumental signatures is shown, and a concept for model based calibration developed. A discussion of advantages for observatory operations, observing, pipeline processing, and data interpretation is accompanied by a briefing about the current status of the Observation Simulation and Instrument Modeling project at the ST-ECF and ESO.

1. Introduction

The increasing importance assigned to tasks circumscribed by the terms "calibration" and "data reduction" reflects the evolution of equipment and the fact that purely morphological investigations have largely been superseded by the demand to turn the last bit of useful information in raw data into significant astrophysical quantities.

Calibration strategies currently adopted are empirical methods aimed at "cleaning" raw data of instrumental and atmospheric effects. Since it is only through the empirical determination of calibration reference data (e.g., dispersion relations, flat fields, sensitivity curves) that both evils—changes of the instrument and changes of atmospheric conditions, respectively—are detected and their effects removed, these calibration strategies often take on the form of a defensive battle repeatedly fought every night.

Space based instrumentation, in particular IUE and HST, has taught that once the effect of the human desire to tweak the instrumental parameters every so often is nullified, pipeline calibration of the raw data can efficiently be implemented. Monitoring and trend analysis of the calibration parameters also shows that predictions made about the instrumental characteristics are usually very reliable. Operational scenarios with pipeline calibration are now also being planned for ground based observatories (e.g., ESO's VLT), and it is easy to show that once instrument stability through configuration control is achieved, all that remains of nightly calibration data taking is to address environmental aspects, e.g., the atmospheric transmission.

[1] Affiliated to the Astrophysics Division of the Space Science Department,of the European Space Agency

Nevertheless, the calibration process remains of the "instrument signature removal" type, even if based on predicted, less noisy calibration data. A long learning process is required to determine the optimum deployment of resources—manpower and observing time—for the calibration task as a whole, and for its subprocesses. This approach also hides the direct link between the engineering parameters of instruments and the differing characteristics in closely related configurations (e.g., the 2-D pattern of echellograms obtained with different grating tilts).

Repeatedly, situations arise where the scrutinizing analysis of science data, in particular when combined with very specific theoretical expectations (e.g., stellar atmospheric models, evolutionary color diagrams), reveals shortcomings or omissions in the "signature removal" calibration. Usually the effects found are neither new to the type of instrument (e.g., non-linearity of a detector) nor unexpected on physical grounds (e.g., scattered light in grating spectrographs). To be fair, almost always the embarrassing situation is not a result of an oversight but rather a consequence of the calibration strategy employed—dealing piecemeal with a highly complex interrelation of physical effects.

Can we do any better? In the following I will lay out the arguments for a radically new approach, highlighting a few examples, and briefly report on the activities in an collaborative ESO/ST-ECF effort to implement observational data calibration and analysis on the basis of physical models.

2. From Observations to Astrophysical Interpretation

Usually the calibration and data analysis process is conceived as a well defined activity, detached from, but linking the observational process with the astrophysical interpretation. The direction is from raw data to better (i.e., "cleaned") data, from counts per pixel to H_0. What really happened during the observation was, however, the application of an operator $\mathbf{O}$ onto a vector $\mathbf{u}$ representing a subvolume of the many-D universe. $\mathbf{O}$ describes the equipment used as well as any other environmental circumstances (e.g., the atmosphere).

Were it not for the ultimate limitation imposed by noise, we could try to just find and apply the inverse operator $\mathbf{C}$ to calibrate, i.e., recover $\mathbf{u}$ from the raw data $\mathbf{d}$, and then go on to interpret $\mathbf{u}$. Eventually the more rewarding "data analysis" strategy is to subject a range of probable $\mathbf{u}'$ to a model of the data taking process $\mathbf{O}'$ and to compare the resultant simulated data $\mathbf{d}'$ with the actual $\mathbf{d}$.

Practically, finding $\mathbf{C}$ empirically is hampered not only by noise, but also by the fact that $\mathbf{O}$ is huge, while we can spend only so much time exploring the parameter space—science exposures are certainly more rewarding. It would help if we had a good idea of what $\mathbf{O}$ looks like. All that is necessary, on paper, is a complete description of the important physical aspects of the instruments. We will see below how far one can go in reality.

In essence, one can implement a two stage process to achieve substantial improvements over the canonical "signature removal" data analysis. Step 1 consists of casting physical principles into code which is capable of simulating observational data. This can be used to generate calibration references that are noise free and contain controllable engineering parameters. At this stage it is straightforward to generate calibration data for a large array of modes that would otherwise have to be covered painstakingly by individual calibration exposures.

After having gained confidence in stage 1, application of optimization techniques, e.g., simulated annealing to simulated data using the physical instrument model, will be step 2 (cf. Rosa 1995).

3. Related Schemes

It is important to note that the process described above differs substantially from many seemingly similar schemes, in that the kernel is an instrument model based on first principles. In observation planning, it is now common practice to simulate the outcome of proposed observations by convolving target models with empirical spectral response curves, adding noise and whatever else the calibration data base can provide. Good examples are the WWW planning tools for the HST instruments. It has also become standard technique now, thanks again to HST, to apply deconvolution techniques and derivatives such as multiple frame analysis (see Hook 1997) to data sets—techniques that usually work best if fed by sophisticated models of the PSF (e.g., TinyTim of HST). Calibration data for HST instrument modes that can not be obtained for reasons of limited time are usually supplied by interpolation between neighboring modes and for all modes forecasted by trend analysis.

The complete physical model of an instrument, in principle, incorporates many of those capabilities. In fact its construction and tuning to actual performance requires a deep understanding based on the experiences gained with many of the tools described above. The big difference is its predictive power.

4. How Far to Go

Exercises such as the FOS scattered light correction (Rosa 1994; Bushouse, Rosa, & Müller 1995) or the FOS dispersion model (Dahlem & Rosa 1997,in preparation) demonstrate that a software model, covering just one particular aspect (here the diffraction and interference properties of gratings and apertures, and the reimaging electron optics of digicons), but going beyond a simple throughput calculation, i.e., correctly describing all relevant physical effects, can be very beneficial in solving problems encountered during the scientific analysis of data. Used in this way, the model appears simply as an additional data analysis tool in support of the calibration process. However, the purpose of this paper is to go one step further, i.e., to advance beyond the "signature removal" calibration strategy currently in use. Had the FOS model been available early on, it would certainly have influenced the specifications for the pipeline, and even earlier the introduction of solar blind detectors.

Obviously, it is necessary to correctly describe all aspects of an instrument to such a degree that the typical percent level accuracy of the canonical calibration can be surpassed. The FOS models mentioned above do not need to incorporate geometric and intensity aspects at once. On the other hand, a model designed to completely cope with the analysis of long slit echelle spectra must be able to predict the geometrical pattern (curvature of orders, of slit images and field distortions), and the blazed sensitivity variations in dispersion direction as well as the spatial intensity profiles (LSF and interorder background).

5. Current Activities

A generic echelle spectrograph model, such as described above, is currently under construction in a collaborative effort between ESO and the ST-ECF. Its immediate application will be to the UVES spectrograph under construction for the VLT observatory and the STIS instrument to be working on HST after the servicing mission in February 1997.

The basis of this effort is a library of C++ classes that provide equipment modules, e.g., grating, mirror, filter, detector, and optical rays (the targets passing through the equipment), as well as modules for other ingredients such as targets, interstellar extinction, and atmospheric properties to compose instrument models and simulate observations (see also Ballester, Banse, & Grosbøl 1997).

References

Ballester, P., Banse, K., & Grosbøl, P. 1997, this volume, 415

Bushouse, H. E., Rosa, M. R., & Müller, Th. 1995, in ASP Conf. Ser., Vol. 77, Astronomical Data Analysis Software and Systems IV, ed. R. A. Shaw, H. E. Payne, & J. J. E. Hayes (San Francisco: ASP), 345

Hook, R. N. 1997, this volume, 147

Rosa, M. R. 1994, in Calibrating Hubble Space Telescope, eds. C. Blades & S. Osmer (Baltimore: STScI), 190

Rosa, M. R. 1995, in Calibrating and Understanding HST and VLT instruments, ed. P. Benvenuti, ESO/ST-ECF Workshop, ESO, 43

Astronomical Data Analysis Software and Systems VI
ASP Conference Series, Vol. 125, 1997
Gareth Hunt and H. E. Payne, eds.

Calibration and Performance Control for the VLT Instrumentation

P. Ballester, K. Banse, and P. Grosbøl

European Southern Observatory, Karl-Schwarzschildstr. 2, D-85748 Garching, Germany

Abstract. The Very Large Telescope will see first light by the end of 1997 and gradually be equipped with up to 14 instruments available on the four telescope units. The framework for data calibration and analysis of the VLT instrumentation is presented, based on the concepts of a calibration plan, pipeline calibration, quality control, and archiving of calibrated data for a ground-based observatory. The role of instrument models for observation preparation and instrument performance control is reviewed.

1. Introduction

The general framework for handling VLT data operations and processing is the Data Flow System, which is composed of a number of subsystems and foundation layers. The Data Flow System (Grosbøl & Peron 1997) includes all services involved, from observation preparation to data analysis. In this paper we present the pipeline and quality control parts of the system.

2. Pipeline Processing

For pipeline processing, the header structure of the data is interpreted by a Data Organizer to associate the individual Frames, identify Reduction Recipes, and create Reduction Blocks. There will be four kinds of pipeline procedures:

- **Calibration Pipeline** will process the data acquired during technical programs and prepare pre-calibrated solutions. Solutions may also be derived from calibration data generated during the observing run. Derived data may individually be submitted to the calibration archive.
- **Reduction Pipeline** performs a quasi real-time calibration of scientific data obtained by supported templates. When possible, the Reduction pipeline applies pre-defined calibrated solutions to guarantee stable performance. When calibration data are taken during an observing program, the quality control pipeline will verify the adequacy of the derived data for this configuration.
- **Quality Control Pipeline** provides quality assessment for instrument performance data and observing conditions, and tests the results of the Reduction Pipeline.

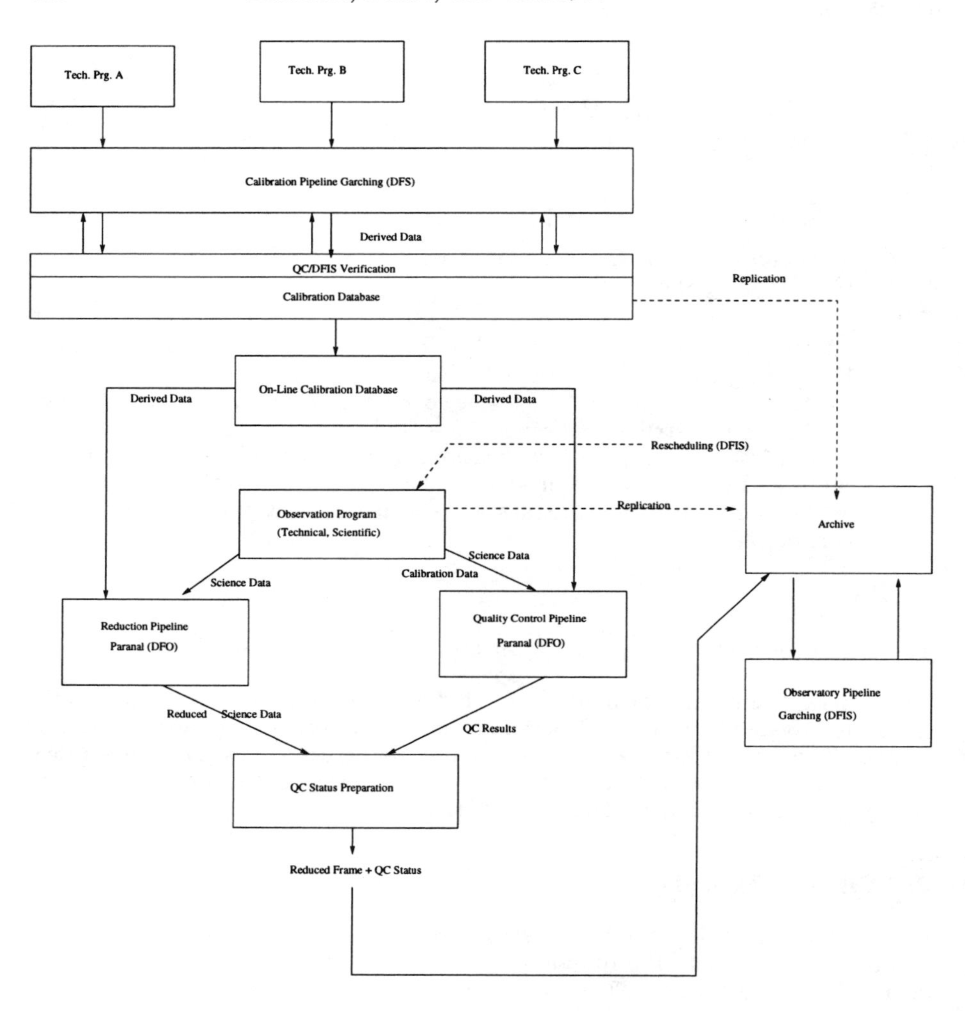

- **Observatory Pipeline** allows re-reduction procedures based on better derived data and algorithms (re-calibration, on-the-fly calibration, final archive preparation, etc.).

3. Quality Control

Quality Control verifies the conditions and the results with which the Reduction Pipeline performs its task. To this end, the Quality Control subsystem verifies the instrumental and observational conditions and compares them to the specifications of the Reduction Pipeline. In addition, Quality Control performs some verification of the data processed by the Pipeline. Instrument performance, observing conditions, and pipeline results contribute to the estimation of a Quality Status attached to each Reduced Frame.

The following levels of testing will be performed in the context of Quality Control. These tasks are presented in a sequence corresponding to the expected level of automation: (*i*) completion testing, (*ii*) routine testing: detector, calibration lamps, arc lines, (*iii*) instrument function testing: focus, instrument response, and (*iv*) specific testing: scattered light, non-linearities. Some of these tasks will be implemented in the Quality Control pipeline and be executed during observations. Others will be implemented as interactive procedures.

An important aspect of Quality Control will be the production and maintenance of instrument models for the purpose of exposure time calculation and instrument control.

4. Instrument Model Framework

The development of an instrument model framework consists of a close collaboration between the ESO Data Management Division (DMD) and Space Telescope European Coordinating Facility (ST/ECF). The VLT and the HST facilities have quite similar demands, because both astronomical observatories are committed to rapid dissemination of data from a variety of instruments to the world-wide community at a large.

One of the VLT instruments under construction is the high resolution echelle spectrograph UVES; first light is planned for 1999. The DMD model for this instrument now succeeds in predicting the geometrical aspects of observational data to better than one resolution element (pixel) of the detector. In parallel, the ST/ECF has produced a computer model for the low-resolution Faint Object Spectrograph (FOS) on HST (Rosa 1995). This software is optimized for simulating internally scattered light, which is a serious nuisance for observations of faint targets.

A direct result of such models is exposure time calculators, which observers can use to estimate the length of each exposure when preparing an observing program. To allow wide access by the scientific community to such tools, the software for these calculators is being made available on the Internet.

The success of these first modeling experiments has led to the definition of a common framework for the development of such models, and to the creation of a versatile software package and associated database. Within this environment, a slight modification of the UVES software was efficiently re-used to model an existing high-resolution spectrograph—CASPEC at the ESO 3.6-metre telescope—and is currently being transformed into a model for the STIS spectrograph on HST.

The next steps will be to provide models for all those instruments that will become operational on the VLT and the HST in the coming years, and to study further the impact of the improved calibrations on new data analysis techniques.

References

Grosbøl, P., & Peron, M. 1997, this volume, 22

Rosa, M. R. 1995, in Calibrating and Understanding HST and VLT instruments, ed. P. Benvenuti, ESO/ST-ECF Workshop, ESO, 43

Astronomical Data Analysis Software and Systems VI
ASP Conference Series, Vol. 125, 1997
Gareth Hunt and H. E. Payne, eds.

The ESO VLT CCD Detectors Software

Antonio Longinotti

European Southern Observatory, Garching bei München, D-85748 Germany, E-mail: alongino@eso.org

Abstract. Charge Coupled Devices (CCD) are currently by far the most widely used type of detector in astronomy. At the ESO Very Large Telescope (VLT), on mount Paranal in Chile, about 40 technical CCD cameras will be in operation for auto-guiding, field viewing, and wavefront sensing, as well as more than ten scientific CCD cameras for instruments working at optical wavelengths. After a brief introduction to the VLT Control System, the VLT CCD Detectors Control Software is presented.

1. Introduction

The VLT Control System is a distributed system consisting of a set of UNIX Workstations, dedicated to high level operations, such as coordination of subsystem activities and interface to the users, and VME-based Local Control Units (LCU), dedicated to the control of sub-systems hardware. The operating system used on the LCUs is VxWorks. All computing units—Workstations and LCUs—are connected through LANs.

The CCD Software Package (Longinotti et al. 1995) is built on top of the VLT common software, which is a layer of services (drivers, libraries, utilities) used by all VLT applications (Raffi 1995). It has been designed as one package to control all CCD cameras, both scientific (SCCD) and technical (TCCD). In this way, costs for software maintenance are reduced and the interface to applications is standardized, in that instrumentation and telescope software interface to all cameras in one and the same way (§4). It became available as part of the VLT Software in October 1995, and four releases have been issued since then. It is currently used at the ESO New Technology Telescope for auto-guiding (two TCCDs), image analysis for active optics (two TCCDs), image quality assessment (two TCCDs), the EMMI instrument (two SCCDs and one TCCD for slit viewing), and the SUSI instrument (one SCCD). It is also used by the VLT FORS instrument (one SCCD), and in ESO labs for SCCD and TCCD new chip and camera tests and verification.

2. System Architecture

In addition to the two standard platforms (Workstation and LCU) described above, a third—called Array Control Electronics (ACE), and based on a transputer network and DSP—is used to control the CCD camera head.

The software running on each platform has the following characteristics:

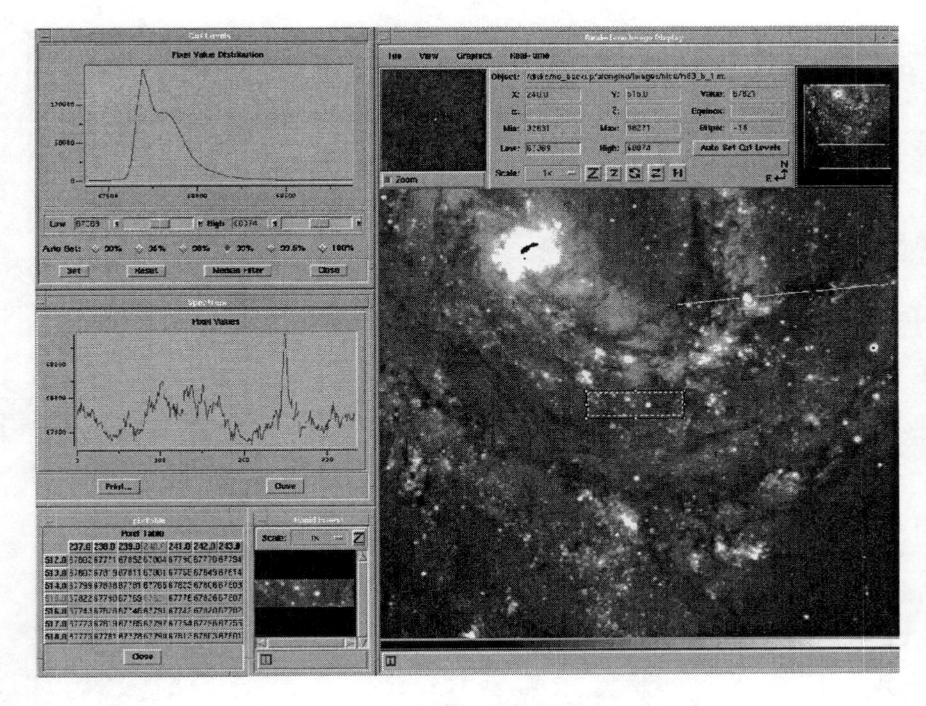

Figure 1. VLT CCD and Real-Time Display.

- **Workstation**. It includes programmatic and interactive user interface.

  The language used is C, with Tcl/Tk used for GUIs.

- **LCU**. The core of the camera system control runs here. Performance is optimized in that operations such as read-out from ACE and image transfer to the Workstation, are performed in parallel whenever possible. Image data are temporarily saved in the LCU memory (e.g., for image re-transmission in case of a network failure).

  The language used is C.

- **ACE**. The transputer and DSP based embedded software runs here. It provides the direct interface to the camera electronics.

  The languages used are Occam (transputers) and C (DSP).

3. Functionality

Compared to the previous generation of ESO CCD systems, in addition to the standard functionality (different readout modes and speeds provided, binning, windowing, execution and control of an exposure, possibly repeated n times, storage of data in FITS files, telemetry, and temperature control), it implements:

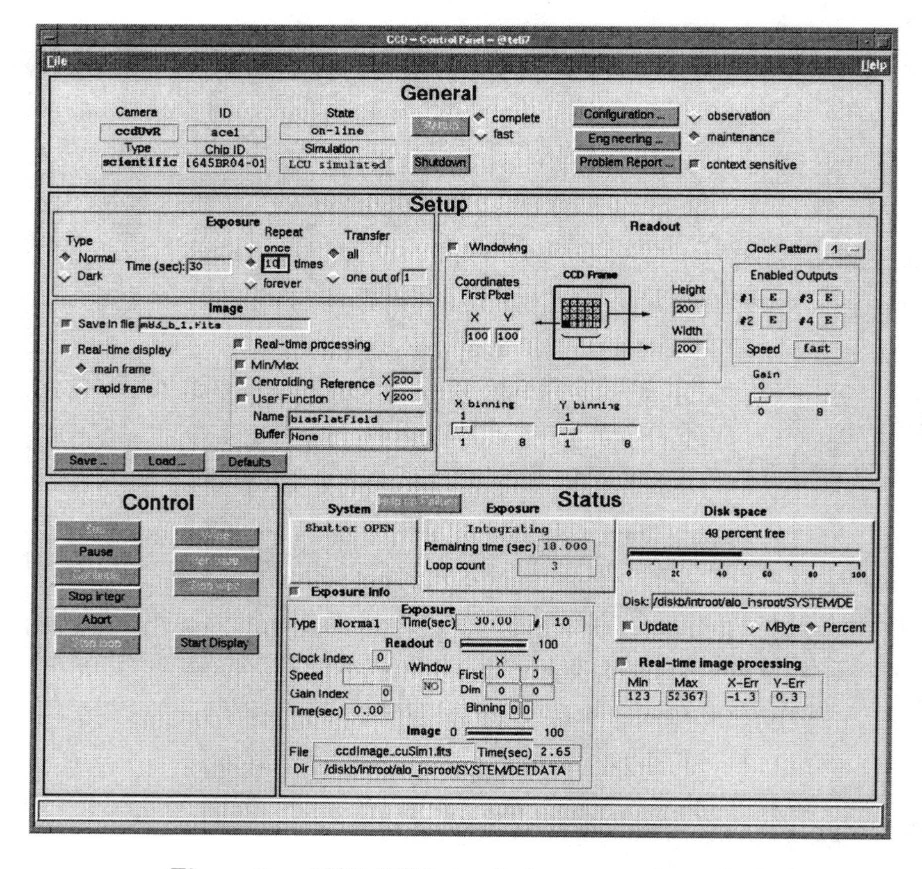

Figure 2. VLT CCD stand-alone control panel.

- **Real-time image processing** on LCU for TCCDs (image quality improvement, centroiding calculation, hook for user function implementing special algorithms).
- **Gain setting** as part of the user setup.
- **Parallel readout from up to four outputs**. The physical image is reconstructed within the LCU.
- **Display of images** while they are being read-out, on any X-11 terminal through the **VLT Real-Time Display** facility (see Figure 1).
- A **Graphical User Interface** (GUI), to control and monitor the status of a CCD camera as a simple stand-alone instrument—extremely useful for laboratory and field testing (see Figure 2 and §5).
- Support for **special control hardware** on LCU, such as the high precision shutter for the FORS instrument.

- Support for **new generation controllers**. About 80% of the LCU software is independent from the particular controller connected to the LCU VME crate, and can therefore be re-used with other controllers.

4. Interface to External Software Packages

The interface to external software is well defined and consists of standard components within the VLT software:

- **Command/Replies**, based on the VLT Message System.
- **Setup files**, containing exposure setup definitions.
- **FITS files**, containing the images produced as the result of exposures.

5. Usage as Stand-alone Instrument

The CCD Software is able to work as a simple stand-alone instrument through a control panel, built using the **VLT panel editor**, which is based on Tcl/Tk (see Figure 2). Among others, it allows the following actions:

- Define a single exposure setup and save it in a setup file, or retrieve already defined setup files.
- Define a sequence of exposures (e.g., for standard calibration operations).
- Start exposures, monitor their status, and possibly stop, pause, or abort them.
- Display images with the VLT Real-time Image Display.
- Interface to VLT data flow.

Acknowledgments. I would like to thank here the people who contributed to the development of the VLT CCD software, in particular, C. Cumani and P. Duhoux.

For more information or questions, please contact A. Longinotti (e-mail to alongino@eso.org).

Information can also be retrieved through anonymous ftp.[1]

References

Raffi, G. 1995, The Messenger, 81 , 5

Herlin, T., Brighton, A., & Biereichel, P. 1995, The Messenger, 81, 6

Longinotti, A., Cumani C., & Duhoux, P. 1995, The Messenger, 82, 7

[1] ftp://te1.hq.eso.org/vlt/pub/doc, files ccd*

Astronomical Data Analysis Software and Systems VI
ASP Conference Series, Vol. 125, 1997
Gareth Hunt and H. E. Payne, eds.

The Observatory Monitoring System: Analysis of Spacecraft Jitter

Peter Hyde, Richard Perrine(CSC), and Kenneth Steuerman

Space Telescope Science Institute, Baltimore, Maryland

Abstract. The Observatory Monitoring System (OMS) is part of the Hubble Space Telescope (HST) ground system at the Space Telescope Science Institute. OMS receives as inputs a file of commands for the HST and files of engineering telemetry data generated by the HST in response to those commands. This paper will discuss how, from its inputs, OMS software provides spacecraft jitter files during observations commanded in the input. Analysis of these jitter files helps astronomers assess the quality of science data obtained during the observations.

1. General Overview

The main purpose of the Observatory Monitoring System(OMS) is to produce files describing the "jitter" of the Hubble Space Telescope as a function of time, so that an astronomer using the telescope will know how much variation there is in the telescope pointing during a science observation. The telescope has its own rectangular coordinate system with three mutually-perpendicular axes—V1, V2, and V3—where V1 coincides with the main axis of the telescope. Since the jitter is the variation in the pointing of the V1 axis, its components can be along the V2 and V3 axes.

The OMS program will compute these components from two inputs. The first is an Event File, which indicates the times of events relating to the telescope's activity. The Event File is generated by parsing a schedule of planned events, which is provided every week to the Operations staff at the Space Telescope Science Institute. The second is a sequence of engineering telemetry files of data sent back by the telescope in response to events commanded (as indicated in the Event File). OMS reads telemetry from the data files and determines what telemetry goes with each observation indicated in the Event File.

The jitter is computed from telemetry data over the time range of the science observations, as indicated in the Event File. For each such observation, OMS produces a jitter table showing the computed jitter as a function of time and a jitter image showing the spread of the jitter during the observation. These will show the astronomer how stable the pointing of the telescope was during a given observation. OMS is run shortly after the science data processing is completed, so that the science data can be evaluated along with the jitter files. The jitter files are often referred to as "observation logs."

OMS can handle several observations in parallel. The present OMS provides for ten instruments, counting each of the three NICMOS detectors as a separate instrument. The only restriction is that OMS may not process two observations at once for the same instrument.

2. Inputs

The main inputs to the OMS program are the Event File and the engineering telemetry files. The Event File provides times for events such as guide-star acquisition, slew, beginning and end of observations, beginning and end of "night," etc. An engineering telemetry file may be in any of several formats, and the format determines the rate of telemetry update and the set of possible telemetry items which may be read from that file. For each such file, the format and start/end times are read from the telemetry file header. Each engineering telemetry file is divided into minor frames, where a minor frame represents a set of telemetry items tagged with the same time.

3. Processing

Some of the telemetry in an engineering telemetry file may be bad, so OMS has a preprocessor which checks the items read from the telemetry and creates an error file to tell the OMS main program which telemetry items should be dropped. Every minor frame must have a valid time. As the data are read from telemetry, the telemetry times are coordinated with the times of the events read in from the Event File, marking beginnings and ends of observations, guide-star acquisitions, etc. At the beginning of each telemetry file, the start time is checked to see whether there is a significant time gap between that time and the last time in the previous engineering telemetry file. If there is, then special processing is done to account for any observations that should have occurred in the time gap, so that no planned observations will be unaccounted for. In addition, there is an array to accumulate the gap time for each observation.

Each minor frame in the telemetry file is processed, unless it is to be dropped (i.e., marked as "bad"). Each telemetry parameter in the minor frames that are not dropped is checked to see whether its value is within range and suitable for use. Some parameters are used to determine guiding status, some to calculate position, some to calculate velocity, etc. As each good parameter is received, the appropriate routines are called in order to calculate or update appropriate items. These include routines handling guide-star acquisition and calculation of the jitter components.

When the guiding status is Fine Lock or Coarse Track, the jitter quantities are calculated and written out to the high-resolution and low-resolution jitter tables of all observations in progress. The high-resolution table has a record of the jitter quantities for every telemetry time during the observation. The low-resolution file has a record only once every three seconds, and some of the values in the record will be averages over the previous three-second interval. At the end of each observation a routine is called to update these files and create the jitter image file.

The jitter image file will be archived for the user. For any observation less than six seconds in length (a "short" observation) the high-resolution jitter table is archived as the jitter table file. For any observation at least six seconds long, the low-resolution jitter table is archived as the jitter table file.

Sometimes there will be minor frames with a valid time but no valid telemetry for computing the v2 and v3 jitter components. Sometimes we have not yet had guide-star acquisition, or a slew or recentering is in progress. In such cases, an item in the jitter table records may not have a valid value and will have "INDEF" entered as its value.

4. Outputs

For each science observation, OMS produces and archives a jitter table file and a corresponding jitter image file. OMS has recently been updated to satisfy the FITS standard, so that both the table files and the image files are being written in FITS format. The jitter image file produced by the OMS analysis program has a primary header followed by an image extension header followed by the image. All of the keywords and values in the .JIH files produced by the old OMS program are either in the primary header or in the image extension header.

For an "internal" observation, or for one which has no valid telemetry data, the image in the jitter image file will be empty. For an "external" observation (as with a normal target) where there is some telemetry, the image is a 64×64 image. In either case, the jitter image file will have complete primary and image extension headers. The jitter image is essentially a histogram of V2 and V3 jitter components. The "brightness" of the image at point (V2, V3) represents how often the telescope jitter value was (V2, V3). The jitter image may be used for PSF analyses or for image deconvolution. The image may be displayed in IRAF with SAOimage, if the IRAF has the FITS kernel. The image header file for each observation contains valuable summary data for that observation.

The jitter table files are in STSDAS binary table format, so that one may use the IRAF commands TREAD or TPRINT to read or copy them. OMS now produces these files in FITS format, so one needs a FITS-kernel version of IRAF to read them.

The jitter table files and jitter image files are archived via the DADS system at the Space Telescope Science Institute. Files no longer considered proprietary may be retrieved by any user. The Institute is creating a new WWW page to facilitate retrieval of these archived files. Meanwhile, these files may be retrieved via Starview. If you want to retrieve the OMS archived files for a science observation [obs-name], you will get the jitter table file and the the jitter image file for [obs-name]. Note that [obs-name] is an eight-character observation name associated with the science data.

The items listed in Table 1 are written to each record of the jitter table file. Some values may be three-second averages. Items marked * have no defined value for short observations. For short observations, items marked ** are current values, not averages.

Acknowledgments. We acknowledge the work of Xuyong Liu (Space Telescope Science Institute) and Bruce Toth (formerly of Space Telescope Science Institute) for developing the algorithms and the original version of the OMS software.

Table 1. Items in Each Time Record of Jitter Table File.

Item	Parameter	Units	Description
1	Time	seconds	time since observation start
2	V2_dom	arcsec	V2 coordinate of dominant FGS
3	V3_dom	arcsec	V3 coordinate of dominant FGS
4	V2_roll	arcsec	V2 coordinate of dominant FGS
5	V3_roll	arcsec	V3 coordinate of dominant FGS
6	SI_V2_AVG	arcsec	avg V2 jitter over 3 seconds **
7	SI_V2_RMS	arcsec	RMS v2 jitter over 3 seconds *
8	SI_V2_P2P	arcsec	Peak-to-peak v2 jitter *
9	SI_V3_AVG	arcsec	avg V3 jitter over 3 seconds **
10	SI_V3_RMS	arcsec	RMS v3 jitter over 3 seconds *
11	SI_V3_P2P	arcsec	Peak-to-peak v3 jitter *
12	RA	degrees	Right Ascension of aperture ref.
13	DEC	degrees	Declination of aperture ref.
14	ROLL	degrees	Angle between North and V3 axis
15	LimbAng	degrees	Angle between Earth Limb & Target
16	TermAng	degrees	Angle between Terminator & Target
17	Latitude	degrees	Latitude of HST subpoint
18	Longitude	degrees	Longitude of HST subpoint
19	Mag_V1	gauss	Magnetic field along V1
20	Mag_V2	gauss	Magnetic field along V2
21	Mag_V3	gauss	Magnetic field along V3
22	EarthMod		Model earth background light
23	SI Specific	varies	Varies with current science instrument
24	DayNight	0 or 1	0 for Day, 1 for Night
25	Recenter	0 or 1	1 if recentering
26	TakeData	0 or 1	1 if taking data
27	SlewFlag	0 or 1	1 if slewing

Astronomical Data Analysis Software and Systems VI
ASP Conference Series, Vol. 125, 1997
Gareth Hunt and H. E. Payne, eds.

Refining the Guide Star Catalog: Plate Evaluations

Oleg M. Smirnov and Oleg Yu. Malkov

Institute of Astronomy, Russian Academy of Sciences, 48 Pyatnitskaya St., Moscow 109017, Russia

Abstract. This is a preliminary report on an investigation of the quality and properties of plates digitized during creation of the Guide Star Catalog (GSC). The results will be vital for reclassifying GSC objects.

1. Introduction

The GSC was created at STScI to support Hubble Space Telescope observations. It contains about 20 million objects, making it the largest all sky photometry source to date. The GSC was created by digitizing 1593 plates and by including bright stars from the HIPPARCOS INCA database. Many objects are measured on more than one plate, and thus have multiple catalog entries. We call such objects multiple-entry objects, or MEOs.

Among the shortcomings of the GSC are: biased magnitudes and incorrect classifications for certain objects, presence of artifact objects. These shortcomings hinder many interesting applications of the GSC, e.g., those involving stellar counts. The Refined GSC (RGSC) project currently in progress at the Center for Astronomical Data, Institute of Astronomy, Moscow,[1] is an attempt to rectify these shortcomings.

2. RGSC: a Refined GSC

The ongoing RGSC project is aimed at creating a new catalog, RGSC, based on the GSC. RGSC will contain all GSC data, plus, for many objects, corrected magnitudes and more detailed and (hopefully) correct classifiers, complete with confidence-of-classification ratings. The primary effort involves verification and reclassification of GSC objects. The following approaches are used:

1. Cross-identification with other catalogs and databases (Malkov & Smirnov 1997). These results are likely to be quite valuable on their own.
2. Multiple-plate analysis (MPA). A large number of objects (MEOs) are registered on more than one plate, and thus have several catalog entries. Each plate has, in effect, its own "opinion" on the nature of a MEO, expressed in that plate's entry for the object. Comparative analysis of multiple entries usually yields insight into the true nature of an object.

---

[1] http://www.inasan.rssi.ru/CAD

3. Probability maps showing the average density of objects of a particular class for given coordinates and magnitude will be built.

4. An expert system will be developed that evaluates results of approaches 1–3 and produces final classifications and confidence estimates.

For any MEO, properties of the individual plates on which the object is registered have at least as much of an effect on the resulting GSC entries as the nature of the object per se. Our initial studies show that these properties vary a great deal from plate to plate. Therefore, detailed analysis of individual plate characteristics is a prerequisite to carrying out accurate multiple-plate analysis of GSC objects.

3. Plate Analysis: Issues and Methods

To determine the likelihood of an object of magnitude M appearing on a given plate, we compute the luminosity function for the plate, calculating limiting magnitude, saturation magnitude, and maximal population magnitude.

To determine the likelihood of an object being misclassified on a given plate, and whether some plates are "special" in that they tend to bias classifications, we calculate $3 \rightarrow 0$ and $0 \rightarrow 3$ tendencies, or $P\{0|3\}$ and $P\{3|0\}$. The former reflects the tendency of a plate to misclassify extended (class 3) objects as stellar (classifier 0), and the latter the reverse. The $3 \leftrightarrow 0$ tendencies are computed using an iterative process. Initial studies demonstrate that $3 \leftrightarrow 0$ tendencies of the majority of plates are a very significant factor in GSC classifications. In particular, as seen in Figure 1, the $N_{non-stellar}$ to $N_{stellar}$ ratio depends on galactic latitude. The best fit to the average ratio as a function of galactic latitude is a sum of a Gaussian and a very small linear component:

$$N_{ns}/N_s = 0.372 \cdot e^{-0.5(|b|+1.44/15.7)^2} + 0.0187 + 2.98 \cdot 10^{-3} \cdot |b|$$

To determine whether the magnitude of objects is biased when measured near a plate edge, we calculate an average magnitude and flux as a function of distance from the plate center (separately for objects of both classes). To determine whether some plates bias magnitudes, for overlapping plates we compute the mean magnitude discrepancy among objects that appear on both plates. To determine whether an object's classification is biased if the object is near a plate edge, we calculate the average density of objects of both classes as a function of distance from the plate center. To determine whether GSC plate quality codes are meaningful, we look for a correlation between plate quality codes and the parameters mentioned above.

We examined whether an object at given coordinates should be expected to appear on a given plate (i.e., determining the area of the sky that the plate really covers). Plate centers are listed in the GSC, and plates are supposed to have a regular (square) shape of a known size. However, most of them have "dead zones"—irregular sections with no objects registered, e.g., clamp marks (evidently the result of scanning technique), broken-off corners, circular or rectangular areas near bright stars and globular clusters (manually removed during GSC production), etc. We produced a "plate atlas" (plots of all the objects registered on a plate), and developed algorithms to determine the "true" boundary of a plate using (a) the plate atlas, (b) actual GSC data, and (c) lists of bright stars, clusters, and other specific objects.

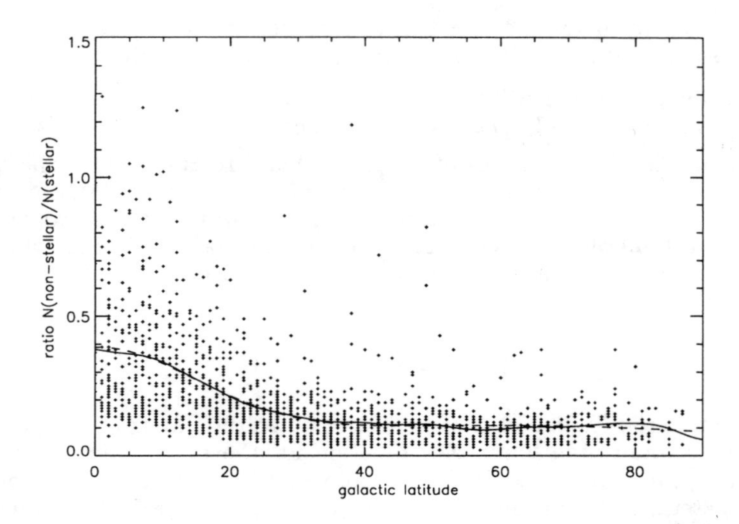

Figure 1. GSC plates: classifier ratios. Solid line is smoothed data, dashed line is the best fit.

4. Plate-related Effects

A number of problems can be traced to the plate-related effects discussed above. A stationary object overlapped by a given plate is not measured on the plate:

Possible causes:	Hit a "dead area."
	Brightness is below the limiting magnitude.
	Brightness is below the saturation magnitude.
Affects:	Multi-plate analysis.

The magnitude of an object (as measured on a given plate) is biased:

Possible causes:	The plate tends to bias magnitudes.
	The object is near the edge of the plate.
Affects:	Stellar counts.

An object is misclassified (and its magnitude is possibly biased):

Possible causes:	The plate biases classifications.
	The object is near the edge of the plate.
Affects:	Multi-plate analysis, stellar counts.

Acknowledgments. This presentation was made possible by financial support from the Logovaz Conference Travel Program. OM is grateful to the Russian Foundation for Fundamental Researches for grant No. 16304.

References

Malkov, O. Yu., & Smirnov, O. M. 1997, this volume, 298

Astronomical Data Analysis Software and Systems VI
ASP Conference Series, Vol. 125, 1997
Gareth Hunt and H. E. Payne, eds.

ZGSC (Compressed GSC) and XSKYMAP

Oleg M. Smirnov and Oleg Yu. Malkov

Institute of Astronomy, Russian Academy of Sciences, 48 Pyatnitskaya St., Moscow 109017, Russia

Abstract. A losslessly compressed version of the Guide Star Catalog has been created by the authors and described in the present paper. Supporting software is also discussed.

1. Introduction

The well-known Guide Star Catalog (GSC) is distributed on two CD-ROMs (about 1.2 GB). This leads to a relatively slow access time; furthermore, it is difficult to keep the GSC completely on-line (it requires a lot of hard drive space, or a jukebox, or at least two CD-ROM drives). Furthermore, the actual data in the catalog is not easily accessible. It is in the form of FITS tables, and the coordinates are given in one standard system (J2000.0). To help PC users solve the problem of data retrieval, we have created the Guide Star Catalog Data Retrieval Software, or GUIDARES (Malkov & Smirnov 1995). This is a user-friendly program which lets one easily produce text samplings of the catalog and sky maps in Aitoff or celestial projections. The main function of GUIDARES is to produce an ASCII table of object entries from a specified region, and, optionally, a graphical sky map of the region. It can handle rectangular and circular regions in four different coordinate systems. GUIDARES is also available under UNIX as a C library, with a graphical interface based on IDL widgets.

2. ZGSC

We have created a compressed version of the GSC, called ZGSC. By using a binary format and an adaptive compression algorithm, the GSC was losslessly compressed by a factor of six, giving the ZGSC a total size of about 200 MB. This makes it possible to store the ZGSC on-line on a hard disk for a dramatic improvement in access time.

An extensive software package was developed to work with the ZGSC. This includes a suite of IDL routines that retrieve data from the ZGSC into IDL arrays, and supporting C libraries for on-the-fly decompression of the catalog. The software facilitates retrieval of circular regions, specified by center and size. Four coordinate systems are supported: equatorial and ecliptic (any equinox), galactic, and supergalactic. The software also allows retrieval of objects of a particular type and/or in a particular magnitude range.

In addition, we have developed a WWW interface to the ZGSC,[1] which allows for quick visualization of data from ZGSC over the WWW. Using a forms-capable WWW browser, the user may define an area in any coordinate system, and receive either a GIF or JPEG image of the selected area reconstructed from the ZGSC, or a ZGSC sampling of the selected area in ASCII, FITS ASCII table, or FITS BINTABLE format.

3. XSKYMAP Software

The XSKYMAP software is an IDL widget application for retrieval, visualization and hard copy of ZGSC samplings. The applications of the XSKYMAP are finder charts, GSC studies (Smirnov & Malkov 1997), etc. XSKYMAP is fully integrated with ZGSC and provides easy access to all retrieval options of the ZGSC. It also provides mouse-based catalog feedback (i.e., click on an object to see the full GSC entry). The software supports mouse operations for zoom in/out and re-center region, and click-and-drag facilities to compute angular separation.

Acknowledgments. This presentation was made possible by financial support from the Logovaz Conference Travel Program. OM is grateful to the Russian Foundation for Fundamental Researches for grant No 16304.

References

Malkov, O. Yu., & Smirnov, O. M. 1995, in ASP Conf. Ser., Vol. 77, Astronomical Data Analysis Software and Systems IV, ed. R. A. Shaw, H. E. Payne, & J. J. E. Hayes (San Francisco: ASP), 182

Smirnov, O. M., & Malkov, O. Yu. 1997, this volume, 426

[1] http://www.inasan.rssi.ru/CAD/zgsc

Astronomical Data Analysis Software and Systems VI
ASP Conference Series, Vol. 125, 1997
Gareth Hunt and H. E. Payne, eds.

Towards Optimal Analysis of HST Crowded Stellar Fields

Peter Linde and Ralph Snel

Lund Observatory, Box 43, S-221 00 Lund, Sweden,
E-mail: peter@astro.lu.se

Abstract. A Nordic group is using the Hubble Space Telescope in a study of stellar populations in the Bar of the Large Magellanic Cloud. Through Strömgren uvby photometry, we determine ages, metallicities, and the luminosity function. We have designed and applied an exposure dithering pattern in order to decrease the effects of undersampling. This also enables a detailed study of detector properties, which is essential for accurate photometry. Careful studies of low level background features are presented. Algorithms developed to analyse the PSF shape reveal variation of this shape with position in the PC field. A comparison verifies improved photometric quality for dithered versus undithered images. The faint end of the luminosity function is studied through application of statistical techniques to very faint background fluctuations.

1. Introduction

A Nordic group (Ardeberg et al. 1997) is using the Hubble Space Telescope in a study of stellar populations in the Bar of the Large Magellanic Cloud (LMC). A total of 35 hours of exposure have been obtained using the WFPC2 camera with Strömgren uvby filters. Through accurate photometry, we determine ages, metallicities, and the luminosity function.

Accurate photometry is essential, and we are investigating various effects affecting the measurements. New algorithms have been developed. Here we present some initial results.

2. Bias Jumps Affecting Image Background

We have noted weak background variations in some of our exposures, typically the shorter ones. Figure 1 shows a set of calibration exposures (with two stars) using the Strömgren uvby filters. In order to see the variations through the noise, the images have been smoothed. The two calibration stars are seen as squares. Typical amplitudes of the background variations are 0.5 ADUs. Aperture photometry results for the v filter, as a function of aperture radius, are shown in Figure 2. The individual frame numbers are given as data points, as well as a standard deviation for each set of points. The spread is larger than expected from photon statistics. A simple attempt to correct for the background variations was made, using an average over all image columns and then reex-

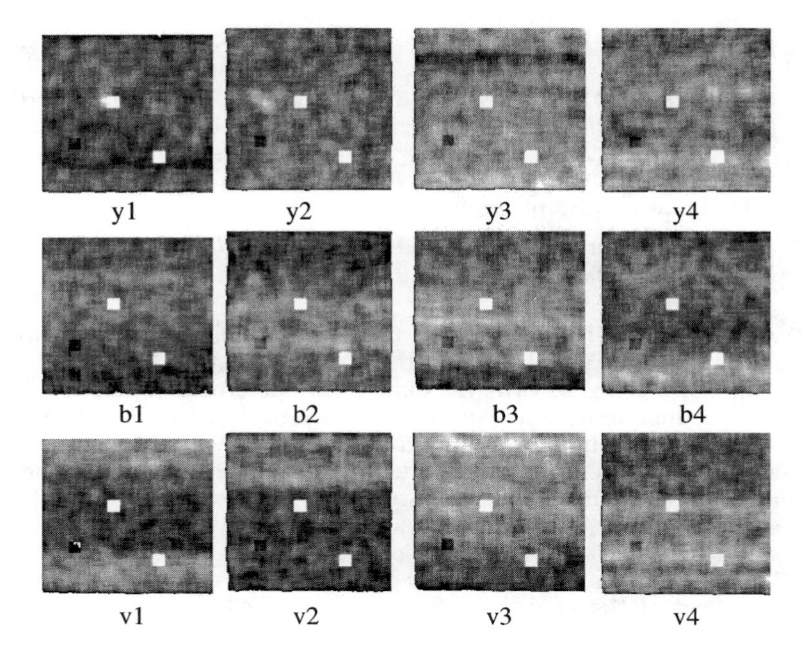

Figure 1. Faint background variations revealed after smoothing. Four images are shown for each filter. The two bright objects are stars.

panding into a background image. In Figure 3, the corresponding results are given. Only a marginal improvement can be noted.

3. Effects of Dithering on Photometry

Sixteen images were obtained of our target LMC field using the F547M (y) filter. For each set of four exposures, consecutive exposures have a quarter pixel offset with respect to the axes of the detector pixel grid. After reconstruction, this effectively improves the sampling of the image.

We have analysed these data using point spread function (PSF) fitting techniques (Daophot/Allstar, Stetson 1987), both as a set of 16 individual exposures and as four sets of dithered images. Figure 5 shows the mean error as a function of magnitude; the dotted line shows non-dithered results and the solid line shows dithered results. A significant improvement in accuracy is noted. The improvement will be more pronounced in our crowded Wide Field Camera (WFC) images, and may be increased further through use of a more sophisticated method of combining the individual dithered images.

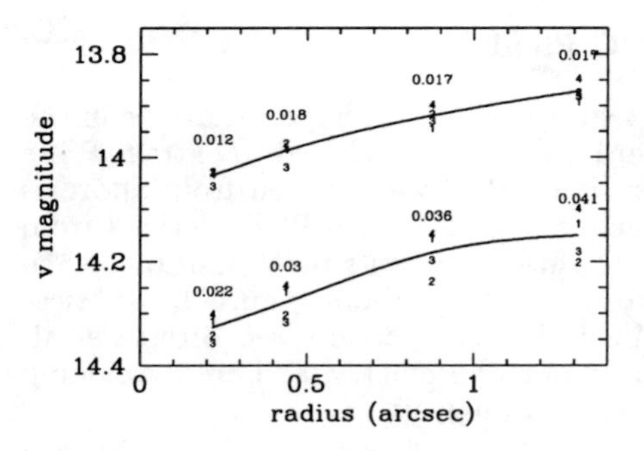

Figure. 2. Aperture photometry magnitude as function of aperture radius.

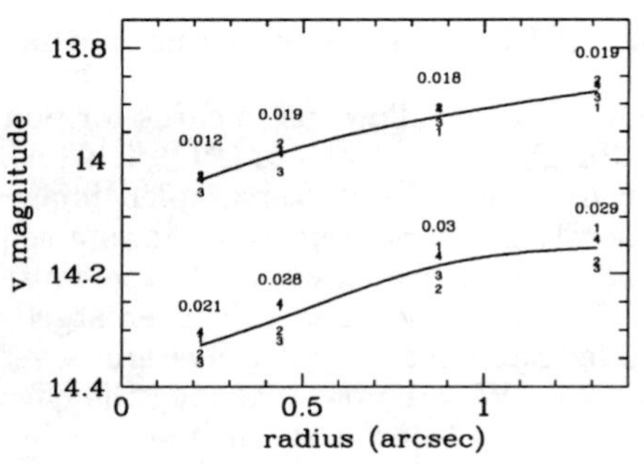

Figure. 3. Same as Fig. 2, but with a correction for background variations.

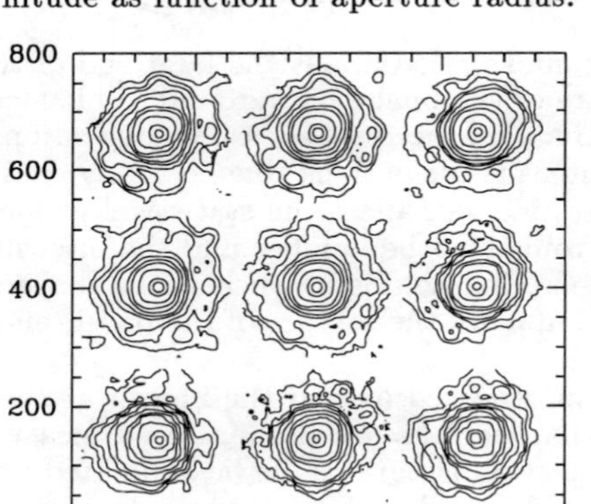

Figure. 4. Variations of PSF shape as function of position on the PC detector.

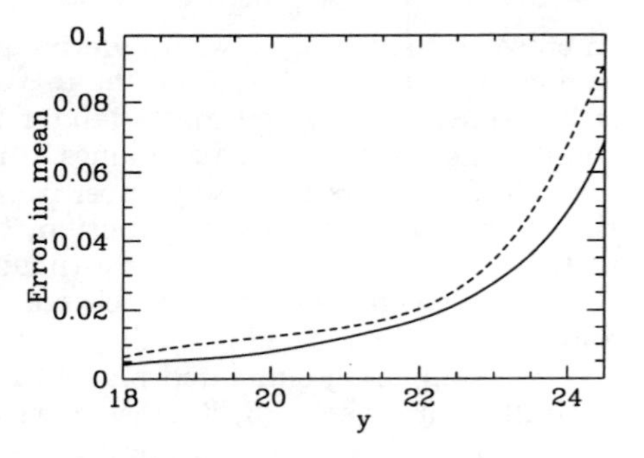

Figure. 5. Dithering effect on photometric accuracy. Full drawn line shows errors derived from 4 dithered images, dashed line from 16 individual images.

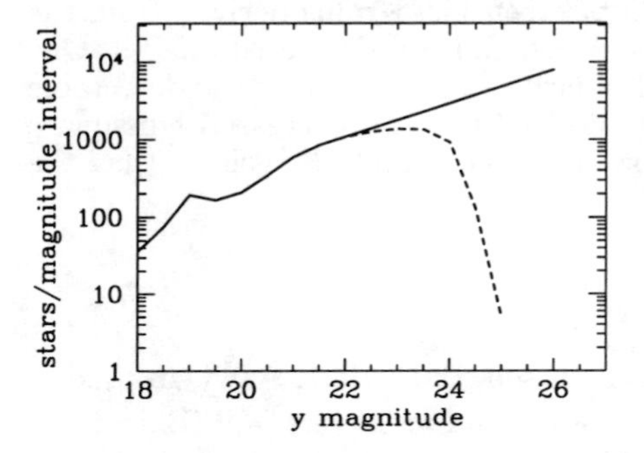

Figure. 6. Luminosity functions. Full drawn line: extrapolated LF. Dashed line: observed, uncorrected LF.

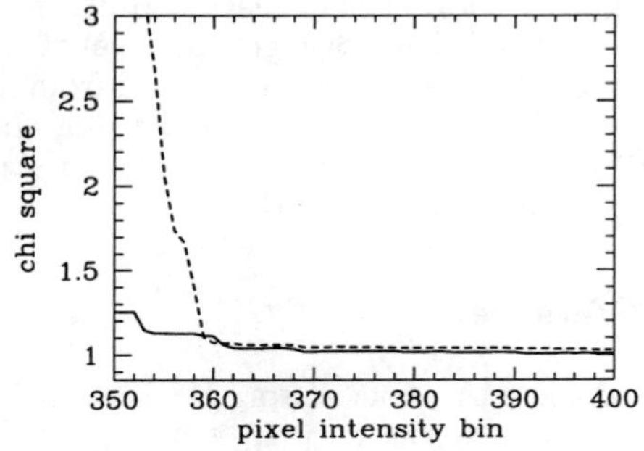

Figure. 7. Goodness of fit for histograms of simulated images, compared to the histogram of the observed image. Full drawn line: extrapolated LF. Dashed line: observed LF.

4. PSF Shape Variations in the PC Field

Our analysis allows us to check for point spread function shape variations in the HST Planetary Camera (PC) field. Algorithms were developed to extract PSFs from dithered and oversampled images. Each PSF was defined from approximately 20 stars. Figure 4 shows a sequence of panels with PSFs derived from different parts of the field. In the figure, the grid represents pixel position on the PC detector, with each PSF enlarged a factor of 20. From Figure 4, a noticeable shape change with position is verified. For comparison, see Biretta et al. (1996). We are now analysing the effect on our photometry, and are developing algorithms to properly include this in the data analysis.

5. The Faint End of the Stellar Luminosity Function

In crowded field images, we are deriving information about the faint end of the luminosity function (LF) from the texture of the image background. This algorithm aims at studying objects fainter than can be reached using conventional completeness estimation techniques. Intensity deviations due to many undetectable faint stars are always positive, which will affect the statistical properties of the pixel intensity distribution. From this, the number and the intensity distribution of faint stars can be estimated. Characteristics such as the shape of the PSF, detector noise properties, etc., influence the texture of the background and have to be modeled carefully.

We compare a simulated image to an observed one. If the images are statistically similar enough, it is assumed that the LFs are similar. As a measure of similarity, the χ^2 measure between the two histograms of the simulated and observed images is used. A lower value of χ^2 indicates a better fit to the histogram. By iteratively adjusting and extrapolating the observed LF, simulated images with increased statistical similarity are created.

Figure 6 shows an example of an observed LF derived from a single WFC image, together with a preliminary model LF. The simple model used has two free parameters: the steepness of the LF and the level of the background. Figure 7 shows the statistical differences between histograms derived from the real image, with a background level of 355 ADU, and two simulated images. One was created using the LF derived from standard measurements of the detectable stars. The other was created using the model LF. The decreased chi-square values for the histogram of the latter shows that the model LF is closer than the observed to the true LF.

References

Ardeberg, A., Gustafsson, B., Linde, P., & Nissen, P.-E. 1997, A&A, in prep.

Biretta, J. A., et al. 1996, WFPC2 Instrument Handbook, vers. 4.0 (Baltimore: STScI)

Stetson, P. B. 1987, PASP, 99, 191

Astronomical Data Analysis Software and Systems VI
ASP Conference Series, Vol. 125, 1997
Gareth Hunt and H. E. Payne, eds.

In-Orbit Calibration of the Distortion of the SOHO/LASCO-C2 Coronagraph

A. Llebaria, S. Aubert, and P. Lamy
Laboratoire d'Astronomie Spatiale-CNRS, Marseille, France

S. Plunkett
Departement of Space Research, University of Birmingham, Birmingham, UK

Abstract. This paper describes distortion calibration procedures for the *SOHO*/LASCO-C2 Coronagraph, based on in-orbit data and extensive image processing methods. It addresses specific problems of externally occulted coronagraphs (obstructed center of field-of-view, strong vignetting, and presence of stray light) and limitations inherent to space-based instrumentation (cosmic rays and limited number of reference points).

1. Introduction

The Large Angle Spectrometric Coronagraph (LASCO) is an instrument aboard the *SOHO* spacecraft, which routinely observes the solar corona in white light (Brueckner et al. 1995). It has now been in successful operation since January 1996 and has produced thousands of images of the corona. LASCO consists of three individual coronagraphs named C1, C2, and C3, each tailored to a specific field-of-view (respectively 3, 6, and 30 solar radii) in order to cope with the tremendous range of brightness of the corona (10^5) and with the problem of rejecting the stray light coming from the Sun. C2 and C3 are externally-occulted coronagraphs, where an external occulter blocks direct solar illumination of the objective. C2 suffers from moderate optical distortion, which must be corrected in order to properly locate the stars crossing its field-of-view (which are subsequently used for photometric calibrations), to retrieve critical attitude parameters, and to insure a correct overlapping of the region common to the C2 and C3 instruments. Practical considerations prevented an accurate calibration of the optical distortion before launch. In-flight calibration has been performed using stars as positional references, and the procedures are described in this paper.

2. Specific Constraints of the LASCO/C2 Instrument

The LASCO/C2 coronagraph has a field-of-view of 1.5° which is imaged onto a 1024×1024 pixels CCD detector so that the pixel field-of-view amounts to 11.5″. It presents a classical barrel distortion, whose amplitude reaches 5 pixels in the corners of the images. The goal of the correction is to restore real positions to within a fraction of a pixel across the whole field-of-view.

Classical correction methods used by the photogrammetry, computer vision, and astronomy communities cannot straightforwardly be implemented for several reasons: (*i*) the lack of extensive ground calibrations (a few images of a grid pattern were obtained but later found unsuitable for analysis), and (*ii*) the central obstruction created by the external occulter which prevents access to the optical axis and the center of distortion.

As a consequence, we decided to rely on stars for absolute positional references. Although C2 is extremely sensitive and detects stars as faint as magnitude 9, a limited number of them (15 in the best cases) is available as reference points for a given image. Therefore, we had to use a large set of images to alleviate this limitation and introduce a sufficient number of reference points.

The calibration of the optical distortion is a particular case of the general calibration of an optical instrument, which consists in determining: (*i*) the so-called internal parameters, which define the correspondence between the ideal and the real images, and (*ii*) the so-called external parameters, which relate the internal coordinates of the ideal images to the outside world.

A favorable point is that we can apply a simple axi-symmetric model to relate the ideal, undistorted images with the real images, because it is able to represent the actual distortion of LASCO-C2 with sufficient accuracy. As a consequence, the shifts due to distortion are constant in moduli over circular annuli and directed to a common center. We can assume that this center coincides with the optical axis.

We adapted the "*Two-Step*" method introduced by Weng, Cohen, & Herniou (1992) as the most appropriate to our problem, given the above constraints. In its general form, this method involves a direct solution for most of the calibration parameters and an iterative solution for the other ones. In our case, we implemented the following procedure:

1. star detection by correlation of successive frames and star identification,
2. initial estimation of the optical parameters (e.g., focal length) and orientation of each individual frame, and
3. fine determination of the intrinsic distortion parameters and estimation of the external parameters from a large set of images.

3. Detection of Stars

Because the method relies on identified stars in image fields as absolute references, we need a reliable detection and identification procedure. Here we define detection as the determination of a star position on the CCD frame and identification as the link between detection (from frame) and star (from catalog).

The actual detection procedure looks for local maxima above local thresholds. However, the CCD images are spoiled with many cosmic ray traces, and the stars are severely sub-sampled, so that it is impossible to reliably detect stars in isolated frames without external information. To validate local maxima as star detections, we overlap detections from successive frames, applying the shift due to their relative displacement which results from the motion of the Sun on the sky. The operational procedure builds up a map of local maxima for each frame. It shifts and adds up five successive maps (centered on the third map). A maximum found in at least three maps is considered a positive detection.

4. Star Identification

Reference stars are obtained from a catalog of star positions limited to the equatorial band spanned by the LASCO-C2 telescope during the year, and limited to visual magnitude 8.5. Typically, 5 to 15 stars are detected in each frame for the period between 2 February 1996 and 29 April 1996.

Each reference map is scaled to the image frame using the initially estimated parameters (internal and external) and identifications are performed using a proximity criterion.

5. First Estimates of Linear Parameters and Distortion

To avoid the mutual interference between distortion and linear parameters, the procedure divides the field (bound by a circle of diameter 24 mm in the focal plane) in four concentric annuli of increasing radius (7, 8, 9, and 10 mm) and fixed width (2 mm) and assumes for each of them a constant equivalent focal length, to be determined later. Then, for each annulus, a minimum linear squares regression (MLSQ) between the reference stars and their associated detections is applied over the full set of images. The process yields estimates of the four equivalent focal lengths, one for each annulus.

The simple model[1] $\delta(\rho) = \rho_m - \rho = a_0\rho + a_1\rho^3$ gives us a first estimate of the radial distortion.

6. Refinement Process

In order to improve the above estimates, we rescale the star map by applying the distortion law and we redo the identification process over the full frame (i.e., without introducing the four annuli). This new association, together with a MLSQ determination of external parameters (residual translation and rotation), allows refining the values of the orbital attitude parameters of LASCO and of the distortion differences between the linearly rescaled references and their related detections. Finally a MLSQ regression of these distortion differences to the polynomial expression $\delta(\rho) = \rho_m - \rho = a_0\rho + a_1\rho^3 + a_2\rho^5 + \ldots$ produces the final improved estimates of the distortion coefficients.

The above method could be repeated iteratively to improve the estimation. In our case this was found unnecessary, as the residuals were small enough (i.e., ≤ 1 pixel) for our purpose.

7. Results

The procedure was applied to a set of 250 images of size 1024×1024 pixels. They were chosen for the same instrumental configuration of parameters (filter, polarizers, size, and exposure time). The initial estimate yielded the pixel[2]

---

[1] note the addition of the term $a_0\rho$ to the classical formula, in order to allow for positive and negative differential distortions with respect to the mean radial distance $\rho = (-a_0/a_1)^{1/2}$ where the equivalent focal length is nominal (i.e., 364 mm)

[2] The bottom left corner pixel has coordinates (0,0)

(512,506) for the optical center and distortion parameters $a_0 = 4.6881.10^{-3}$ and $a_1 = -1.084.10^{-4}$ (where ρ is in mm).

The evolution of the estimation for successive refinements and for different data sets are given in the following table:

Period	Itera tion	a_0	a_1	a_2	Number of references
Apr 1996	0	$4.6881.10^{-3}$	$-1.0840.10^{-5}$	-	1589
Apr 1996	1	$6.5404.10^{-3}$	$-1.4441.10^{-5}$	$2.2103.10^{-7}$	1589
Apr 1996	2	$6.0619.10^{-3}$	$-1.4672.10^{-5}$	$2.0899.10^{-7}$	1589
Mar 1996	1	$5.1344.10^{-3}$	$-1.2234.10^{-5}$	$1.0979.10^{-7}$	1115
Feb 1996	1	$4.4836.10^{-3}$	$-1.1276.10^{-5}$	$0.5996.10^{-7}$	1477

8. Conclusion

In this article, a specific method for determining the distortion parameters of an externally occulted coronagraph from in-orbit images has been described. Good sensitivity and stable pointing over a long period of time, in order to get a significant number of reference points, were of critical importance. The method has been successful in correcting the images with an accuracy better than one pixel.

References

Brueckner, G. E., et al. 1995, Solar Physics, 162, 357

Weng, J., Cohen, P., & Hermion, M. 1992, IEEE Transactions on Patt. Anal. and Mach. Intell., 14, N10

Astronomical Data Analysis Software and Systems VI
ASP Conference Series, Vol. 125, 1997
Gareth Hunt and H. E. Payne, eds.

NICMOS Calibration Pipeline: Processing Associations of Exposures

H. Bushouse, J. MacKenty, C. Skinner

Space Telescope Science Institute, Baltimore, MD 21218

E. Stobie

University of Arizona, Steward Observatory, Tucson, AZ 85721

Abstract. The Hubble Space Telescope (HST) Near Infrared Camera and Multi-Object Spectrometer (NICMOS) will often produce multiple exposures of a given target, usually for the purposes of measuring and removing thermal background emission and for rejecting cosmic rays. The processing of these associated images presents new challenges to the HST data processing pipelines, which heretofore operated only on single images. In this paper we describe the concept of associated observations and how they will be processed collectively by the NICMOS calibration pipeline.

1. Introduction

The basic or "atomic" element within the HST ground system has been the "exposure." In this mode, a single exposure results in a single "dataset," which is given a unique name and is pipeline processed, calibrated, and archived separately from all other datasets. The second-generation HST instruments present many instances in which the combination of data from two or more exposures is necessary to create a scientifically useful data product. For NICMOS, for example, the HST thermal background is expected to vary temporally, therefore multiple exposures (dithered for small targets and "chopped" onto blank sky for larger targets) will be necessary to measure and remove this background. Multiple exposures will also be used to reject cosmic rays, as well as for constructing mosaics of large angular-sized targets. While this has been standard practice in existing data reduction schemes for ground-based IR observations, it is a new paradigm for the HST ground system.

2. What is an Association?

An association is a means of identifying a set of exposures as belonging together and being, in some sense, dependent upon one another. The association concept permits these exposures to be calibrated, archived, retrieved, and reprocessed as a set rather than as individual exposures. Associations are defined by observers in their proposals. Typical usage will be:

- to obtain multiple exposures at a single sky position for the purpose of identifying and rejecting cosmic ray events,
- to obtain a sequence of slightly offset (dithered) exposures to improve flat fielding, avoid bad pixels, build a mosaic for large angular-sized targets, and, for sufficiently compact targets, to remove the background illumination, and
- to obtain a sequence of observations in which the field of view is chopped between the target and one or more offset sky regions to remove the background illumination for large angular-sized targets.

The individual exposures are logically linked through the use of an association table, which contains information on each exposure and is used by the data processing system to identify the contents of an association.

A set of predefined observing patterns are provided for NICMOS dithered, chopped, and combined dither-chop observations. All patterns are open ended; the observer specifies the total number of desired positions to be executed. The observer also specifies the size of the dither and/or chop movements to be executed between successive exposures within the pattern.

3. NICMOS Data Calibration

The NICMOS data calibration process is divided into two stages. Stage 1 (CALNICA) calibrates individual exposures without information from other members of the association. It performs the typical instrumental calibration steps such as dark-current subtraction, non-linearity corrections, and flat-fielding. Stage 2 of calibration (CALNICB) only operates on associated images that have previously gone through CALNICA processing, and creates combined images for each distinct (non-overlapping) pointing within the association. The CALNICB output images are generically referred to as "mosaics."

All stages of calibration propagate and update (as appropriate) data arrays containing statistical errors, data quality flags, number of samples, and exposure time per pixel that accompany the actual science image.

The CALNICA and CALNICB software is written in ANSI C. Data I/O is accomplished using IRAF I/O library functions. A set of C-to-IRAF interface routines (developed at STScI) provides the linkage between the calibration software and the IRAF libraries.

4. CALNICB Processing

The operation of CALNICB is driven by an association table, which is a FITS binary table that contains a list of the individual exposures that make up the association. This design allows observers to modify the contents of an association by adding, deleting, or editing individual exposures and then reprocessing. This is easily accomplished by modifying the association table and (if desired) the files containing the individual exposures.

CALNICB processing is divided into the following major steps:

- read the list of input images from the association table and load all data,

- determine processing parameters from header keywords,
- combine multiple images (if any) at individual pattern positions,
- measure and remove background illumination,
- create mosaic images from overlapping pattern positions, and
- create an updated output association table.

4.1. Determining Processing Parameters

CALNICB processing parameters are determined from FITS header keyword values in the input images. If none of the predefined observing patterns was used, then all the images in the association are multiple exposures at a single pointing. If a pattern was used, CALNICB determines how many and which images were taken at each pattern position, and which of the positions correspond to target or sky observations. The type of pattern used (if any) also determines the number of output mosaic images that will be produced. If either no pattern or a pure dither pattern was used, there will be a single output mosaic image made up of either the multiple images at a single pointing (the no pattern case) or a true mosaic of the entire set of overlapping dither images. Pure chop patterns result in one mosaic image for the target position, as well as one mosaic image for each of the offset sky (chop) positions. Combined dither-chop patterns result in one mosaic for the target dither sequence and one for the sky dither sequence.

4.2. Combining Multiple Exposures

In the event that the observer requests multiple exposures at each pattern position, the individual images are combined. Cross-correlation techniques are used to measure and correct for any misregistration between images that may be present due to, for example, small amounts of telescope drift. The combining procedure computes the average value of the n samples for each image pixel, excluding flagged pixels, and uses iterative sigma-clipping to reject deviant values. The number of samples used, as well as the total integration time, at each pixel are updated in their respective data arrays.

4.3. Background Estimation and Removal

Background removal is accomplished in two steps: first, a scalar background level is determined from the images in the association and subtracted from them; second, a reference image containing measured spatial variations in the HST thermal background is subtracted from each association image.

The scalar background level is first determined for each association image, and then a final value is determined by averaging the individual results. Iterative sigma clipping is used at both levels of averaging to reject deviant values that may be the result of cosmic ray events or sources contained within the images. For dithered patterns, all positions are used in the scalar background computation. For chopped or combined dither-chop patterns, only the off-target positions are used.

4.4. Mosaicking

A mosaic image is constructed by:

- determining the relative offsets of each image (using cross-correlation techniques),
- using the relative offsets to determine the total size of the mosaic and the position of each image within the mosaic,
- resampling (via bi-linear interpolation) each image into its appropriate location within the mosaic stack, and
- combining (collapsing the stack) the samples for each pixel in the mosaic.

The process of combining the samples for each mosaic pixel is identical to that used earlier for combining multiple images at a given pattern position; the average pixel value is computed using iterative sigma-clipping to reject outliers, as well as excluding flagged pixels. Values for the number of samples and integration time arrays are updated as appropriate.

4.5. Output Table

The output association table contains a copy of all the information from the input table plus new columns of information derived during processing. The new information includes a flag to indicate whether or not a particular image was used in the background calculation, the background level computed for the image, and the x and y pixel offsets of each image relative to its reference image.

CALNICB will allow an observer to reuse an output table as input to the program in order to reprocess a given association using different parameters. The observer can modify the table to indicate whether or not certain images should be used in the background calculation, manually set the background level and skip the run-time calculation, or set specific image offset values and skip the run-time calculation of offsets. An example association table for a five position spiral-dither pattern is shown in Table 1. Note that the last row of the table pertains to the output mosaic image.

Table 1. Association Table.

Name	Type	Prsnt	BkgImage	MeanBkg (DN/sec)	XOffset (pixels)	YOffset (pixels)
N2RUSG01T	EXP-TARG	yes	yes	0.293	0.00	0.00
N2RUSG02T	EXP-TARG	yes	yes	0.292	36.20	0.00
N2RUSG03T	EXP-TARG	yes	yes	0.292	36.20	36.20
N2RUSG04T	EXP-TARG	yes	yes	0.293	0.00	36.20
N2RUSG05T	EXP-TARG	yes	yes	0.293	−36.20	36.20
N2RUSG010	PROD-TARG	yes	no	INDEF	INDEF	INDEF

Astronomical Data Analysis Software and Systems VI
ASP Conference Series, Vol. 125, 1997
Gareth Hunt and H. E. Payne, eds.

NICMOS Related Software Development at the ST-ECF

R. Albrecht,[1] W. Freudling,-[2] A. Caulet,[1] R. A. E. Fosbury,[1] R. N. Hook,[2] H.-M. Adorf,[2] A. Micol,[1] R. Thomas[3]

Space Telescope – European Coordinating Facility

Abstract. In collaboration with the STScI and with the NICMOS Investigation Definition Team we have been developing software to predict aspects of NICMOS performance and to analyze NICMOS grism data. Quick look analysis software has been developed to extract spectra from NICMOS grism images and matching direct images to support ground testing and in-orbit verification. The TinyTim point spread function generation package has been modified to produce NICMOS PSFs for various instrumental modes. Software for creating simulated data has been developed in order to test various analysis procedures. A "grism spectra extraction pipeline" has been produced which allows the location, extraction, and calibration of grism spectra from NICMOS frames with a minimum of human intervention. Tools to convert between units and photometric systems are available.

1. Introduction

NICMOS (Near Infrared Camera and Multi-Object Spectrometer) will be installed in the Hubble Space Telescope (HST) during the Second Servicing Mission in February 1997.

Software is being developed at the STScI to ensure the proper calibration of data obtained with NICMOS. Software is also being developed within the NICMOS Investigation Definition Team (IDT, PI Rodger Thompson) to support the IDT science program.

Among the operating modes of NICMOS is a grism spectrum mode. While this mode is not being considered one of the primary operating modes, it nonetheless provides a very important scientific capability, and has the highest potential for serendipitous discovery. A strategy for a routine parallel grism survey has thus been developed.

In collaboration with the NICMOS IDT and with the STScI, the ST-ECF has been developing software to support the NICMOS grism mode from ground testing to routine pipeline calibration. This includes interactive tools for the examination of grism and associated direct frames, for computing point spread

---

[1]European Space Agency, Space Science Department, Astrophysics Division

[2]European Southern Observatory, Garching bei München, Germany

[3]Free-lance

functions, and a component of pipeline calibration for the automatic detection and extraction of grism spectra. A Web-based tool has been produced for magnitude/flux conversion.

2. NICMOSLook

NICMOSLook is an interactive tool to extract spectra for individual sources on a NICMOS grism image. A matching direct image of the same field is used to determine the location of sources and, therefore, the zero point for the wavelength scale for each individual spectrum.

The tool is implemented as an IDL widget. After loading the grism image and the direct image, both images can individually be displayed and manipulated. The user can locate objects in various ways: with the cursor, by supplying object coordinates, or by setting a threshold for an automatic object search.

Spectra can be extracted for any user-selected object or for all objects at once. The dispersion and distortion spectra are read from a database. There are several options for the spatial weighting of the spectra. The weights can be supplied by the user. Predefined weight functions are equal weights or weights computed from simulated point spread functions (see below).

The output of NICMOSlook are wavelength calibrated spectra, which are plotted on the screen and can be saved as PostScript files or ASCII data files.

3. TinyTim for NICMOS

The detection of objects on grism images, weighted extractions of grism spectra, deconvolution of images and spectra, and simulations of imaging data all require at least approximate knowledge of PSFs. One part of the ST-ECF NICMOS project was therefore to develop the capability to easily create PSFs for the NICMOS filters and grisms.

The TinyTim software is well known to anybody who ever worked on pre-COSTAR data. Originally written by John Krist at the STScI, it can generate theoretical point spread functions for the FOC and the WF/PC, both before and after the first refurbishment mission in late 1993, supporting image restoration work and other analysis for which knowledge of the PSF is needed (Krist 1994).

TinyTim has been modified at the ST-ECF to produce PSFs for the various operating modes of NICMOS. These PSFs are being used by NICMOSLook (see above) and by the grism spectrum extraction pipeline (see below) to extract spectra from grism frames, in particular for spectra in crowded fields with overlapping objects.

The modifications made include the addition of the NICMOS optical system geometry (focal ratios, pixel sizes, position of the NICMOS cameras in the HST focal plane, etc.) and the incorporation of the additional pupil obscuration introduced by the "cold masks."

The original TinyTim used stellar spectral type as a simple way of specifying the spectral energy distribution of the source. This is clearly not appropriate for an infrared system and, after discussion with the NICMOS PI, it is planned to incorporate three ways of specifying SEDs: a blackbody of a given temperature, a powerlaw of a given slope, and a user-specified spectrum.

4. Simulated Data

It is obvious that for the testing of any data analysis software, a good set of representative data is required.

There are two distinctly different aspects to this exercise: on the one hand the data have to be similar to, if not identical with, the expected actual data in terms of structure, header, keywords, etc. Although we made considerable efforts to anticipate the expected configuration, we expect to have to make considerable changes.

On the other hand, the simulated data have to be representative in terms of the pixel content. Different data sets corresponding to different configurations of NICMOS, different roll angles of the HST, and different degrees of crowding in the field were constructed.

5. The Grism Spectrum Extraction Pipeline

The interactive grism extraction tool NICMOSlook described above is a convenient tool to extract spectra from small numbers of grism images. However, a routine extraction of spectra from large number of NICMOS grism images requires a tool which extracts spectra without human interaction. Such a need may arise from large survey-type grism programs, for archival research, or for programs where grism images contain a larger number of objects which are not the primary object of interest. Ideally, such a grism extraction can be run at the same time as the STScI pipeline data reduction.

Establishing the requirements for such a software package, we quickly found that we did not have the resources to write it from scratch. On the other hand, it soon became evident that most of the individual steps needed for the reduction have been coded before in one way or the other. The challenge was to harvest as much as possible of this available functionality and to combine it into a package which would meet the requirements of minimum human intervention and operational resilience. Currently, we are implementing such a grism extraction "pipeline" (called calnicc as it requires the STScI developed calnica and calnicb pipeline steps) as an IDL program, which calls C modules where necessary.

The capabilities of the program are the following. Objects are identified and classified as stars or galaxies using a neural network approach implemented as the SExtractor program (Bertin & Arnouts 1996). The wavelength calibration of the extracted spectra are performed using the position of the objects as determined by the SExtractor program as the zero point, and using parameterized dispersion relations. After extractions of the spectra, they are corrected for the wavelength dependence of the quantum efficiency of the detector. The flux scale is then computed using the standard NICMOS flux calibration data. The extracted spectra will automatically be checked for artifacts from bad data and contamination from nearby objects. All spectra will be automatically searched for emission and absorption lines. In addition, the continuum emission will automatically be determined. The final data products are binary FITS tables with the spectra, error estimates, object parameters derived from the direct imaging, and details of the spectrum extraction process.

6. NICMOS Tools

The near infrared is the region where the Magnitudes meet the Janskys. In order to support the work on the NICMOS related software, and to provide tools for the community to construct observing programs for NICMOS, the ST-ECF developed a Web-based unit converter.

Taking advantage of recent developments among Web browsers, in particular of the emergence of Java (Arnold & Gosling 1996) and Javascript, we decided to implement the tool in Javascript. This has the advantage that the functionality can be invoked through the Web, but that the computational load is exported to the client, thus offloading the host. The tool is available at: http://ecf.hq.eso.org/nicmos/tools/nicmos_units_tool.html. Details of design and implementation are given in another poster paper (Micol et al. 1997).

When invoked from a Web browser capable of handling Javascript, the user is presented with a form-type interface.

Three are the input fields: Magnitude, Flux, or NICMOS. Correspondingly the user can select the associated units:

- Photometric system (CIT or UKIRT) and band (from V to Q) for the Magnitude;
- Some of the most used units for the Flux;
- Janskys and $Jy/arcsec^2$ are the units for NICMOS.

As in a spread sheet, whenever an input field is changed, the other two will be automatically computed according to the chosen units.

For better user-friendliness, two fill-in forms are provided: one for point sources and the other for extended sources. Both the forms have the same functionality and layout.

References

Arnold, K., & Gosling, J. 1996, The Java Programming Language (Reading, MA: Addison-Wesley)

Bertin, E., & Arnouts, S. 1996, A&AS, 117, 393

Krist, J. 1994, The Tiny Tim User's Manual (Baltimore, MD: Space Telescope Science Institute)

Micol, A., Albrecht, R., & Pirenne, B. 1997, this volume, 104

Astronomical Data Analysis Software and Systems VI
ASP Conference Series, Vol. 125, 1997
Gareth Hunt and H. E. Payne, eds.

IDL Library Developed in the Institute of Solar-Terrestrial Physics (Irkutsk, Russia)

S. K. Konovalov, A. T. Altyntsev, V. V. Grechnev, E. G. Lisysian, and G. V. Rudenko

Institute of Solar-Terrestrial Physics, Lermontov St. 126, Irkutsk, Russia 664033, E-mail: root@sitmis.irkutsk.su

A. Magun

Institute of Applied Physics, University of Bern, Switzerland, 3012 Bern Sidlerstrasse 5, E-mail: magun@sun.iap.unibe.ch

Abstract. We process and analyze data provided by the SSRT interferometer using IDL. Thanks to the well-known capabilities of IDL, this has expedited our research. Special requirements, convenience, and efficiency led us to the creation of an expanding library of IDL routines and programs, described herein.

1. Introduction

We have been processing and analyzing data provided by the Siberian Solar Radio Telescope (SSRT) (Smolkov et al. 1986). Much of these data required development of special techniques for processing and previewing. Using the well-known capabilities of IDL, we were able to speed up our research. However, a variety of routines had to be developed due to the specific needs of our research and the features of our instrument, as well as our desire to make processing still more convenient and efficient. Having started this work in collaboration with the Institute of Applied Physics (Bern), we continued it in Irkutsk. Coming up with our own IDL installation allowed us to expedite our work. As a result, a whole set of routines and programs supporting our programming has emerged (Konovalov et al. 1997). This set contains special routines for our use, as well as routines which could be of common interest. This library is being currently expanded.

2. Overview

In response to inherently specific features of our instrument, the SSRT, and to our research needs and preferences, we have developed, supplemented, and corrected some routines from the standard IDL library. Since our IDL library contains more than 100 routines and functions, we can present here only an overview of these routines, classifying them by category, and briefly describing their capabilities. The routines can be used on UNIX and MS Windows platforms, and all can be run under IDL 3.0.1 or later versions.

3. General Routines and Functions

These routines and functions were written for specific needs, but could be generally useful:

String-type variables manipulation: We have added functions which allow (*i*) finding a given substring in a string, (*ii*) replacing it with a given model of arbitrary length, and (*iii*) splitting a string containing any delimiter (e.g., whitespace) into substrings.

Files and file names manipulation: These routines provide: (*i*) conversion of text files between UNIX and MS Windows; (*ii*) adding an End-Of-File marker; (*iii*) extracting a short file name, extension, file name without extension, and directory name from the full path name; (*iv*) automatically recognizing the type of some graphics files (GIF, TIFF, BMP, or FITS); (*v*) creating a new file name, given a wildcarded pattern; (*vi*) an interactive tool for deleting a file, etc.

Mathematical functions: We have developed routines for rapidly evaluating some often-required functions: (*i*) SIGN and SINC ($\sin x/x$); (*ii*) recognizing even numbers; and (*iii*) calculating the Gaussian function. Furthermore, some operations for array manipulation have been implemented: (*i*) conversion of 1-D subscripts into 2-D; (*ii*) extracting subarrays; and (*iii*) searching for coinciding elements in different arrays.

Graphics windows manipulation: We have developed routines expanding the standard IDL tools with the following capabilities: (*i*) bringing all existing graphics windows to the front; (*ii*) deleting all existing graphics windows; (*iii*) displaying current cursor coordinates within a selected graphics window; (*iv*) saving and restoring scaling of the axes in a given window; and (*v*) converting a cursor into another shape.

Plotting routines: We have developed the routines which: (*i*) plot a graph versus time expressed in hours, minutes, and seconds; (*ii*) plot a graph of a fragmented array; (*iii*) plot a straight line of a given slope crossing a given point; (*iv*) plot a given arc or circle; (*v*) overplot a triangle-shaped marker onto a plot or image; (*vi*) display an array in the brightness representation enclosed by axes; and (*vii*) provide a box-shaped cursor for widget-based programs, etc.

Curve analysis: The developed routines allow (*i*) finding (and, if needed, marking on the plot) local extremes; (*ii*) selecting a local peak; (*iii*) calculating its width; (*iv*) finding the circle passing through three given points; and (*v*) finding a line bounding a given fast-oscillating curve (similar to the detection of a radio signal).

Graphics file manipulation and image processing: We use FITS files, so we have routines to handle the headers (searching for a keyword and returning the corresponding value, extracting of time and date from the header, etc.). We can automatically recognize the type of some graphics files (GIF, TIFF, BMP, or FITS) and display them on the screen. Routines were developed to interactively measure the coordinates of the solar disk's center and radius, and to save the image contained in a graphics file into a FITS file. There is also a routine converting a plot obtained from an image into a digital array, which can be saved for further processing.

Viewers: Some viewers have been developed for the SSRT data, as well as for the standard format graphics or digital files. There is an interactive array viewer which allows looking at data in various representation (wire-mesh surface, shaded surface, halftone image, contour, and their combinations), measuring profiles in two directions, and reading pixel values. Contour levels can be selected interactively. There is also a viewer for text files or string-type arrays entered on the command line (similar to XDISPLAY-FILE).

Date and time formats and conversion: A few routines and functions have been developed which are used in reading data records, in calculations, and in displaying date and time in a suitable form.

Astronomical calculations: Our routines: (*i*) compute an hour angle, a declination, and a radius of the Sun; (*ii*) calculate the heliographic, Carrington's, and plane coordinates, and perform transformation between them; (*iii*) compute the time interval between two events given in different calendar formats; (*iv*) transform the coordinates on the Sun according to differential rotation; and (*v*) rotate a flat image of the Sun around the polar axis.

Service widgets: They are used for inputting a date and time, a brief text, or coordinates. There are widgets for issuing a message, for obtaining an answer to a question, and for selection of one item from the list, etc.

Other: The following routines we have not classified: (*i*) conversion of a bit-serial array into a byte array; (*ii*) creation of a new system variable !UC containing universal constants; (*iii*) selection and preparing of the graphics device required for the next output (including setting of sizes, thickness, color, etc.), and closing them; and (*iv*) deallocation of all possible file units (convenient for debugging.)

4. Routines and Functions for SSRT Data Users

These routines are oriented to SSRT data processing.

SSRT: instrumentation, data processing: A range of routines is used to calculate the parameters of the SSRT beam and response as well as astrometric values; to read and convert the information contained in the original records; to execute some actions with original files; to perform data processing; and to simulate instrumental characteristics of the SSRT and its response.

Special routines: We have also some routines completely for our own convenience, such as converting ASCII codes from Roman into Russian, etc.

5. Supplemented and Corrected Routines

We have modified the routine CONGRID to transform small arrays correctly. We have also supplemented the routine IMAGE_CONT with various keywords, and edited it to display small arrays correctly. The routine READ_BMP has been enhanced to handle monochrome images.

We welcome any interest and cooperation.

Acknowledgments. We are grateful to Drs. B. Kliem (AIP, Germany), Yu. M. Rosenraukh, T. A. Treskov, S. V. Lesovoi, V. G. Miller, A. V. Bulatov, and B. I. Lubyshev (ISTP, Russia) for the helpful discussion and the assistance in providing us with the data.

This work was supported by grants International Science Foundation (ISF) RLD000 and RLD300, ESO C&EE Programme A-03-049 and A-05-013, Swiss National Science Foundation 20 29871.90, and INTAS (INTAS-94-4625). Our special thanks to ISF for supporting our participation at this conference.

References

Smolkov, G. Ya., Pistolkors, A. A., Treskov, T. A., Krissinel, B. B., & Putilov, V. A. 1986, Ap&SS, 119, 1

Konovalov, S. K., Altyntsev, A. T., Grechnev, V. V., Lisysian, E. G., & Magun, A. 1997, this volume, 100

Astronomical Data Analysis Software and Systems VI
ASP Conference Series, Vol. 125, 1997
Gareth Hunt and H. E. Payne, eds.

The Data Handling System for the NOAO Mosaic

Doug Tody

IRAF Group, NOAO,[1] PO Box 26732, Tucson, AZ 85726

Abstract. This paper presents the data handling system being built for the NOAO Mosaic by the IRAF group at NOAO. This system consists of a Data Capture Agent which assembles and saves the disk images, a Real Time Display and Mosaic viewer used to view the very large (134 MB) Mosaic images during and after readout, and IRAF software for CCD processing, quick look, and general data interaction. The system architecture is based on general message bus and distributed shared object facilities. The Real Time Display and Mosaic viewer is a general, user extensible image viewer based on the existing Ximtool and SAOtng display software from NOAO and SAO. Companion papers describe the Mosaic data format and data reductions. Like IRAF itself, all of this software is portable and controller independent hence suitable for use by other observatories, particularly if they already use IRAF in the observing environment.

1. Introduction

The NOAO Mosaic is a large-field detector consisting of eight 2K×4K CCDs arranged in a 4×2 mosaic, for a total detector size of 8K×8K. At 16 bits per pixel, raw images are 134 MB in size. The instrument will be used on the 4-meter and 0.9-meter telescopes on Kitt Peak in the northern hemisphere and on the 4-meter telescope on Cerro Tololo in Chile. The field of view of the Mosaic ranges from 36′ to 1° depending on the telescope, with a scale of 0.26″/pixel at the 4-meter. Because the field is so large, optical field correctors are required. Similar instruments are being built at a number of other observatories, e.g., CFHT/U.Hawaii, Keck/Lick, and McDonald. The software described in this paper is being developed in collaboration with these and other groups.

While there is little fundamentally new about Mosaic data handling, the overall system accentuates old problems to a degree rarely before seen in ground based telescopes. The use of multiple CCDs causes problems with misaligned grids and gaps between the CCDs, requiring interpolation, image combination, and dithering to rectify the data. The CCDs can have different bias and flat characteristics requiring calibration before they can be viewed together on the display. The large field and the use of optical correctors mean that field distortions are significant, and combined with the misalignment of the CCDs this complicates coordinate determination and astrometry. The use of multiple CCDs

[1]National Optical Astronomy Observatories, operated by the Association of Universities for Research in Astronomy, Inc. (AURA) under cooperative agreement with the National Science Foundation.

requires that data be read out simultaneously from all CCDs, hence the raw data is interleaved as it arrives from the controller and must be "unscrambled" before being written to disk or displayed. Finally, the images are very large. A powerful computer system and efficient software is required to be able to handle such large images. Even viewing the data is difficult since the image is a composite of a number of smaller images, and at 8K×8K or 64 megapixels, the area of the full Mosaic is about 50 times that of the typical workstation screen.

2. The Mosaic Data Handling System

The Mosaic data handling system takes the raw data as they are generated by the CCD controller during frame readout and does all subsequent processing of the data, including capture to disk, real time image display, quick look and data quality assessment, pipeline data reductions, taping, queueing of data to the data archive, and, if desired, re-reduction of the data at the observer's home institution using IRAF.

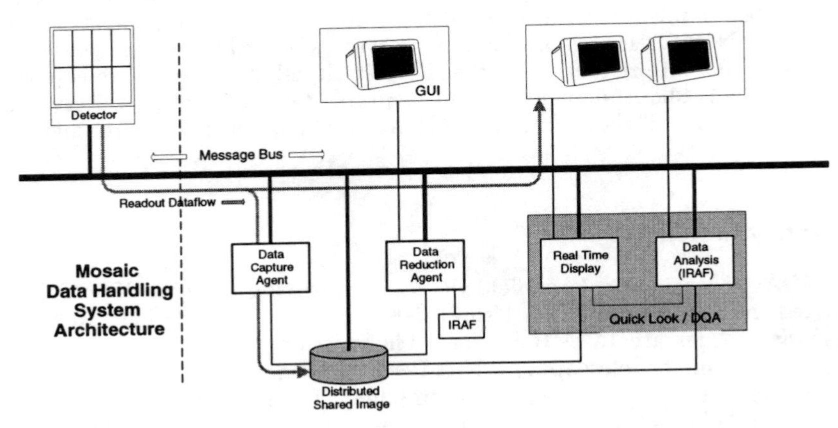

Figure 1. Data Handling System Architecture.

Figure 1 illustrates the software architecture of the Mosaic Data Handling System (DHS). As the Mosaic is read out, pixel and header data packets are written to the *message bus* which connects all elements of the data system. The *Data Capture Agent* (DCA) captures these data packets and builds an observation file on disk. At the same time data is sent to the *Real Time Display* (RTD) which displays the Mosaic during frame readout. Quick look is provided by the RTD and by IRAF, which can interact with the RTD during and after frame readout. The *Data Reduction Agent* (DRA) directs the post-processing of each observation file, applying standard calibrations and writing the data to tape and to the data archive.

2.1. Messaging and Shared Data Access

The heart of the Mosaic data handling system is the message bus, which connects all data system components. The message bus provides flexible and efficient facilities for components to communicate with each other. The message bus (which is a software facility) supports both distributed and parallel computing, connecting multiple host computers or multiple processors on the same host.

The message bus provides two methods for components to communicate with each other. *Producer/consumer events* allow components to listen for (consume) asynchronous event messages produced and broadcast by other components. *Requests* allow synchronous or asynchronous remote procedure calls (method invocations) to be directed to services or data objects elsewhere on the message bus. Discovery techniques can be used to determine what services are available and to query their methods. Host computers and components can dynamically connect or disconnect from the bus. The bus can automatically start services upon request; or services and other components can be started by external means, connecting to the message bus during startup.

An important class of message bus component is the *Distributed Shared Object* (DSO). DSOs allow data objects to be concurrently accessed by multiple clients. The DSO provides methods for accessing and manipulating the data object and locking facilities to ensure data consistency. DSOs are distributed, meaning that clients can be on any host or processor connected to the message bus. In the case of the Mosaic DHS, the principal DSO is the *distributed shared image* which is used for data capture, to drive the real time display, and for quick look interaction from within IRAF. The distributed shared image uses shared memory for efficient concurrent access to the pixel data, and messaging to inform clients of changes to the image.

The Mosaic DHS uses a custom message bus API which is layered upon some lower level messaging system. At present we are using the Parallel Virtual Machine (PVM) facility; in the future we might use other facilities such as CORBA. The use of a custom API provides isolation from the underlying messaging facility and aids development of a standard framework and set of services for integrating a set of applications.

2.2. Data Capture

When the Mosaic is read out the controller writes a stream of message packets onto the message bus. These take the form of requests to the *Data Capture Agent* (DCA), which operates as a service on the message bus. The DCA sits in a loop handling requests from the message bus. Incoming header data is buffered internally within the DCA. Incoming pixel data blocks are unscrambled and written directly to the output image using the distributed shared image facility. When the readout is finished a table driven keyword translation module (implemented as a configurable Tcl script) transforms the input device dependent detector keywords as necessary to conform to the data format required by the DHS. Ultimately a new observation file is written to disk and passed off to the DRA for post-processing. The observation file is a multi-extension FITS file containing one IMAGE extension for each amplifier of the Mosaic. The DCA can handle multiple simultaneous readouts from different clients; the clients can be on any host computer connected to the message bus.

2.3. Real Time Display and Quick Look

The primary function of the *Real Time Display* (RTD) is to display the Mosaic in real time during readout; pixels appear in the display as soon as readout begins. As noted above the DCA does not just write to a disk image, it writes to a *distributed shared image*, a type of DSO. At the same time that the DCA is writing to the DSO, the RTD is reading from it and displaying the incoming data. The DCA receives an incoming write-pixel request on the message bus, obtains locks on a set of regions in the output image (e.g. 16 regions for 8 CCDs with 2 amps each), copies the input data to the output regions, and then frees the regions. This causes the DSO to send messages to all clients, such as the RTD, which want to be informed of changes to the image. The RTD then performs any on-the-fly calibration or other processing and updates the displayed image.

To the user the RTD is an image browser displaying to one or more workstation screens, two in the case of the NOAO Mosaic. One screen shows the full mosaic dezoomed at 50-to-1. The second screen shows a zoomed up region of the mosaic. Multiple zoom windows can be active on multiple screens.

The RTD is not just a real time display, it is a fully functional image viewer with extensive builtin functions for quick look image analysis. Additional functionality (possibly quite extensive) can be added via a dynamic "plug-in" facility, which allows users or projects to easily customize the display or tailor it for existing data systems. Finally, extensive image analysis is available via IRAF or any other external image analysis system which interfaces with the RTD and DSO. IRAF sees the DSO as if it were a conventional disk image, allowing any IRAF task to be used. This allows tight integration of IRAF quick look or analysis tasks with the RTD. It is even possible to use an IRAF task to operate upon the incoming image during readout, before readout has completed.

2.4. Data Processing

The Mosaic DHS includes a full pipeline data reduction capability, plus facilities for taping, archiving, viewing and managing the data set. All data reduction is performed by IRAF tasks under the direction of the *Data Reduction Agent* (DRA). The DRA is driven by a device dependent script hence is user configurable and easily adapted for new instrument configurations. Companion papers by Frank Valdes (Valdes 1997a, 1997b) describe the Mosaic data format and data reductions.

3. Summary

The primary function of the Mosaic data handling system is to process data from the NOAO Mosaic. The significance of the project is much greater however. The DHS itself is applicable to any type of data and when completed will be used for general data acquisition within NOAO and at other observatories. The message bus, DSO, and plug-in image display (RTD) technology used by the Mosaic is being developed as a more general facility for use in the IRAF system and by other projects. This work is supported in part by grants from the NASA Astrophysics Data Program and from the NASA Applied Information Systems Research program. See the Mosaic project Web page (http://iraf.noao.edu/projects/mosaic) for additional information on these efforts.

References

Valdes, F. 1997a, this volume, 455

Valdes, F. 1997b, this volume, 459

Astronomical Data Analysis Software and Systems VI
ASP Conference Series, Vol. 125, 1997
Gareth Hunt and H. E. Payne, eds.

IRAF Data Reduction Software for the NOAO Mosaic

Francisco Valdes

IRAF Group, NOAO,[1] *PO Box 26732, Tucson, AZ 85726*

Abstract. NOAO is building a large format (8K×8K pixels) camera using a mosaic of eight 2048×4096 CCDs with eight amplifier readouts initially, and eventually up to 16. This paper describes the data reduction architecture and software being developed for this instrument. The real time display and data capture are described by Tody (1997) and the data format for recording data is described in separate paper by Valdes (1997).

Figure 1 illustrates the components and data flow of the NOAO CCD Mosaic Software System.[2] The data reduction components are highlighted here.

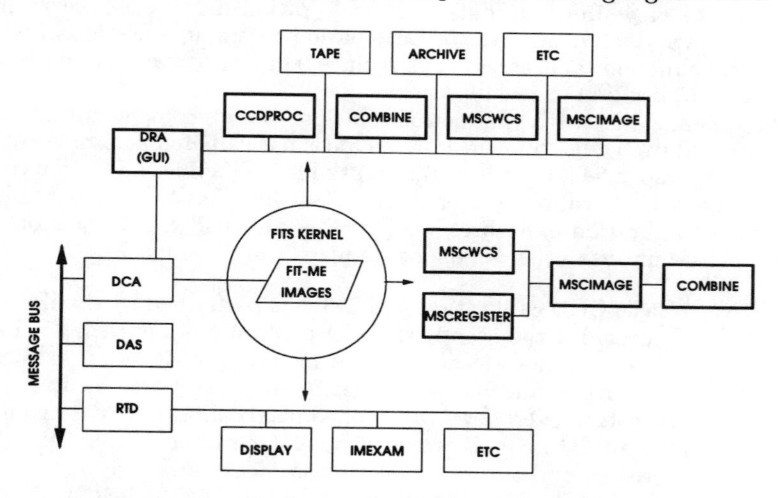

Figure 1. The NOAO CCD Mosaic Software System.

The *data acquisition system* (DAS) sends pixel and descriptive information through a *message bus* to subscribing clients. Two major clients are the *real time display* (RTD) and the *data capture agent* (DCA). The RTD displays the image data they are being received and interacts with other processes such as the IRAF task IMEXAMINE. The DCA writes observation information to a *FITS*

[1]National Optical Astronomy Observatories, operated by the Association of Universities for Research in Astronomy, Inc. (AURA) under cooperative agreement with the National Science Foundation.

[2]http://iraf.noao.edu/projects/ccdmosaic/Reductions/reductions.html

Multiextension file (FITS-ME) in the Mosaic data format, and sends observation file status messages to the data reduction agent. The message bus, RTD, DCA, and Mosaic data format are described elsewhere (see Tody 1997; Valdes 1997).

DRA: The *data reduction agent* (DRA) operates on the observation files. It is a high level task, with a sophisticated graphical user interface (GUI), to perform pipeline calibrations, reductions, data quality assessment, archiving and possibly other functions, by communicating with IRAF tasks which access the observation files through the IRAF *FITS Image Kernel* (Zarate & Greenfield 1995).

The DRA is a continuously running, event-driven process, triggered when the data capture agent finishes writing an observation to disk and when the user initiates an action via the GUI. The first case provides automatic processing and archiving. The second case allows manual calibrations or recalibrations of the automatically processed data when additional or improved calibration data become available. For example, automatic processing can proceed using calibration data from the start of the night, and recalibration can be done after calibrations at the end of the night are obtained.

The DRA is controlled by a GUI. This interface provides a browsing tool for the observations, processing status information, quality assessment information, a tool to view and manipulate the calibration data base, the ability to delete or exclude data, control of the automatic processing, selection of pipeline calibration, reduction, and assessment recipes, control of parameters, recall of raw data, and initiation of recalibrations.

The pipeline calibration, reduction, and quality assessment are defined by *recipes* selected from a list of recipes. A recipe is basically a *macro* or *script* that is executed on a specified disk file or set of disk files. Pipeline calibration consists of the standard CCD calibration operations, setting the WCS, and combining sequences of calibration exposures with scaling and bad pixel rejection. The output includes uncertainties, an exposure map, and pixel masks.

CCDPROC: Basic CCD calibration processing is performed by the IRAF task CCDPROC. It provides the standard CCD calibrations for each of the amplifier/CCD readouts of the Mosaic. These include pre/overscan calibration, trimming of pre/overscan and bad edge regions, bad pixel and saturated pixel masking and replacement, zero level (also called bias) calibration, dark count calibration, flat field calibration, and propagation of uncertainties from the detector readout characteristics and the calibration data. This task may also combine calibrated amplifier images into a single image for the CCD. The output Mosaic format then consists of multiple extensions for the CCDs.

The Mosaic version of CCDPROC is actually a relatively simple task that understands the details of the Mosaic data format. It extracts the individual amplifier images and associated data, such as pixel masks and uncertainties, and passes them to a lower level task to do the actual processing. It then takes the calibrated data and updated associated data and puts them back into the Mosaic data format. The lower level task is written to process individual images and associated pixel masks and uncertainties from an input to an output and has no knowledge of the details of the Mosaic data format.

COMBINE (Calibration Images): Calibration observations, such as flat fields, generally include many exposures to minimize noise. The exposure sequences are identified by the DRA which uses COMBINE to create master calibrations.

The DRA keeps track of the master calibrations and applies the appropriate one to new science exposures.

COMBINE combines the individual elements of the Mosaic matched by amplifier or CCD identification. The combining is done pixel-by-pixel within each amplifier/CCD image. It also propagates combined bad pixel masks, variance images, and exposure maps. The Mosaic version of COMBINE is a relatively simple task that understands the details of the Mosaic data format. It extracts the individual amplifier/CCD images and associated data, such as pixel masks and uncertainties, and passes them to a lower level task to do the actual combining. The combined data and updated associated data are then put back into the Mosaic data format.

COMBINE (Dithered or Rastered Science Images): The combining of calibration exposures is straightforward in the sense that there does not need to be any interpolation, shifting, and coordinate manipulation. The combining of dithered or rastered science exposures is more complex, particularly with regard to coordinate systems. Such data are first resampled into a single image in a celestial coordinate system that can be shifted by integer amounts along both image axes before combining. This is done by MSCIMAGE. COMBINE uses the coordinate system produced by MSCIMAGE to shift and then combine dithered or rastered observations.

MSCWCS: The Mosaic World Coordinate System (WCS) maps the image pixels to celestial coordinates on the sky. The mapping is stored in the headers for each amplifier/CCD image. The WCS is defined in two stages. The first stage applies a predetermined calibration and the second stage adjusts this calibration based either on a catalog of sources in the field of the exposure or registers the WCS in multiple overlapping exposures based on common objects in the images.

The WCS calibration file consists of *plate solutions* for each amplifier/CCD determined from calibration exposures. The plate solution is then applied to observations by adding the telescope pointing and, possibly, instrument position angle. In other words, the WCS is determined once at some telescope pointing reported by the telescope control system. This WCS is used for other telescope pointings with a zero point offset set by the difference in reported telescope coordinates between the calibration and the observation. If the detector may be rotated then the calibration also includes a rotation axis origin determination and uses the difference in instrument position angles to adjust the WCS.

The first stage of setting the WCS for an observation using a calibration file and the telescope pointing is a basic calibration operation performed by MSCWCS. Note that if the WCS is set at an earlier stage by the data acquisition system or the data capture agent then this option of MSCWCS might not be needed.

The WCS set by the first stage is likely to be off by a small amount due to errors in the telescope pointing and instrument flexure. The second stage is to use objects in the image to adjust the WCS. This second stage may use many objects and a full astrometric catalog to make a new calibration. However it is more likely that there are only a few objects and possibly no source catalog. In that case the few objects can be used to make small zero point and rotation adjustments in either an absolute sense if the objects have known celestial coordinates or a relative sense if common objects in multiple exposures are used to register the exposures.

The adjustment of the WCS using a catalog of sources in the field of observation is also performed by the task MSCWCS. It assumes that the existing WCS is fairly close. It takes each source in an input source catalog and searches near the expected position in the image for an object. The object position is determined using a centering algorithm. Once a set of measured pixel positions and catalog celestial coordinates is determined the WCS can be adjusted for an offset and rotation or possibly a new plate solution can be computed.

MSCWCS can be run automatically given a good first WCS and a catalog of sources. If the observer supplies the source catalog or the data reduction agent can automatically obtain a catalog (say by using the telescope coordinates and a *catalog server*) then this second stage WCS calibration performed by MSCWCS can be part of the basic calibration performed by the DRA.

MSCREGISTER: The task MSCREGISTER uses objects in a set of Mosaic observations to adjust the world coordinate system (WCS) for each observation to best *register* the objects. This means that overlapping objects will have nearly the same coordinates subject to the limitations set by the form of the WCS description. The set of objects need not appear in all observations but there must be some reasonable overlap so that each observation has common objects with one other observation and all the observations form a single contiguous region.

Several algorithms are required. The objects in each amplifier/CCD image must be cataloged. Then common objects between the many catalogs must be identified. Finally the set of WCS must be registered in some "best" way.

MSCIMAGE: In the basic processing data from the individual amplifiers and CCDs are kept separate except that data from multiple amplifiers from a single CCD may be merged together into a single unit for the CCD. The observer may analyze the calibrated exposures keeping the readouts separate. This avoids any resampling of the pixel data. However, the observer may wish to resample the elements of the Mosaic into a single large image. MSCIMAGE uses the WCS to resample the pixels into a uniform grid on the sky. This corrects alignment errors in the detector and optical distortions.

Basically, a uniform sky grid of equal sized pixels about some point in the sky is defined and the observed pixels are interpolated to this grid. By using the same grid for dithered or rastered sets of observations, the images can then be combined using only integer pixel shifts in the two image axes. The goal is to require only a single interpolation of the data.

References

Tody, D. 1997, this volume, 451

Valdes, F. 1997, this volume, 459

Zarate, N., & Greenfield, P. 1996, in ASP Conf. Ser., Vol. 101, Astronomical Data Analysis Software and Systems V, ed. G. H. Jacoby & J. Barnes (San Francisco: ASP), 331

Astronomical Data Analysis Software and Systems VI
ASP Conference Series, Vol. 125, 1997
Gareth Hunt and H. E. Payne, eds.

Data Format for the NOAO Mosaic

Francisco Valdes

IRAF Group, NOAO,[1] PO Box 26732, Tucson, AZ 85726

Abstract. NOAO is building a large format (8K×8K pixels) camera using a mosaic of eight 2048×4096 CCDs with eight amplifier readouts initially, and eventually up to 16. This paper outlines the data format and keyword definition methodology for data from this instrument. The software system is described in other papers (see Tody 1997; Valdes 1997).

1. The Data File Structure

NOAO CCD Mosaic data is stored as a single FITS file for each observation. The FITS file contains a *primary header* with no associated data and a number of *extensions.* The primary header describes the file contents, and contains *global* keyword information applicable to all extensions. The extensions include the image data from each amplifier, pixel masks, uncertainty arrays, exposure maps, auxiliary tables, etc. The image data extensions are always present, and other information is added during the reductions.

Figure 1 shows the data file structure, and illustrates how the *inheritance* convention (Zarate & Greenfield 1995) defines the header for each image extension as the combination of the global and individual header keywords.

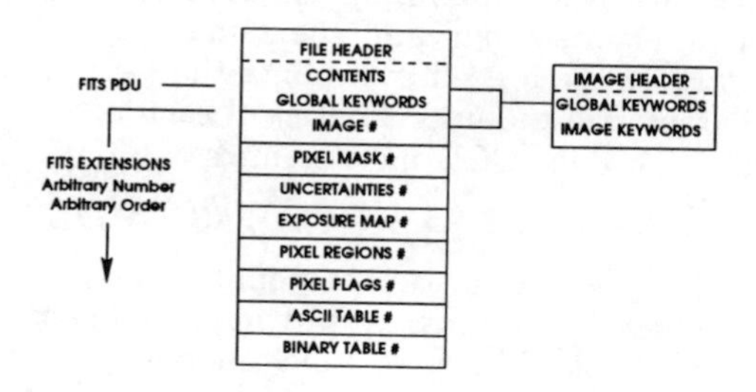

Figure 1. Mosaic Data File Structure.

The Image Data The basic image data consist of separate FITS Image Extensions for each amplifier. Each has an *extension name* used by the IRAF software to refer to the image through the *IRAF FITS Image Kernel* (Zarate

[1] National Optical Astronomy Observatories, operated by the Association of Universities for Research in Astronomy, Inc. (AURA) under cooperative agreement with the National Science Foundation.

& Greenfield 1995) (e.g., "`obs123[im3]`" refers to the image data for the third amplifier). The pixel masks, uncertainty arrays, and other array type extensions are also accessed through the FITS kernel by extension name.

Pixel Masks *Pixel masks* assign an non-negative integer value to each pixel in an image. The meaning of the mask value depends on the purpose of the mask—there may be more than one assigned to an image—and the application that will use it. Because it is often the case that most pixels have the same mask value IRAF provides a special representation called a *pixel list*. This representation is very compact. Some types of real data, such as the uncertainty array (described in the next section), may also consist of regions of constant value or be usefully mapped to integers using something like the BSCALE and BZERO method in FITS. The pixel list is stored as a FITS extension in a form still to be defined but most likely based on a binary table. The FITS kernel will be able to convert this type of extension to a standard IRAF pixel mask so applications may use these without knowledge of the FITS representation.

The types of integer pixel masks that might be used are identification of good and bad pixels with a set of code values for the type of bad pixel (CCD defect, saturated, etc.), the number of input pixels contributing to a pixel in a combined image, data quality flags, and marking regions for various purposes. The types of real pixel masks are uncertainty values and exposure maps showing the accumulated exposure time contributing to a pixel.

Uncertainties An important aspect of the image data is the uncertainties. Many of the concepts are reasonably well understood such as the characterization of the uncertainties in the raw CCD data in terms of a readout noise and Poisson statistics and how uncertainties are propagated when combining pixels with independent errors. Others are less well understood such as what happens with resampling. The biggest dilemma has been how to maintain the uncertainty information without doubling the data volume by using an associated data array of uncertainty values of the same size as the image data. In terms of the data structure we need something that will be compact yet offer the flexibility to characterize the uncertainties of each pixel.

The model we propose for CCD uncertainties is

$$\sigma^2(i,j) = A + (B + I(i,j))) * f(U(i,j)) \tag{1}$$

where $I(i,j)$ is the data, A and B are constants, $U(i,j)$ is an array of values, and f is a mapping function. In order to provide a compact description $U(i,j)$ is represented as a pixel list of integers which, hopefully, have large regions of constant value. The use of integers means that the variances will be quantized at some precision. The mapping function f can be defined to adjust the resolution at different levels. Note that there is already a mapping relative to the pixel sigmas because of the definition in terms of the variance.

This model allows easy propagation of errors in the common cases. The A value is a constant noise term. Typically this would be the CCD readout noise. When adding or subtracting two images, the corresponding A terms add. The B term is used when adding or subtracting constant values from images. For raw CCD data this value is zero. The usefulness and compactness of this model, that is how well the idea of largely constant areas in the U array will work in practice, still needs to be investigated. Preliminary experiments show promise that this approach will work effectively.

2. The Observation Information

The basic observational data consist of CCD pixel values and descriptive or documentary information associated with the observation. Pixel values are recorded as separate FITS image rasters for each amplifier. Observation information is recorded in FITS headers, both the primary global header and the extension image headers. This section outlines the method used to identify the interesting observation information and translate it into FITS keywords. Complete details are given on the NOAO Keyword Definitions[2] page.

There are three steps to defining the content of the observation information to be stored in the data file: (*i*) define a *logical model* of the observation information that includes **all** relevant or interesting items, (*ii*) translate the information into well defined and documented FITS keywords, and (*iii*) to collect the information and place it into the observation data file using the keyword definitions. The information actually recorded will be a subset of all the logical observation information identified in the first step.

The Logical Model The logical model of an observation attempts to identify **all** the information about an observation in a systematic manner. This model and framework is general for all ground-based optical and near-IR data and may be adopted and extended by other observatories.

The logical model analyzes an observation into a hierarchy of *classes* modeled after the logical components of the astronomical sources, the instrumentation, the data format, and archiving. A class consists of information elements which are either individual pieces of information or instances of another class. An element may also be an array of one or more instances such as, for example, information about multiple objects in the field of view.

In this short paper we can only present a brief example to give a flavor of this methodology. The root class is OBSERVE. It consists mostly of other classes such as OBJECT[n], TELESCOPE, INSTRUMENT, and DETECTOR. Note that the OBJECT class can have multiple instances. Many classes, such as the those previously mentioned, include general subclasses such as COORDINATE. A COORDINATE class has elements such as right ascension, declination, and system. This not only allows multiple coordinates for things like objects and telescope pointings but also different coordinate system types and equinoxes.

The COORDINATE class example shows one of the powers of this methodology. In the class we define an element such as equinox. Then we can be assured that any coordinate included as an element of some other class will have the equinox explicitly included and not forget that a coordinate must include this to be complete.

An information element has a hierarchical identifier. Examples of these are "`Object[n].Coordinate.ra`" and "`Detector.Ccd[n].Amp[n].Exp.darktime`." In words the latter example says there is a piece of exposure information that applies to a particular amplifier, in a particular CCD, in a detector which gives the effective dark current time. The capitalized words are instances of classes and shows that darktime is a node element of the Exp class which is a subclass element of the Amp class and so on.

---

[2] http://iraf.noao.edu/docs/keywords/

In our draft logical model there are 45 classes. The classes have anywhere from two to ten elements many of which are instances of another class. This is a manageable description even though if we expand out all possible elements as identifiers we get a very large number of elements. This leads us to be confident that we have identified all the pieces of information which we would not have if we started by trying to define all the leaf elements directly.

Clearly it is not possible to define all the information for every instrument and type of observation. However, the logical class model can be extended in a systematic way. This can be done by adding additional elements to a class or adding new classes. Instrument or system specific classes, such as for a particular instrument or array controller, may be added to define parameters which do not fit the general observation model. After the logical model is extended then the mapping to FITS keywords can be made.

FITS Keywords This section outlines the mapping of the logical observation identifiers to FITS keywords. The logical model is very general and could be used by many observatories. The mappings to FITS keywords could be more observatory specific though the same mappings could also be used by different observatories.

Every piece of information identified by the logical model has both a logical name and a FITS keyword. The mapping is given in a *keyword dictionary*. The keyword dictionary not only defines the keywords but acts as additional and more detail documentation about the meanings and assumptions for the information recorded in the image headers.

The keyword dictionary gives the logical identifier, the FITS keyword, a substitute keyword, the units, comment string, definition, etc. It is the substitute keyword that allows mapping the very large number of logical elements to a far few number of keywords. Note that there is no requirement that all the defined keywords appear in the FITS header. There are several reasons why items will not appear. Some items do not make sense for particular instruments, some items may not be available to the data acquisition system, and items with identical values may be mapped to a common keyword.

While the logical header identifies each possible item separately, many items will have the same value. These can be mapped to a single FITS keyword through the substitute keyword entry of the dictionary. An example of this is the coordinate system identification which may be the same for all coordinates; i.e., all coordinates are given in FK5 with equinox J2000. Items may also map to the same keyword because there is no precise value but a related value is approximately correct. An example of this is if the location of the center of the detector on the sky is not known then the telescope position my be substituted.

References

Tody, D. 1997, this volume, 451

Valdes, F. 1997, this volume, 455

Zarate, N., & Greenfield, P. 1995, in ASP Conf. Ser., Vol. 101, Astronomical Data Analysis Software and Systems V, ed. G. H. Jacoby & J. Barnes (San Francisco: ASP), 331

Part 11. AXAF

Astronomical Data Analysis Software and Systems VI
ASP Conference Series, Vol. 125, 1997
Gareth Hunt and H. E. Payne, eds.

ASC Data Analysis Architecture

M. Conroy, W. Joye, J. Herrero, S. Doe

AXAF Science Center, Smithsonian Astrophysical Observatory, 60 Garden Street, Cambridge, MA 02138

A. Mistry

AXAF Science Center, TRW, 60 Garden Street, Cambridge, MA 02138

Abstract. The AXAF Science Center (ASC) is using an "open architecture" approach to develop its data analysis environment. The system is a loosely coupled environment consisting of several major applications: visualizer, browser, fitter/modeler, as well as the data analysis tool-box. The ASC Data Model and Interprocess Communications (IPC) provide the data interface between applications and tools. The Navigator, CLI, and Profile Editor provide the user with different control methods to access these components. The modular design provides a flexible, configurable environment in which the user can create customized applications from the standard components.

1. Introduction

The ASC must develop, distribute, and support a portable data analysis package to provide observers world-wide with the ability to analyze AXAF data. The "open architecture" approach adopted for the ASC Data Analysis System allows maximum reuse of existing software, while providing an environment that can be customized and adapted for user-specific or future project needs. The architecture defines loosely coupled interfaces to facilitate customization of the environment by replacement of components.

2. Data Analysis Environment

A diagram of the data analysis environment appears in Figure 1. The components of the environment can be categorized as follows:

- **Control Mechanisms** A Graphical User Interface (GUI) and a simple command line interface (CLI) are required for data analysis tools and applications; this allows both interactive and batch mode control of the components. The Pipeline Controller provides a mechanism for configuring the components for automated processing and monitoring.
- **Tools and Applications**. Tools constitute the core of the data analysis system. They are built using layered libraries, and are used in many roles by the ASC. For automated pipeline processing, they are configured in

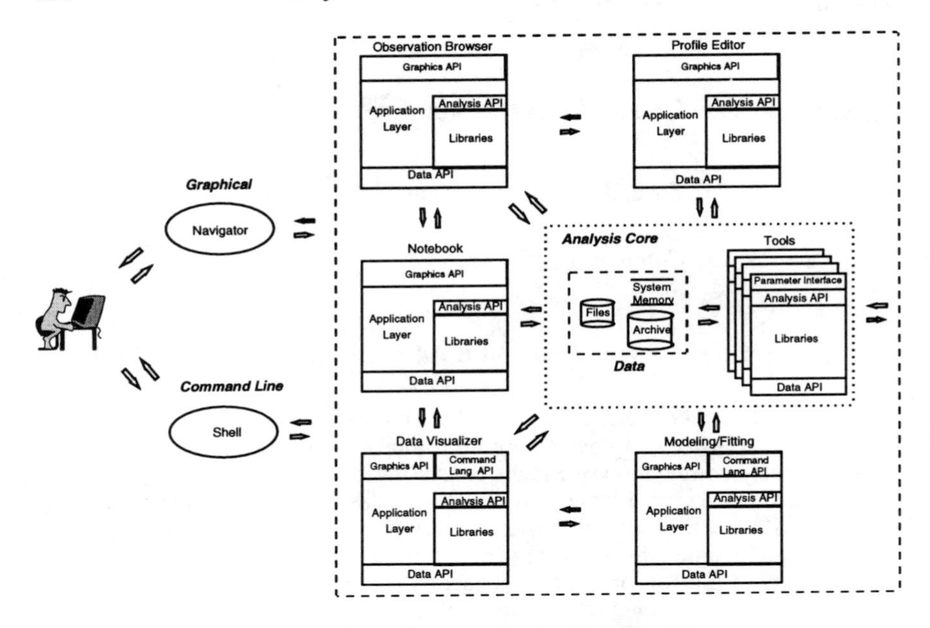

Figure 1. Data analysis environment.

pipeline profiles and invoked by the pipeline controller. They can be run interactively by the user from the command line or built into simple shell scripts or user-defined pipelines. Analysis applications, are built from the basic tool-box but are provided with interactive visualization and control features to facilitate the complex operations.

- **Visualization** components consist of image display, plot (line graphics) display, and browser displays. The Data Browser and Observation Browser provide the standard browsing capabilities and provide both ASCII screen displays and 1-D, 2-D, and 3-D image and line graphic representations.

- **Infrastructure** components provide the mechanisms to connect the components to each other and to the user. Inter Process Communication (IPC) is used to pass messages between tools and applications in the environment. Helper tools encapsulate IPC calls to provide command-line access to services and status information in the applications. The SAO IRAF compatible parameter interface provides a flexible mechanism for invoking tools either interactively or in batch mode. The traditional parameter interface has been extended to recognize and interpret both stand-alone tools that compute parameter settings and IPC commands and dataset parameters that allow tools and applications to share common parameter settings.

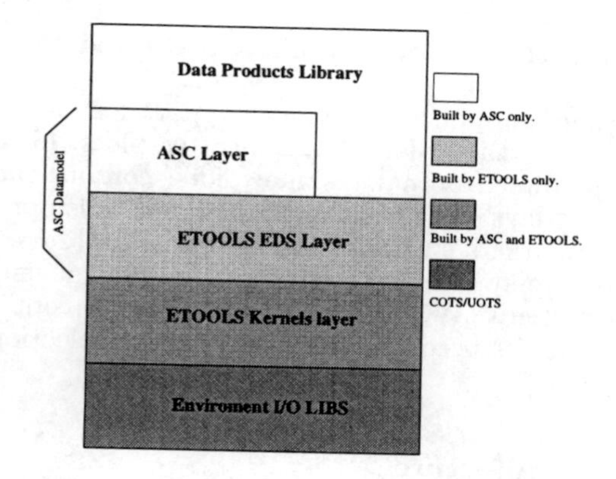

Figure 2. ASC Data API and Data Model.

3. Data Model

The data model provides a high-level application interface (API) through which all tools and applications can interact with the data. The Data Products layer defines the data API for the tools and applications. The data products consist of the standard high-energy astrophysics data products such as event-lists, spectra, and light-curves. This interface is independent of the physical storage formats (Herrero, Oberdorf, & Conroy 1997). Figure 2 shows the layered architecture of the Data API, the data model, and the dynamic data formatting(DDF).

4. Tool Architecture

The tools represent stand-alone executables that perform single functions. Each tool is invoked with user-specified parameters and transforms input data into output data. They are designed using an "open architecture," and are built using layered libraries (Conroy et al. 1996). The SAO Parameter Interface provides the mechanism by which these tools may be configured into the analysis environment when desired. The parameter values may be specified directly or via helper tools that invoke IPC calls to other parts of the environment or to a shared dataset parameter file. In this way, it is possible for a tool to perform analysis on an image region indicated by the Data Visualizer. Similarly, user scripts and pipelines can produce PostScript graphics plots when running in a batch environment, by accessing services without user interaction.

5. Visualizer Architecture

The visualization architecture also employs a layered approach which allows the graphics API to be defined independently of the image and graphics engines used to implement the functions. This layered design will allow the Data Visualizer to be built with either public domain display engines or commercially licensed engines, depending on the availability of the products. For instance, SAOtng (Mandel & Tody 1995) is the targeted public domain imaging engine, whereas

IDL represents a popular commercial product that could be linked to the Data Visualizer.

The visualization component provides support for the standard types of displays: 2-D and 3-D image plots, 2-axis scatter plots, histogram (1-D line) plots, strip-plots, predicted/actual/residual plots, contour plots, and sky-grid plots with catalog overlays. The Data Visualizer must also provide the interactive ability to draw markers such as circle, ellipses, contours, masks, polygons and boxes, and to display cursor coordinate read-outs in any of the support World Coordinate Systems (WCS). The IPC mechanism coupled with the SAO parameter interface can be used to communicate these selections to the analysis tools.

6. Application Architecture

The Modeling/Fitting Application (Doe et al. 1997) is an example of a high-level application that provides access to the model building, fitting, and visualization tools in an interactive application. One of the motivations for building this application is to allow graphical monitoring of the progress of the fitting process. The modeling portion of the application is controlled via a mini-language that provides natural language constructs to define the model components, operations, and the interrelationship between the modeling parameters.

7. Control Architecture

Navigator provides a graphical user interface (GUI) to the entire ASC system. The Data Analysis GUI represents the components of the Navigator that interact with the portable Data Analysis system. The GUI provides a powerful and intuitive interface for interactive analysis. The command-line interface allows access to the analysis components via a simple interface. This is particularly important to support semi-automatic user processing through scripts. The pipeline controller is the third control component. It can be invoked on pre-defined or user-defined pipeline profiles that execute complex sequences of operations.

Acknowledgments. This project is supported by NASA contract NAS8-39073 (ASC). We also gratefully acknowledge the many fruitful conversations with the other members of the ASC Data System and Science Data System.

References

Conroy, M., Doe, S., & Herrero, J. 1996, in ASP Conf. Ser., Vol. 101, Astronomical Data Analysis Software and Systems V, ed. G. H. Jacoby & J. Barnes (San Francisco: ASP), 285

Doe, S., Sieminginowska, A., Joye, W., & McDowell, J. 1997, this volume, 492

Herrero, J., Oberdorf, O., & Conroy, M. 1997, this volume, 469

Mandel, E., & Tody, D. 1995, in ASP Conf. Ser., Vol. 77, Astronomical Data Analysis Software and Systems IV, ed. R. A. Shaw, H. E. Payne, & J. J. E. Hayes (San Francisco: ASP), 15

Astronomical Data Analysis Software and Systems VI
ASP Conference Series, Vol. 125, 1997
Gareth Hunt and H. E. Payne, eds.

Implementation Design of the ASC Data Model

J. Herrero, O. Oberdorf, M. Conroy, J. McDowell

AXAF Science Center, Smithsonian Astrophysical Observatory, 60 Garden Street, MS 81, Cambridge, MA 02138

Abstract. The ASC data model provides an abstract description of *AXAF* datasets. Through the data model API, tools and applications will have efficient and transparent access to heterogeneous disk formats. To accomplish this, the data model will use a layered design.

1. Requirements

Previous requirements for the ASC data model stressed flexibility, extensibility, and economy (Conroy et al. 1995). These requirements can be further quantified so that an implementation design can be easily judged. What follows is a list of the quantified requirements that the implementation design had to meet:

Open Architecture: The system shall be implemented so that support for new data formats or new functionality can easily be added, without disrupting previous code or already developed tools and applications.

Standard API Independent of Data Format: The system shall be implemented so that different disk formats can be accessed through the same API.

Code Reuse: The system shall be implemented atop existing libraries that will do the actual low level file I/O.

Support ASC Needs: The system shall support the need of the ASC (*AXAF* Science Center) data analysis environment, so that tools and applications can easily be developed.

2. Implementation Approach

In order to to meet the requirement of support for multiple file formats through the same API, the library uses a layered approach. The bottom layer provides a standard API to multiple file formats. This API provides the other levels with a set of generic objects, such as tables and images, and will be implemented by adopting the ETOOLS package currently under development by a collaboration of CEA and SAO (Abbott et al. 1995; Abbott et al. 1997). The middle layer will add "meaning" to the ETOOLS data objects so that ASC applications can be designed easily and implemented in an consistent way. It will also provide scientific functions to manipulate the data. The top layer will consist of the

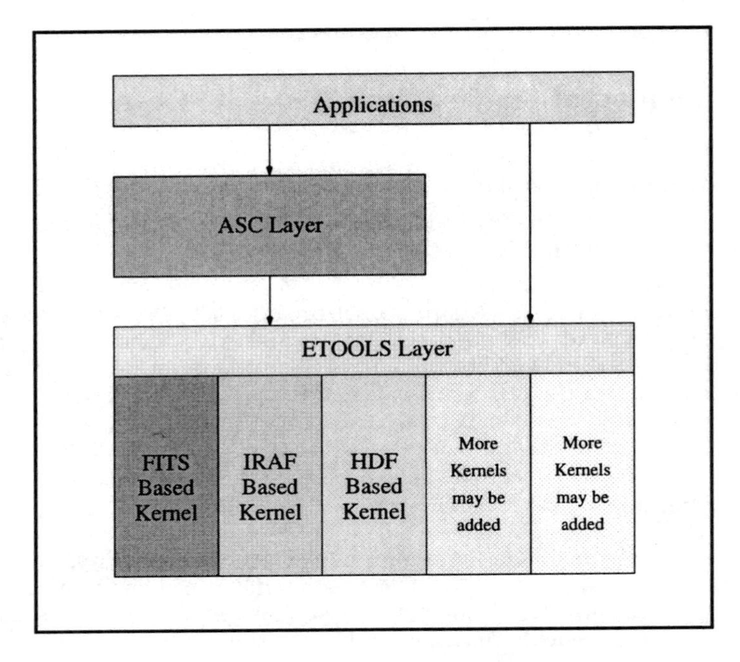

Figure 1. Layered Design of the ASC Data Model.

actual ASC tools and applications. Please refer to Figure 1 to see a diagram of the different layers in our implementation design.

3. ASC Layer

The ASC Data Library layer will provide the required *AXAF* science functionality to the generic data. It will be responsible for providing scientifically meaningful data objects, constructed from ETOOLS primary constructs, and data manipulation functions.

3.1. Dataset Layer

The Dataset Layer gives scientific meaning to the underlying generic objects. This layer will enhance or group the generic objects to produce specialized objects. For example, tools that need to operate on a "Light Curve" object would go through this interface to access their source or target objects. Here are two examples of new objects that the Dataset Layer would provide:

ASC Light Curve Object consists of a table with a binned time column and other columns containing either count rate or flux information and other auxiliary information. It is compatible with HEASARC rate files, and is interpreted as a light curve of count rate or flux versus time.

ASC Source List Object consists of a table with a column including positional information, a column containing source name identifiers (optional), and other optional columns. It is interpreted as a list of sources, and is used as input to programs which calculate derived source properties.

3.2. Science Layer

The Science Layer lies just above the Generic Layer. It contains routines to add additional meaning and functionality to the generic objects. For example, this layer will allow a column to have associated error values. Some examples of Science Layer functionality:

Error Columns: The library will keep track of columns which are error values for other columns in a table object. Science Layer function calls will automatically detect and handle cases where error values are provided.

Coordinate Transforms: The library will keep track of values which are actually coordinates. Science layer function calls will automatically perform any required coordinate transformations.

4. Generic Layer

The decision to adopt the ETOOLS package as the generic layer of the implementation of the ASC data model was made because it meets the implementation requirements. The most important way in which ETOOLS meets these requirements is by providing a uniform interface through which different data file formats can be accessed. This is done by a kernel mechanism, in which there exists one kernel for each data file format desired (e.g., one kernel for FITS files, one kernel for IRAF files, etc.). The following list outlines all the ways in which the ETOOLS package meets the ASC data model implementation requirements:

Open Architecture: The ETOOLS kernel mechanism allows more file formats to be supported by adding more kernels.

Standard API Independent of Data Format: Because of the ETOOLS kernel approach, tools and applications designed and implemented on top of the ETOOLS package can access different disk file formats through the same API.

Code Reuse: The ETOOLS project is not inventing any new file formats. Instead, it provides a standard way to use existing file formats and libraries.

Support ASC Needs: The ETOOLS project provides enough primary constructs to support the ASC tools and application needs, including support for tables, images, filters, and WCS transforms.

4.1. Supported Kernels

The ASC will write at least two ETOOLS kernels to implement its Data Model, the first will be an IRAF based kernel and the second kernel will be a FITS based kernel. Other kernels could be implemented, like an HDF based kernel.

IRAF Based Kernel would use IRAF formats to store the disk data. Event tables will be stored using QPOE files, images will be stored using IRAF image files, and filtering would be done by the QPOE/QPEX filter library.

FITS Based Kernel would use FITS to store the disk data. Event tables will be stored as FITS table extensions, images will be stored as FITS images, and filtering will be done by an ETOOLS provided filter engine.

HDF Based Kernel would use HDF file formats to store the disk data. At the moment this (possible) kernel has not yet been designed.

4.2. Kernel Example, an IRAF QPOE Kernel

The ASC IRAF kernel will use IRAF files to store data on disk. One requirement for this kernel is that the object files (and especially the event tables) can be copied out of the ETOOLS environment, and still be valid IRAF files that can be analyzed in the IRAF environment.

Tables are one of the more important objects in the analysis of event data. Event tables will be stored using QPOE files.

Images will be implemented using IRAF image files.

Filtering will reuse the power and flexibility of the QPOE/QPEX filtering mechanism.

Object Properties will be implemented using either QPOE headers or image headers, depending on the situation.

Acknowledgments. This project is supported by NASA contract NAS8-39073 (ASC). We also gratefully acknowledge the many fruitful conversations with the other members of the ASC Data System and Science Data System.

References

Abbott, M., Kilsdonk, T., Olson, E., Christian, C., Conroy, M., Brissenden, R., & Herrero, J. 1997, this volume, 96

Abbott, M., Kilsdonk, T., Olson, E., Christian, C., Conroy, M., Brissenden, R., Van Stone, D., & Herrero, J. 1996, in ASP Conf. Ser., Vol. 101, Astronomical Data Analysis Software and Systems V, ed. G. H. Jacoby & J. Barnes (San Francisco: ASP), 57

Conroy, M., Doe, S., & Herrero, J. 1996, in ASP Conf. Ser., Vol. 101, Astronomical Data Analysis Software and Systems V, ed. G. H. Jacoby & J. Barnes (San Francisco: ASP), 285

Van Stone, D., Conroy, M., & McDowell, J. 1996, in ASP Conf. Ser., Vol. 101, Astronomical Data Analysis Software and Systems V, ed. G. H. Jacoby & J. Barnes (San Francisco: ASP), 199

Astronomical Data Analysis Software and Systems VI
ASP Conference Series, Vol. 125, 1997
Gareth Hunt and H. E. Payne, eds.

ASC Coordinate Transformation—The Pixlib Library

H. He, J. McDowell, M. Conroy

Smithsonian Astrophysical Observatory, 60 Garden Street, Cambridge, MA 02138, E-mail: hhe@cfa.harvard.edu

Abstract. We describe a coordinate library for *AXAF* data analysis. The library handles transformations between celestial coordinates and instrumental (mirror, focal plane, detector pixel) coordinate systems. The need for careful transformations is driven by the accuracy of the detectors and the attitude determination system. The coordinate systems are characterized by parameter files generated from experimental and calibration data. Transformation calculations are performed by matrix-representation routines for maximum flexibility. This library is implemented in ANSI C, and uses the SAO IRAF-compatible parameter interface.

1. Introduction

Pixlib library is intended to perform *AXAF* Science Center (ASC) coordinate system data analysis. The library facilitates space telescope simulations dealing with five science instruments (SI) or detectors (e.g., ACIS, HRC-I), each constructed up to ten chips. The pixlib is designed to be a structured system, to promote system extensibility and maintainability. The fundamentals of the program are provided by a number of modules which rely on external parameter interface files. The parameter files determine the essential characteristics of the ASC coordinate systems. The coordinate transformation computation utilizes matrix arithmetic algorithms for coding efficiency and software reusability. The integrated library provides a comprehensive collection of functions for user applications, and these functions are well documented, easy to use, and FORTRAN-binding compatible. At the present, the library affords more than 50 transformations among ~20 ASC coordinate systems.

2. System Design

Figure 1 lists the most important ASC coordinate systems and sketches the transformation paths among them. Items threaded by line arrows denote coordinate systems, e.g., **CHIP** for 2-D chip pixel coordinates, **CPC** 2-D chip physical coordinates in mm, **TDET** 2-D tiled detector plane pixel coordinates, and **LSI** for local SI coordinates. The line arrows connecting a pair of coordinate systems indicate the coordinate transformation is both forward and backward, and the aim of the transformation is explained at the side. For instance, transformation of **CPC** to **LSI** is to account for chip orientation in space, and **N** to **FP** is to convert the 3-D HRMA nodal position to the 2-D pixel focal plane co-

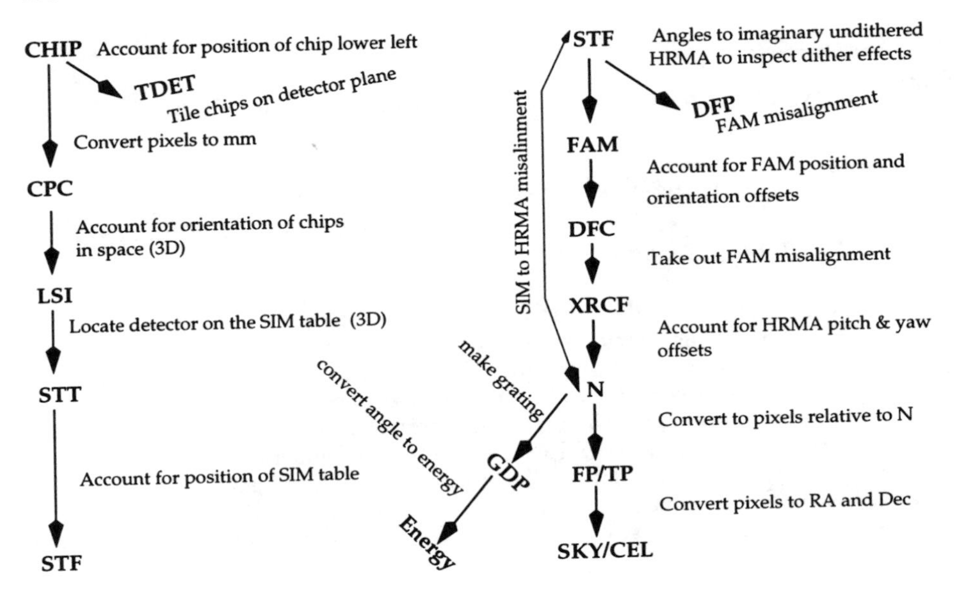

Figure 1. ASC coordinate systems and the transformations.

ordinates. Details on the definitions and determinations of the ASC coordinate systems are discussed in a memo (McDowell 1996).

The goal of pixlib is to build up a software framework for the coordinate systems outlined above, and provide functions to perform the coordinate transformations between any two systems. Borrowing from the C++ class concept, the pixlib distinguishes "private" (lower-level) data/functions and "public" (upper-level) routines. The private section, composed of more than five independent modules, is constructed on a set of external parameter files that hold the definitions and aspects of the ASC coordinate systems (to be discussed in next section). A struct member is defined in each module in order to handle the interface with the external files. The initiated data members of the struct are then passed to the function members of the module that are to be invoked in the upper-level applications. Several important modules of the telescope simulator are briefed below. The function of module `pix_detplan.c` is to layout chips on a **TDET** plane following the specifications (e.g., chip lower-left corner position and orientation, etc.) given in relevant parameter files. The motivation for introducing **TDET** is that the *AXAF* detector chips do not lie in a plane, making it difficult to inspect simultaneously both individual chip pixels and an overall geometry of the detector. To retain this information, chips are tiled in an approximately correct relative orientation, but the chip edges are paralleled and the gaps between the chips are edged. The `pix_grating.c` subsystem focuses on grating observation of a dispersed spectrum, determining the photon energy given a zero-order position for the undispersed photon trajectory, or traces the dispersed photon trajectory for a given energy. Module `pix_chip2stf.c` sets up **CHIP**, **CPC**, **LSI**, **STT**, and **STF** coordinate systems, and provides methods for the system transformations. Similarly, `pix_stf2tpc.c` and `pix_tpc2src.c` initiate the rest of coordinate systems and complete the transformations between **STF** and **CEL**.

By providing a rich collection of transformation functions, the "public" module hides the details of the implementations by wrapping one or more lower-level method(s). Unlike the lower-level programs, which are especially heavy on pointer manipulation, the upper-level functions are controlled by arguments passed by value. This type of manipulation was particularly desirable, since it allows FORTRAN-binding compatibility. Once the coordinate framework is built (or initiated), a requested transformation call from, say, **CPC** to **FP**, is able to output coordinates in **CPC** for given coordinates in **FP** without asking for other inputs.

3. Data Structure

The pixlib adopts SAO IRAF-compatible parameter file(s) as data I/O structure. There are six parameter files which are generalized from experimental or calibration data (McDowell 1996). These files specify several primary coordinate origins, relevant orientations, and others. Those are described as follows.

- `pix_origin_table.par` sets eleven *AXAF* primary coordinate system origins in spacecraft and HRMA nodal coordinates,
- `pix_tdet_planesys.par` defines nine **TDET** plane systems (dimension and center) of the five detectors,
- `pix_tdet_nomfocus.par` determines nominal focus points of **TDET** plane systems,
- `pix_tdet_refcoords.par` specifies the lower-left corner positions in **TDET** of 17 chips,
- `pix_corner_lsi.par` maps 2-D chip corner positions to 3-D **LSI** coordinates,
- `pix_sim_table.par` defines seven aim-points, the offsets from SIM transformation table (**STT**), of detectors, and
- `pix_size_cntr.par` constitutes 2-D instrumental pixel systems (i.e., **SKY**, **FP**, **TP**, and **GDP**) and grating properties.

The data are entered in the parameter files as a string, such as "(x, y, z)", in order to retain clarity. The string is interpreted back into the individual numbers when the file is opened. These coordinate system definition files are hidden from the user and looked up internally, once per run, in lower-level modules during the pixlib initiation. On the other hand, an additional parameter file is created specially for user interface. It allows the user to specify the chip of a detector, aim-point, focal plane system, etc. for performing a telescope simulation or data analysis.

4. Matrix Algorithm

The pixlib coordinate transformation computations are performed with matrix arithmetic. A portable matrix library package provides utilities to handle matrix algebra (i.e., vector-vector cross/dot/addition/subtraction, matrix(vector)-matrix multiple, matrix transpose/compose). Specifically, we use homogeneous coordinates (Salmon & Slater 1987) to simplify further matrix representation of coordinate transformations. Application of homogeneous coordinates to the ASC coordinate systems is briefly described below.

Coordinate transformation from frame A to frame B can be concisely expressed as $\wp(B) = F(A,B)\wp(A)$. Where $\wp$ is a 4×1 homogeneous coordinate vector, defined as $\wp \equiv \begin{pmatrix} P_1 & P_2 & P_3 & 1 \end{pmatrix}^T$ for a conventional 3-D spatial vector $\begin{pmatrix} P_1 & P_2 & P_3 \end{pmatrix}^T$. $F(A,B)$ is a 4×4 transformation matrix expressed in terms of rotation and translation matrices,

$$F(A,B) = \begin{pmatrix} R(A,B) & A_0(B) \\ O & I \end{pmatrix}_{4\times4}$$

$R(A,B)$ above denotes a 3×3 A-to-B rotation matrix, $A_0(B)$ a 3×1 translation (column) matrix of frame A origin in frame B. O is 1×3 zero matrix and I 1×1 unit matrix. To obtain the backward B-to-A transfer, $\wp(A) = F(B,A)\wp(B)$, the corresponding rotation and translation matrices for the $F(B,A)$ are derived, assuming $R(A,B)$ orthogonal, as

$$R(B,A) = R^T(A,B) \quad \text{and} \qquad B_0(A) = -R^T(A,B)A_0(B).$$

We use a transformation struct to compose the matrices, or arrays of $F[4][4]$, $R[3][3]$, and $A_0[3]$, for reusability. The struct is dynamically allocated and instantiated when a upper-level request is called. The values of A_0 and R elements are either taken from the parameter files or derived by parsing arguments, and the F are then composed with the known A_0 and R. The requested coordinate transformation from A to B is thus completed by a matrix-vector multiple performance.

Acknowledgments. We are grateful to SAO colleagues for offering discussions and knowledge regarding the paper. This project is supported from the *AXAF* Science Center (NAS8-39073).

References

McDowell, J. 1996, ASC Coordinates, Revision 4.0, SAO/ASCDS

Salmon, R., & Slater, M., 1987, Computer Graphics: Systems and Concepts (Reading, MA: Addison-Wesley)

Astronomical Data Analysis Software and Systems VI
ASP Conference Series, Vol. 125, 1997
Gareth Hunt and H. E. Payne, eds.

Simulated AXAF Observations with MARX

Michael W. Wise, David P. Huenemoerder, and John E. Davis

MIT Center for Space Research, AXAF Science Center
70 Vassar St. Building 37-644, Cambridge, MA 02139-4307

Abstract. The AXAF Science Center group at MIT has developed an end-to-end simulator of the AXAF satellite called **MARX**. **MARX** includes models of the AXAF mirrors, the low- and high-energy transmission grating assemblies (LETG and HETG), and the HRC and ACIS focal plane detectors. We discuss the role of **MARX** within the ASC Data System and present sample simulated images and spectra for two typical cosmic X-ray sources.

1. Introduction

The AXAF Science Center (ASC) Data Analysis System will include the ability to simulate the detailed response of the AXAF satellite to X-ray sources. These simulations will be used for a variety of purposes including: development of processing algorithms; ground-based and on-orbit performance prediction; testing of the standard processing pipelines; and scientific observation planning and prediction. As part of this modeling effort, the AXAF Science Center group at MIT has developed a modular, portable, stand-alone simulator **MARX** (**M**odel of **A**XAF **R**esponse to **X**-rays). In this paper, we will briefly discuss some of the motivations which have driven **MARX**'s development as well the simulator's current level of functionality. To demonstrate its capabilities, we present simulated AXAF images and spectra for two cosmic sources.

2. Motivations

MARX was originally developed to provide sample AXAF data with an *exactly* known instrument response in order to develop and test spectral extraction and analysis algorithms. It has since developed into an accurate and flexible model of the in-flight capabilities of AXAF and provides a fast and portable alternative to other simulation methods. Here we list some of the uses **MARX** simulations serve within the ASC Data System:

- **Development of data analysis software**
  MARX provides realistic data incorporating a variety of physical effects which characterize AXAF. These data can be used to develop new algorithms, such as event-based spectral extraction of HETG and LETG spectra, or deconvolution of overlapping orders in the case of the LETG. **MARX** simulations provide knowledge of the exact solution for comparison with the reconstructed data.

- **Fitting and Data Analysis**
  As part of the ASC Data System, a fitting application is being designed to allow comparison of models with calibration and flight data (see Doe et al. 1997). This Fit Engine will be capable of driving **MARX** to produce high-quality simulations as part of the fitting process.

- **Calibration**
  Ground calibration of AXAF will utilize detailed predictions to plan the tests and as an aid to the interpretation and digestion of resulting test data. Many of these predictions are being compiled using **MARX**.

- **Flight Observation planning**
  MARX provides a realistic model of the AXAF response to cosmic X-ray sources and can be used for detailed planning of anticipated science programs. One may investigate issues related to sensitivity, spatial resolution, or spectral resolution, for instance. **MARX** will be released to the community as part of the proposal planning software package.

3. Functionality

MARX provides the capability to simulate the various combinations of scientific instruments on-board the AXAF satellite and includes models of the AXAF mirrors, the low- and high-energy transmission grating assemblies (LETG and HETG), and the HRC and ACIS focal plane detectors. For several components of the system, the user may choose between several simulation modes depending on their needs. The mirrors, for example, can be simulated using a simple effective area model or via a full ray-trace if desired. A number of simple source models (extended or point-like) are supported and users may extend the capabilities by incorporating their own models. Sources may have arbitrary spectral energy distributions which are specified via an input ASCII file. **MARX** also provides support for simulations of ground-based calibration tests at XRCF. The HRMA shutter assembly is modeled as well as sources at a finite distance.

4. Implementation

MARX has been coded entirely in C and consists of a single executable. Control of the program is accomplished through a parameter file using the IRAF parameter interface library. The simulator is reasonably compact (requiring only 6 MB for the entire distribution) and fast (3,000 photons/sec on a SPARC 20). It has been successfully compiled under SunOS, Solaris, and Linux. Output from **MARX** can take several forms: a directory of simple binary vectors for each photon property, a FITS photon event list from which images and spectra may be extracted, an SAOSAC DPDE rayfile, or even ASCII files.

5. Sample Simulations

5.1. UX Ari Simulation

UX Ari is a bright RS CVn binary system—a prototypical coronal source. A continuous emission measure model was constructed based on the published

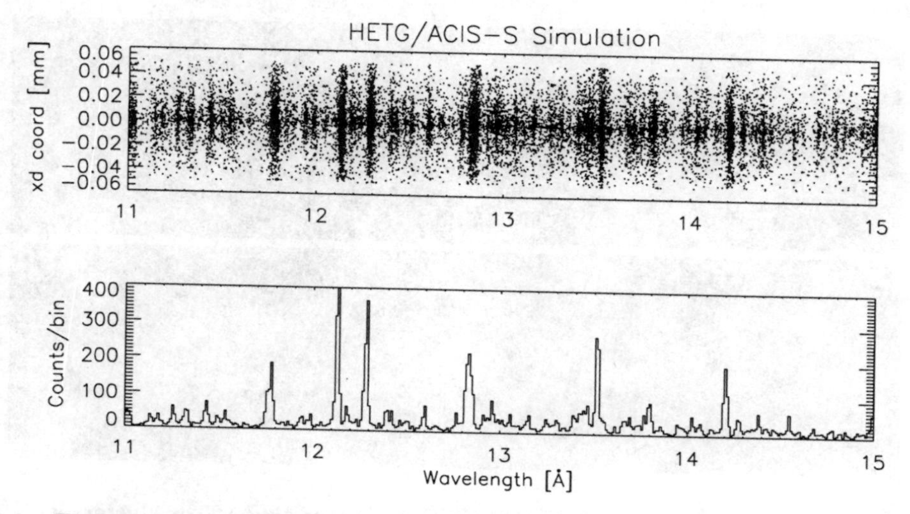

Figure 1. A simulation of a stellar coronal plasma as detected by the HETG with the ACIS-S detector. In the upper panel, an expanded view of the detected photon distribution is shown. Spectral lines show up as dark, vertical bands. The lower panel gives the total counts per pixel for the region of the spectrum shown in the upper panel.

EUVE data (Dupree 1996); the peak emission occurred near 10^7 K. Simulations were run for both the HETG and LETG

with the same input spectrum for an exposure time of approximately 40 ksec. The flux for the simulations was set at 4×10^{-11} ergs cm^{-2} s^{-1} in the range 0.1–10 keV. For the HETGS Simulation, one million input photons gave 124,000 output photons for an equivalent exposure time of 36 ksec. In the extraction rectangle, there were 14,000 first order HEG photons and 50,000 MEG. The high-order contribution was about 4% of the first order, and about 50,000 photons went into the zeroth order. With the LETGS, four million input photons resulted in 85,000 detected photons for a 39 ksec equivalent exposure time. The zeroth order had 33,000 photons, and the first order 46,000, with high order contribution about 10% of the first order. Figure 1 shows an expanded view of a portion of the detected HETG image as well as the extracted spectrum.

5.2. Abell 2029 Simulation

The Abell 2029 cluster of galaxies is a bright, nearby (z=0.0767) cluster which is bright in X-rays and contains a strong cooling flow. Such cooling plasmas should be rich in X-ray line emission and make interesting targets for the HETGS. A2029 is, however, an extended object making analysis of its spectrum more involved. The spectrum for the cluster emission was taken from XSPEC fits to ASCA data for this object (Wise & Sarazin 1997, in preparation). The flux for the simulations was taken to be 7.5×10^{-11} ergs cm^{-2} s^{-1} in the range 2–10 keV. The spectral emission was assumed to consist of two components: isothermal cluster emission from a T=7.45 keV Raymond-Smith thermal plasma with an abundance of 0.4 solar and a central cooling flow with $\dot{M}_X=300\,M_\odot\,yr^{-1}$.

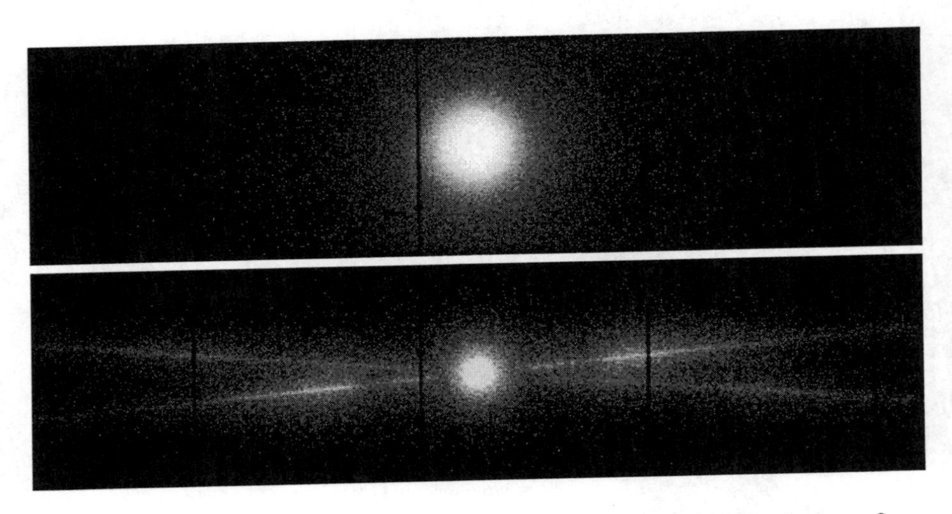

Figure 2. A simulated AXAF observation of the A2029 cluster of galaxies. The lower image depicts an observation utilizing the HETG on-board AXAF while the upper image shows the cluster without the HETG in place.

The cluster emission was assumed to be distributed in a standard spherically symmetric, isothermal beta surface brightness distribution with a core radius of r_c=200 kpc (~100″) and β=0.73 (Sarazin 1986). The cooling flow emission was modeled spatially as a radially symmetric Gaussian with σ=100 kpc (~50″) corresponding to the cooling radius of the flow. A 100 ksec simulated HETG observation yielded 78,000 detected events. The same exposure without the grating in place yielded 218,000 detected events. Figure 2 shows the resulting images from the two simulations.

Acknowledgments. We gratefully acknowledge the aid and support of various members of the AXAF project including the Mission Support Team, the Calibration group, and the Data Systems group.

References

Doe, S., Siemiginowska, A., Joye, W., & McDowell, J. 1997, this volume, 492

Dupree, A. K. 1996, in ASP Conf. Ser., Vol. 109, Cool Stars 9, ed. R. Pallavicini & A. K. Dupree (San Francisco: ASP), 237

Astronomical Data Analysis Software and Systems VI
ASP Conference Series, Vol. 125, 1997
Gareth Hunt and H. E. Payne, eds.

The AXAF Science Center Performance Prediction and Calibration Simulator

R. A. Zacher, A. H. MacKay, B. R. McNamara, and L. P. David

SAO/ASC, Cambridge, MA 02138

Abstract. We are developing and integrating software to simulate the focal plane detectors, shutters, and gratings for the Advanced X-ray Astrophysical Facility (AXAF). AXAF is one of four observatories in the NASA "Great Observatory" series, scheduled for launch in 1998. AXAF will offer unprecedented spatial and spectral resolution in the X-ray band ranging from 0.1–10 keV. The path of X-ray photons is simulated from the exit of the telescope mirrors to the focal plane. Each major functional element of the simulation is represented by an independent module. Module execution and inter-module communication is accomplished within a pipeline architecture. The software is written in C/UNIX and utilizes a number of existing astronomical software libraries. Detailed models are being developed for the two focal plane instruments. These instruments are ACIS, which is a CCD camera, and HRC, which is a microchannel plate detector. Realistic detector output files are generated in a variety of formats. The simulations are currently being used for planning calibration activities, on-orbit performance prediction and for testing the analysis and telemetry software.

1. Introduction

We are developing computer models to simulate the focal plane detectors of the Advanced X-ray Astrophysical Facility (AXAF). The models, being developed as part of the AXAF Science Center, are being used to aid in calibration planning and to characterize the performance of the AXAF observatory. Scripts have been developed to configure and run the simulations automatically from a test database. Depending on the output mode selected, the results can be viewed directly, sent through telemetry processing, or fed into higher-level analysis pipelines in the Data System.

Much of our work has been focused on developing high-fidelity simulations of the two main Scientific Instruments located in the focal plane of the telescope. These are the AXAF CCD Imaging Spectrometer (ACIS) and the High Resolution Camera (HRC). In addition to these, we have integrated gratings modules and we use output from other simulators which model the sources and telescope mirrors (see Figure 1).

1.1. AXAF CCD Imaging Spectrometer (ACIS)

ACIS is a charge-coupled device optimized for X-ray detection. Its 24μm pixel size offers $^1/_2''$ resolution in the AXAF focal plane. The field of view is $16\times16'$ for

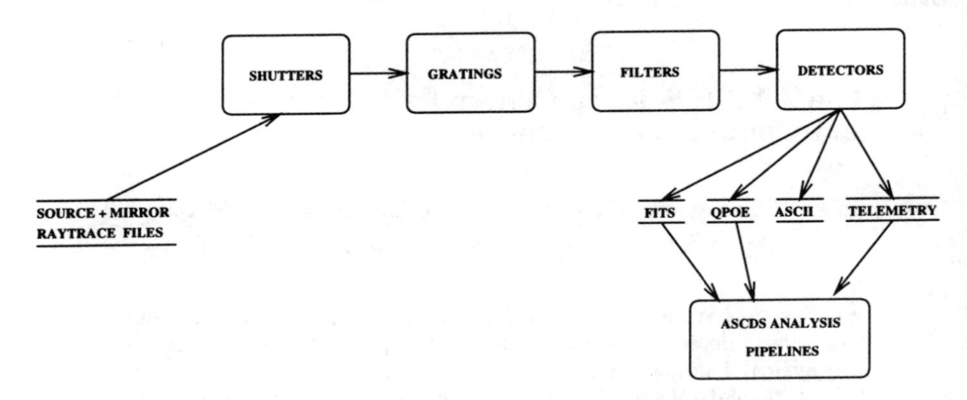

Figure 1. **Simulation Schematic.** The SHUTTERS module simulates 16 separately configurable shutters behind the mirror assembly. The GRATINGS module simulates the High, Medium, and Low Energy Transmission Gratings. The FILTERS module simulates optical blocking filters in front of the detectors. The detectors are ACIS (a charge coupled device) and HRC (a microchannel plate).

the imaging array and $8\times48'$ for the spectroscopic array. The chip is modeled as a multilayer structure. The incident X-ray photons create a charge cloud whose position and size are determined by the silicon absorption depth, the photon energy, the photon position and the dopant concentration. The charge cloud then drifts to the surface of the chip under the influence of a layer dependent electric field, which we model using a Monte Carlo method. At the surface, the charge is mapped onto a 3×3 pixel array. The functional dependence for the number of electrons (n_e) created by an incident X-ray of energy E_x is given by $n_e{\sim}n_x(E_x/\Delta E)$. Here, n_x is a function calculated by the Monte Carlo program and Δ=3.65 eV is the energy required to liberate a charge carrier. Additional features modeled include read noise, charge transfer inefficiency, bias, gain, and layer thicknesses. The algorithms used in the CCD simulation are based on the program XRAYSIM developed by Lumb et al. (1994), which in turn was based on analytical calculations by Janesick (1987, 1988) and Hopkinson (1987).

The simulator can be operated in two modes. The *Event List Mode* outputs the abovementioned 3×3 pixel array to a FITS event list. This mode is designed for high throughput and does not model effects which arise when two photons hit the same location on the chip in a given integration period. The *Full Frame Mode* embeds the 3×3 pixel array in a much larger rectangular array in memory. This mode includes the effects which arise when multiple photons hit the same location in a given integration period. The *Full Frame Mode* has two output formats available. The events can be extracted from the array in memory using the same algorithm used to detect events in real ACIS frames. The extracted events are output to a FITS event list. Alternatively, the full arrays can be written to FITS image files, one for each integration period. These image files are similar to those produced by the physical chips before event extraction. Event detection in the array in memory yields substantial performance gains over detection of the events in the FITS image files.

1.2. High Resolution Camera (HRC)

The HRC is a microchannel plate (MCP) detector that provides a spatial resolution of less than $^1/_2''$. The field of view is $31\times31'$ for the imaging array and $7\times97'$ for the spectroscopic array. Resolving power is limited, with $E/\Delta E \sim 1$. The simulation models the UV Ion Shield (UVIS), the MCP itself, and the wire charge grid. The UVIS is modeled using a generalized filter program that statistically simulates photon absorption by applying a transmission curve to the input photon energy. The MCP surface is modeled as a surface of circular pores with a diameter of 0.0125 mm and spacing of 0.015 mm. A model of quantum efficiency as a function of incident angle is also applied. The wire grid charge resulting from the charge cloud produced by the MCP is modeled by a scaled Lorentz function. Events are passed into a telemetry simulator that models dead time induced by telemetry bandwidth limitations. Output modes are raw telemetry, FITS event list, and QPOE image formats. The HRC simulator has also been adapted to simulate a similar instrument called the High Speed Imager (HSI) which is used for telescope mirror calibration.

1.3. Mirror and Grating Modules

The mirror simulation's raytrace output can be projected directly on to the model detectors, or diffracted by the gratings module before projection on to the model detectors. The dispersed gratings spectrum provides a resolving power of $\mathrm{E}/\Delta\mathrm{E} \sim 10^{2\rightarrow 3}$. The High Energy Transmission Grating is typically used in conjunction with the ACIS detector and the Low Energy Transmission Grating with the HRC.

2. Architecture

The simulator control hierarchy is depicted in Figure 2. The simulators run as a set of UNIX processes, each of which represents a physical component being modeled. These processes are started, monitored, and stopped by the ASCDS Pipeline Controller. The simulators utilize common ASCDS libraries where possible, such as the IRAF parameter interface. Events (photons) are passed from one process in the pipeline to the next with each process performing some necessary action on an event before passing it along. The action may be to alter the event or to decide not to propagate it. The simulators are implemented primarily in C, with some supporting code written in Perl.

3. Conclusions

The software is designed to easily accommodate modifications and enhancements. The modular pipeline approach has facilitated the interchange of modules as we continue to upgrade the fidelity and capabilities of the simulations. In order to further improve flexibility and provide access to data in the simulator pipelines, a C++ Application Program Interface to the simulator data stream is being prototyped.

Acknowledgments. This project is supported by NASA contract NAS8-39073 (ASC). We would like to thank the following individuals for their contributions to the simulations. Terry Gaetz, Diab Jerius, and Dan Nguyen developed

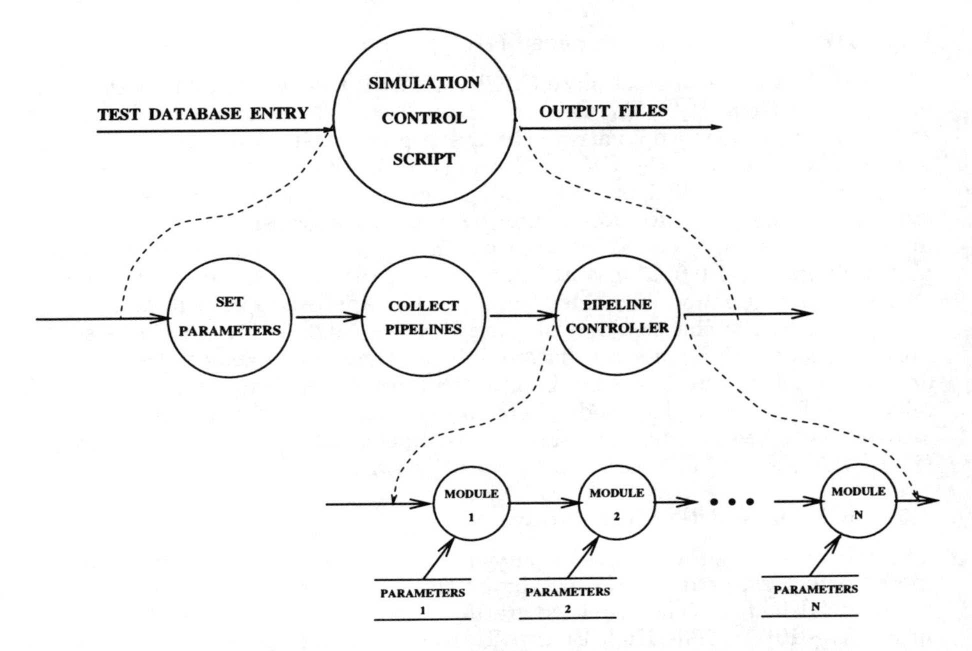

Figure 2. **Simulation Control Hierarchy.** The hierarchy of program control is depicted. Using database entries describing the test to be performed, parameters describing the configuration are set. The ASCDS pipeline controller initiates the raytrace and monitors program execution.

the mirror and source models. John Davis, Dan Nguyen, and Mike Wise developed the gratings model. Diab Jerius developed the HRC pore surface model. Dave Plummer developed the HRC Telemetry dead time simulation. Adam Dobrzycki developed the HRC charge grid algorithm.

References

Janesick, J. R., Elliot, T., Collins, S., Taher, D., Campbell, D., & Garmire, G. 1987, Optical Engineering, 26, 2, 156

Janesick, J. R., Elliot, T., Bredthauer, R., Chandler, C., & Burke, B. 1988, SPIE, 982, 70

Hopkinson, G. R. 1987, Optical Engineering, 26, 8, 766

Lumb, D., Townsley, L., Nousek, J., Burrows, D., & Corbet, R., 1994, personal communication

Astronomical Data Analysis Software and Systems VI
ASP Conference Series, Vol. 125, 1997
Gareth Hunt and H. E. Payne, eds.

Modeling AXAF Obstructions with the Generalized Aperture Program.

D. Nguyen, T. Gaetz, D. Jerius, and I. Stern

Smithsonian Astrophysical Observatory, Cambridge, MA 02138

Abstract. The generalized aperture program is designed to simulate the effects on the incident photon stream of physical obstructions, such as thermal baffles and pre- and post-collimators. It can handle a wide variety of aperture shapes, and has provisions to allow alterations of the photons by the apertures. The philosophy behind the aperture program is that a geometrically complicated aperture may be modeled by a combination of geometrically simpler apertures. This is done by incorporating a language, *lua*, to lay out the apertures. User provided call-back functions enable the modeling of the interactions of the incident photon with the apertures.

This approach allows for maximum flexibility, since the geometry and interactions of obstructions can be specified by the user at run time.

1. Introduction

The Advanced X-ray Astrophysics Facility (AXAF), due to be launched in 1998, is the third of NASA's four Great Space Observatories. AXAF will provide unprecedented capabilities of high resolution imaging and spectroscopy over the X-ray range of 0.1–10 keV.

As part of our efforts to support the AXAF program, the SAO AXAF Mission Support Team has developed a software suite to simulate AXAF images generated by the flight mirror assembly. One of the tasks of this system is to simulate the physical obstructions in front of and behind the AXAF mirrors.

The generalized *aperture* program is designed to simulate the effects on the incident photon stream of apertures in physical obstructions, such as the X-ray and thermal baffles. It can handle a wide variety of aperture shapes, and has provisions to allow alteration of the photons by the apertures. Apertures can simply pass, block, redirect, modify, or generate new (e.g., fluorescent) photons.

2. The Generalized Aperture Description

The conventional approach to simulating physical obstructions is to assign absolute positions and orientations to each aperture. This approach would be tedious, error prone, and inflexible to changes in design later on.

The modeling of apertures as combinations of sub-apertures, and recognizing that the bookkeeping involved in tracking the photons through the apertures is independent of any particular type of aperture, led structuring the program into three components: the *front end*, *central engine*, and the *back end*.

The *front end*, which parses the description of the openings and generates lists of apertures against which to check the photons; the *central engine*, which reads the photons and checks them against the apertures in the lists, and the *back end*, composed of modules which model each of the apertures, are called upon by the front end and central engine.

The three sections of the program are encapsulated so that a minimum of information is exchanged between them, permitting integration of new modules into the back end without affecting the rest of the program.

2.1. The Front End

The user interface The *aperture* program's user interface is a programming language, *lua*. Embedding an interpreted language such as *lua* in the front end of the *aperture* program enables the application program *aperture* to be programmable at run time. The *lua* language is a small language with flow control (*if..then, while..do..end*, etc.), functions, floating point arithmetic, multi-dimensional arrays and structures, and implicit dynamical memory allocation.

Aperture Positions and Orientations The *aperture* program uses the concept of a local coordinate system in which each aperture is placed. It keeps track of the transformations required to map between the external coordinate system in which the input photon positions and directions are specified, and the local coordinate systems of the apertures. It provides for the hierarchical layering of coordinate systems:

- A *global* coordinate system, relative to which assemblies are specified.
- An *assembly* coordinate systems, relative to which either sub-assemblies or apertures are specified.
- A *sub-assembly* coordinates systems, relative to which either nested sub-assemblies or apertures are specified.

At each level, changes to the current coordinate system do not affect the higher level coordinate systems.

2.2. The Central Engine

The generalized aperture program takes a list of assemblies and apertures created by the front end, and compares each input photon to the apertures, in turn. It calls functions provided by the back end modules, which do the actual checking of the photons. It provides the logic to move the photon to the next assembly, should an aperture accept a photon and pass it along.

2.3. The Back End

Each back end module consists of two components. The first is called by the user's aperture definition script, and creates an instance of an aperture with a given set of aperture specific parameters, attaching it to the list of apertures for the current assembly. The second component contains the logic necessary to determine if a photon falls within it and is affected by it. Apertures can simply pass, block, redirect, modify, or generate new photons (e.g., fluorescent).

3. Detailed Descriptions of the Aperture Program

The *lua* program creates assemblies and sub-assemblies, the assemblies are represented internally as a list of *Assembly* structures. The sub-assemblies are temporary constructs which exist only to allow stacking of the current transformation matrix.

The back end modules called from the *lua* program create instances of apertures by creating aperture-specific data objects which contain the information necessary to fully describe the aperture. For example, an annulus is described by its inner and outer radius; a rectangle by its height and width. The module itself is described by an aperture module structure, which contains pointers to standard routines provided by the module (aperture instantiation, initialization, photon processing, and cleanup). This structure and the data object are passed to the front end utility routine, which encapsulates them inside an abstract object and appends that object to the list of apertures for the current assembly.

The processing of photons begins by reading in a photon. This is usually done from the specified input stream, but may be done from a special internal stack if any photons have been generated by an aperture. It then determines the first valid assembly with which the photon will interact. Usually this is the first assembly, but if the photon is generated by an aperture, the first valid assembly is determined by the assembly to which the generating aperture belongs.

Beginning with this first valid assembly, the central engine simulates the interactions of the photon with the apertures by an in-order traversal of the assembly's aperture list. At each visit of an aperture, the module associated with the aperture is passed the data object specific to the aperture and a copy of the photon data packet. It uses these data to determine whether it can process the photon, and returns a code to the central engine indicating whether or not it has accepted the photon for processing and the state of the photon after processing. Normally, the module does not alter the contents of the photon data packet; passing it a copy is a safety measure. In the event that it has done so purposefully, it signals the central engine (via the return code) to transfer the contents of the copied photon data packet to the original, so that the remaining apertures interact with the modified photon.

Depending upon the returned code, the central engine may discard the photon and read a new one, begin processing the next assembly in line, or restart the traversal of the current assembly. The last action is taken if the option *loop-mode* is enabled and photons have been redirected or generated. It is the only means by which a photon can successfully interact with more than one aperture in an assembly.

References

Du Bois, P. F. 1994, Computers in Physics, 8,

Ierusalimschy, R., de Figueiredo, L. H., & Filho, W. C. 1995, Reference Manual of the Programming Language Lua 2.1[1]

---

[1] ftp://ftp.inf.puc-rio.br/pub/docs/techreports/95_12_ierusalimschy.ps.gz

Astronomical Data Analysis Software and Systems VI
ASP Conference Series, Vol. 125, 1997
Gareth Hunt and H. E. Payne, eds.

The AXAF Ground Aspect Determination System Pipeline

M. Karovska, T. Aldcroft, R. A. Cameron, and J. DePonte

Harvard-Smithsonian Center for Astrophysics, Cambridge, MA 02138, USA, E-mail: karovska@cfa.harvard.edu

M. Birkinshaw

Department of Physics, University of Bristol, Bristol, BS8 1TL, UK, and Harvard-Smithsonian Center for Astrophysics, Cambridge, MA 02138, USA

Abstract. This article describes the *AXAF* aspect determination pipeline and its components, and highlights a procedure for determining centroids—a crucial step in deriving the post-facto aspect solution.

1. Introduction

The aspect determination system (ADS) is responsible for the collection of data that allow the pointing direction and roll angle of the *AXAF* Observatory to be reconstructed from the telemetry. The major parts of this subsystem are:

1. the aspect camera assembly (ACA) with its stray light shade,
2. the gyroscopes (inertial reference units—IRUs),
3. the fiducial light assembly, and
4. the fiducial transfer system.

The *AXAF* aspect camera is a 0.11 m diameter Ritchey-Chrétien telescope with a 3-element refractive corrector, which images about 2 square degrees of sky into one of two red-sensitive 1024×1024 CCDs. The optics of the camera are deliberately defocused, so that the FWHM of star image is about 9″, well spread out relative to the CCD scale of 5″/pixel. The aspect camera directly views the optical sky, and also views fiducial lights arranged around an X-ray detector, imaged via the fiducial transfer system consisting of a retroreflector/collimator and periscope (see Birkinshaw & Karovska 1995). Up to eight images (normally five guide stars and three fiducial lights) are tracked by the aspect camera, and their astrometry provides a history of the celestial pointing coordinates and roll angle for each X-ray observation.

2. Aspect Determination Pipeline

The Ground Aspect Determination System (GADS) pipeline is a collection of tools which process aspect telemetry and generate aspect data products. It

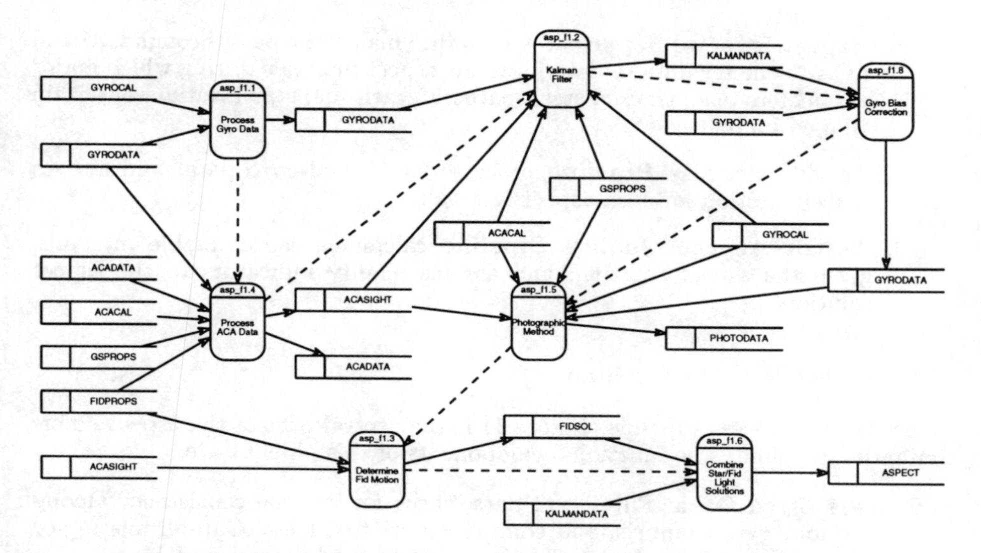

Figure 1. The Aspect Process Pipeline and its components.

produces a three-axis aspect solution in J2000.0 celestial coordinates, to support both image reconstruction (with error less than 0.5″) and celestial location for X-ray data (with error less than 1″).

The aspect solution is calculated from telemetry data (ACADATA and GYRODATA on Figure 1), generated by the aspect camera assembly and the inertial reference units (gyroscopes), using their associated calibration and alignment data (GYROCAL, ACACAL, GSPROPS, and FIDPROPS in Figure 1). The ACA telemetry contains pixelated images of stars between 10th and 6th magnitude and fiducial lights, nominally at 7th magnitude apparent brightness, at intervals of 1.025, 2.050, or 4.100 s (corresponding respectively to 4×4, 6×6, and 8×8 pixel images). The *AXAF* IRUs provide four axes of integrated angular rate telemetry at 0.25625 s intervals, from which three orthogonal axis rate data are derived.

The aspect camera data are analyzed to determine the centroid of each star or fiducial light image. The star centroids are combined with the IRU rate data in a Kalman filter to provide a time-varying optimal aspect solution and associated covariance matrix, assuming known noise models for the aspect camera and IRU data. The relative offsets between the star centroids and the fiducial light centroids are then calculated. Finally, the boresight calibration is applied to the relative offset, to transform the optical celestial aspect solution to the X-ray focal plane data. The GADS pipeline also performs a simple least-squares simultaneous fit of ACA position data and IRU rate data, and generates "photographs" of the ACA guide stars, for diagnostic and simple quality-assurance purposes. The pipeline also determines intervals during which the aspect solution is stable, assigns aspect quality indicators, and updates databases and the *AXAF* Optical Star Selection (AOSS) catalog.

The aspect pipeline consists of three components:

1. **Aspect Interval Separation Pipeline** makes one pass through IRU and ACA telemetry data to determine the aspect intervals during which major parameters (e.g., tracked star, spacecraft attitude) stay invariant or within allowed limits.
2. **Aspect Process Pipeline** processes ACA and gyro data, and derives aspect solution for each aspect interval.
3. **Assign Aspect Quality Pipeline** calculates aspect stable intervals, given the aspect solution, and assigns quality indicators to the aspect solution.

3. Aspect Process Pipeline

The Aspect Process Pipeline (Figure 1) is the central part of the aspect determination pipeline. The functional components of this pipeline are:

Process Gyro Data: Filter gyro data, check for internal consistency among the four gyro channels, and convert counts to angles. Gap-fill missing or filtered data, and calculate 3-axis spacecraft body angular rate.

Process ACA Data: Process aspect camera data, apply corrections, and calculate centroid coordinates (ACA sightings).

Kalman Filter: Optimally combine ACA star centroids and IRU data to determine ACA celestial location and image motion, using a Kalman filter and smoother. The Kalman filter and smoother also provides error estimates for the position and rate estimates in each axis, in the form of a covariance matrix.

Gyro Bias Correction: Correct bias drift rate using the estimate from the Kalman filter.

Photographic Method: Use photographic method for solving for the absolute celestial location and roll.

Determine Fiducial Motion: Perform a time-averaged solution of the fiducial light positions, calculate the fiducial light field centroid, and derive field motion as a function of time.

Combine Star/Fid Light Solutions: Calculate offset between fiducial light solutions and filtered star solutions, and apply boresight calibration, to generate image motion and celestial location at the focal plane science instrument.

4. Centroid Determination

We plan to use a PSF-fitting routine to calculate centroids and fluxes for each star and fiducial light image, where the PSFs obtained by interpolating from a database of theoretical PSFs of the optical system. Each model PSF is defined on a pixel grid that is finer than that of the aspect camera itself, and is calculated using the parameters of the optical system as measured before launch or on orbit.

Using a subset of ACA pixel data from the Aspect Interval, we identify poorly-fitting PSFs and find spoiler stars in star and fiducial light images. (Spoilers are faint stars in the vicinity of guide stars or fiducial lights). We then create new "effective" PSFs incorporating any spoilers, for image centroiding. We apply multidimensional χ^2 minimization routines (e.g., Powell minimization routine) to derive the optimal values for the coordinates and brightness of the image and the local sky background level.

Acknowledgments. *AXAF* Science Center and SAO staff that contributed to this work include: Daniel Schwartz, Martin Elvis, Jonathan McDowell, Gerald Cardillo, and Peter Daigneau. We thank Nanci Kascak from TRW for providing us with the Kalman filter and gyro model code. This work was supported by NASA Contract No. NAS8-39073.

References

Birkinshaw, M., & Karovska, M. 1995, AXAF News, No. 3

Astronomical Data Analysis Software and Systems VI
ASP Conference Series, Vol. 125, 1997
Gareth Hunt and H. E. Payne, eds.

Fitting and Modeling in the ASC Data Analysis Environment

S. Doe, A. Siemiginowska, W. Joye, and J. McDowell

Smithsonian Astrophysical Observatory, 60 Garden Street, MS 81, Cambridge, MA 02138

Abstract. As part of the AXAF Science Center (ASC) Data Analysis Environment, we will provide to the astronomical community a Fitting Application. We present a design of the application in this paper. Our design goal is to give the user the flexibility to use a variety of optimization techniques (Levenberg-Marquardt, maximum entropy, Monte Carlo, Powell, downhill simplex, CERN-Minuit, and simulated annealing) and fit statistics (χ^2, Cash, variance, and maximum likelihood); our modular design allows the user easily to add their own optimization techniques and/or fit statistics. We also present a comparison of the optimization techniques to be provided by the Application. The high spatial and spectral resolutions that will be obtained with AXAF instruments require a sophisticated data modeling capability. We will provide not only a suite of astronomical spatial and spectral source models, but also the capability of combining these models into source models of up to four data dimensions (i.e., into source functions $f(E, x, y, t)$). We will also provide tools to create instrument response models appropriate for each observation.

1. Introduction

Fitting models to data is a vital part of the analysis of astronomical data. As part of the ASC Data Analysis Environment, we have designed a Fitting Application. Although other fitting packages (e.g., XSPEC; see Arnaud 1996) exist, the high resolution and sensitivity of AXAF data present new challenges to the modeling and fitting of data; fitting models of the form $f(E, x, y, t)$ is a requirement for our software, and so we have been compelled to design our own Fitting Application. This paper presents a design of the flight version (Release 3) of our Fitting Application. We also discuss a preliminary test of the performance of the X-Ray Calibration Facility (XRCF), Release 1 version of our fitting software (Doe, Conroy, & McDowell 1996).

2. Design of the Fitting Application

The design of our Fitting Application is shown in Figure 1. The Application is controlled through a GUI, the Fit Monitor/Navigator. (The modules and tools discussed below may also be run from outside, without invoking the Navigator.) As a Monitor, it monitors the progress of the Fitting Engine through parameter

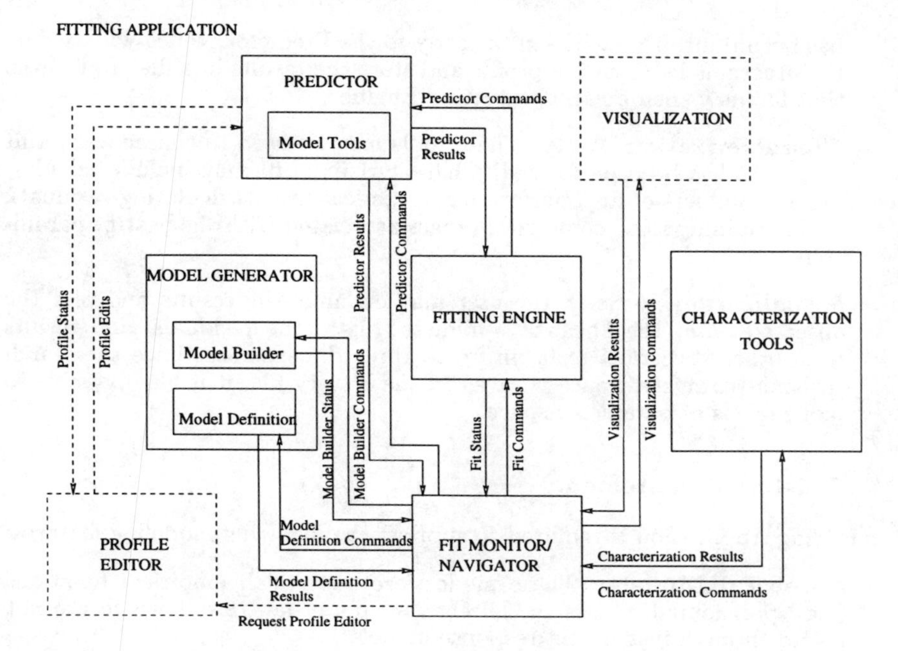

Figure 1. The Fitting Application.

space, and can halt the engine when necessary. As a Navigator, it allows the user to invoke the following utilities:

- **Fitting Engine.** This is the main component, responsible for searching parameter space. Given data, a parameterized model, and some convergence criteria, the Engine calculates predicted data values, compares them to the observed data, and searches parameter space for the parameters that yield a "best fit." The history of the search is recorded in a history file; the "best predicted data" are also calculated with the best-fit parameters. The Engine supports a variety of fit optimization algorithms, fit statistics, and convergence criteria.

- **Model Generator and Predictor.** The Model Generator has two functions. The Generator allows the user to *define* a model, by combining simpler models according to the rules imposed by the modeling language, and by setting the parameters of the model. The Generator also allows the user to *build* a model, by parsing the modeling language expression and storing the result in a format used by the Predictor. The Predictor may then take a model built by the Model Generator, and *calculate* predicted data over some data (sub)space. This may involve calls to pre-defined functions (e.g., a Gaussian or a power-law), or the execution of a series of external programs, which have been defined in a profile built with the Profile Editor.

- **Profile Editor.** With the Profile Editor, a *profile*, listing a number of programs, and their order of execution, may be built. Once such a profile has been built, it may be sent directly to the Predictor, which will execute the programs listed in the profile and store the results in a file. Data from that file may then be sent to Fitting Engine.

- **Characterization Tools.** These tools may be used to characterize a fit after a fit has been performed. Characterizing a fit may include calculating a "goodness-of-fit," performing statistical tests, calculating residuals, or determining some confidence ranges associated with the best-fit parameters.

- **Visualization.** Finally, the user may examine the results and plot the observed data, the "best-fit" predicted data, the residuals, and results from other statistical tests, in up to three dimensions. Since the search through parameter space is saved in the history file, it is also possible to plot regions of parameter space.

3. Modeling Requirements

The Fitting Application is required to support the following modeling features:

- **Empirical Models.** These models are analytical, empirical functions (e.g., polynomial, Gaussian, Lorentzian, power-law, etc.), which are not folded through instrument response models.

- **Astronomical Source Models.** These models include a variety of spatial, spectral, and temporal models of astronomical X-ray sources. Spectral models from the XSPEC package will also be available. However, the user may combine spatial, spectral, and temporal models into models capable of modeling a data space of up to four dimensions (e.g., $f(E, x, y, t)$).

- **AXAF Instrument Response Models.** Astronomical source models may be folded through AXAF instrument and mirror response models. From the instrument responses provided, a response model appropriate for a given AXAF observation may be generated. Response models appropriate for other missions (e.g., *Einstein*, *ROSAT*) may also be used.

- **Modeling Language.** Due to the high spatial and spectral resolutions of the AXAF instruments, it is highly desirable to be able to, e.g., combine models in "joint" spatial-spectral modes, which requires a modeling language sophisticated enough to permit users to build such "joint" models.

4. Comparison of Optimization Algorithms

We have implemented a Release 1 (XRCF) version of the Fitting Engine; this implementation includes the optimization algorithms listed in the table below. The implementation of the Levenberg-Marquardt algorithm is that contained in Numerical Recipes (1992); for the other algorithms, we have used the OPTIM library (Birkinshaw 1995). In this table, we present the execution time of the Engine, relative to the execution time of the Engine when the simplex algorithm has been selected. (At present, we are exploring ways to optimize the implementation of these algorithms, particularly the Powell and Levenberg-Marquardt routines.) In each run of the Engine, a 2-D Gaussian was fit to an array of 900 data points. We also present the number of lines of code for the implementation of each algorithm.

Table 1. Comparison of Optimization Algorithms.

Algorithm	Execution Time [a]	SLOCs [b]
grid search	1.6	233
grid search + Powell	2176.0	1058
Levenberg-Marquardt	3.0	332
Monte Carlo	2.7	163
Monte Carlo + Powell	477.0	977
Powell	13.3	814
simulated annealing (1)	249.6	259
simulated annealing (2)	389.4	265
simulated annealing (1) + Powell	272.4	1073
simulated annealing (2) + Powell	202.2	1079
simplex	1.0	368

[a]Relative to execution time for simplex.

[b]Source lines of code.

Acknowledgments. This project is supported by NASA contract NAS8-39073 (ASC). We would like to thank Mark Birkinshaw for making his OPTIM library available at the ASC; we also thank Michael Wise and Antonella Fruscione for many fruitful discussions.

References

Arnaud, K. A. 1996, in ASP Conf. Ser., Vol. 101, Astronomical Data Analysis Software and Systems V, ed. G. H. Jacoby & J. Barnes (San Francisco: ASP), 17

Birkinshaw, M. 1995, CfA internal memo

Doe, S., Conroy, M., & McDowell, J. 1996, in ASP Conf. Ser., Vol. 101, Astronomical Data Analysis Software and Systems V, ed. G. H. Jacoby & J. Barnes (San Francisco: ASP), 155

Press, W. H., Teukolsky, S. A., Vetterling, W. T., & Flannery, B. P. 1992, Numerical Recipes, 2nd ed. (Cambridge, Cambridge University Press), 387

Author Index

Index

Colophon

These Proceedings were prepared in nearly the same manner as the ADASS IV volume (ASP Conference Series, Volume 77), as a single LaTeX document using a style based on the ASP conference style. A skeleton document was constructed to incorporate the 117 separate contributions, as chapters, along with the front and back matter. The table of contents, indices, and cross-references between contributions were done with the standard LaTeX tools.

The few differences from the ADASS IV volume anticipate the conversion of the volume into electronic form for the Web. The ADASS V volume (ASP Conference Series, Volume 101), edited by George Jacoby and Jeannette Barnes, contained links to the NASA Astrophysics Data System for as many of the references as possible—we have tried to do the same. The HTML 3.2 standard contains a few new special symbols that allowed us to reduce the number of bitmaps required to reproduce the math in the papers. This should reduce download times.

Again, pages were typeset in 10 point Computer Modern fonts on pages approximately the same size as those in the final product. Tomas Rokicki's *dvips* program was used to scale the output up to 11 point for the final proofs, which were printed at NRAO on an HP LaserJet 4M, using 600 dpi Computer Modern fonts constructed especially for this project. After being reduced by the publisher, the final result is true 10 point Computer Modern fonts at an effective resolution of 660 dpi.

All authors submitted their manuscripts electronically to the editors. One was submitted by electronic mail, the rest were deposited in a designated anonymous `ftp` area. All were submitted with LaTeX markup.

The 117 individual contributions were accompanied by 118 encapsulated PostScript (EPS) files, all of which were used in the final product. Only one figure was submitted on paper. About 60% of the EPS files required manual editing, either to repair an incorrect BoundingBox, or to rotate figures submitted in landscape orientation back to portrait orientation. The lone paper figure was scanned for inclusion in the volume.

The proceedings of this conference are once again being made available on the World Wide Web thanks to the permission of the Editors of the Astronomical Society of the Pacific. This is available from NRAO[1] and STScI.[2] The conversion was done by feeding the entire LaTeX document to a modified version of Nikos Drakos' **LaTeX2`html`**[3] program.

---

[1] http://www.cv.nrao.edu/adass/adassVI

[2] http://www.stsci.edu/stsci/meetings/adassVI

[3] http://cbl.leeds.ac.uk/nikos/tex2html/doc/latex2html/latex2html.html

ROND-POINT 2

Méthode de français basée sur l'apprentissage par les tâches

Catherine Flumian
Josiane Labascoule
Corinne Royer

AVANT-PROPOS

ROND-POINT s'adresse à des apprenants grands adolescents et adultes et comprend trois niveaux (débutant, intermédiaire et avancé) qui couvrent les niveaux A1 et A2 (**ROND-POINT 1**), B1 (**ROND-POINT 2**) et B2 (**ROND-POINT 3**) du *Cadre européen commun de référence pour les langues*. Ce deuxième niveau aide aussi à la préparation du DELF B1, en vigueur depuis septembre 2005.

LA PERSPECTIVE ACTIONNELLE ET L'APPRENTISSAGE PAR LES TÂCHES

Le *Cadre européen commun de référence* (CECR) établit les bases théoriques et fournit les outils méthodologiques nécessaires pour surmonter les carences des approches dites communicatives. Dans ce but, le CECR formule une proposition méthodologique cohérente et privilégie ce qu'il appelle une perspective actionnelle. Cela signifie que les usagers et les apprenants d'une langue sont, avant tout, considérés « comme des acteurs sociaux ayant à accomplir des tâches dans des circonstances et un environnement donnés... ». C'est dans ce sens que **ROND-POINT** est la première méthode de français basée sur l'apprentissage par les tâches.

UN ENSEIGNEMENT CENTRÉ SUR L'APPRENANT

Les situations proposées en classe sont trop souvent éloignées de l'environnement de l'apprenant. L'apprentissage par les tâches surmonte cette difficulté en centrant sur l'élève les activités réalisées en classe.
À partir de sa propre identité et en s'exprimant selon ses propres critères, l'élève développe de manière naturelle ses compétences communicatives dans la langue cible.

DES PROCESSUS AUTHENTIQUES DE COMMUNICATION

La mise en pratique de la perspective actionnelle, telle que nous l'avons conçue pour **ROND-POINT**, entraîne de profondes modifications. La communication qui s'établit au cours de l'exécution des tâches est enfin authentique et la classe — cet espace partagé dans le but d'apprendre (et d'utiliser) une langue réelle — devient un lieu où chacun vit des expériences de communication aussi riches et authentiques que celles que les apprenants vivent en dehors de la classe.

LES COMPOSANTS DE ROND-POINT

Chaque niveau de la méthode comprend un livre de l'élève (avec CD inclus), un cahier d'exercices (avec CD inclus) et un guide pédagogique. Chaque unité du ***Cahier d'exercices*** offre des activités spécialement conçues pour consolider les compétences linguistiques développées dans le ***Livre de l'élève*** et pour entraîner les apprenants aux examens du DELF.
Le ***Guide pédagogique*** explique les concepts méthodologiques sous-jacents et suggère différentes exploitations pour les activités du ***Livre de l'élève***.

Avec l'approche actionnelle, la méthodologie de l'enseignement-apprentissage du français langue étrangère prend un tournant radical. Il ne s'agit pas de rejeter pour autant les apports de l'approche communicative — qui y songerait ? — mais l'éclairage porte dorénavant sur une composante essentielle de la communication, à savoir : l'action.

Dans cette optique, l'apprenant est d'abord considéré comme « acteur social » et agit ou, mieux encore, interagit socialement en vue de maîtriser la langue cible. L'accent est donc placé sur la réalisation de tâches, tâches qui s'exécutent en commun.

L'enseignant et ses apprenants se fixent donc, séance après séance, des objectifs qui relèvent toujours d'un « faire », d'une action à entreprendre en commun.

Dans **ROND-POINT 2**, par exemple, les auteurs proposent des tâches qui font appel au jeu et à la créativité : chercher une personne avec qui partager un appartement, organiser un week-end avec des amis, mettre au point un produit qui va nous faciliter la vie, élaborer un test de personnalité, débattre sur des sujets quotidiens ou encore raconter des anecdotes.

Bien sûr, pour atteindre ces objectifs, l'apprenant aura besoin de « ressources » (connaissances culturelles et langagières) afin d'accomplir la tâche fixée.

Dans **ROND-POINT 2**, l'organisation des unités révèle clairement le plan stratégique à exécuter afin que l'apprenant puisse atteindre les objectifs :

1. Visualiser les compétences à atteindre, la tâche à réaliser et avoir un premier contact avec le vocabulaire.
2. Entrer dans le contexte, en réalisant des activités d'apprentissage propres aux différentes compétences.
3. Acquérir les ressources, en incorporant les « instruments » indispensables pour agir efficacement en société ; c'est-à-dire, être capable d'appliquer les « normes » et « tonalités culturelles, sociales et linguistiques requises », selon le « genre du contexte ».
4. Réaliser la tâche en déployant ses compétences et son aptitude d'« acteur social confirmé ».
5. Découvrir des éléments de la culture francophone et comparer avec la culture d'origine de l'apprenant.

On reconnaîtra facilement, grâce aux termes entre guillemets, les composantes du modèle communicationnel de Dell Hymes, père de l'ethnographie de la communication, discipline qui place l'action sociale au centre de ses préoccupations sociolinguistiques.

À la question cruciale que se posent bien des didacticiens, dont Philippe Meirieu, dans son ouvrage : *Apprendre... Oui. Mais comment ?*, **ROND-POINT 2**, par son ancrage consciencieux et volontaire dans l'approche actionnelle, apporte une réponse pragmatique efficace.

Geneviève-Dominique de Salins
Professeur émérite

DYNAMIQUE DES UNITÉS :

Les unités de **ROND-POINT** sont organisées en cinq doubles pages qui vous apportent progressivement le lexique et les ressources grammaticales nécessaires à la communication :

- La rubrique ANCRAGE offre un premier contact avec le vocabulaire et les thèmes de l'unité. On y présente les objectifs, le contenu grammatical de l'unité et la tâche que vous devrez réaliser sous la rubrique TÂCHE CIBLÉE.
- La rubrique EN CONTEXTE propose des documents et des activités proches de la réalité hors de la classe. Ces documents vont vous permettre de développer une capacité de compréhension réelle.
- La rubrique FORMES ET RESSOURCES vous aide à systématiser les aspects de la grammaire nécessaires à la réalisation de la tâche ciblée.
- La rubrique TÂCHE CIBLÉE crée un contexte de communication où vous allez réutiliser tout ce que vous avez appris dans les étapes antérieures.
- La rubrique REGARDS CROISÉS fournit des informations sur le monde francophone et vous invite à réfléchir aux contrastes des cultures en contact.

À la fin du livre, un MÉMENTO GRAMMATICAL réunit et développe tous les contenus abordés dans chaque unité, et notamment ceux présentés dans la rubrique FORMES ET RESSOURCES.

ANCRAGE

COMMENT EXPLOITER CES PAGES

- En général on vous propose de petites activités de découverte du vocabulaire.
- L'image est très importante, elle va vous aider à comprendre les textes et le vocabulaire.
- Vos connaissances préalables dans d'autres domaines (autres langues, autres matières) et votre expérience du monde sont aussi des ressources utiles pour aborder l'apprentissage du français. Utilisez-les !

EN CONTEXTE

COMMENT EXPLOITER CES PAGES

- Dès le début vous allez être en contact avec la langue française telle qu'elle est dans la réalité.
- Ne vous inquiétez pas si vous ne comprenez pas chaque mot. Ce n'est pas nécessaire pour mener à bien les activités proposées ici.
- *Les textes en rouge* offrent des exemples qui vont vous aider à construire vos propres productions orales.
- *Les textes en bleu* sont des modèles de productions écrites.

FORMES ET RESSOURCES

COMMENT EXPLOITER CES PAGES

- Vous allez presque toujours travailler avec une ou plusieurs personnes. Ceci va vous permettre de développer vos capacités d'interaction en français.
- Dans plusieurs activités on vous demande de réfléchir et d'analyser le fonctionnement d'une structure. Ce travail de réflexion vous aidera à mieux comprendre certaines règles de grammaire.
- Vous trouverez regroupées dans une colonne centrale toutes les ressources linguistiques mises en pratique. Cette fiche de grammaire vous aidera à réaliser les activités et vous pourrez la consulter autant de fois que vous le voudrez.

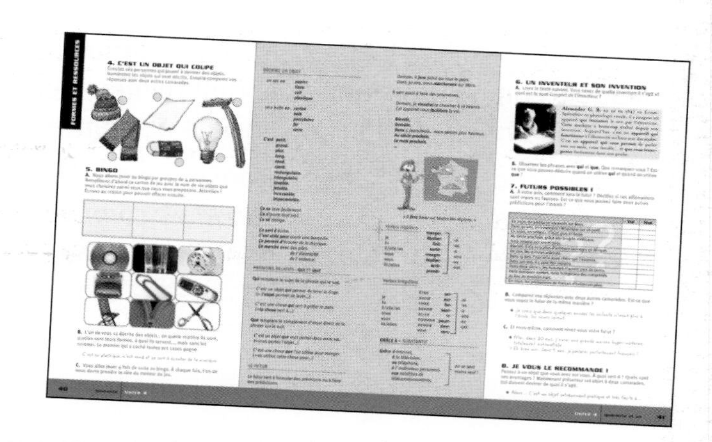

COMMENT EXPLOITER CES PAGES

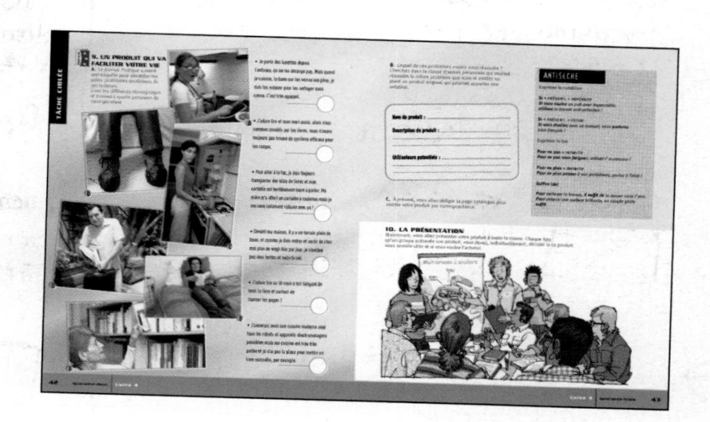

- L'aisance et l'efficacité communicatives sont ici essentielles.
- Vous allez réaliser cette tâche en coopération : vous allez résoudre un problème, échanger des informations et des opinions, négocier des solutions, élaborer des textes, etc.
- La phase de préparation est très importante. C'est l'occasion de mobiliser efficacement ce que vous avez appris. Mais, c'est aussi l'occasion de vous montrer créatif et autonome. Pour cela vous devez être capables d'évaluer vos besoins ponctuels en vocabulaire et en grammaire.
- Vous pouvez chercher les ressources dont vous avez besoin dans le livre, dans un dictionnaire ou dans l'« Antisèche », une petite fiche qui vous fournit de nouvelles ressources langagières. Vous allez discuter avec les membres de votre groupe à propos de la manière de réaliser la tâche et vous pouvez aussi solliciter ponctuellement l'aide de votre professeur.

Cet icone indique les activités que vous pouvez classer dans votre Portfolio européen des langues.

COMMENT EXPLOITER CES PAGES

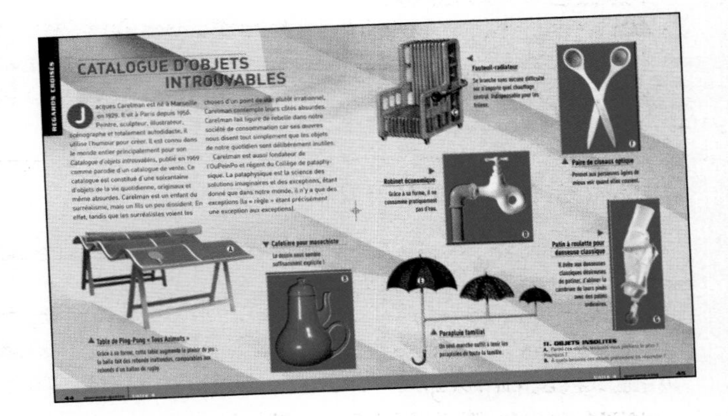

- Vous allez trouver dans ces pages des informations qui vont vous permettre de mieux connaître et mieux comprendre les valeurs culturelles, les comportements et la vie quotidienne dans différents pays où l'on parle français.
- Très souvent on vous demandera de réfléchir à votre propre identité culturelle, à vos propres expériences de la vie pour mieux comprendre ces nouvelles réalités culturelles.
- Certains documents peuvent vous sembler complexes, ne vous inquiétez pas : ce sont des « échantillons » de culture qui sont là pour vous montrer une autre réalité. Ce ne sont pas des textes à reproduire.

COMMENT EXPLOITER CES PAGES

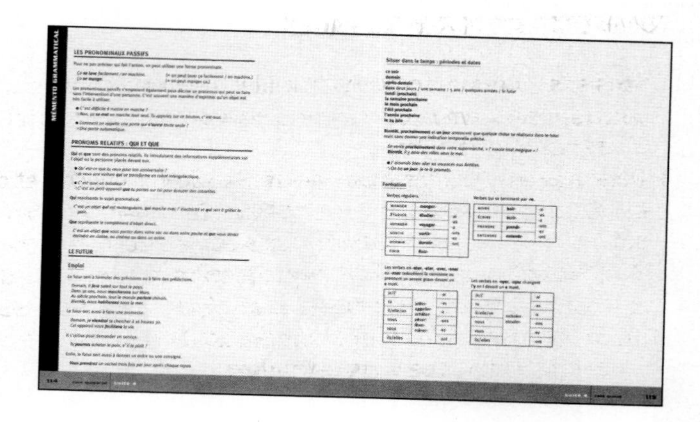

À la fin du livre, le MÉMENTO GRAMMATICAL développe les explications contenues dans la fiche de grammaire de la rubrique FORMES ET RESSOURCES.

- Vous pourrez consulter cet outil à tout moment de votre apprentissage.
- Il vous aidera dans les activités centrées sur la découverte et la conceptualisation d'aspects formels et sera un appui pour le développement de votre autonomie.

ANCRAGE

CHERCHE COLOCATAIRE

Nous allons chercher dans la classe une personne avec qui partager un appartement.

Pour cela nous allons apprendre à :

- **parler de nos goûts, de notre manière d'être et de nos habitudes**
- **décrire l'endroit où nous habitons**
- **exprimer des ressemblances, des différences et des affinités**
- **nous orienter dans l'espace**
- **communiquer nos impressions et nos sentiments**

Et nous allons utiliser :

- ***adorer, détester, ne pas supporter***
- ***(m') intéresser, (m') ennuyer, (me) déranger...***
- **le conditionnel : *moi, je préférerais...***
- **les prépositions de localisation dans l'espace : *à droite, à gauche, en face de...***
- **les questions : *quand, où, à quelle heure...***
- ***avoir l'air* + adjectif qualificatif**
- **les intensifs : *si, tellement***

1

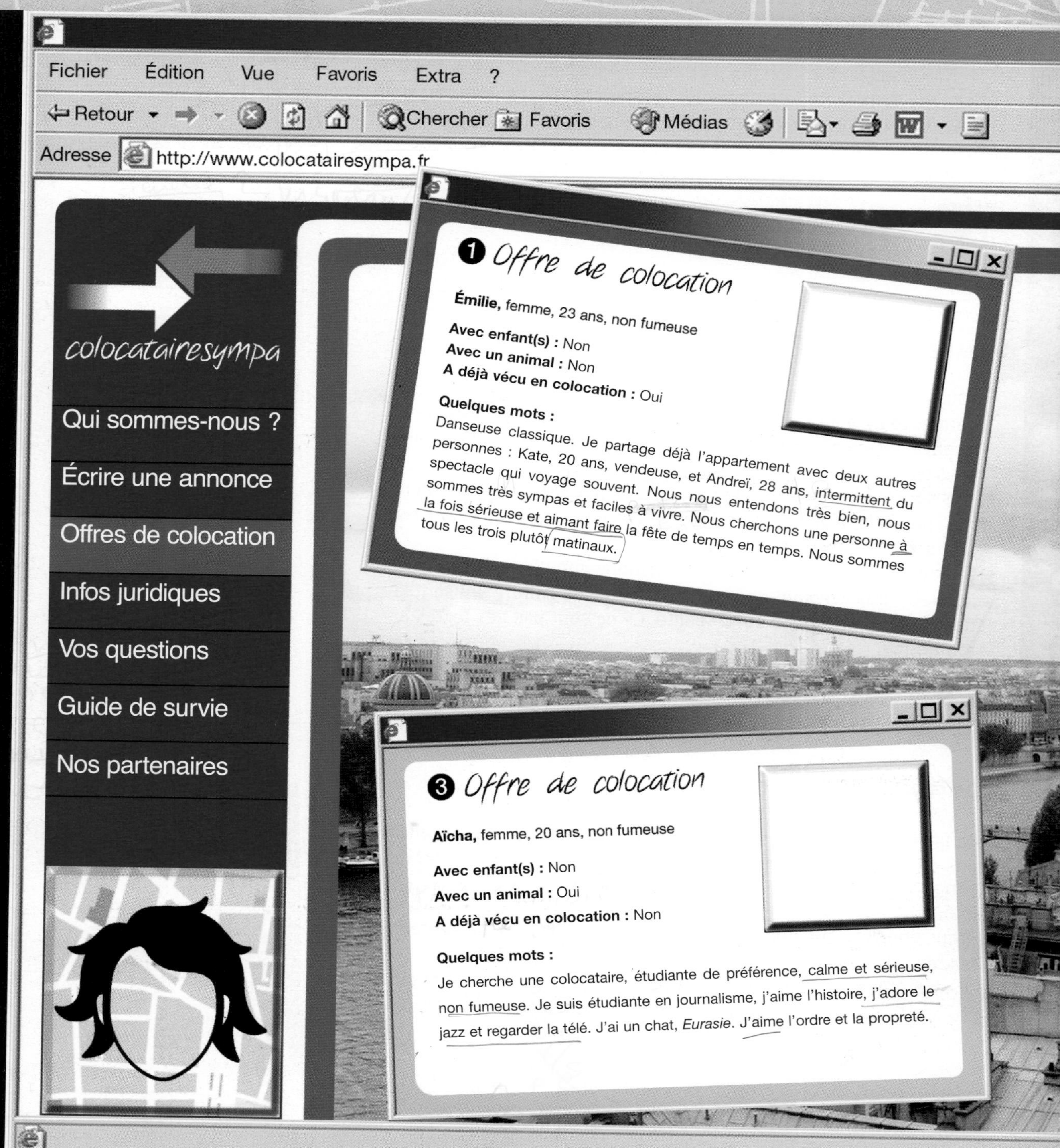

Fichier Édition Vue Favoris Extra ?
Retour
Chercher
Favoris
Médias
Adresse http://www.colocatairesympa.fr
colocatairesympa
Qui sommes-nous ?
Écrire une annonce
Offres de colocation
Infos juridiques
Vos questions
Guide de survie
Nos partenaires
❶ Offre de colocation
Émilie, femme, 23 ans, non fumeuse
Avec enfant(s) : Non
Avec un animal : Non
A déjà vécu en colocation : Oui
Quelques mots :
Danseuse classique. Je partage déjà l'appartement avec deux autres personnes : Kate, 20 ans, vendeuse, et Andreï, 28 ans, intermittent du spectacle qui voyage souvent. Nous nous entendons très bien, nous sommes très sympas et faciles à vivre. Nous cherchons une personne à la fois sérieuse et aimant faire la fête de temps en temps. Nous sommes tous les trois plutôt matinaux.
❸ Offre de colocation
Aïcha, femme, 20 ans, non fumeuse
Avec enfant(s) : Non
Avec un animal : Oui
A déjà vécu en colocation : Non
Quelques mots :
Je cherche une colocataire, étudiante de préférence, calme et sérieuse, non fumeuse. Je suis étudiante en journalisme, j'aime l'histoire, j'adore le jazz et regarder la télé. J'ai un chat, Eurasie. J'aime l'ordre et la propreté.

1. WWW.COLOCATAIRESYMPA.FR

A. Ces quatre jeunes femmes habitent Paris et cherchent un colocataire pour partager le loyer. Lisez les messages qu'elles ont laissés sur Internet et regardez les photos. Est-ce que vous pouvez retrouver qui est qui ?

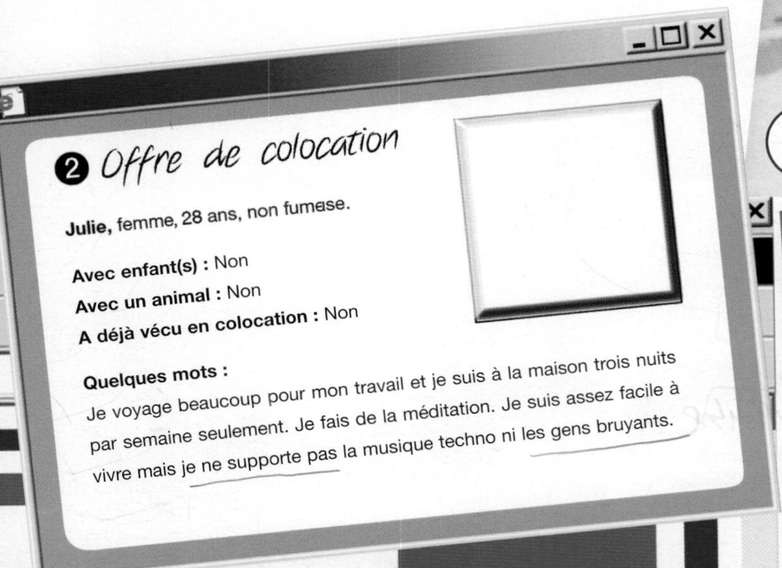

❷ Offre de colocation

Julie, femme, 28 ans, non fumeuse.

Avec enfant(s) : Non
Avec un animal : Non
A déjà vécu en colocation : Non

Quelques mots :
Je voyage beaucoup pour mon travail et je suis à la maison trois nuits par semaine seulement. Je fais de la méditation. Je suis assez facile à vivre mais je ne supporte pas la musique techno ni les gens bruyants.

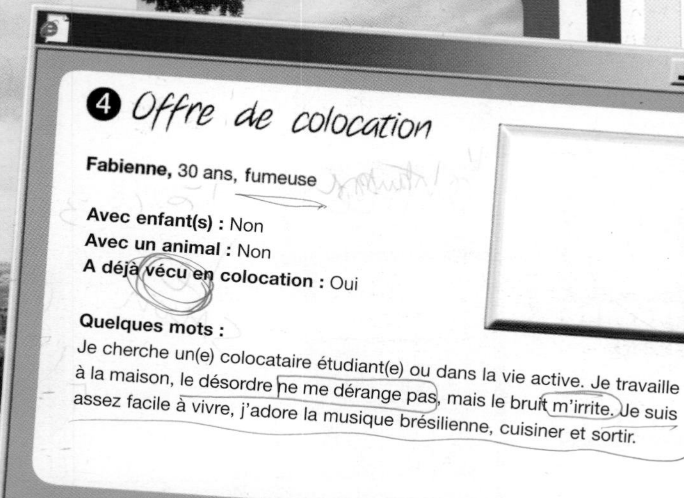

❹ Offre de colocation

Fabienne, 30 ans, fumeuse

Avec enfant(s) : Non
Avec un animal : Non
A déjà vécu en colocation : Oui

Quelques mots :
Je cherche un(e) colocataire étudiant(e) ou dans la vie active. Je travaille à la maison, le désordre ne me dérange pas, mais le bruit m'irrite. Je suis assez facile à vivre, j'adore la musique brésilienne, cuisiner et sortir.

B. Quelle impression vous font ces personnes ?

- Émilie a l'air très sociable, très tolérante.
- Oui, et elle a l'air assez sympathique.

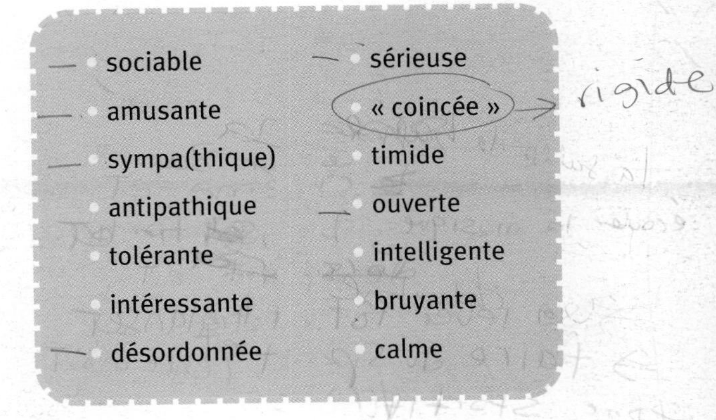

sociable	sérieuse
amusante	« coincée »
sympa(thique)	timide
antipathique	ouverte
tolérante	intelligente
intéressante	bruyante
désordonnée	calme

C. Imaginez que vous allez vivre à Paris une année. Vous cherchez une chambre en colocation. Lisez de nouveau ces annonces, avec qui préféreriez-vous habiter ?

- Moi je préférerais habiter avec Aïcha, parce qu'elle est étudiante comme moi et non fumeuse.
- Moi non ! Je préférerais habiter avec Émilie parce qu'elle a l'air très ouverte et sociable.

2. DES APPARTEMENTS À LOUER

A. Regardez cette annonce immobilière. Pouvez-vous identifier chaque pièce ?

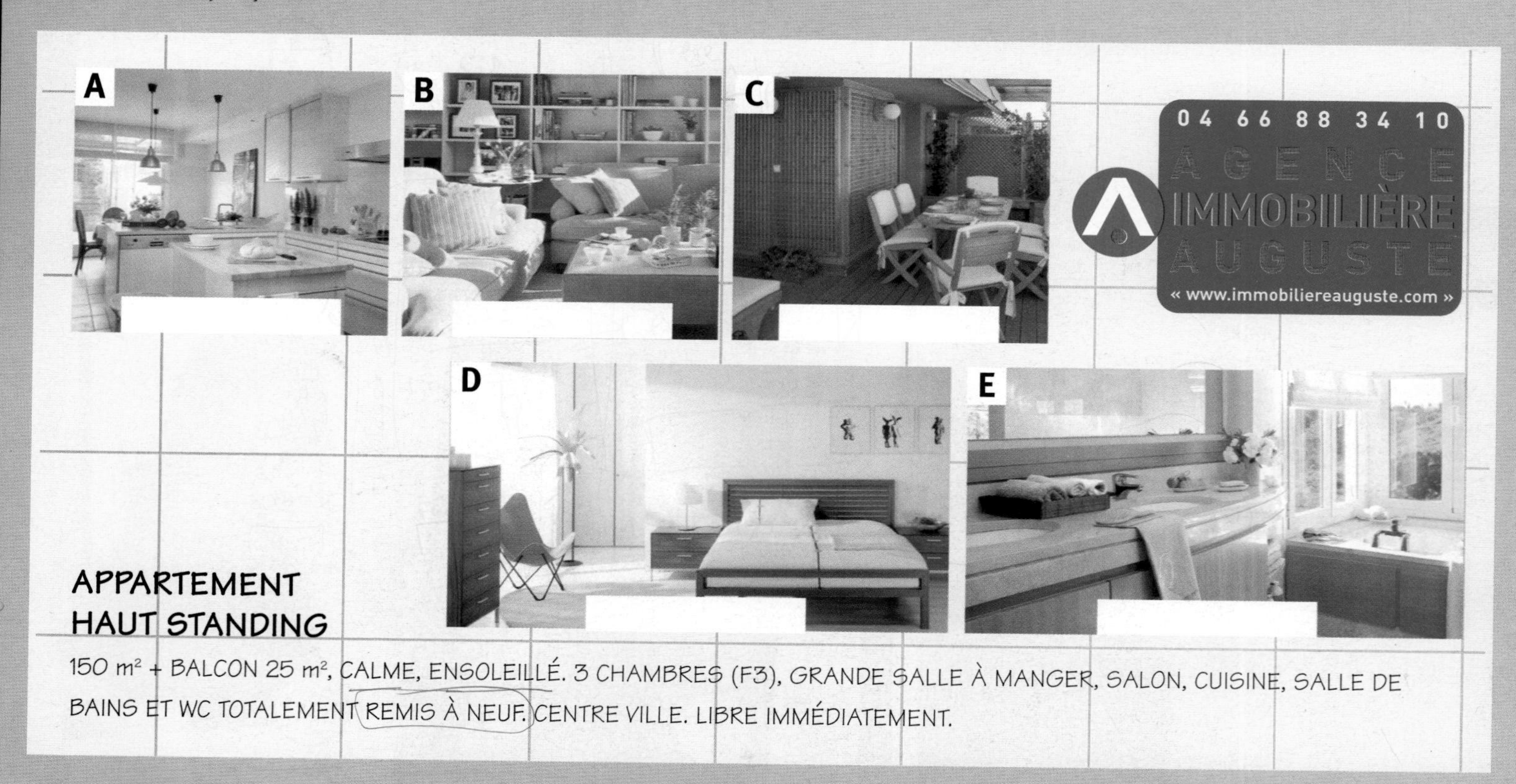

B. L'agent immobilier fait visiter l'appartement ci-contre. Écoutez la conversation. Où se trouve chaque pièce ?

- à côté (de)
- à gauche
- à droite
- au fond (de)
- en face (de)

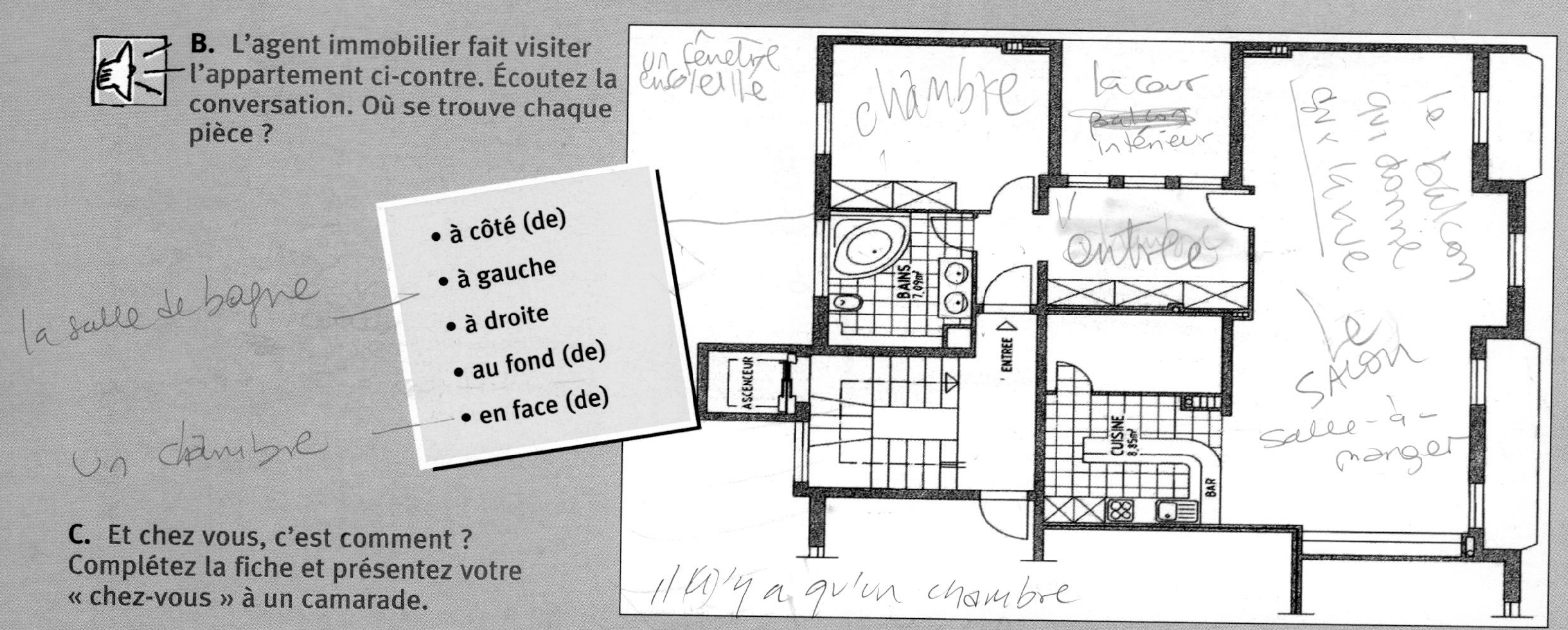

C. Et chez vous, c'est comment ? Complétez la fiche et présentez votre « chez-vous » à un camarade.

Nombre de chambres : • studio • 1 chambre (F1) • 2 chambres (F2) • 3 chambres (F3) • 4 chambres (F4) • 5 chambres (F5)

grand	petit	au centre ville	loin du centre ville
calme	bruyant	ensoleillé	sombre

- Chez moi, c'est plutôt grand. Il y a 6 pièces : la cuisine, la salle à manger, 2 chambres et un salon, une salle de bains et des toilettes, bien sûr. C'est un appartement...

3. VOTRE VEDETTE AU QUOTIDIEN

A. Lisez cette interview de la chanteuse Lara Garacan et complétez sa fiche.

Interview de la chanteuse Lara Garacan

Bonjour Lara !
Bonjour !
Lara, quand vous n'êtes pas sur scène, que faites-vous ?
Eh bien, vous savez, dans ce métier, on a besoin de se retrouver seul avec soi-même. Il faut se protéger de la surmédiatisation. Donc, quand j'ai du temps pour moi, je m'occupe d'une ferme que j'ai dans le Gers, et puis, j'ai une grande passion pour l'eau, la mer, le soleil. Dès que je peux, je vais voir la mer.
Qu'est-ce que vous aimez particulièrement ?
J'aime la vie de famille, les enfants, cuisiner, accueillir des amis autour d'un bon plat. Il n'y a rien de meilleur qu'une bonne table avec des amis et des rires.
On dit que vous êtes une révoltée... Qu'en pensez-vous ?
Oui, je me sens révoltée, je supporte mal le système qu'on nous impose, alors j'écris beaucoup.
Écrire, ça vous permet d'exprimer votre révolte ?
Oui, j'en ai besoin pour exprimer mes révoltes. C'est pour ça que j'aime les sports à risques, je pense. Je fais du deltaplane et du ponting, J'adore les sensations fortes. Je suis un peu impatiente parfois et nerveuse, j'ai besoin de dépenser mon énergie.
Qu'est-ce que vous détestez ?
Je n'aime pas les hypocrites, je ne supporte pas qu'on me donne des ordres et il y a plein de petits détails qui me dérangent.
Comme quoi par exemple ?
La fumée, le bruit de la circulation en ville ou les chiens de mon voisin.

- Elle adore : ....
- Elle supporte mal : ....
- Elle déteste : ....
- .... la dérange/nt (beaucoup)
- Elle fait (souvent) : ....

B. Quelle impression vous fait Lara Garacan ?

- Elle a l'air facile à vivre.
- Je ne sais pas. En tout cas, elle a l'air très dynamique.

C. Est-ce que vous avez des points communs avec Lara Garacan ? Parlez-en avec un camarade.

- Moi aussi j'aime la mer et le soleil.
- Moi aussi...

4. MON APPART

A. Individuellement, faites le plan de votre appartement ou de votre maison en écrivant le nom des pièces.

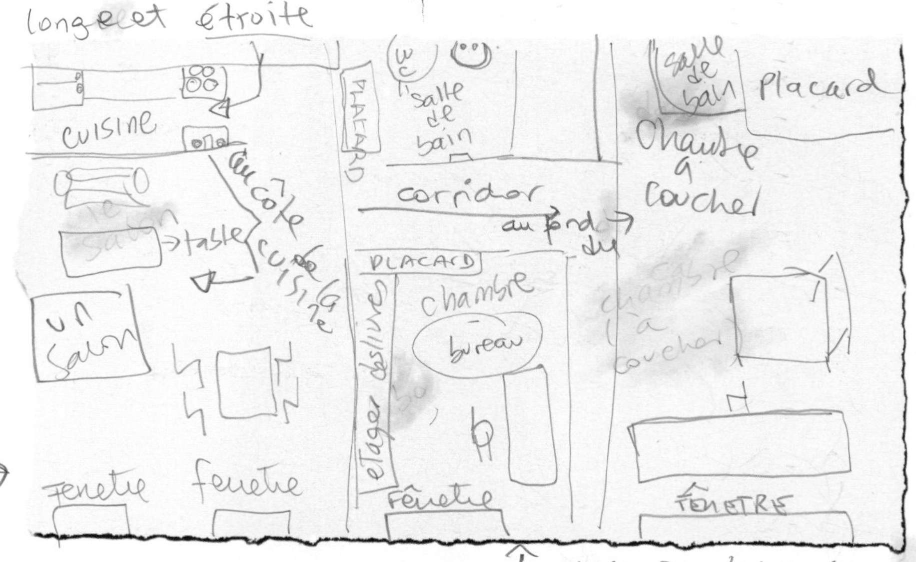

B. Maintenant expliquez à un camarade la disposition des pièces chez vous. Celui-ci devra dessiner le plan de votre appartement ou maison en suivant vos indications.

- ● Alors, quand tu rentres, à droite, il y a une chambre.
- ○ Comme ça ?
- ■ Oui voilà, et à côté, il y a la salle de bains.

5. MOI, JE M'ENTENDRAIS BIEN AVEC...

Complétez cette fiche avec la description d'une personne de votre entourage (un ami, un cousin, une sœur, un frère...). Ensuite, par groupes de quatre, lisez cette description. Chacun doit décider s'il s'entendrait bien avec cette personne ou non.

Nom : .... Âge : ....

Il/elle adore : ....
....
Il/elle déteste : ....
....
....

- danser
- l'ordre
- sortir
- inviter
- se lever tôt
- la télé
- nager
- le cinéma
- le sport
- la musique
- lire
- autre :

- les chats
- le désordre
- le bruit
- la saleté
- la fumée
- autre :

....
.... le/la gêne/ent beaucoup.
....
.... le/la dérange/ent beaucoup.

Il/elle fait souvent ....
....

- du sport
- de la musique
- la sieste
- des voyages
- du vélo
- autre :

EXPRIMER DES IMPRESSIONS

Avoir l'air + adjectif

*Elle **a l'air ouverte.*** (= elle semble être ouverte)

Dans cette expression, l'adjectif s'accorde avec le sujet.

***Il a l'air** sérieux.*
***Elle a l'air** sérieuse.*

Trouver + adjectif

*Je le **trouve** plutôt antipathique.* (= il me semble...)

*Elle est belle, mais je la **trouve** un peu froide.*

EXPRIMER DES SENTIMENTS

IRRITER : *Le bruit **(, ça) m'irrite.***
DÉRANGER : *La fumée **(, ça) me dérange.***
GÊNER : *La pollution **(, ça) me gêne.***
AGACER : *Le maquillage **(, ça) m'agace.***
ÉNERVER : *La tranquillité **(, ça) m'énerve.***
PLAIRE : *La danse **(, ça) me plaît.***

Ces verbes peuvent tous s'employer avec la forme impersonnelle **ça me**, qui se conjugue à la 3e personne du singulier.

Si le sujet est pluriel, le verbe se conjugue à la 3e personne du pluriel.

*Tous ces bruits **m'irritent.***

EXPRIMER L'INTENSITÉ FORTE

***Qu'est-ce que** c'est sombre !*
*Je le trouve **tellement** beau !*
*Elle est **si** belle !*

CONDITIONNEL

Les terminaisons du conditionnel sont les mêmes pour tous les verbes.

AIMER
j'aimer**ais**
tu aimer**ais**
il/elle/on aimer**ait**
nous aimer**ions**
vous aimer**iez**
ils/elles aimer**aient**

FINIR
je fini**rais**
tu fini**rais**
il/elle/on fini**rait**
nous fini**rions**
vous fini**riez**
ils/elles fini**raient**

Ce temps sert à exprimer un désir.

*Je **préférerais** habiter avec Sonia.*

***J'aimerais** dîner en tête à tête avec Johnny Depp.*

*Je **passerais (volontiers)** une semaine de vacances avec Eminem.*

Il sert aussi à faire une suggestion, une proposition.

*On **pourrait** chercher un troisième colocataire.*
*Il **pourrait** dormir dans la salle à manger.*

ORIENTATION DANS L'ESPACE

à droite (de)

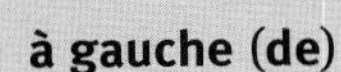

à gauche (de)

au coin (de)

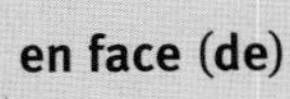

en face (de)

au fond (de)

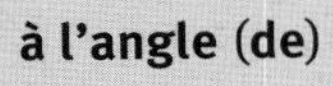

à l'angle (de)

derrière

devant

6. COLIN-MAILLARD

Règle du jeu : sur un papier, dessinez un parcours (vous pouvez vous inspirer de l'exemple). Votre camarade devra faire ce parcours avec un stylo les yeux fermés en essayant de suivre vos indications. Il doit sortir du parcours le moins possible.

- Alors, va tout droit, encore un peu. Arrête ! Maintenant tourne à droite...

7. TELLEMENT SÉDUISANT !

Pensez à des personnes, célèbres ou non,

- que vous aimeriez rencontrer.
- avec qui vous passeriez volontiers une semaine de vacances.
- avec qui vous dîneriez en tête-à-tête.
- avec qui vous partiriez en voyage à l'aventure.
- avec qui vous aimeriez sortir un soir.
- que vous inviteriez chez vous le jour de Noël.

- Moi, j'aimerais rencontrer Viggo Mortensen. Il est si séduisant !
- Et bien moi, je passerais volontiers une semaine...

TÂCHE CIBLÉE

8. À LA RECHERCHE D'UN APPARTEMENT

A. Vous cherchez un logement avec deux autres camarades, voici trois plans d'appartements. Il y a un F2 très grand et clair, un F3 moyen et un F4 avec des chambres assez petites et plus sombres. Mettez-vous d'accord sur l'appartement que vous allez choisir.

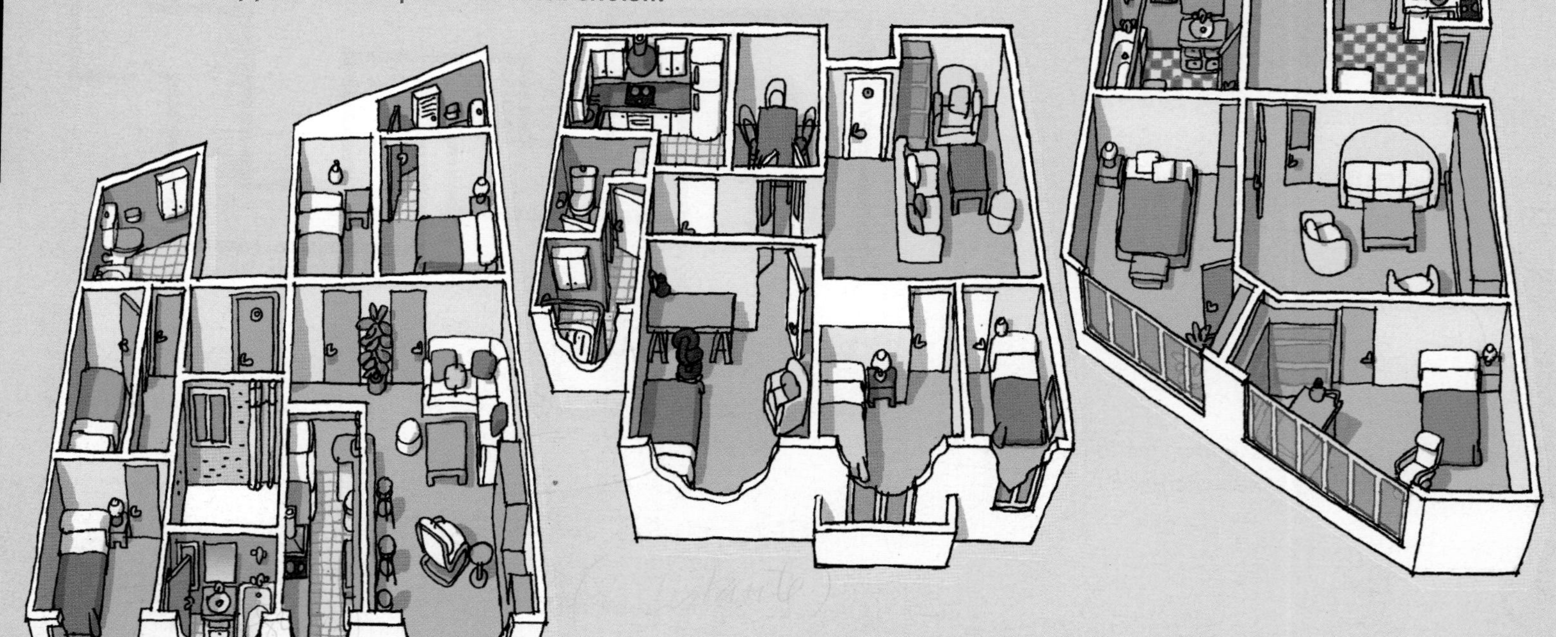

- On pourrait prendre l'appartement de 4 chambres.
- Non, les chambres sont trop petites, on prend... ?
- Non, je trouve que...

B. Vous venez de louer cet appartement avec vos deux camarades. Décidez comment vous allez partager l'espace et organiser votre cohabitation.

- Moi je prends cette chambre.
- Oui, et moi celle-là.
- Non, je ne suis pas d'accord...

9. LE/LA QUATRIÈME COLOCATAIRE

A. Le loyer de votre appartement a beaucoup augmenté. Vous décidez de chercher un quatrième colocataire. Où va-t-il dormir ?

- On pourrait partager une chambre ?
- Non, je crois qu'on pourrait...

B. Vous avez passé une petite annonce dans la presse et quelques personnes vous ont contactés. Qu'est-ce que vous allez leur demander pour savoir si vous allez pouvoir vous entendre ? Chacun de vous va interviewer un candidat. Préparez ensemble les questions que vous allez lui poser.

1. Est-ce que tu fumes ?
2. À quelle heure tu te lèves normalement ?
3.
4.

...

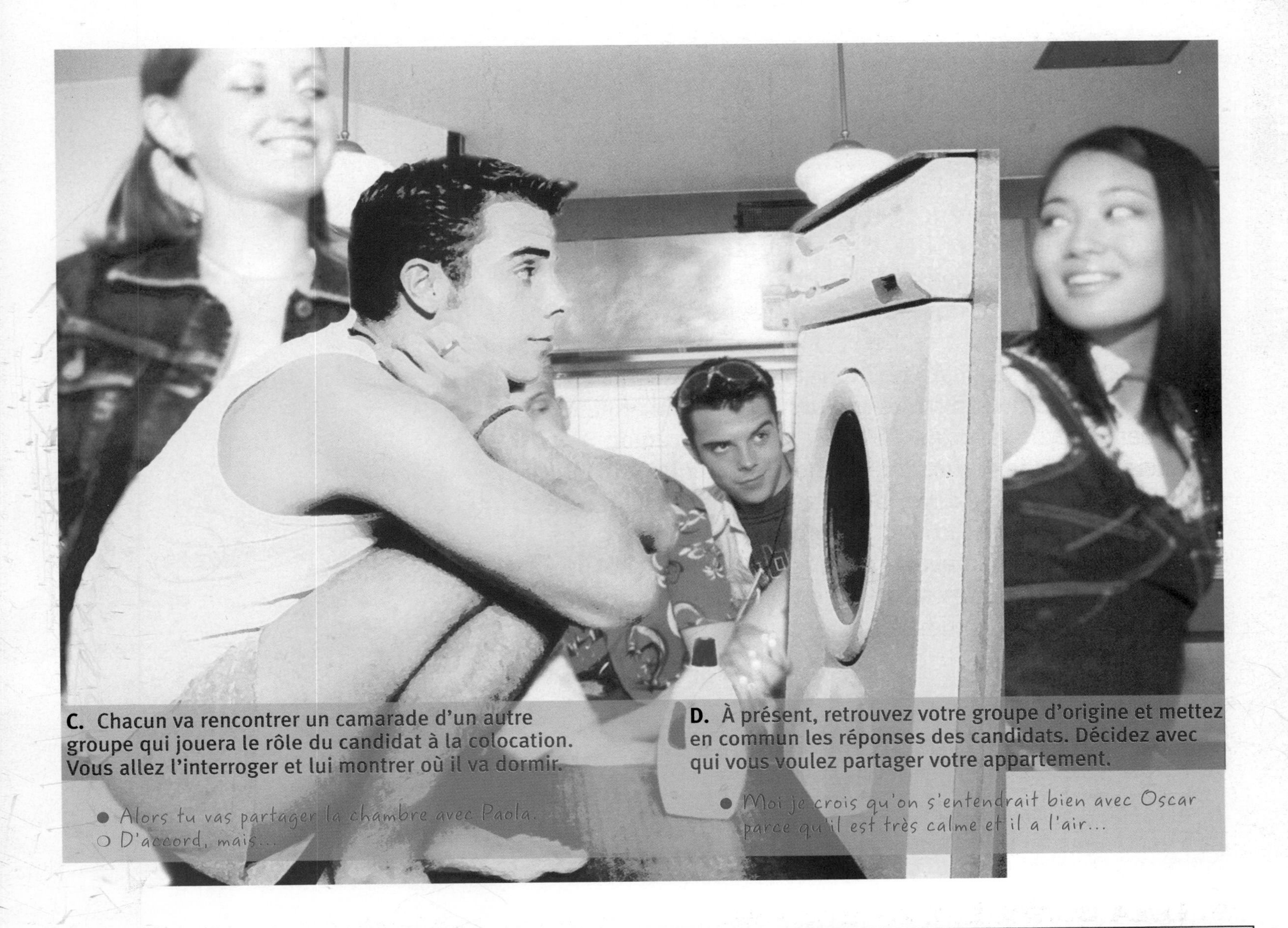

C. Chacun va rencontrer un camarade d'un autre groupe qui jouera le rôle du candidat à la colocation. Vous allez l'interroger et lui montrer où il va dormir.

- Alors tu vas partager la chambre avec Paola.
- D'accord, mais...

D. À présent, retrouvez votre groupe d'origine et mettez en commun les réponses des candidats. Décidez avec qui vous voulez partager votre appartement.

- Moi je crois qu'on s'entendrait bien avec Oscar parce qu'il est très calme et il a l'air...

10. UN COURRIEL POUR LE QUATRIÈME COLOCATAIRE

Vous avez décidé qui sera le quatrième colocataire ? Bien, alors vous prévenez par courriel la personne que vous avez choisie.

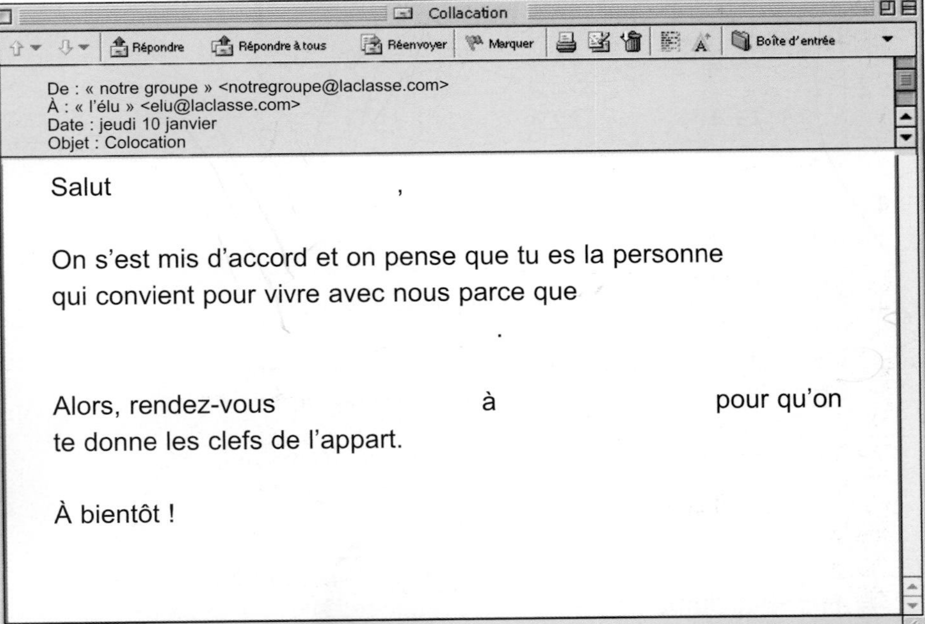

Collacation

De : « notre groupe » <notregroupe@laclasse.com>
À : « l'élu » <elu@laclasse.com>
Date : jeudi 10 janvier
Objet : Colocation

Salut ,

On s'est mis d'accord et on pense que tu es la personne qui convient pour vivre avec nous parce que .

Alors, rendez-vous à pour qu'on te donne les clefs de l'appart.

À bientôt !

LES FRANCILIENS RESTENT À LA MAISON

En Île-de-France, les jeunes restent un peu plus longtemps chez leurs parents que dans les autres régions françaises. Mais pourquoi cette tendance des jeunes à rester chez leurs parents s'accentue-t-elle en région parisienne ? Cette différence avec le reste de la France s'explique essentiellement par la proximité des universités. En effet, il y a beaucoup de facultés à Paris et dans sa région. Par conséquent, les jeunes de l'Île-de-France qui décident de faire des études universitaires ne sont pas obligés de quitter le domicile familial. En province, par contre, les universités sont souvent éloignées du domicile des parents et les jeunes doivent quitter leur famille pour poursuivre leurs études.

11. ENCORE CHEZ LEURS PARENTS

A. Lisez le texte. Est-ce que c'est aussi comme ça dans votre pays ?

B. Regardez ce tableau. C'est le pourcentage de jeunes qui vivent encore chez leurs parents dans la région parisienne. Qu'est-ce que vous remarquez ?

	HOMMES	FEMMES
20-24 ans	71%	56%
25-29 ans	27%	15%

C. D'après vous, pourquoi les filles partent-elles plus jeunes de la maison ? Choisissez une réponse.

a. Les filles ont plus souvent des conflits avec leurs parents au sujet des relations amoureuses.

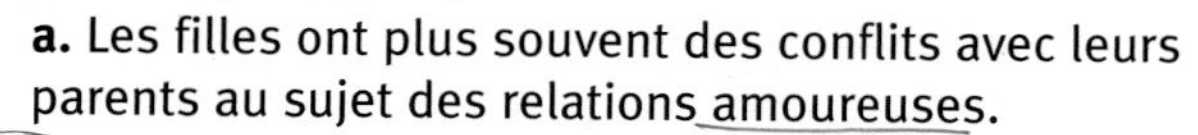

b. Les filles étudient moins longtemps et commencent à travailler avant les garçons. Comme elles ont un salaire régulier, elles peuvent s'émanciper.

c. Les filles sont mieux préparées pour tenir une maison. Elles savent cuisiner, faire la vaisselle, le ménage et les courses.

D. Maintenant écoutez le sociologue Philippe Douchard et vérifiez votre réponse.

12. TÉMOIGNAGES

Lisez ces témoignages de jeunes Français.
Est-ce qu'il y a des points communs avec ce qui se passe dans votre pays ?

Les filles sont plus souvent pressées de partir que les garçons

VANESSA

J'ai 24 ans et je suis aide-soignante. Je loue un studio depuis deux ans. Le loyer est un peu cher mais je préfère mon indépendance. Je n'ai pas besoin de dire à mes parents avec qui je sors, où je vais et à quelle heure je rentre.

Quand les jeunes s'émancipent, ils vont souvent habiter à moins de 5 kilomètres de chez leurs parents

STEVEN

J'ai 23 ans et je travaille en intérim. C'est un peu dur, je ne gagne pas beaucoup d'argent et mon loyer représente 40% de mon salaire. J'habite pas très loin de chez mes parents, dans le quartier où j'ai grandi. Comme ça, le week-end, je me retrouve avec mes copains de toujours, on s'entend bien ! C'est important de se soutenir mutuellement.

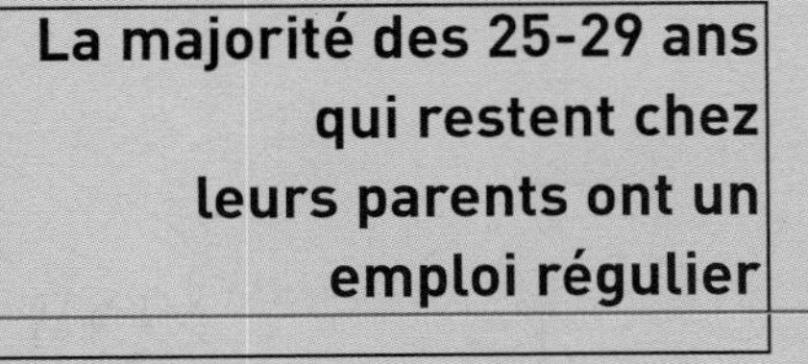

La majorité des 25-29 ans qui restent chez leurs parents ont un emploi régulier

THIERRY

J'ai 27 ans, je suis mécanicien et j'ai un emploi stable depuis un an. Je vis encore chez mes parents et ça ne pose aucun problème. Mes parents sont très compréhensifs et tolérants, ils ont très bien accepté la présence de Johanne, ma petite amie. Nous avons l'intention de faire quelques économies et dans deux ou trois ans, nous nous installerons chez nous.

ANCRAGE

SI ON ALLAIT AU THÉÂTRE ?

Nous allons organiser un week-end dans notre ville pour des amis français.

Pour cela nous allons apprendre à :

- exprimer nos préférences en matière de loisirs
- faire part des expériences
- faire des suggestions et exprimer une envie
- inviter quelqu'un
- accepter ou refuser une invitation et à fixer un rendez-vous (lieu et date)

Et nous allons utiliser :

- *c'était (très)* + adjectif ; *il y avait plein de* + nom
- *ça te dit de* + infinitif
- *si on* + imparfait
- *avoir envie de* + infinitif
- le futur proche
- les jours de la semaine, les moments de la journée et l'heure
- des prépositions et locutions de localisation : *au centre de, (pas) loin de, (tout) près de*

2

1
NUITS BLANCHES
La meilleure musique de tes nuits blanches Dj Jean-Jean
12, rue du Blé
danser
écouter
techno
2
POLYGONE
ANTIGONE
3
EUROMOTEUR
Foire internationale de l'automobile
Palais des expositions, Pavillons X et XI
4
Finale de la Ligue des Champions
Olympique de Marseille-Milan
Samedi 20 heures
Stade Municipal
5
KRONOS LE JEU DE RÔLE
2000 parties de KRONOS simultanées
Inscrivez-vous dès aujourd'hui
Antechrone
COURS BASIQUE DE PHOTO NUMÉRIQUE

1. À FAIRE

A. Regardez les images. Quelles sont les activités qu'on peut faire dans cette ville ce week-end ?

B. Imaginez que vous passez le week-end dans cette ville. Lesquelles de ces activités vous intéressent ? Parlez-en avec deux autres camarades.

- Moi, j'aimerais bien prendre un bain dans un hammam et suivre un cours de photo numérique.
- Et bien moi, j'aimerais...

- ○ suivre un cours de photo numérique
- ○ aller en boîte
- ○ prendre un bain au hammam
- ○ aller au cirque
- ○ visiter un salon ou une exposition
- ○ faire une partie de jeu de rôle avec des internautes du monde entier
- ○ faire du patin à glace
- ○ voir un match de football
- ○ faire du shopping

2. ÇA TE DIT ?

A. Maintenant, écoutez ces conversations entre des amis qui parlent de ce qu'ils vont faire ce week-end. Où vont-ils ?

1. Mario et Lucas : ....
2. Sonia et Nathanaël : ....
3. Lise et Katia : ....

B. Et vous ? Qu'est-ce que vous faites normalement le vendredi ou le samedi soir ? Parlez-en avec deux ou trois camarades.

	souvent	quelquefois	jamais
Je vais danser / en boîte...			
Je sors dîner.			
Je joue avec des amis à un jeu de société / jeu de rôle...			
Je « chate » avec mes amis.			
Je fais du théâtre, de la danse...			
Je vais au cinéma.			
J'organise des soirées DVD avec des amis.			
Autres :			

- Moi, le samedi soir je vais souvent danser avec les copains.
- Moi, le vendredi soir, je vais quelquefois au cinéma ou bien j'organise un repas à la maison.

3. VIVEMENT LE WEEK-END !

A. Tous ces gens parlent de leurs projets pour le week-end. Luc, par exemple, aimerait sortir avec Roxane, mais est-ce qu'elle va accepter ? Regardez ces illustrations, dans chacun des extraits une phrase manque. Replacez les phrases ci-contre dans le dialogue correspondant.

- Si on allait voir « Spiderman » ?
- Ça te dit de venir avec moi ?
- Moi, j'ai très envie d'aller danser.
- Euh, je suis désolé mais je ne suis pas libre samedi !

1

- ○ Allô !
- ● Bonjour Roxane ! C'est Luc !
- ○ Ah, bonjour Luc !
- ● Dis-moi, est-ce que tu es libre ce week-end ?
- ○ Euh... oui, pourquoi ?
- ● Et bien, j'ai deux entrées pour le concert de Björk samedi soir.
- ○ Ah oui ? Génial !
- ● ....
- ○ Oui, merci pour l'invitation !

2

- ○ Qu'est-ce que tu fais, toi, ce week-end ?
- ■ .... Avec Samuel on va au Macadam Pub vendredi soir. Y'a des soirées salsa tous les vendredis. L'ambiance est très très sympa. Et toi ?
- ○ Moi je sors avec qui tu sais !!
- ■ Avec Luc ? C'est pas vrai !!
- ○ Si si ! Il m'a invité au concert de Björk.
- ■ Super !!

3

Nadège — Thomas — Yasmine

- ❑ Qu'est-ce qu'on fait samedi soir ?
- ▲ ....
- ▼ Ah je l'ai vu, c'est pas terrible !
- ❑ Oui, et puis moi, les films d'action, c'est pas mon truc.
- ▲ Et si on allait voir « Désirs et murmures ». Il paraît que c'est super bon !
- ▼ Ouais, moi je suis d'accord ! Et toi, Thomas ?
- ❑ Ouais, pour moi, c'est d'accord. On prévient Luc ?
- ▼ Ok, je m'en charge.

4

- ● Allô ?
- ▲ Allô Luc ?
- ● Ah salut Yasmine !
- ▲ Écoute, samedi soir on sort avec les copains. Tu veux venir ?
- ● ....

B. Écoutez les dialogues complets et vérifiez puis résumez ce qu'ils vont faire ce week-end.

Luc va sortir .... avec .... et ils vont aller ....
Sandra va sortir .... avec Samuel et ils vont ....
Yasmine va sortir .... avec .... et ils vont ....

C. Écoutez de nouveau. Vous avez remarqué comment... ?

- On propose de faire quelque chose.
- On exprime un désir.
- On accepte une proposition.
- On refuse une invitation.

4. TROIS CINÉPHILES

A. Lisez ces synopsis, est-ce que vous savez à quel titre correspond chacun ?

1

Film néo-zélandais/américain (2003)

▸ Les armées de Sauron ont attaqué Minas Tirith, la capitale de Gondor. Les Hommes luttent courageusement. Tous les membres de la communauté font tout pour détourner l'attention de Sauron et donner à Frodon une chance d'accomplir sa mission. Voyageant à travers les terres ennemies, Frodon doit se reposer sur Sam et Gollum, tandis que l'anneau continue de le tenter.

2

Film français/américain (1988)

▸ Ce film raconte l'histoire de deux enfants en Grèce, passionnés par la plongée en apnée. Leur rivalité se poursuivra dans leur vie d'adulte. Lequel des deux descendra le plus loin ? Vous découvrirez leurs amours, leurs amitiés, avec les humains et avec les dauphins.

3

Film américain (1990)

▸ Clarisse Sterling, une jeune stagiaire du FBI, est chargée d'enquêter sur une série de meurtres épouvantables commis dans le Middle West par un psychopathe connu sous le nom de « Buffalo Bill ».

B. Marc, Léna et Stéphane sont trois jeunes qui « chatent » parfois sur Internet. Voici un fragment d'un de leurs « chats » se rapportant aux films ci-contre. Qu'est-ce que chacun a vu ? Quand ? Qu'est-ce qu'ils en ont pensé ?

Adresse http://www.chatjeune.com

Marc-19 dit :
Salut les gars !
Je suis allé au ciné club la semaine dernière j'ai vu un bon film, vraiment je vous le recommande.
Stephparis dit :
Ah ! Ouais ! C'est quoi ?
Marc-19 dit :
Je me souviens plus du titre, c'est pas très récent, c'ést en anglais sous-titré.
Lena-na dit :
Et ça raconte quoi ?
Marc-19 dit :
C'est un film policier, mais avec beaucoup de suspense. C'est génial !
Stephparis dit :
Ben moi aussi je suis allé au ciné, hier au soir et j'ai vu « Le Retour du roi ».
Lena-na dit :
Tu as aimé ?
Stephparis dit :
Euh ! Oui, j'ai bien aimé, c'est pas mal mais je m'attendais à autre chose. Avec toute la publicité qu'il y a eu, je pensais voir quelque chose d'extraordinaire.
Marc-19 dit :
Moi, j'ai lu le livre et c'est super !
Lena-na dit :
Eh ben moi ce week-end j'ai revu un film pour la troisième fois. C'est mon film préféré.
Stephparis dit :
Trois fois ! ? C'est quoi cette merveille ?
Lena-na dit :
C'est un vieux film, très poétique. Les images sont superbes. C'est sur la mer, les dauphins... J'adore.

	Quel film ?	Quand ?	Appréciation positive	Appréciation négative
Marc				
Léna				
Stéphane				

C. Aimez-vous le cinéma ? Pensez à un film qui vous a plu : recommandez-le à vos camarades.

● *Vous avez vu « Vatel ». C'est un film génial de...*

5. QU'EST-CE QUE VOUS AVEZ FAIT CE WEEK-END ?

A. Vous allez découvrir ce que les autres personnes de la classe ont fait ce week-end. Mais, d'abord, remplissez vous-même ce questionnaire.

Je suis resté(e) chez moi.

Je suis allé/e
- au cinéma.
- à un concert.
- en discothèque.
- chez des amis.
- ailleurs : ....

J'ai fait
- du football.
- du skate.
- une partie de cartes.
- autre chose : ....

J'ai vu
- un film.
- une exposition.
- autre chose : ....

B. Parlez maintenant avec deux ou trois camarades de vos activités du week-end.

Bonito

C'était	génial. chouette. sympa. nul. ennuyeux.

Je me suis	vraiment vachement bien beaucoup pas mal	amusé/e. ennuyé/e.

- ● Qu'est-ce que tu as fait ce week-end ?
- ❍ Je suis allée à un concert de musique classique...

C. Quelles sont les trois activités les plus fréquentes dans votre classe ?

D. Vous avez déjà des projets pour le week-end prochain ? Parlez-en avec deux autres camarades.

- ● Moi, le week-end prochain je vais aller au cinéma.
- ❍ Eh bien moi, je vais peut-être sortir en boîte. Et toi ?
- ■ Moi, je ne sais pas encore.

6. J'AI A-DO-RÉ !

A. Mettez-vous par groupes de trois et parlez d'un lieu où vous êtes allés et que vous avez adoré ou détesté. Vous pouvez utiliser un dictionnaire ou demander de l'aide à votre professeur.

- ● Un lieu que tu as adoré ?
- ❍ Moi, la Sicile, c'est dans le sud de l'Italie. C'était très joli et il faisait très très beau. C'était très très bien. Vraiment !
- ● Et un lieu que tu as détesté ?

DÉCRIRE ET ÉVALUER (UN SPECTACLE, UN WEEK-END...)

- ● *Tu es allé au cinéma ?*
- ❍ *Oui, j'ai vu Spiderman III. **C'était génial !***

C'était (vraiment)	**super.** **génial.**
C'était (très très*)	**joli.** **mauvais.** **intéressant.** **bien.** **sympa.**

* À l'oral, on répète souvent **très** pour donner plus de force à l'appréciation.

C'est un journal d'informations / un film / un documentaire / un jeu concours / un « reality show » / une retransmission sportive / une série télévisée / un magazine

IL Y AVAIT PLEIN DE + SUBSTANTIF

Plein de s'utilise en code oral dans le sens de « beaucoup de ».

- ● *En boîte hier, il y avait **plein de** fumée. Impossible de respirer !*
- ❍ *Oui, mais il y avait aussi **plein de** gens intéressants !*

* À l'oral, **il y avait** est souvent prononcé **y'avait.**

PROPOSER, SUGGÉRER QUELQUE CHOSE

Ça me/te/vous/... dit de/d' + INFINITIF

***Ça ne me dit rien de visiter** un musée.*

***Ça te dit d'aller** en boîte ?*

***Ça vous dit de voir** un film ?*

- ● ***Ça te dit de manger** un couscous ?*
- ❍ *Non, **ça me dit** rien du tout.*

(Et) Si on + IMPARFAIT ?

Et si on **allait** au cinéma ce soir ?
mangeait au restaurant ?
regardait un film à la télé ?
allait en boîte ?

AVOIR ENVIE DE

Avoir envie de sert à exprimer un désir.

J'ai envie de danser / vacances.

ACCEPTER OU REFUSER UNE PROPOSITION OU UNE INVITATION

Vous acceptez : **Volontiers !**
D'accord !
Entendu !

Vous refusez poliment une invitation.

(Je suis) désolé/e, mais je ne suis pas libre.
je ne suis pas là.
je ne peux pas.
j'ai beaucoup de travail.
je n'ai pas le temps.

FIXER L'HEURE D'UN RENDEZ-VOUS

- ***On se retrouve à quelle heure ?***
- ○ ***À** vingt heures.*
- *D' accord.*

LES MOMENTS DE LA JOURNÉE ET L'HEURE

samedi (matin) / lundi (soir) / vendredi **à**
dix heures
dix heures cinq
dix heures **et quart**
dix heures vingt-cinq
dix heures **et demie**
dix heures trente-cinq
onze heures **moins** vingt
onze **moins le quart**
onze heures **moins** cinq

INDIQUER UN LIEU

- *Je vous recommande la pizzeria « Chez Geppeto ».*
- ○ *Ah! oui, **c'est où ?***
- ***Tout près de** chez moi, **place** de la Claire Fontaine.*

Dans le sud/l'est/l'ouest/le nord **de** l'Espagne.
Au sud, au nord, à l'est, à l'ouest **de** Paris.
À Berlin
Au centre de Londres
Dans mon quartier
Pas loin de chez moi
(Tout) près de la fac / **du** port
(Juste) à côté de la gare/ **du** stade
Devant le restaurant L'eau vive
Sur la place du marché
À la piscine / **Au** Café des sports
Au 3, rue de la Précision

7. CE SOIR À LA TÉLÉ

A. À deux : regardez le programme de quatre chaînes de télévision. De quel type d'émissions s'agit-il ?

TF1	La 2	La 3	Canal +
20h30 **Le millionnaire** Ce soir deux candidats de la région parisienne vont s'affronter pour gagner des millions.	20h55 **La CRIM** Suite de l'enquête de l'inspecteur Rive sur une étrange affaire d'enlèvement.	20h55 **Thalassa** Escale au Pérou. Thalassa nous emmène ce soir sur les côtes du Pacifique pour découvrir Paracas : un oasis au milieu de la mer.	21h00 **From Hell (2001)** Sur les traces de Jack l'éventreur, avec Johnny Depp.
21h40 **Le loftstory** Retransmission en direct de la vie au quotidien de nos 8 amis enfermés maintenant depuis 5 semaines dans une maison de la région parisienne. Qui devra s'en aller demain ?	23h50 **Contre-courant** Les passagers clandestins. Notre envoyé spécial a filmé ces jeunes qui mettent leur vie en danger pour partir vers un avenir meilleur.	00h55 **Le Journal et La Météo**	23h15 **Championnat NBA**
			02h30 **Tradition et Folk** Concert de l'Ensemble de musique traditionnelle tunisienne.

B. Est-ce que cette programmation ressemble à celle de la télévision dans votre pays ? Quelles émissions aimez-vous voir ?

- Moi, le jeudi soir, je regarde « Acoustic » sur TV5, c'est super !
- ○ C'est à quelle heure ?

8. ON PREND RENDEZ-VOUS

À deux, imaginez que ce sont les activités proposées cette semaine dans votre ville. Mettez-vous d'accord pour en choisir une ensemble.

- Ça te dit d'aller voir « Matrix » dimanche ?
- ○ Oui, à quelle heure ?
- À vingt heures.
- ○ Ah non, c'est trop tard.
- Et à dix-sept heures trente ?

MICROPOLIS La cité des insectes
Au moyen d'images, de sons d'une architecture et d'une scénographie spectaculaires, originales et ludiques, ce musée transporte le visiteur dans l'univers des insectes.
HORAIRES
De mars à décembre : de 10 h à 16 h
De juin à septembre : de 9 h à 19 h

GRAND CONCERT DE BRUCE SPRINGSTEEN
Stade des Étoiles
Samedi à 21 h

BOWLING
Centre commercial des 2 Ponts
13 pistes
Tous les jours de 11 h à 1 h du matin

RÉSERVE ANIMALIA
Dans notre réserve vous pouvez voir de près des animaux sauvages en pleine liberté, comme dans leur habitat naturel.
Ouvert tous les jours, toute l'année.

Cinéma FOX
MATRIX
Séances :
14 h 30, 17 h 30, 20 h et 22 h

9. CE WEEK-END, ON SORT !

A. Imaginez qu'un ami vient passer le week-end dans votre ville. Vous pouvez lui recommander des lieux où aller ?

Lieux à visiter dans les environs :

Musées et monuments :

Endroits où manger :

Bars, pubs et discothèques :

Autres activités à faire :

B. Comparez vos recommandations avec celles de deux autres camarades. Vous connaissez tous ces lieux ?

- ● La Pizzeria 4 Staggioni, c'est bien ?
- ○ Oui, on y mange bien et c'est pas cher.
- ● Et c'est où ?
- ○ Dans le centre, près de chez moi.

10. CHARLINE, RACHID ET SARAH

PORTFOLIO

A. Charline, Rachid et Sarah cherchent des correspondants. Lisez les messages qu'ils ont laissés sur Internet. Est-ce qu'ils aiment faire les mêmes choses que vous ? Cherchez les points communs que vous avez avec eux et commentez-les avec un ou deux camarades.

- ● Rachid aime le football et moi aussi.
- ○ Charline est comme moi, elle adore la musique.

Nom : Loiseau · **Prénom :** Sarah
Courriel : ssarah@mot.com

J'aimerais correspondre avec des jeunes qui, comme moi, aiment la nature (je fais de la randonnée et j'ai une super collection de scarabées). Je fais partie d'une association écologique et je veux devenir vétérinaire ou biologiste. J'aime aussi lire, surtout des romans de voyages (je suis une authentique fan de Jules Verne) et je rêve de voyager dans le monde entier.

Nom : Agili

Prénom : Rachid

Courriel : rachidagili@prop.com

Si tu aimes le football (je suis supporter du Paris Saint-Germain), le cinéma, (je fais des courts-métrages avec des copains) la B.D. (Obélix et Astérix, Tintin,Titeuf...) et les parcs de loisirs (j'en ai visité 5 jusqu'à présent), eh bien, je suis le correspondant idéal. Alors, j'attends ta réponse.

Nom :
Boudou

Prénom :
Charline

Courriel :
charline@wanadoo.fr

Salut, je suis une jeune Parisienne et je cherche des correspondants de tous pays. J'adore la musique (je joue de la guitare avec un groupe de copains) et les sports (natation, VTT, courses de motos). Je suis très ouverte, curieuse de tout et j'adore faire la fête entre amis.

B. Imaginez que ces trois jeunes viennent passer le week-end prochain dans votre ville. Est-ce qu'il y a des endroits que vous pourriez leur recommander en fonction de leurs goûts ? Est-ce qu'il y a actuellement des événements (spectacles, expositions, concerts, etc.) qui pourraient les intéresser ? Parlez-en avec deux camarades.

- *Sarah aime la nature. Elle pourrait visiter le Jardin botanique. Il y a...*

C. Votre école et celle de Charline, Rachid et Sarah ont organisé un échange. Ils arrivent vendredi soir ! Avec deux autres camarades, décidez lequel des trois vous voulez accompagner ce week-end puis mettez au point le programme.

D. Vous allez maintenant présenter oralement votre programme.

SAMEDI

8 ______

10 ______

12 ______

14 ______

16 ______

18 ______

20 ______

22 ______

DIMANCHE

8 ______

10 ______

12 ______

14 ______

16 ______

18 ______

20 ______

22 ______

11. DEUX GÉNÉRATIONS DE FRANÇAIS ET LEURS LOISIRS

A. Lisez ces deux petits textes à propos des loisirs de deux générations de Français. Est-ce que les 11-20 ans et les 25-35 ans se comportent de la même manière dans votre pays ? Est-ce qu'ils ont aussi de l'argent de poche ? Échangez vos impressions avec deux autres personnes.

B. Quel est votre âge ? Est-ce que vous appartenez à l'une de ces deux générations ? Si c'est le cas, est-ce que vous vous reconnaissez dans leur description ? Vous pouvez rédiger un petit texte où vous décrirez le mode de vie de votre génération dans votre pays.

ARGENT DE POCHE

- Les 11-14 ans reçoivent en moyenne 17 euros d'argent de poche par mois.
- Les 15-17 ans disposent en moyenne de 46 euros par mois.
- Les 18-20 ont 99 euros par mois.

LA GÉNÉRATION LOFT*

Ils ont entre 11 et 20 ans et, en 2002, ils représentent seulement 13,1% de la population française. Les amis ont, à leurs yeux, beaucoup d'importance. Cette génération s'exprime par SMS, visite les forums sur internet et « chate » avec des amis du monde entier rencontrés sur le réseau (Internet).

Les garçons comme les filles donnent beaucoup d'importance à leur look et faire du shopping est l'un de leurs loisirs préférés. En matière de mode, les garçons aiment les marques de sport et les filles achètent des marques peu chères qui leur permettent de changer de look fréquemment.

* Ce nom vient d'une émission de télé réalité « Loft story » où des jeunes gens devaient vivre pendant plusieurs mois enfermés dans un grand appartement ou loft.

LES TRENTENAIRES

Une nouvelle tribu est née. Ils ont entre 25 et 35 ans et ils mènent une vie professionnelle dure et trépidante. On les appelle aussi les « adulescents » (adulte + adolescent) parce qu'ils ont une famille, un travail et des responsabilités mais ils ne veulent pas vieillir. Ils aiment retrouver les sensations, les émotions et les jeux de leur enfance. Le week-end, ils vont à des soirées spéciales où ils chantent et dansent sur les airs de leur enfance tandis que sur les écrans géants, défilent des extraits de Goldorak, Capitaine Flam, Candy ou Casimir. La nostalgie de ces années-là est devenue un style de vie !

12. TOUS LES JEUNES FONT LES MÊMES CHOSES ?

Écoutez ces interviews de trois jeunes provenant de trois pays différents. Notez, suivant le modèle du tableau ci-dessous, ce qu'ils font pendant leur temps libre.

	Quand ?	Quoi ?	Avec qui ?
1. Rebecca (Suisse)			
2. Valérie (Québec)			
3. Olivier (France)			

C'EST PAS MOI !

allons mettre au
un alibi et justifier
emploi du temps.

ela nous allons
ndre à :

- nter des événements
  nformant de leur
  cession dans le temps
- rire un lieu, une personne,
  circonstances
- nander et donner des
  rmations précises
  eure, le lieu, etc.)

s allons utiliser :

- parfait et le passé
  posé
- *bord, ensuite, puis,*
  ès, enfin
- *nt* + nom, *avant de* +
  nitif
- *ès* + nom/infinitif passé
- *en train de*
  'imparfait)
- *appeler* au présent
- *ne semble que*
- exique des vêtements
  uleurs et matières)
- escription physique
- marqueurs temporels :
  soir, dimanche dernier,
  nt-hier vers 11 heures
  te...

1. GRANDS ÉVÉNEMENTS

A. Regardez ces photos. De quels événements s'agit-il ?

LANCE ARMSTRONG A GAGNÉ SON SIXIÈME TOUR DE FRANCE !

COUPE DU MONDE : LE BRÉSIL A REMPORTÉ LA VICTOIRE CONTRE L'ALLEMAGNE

CHUTE DU MUR DE BERLIN : DES FAMILLES SE SONT RETROUVÉES DANS LA PLUS GRANDE ÉMOTION 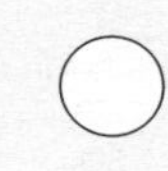

11 OSCARS POUR « LE SEIGNEUR DES ANNEAUX » : LE RECORD DE TITANIC ET BEN-HUR A ÉTÉ ATTEINT ! 

INCROYABLE ! ON A MARCHÉ SUR LA LUNE 

LE PRÉSIDENT FRANÇOIS MITTERRAND ET ÉLISABETH II ONT INAUGURÉ LE TUNNEL SOUS LA MANCHE 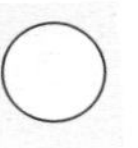

CLONAGE : LA BREBIS DOLLY EST NÉE 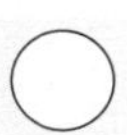

BECKHAM ET VICTORIA SE SONT MARIÉS

B. C'était en quelle année ?

• 1969 • 1989 • 1994 • 1996 • 1999 • 2002 • 2004 • 2004

- ● *Je pense que le Brésil a remporté la Coupe du Monde de football contre l'Allemagne en 2000.*
- ○ *Non, moi, je crois que c'était en 2002.*

C. Vous rappelez-vous certains de ces événements ? Quel âge aviez-vous cette année-là ?

- ● *Moi, en 1969, je n'étais pas né !*
- ○ *Moi non plus !*
- ■ *Moi, je me rappelle la Coupe du Monde de 2002. J'avais 12 ans.*

D. Est-ce qu'il y a des événements sportifs, culturels, scientifiques ou autres qui vous ont marqué ? C'était en quelle année ?

2. UN BON ALIBI

Essayez de répondre : où étiez-vous ? Dans le tableau ci-dessous, voici quelques suggestions pour vous aider.

Où étiez-vous...

- dimanche dernier à 14 heures ?
- hier soir à 19 heures ?
- le 31 décembre à minuit ?
- le jour de votre dernier anniversaire à 22 heures ?
- avant-hier à 6 heures du matin ?
- ce matin à 8 heures 30 ?

(Moi) j'étais...

- en train de regarder la télévision.
- en train de prendre un bain.
- en train de manger.
- en train de dormir.
- en train d'étudier.
- en train de...

- en classe.
- chez des amis.
- en voyage.
- au cinéma.
- chez moi.
- au travail.

- avec un/e ami/e.
- avec ma femme/mon mari.
- avec ma famille.
- avec mon/ma petit/e ami/e.

- je ne me rappelle pas bien.
- il me semble que...
- je crois que...

- ● *Moi, hier à 17 heures 20, j'étais chez moi en train de regarder la télévision.*
- ○ *Moi, j'étais au cinéma avec un ami.*
- ■ *Moi, je ne me rappelle pas bien, mais il me semble que...*

3. LE COMMISSAIRE GRAIMET MÈNE L'ENQUÊTE

A. Voici un extrait d'un roman policier où le commissaire Graimet mène une enquête sur un vol qui a eu lieu la semaine dernière. Lisez ce dialogue entre le commissaire Graimet et l'un des témoins du *hold-up*. Pouvez-vous identifier les deux gangsters parmi ces cinq suspects ?

L'inspecteur Graimet allume sa pipe et commence à poser des questions :

– Alors, qu'est-ce que vous avez vu ?

– Eh bien, hier matin, à 9 heures, je suis allé à ma banque pour retirer de l'argent… Je faisais la queue au guichet quand…

– Il y avait beaucoup de monde ?

– Oui, euh, il y avait cinq personnes devant moi.

– Est-ce que vous avez remarqué quelque chose de suspect ?

– Oui, euh, juste devant moi, il y avait un homme...

– Comment était-il ?

– Grand, blond, les cheveux frisés.

– Comment était-il habillé ?

– Il portait un jean et un pull-over marron.

– Et alors ? Qu'est-ce qui était suspect ?

– Eh bien, il avait l'air très nerveux. Il regardait souvent vers la porte d'entrée.

– Bien, et qu'est-ce qui s'est passé ?

– Soudain, un autre homme est entré en courant et…

– Comment était-il ?

– Euh, eh bien il était plutôt de taille moyenne, roux, les cheveux raides… Il avait l'air très jeune. Ah ! Et il portait des lunettes.

– Et, à ce moment-là, qu'est-ce qui s'est passé ?

– L'homme qui était devant moi a sorti un revolver de sa poche et il a crié « Haut les mains ! C'est un hold-up ».

– Alors, qu'est-ce que vous avez fait ?

– Moi ? Rien ! J'ai levé les bras comme tout le monde.

12

B. Regardez les verbes employés dans ce dialogue, ils sont tous au passé mais ils sont conjugués à deux formes différentes. Vous pouvez les séparer en deux groupes ? Quelles remarques pouvez-vous faire sur leur construction ?

Vous avez vu	J'étais

C. Lisez de nouveau le dialogue et essayez d'expliquer à quoi servent l'une et l'autre forme verbale.

D. Pensez à l'un de ces personnages et décrivez-le à votre camarade qui devra deviner de qui il s'agit.

● *Il est plutôt grand, il a les cheveux raides... et il porte un blouson marron.*
○ *C'est celui-ci !*

Il porte
- une veste
- un blouson
- un pull-over
- une chemise
- une cravate
- un pantalon
- un jean
- des chaussures
- une casquette
- des lunettes
- une moustache
- la barbe

Il a les cheveux
- courts
- raides
- bruns
- roux
- longs
- frisés
- blonds
- châtains

Il est chauve

Il a les yeux
- bleus
- noirs
- gris
- verts
- marron

Il est (plutôt)
- grand
- de taille moyenne
- petit
- gros
- mince
- maigre

4. FAIT DIVERS

A. Olivier Debrun a été victime d'un vol. L'agent de police qui l'a interrogé a pris des notes sur son carnet. Lisez ses notes, puis essayez avec un camarade d'imaginer ce qui est arrivé à Olivier Debrun.

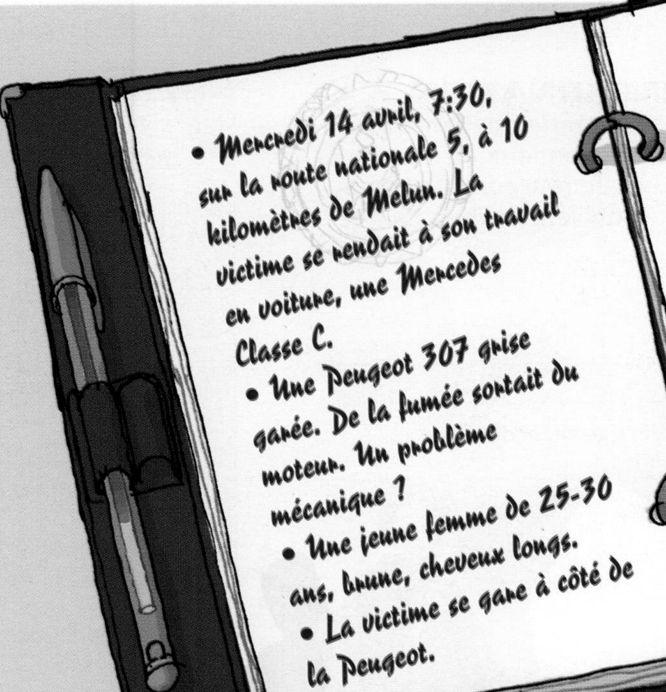

• Mercredi 14 avril, 7:30, sur la route nationale 5, à 10 kilomètres de Melun. La victime se rendait à son travail en voiture, une Mercedes Classe C.
• Une Peugeot 307 grise garée. De la fumée sortait du moteur. Un problème mécanique ?
• Une jeune femme de 25-30 ans, brune, cheveux longs.
• La victime se gare à côté de la Peugeot.
• Un homme d'environ 35 ans, grand, châtain, mal rasé apparaît et menace la victime avec une arme à feu.
• La femme demande les clefs de contact de la Mercedes, met le moteur en marche.
• Ils emportent son téléphone portable ; 3 cartes de crédit ; 200 Euros.
• Malfaiteurs partis en direction de Fontainebleau.

SOCIÉTÉ

Vol de voiture à main armée sur la N5

Mercredi matin, un homme a été victime d'un couple de malfaiteurs sur la nationale 5, près de Melun.

Olivier Debrun .... comme d'habitude quand il a vu .... arrêtée sur le bord de la route nationale 5 .... faisait signe aux automobilistes de s'arrêter. « .... raconte Olivier Debrun, alors j'ai pensé qu'elle avait un problème mécanique et je me suis arrêté pour l'aider. » À ce moment-là, le complice de la jeune femme, qui était caché dans la Peugeot, est sorti et a menacé la victime avec .... Olivier Debrun a été contraint de donner .... ainsi que .... , et .... qu'il portait sur lui.
Les deux complices se sont enfuis ....

B. Écoutez les déclarations de la victime. Est-ce que cela correspond à ce que vous aviez imaginé ? Maintenant, en vous appuyant sur les déclarations de la victime et sur les notes de police, complétez l'article ci-contre qui doit être publié dans la presse.

5. JE PORTAIS

A. Mettez-vous deux par deux. Vous rappelez-vous comment votre camarade était habillé à la classe précédente ?

● Je me rappelle que tu portais un pull-over bleu marine.
❍ Non, pas du tout ! J'avais une chemise blanche !

B. Maintenant, individuellement, écrivez sur une feuille comment vous étiez habillé(e) à la dernière classe de français. Puis, donnez votre description anonyme au professeur qui l'affichera au tableau.

C. Formez des groupes de quatre personnes, lisez les descriptions au tableau et essayez de vous rappeler qui était habillé de cette manière. Le groupe qui a reconnu le plus de personnes a gagné.

6. CROYEZ-VOUS À LA RÉINCARNATION ?

Imaginez que vous vous souvenez d'une vie antérieure. Pensez à un métier d'autrefois, vous pouvez utiliser un dictionnaire ou demander à votre professeur. Ensuite, expliquez à tour de rôle comment vous étiez et ce que vous faisiez. Vos camarades doivent deviner de quel métier il s'agit.

- Je vivais en/au/à ....
- Je portais ....
- J'étais ....
- On me/m' respectait/aimait/craignait/écoutait ....
....
- J'avais ....
- Je faisais ....
- J'aimais ....

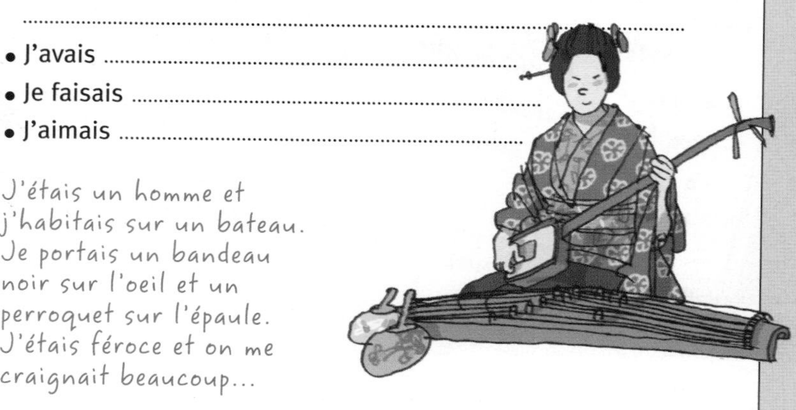

● J'étais un homme et j'habitais sur un bateau. Je portais un bandeau noir sur l'oeil et un perroquet sur l'épaule. J'étais féroce et on me craignait beaucoup...

7. DEVINEZ CE QU'IL A FAIT AVANT ET APRÈS !

A. Qu'est-ce que vous avez fait hier ? Complétez les phrases suivantes.

- Hier, avant de sortir de chez moi, je/j' ....
- Après le déjeuner, je/j' ....
- Aussitôt après avoir dîné, je/j' ....
- Juste avant de me coucher, je/j' ....

B. À deux, faites maintenant des suppositions sur ce que votre camarade a fait hier.

● Hier, avant de sortir de chez toi, tu as allumé ton portable ?
❍ Non, avant de sortir de chez moi, j'ai bu un café.

L'IMPARFAIT

Pour former l'imparfait, on prend la base du présent à la première personne du pluriel et on ajoute les terminaisons de l'imparfait.

SE LEVER
je me lev**ais** [e]
tu te lev**ais** [e]
il/elle/on se lev**ait** [e]
nous nous lev**ions** [iõ]
vous vous lev**iez** [ie]
ils/elles se lev**aient** [e]

ÊTRE:
j'**é**tais
tu **é**tais
il/elle/on **é**tait
nous **é**tions
vous **é**tiez
ils/elles **é**taient

L'imparfait situe une action dans le passé sans signaler ni le début ni la fin de l'action. Il sert à parler de nos habitudes dans le passé.

À cette époque-là, elle ***se levait*** *tous les matins à 6 heures.*

L'imparfait sert aussi à décrire une action en cours.

● *Que* ***faisiez-vous*** *samedi soir dernier ?*
❍ *Moi, je* ***regardais*** *la télévision.*

Ou bien à décrire les circonstances qui ont entouré un événement.

Il n' est pas venu en classe parce qu' il ***était*** *malade et* ***avait*** *de la fièvre.*

LE PASSÉ COMPOSÉ

Le passé composé sert à raconter l'événement.

● *Qu' est-ce qui s' est passé ?*
❍ *Deux malfaiteurs* ***ont attaqué*** *la banque.*

● *Qu' est-ce que* ***vous avez fait*** *hier soir ?*
❍ ***Je suis allé*** *au cinéma.*

VOIR		ALLER	
j'**ai**	**vu**	je **suis**	**allé/e**
tu **as**		tu **es**	
il/elle/on **a**		il/elle/on **est**	
nous **avons**		nous **sommes**	**allé/s/es**
vous **avez**		vous **êtes**	
ils/elles **ont**		ils/elles **sont**	

Tous les verbes pronominaux (**se lever, s'habiller** etc.) et les verbes **entrer, sortir, arriver, partir, passer, rester, devenir, monter, descendre, naître, mourir, tomber, aller, venir** se conjuguent avec l'auxiliaire **être.** Dans ce cas, le participe s'accorde avec le sujet.

LES PARTICIPES PASSÉS

Il existe 8 terminaisons pour les participes passés.

-é → étudié	-i → fini	-it → écrit
-is → pris	-ert → ouvert	-u → vu
-eint → peint	-aint → plaint	

Attention à la place des adverbes !

*Il a **beaucoup** dormi.*
*Nous avons **bien/mal** mangé.*

EXPRESSION DE LA NÉGATION

*Je **ne** suis **pas** allé au cinéma.*

En français oral, **ne** disparaît souvent.

● *Et Pierre, il est **pas** venu ?*
❍ *Non, je l' ai **pas** vu.*

SITUER DANS LE TEMPS

Hier,
Hier matin,
Hier après-midi, — je suis allé au cinéma.
Hier soir,
Avant-hier,

Ce matin,
Cet après-midi, — j'ai fait de la gymnastique.

Dimanche, lundi, mardi... j'ai joué au football.
Vers 7:30
À 20:00 **environ**

LA SUCCESSION DES ÉVÉNEMENTS

D'abord, *j' ai pris mon petit déjeuner.*
Ensuite, *je me suis douché.*
Puis, *je me suis habillé.*
Après, *je suis sorti.*
Et puis, *j' ai pris l' autobus.*
Enfin, *je suis arrivé au travail.*

Un moment antérieur

Avant + NOM
Avant les examens, *j' étais très nerveuse.*

Avant de + INFINITIF
Avant de me coucher, *je me suis douché.*

Un moment postérieur

Après + NOM
Après le déjeuner, *ils ont joué aux cartes.*

Après + INFINITIF PASSÉ
Après avoir déjeuné, *ils ont joué aux cartes.*

8. « JE » EST MULTIPLE

Par groupes de quatre. Chacun réécrit les phrases ci-dessous sur une feuille, puis vous finissez la première phrase comme vous le souhaitez. Ensuite, vous pliez la feuille pour que l'on ne voie pas ce que vous avez écrit et vous la faites passer à votre voisin de droite qui complètera la 2e phrase, pliera la feuille et la fera passer à son voisin. La feuille doit circuler jusqu'à ce que le texte soit complet.

Samedi matin, à 8 heures, je/j' ....

Ensuite, je/j' ....

Après, vers 11h30, je/j' ....

L'après-midi, entre 14 heures et 16 heures, je/j' ....
et je/j' ....

Comme il faisait beau, je/j' ....

et puis je/j' ....

À 18 heures, je/j' ....

Enfin, je/j' .... C'était une bonne journée !

9. C'EST LA VIE !

A. Écoutez Damien qui raconte à une amie ce qui a changé dans sa vie depuis quelques années. Quels sont les thèmes dont il parle ? Notez-les dans le tableau ci-dessous dans leur ordre d'apparition.

LOISIRS | ASPECT PHYSIQUE | AMIS | LIEU D'HABITATION/LOGEMENT

Thèmes de conversation	Changement
① aspect physique	
②	
③	
④	

B. Écoutez de nouveau leur conversation et notez les changements dont parle Damien. À votre avis, ce sont des changements positifs ou négatifs ?

C. Maintenant, pensez à deux changements dans votre vie, complétez les phrases ci-dessous puis parlez-en avec deux autres camarades.

Avant .... Aujourd'hui, ...

Quand j'étais petit(e)/plus jeune, ...

Mais aujourd'hui, ...

Il y a .... ans, ... Mais aujourd'hui, ...

....

10. QU'EST-CE QUI S'EST PASSÉ ?

Écoutez cette information retransmise par une radio locale et remettez les dessins dans l'ordre chronologique.

11. INTERROGATOIRE

A. La police soupçonne certains membres de votre classe d'être les auteurs de cet étrange cambriolage. Elle veut les interroger à propos de leur emploi du temps, hier soir entre 19 heures et 23 heures. Divisez votre classe en deux groupes : un groupe d'enquêteurs et un groupe de suspects. Pendant que les enquêteurs, deux par deux, vont préparer un questionnaire, les suspects, deux par deux, vont élaborer leur alibi. Les éléments du tableau ci-contre peuvent également vous aider.

- Où étaient les suspects entre .... et .... ?
- Qu'ont fait les suspects entre .... et .... ?
- Comment étaient-ils habillés ?
- Où sont-ils allés après .... et avant .... ?
- Où ont-ils dîné ? Avec qui ?
- À quelle heure se sont-ils séparés ?
- Autres :

B. Chaque suspect est maintenant interrogé individuellement par un policier. Pour essayer de trouver des failles dans les alibis de vos suspects, vous pouvez leur demander de décrire des lieux, des personnes et les questionner sur de petits détails supplémentaires.

- *Où étiez-vous entre 19 heures et 23 heures ?*
- *Je suis allée au restaurant.*
- *Avec qui ?*
- *Avec Aleksandra ?*
- *Comment était le serveur ?*

C. Après les interrogatoires, les enquêteurs comparent les deux déclarations et décident si leurs suspects sont coupables ou non.

- *Aleksandra et Nadia sont coupables : Aleksandra dit qu'elle est allée avec Nadia au restaurant à 21 heures, mais Nadia affirme...*

12. GENTLEMAN OU CAMBRIOLEUR ?

A. Lisez ce texte. Comment imaginez-vous Arsène Lupin ? Et dans la littérature de votre pays, existe-t-il un voleur aussi connu que celui-ci ? Comment est-il ?

ARSÈNE LUPIN

Le personnage d'Arsène Lupin est né le jour où Pierre Laffitte, un grand éditeur qui venait de lancer le magazine *Je sais tout*, a demandé à Maurice Leblanc d'écrire une nouvelle policière dont le héros serait l'équivalent français de Sherlock Holmes. C'est donc dans ce contexte, en 1905, que le premier Arsène Lupin s'impose immédiatement. Il est en fait très différent de Sherlock Holmes. D'abord, ce n'est pas un détective, mais un brillant cambrioleur. Et tandis que Holmes a un côté obscur, dans la vie d'Arsène Lupin tout est clair. Il est de nature gaie et optimiste.

S'il y a eu une disparition ou un vol, on sait immédiatement que le coupable ne peut être qu'Arsène Lupin, ce vif, audacieux et impertinent cambrioleur qui se moque de l'inspecteur Ganimard. Il ridiculise les bourgeois et porte secours aux faibles, Arsène Lupin est, en quelque sorte, un Robin des Bois de la « Belle Époque ».

Il ne se prend pas au sérieux et n'a comme arme que les jeux de mots. C'est un anarchiste qui vit comme un aristocrate ; il n'est jamais moralisateur ; il ne donne pas son cœur à la femme de sa vie, mais aux femmes de ses vies. Il symbolise la double vie, la loi transgressée dans l'élégance et la séduction.

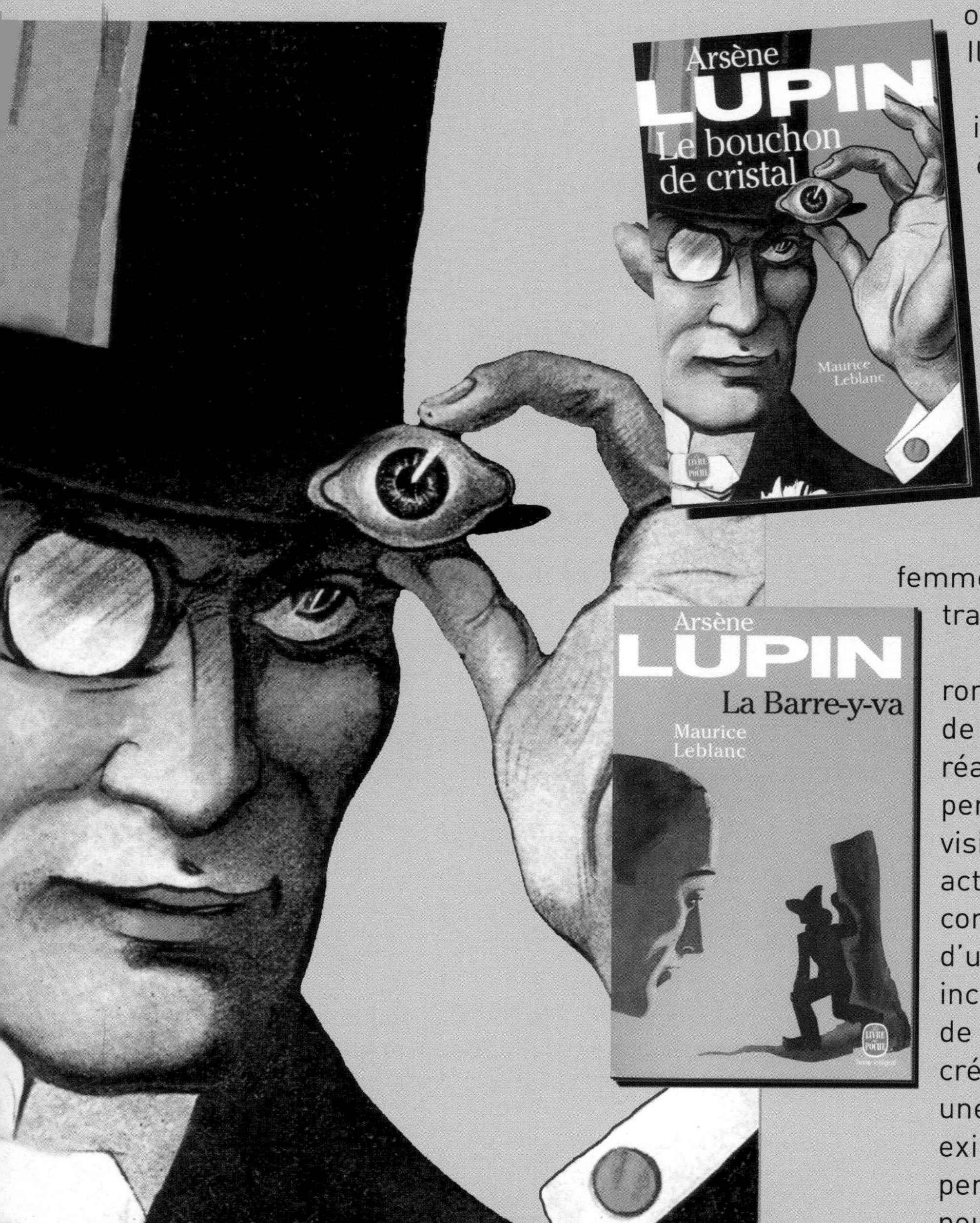

Les aventures de Lupin s'étalent sur seize romans, trente-sept nouvelles et quatre pièces de théâtre entre 1905 et 1939. Certains réalisateurs français ou étrangers ont porté le personnage du gentleman-cambrioleur à la télévision et au cinéma. Malgré les performances des acteurs, aucune des réalisations n'a pu rendre compte de l'essence du personnage. Les traits d'un acteur, quel qu'il soit, ne peuvent en effet incarner un personnage qui, justement, n'a pas de visage et que personne, si ce n'est son créateur, ne peut reconnaître. Arsène lupin est une figure sans traits physiques et qui ne peut exister finalement que par les mots, un personnage que chacun dans son imaginaire peut modeler à sa guise.

B. Écoutez cette chanson : elle raconte l'histoire de deux autres voleurs. Lesquels ?

Vous avez lu l'histoire
De Jesse James
Comment il a vécu
Comment il est mort
Ça vous a plus hein
Vous en d'mandez encore
Eh bien
Ecoutez l'histoire
De Bonnie and Clyde

Alors voilà
Clyde a une petite amie
Elle est belle et son prénom
C'est Bonnie
À eux deux ils forment
Le gang Barrow
Leurs noms
Bonnie Parker et Clyde Barrow

Bonnie and Clyde
Bonnie and Clyde

Moi lorsque j'ai connu Clyde
Autrefois
C'était un gars loyal
Honnête et droit
Il faut croire
Que c'est la société
Qui m'a définitivement abîmé

Bonnie and Clyde
Bonnie and Clyde

Qu'est-c' qu'on n'a pas écrit
Sur elle et moi
On prétend que nous tuons
De sang-froid
C'est pas drôl'
Mais on est bien obligé
De fair' tair'
Celui qui se met à gueuler

Bonnie and Clyde
Bonnie and Clyde

Chaqu'fois qu'un polic'man
Se fait buter
Qu'un garage ou qu'un'
banque
Se fait braquer
Pour la polic'
Ça ne fait d'myster'
C'est signé Clyde Barrow
Bonnie Parker

Bonnie and Clyde
Bonnie and Clyde

Maint'nant chaqu'fois
Qu'on essaie d'se ranger
De s'installer tranquill's
Dans un meublé
Dans les trois jours
Voilà le tac tac tac
Des mitraillett's
Qui revienn't à l'attaqu'

Bonnie and Clyde
Bonnie and Clyde

Un de ces quatr'
Nous tomberons ensemble
Moi j'm'en fous
C'est pour Bonnie que
je tremble
Quelle importanc'
Qu'ils me fassent la peau
Moi Bonnie
Je tremble pour Clyde Barrow

Bonnie and Clyde
Bonnie and Clyde

D'tout'façon
Ils n'pouvaient plus s'en
sortir
La seule solution
C'était mourir
Mais plus d'un les a suivis
En enfer
Quand sont morts
Barrow et Bonnie Parker

Bonnie and Clyde
Bonnie and Clyde

ÇA SERT À TOUT !

s allons mettre au
t un produit qui
itera notre vie.

cela nous allons
endre à :

- ommer et présenter des
jets
- écrire et expliquer le
nctionnement d'un
jet
- ractériser des objets et
vanter leurs qualités
- nvaincre

ous allons utiliser :

- lexique des formes et
s matières
- s pronominaux passifs :
se casse, ça se boit...
- elques expressions
ec des prépositions :
re facile à / utile pour,
rvir à, permettre de...
- s pronoms relatifs *qui*
que
- futur simple
- *âce à...*
- + présent
- *ur/pour ne pas/*
ur ne plus + infinitif

4

1. À QUOI ÇA SERT ?

A. Regardez ces objets. Vous savez comment ils s'appellent ?

- un grille-pain
- des lunettes de soleil
- une machine à laver
- un ouvre-boites
- un sac à dos
- un casque de vélo
- un sèche-cheveux
- un antivirus d'ordinateur
- une machine à calculer

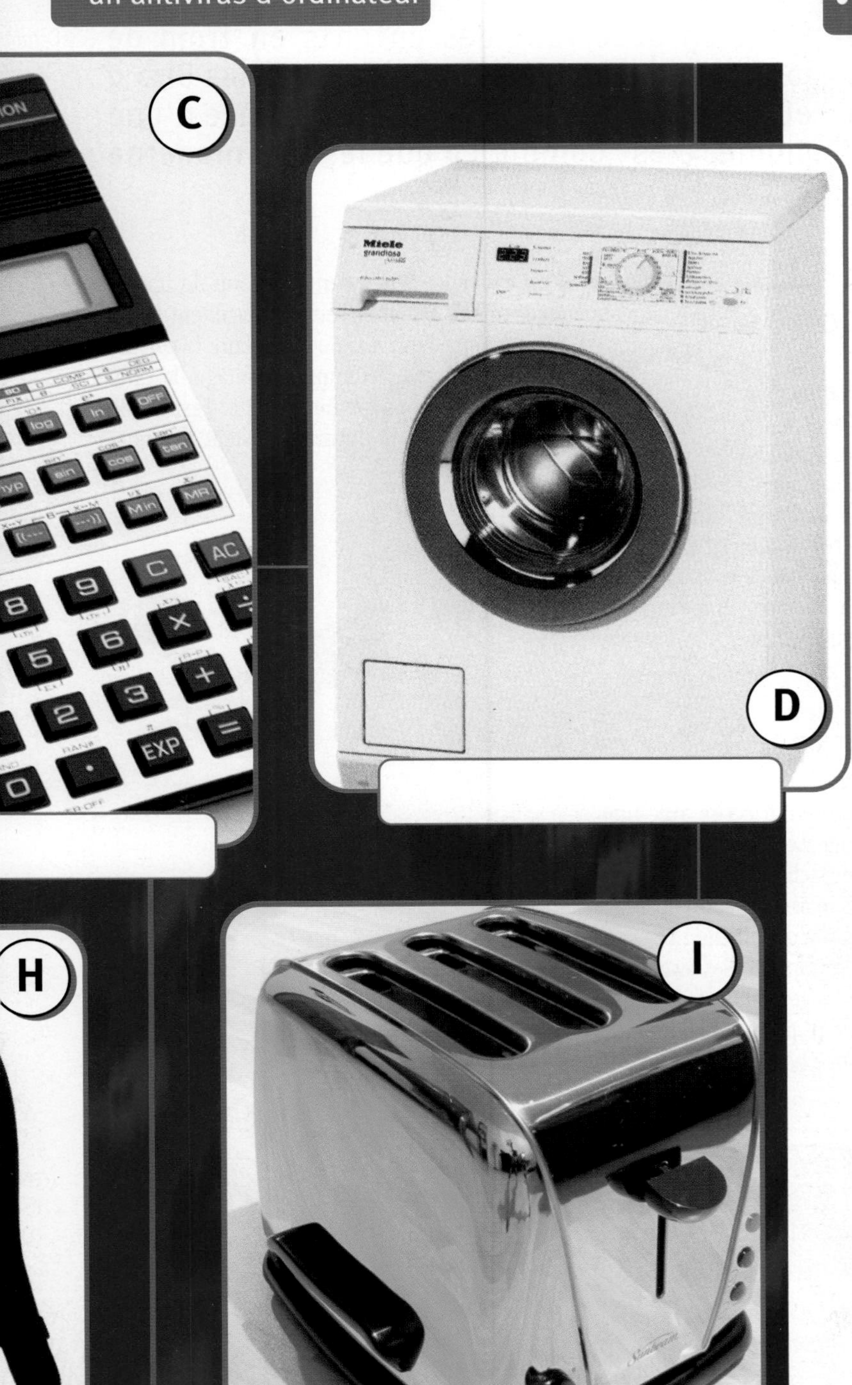

B. À quoi servent ces objets ?

❍ Ça sert à ouvrir une boîte de conserve.
❍ Ça permet de laver le linge.
❍ Ça sert à calculer.
❍ Ça sert à griller le pain.
❍ C'est utile pour se protéger contre les virus informatiques.
❍ C'est utile pour se protéger la tête quand on roule à vélo.
❍ Ça permet de se sécher les cheveux.
❍ Ça sert à se protéger du soleil.
❍ C'est utile pour voyager.

C. Lesquels de ces objets n'utilisez-vous jamais et lesquels utilisez-vous souvent ?

● Moi, je n'utilise jamais de sèche-cheveux.
○ Moi si, je me sers souvent d'un sèche-cheveux.

D. Regardez comment sont construits ces noms. Que remarquez-vous ? Pouvez-vous les classer selon leur structure ?

antivirus	ouvre-boites
sac à dos	casque de vélo

E. Est-ce que vous pouvez penser à d'autres mots construits de la même manière ? Cherchez dans le dictionnaire ou demandez de l'aide à votre professeur.

2. DES INVENTIONS QUI ONT CHANGÉ NOTRE VIE

A. Lisez cet article. Savez-vous en quelle année sont apparus ces objets ? **1925, 1945, 1953, 1974, 1981** ?

B. Lesquelles de ces inventions sont pour vous les plus utiles ? Qu'ajouteriez-vous à cette liste d'inventions qui ont changé notre vie ? Parlez-en avec un camarade.

7 INVENTIONS DU XXe SIÈCLE

Selon Roland Moreno, l'inventeur de la carte à puce, on ne peut pas créer à partir de rien. Toute nouveauté dépend de la capacité à associer deux concepts qui existent déjà. Par exemple, la plume pour écrire et la bille. C'est en observant des enfants en train de jouer aux billes que l'ingénieur Ladislas Biro a eu l'idée de mettre une bille à la pointe d'une plume. C'est comme ça que le stylo moderne est né !

L'ORDINATEUR PERSONNEL (PC)

L'ordinateur a profondément transformé la manière de communiquer, la manière d'apprendre et la manière de jouer. Il y a à peine deux décennies, pour acheter un billet d'avion ou de train, on était toujours obligé de se déplacer et les lettres d'Amérique mettaient des semaines à arriver. Aujourd'hui, grâce à un ordinateur connecté à Internet, vous pouvez écrire votre courrier et l'envoyer en quelques minutes, écouter des CD ou la radio, lire les informations, réserver un billet d'avion, consulter un médecin, acheter un livre ou une voiture sans bouger de chez vous.

LA CARTE À PUCE

C'est l'ingénieur français Roland Moreno qui a inventé la carte à puce. Autrefois, les gens n'avaient pas de compte bancaire et ils gardaient leur argent chez eux, sous le matelas ou dans une chaussette. La carte à puce a permis le développement de la monnaie électronique : télécartes, cartes bancaires, porte-monnaie électroniques. Aujourd'hui, qui n'a pas deux ou trois cartes à puce dans son portefeuille ?

LE RUBAN ADHÉSIF (SCOTCH)

Avant l'invention du ruban adhésif, toutes les petites réparations étaient très compliquées et exigeaient l'intervention d'un spécialiste. Aujourd' hui, tout le monde se sert de ce ruban adhésif et imperméable. Grâce à lui, vous réparez vous-même et très facilement un livre déchiré, un robinet qui fuit, une boîte qui ferme mal…

LE STYLO BIC

Il a amélioré la vie des écoliers. Autrefois, les écoliers apprenaient à écrire avec une plume et un encrier. Le soir, ils rentraient chez eux couverts de taches d'encre sur leurs vêtements, sur leurs mains… L'apprentissage de l'écriture était bien plus difficile ! Le stylo bic a été une petite révolution. Ses avantages sont nombreux : il est léger, solide, facilement transportable, durable (il permet 3 kilomètres d'écriture !), propre et très bon marché.

LES PLATS SURGELÉS

Avant, la préparation d'un bœuf-bourguignon ou d'un cassoulet demandait beaucoup de travail. Aujourd'hui, les plats congelés facilitent la vie des personnes qui n'aiment pas faire la cuisine ou qui n'ont pas le temps. L'avantage de cette technique de conservation est que l'aspect, le goût et les qualités nutritives restent pratiquement intacts.

L'AVION

LES MOUCHOIRS EN PAPIER

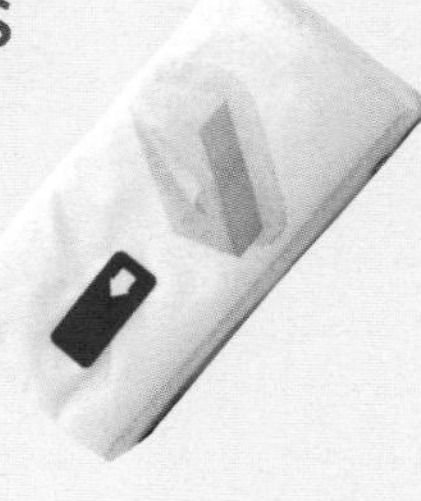

C. En quoi est-ce que l'avion et les mouchoirs en papier ont changé nos vies ? Avec un camarade, écrivez les textes correspondants.

3. JE CHERCHE QUELQUE CHOSE

A. Regardez ce catalogue de vente par correspondance. Vous avez déjà utilisé ces objets ? Vous croyez qu'ils peuvent vous être utiles ? Parlez-en avec un camarade.

● Tu as déjà utilisé un rasoir à peluches ?
○ Non, jamais, et toi ?

■ La bombe lacrymogène peut m'être utile, parce que je rentre souvent seule le soir et...

BOMBE LACRYMOGÈNE
Neutralisant incapacitant
SA
ANTI-AGR
ACTION
30
NEUTRALISANT
Aérosol de défense hyper efficace. Son jet puissant neutralise tout agresseur sans risque pour l'environnement.
6,86 €
• Cont. : 75 ml

Le rasoir à peluches
Indispensable !
Le rasoir à peluches élimine les bouloches des tissus en laine. Vos pull-overs retrouveront un aspect neuf. Grâce à sa petite taille, il est très maniable et vous pourrez l'amener en voyage.
AVANT
14 € 00
avec reservoir
APRÈS
• Fonctionne avec 2 piles LR6
• Livré avec brosse de nettoya-

La brosse anti-peluches

Génial !

Pratique et efficace pour enlever les peluches, poils d'animaux, cheveux... Grâce à ses feuilles autocollantes, vos vêtements, canapés, fauteuils et tapis seront toujours propres !

La tente randonnée

Seulement 30,34€ !

Grâce à cette tente particulièrement confortable, vous apprécierez les joies du camping. Entièrement doublée en polyester imperméable, son armature est en fibre de verre légère. Livrée avec sac de rangement.

B. Écoutez maintenant la conversation d'Emma avec un vendeur. Qu'est-ce qu'elle cherche ? Elle devra finalement acheter ce produit par correspondance. Vous pouvez remplir pour elle le bon de commande ?

BON DE COMMANDE

Quantité	Référence	Nom de l'article	Prix unitaire	Total
Frais de port et d'emballage :				7,50€
TOTAL				

4. C'EST UN OBJET QUI COUPE

Écoutez ces personnes qui jouent à deviner des objets. Numérotez les objets qui sont décrits. Ensuite comparez vos réponses avec deux autres camarades.

5. BINGO

A. Nous allons jouer au bingo par groupes de 4 personnes. Remplissez d'abord ce carton de jeu avec le nom de six objets que vous choisirez parmi ceux que nous vous proposons. Attention ! Écrivez au crayon pour pouvoir effacer ensuite.

B. L'un de vous va décrire des objets : en quelle matière ils sont, quelles sont leurs formes, à quoi ils servent..., mais sans les nommer. Le premier qui a coché toutes ses cases gagne.

C'est en plastique, c'est rond et ça sert à écouter de la musique.

C. Vous allez jouer 4 fois de suite au bingo. À chaque fois, l'un de vous devra prendre le rôle du meneur de jeu.

DÉCRIRE UN OBJET

*un sac **en*** ***papier***
tissu
cuir
plastique

*une boîte **en*** ***carton***
bois
porcelaine
fer
verre

C'est ***petit.***
grand.
plat.
long.
rond.
carré.
rectangulaire.
triangulaire.
lavable.
jetable.
incassable.
imperméable.

Ça se *lave facilement.*
Ça s'*ouvre tout seul.*
Ça se *mange.*

Ça sert à *écrire.*
C'est utile pour *ouvrir une bouteille.*
Ça permet d'*écouter de la musique.*
Ça marche avec *des piles.*
de l' électricité.
de l' essence.

PRONOMS RELATIFS : QUI ET QUE

Qui remplace le sujet de la phrase qui le suit.

*C' est un objet **qui** permet de laver le linge.*
(= **l'objet** permet de laver...)

*C' est une chose **qui** sert à griller le pain.*
(=**la chose** sert à...)

Que remplace le complément d'objet direct de la phrase qui le suit.

*C' est un objet **que** vous portez dans votre sac.*
(=vous portez l'objet...)

*C' est une chose **que** l'on utilise pour manger.*
(=on utilise cette chose pour...)

LE FUTUR

Le futur sert à formuler des prévisions ou à faire des prédictions.

*Demain, il **fera** soleil sur tout le pays.*
*Dans 30 ans, nous **marcherons** sur Mars.*

Il sert aussi à faire des promesses.

*Demain, je **viendrai** te chercher à 16 heures.*
*Cet appareil vous **facilitera** la vie.*

Bientôt,
Demain,
***Dans** 5 jours/mois... nous serons plus heureux.*
Au siècle prochain,
Le mois prochain,
...

*« Il **fera** beau sur toutes les régions. »*

Verbes réguliers

	manger-	
je	**étudier-**	-ai
tu	**finir-**	-as
il/elle/on	**sortir-**	-a
nous	**manger-**	-ons
vous	**étudier-**	-ez
ils/elles	**écrir-**	-ont
	prendr-	

Verbes irréguliers

	ÊTRE	**ser-**	
je	AVOIR	**aur-**	-ai
tu	FAIRE	**fer-**	-as
il/elle/on	SAVOIR	**saur-**	-a
nous	ALLER	**ir-**	-ons
vous	POUVOIR	**pourr-**	-ez
ils/elles	DEVOIR	**devr-**	-ont
	VOIR	**verr-**	

GRÂCE À + SUBSTANTIF

***Grâce à** Internet,*
***à** la télévision,*
***au** téléphone,*
***à** l'ordinateur personnel,*
***aux** satellites de télécommunications,*
— on se sent moins seul !

6. UN INVENTEUR ET SON INVENTION

A. Lisez le texte suivant. Vous savez de quelle invention il s'agit et quel est le nom complet de l'inventeur ?

Alexander G. B. est né en 1847 en Écosse. Spécialiste en physiologie vocale, il a imaginé un appareil **qui transmet** le son par l'électricité. Cette machine a beaucoup évolué depuis son invention. Aujourd'hui, c'est un **appareil qui fonctionne** à l'électricité ou bien avec des ondes. C'est un **appareil qui vous permet** de parler avec vos amis, votre famille... et **que vous transportez** facilement dans une poche.

B. Observez les phrases avec **qui** et **que.** Que remarquez-vous ? Est-ce que vous pouvez déduire quand on utilise **qui** et quand on utilise **que** ?

7. FUTURS POSSIBLES !

A. À votre avis, comment sera le futur ? Décidez si ces affirmations sont vraies ou fausses. Est-ce que vous pouvez faire deux autres prédictions pour l'avenir ?

	Vrai	Faux
En 2050, on partira en vacances sur Mars.		
Dans 30 ans, on traversera l'Atlantique sur un pont.		
En 2020, les enfants n'iront plus à l'école.		
Au siècle prochain, grâce aux progrès médicaux, nous vivrons 120 ans et plus.		
Bientôt, il n'y aura plus d'animaux sauvages en Afrique.		
Un jour, les voitures voleront.		
Dans 15 ans, l'eau sera aussi chère que l'essence.		
Dans 100 ans, il y aura des mutants.		
Dans deux siècles, les hommes n'auront plus de dents.		
Dans quelques années, nous mangerons des comprimés au lieu de produits frais.		
En 2090, les professeurs de français n'existeront plus.		

B. Comparez vos réponses avec deux autres camarades. Est-ce que vous voyez le futur de la même manière ?

- *Je crois que dans quelques années les enfants n'iront plus à l'école, les cours seront...*

C. Et vous-même, comment rêvez-vous votre futur ?

- *Moi, dans 20 ans, j'aurai une grande maison hyper-moderne, totalement automatisée.*
- *Eh bien moi, dans 5 ans, je parlerai parfaitement français !*

8. JE VOUS LE RECOMMANDE !

Pensez à un objet que vous avez sur vous. À quoi sert-il ? Quels sont ses avantages ? Maintenant présentez cet objet à deux camarades, qui doivent deviner de quoi il s'agit.

- *Alors... C'est un objet extrêmement pratique et très facile à...*

TÂCHE CIBLÉE

9. UN PRODUIT QUI VA FACILITER VOTRE VIE

A. Le journal *Pratique* a mené une enquête pour identifier les petits problèmes quotidiens de ses lecteurs.
Lisez les différents témoignages et trouvez à quelle personne ils correspondent.

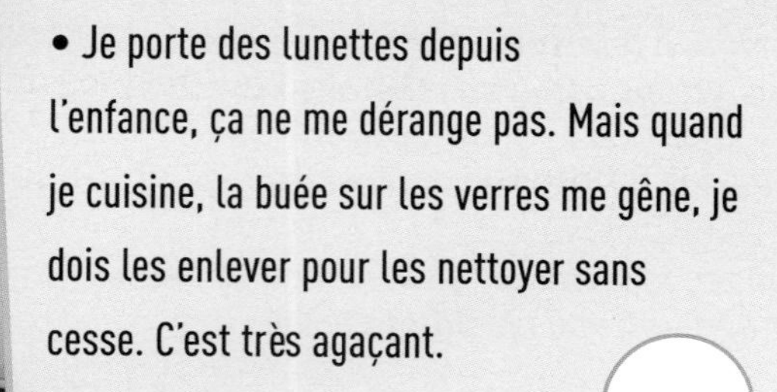

- Je porte des lunettes depuis l'enfance, ça ne me dérange pas. Mais quand je cuisine, la buée sur les verres me gêne, je dois les enlever pour les nettoyer sans cesse. C'est très agaçant.

- J'adore lire et mon mari aussi, alors nous sommes envahis par les livres, nous n'avons toujours pas trouvé de système efficace pour les ranger.

- Pour aller à la fac, je dois toujours transporter des kilos de livres et mon cartable est terriblement lourd à porter. Ma mère m'a offert un cartable à roulettes mais je me sens tellement ridicule avec ça !

- Devant ma maison, il y a un terrain plein de boue, et comme je dois entrer et sortir de chez moi plus de vingt fois par jour, je n'enlève pas mes bottes et salis le sol.

- J'adore lire au lit mais c'est fatigant de tenir le livre et surtout de tourner les pages !

- J'aimerais avoir une cuisine moderne avec tous les robots et appareils électroménagers possibles mais ma cuisine est très très petite et je n'ai pas la place pour mettre un lave-vaisselle, par exemple.

B. Lequel de ces problèmes voulez-vous résoudre ? Cherchez dans la classe d'autres personnes qui veulent résoudre le même problème que vous et mettez au point un produit original qui pourrait apporter une solution.

Nom du produit : ....

Description du produit : ....

....

....

Utilisateurs potentiels : ....

....

....

C. À présent, vous allez rédiger la page catalogue pour vendre votre produit par correspondance.

ANTISÈCHE

Exprimer la condition

Si + PRÉSENT, + IMPÉRATIF
Si vous voulez *un pull-over impeccable,* ***utilisez*** *la brosse anti-peluches !*

Si + PRÉSENT, + FUTUR
Si vous étudiez *avec ce manuel, vous* ***parlerez*** *bien français !*

Exprimer le but

Pour ne pas + INFINITIF
Pour ne pas vous fatiguer, *utilisez l'ascenseur !*

Pour ne plus + INFINITIF
Pour ne plus penser *à vos problèmes, partez à Tahiti !*

Suffire (de)

Pour *nettoyer la brosse,* ***il suffit de*** *la passer sous l'eau.*
Pour *obtenir une surface brillante, un simple geste* ***suffit****.*

10. LA PRÉSENTATION

Maintenant, vous allez présenter votre produit à toute la classe. Chaque fois qu'un groupe présente son produit, vous devez, individuellement, décider si ce produit vous semble utile et si vous voulez l'acheter.

CATALOGUE D'OBJETS INTROUVABLES

Jacques Carelman est né à Marseille en 1929. Il vit à Paris depuis 1956. Peintre, sculpteur, illustrateur, scénographe et totalement autodidacte, il utilise l'humour pour créer. Il est connu dans le monde entier principalement pour son *Catalogue d'objets introuvables*, publié en 1969 comme parodie d'un catalogue de vente. Ce catalogue est constitué d'une soixantaine d'objets de la vie quotidienne, originaux et même absurdes. Carelman est un enfant du surréalisme, mais un fils un peu dissident. En effet, tandis que les surréalistes voient les choses d'un point de vue plutôt irrationnel, Carelman contemple leurs côtés absurdes. Carelman fait figure de rebelle dans notre société de consommation car ses œuvres nous disent tout simplement que les objets de notre quotidien sont délibérément inutiles.

Carelman est aussi fondateur de l'OuPeinPo et régent du Collège de pataphysique. La pataphysique est la science des solutions imaginaires et des exceptions, étant donné que dans notre monde, il n'y a que des exceptions (la « règle » étant précisément une exception aux exceptions).

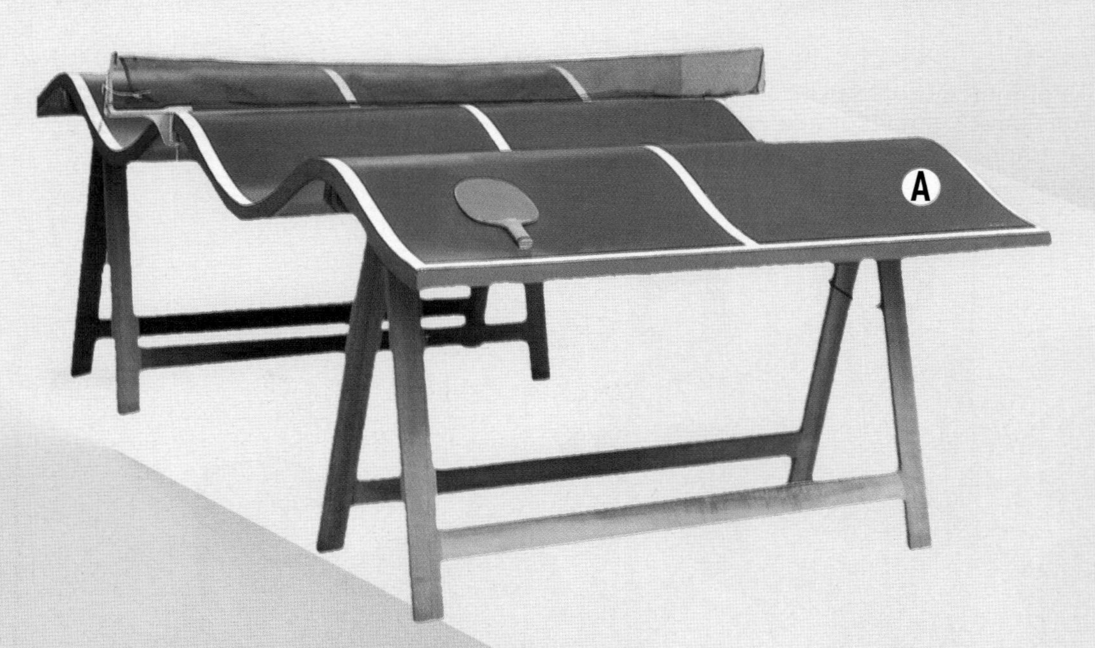

▲ Table de Ping-Pong « Tous Azimuts »

Grâce à sa forme, cette table augmente le plaisir du jeu : la balle fait des rebonds inattendus, comparables aux rebonds d'un ballon de rugby.

▼ Cafetière pour masochiste

Le dessin nous semble suffisamment explicite !

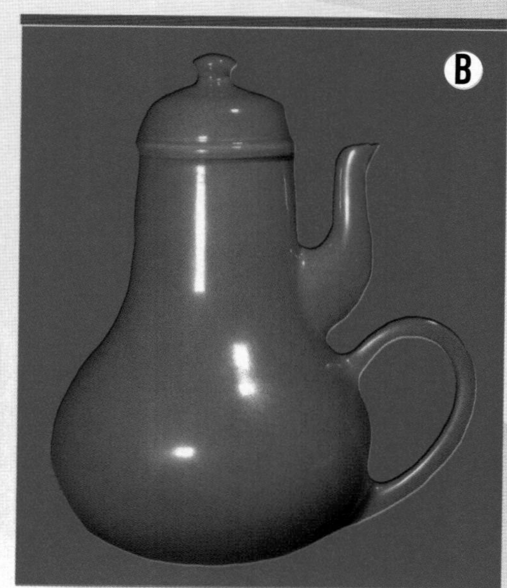

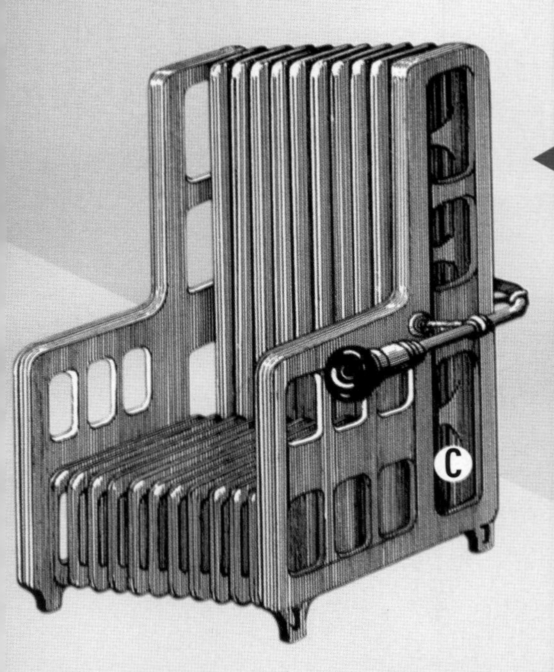

Fauteuil-radiateur

Se branche sans aucune difficulté sur n'importe quel chauffage central. Indispensable pour les frileux.

Paire de ciseaux optique

Permet aux personnes âgées de mieux voir quand elles cousent.

Robinet économique

Grâce à sa forme, il ne consomme pratiquement pas d'eau.

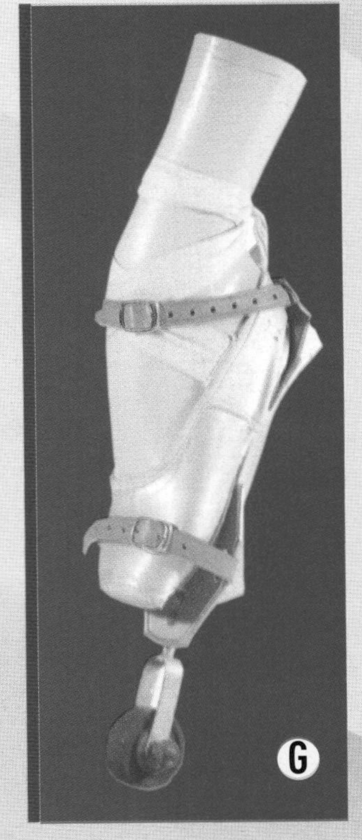

Patin à roulette pour danseuse classique

Il évite aux danseuses classiques désireuses de patiner, d'abîmer la cambrure de leurs pieds avec des patins ordinaires.

E

Parapluie familial

Un seul manche suffit à tenir les parapluies de toute la famille.

11. OBJETS INSOLITES

A. Parmi ces objets, lesquels vous plaisent le plus ? Pourquoi ?
B. À quels besoins ces objets prétendent-ils répondre ?

JE SERAIS UN ÉLÉPHANT

s allons élaborer un
de personnalité et
arer un entretien
ecrutement.

cela nous allons
endre à :

- aluer des qualités
ersonnelles
- rmuler des
ypothèses
- oisir des manières de
adresser à quelqu'un
lon la situation et les
lations entre les
ersonnes
- mparer et justifier nos
oix

us allons utiliser :

- *oir du, de la , des* + sub-
antif
- *anquer de* + substantif
- ypothèse
+ imparfait, conditionnel
- s pronoms compléments
rects et indirects : *me,*
le, la, les, lui, leur, etc.
- tutoiement et le
uvoiement
- comparaison

5

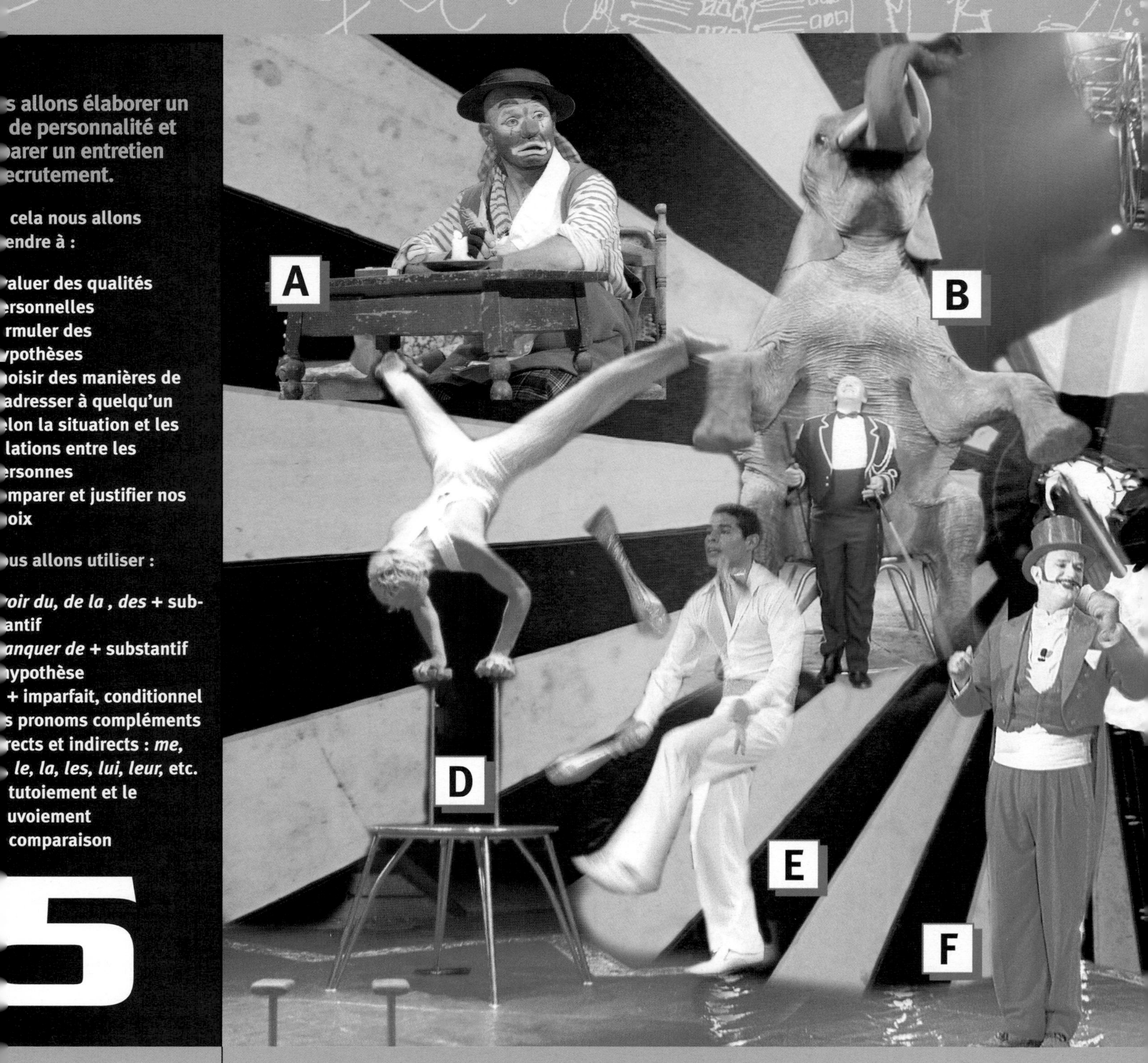

1. LE PLUS GRAND SPECTACLE DU MONDE

A. Regardez ces photos, vous pouvez identifier le métier de ces personnes ?

- ❑ des trapézistes
- ❑ un clown
- ❑ un présentateur
- ❑ un cracheur de feu
- ❑ un jongleur
- ❑ un équilibriste
- ❑ des acrobates
- ❑ un dresseur d'éléphants

C

G

H

B. À votre avis, quelles sont les qualités essentielles pour exercer ces métiers ?

Être
adroit / créatif / souple / agile / téméraire...

Avoir
de l'imagination / des réflexes / du sang-froid / un bon sens de la coordination / de l'équilibre / une grande capacité de concentration / une grande confiance en l'autre...

Ne pas avoir
le vertige / peur des animaux sauvages / peur du ridicule...

Aimer
les enfants / les animaux / le risque / faire rire / le contact avec le public...

Savoir
parler en public / se faire obéir par les animaux / raconter des histoires / jouer de la flûte...

- ● *Pour être clown, il faut aimer faire rire.*
- ○ *Oui, et il ne faut pas avoir peur du ridicule.*

C. Lequel de ces métiers pourriez-vous exercer ? Lequel est-ce que vous ne pourriez pas faire ? Parlez-en avec deux autres camarades.

- ● *Moi, je crois que je pourrais être clown. J'aime bien faire rire et j'aime beaucoup les enfants.*
- ○ *Moi, je ne pourrais pas ! Je manque totalement d'imagination et j'ai peur du ridicule.*

2. QUALITÉS ANIMALES ?

A. Les animaux sont souvent associés à certaines qualités ou à certains défauts. Vous savez quelles sont les associations que l'on fait en français ? Essayez de deviner !

fort		un agneau
rusé		une mule
malin		un singe
doux		une pie
têtu		une tortue
bavard	**comme**	un bœuf
lent		une limace
fainéant		une taupe
muet		un renard
myope		une carpe

B. Est-ce que l'on fait les mêmes associations dans votre culture ? Est-ce qu'il y en a d'autres ?

3. ÊTES-VOUS SOCIABLE OU MISANTHROPE ?

A. Répondez à ce test puis, avec un camarade, comparez vos réponses. Lequel d'entre vous est le plus sociable ? Pourquoi ?

- *Moi, je pense que tu es plus sociable que moi parce que tu as choisi l'agneau comme animal...*

SOCIABLE OU MISANTHROPE ?

1. Vous bloquez la rue avec votre voiture et un automobiliste klaxonne...
- A. Vous lui souriez sans bouger.
- B. Vous vous excusez et vous partez immédiatement.
- C. Vous l'ignorez complètement.

2. Quel animal aimeriez-vous être ?
- A. Un chimpanzé.
- B. Un ours.
- C. Un agneau.

3. Un ami vous a appelé, vous étiez absent...
- A. Vous attendez qu'il rappelle.
- B. Vous l'appellerez après dîner.
- C. Vous le rappelez immédiatement.

4. Si vous deviez partir vivre ailleurs vous iriez à...
- A. New York.
- B. Oulan-Bator en Mongolie.
- C. México.

5. Un touriste vous demande son chemin...
- A. Vous répondez « *Sorry, I don't speak English* ».
- B. Si vous avez le temps, vous l'accompagnez jusqu'à sa destination.
- C. Vous lui recommandez de prendre un taxi.

6. Si vous ne deviez pas travailler, qu'est-ce que vous feriez ?
- A. Vous iriez tous les soirs en boîte.
- B. Vous iriez vivre dans un petit village perdu.
- C. Vous feriez du bénévolat dans une ONG.

7. Si vous vous inscriviez à une activité de loisir, ce serait...
- A. Du basket-ball.
- B. De la natation.
- C. De la salsa.

8. Si vous pouviez vivre la vie d'un personnage de roman, qui aimeriez-vous être ?
- A. Robinson Crusoé.
- B. D'Artagnan.
- C. Arsène Lupin.

9. Si vous invitiez cinq personnes à dîner ce soir et qu'il vous manquait trois chaises, qu'est-ce que vous feriez ?
- A. Vous iriez demander trois chaises à un voisin.
- B. Vous organiseriez un buffet froid.
- C. Vous annuleriez le repas et passeriez la soirée seul(e).

10. Si vous voyiez un aveugle sur le point de traverser un carrefour dangereux, qu'est-ce que vous feriez ?
- A. Vous l'observeriez pour intervenir si c'était nécessaire.
- B. Vous le prendriez par la main pour l'aider à traverser.
- C. Vous penseriez qu'il doit être habitué à traverser ce carrefour et vous continueriez votre chemin.

B. Faites vos comptes. Si vous obtenez le même nombre de réponses pour deux symboles, cela signifie que votre personnalité oscille entre deux portraits.

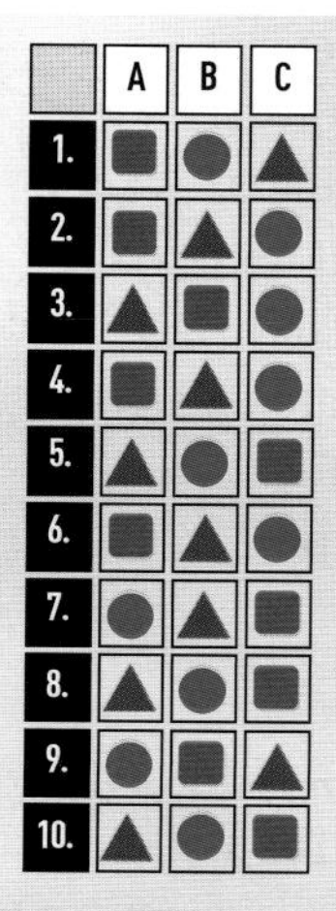

	A	B	C
1.	■	●	▲
2.	■	▲	●
3.	▲	■	●
4.	■	▲	●
5.	▲	●	■
6.	■	▲	●
7.	●	▲	■
8.	▲	●	■
9.	●	■	▲
10.	▲	●	■

Vous avez une majorité de ●

Sociable et généreux !

Vous êtes quelqu'un de très sociable. Vous êtes toujours attentif aux besoins des autres et vous avez bon caractère. Par contre, vous manquez d'agressivité et certaines personnes autour de vous ont tendance à abuser de votre gentillesse.

Vous avez une majorité de ■

La société représente pour vous le confort !

Vous êtes sociable par intérêt. En effet, vous préférez les avantages que vous offrent la vie en société et vous êtes quelqu'un de fondamentalement urbain. Pas question pour vous de vous exiler au fond de la forêt amazonienne, car vous pensez que vous n'avez rien à y faire.

Vous avez une majorité de 

Vous manquez de confiance en la société !

Vous avez une tendance misanthrope. Vous avez besoin d'être seul pour vous détendre réellement et être capable d'affronter le stress de la vie en société. Vous n'avez pas le sens de l'humour et vous vous sentez parfois attiré par les expériences mystiques.

4. INTRIGUES AMOUREUSES

A. Voici deux extraits de roman-photo. Des phrases manquent, à vous de les identifier pour compléter les dialogues. Ensuite, écoutez et vérifiez.

a. Asseyez-vous, je vous en prie !
b. Mais asseyez-vous donc !
c. Entrez, je vous en prie !
d. Je te présente ma collègue.
e. Ben, entre !
f. Assieds-toi, si tu veux.
g. Entrez, entrez !

B. Quand est-ce qu'on dit **vous** ? Cochez les bonnes réponses.

Quand on parle...

❑ à plusieurs personnes à la fois.	❑ à un ami.	❑ à quelqu'un de notre famille.
❑ à une personne que l'on ne connaît pas.	❑ à une personne âgée.	❑ à un camarade de classe.
❑ à une personne avec qui on a des contacts superficiels (un voisin, un commerçant...).	❑ à un supérieur hiérarchique.	❑ à un collègue de même niveau hiérarchique.
	❑ au professeur.	❑ aux parents de son/sa petit/e ami/e.

5. DEVINETTES

A. Lisez ces phrases, savez-vous ce qu'elles définissent ?

1. On **le** considère comme le meilleur ami de l'homme.
2. On **la** change deux fois par an.
3. On **les** ouvre le matin et on **les** ferme le soir.
4. On **lui** écrit pour avoir des cadeaux.
5. Les marins sont perdus s'ils **les** entendent chanter.
6. On **leur** offre un cadeau en mai.
7. On **leur** envoie des messages sans savoir s'ils existent.
8. Les enfants **lui** confient leurs dents de lait quand elles tombent.

B. Observez les pronoms en caractère gras dans les devinettes. Pouvez-vous compléter le tableau suivant ? Quelle différence observez-vous entre les compléments d'objet direct (COD) et les compléments d'objet indirect (COI) ?

COD	le	MASCULIN SINGULIER	On considère le chien comme le...
	la		
		MASC. ET FÉM. PLURIEL	
COI	lui		

C. Avec un camarade, faites correspondre les termes suivants avec leur définition : **les enfants, la Terre, la télévision, l'eau, les amis, l'amour, l'Univers**. Utilisez le pronom qui convient.

- On le
- On la
- On les
- On lui
- On leur

attribue 4,5 milliards d'années : ....
demande souvent des conseils : ....
croit en expansion : ....
raconte des histoires avant de dormir : ....
cherche toute la vie : ....
regarde en moyenne 2 heures par jour : ....
doit la vie : ....

6. TROIS ANIMAUX

Voici un test (très sérieux). D'abord, réfléchissez pour y répondre, puis parlez-en avec un camarade. Après, votre professeur vous donnera des clés afin d'interpréter vos réponses.

- Si vous étiez un animal, quel animal seriez-vous ? Pourquoi ?
- Si vous ne pouviez pas être cet animal, quel autre animal seriez-vous ? Pourquoi ?
- Si vous ne pouviez pas être ce deuxième animal, quel troisième animal seriez-vous ? Pourquoi ?

● Moi, si j'étais un animal, je serais un éléphant.
○ Pourquoi ?
■ Parce que les éléphants sont très forts...,
● Et si tu ne pouvais pas être un éléphant, qu'est-ce que tu serais ?

LES PRONOMS COD ET COI

Quand on parle de quelqu'un ou de quelque chose qui a déjà été mentionné ou bien est identifiable grâce au contexte, on utilise les pronoms compléments d'objet direct (COD) et compléments d'objet indirect (COI), afin de ne pas le répéter.

Le COD

Le COD représente la chose ou la personne sur laquelle s'exerce l'action exprimée par le verbe.

● *Tu regardes beaucoup **la télévision** ?*
○ *Non, je **la** regarde surtout le week-end.*

● *Tu écoutes **le professeur** quand il parle ?*
○ *Bien sûr que je **l'** écoute !*

Pour identifier le COD dans une phrase, on peut poser la question avec **quoi** ou **qui**.

*Tu regardes **quoi** ?* (**la télévision**)
*Tu écoutes **qui** ?* (**le professeur**)

Le COI

Le COI est la chose ou la personne qui reçoit indirectement l'action que fait le sujet. Il est introduit par une préposition.

● *Qu'est-ce que vous offrez **à Charlotte** pour son anniversaire ?*
○ *On **lui** offre un pull-over.*

● *Est-ce que tu as téléphoné **à tes parents** ?*
○ *Oui je **leur** ai téléphoné ce matin.*

Pour identifier le COI dans une phrase, on peut poser la question **à qui**.

*Tu as téléphoné **à qui** ?* (**à tes parents**)

Les pronoms COD et COI se placent normalement devant le verbe dont ils sont compléments.

● *Et ton travail ?*
○ *Je peux **le** faire demain.* ~~*Je le peux*~~ *faire.*

● *Tu as parlé à Marie-Laure ?*
○ *Je vais **lui** parler ce soir.* ~~*Je lui vais*~~ *parler.*

À l'oral, on utilise souvent les pronoms COD et COI avant même d'avoir mentionné l'élément auquel ils se réfèrent.

● *Alors, tu **les** as faits **tes devoirs** ?*

○ *Qu'est-ce que tu **lui** as acheté, à **maman** ?*

FAIRE UNE HYPOTHÈSE AU PRÉSENT

Pour exprimer une hypothèse au présent, on utilise la forme :

Si + IMPARFAIT, CONDITIONNEL PRÉSENT

- ● ***Si vous gagniez*** *beaucoup d'argent à la loterie, qu'est-ce que* ***vous feriez*** *?*
- ❍ ***Je ferais*** *le tour du monde.*
- ● ***Si vous étiez*** *un animal, quel animal* ***seriez-vous*** *?*
- ❍ *Moi,* ***je serais*** *un éléphant.*

LE CONDITIONNEL

Pour former le conditionnel, on prend la base du futur simple et on ajoute les désinences de l'imparfait :

ÊTRE	Futur	Conditionnel	
	ser-	je ser**ais**	[e]
		tu ser**ais**	[e]
		il/elle/on ser**ait**	[e]
		nous ser**ions**	[iõ]
		vous ser**iez**	[ie]
		ils/elles ser**aient**	[e]

AVOIR	→	**aur-**	
FAIRE	→	**fer-**	
SAVOIR	→	**saur-**	
ALLER	→	**ir-**	-ais
POUVOIR	→	**pourr-**	-ais
DEVOIR	→	**devr-**	-ait
VOIR	→	**verr-**	-ions
VOULOIR	→	**voudr-**	-iez
VENIR	→	**viendr-**	-aient
ÉTUDIER	→	**étudier-**	
AIMER	→	**aimer-**	
DORMIR	→	**dormir-**	

TU OU VOUS ?

Tu suppose une relation de familiarité et s'utilise pour parler aux enfants, aux membres de la famille, aux amis et, dans certains secteurs professionnels, aux collègues de même niveau hiérarchique. **Vous** s'utilise pour marquer le respect ou la distance. Les statuts des interlocuteurs ne sont pas toujours égaux et souvent l'un des interlocuteurs tutoie tandis que l'autre vouvoie. Quand des locuteurs francophones veulent passer au tutoiement, ils le proposent clairement.

- ● *On se tutoie ?*
- ❍ *Tu peux me tutoyer, si tu veux.*

7. DANS LE DÉSERT

A. Vous allez entendre une histoire. Fermez les yeux et imaginez que vous partez en voyage, dans le désert du Sahara. En visualisant ce que l'on vous dit, répondez mentalement aux questions qui vous seront posées ou écrivez vos réponses en profitant de chaque pause. Votre professeur peut vous donner une fiche à remplir.

B. Maintenant, ouvrez lentement les yeux et regardez ce que vous avez écrit. Le professeur va vous donner des clefs pour interpréter vos réponses. Est-ce que vous êtes d'accord avec ces interprétations ? Parlez-en avec un camarade de classe.

8. SI C'ÉTAIT...

Vous connaissez le jeu du portrait ? Toute la classe va élaborer une liste avec 10 personnes connues de tous. Puis, individuellement, vous allez en définir une en répondant aux questions qui vous sont posées.

- Si c'était un animal, ce serait .... ?
- Si c'était un objet, ce serait .... ?
- Si c'était une profession, ce serait ...... ?
- Si c'était une couleur .... ?
- Si c'était un pays .... ?
- Si c'était une pièce de la maison
- .... ?
- Si c'était ....

- ● Si c'était un animal, qu'est-ce que ce serait ?
- ❍ Si c'était un animal, ce serait un hippocampe.

9. COMMENT RÉAGISSEZ-VOUS ?

Comment réagissez-vous dans les situations suivantes ? Comparez vos réactions avec celles d'un camarade.

Qu'est-ce que vous faites

- si vous trouvez un animal abandonné ?
- quand vous avez vu un film qui vous a beaucoup plu ?
- quand vous voyez un pantalon qui vous plaît dans une vitrine ?
- si un ami vous demande de lui prêter votre voiture/moto/vélo ?
- quand vos parents vous donnent un conseil ?
- si un ami a un problème ?
- si vous devez de l'argent à un ami/collègue... ?
- quand vous n'avez pas compris ce que le professeur vient d'expliquer ?
- quand vous n'aimez pas le livre que vous venez de commencer ?
- si on vous demande de garder un secret ?

- ● Moi, quand j'ai vu un film qui m'a plu, je le recommande à mes amis.
- ❍ Et bien moi, je le recommande à...

10. NE TUTOYEZ PAS VOTRE INTERLOCUTEUR !

A. Cet article donne quelques conseils pour réussir un entretien de recrutement. En le lisant, vous découvrirez ce qu'il faut faire ou ce qu'il ne faut pas faire, dans cette situation. À votre avis, le comportement du candidat de la photo est-il adéquat ? Rédigez la règle correspondante !

B. Est-ce que ces règles sont valables dans votre pays ? Si vous découvrez des règles spécifiques à votre pays, rédigez-les.

- *Dans mon pays, on ne doit pas serrer la main de son interlocuteur.*

RÈGLES D'OR DE L'ENTRETIEN DE RECRUTEMENT

Tous les conseils pour réussir l'étape finale de votre recherche d'emploi : l'entretien de recrutement

À éviter :

- Manquer de ponctualité.
- Négliger votre aspect vestimentaire.
- Vous approcher à moins de 90 centimètres de votre interlocuteur.
- Serrer mollement la main de votre interlocuteur (votre poignée de main doit être ferme).
- Prendre l'initiative : attendez que votre interlocuteur vous invite à vous asseoir.
- Fuir le regard de votre interlocuteur (regardez-le dans les yeux, mais sans le fixer).
- Lui proposer de vous tutoyer.

Vous êtes jeune diplômé et vous allez vous présenter à votre premier entretien d'embauche ? Vous êtes étranger et vous cherchez du travail en France ? Voici les règles d'or de l'entretien de recrutement.

Sachez que la première impression que vous produirez sera déterminante. Alors, soignez votre apparence, soyez courtois et souriant. Montrez-vous sûr de vous mais pas arrogant. Faites attention à votre gestuelle et à vos mimiques. En effet, selon certaines études, 90% de la communication est constitué sur le langage non-verbal !

11. MÉTIER INSOLITE

A. Par groupes de trois, pensez à un métier peu courant (le professeur peut vous aider), puis définissez quelles sont, à votre avis, les qualités essentielles pour exercer ce métier.

- *Je propose « professeur de tai-chi ». Un professeur de « tai-chi » doit avoir de l'équilibre et un bon sens de la coordination...*
- *Moi, je propose souffleur au théâtre ou psychologue pour animaux.*

B. Maintenant, toujours en groupes, vous devez préparer un test (en trois exemplaires) : six à huit questions qui permettent d'identifier si une personne a les qualités requises pour exercer ce métier insolite. Attention : dans votre test, ne faites pas référence au métier auquel vous pensez ! Voici quelques exemples de questions.

1. **Lequel de ces trois objets aimeriez-vous être ? Pourquoi ?**
   a. une montre
   b. un vase
   c. un marteau
2. **Si vous deviez travailler ...., que feriez-vous ?**
   a. ....
   b. ....
   c. ....
3. **Aimeriez-vous travailler avec .... ?**
4. **Qu'est-ce que vous feriez si vous pouviez/alliez................ ?**

SOURCIER

LAVEUR DE VITRES

CONDUCTEUR DE TRAIN

C. Vous êtes directeur/trice de ressources humaines et vous avez publié dans la presse une petite annonce de recrutement. À présent, chacun d'entre vous va simuler un entretien d'embauche. Vous allez interviewer une personne d'un autre groupe et noter ses réponses. Attention : ne révélez pas à votre candidat pour quel métier il postule !

ENTREPRISE
sérieuse recrute :
Femme ou homme,
SALAIRE INTÉRESSANT,
PERSPECTIVES D'ÉVOLUTION
Pour un entretien individuel
avec la société, contactez-nous :
Mél : carriere@facil.com
Tél. : 01 56 44 44 44

ANTISÈCHE

- *Qu'est-ce que vous pensez de Yannick ?*
- *À mon avis, il manque de patience pour ce travail.*

COMPARAISON

*Lucien est **plus** patient que Philippe.*
*Vincent est **moins** dynamique **que** Nathalie.*
*Kevin est **aussi** sérieux **que** Sybille.*

*Paul a **plus/autant/moins** de patience que Luc.*

Martin aime la solitude.
***Par contre,** il ne supporte pas la chaleur.*

D. Maintenant, reformez le groupe d'origine et comparez les réponses que vous avez obtenues. Quel est le candidat que vous choisissez et pourquoi ? Ensuite, expliquez votre choix au reste de la classe.

- Nous choisissons Paul pour être ... parce qu'il est ... et aussi parce que ...

12. TRAVAIL, TRAVAIL

Légalement, une semaine de travail représente combien d'heures dans votre pays ? Cela vous semble-t-il beaucoup ? Peu ? Quels seraient les avantages et quels seraient les inconvénients si une loi comme celle des 35 heures en France était appliquée dans votre pays ?

TEMPS DE TRAVAIL ET TEMPS DE LOISIRS

Tout au long du vingtième siècle, de grandes lois sociales ont rythmé la vie des Français au travail. En 1906, le repos dominical est imposé et en 1919 on passe à la journée de 8 heures et à la semaine de 48 heures.

Mais la date qui reste dans la mémoire collective correspond à la loi sur les congés payés. Elle est associée au gouvernement du Front populaire. C'est en effet en 1936, sous le gouvernement de Léon Blum, que deux lois sociales instaurent la semaine de 40 heures et le droit pour tous les salariés à 12 jours de congés payés par an.

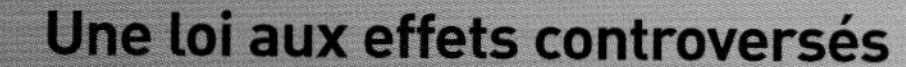

Une loi aux effets controversés

À partir de 1982, on assiste à une augmentation du temps libre et le droit aux loisirs est finalement mieux respecté que le droit au travail. Cependant beaucoup de Français s'interrogent sur ce temps « libéré », car ils ont le sentiment que le niveau de stress au travail est supérieur. Pour beaucoup de gens, le travail est depuis toujours un moyen de donner un sens à la vie, et ils se demandent si le travail n'a pas été dévalorisé au profit des loisirs.

Avant la mise en application des 35 heures en 2000, les Français se montraient favorables à son principe. Actuellement, 28% des Français concernés par cette loi pensent qu'elle a diminué la qualité de leur vie quotidienne. La principale explication est l'accroissement de la pression exercée sur les salariés concernés, tant en matière de productivité que de flexibilité. Pourtant, dans la majorité des cas, le salaire antérieur a été maintenu. Les salariés ont peur aussi d'une baisse du pou-

1906	• Le repos dominical est imposé.
1919	• Passage à la journée de 8 heures et à la semaine de 48 heures.
1936	• les Français ont droit à 2 semaines de congés payés (réformes du Front populaire), semaine de 40 heures de travail.
1956	• 3ème semaine de congés payés.
1965	• 4ème semaine de congés payés.
1982	• 5ème semaine de congés payés, semaine de 39 heures et retraite à 60 ans.
2000	• application de la loi sur les 35 heures par semaine

voir d'achat compte tenu de la limitation des heures supplémentaires et de la modération des hausses de salaires. Enfin, cette loi n'a pas tenu ses promesses en matière de création d'emploi, d'où une grande déception de la part des salariés.

Travail « à la carte »

Les grands principes de la RTT (réduction du temps de travail) étaient de lutter contre le chômage, en acceptant de répartir davantage le temps de travail. Malheureusement, il semble qu'elle a accentué les inégalités entre, d'une part, les salariés qui travaillent moins et bénéficient de leur salaire antérieur et, d'autre part, les indépendants qui continuent de travailler beaucoup plus de 35 heures par semaine.

Finalement, les Français rêvent d'un travail et d'une retraite « à la carte » plutôt que d'un menu imposé.

(inspiré de Francoscopie 2003)

13. TRAVAILLER OU NE PAS TRAVAILLER ?

A. Vous allez écouter une chanson intitulée « Le travail c'est la santé », écrite par Boris Vian. Écoutez-la en lisant les paroles. Quelle est, selon vous, l'opinion de l'auteur sur le travail ? Mettez-vous d'accord avec votre camarade.

paroles : Boris Vian
musique : Henri Salvador

Le travail c'est la santé
Rien faire c'est la conserver
Les prisonniers du boulot
Font pas de vieux os

Ces gens qui courent au grand galop
En auto, métro ou vélo
Vont-ils voir un film rigolo ?
Mais non, ils vont à leur boulot

Refrain

Ils bossent onze mois pour les vacances
Et sont crevés quand elles commencent,
Un mois plus tard, ils sont costauds
Mais faut reprendre le boulot

Refrain

Dire qu'il y a des gens en pagaille
Qui courent sans cesse après l'travail
(faut être fou)

Moi, le travail me court après.
Il n'est pas près d'me rattraper! Ah ah ah

Refrain

Maintenant, dans le plus p'tit village,
Les gens travaillent comme des sauvages
(j'les ai vus)
Pour se payer tout le confort
Quand ils ont tout, ben, ils sont morts !

Refrain

B. En vous aidant des informations sur les congés payés des Français, pouvez-vous dire à quelle époque cette chanson a probablement été écrite ?

C. Y a-t-il des chansons sur le thème du travail dans votre langue ?

JE NE SUIS PAS D'ACCORD !

s allons organiser
ébat sur la télé et
irer des conclusions
d'améliorer la
grammation de la
vision.

cela nous allons
endre à :

poser notre point de
e et à le défendre
endre la parole
formuler des
guments
liser des ressources
ur le débat

us allons utiliser :

subjonctif après les
pressions d'opinion à la
me négative : *je ne crois
s que / pense pas que
sait que, il est vrai
e... mais, par rapport à,
st-à-dire, d'ailleurs, en
fet, car, par conséquent,
une part... d'autre part,
ême si, par contre*
pronom relatif *dont*

1. TÉLÉSPECTATEUR : UN PEU, BEAUCOUP, PAS DU TOUT ?

A. Quels types d'émissions préférez-vous regarder ?

- ❑ les journaux télévisés
- ❑ les documentaires
- ❑ les reportages
- ❑ les films
- ❑ les téléfilms, les séries et les feuilletons
- ❑ les dessins animés
- ❑ les jeux
- ❑ les émissions de télé réalité
- ❑ les débats télévisés
- ❑ les émissions culturelles
- ❑ les programmes pour enfants

● Moi, je regarde les reportages ; c'est intéressant, ça permet de connaître d'autres pays, d'autres cultures. Et toi ?
○ Moi, je préfère les films, ça me relaxe.

B. Le magazine *Télé pour tous* fait une enquête, répondez-y individuellement.

VOUS, TÉLÉSPECTATEUR

LES CHAÎNES	OUI	NON
Êtes-vous un/e adepte du zapping ?		
Avez-vous une chaîne préférée ?		
Si oui, laquelle ?		

LES JOURNAUX TÉLÉVISÉS	OUI	NON
Choisissez-vous votre journal télévisé en fonction du/de la présentateur/trice ?		
En fonction du point de vue adopté pour présenter l'information ?		

AVEC QUI ?	OUI	NON
Aimez-vous regarder la télé avec des amis ?		
Préférez-vous être seul(e) pour regarder la télé ?		

COMBIEN DE TEMPS?	
Combien d'heures par jour regardez-vous la télévision ?	
moins de 2 heures par jour	
entre 2 heures et 5 heures par jour	
plus de 5 heures par jour	

INFLUENCES	OUI	NON
La télévision modifie-t-elle votre façon de voir les choses ?		
Y a-t-il des émissions dont vous parlez avec vos amis ?		
Si oui, lesquelles ?		

TÉLÉ MANIAQUE	OUI	NON
Si vous avez des insomnies, vous mettez-vous devant le petit écran ?		
Quand vous ne savez pas quoi faire, regardez-vous la télévision ?		
Allumez-vous de manière systématique la télévision, même si vous ne la regardez pas nécessairement ?		
Mangez-vous devant la télévision ?		

C. Après avoir répondu au questionnaire, regardez vos réponses et comparez-les avec celles d'un camarade. Quel type de téléspectateur pensez-vous être ?

a. un « accro » de télévision
b. un amateur de télé
c. un antitélé

● Moi, je suis un « accro » de télévision, je la regarde environ 3 heures tous les jours et je zappe beaucoup, et toi ?
○ Moi, ...

2. TÉLÉ RÉALITÉ

A. Lisez ces deux opinions sur la télé réalité, puis retrouvez les arguments pour et les arguments contre la télé réalité et placez-les dans le tableau.

	Pour	Contre
Catherine		
Alex		

La télé réalité ? Pour ou contre ?

Catherine Pasteur, peintre

Tout dépend de l'émission de télé réalité dont on parle. Prenons, par exemple, l'émission « Star Academy », l'école des jeunes chanteurs. Cette émission permet à de jeunes talents de perfectionner leur technique de chant et les finalistes peuvent enregistrer un ou plusieurs disques. Cette émission offre donc la possibilité à de jeunes talents de réaliser leur rêve.

Mais il y a aussi des émissions de télé réalité comme, par exemple, « Loft Story » où plusieurs personnes sont enfermées dans un appartement pendant trois mois. Dans ce type d'émissions de télé réalité, l'idée est toujours la même : des gens sont filmés dans leur quotidien et les spectateurs les regardent comme au zoo. Il n'y a aucune créativité, aucune esthétique.

Il est vrai que l'une des fonctions de la télévision est de distraire les téléspectateurs, mais quel est l'intérêt de regarder des gens qui se disputent ou qui parlent de leur intimité ? Je ne crois pas que cela soit une distraction saine.

Ce qui me choque aussi, c'est que le scénario de la plupart de ces émissions de télé réalité est écrit à l'avance. En effet, même si le téléspectateur a l'impression que tout est improvisé et naturel, il s'agit en réalité d'un montage. Tout est décidé à l'avance. On trompe le téléspectateur, on lui fait croire qu'il peut décider alors que c'est complètement faux.

En conclusion, je crois que ces émissions n'apportent rien au téléspectateur, elles n'ont qu'un but : faire gagner beaucoup d'argent aux producteurs et à la chaîne télévisée.

Alex Lecocq, psychologue

Pourquoi est-ce que la télé réalité remporte un tel succès auprès des téléspectateurs en France ? Probablement parce que la principale caractéristique de ces émissions est de présenter à l'écran des gens ordinaires avec lesquels les téléspectateurs s'identifient facilement.

C'est bien ou c'est pas bien ? D'un côté, cela représente une démocratisation de la télévision. Tout le monde peut passer à la télévision. Les émissions de télé réalité alimentent aussi les conversations : on en parle en famille, entre amis, avec les collègues et c'est une conversation qui ne provoque pas de conflits.

D'un autre côté, les protagonistes de ces émissions sont des personnes sans mérite particulier, qui deviennent rapidement riches et célèbres grâce à la télévision. Je pense qu'on peut y voir une justification de la médiocrité.

En conclusion, je ne crois pas qu'il faille condamner ce type d'émissions. L'une des fonctions de la télévision est de distraire le spectateur et ces émissions jouent bien ce rôle. Par contre, je crois qu'il faut limiter leur nombre. Il y a, actuellement, trop d'émissions de ce type et c'est plutôt la diversité à la télé qui est en danger.

B. Partagez-vous les opinions de Catherine et d'Alex ? Ajouteriez-vous d'autres arguments ? Commentez-le avec un camarade.

- *Je suis d'accord avec Alex. Moi aussi je pense que la télé réalité...*

3. LE PIERCING ET LES TATOUAGES

A. Lisez la transcription d'un débat radiophonique au sujet du piercing et des tatouages et ajoutez, là où ils manquent, les mots et expressions du tableau ci-dessous.

• en tant que	On situe une opinion à partir d'un domaine de connaissance ou d'expérience.
• d'ailleurs	On justifie, développe ou renforce l'argument ou le point de vue qui précède en apportant une précision.
• il est vrai que... mais	On reprend un argument et on ajoute une idée qui le nuance ou le contredit.
• car	On introduit une cause que l'on suppose inconnue de l'interlocuteur.
• je ne partage pas l'avis de/d'	On marque le désaccord avec l'opinion de quelqu'un.
• par conséquent	On introduit la conséquence logique de quelque chose.
• on sait que	On présente un fait ou une idée que l'on considère admis par tout le monde.

Présentateur : Evelyne Jamel, en tant que sociologue, que pensez-vous du phénomène du piercing et du tatouage chez les jeunes?
Evelyne Jamel : Le piercing comme le tatouage existent depuis très très longtemps dans certaines civilisations. En Afrique, en Océanie ou au Japon le piercing ou le tatouage sont des rites. Mais dans notre société, ils correspondent à deux phénomènes : **d'une part** c'est un phénomène de mode ; on porte un piercing ou un tatouage pour des raisons esthétiques. .... beaucoup de piercings ou de tatouages sont de faux piercings ou de faux tatouages.
P. : Comment ça, de faux piercings et de faux tatouages !?
E. J. : Oui, **c'est-à-dire** qu'ils ne sont pas permanents.
P. : Et d'autre part ?
E. J. : Eh bien, **d'autre part**, il s'agit d'un phénomène de contestation. C'est le mouvement punk qui les a mis à la mode il y a une trentaine d'années. C'est une façon de se révolter ou de montrer que l'on appartient à un groupe.
P. : Est-ce qu'il y a beaucoup de jeunes qui portent un tatouage ou un piercing ?
E. J. : En France, 8% des jeunes de 11 à 20 ans ont un piercing et 1% portent un tatouage.
P. : Albert Lévi, qu'en pensez-vous ?
Albert Lévi : Bien, ... médecin, je dois mettre en garde contre les risques du piercing ou du tatouage. Un piercing au nombril avant 16 ans n'est pas du tout recommandable .... les adolescents peuvent encore grandir et la peau peut éclater.
Le piercing représente .... un risque pour la santé.
P. : Et est-ce que les tatouages sont moins dangereux ?
A. L. : C'est pareil. Le matériel de tatouage doit être parfaitement désinfecté et je ne pense pas que ces règles d'hygiène élémentaires soient toujours respectées.
P. : Donc, à votre avis, est-ce que ces pratiques devraient être interdites ?
A. L. : **En effet**, interdire pourrait être une solution.
P. : Evelyne Jamel, êtes-vous d'accord ?
E. J. : Mais non, pas du tout ! .... du docteur Lévi, **même si** ses inquiétudes par rapport à ces pratiques sont justifiées. .... le piercing ou le tatouage comportent des risques .... interdire n'est pas la solution. .... si l'on interdit à un adolescent de se faire un piercing, il s'en fera deux ! **Par contre,** les parents peuvent expliquer à leurs enfants les risques du piercing ou du tatouage et...

B. Maintenant, écoutez le débat et vérifiez vos réponses.

C. Les invités de l'émission utilisent, pour exposer et défendre leur point de vue, des expressions typiques en situation de débat. Elles sont en caractères gras dans le texte. Comprenez-vous à quoi elles servent ? Trouvez-vous des équivalents dans votre langue ?

4. TOUS CEUX DONT ON PARLE

A. Disposez des chaises en cercle. Chacun s'assoit sur une chaise sauf un élève qui reste au centre. Cet élève lit une des phrases suivantes ou en invente une. Ceux que la phrase concerne se lèvent et, le plus vite possible, essaient de se rasseoir sur une autre chaise. Celui qui est au milieu essaie aussi de s'asseoir. La personne qui reste sans chaise doit donner l'ordre suivant.

- Toutes les personnes dont les yeux sont marron/bleus... se lèvent.
- Toutes les personnes dont les chaussures sont noires/marron... se lèvent.
- Toutes les personnes qui ont des lunettes se lèvent.
- Toutes les personnes dont le prénom commence par D/L/M... se lèvent.
- Toutes les personnes qui portent un pantalon se lèvent.
- Toutes les personnes qui font de la natation se lèvent.
- Toutes les personnes dont les cheveux sont blonds se lèvent.
- Toutes les personnes dont le nom de famille comprend un S se lèvent.

B. Maintenant, observez les phrases avec **dont**. Comprenez-vous comment il s'utilise ?

5. D'ACCORD OU PAS D'ACCORD ?

A. Lisez ces phrases. Êtes-vous d'accord ?

- ❑ Tous les hommes sont égaux.
- ❑ Il y a beaucoup de différences entre les hommes et les femmes.
- ❑ Les voitures doivent être interdites dans les grandes villes.
- ❑ Tout le monde peut choisir ce qu'il veut faire.
- ❑ Les voyages forment la jeunesse.
- ❑ Les extraterrestres existent.
- ❑ La vie est plus facile pour les hommes que pour les femmes.
- ❑ Le français est plus facile que l'anglais.
- ❑ Les examens sont totalement indispensables.

• Moi, je ne crois pas que tous les hommes soient égaux...

B. Quand vous utilisez les expressions **je ne crois pas que** ou **je ne pense pas que**, quelles remarques pouvez-vous faire à propos du verbe qui suit ?

EXPRIMER SON POINT DE VUE

À mon avis,
D'après moi,
Je pense que
Je crois que
+ INDICATIF
*Martin **est** un bon candidat.*

Je ne pense pas que
Je ne crois pas que
+ SUBJONCTIF
*Martin **soit** un bon candidat.*

● ***Je pense que** le français est plus facile que l'anglais. Tu ne crois pas ?*
❍ *Mais non! **Je ne pense pas** que le français **soit** plus facile...*

LE SUBJONCTIF

Le présent du subjonctif est construit à partir du radical du verbe à la troisième personne du pluriel du présent de l'indicatif (pour **je, tu, il** et **ils**) ainsi que des formes **nous** et **vous** de l'imparfait.

	DEVOIR
Ils **doiv**ent	que je doi**ve**
	que tu doi**ves**
	qu'il/elle/on doi**ve**
	qu'ils/elles doi**vent**
Nous **devions**	que nous **devions**
Vous **deviez**	que vous **deviez**

Les verbes **être, avoir, faire, aller, savoir, pouvoir, falloir, valoir** et **vouloir** sont irréguliers.

ÊTRE	AVOIR
que je **sois**	que j'**aie**
que tu **sois**	que tu **aies**
qu'il/elle **soit**	qu'il/elle **ait**
que nous **soyons**	que nous **ayons**
que vous **soyez**	que vous **ayez**
qu'ils/elles **soient**	qu'ils/elles **aient**

FAIRE
que je **fasse**
que tu **fasses**
qu'il/elle **fasse**
que nous **fassions**
que vous **fassiez**
qu'ils/elles **fassent**

LES CONNECTEURS

***On sait que** la télévision est un moyen de communication très important.*	On présente un fait ou une idée que l'on considère admis par tout le monde.
***En tant que** psychologue pour enfant, je dois dire que...*	On présente un point de vue à partir d'un domaine de connaissance.
***Par rapport à** la violence très présente dans les films, je trouve que...*	On signale le sujet ou le domaine dont on veut parler.
***D'une part**, les parents ne surveillent pas suffisamment leurs enfants, **d'autre part**...*	On présente deux aspects d'un sujet.
***D'ailleurs**, nous ne pouvons pas prétendre que la télé est coupable de...*	On justifie, développe ou renforce l'argument ou le point de vue qui précède.
***C'est-à-dire que** les parents et la société en général sont aussi responsables de l'éducation...*	On explicite en développant l'idée qui précède.
***En effet**, la télévision n'est pas la seule responsable de...*	On confirme et on renforce l'idée qui vient d'être présentée.
***Car** la violence est présente aussi dans d'autres aspects de la vie des enfants.*	On introduit une cause que l'on suppose inconnue de l'interlocuteur.
***Je ne partage pas l'avis de** M. Delmas.*	On marque son désaccord avec l'opinion de l'interlocuteur.
***Il est vrai que** les parents doivent surveiller leurs enfants, **mais** la télévision est un service public et...*	On reprend l'argument de l'interlocuteur et on ajoute une idée qui le nuance ou le contredit.
***Par contre**, certaines télévisions ne comprennent pas qu'elles ont un rôle...*	On introduit une idée ou un fait qui contraste avec ce qu'on a dit précédemment.

LE PRONOM RELATIF **DONT**

Dont peut être complément du nom.

- *Je connais un gars **dont** le père est animateur à la télé. (= son père est animateur)*

Il peut être aussi complément prépositionnel d'un verbe qui impose l'utilisation de la préposition **de**.

- *C'est une chose **dont** on parle souvent. (= on parle de la télévision)*

6. MESDAMES, MESSIEURS, BONSOIR !

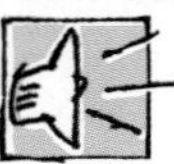

A. Christian Laurier anime un débat télévisé. Il introduit chaque sujet grâce à une sorte de petite énigme. Écoutez ses introductions et notez les mots-clés que vous comprenez. Ensuite, avec un camarade, faites des hypothèses sur les sujets abordés. Voici des indices pour vous aider.

	Indice	Mots-clés	Sujet abordé
1.	C'est un moment dont nous rêvons toute l'année.		
2.	C'est un thème dont les Français se préoccupent beaucoup.		
3.	C'est un gaz dont la Terre a besoin pour se protéger contre les rayons du soleil.		
4.	C'est un engin dont on se sert trop.		

B. Avec un ou deux camarades, essayez à votre tour de préparer une petite introduction sur un sujet de votre choix. Puis lisez-la à voix haute. Vos camarades vont essayer de deviner quel sujet vous voulez introduire.

7. CHIEN OU CHAT ?

A. Par groupes de quatre, choisissez un des sujets suivants. Vous pouvez en ajouter d'autres si vous voulez. Dans le même groupe, deux d'entre vous vont prendre parti pour une option et les deux autres pour l'option contraire.

Prendre la voiture	ou	les transports publics.
Vivre en ville	ou	à la campagne.
Étudier le français	ou	une autre langue.
Travailler à l'étranger	ou	dans son propre pays.
Les vacances à la plage	ou	à la montagne.
Regarder un film à la télé	ou	aller au cinéma.
....	ou	....

B. Chaque binôme doit préparer son argumentation, et défendre brièvement son point de vue en réagissant aux opinions des autres.

- *C'est vrai que les vacances à la montagne sont plus tranquilles que les vacances à la mer, mais ...*

8. ON EN DISCUTE

Jean-Philippe Cuvier présente chaque semaine l'émission « On en discute ». Le thème abordé ce soir est « Pour ou contre la télévision aujourd'hui ». Écoutez-le présenter ses invités et complétez ces fiches de présentation avec l'argument principal de chaque invité.

TV 22

Coralie

- Lycéenne
- 18 ans

❏ pour / ❏ contre

- Argument :

....
....

TV 22

Pascal Lumour

- Association « Front de libération télévisuelle »
- 38 ans

❏ pour / ❏ contre

- Argument :

....
....

TV 22

Valérie Toubon

- Professeur de mathématiques, collège Henri IV à Poitiers
- 48 ans
- 22 ans d'expérience dans l'enseignement

❏ pour / ❏ contre

- Argument :

Les enfants passent trop de temps devant la télévision.

TV 22

Denis Lambert

- Psychologue pour enfants
- 52 ans

❏ pour / ❏ contre

- Argument :

....
....

9. ÊTES-VOUS POUR OU CONTRE ?

A. Que pensez-vous des différents arguments présentés par ces personnes ? Choisissez votre camp : pour ou contre, puis formez des groupes avec ceux qui partagent votre opinion. Vous pouvez ensemble ajouter de nouveaux arguments.

- Moi je suis plutôt d'accord avec Pascal Lumour, je trouve que...
- ○ Moi aussi...

B. Maintenant, dans chaque groupe, préparez le débat. Justifiez vos points de vue et vos arguments en cherchant des exemples pris dans la programmation de la télévision de votre pays.

- Bon, nous pouvons dire, en tant que téléspectateurs, que les chaînes de TV nous semblent toutes pareilles, n'est-ce pas ?
- ○ Oui, par exemple, les émissions sportives sont toujours consacrées au football et au...

TV 22

Gérard Rhodes

• Cinéaste

• 28 ans

❏ pour / ❏ contre

• Argument :

....

....

Raymonde Pariot

• Sociologue et historienne

• 67 ans

❏ pour / ❏ contre

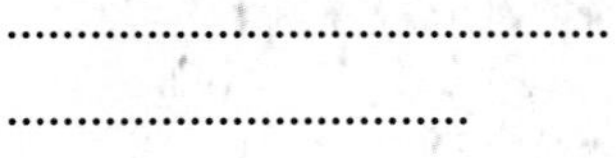

• Argument :

....

....

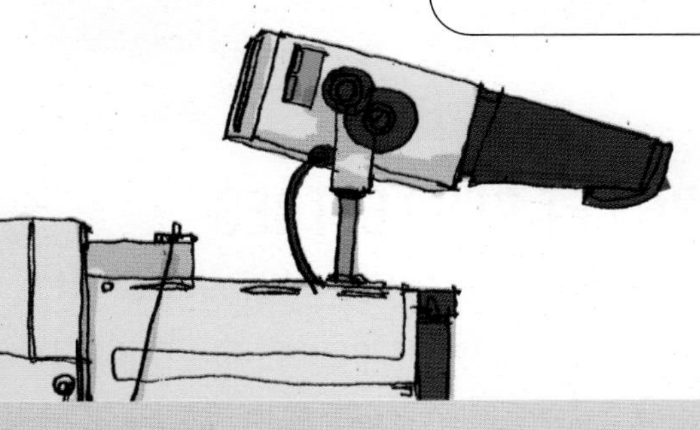

C. Vous allez ensuite débattre avec un autre groupe qui a choisi le camp adverse. Attention, vous devez tirer des conclusions et parvenir à un accord afin d'améliorer la qualité de la télé de votre pays en formulant une série de propositions.

ANTISÈCHE

QUAND ON NE COMPREND PAS QUELQUE CHOSE OU QUAND ON VEUT DEMANDER DES EXPLICATIONS

Tu peux répéter/ Vous pouvez répéter, s' il te/vous plaît ?
Je ne comprends pas ce que tu dis/vous dites.
Pardon, mais je ne sais pas si j' ai bien compris.

QUAND ON NE SAIT PAS SI ON A ÉTÉ COMPRIS

Je ne sais pas si je me fais comprendre...
Tu vois/comprends ce que je veux dire ?
Vous voyez/comprenez ce que je veux dire ?

10. PROGRAMMES DU PETIT ÉCRAN

A. Voici quatre émissions emblématiques de la télévision française. Lisez le résumé de ces émissions. En avez-vous déjà entendu parler ? Laquelle ou lesquelles de ces émissions aimeriez-vous voir ?

THALASSA

Ce magazine existe depuis 1975. Il se compose de reportages divers et très complets, toujours en relation avec la mer. Chaque reportage est suivi d'un débat entre les présentateurs et des personnalités invitées sur le plateau.

DES CHIFFRES ET DES LETTRES

Ce jeu télévisé a été créé en 1965 et il s'est très vite popularisé car, d'emblée, les téléspectateurs ont pris le réflexe de jouer en même temps que les candidats. Cette émission s'est convertie en un véritable phénomène social et les grandes finales attirent de huit à douze millions de téléspectateurs. Plusieurs centaines de clubs existent en France et même à l'étranger. Le décor de l'émission est très sobre et inspire le calme. On n'entend ni rires, ni applaudissements, contrairement aux autres émissions de jeux. L'objectif du jeu est, d'une part, de découvrir le mot le plus long à partir de 9 lettres tirées au sort, et, d'autre part, de parvenir au chiffre annoncé en faisant diverses opérations de calcul.

DICOS D'OR ou LA DICTÉE DE PIVOT

Imaginez des millions de téléspectateurs, stylo en main, en train d'écrire la dictée lue par le célèbre présentateur Bernard Pivot. La dictée de Pivot est retransmise à la télévision chaque année depuis 1985 et elle remporte un succès énorme. En 2004, ce sont 15 000 adultes et 500 000 enfants qui se sont inscrits à ce concours télévisé d'orthographe française.

Les mots-valises

Que les Libanais se soient déjà adonnés par deux fois à la perverse jouissance de la dictée, n'étonne pas les Français. Ils savent, ne serait-ce que par des on-dit, que les Beyrouthins aiment et pratiquent la langue de La Fontaine, de Chateaubriand et de Mérimée, et qu'ils ne se sont jamais laissé décourager par les règles des participes passés des verbes pronominaux. Ils se sont entraînés à en déjouer les pièges. Ils se sont même amusés à conjuguer des verbes au subjonctif imparfait. Quel courage !

FAUT PAS RÊVER

Cette émission a été créée en 1990 par Georges Pernoud (le créateur de l'émission THALASSA). On y découvre deux reportages tournés à l'étranger et un troisième tourné en France. Ces reportages sont souvent insolites. Par exemple, vous pouvez partager le quotidien d'une tribu du Kenya, découvrir comment on devient geïsha aujourd'hui, ou encore découvrir de surprenants accents régionaux en France. Après chaque reportage, une personnalité est invitée à réagir et à commenter ce qu'elle vient de voir.

B. Avez-vous des émissions semblables dans votre pays ? Quelles sont pour vous les émissions les plus emblématiques de votre pays ?

QUAND TOUT À COUP...

s allons raconter
anecdotes
onnelles.

cela nous allons
endre à :

conter au passé
ganiser les
énements dans un récit
crire les
constances qui
tourent le récit

us allons utiliser :

nparfait, le plus-
e-parfait et le passé
mposé dans un récit
elques marqueurs
nporels : *l'autre jour,
y a, ce jour-là, tout à
up, soudain, au bout
, à ce moment-là, la
ille, le lendemain,
alement*
forme passive

7

1. SOUVENIRS, SOUVENIRS

A. Jean-Paul nous montre son album de photos. Regardez les photos et, avec un camarade, retrouvez les titres qui, à votre avis, correspondent à chacune. Vous pouvez deviner où est Jean-Paul sur chaque photo ?

1. Vacances en Bretagne
2. Tout bronzé au Brésil
3. Champions de la ligue, mon premier (et dernier) grand exploit sportif !
4. Dans le camion de tonton
5. Super chic pour le mariage de Denise
6. À trottinette, j'ai toujours aimé la vitesse !!

B. Maintenant, écoutez Jean-Paul qui commente les photos. Pouvez-vous identifier certains membres de sa famille et dire quand et où la photo a été prise ? Notez vos réponses.

	AVEC QUI ?	QUAND / À QUEL ÂGE ?	OÙ ?
1			
2			
3			
4			
5			
6			

2. LES AVENTURES DE LA PETITE JO

A. Avez-vous organisé une course d'escargots quand vous étiez enfant ? C'est très facile ! Sur la liste ci-dessous, il y a tout ce qu'il faut. Avec un camarade, essayez d'imaginer comment ça s'organise.

● D'abord on marque chaque escargot.
○ Oui, puis...

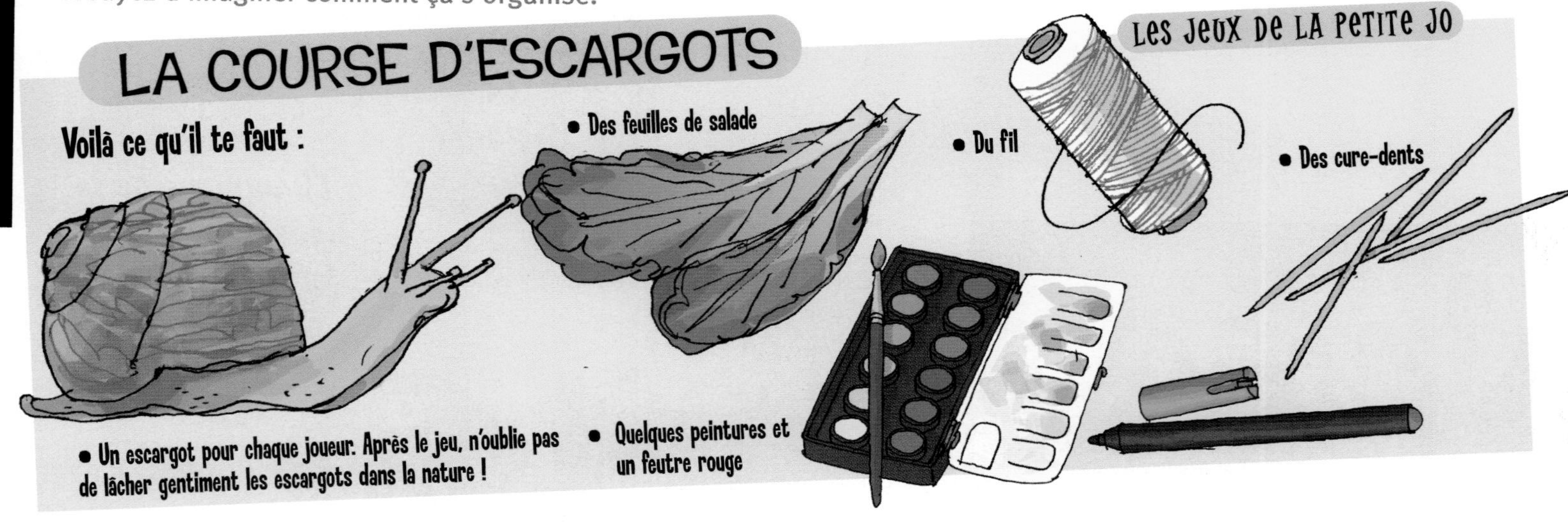

LA COURSE D'ESCARGOTS
Les jeux de la Petite Jo
Voilà ce qu'il te faut :
• Des feuilles de salade
• Du fil
• Des cure-dents
• Un escargot pour chaque joueur. Après le jeu, n'oublie pas de lâcher gentiment les escargots dans la nature !
• Quelques peintures et un feutre rouge

B. Lisez maintenant le récit de la Petite Jo. Est-ce que la course d'escargots s'organise comme vous l'aviez pensé ?

Lundi dernier, avec les copains, on est allé jouer derrière la maison du voisin. **C'était** le premier mai, alors on **n'avait** pas école. Là, **on a trouvé** beaucoup d'escargots. C'est vrai qu'**il avait plu** toute la nuit et les escargots adorent la pluie. Alors, **on a pris** quelques escargots et puis, avec tous les copains, Paul, Eric, Lulu et Frédo **on a organisé** une course d'escargots. C'est drôlement marrant les courses d'escargots. **On leur a peint** la coquille avec quelques peintures de différentes couleurs et **on a écrit** des numéros dessus,

puis **on les a tous placés** sur la ligne de départ. **On avait** chacun un cure-dents avec un fil et une feuille de salade attachée au bout du fil. Oui, parce que les escargots, ça aime beaucoup la salade ! **Ça a duré** des heures !!. 50 centimètres, c'est long, c'est drôlement long pour un escargot ! Finalement, **mon escargot a gagné** !!! **J'étais** drôlement contente car normalement je ne gagne jamais les courses d'escargots ! La dernière fois, c'est l'escargot de Paul qui **avait gagné** et, la fois d'avant, c'est celui de Lulu qui **avait gagné** !

C. La Petite Jo utilise trois temps du passé pour raconter son histoire. Ces verbes sont signalés en caractères gras. Est-ce que vous pouvez les distinguer ? À quels temps sont-ils ?

D. Quelles remarques pouvez-vous faire sur la construction du plus-que-parfait par rapport à l'imparfait et au passé composé ?

Le passé composé sert à raconter les événements successifs d'une histoire. L'histoire que la Petite Jo raconte se passe lundi dernier.

L'imparfait ne signale ni le début ni la fin d'une action et sert à expliquer les circonstances de l'histoire qu'on raconte.

Le plus-que-parfait sert à évoquer des événements qui se sont passés avant l'histoire qu'on raconte et qui sont utiles pour mieux comprendre cette histoire.

3. HISTOIRE D'UNE VIE

A. Lisez cette biographie de Mata-Hari. Vous trouvez qu'elle a eu une vie intéressante ?

LES FEMMES QUI ONT CHANGÉ L'HISTOIRE

Mata-Hari

Qui était Mata-Hari ?

Margaretha Geertruida Zelle, alias Mata-Hari, est née aux Pays-Bas. C'était la fille d'un marchand de chapeaux. À 19 ans, elle s'est mariée avec un militaire en poste sur l'île de Java. Après plusieurs années passées aux Indes néerlandaises, elle a divorcé et elle est partie vivre à Paris. C'étaient les années folles et, en ce début de siècle, Paris s'amusait, la ville était avide de modernité et de distractions. Mata-Hari se faisait alors passer pour une danseuse javanaise et Émile Guimet, un grand collectionneur d'art oriental, l'a invitée à danser dans son musée. C'est à lui qu'elle doit son nom de scène : « Mata-Hari » ce qui signifie « soleil levant » en javanais. Mata-Hari ne savait pas danser mais elle était belle et savait se déshabiller avec talent et mouvoir un long corps mince et bronzé. Le spectacle a eu un grand succès et Mata-Hari est partie danser ensuite à Madrid, à Monte Carlo, à Berlin, à Vienne... Elle y a fréquenté des aristocrates, des diplomates, des banquiers, des industriels, des militaires...

Quand la Grande Guerre a éclaté, Mata-Hari, a continué à voyager librement à travers l'Europe. Les services secrets français ont alors voulu profiter de son exceptionnelle facilité de déplacement et du fait qu'elle parlait plusieurs langues. Le capitaine Ladoux, chef des services du contre-espionnage, lui a demandé d'aller espionner le Kronprinz (le prince héritier de l'Empire allemand) dont elle avait fait la connaissance. Mais un peu plus tard, des télégrammes ont été interceptés ; ils apportaient des détails sur les déplacements de l'agent allemand H21, des détails qui correspondaient aux déplacements de Mata-Hari.

Cette année-là, la tension était très forte en France. La guerre durait depuis 3 ans et elle avait fait beaucoup de victimes. La société française inquiète, commençait à douter de la victoire et l'on voyait des espions partout. Les services secrets ont donc pris la décision d'arrêter Mata-Hari. Elle a été jugée rapidement et condamnée à mort par un tribunal militaire. Un jour d'automne, à l'aube, Mata-Hari a été conduite face au peloton d'exécution. Fière et le sourire aux lèvres, elle a refusé d'avoir les yeux bandés, quand l'officier a crié « feu », elle a envoyé, du bout des doigts, un dernier baiser. Elle avait seulement 41 ans.

L'histoire de France en quelques dates

481 : Clovis, roi des Francs, conquiert la Gaule.
1661 : Louis XIV devient roi de France.
1789 : Révolution Française.
1804 : Napoléon Bonaparte devient empereur et part à la conquête de l'Europe.
1870 : La France et l'Allemagne entrent en guerre.
1871 : Victoire de l'Allemagne.
1914 : L'assassinat, à Sarajevo, de l'archiduc d'Autriche déclenche la Première Guerre Mondiale.
1918 : Les États-Unis entrent en guerre aux côtés de la France, l'Angleterre et la Russie.
1939 : La Grande Bretagne et la France déclarent la guerre à l'Allemagne qui suit une politique d'expansion en Europe.
1940 : La France est contrôlée par l'Allemagne.
1944 : Débarquement de Normandie.
1945 : Libération de la France et défaite de l'Allemagne.
1957 : la République Fédérale d'Allemagne, la Belgique, la France, l'Italie, le Luxembourg et les Pays-Bas créent la CEE.
2004 : L'Union Européenne accueille sept nouveaux pays.

B. L'histoire de Mata-Hari est liée à l'histoire de la France. À quel moment de l'histoire se sont passés les événements racontés dans ce texte ?

C. Imaginez la phrase que Mata-Hari aurait pu dire en lançant son dernier baiser.

4. C'EST COMME ÇA, LA VIE

A. Séverine raconte dans un magazine ce qui lui est arrivé alors qu'elle était au restaurant avec un ami. Lisez ce texte et introduisez les marqueurs qui manquent.

• au bout de • finalement • l'autre jour • la veille • Le lendemain • soudain • tout à coup

C'EST COMME ÇA, LA VIE

« Je ne comprenais pas pourquoi on m'apportait des fleurs»

Romantiques, drôles ou surprenantes, rien ne vaut les histoires vécues. Nous t'offrons cette page pour que tu puisses raconter les tiennes. Aujourd'hui, Séverine nous explique ce qui lui est arrivé alors qu'elle dînait au restaurant.

.............., je suis allée au restaurant avec mon ami parce que ...., j'avais terminé tous les examens et je voulais fêter ça. Nous sommes allés dans un restaurant assez chic dans mon quartier. Nous nous sommes installés et nous avons choisi notre menu. À la table à côté de nous, il y avait un couple. Nous en étions au dessert quand ...., le serveur est venu vers moi avec un grand bouquet de roses. Je ne comprenais pas pourquoi on m'apportait des fleurs et surtout qui me les envoyait. Alors, comme mon ami affirmait que ce n'était pas lui qui m'avait fait parvenir ce bouquet, nous avons commencé à nous poser des questions. J'allais appeler le serveur quand ...., un homme s'est dirigé vers moi avec un violon à la main. Mais avant même qu'il arrive jusqu'à moi, notre voisin de table furieux s'est levé et a dit au violoniste de venir devant la jeune femme qui l'accompagnait. .... quelques minutes, le serveur est arrivé, très gêné. Il s'est excusé et m'a dit qu'il s'était trompé de table et que les roses étaient pour la dame de la table voisine. ...., tout s'est arrangé.
.................., par hasard, j'ai croisé dans la rue notre voisin de table qui m'a invitée à prendre un café.

Séverine, Strasbourg

B. Vous allez écouter trois personnes qui commencent à raconter une anecdote. Comment croyez-vous que ces histoires finissent ? Parlez-en avec deux autres camarades.

C. Maintenant écoutez et vérifiez.

5. UN ÉVÉNEMENT ET SON CONTEXTE

A. En groupe, faites la liste de ce que chacun a fait de spécial récemment : des choses amusantes, importantes ou futiles, prévues ou imprévues... Précisez, si vous le pouvez, la date et l'heure.

● Samedi dernier, à midi, je suis allé manger chez un copain...
○ Mercredi soir, j'ai vu un match de football.
■ Ben moi, j'ai...

B. Maintenant, choisissez un de ces événements et expliquez certaines circonstances qui ont entouré cet événement.

● Samedi dernier, je suis allé manger chez un copain. C'était son anniversaire, sa mère avait fait un gâteau délicieux...

RACONTER UNE HISTOIRE, UN SOUVENIR, UNE ANECDOTE

Nous pouvons raconter une histoire sous la forme d'une succession d'événements au passé composé.

*Il **a versé** le café dans la tasse. Il **a mis** du sucre dans le café. Avec la petite cuiller, il **a remué**. Il **a bu** le café.*

Pour chaque événement, nous pouvons expliquer les circonstances qui l'entourent. On utilise alors l'imparfait.

*Il a versé le café dans la tasse. Il **était** très fatigué et **avait** très envie de prendre quelque chose de chaud. Il a mis du sucre, **c'était** du sucre brun...*

Pour parler des circonstances qui précèdent l'événement, on utilise le plus-que-parfait.

***Il avait mal dormi** et il était très fatigué, alors il a pris une bonne tasse de café. C' était du café brésilien **qu'il avait acheté** la veille.*

LE PLUS-QUE-PARFAIT

Pour former le plus-que-parfait d'un verbe, on met l'auxiliaire **être** ou **avoir** à l'imparfait et on le fait suivre du participe passé :

J'**avais**	**dormi**
Tu **avais**	
Il/elle/on **avait**	
Nous **avions**	
Vous **aviez**	
Ils/elles **avaient**	

Quand un verbe est conjugué avec l'auxiliaire **être**, il s'accorde en genre et en nombre avec le sujet.

*Cette année, Claire et Lulu sont parties en Italie, l' été dernier, elles étaient allé**es** en Espagne.*

SITUER DANS LE TEMPS

On peut raconter une anecdote, en la situant dans le passé, mais sans préciser quand.

***L'autre jour,** je suis allé à la plage avec des amis.*

On peut raconter une anecdote en la situant d'une manière plus précise dans le passé.

Il y a un mois environ,	*je suis allée au restaurant avec Louis.*
Lundi dernier,	
Samedi soir,	

On peut introduire des circonstances qui entourent ou précèdent l'événement.

__À cette époque-là,__ j' habitais dans le centre-ville.

Lundi dernier je suis allée au centre-ville. __Ce jour-là,__ il pleuvait des cordes.

Samedi soir, j' ai vu une chose étrange. __Ce soir-là,__ j' avais décidé de faire une balade au bord de la rivière...

Certains marqueurs temporels introduisent les circonstances qui ont précédé l'événement.

La police a arrêté jeudi un homme qui, __quelques jours auparavant/deux jours avant/ la veille,__ avait cambriolé la bijouterie de la place Tiers.

D'autres marqueurs indiquent la durée entre deux événements.

Ils se sont mariés en 2001 et, __au bout de__ quelques années, Charline est née.

D'autres marqueurs peuvent signaler le déclenchement d'un événement imprévu.

J' étais en train de regarder la télé __quand soudain/tout à coup__ la lumière s' est éteinte.

Pour conclure le récit et indiquer la conséquence de l'histoire.

Je me suis levée tard, le téléphone a sonné, ma voiture ne voulait pas démarrer. __Finalement,__ je suis arrivée en retard.

LA VOIX PASSIVE

Dans une phrase normale, à la voix active, le sujet du verbe fait l'action.

André Le Nôtre a dessiné les jardins de Versailles.

Dans une phrase à la voix passive, le sujet du verbe ne fait pas l'action.

Les jardins de Versailles __ont été dessinés__ par André Le Nôtre.

La voix passive se construit avec **être** + participe passé. Le temps verbal est indiqué par l'auxiliaire **être** et le participe s'accorde en genre et en nombre avec le sujet.

L' aéroport __sera dessiné__ par Jean Nouvel.
L' église __a été dessinée__ par un grand architecte.
Les ponts __sont dessinés__ par des ingénieurs.

6. PAR QUI ?

A. Vous savez qui a été acteur dans les événements suivants ? Travaillez avec un camarade.

La Bataille de Waterloo a été gagnée		le peuple français le 21 janvier 1793.
La guillotine a été inventée		Gutenberg vers 1440.
L'Amérique a été découverte		les frères Lumière en 1895.
La Joconde a été peinte		Alexander Fleming en 1928.
La Tour Eiffel a été construite		Jules Vernes en 1873.
Louis XVI a été guillotiné		Léonard de Vinci au seizième siècle.
La Bataille de Waterloo a été perdue	par	Gustave Eiffel au dix-neuvième siècle.
La Gaule a été conquise		Christophe Colomb au quinzième siècle.
La presse à imprimer a été créée		le duc de Wellington le 18 juin 1815.
« Le Tour du monde en quatre-vingts jours » a été écrit		Vincent Van Gogh en 1888.
La pénicilline a été découverte		Napoléon Bonaparte le 18 juin 1815.
Le cinéma a été inventé		Jules César au premier siècle avant Jésus Christ.
« Les Tournesols » ont été peints		Joseph Ignace Guillotin au dix-huitième siècle.

B. Observez comment ces verbes sont construits. Qu'est-ce que vous remarquez ?

7. LES TITRES À LA UNE

Par deux, vous allez jeter deux fois un dé ou choisir deux numéros entre 1 et 6. Chaque numéro renvoie à une moitié du titre d'un journal. Rédigez ensuite l'article correspondant.

① La célèbre Madame Soleil
② Un homme arrêté plusieurs fois pour infraction au code de la route
③ Le maire de la plus petite commune française
④ Un professeur de chimie à la retraite
⑤ Un groupe d'élèves d'une classe de français
⑥ Astérix

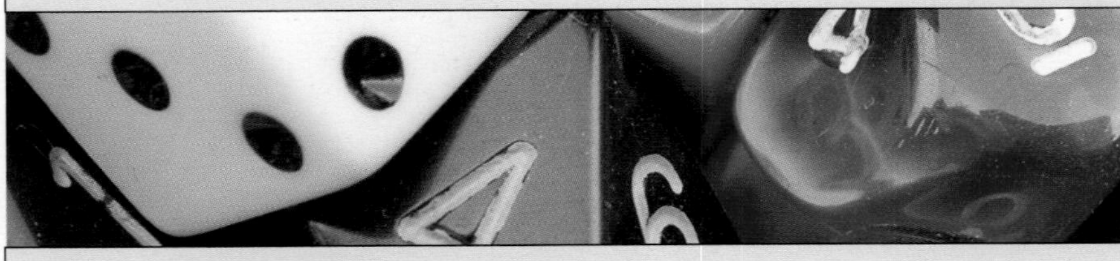

① a décidé de participer au prochain tour de France.
② vient de publier sa biographie.
③ a été kidnappé(e).
④ a été proposé(e) pour le prix Nobel de la paix.
⑤ est tombé(e) d'un train en marche.
⑥ a gagné 200 millions d'euros au loto.

8. LA PREMIÈRE FOIS

A. On se rappelle souvent des « premières fois ». Avec un camarade essayez de vous souvenir de la première fois que vous avez/êtes...

- fait de la bicyclette
- participé à un spectacle
- conduit une voiture
- fait du ski
- fait du cheval
- allé(e) à l'école
- allé(e) à l'étranger
- monté(e) dans un avion
- ....
- ....

● *La première fois que je suis allé à l'étranger, j'avais 12 ans. C'était pour les vacances d'été et mon père voulait nous faire une surprise. Alors, nous sommes allés à Amsterdam. C'était super parce que...*

B. Vous souvenez-vous de la dernière fois que vous avez fait ces choses-là ?

9. C'EST ARRIVÉ À QUI ?

A. Vous avez sûrement vécu des moments de grande émotion. Regardez la liste proposée. Essayez ensuite de vous rappeler les circonstances et les détails. Puis, individuellement, complétez le tableau. Bien sûr, vous pouvez aussi inventer si vous en avez envie !

Une personne célèbre que j'ai rencontrée	
Un lieu où je me suis perdu(e)	
Un avion/un train/un bus que j'ai raté	
Un plat insolite que j'ai mangé	
Une mauvaise rencontre que j'ai faite	
De l'argent que j'ai perdu	
Un jour où j'ai eu très peur	
Une soirée inoubliable	
Une expérience amusante	
Une expérience embarrassante	
Autre :	

B. Maintenant, mettez-vous par groupe de trois et racontez entre vous les aventures les plus curieuses ou les plus intéressantes. Décidez si c'est vrai ou pas.

- Moi, un jour, j'ai croisé Gérard Depardieu à l'aéroport. Je partais à...

C. Parmi les histoires vraies que l'on vous a racontées, choisissez la plus intéressante. Celui qui a vécu cette histoire la raconte de nouveau aux autres dans les moindres détails. Les deux autres peuvent prendre des notes et poser des questions. Préparez-vous bien car, ensuite. vous allez devoir raconter cette histoire comme si vous l'aviez vécue !

- ● Il y a quelques années, un jour, pendant l'été...
- ○ Quel âge tu avais ?
- ● J'avais 14 ans.
- ...
- ❑ La veille, mon grand frère avait eu son permis de conduire et il voulait...

D. Chaque groupe raconte son anecdote devant la classe. À tour de rôle, chacun des trois membres du groupe raconte l'histoire à la première personne, comme s'il l'avait vécue. Ensuite, le groupe-classe a le droit de poser des questions pour essayer de deviner lequel des trois a réellement vécu cette anecdote.

- Moi je crois que c'est William qui a vécu cette aventure parce que...

LES ANTIHÉROS DE LA BD

Au début du siècle, les Français dessinaient beaucoup pour les enfants puis, dans les années 50 et 60, pour les adolescents. Mais depuis les années 70, il y a des BD pour tous les goûts et tous les âges. À côté des héros positifs comme Astérix, on trouve des héros beaucoup moins brillants et parfois même carrément antipathiques. Par exemple, en 1957, Franquin invente Gaston Lagaffe qu'il surnomme « le héros sans emploi ». En 1977, Christian Binet publie les premières péripéties des Bidochons et, en 1982, Jean-Marc Lelong présente Carmen Cru, une mamie misanthrope.

On dit qu'elle s'est fait arracher les dents pour les vendre et acheter son vélo.

On dit qu'elle attrape les petits enfants, les coupe en petits morceaux et les fait brûler dans ses cabinets.

Gaston Lagaffe

Gaston Lagaffe travaille à la rédaction d'un journal, mais il est totalement inutile et incompétent. Il est lent, maladroit, mou et dangereux pour son entourage. Son plus grand plaisir, quand il ne dort pas au travail, c'est de faire de la musique avec des instruments de sa création, ce qui est très désagréable pour ses collègues. Son obsession : NE PAS TRAVAILLER. Il aime aussi les animaux qu'il garde au bureau car son appartement est trop petit. Comme son nom l'indique, Gaston fait sans arrêt des « gaffes ». Il est ingénieux, mais provoque toujours des catastrophes. Gaston fait aussi de nombreuses blagues à l'agent Longtarin : il se venge ainsi de toutes les amendes qu'il doit payer à cause de ce policier.

Carmen Cru

Carmen Cru est une grand-mère insupportable. Elle vit seule dans une petite maison et les promoteurs l'entourent progressivement d'immeubles en béton. Ils veulent la faire partir pour pouvoir construire d'autres immeubles, mais Carmen Cru résiste et se défend. Elle circule sur un vieux vélo, se soigne avec une boîte de médicaments qui date d'avant la guerre et dans son quartier, tout le monde l'appelle « la vieille ».

Les Bidochons

Raymonde et Robert Bidochon sont une caricature du couple de Français moyens. Robert s'imagine qu'il est beaucoup plus intelligent que sa femme. Quant à elle, elle accepte tous les caprices de son mari. Les Bidochons sont victimes de la société : à l'hôpital, on leur dit qu'ils n'existent pas dans les fichiers ; en voyage organisé, ils n'ont pratiquement pas le temps de descendre de l'autobus ; quand ils achètent un téléphone portable, ils ne savent pas le faire fonctionner...

10. TROIS ANTIHÉROS

A. Lisez le texte sur la bande dessinée et observez ces personnages. Les reconnaissez-vous ?

B. Aimeriez-vous lire les aventures de ces personnages ? Quels sont ceux qui vous intéressent le plus ?

C. Quels autres personnages de la bande dessinée française ou belge connaissez-vous? Il s'agit de héros ou d'antihéros ? Les aimez-vous ?

ANCRAGE

IL ÉTAIT UNE FOIS...

Nous allons raconter un conte traditionnel en le modifiant.

Pour cela nous allons apprendre à :

- exprimer des relations logiques de temps, cause, finalité et conséquence
- raconter une histoire

Et nous allons utiliser :

- passé simple
- *pour que* + subjonctif
- *afin de* + infinitif
- *si/tellement... que*
- *lorsque*
- le gérondif
- *pendant que, tandis que*
- *pourtant*
- *puisque*

1. QUE TU AS DE GRANDES OREILLES !

A. Lisez bien les phrases du tableau. Elles proviennent de contes traditionnels. À quel conte croyez-vous qu'elles correspondent ?

	A	B	C	D	E	F	G	H	I	J	K	L	M
Cendrillon													
Le loup et les sept chevreaux													
Le vilain petit canard													
Blanche-Neige													
Le Petit Chaperon rouge													
Le Petit Poucet													

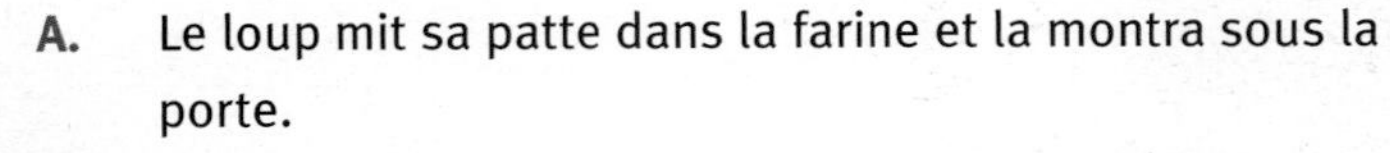

A. Le loup mit sa patte dans la farine et la montra sous la porte.

B. « Miroir, mon gentil miroir, qui est la plus belle en ce royaume ? »

C. Il était tellement petit qu'on l'appelait Petit Poucet.

D. Grand-mère, que tu as de grandes oreilles !

E. « Emmène-la dans la forêt et arrange-toi pour qu'elle n'en sorte pas vivante » ordonna la reine au chasseur.

F. La fée passa sa baguette magique sur la citrouille qui se transforma rapidement en un superbe carrosse.

G. « Montre tes pattes pour que nous puissions voir si tu es vraiment notre chère maman » dirent-ils en cœur.

H. Elle descendit l'escalier tellement vite qu'elle perdit une de ses pantoufles de vair.

I. Pendant que les sept nains travaillaient à la mine, elle s'occupait de la maison.

J. Lorsque la mère canard vit qu'il était si laid, elle dit à ses frères de s'éloigner de lui.

K. « Toi, tu passes par ici, et moi, je passe par là, on verra qui arrivera le premier chez ta grand-mère » dit le loup.

L. Grâce aux bottes magiques de l'ogre, il parcourut sept lieues d'un pas.

M. « Puisque tu es si laid, tu ne joueras pas avec nous » lui crièrent ses frères.

B. Maintenant comparez vos réponses avec celles d'un camarade.

- Qu'est-ce que tu as mis ?
- Cendrillon : b...

2. QU'EST-CE QU'UN CONTE ?

A. Qu'est-ce que vous savez sur les contes ? Lisez les affirmations suivantes et répondez individuellement : vous semblent-elles plutôt vraies, plutôt fausses ou vous ne le savez pas ?

1. Les contes n'existent pas dans certaines cultures.
2. Les contes sont souvent des histoires racontées aux enfants pour qu'ils s'endorment.
3. Dans les contes européens, le rôle du méchant est très souvent représenté par un dragon.
4. Les contes sont des histoires pour enfants, mais aussi pour adultes.
5. Les contes commencent par une formule spéciale.
6. Les contes ne se terminent pas toujours bien.

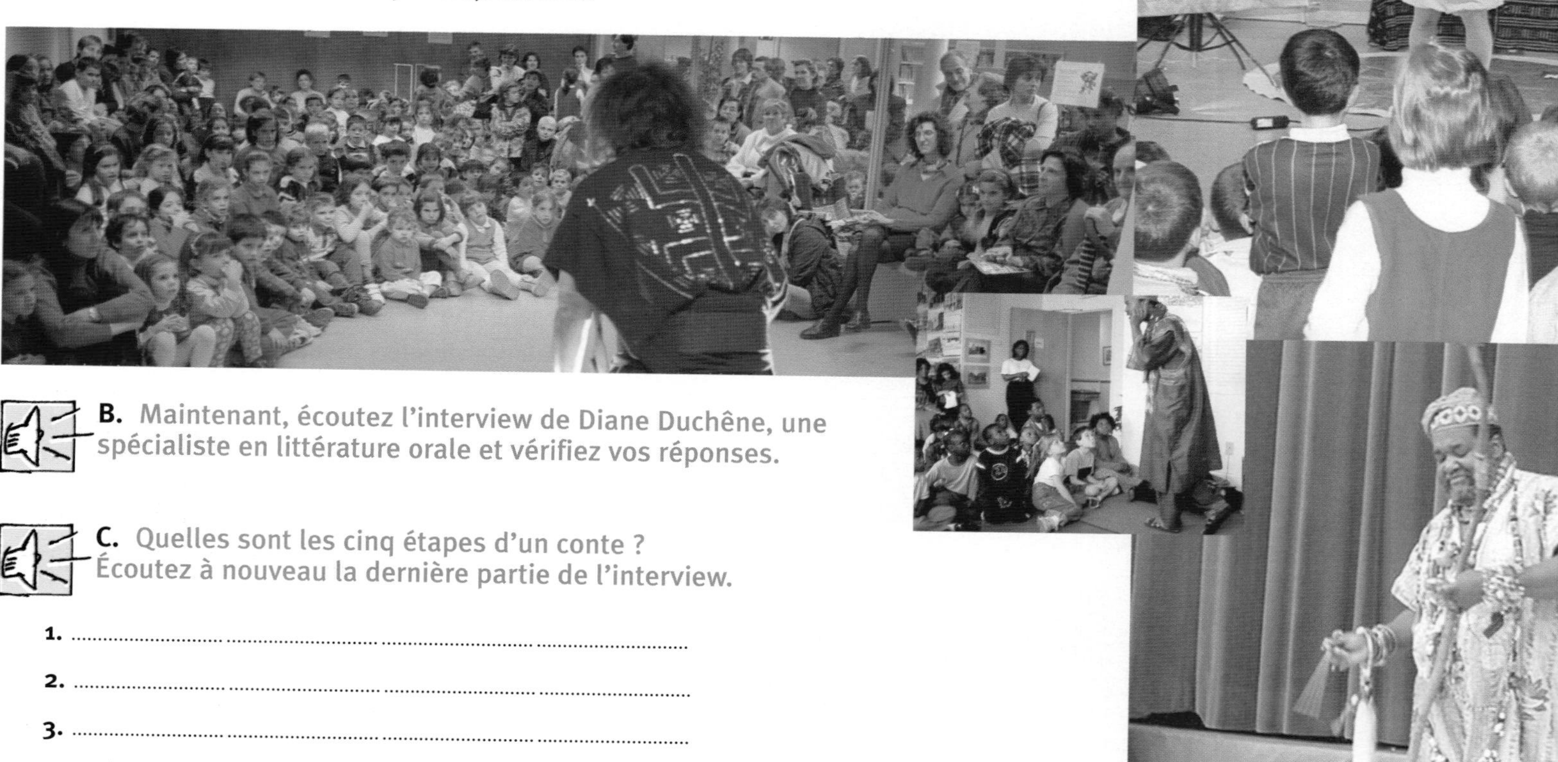

B. Maintenant, écoutez l'interview de Diane Duchêne, une spécialiste en littérature orale et vérifiez vos réponses.

C. Quelles sont les cinq étapes d'un conte ? Écoutez à nouveau la dernière partie de l'interview.

1. ....
2. ....
3. ....
4. Le héros trouve une bonne solution.
5. ....

3. LE PETIT POUCET

A. Vous connaissez le conte du Petit Poucet, n'est-ce pas ? Voici neuf extraits de ce conte, à vous de les remettre dans l'ordre avec l'aide de deux camarades.

Il était une fois un bûcheron et sa femme qui avaient sept fils. Le plus jeune était si petit que tout le monde l'appelait Petit Poucet. Le Petit Poucet était un petit garçon très intelligent et très attentif.

Ce soir-là, après avoir mangé un petit morceau de pain, les enfants allèrent se coucher. Le Petit Poucet qui n'avait pa faim garda le pain dans sa poche. Le lendemain, le bûcheron dit à sa femme : « Ma femme, nous n'avons plus rien à mang Nous devons abandonner nos enfants. » Un peu plus tard, le bûcheron partit avec ses enfants dans la forêt. Tout au long d chemin, le Petit Poucet jeta des miettes de pain.

Une minute et cent mille lieues plus loin, le Petit Poucet arriva devant le palais royal. Il entra dans la salle où le roi tenait un conseil de guerre. « Que viens-tu faire ici, mon garçon ? », demanda le roi. « Je suis venu pour gagner beaucoup d'argent », expliqua le Petit Poucet. « Sire, je peux vous aider à gagner la guerre grâce à mes bottes de sept lieues. »

Effectivement, grâce au Petit Poucet et aux bottes de sept lieues, le roi gagna la guerre et donna au Petit Poucet un grand sac d'or.

C'est fatigant de parcourir sept lieues à chaque pas et l'ogre, qui n'avait pas vu les enfants, s'arrêta pour se reposer. Mais il était tellement fatigué qu'il s'endormit aussitôt.

Alors, le Petit Poucet chuchota à ses frères : « Partez, courez jusqu'à la maison pendant qu'il dort. »

Ensuite, le Petit Poucet s'approcha de l'ogre qui dormait et lui retira les bottes. Puis il les enfila et partit en courant vers le palais royal.

Le Petit Poucet remercia le roi et rentra bien vite chez lui. Depuis ce jour, lui et sa famille ne connurent plus jamais la misère et ils vécurent heureux très longtemps.

3 Le Petit Poucet, qui avait tout entendu, sortit dans le jardin et remplit ses poches de petits cailloux blancs. Ensuite, il alla se coucher. Le lendemain, quand le bûcheron emmena ses enfants dans la forêt, le Petit Poucet jeta tout au long du chemin les petits cailloux. Lorsque le père les abandonna, les enfants se mirent à pleurer. Le Petit Poucet consola ses frères et leur dit qu'il suffisait de suivre les petites pierres pour rentrer à la maison. Lorsque les enfants arrivèrent chez eux, leur mère les embrassa en pleurant de joie.

Le lendemain matin, quand l'ogre alla chercher les enfants, il ne trouva personne. Les enfants s'étaient échappés. « Ces petits vauriens se sont échappés !... » hurla l'ogre. « Femme, donne-moi mes bottes de sept lieues ! » Les enfants étaient déjà assez loin. Ils étaient sortis de la forêt et ils voyaient déjà la maison de leurs parents lorsque, tout à coup, l'ogre apparut. Le Petit Poucet et ses frères se cachèrent derrière un arbre.

Un soir, alors que les enfants dormaient, sauf le Petit Poucet qui s'était caché sous une chaise, le bûcheron dit à sa femme : « Ma femme, nous ne pouvons plus nourrir nos garçons. Nous devons les abandonner dans la forêt. »

« Abandonner nos chers enfants ! Je ne veux pas ! » dit sa femme en pleurant.

« Il n'y a pas d'autres solutions » répondit le bûcheron.

Le Petit Poucet croyait retrouver son chemin grâce aux miettes de pain. Malheureusement, il n'en retrouva pas une seule car les oiseaux avaient tout mangé. Alors, avec ses frères, ils marchèrent pendant des heures dans la forêt. Tout à coup, ils virent une maison et le Petit Poucet alla frapper à la porte. Mais c'était la maison d'un ogre très méchant qui mangeait les enfants. L'ogre les enferma pour les manger le jour suivant.

B. Regardez les verbes dans l'extrait nº 3. Un nouveau temps du passé a été introduit. Soulignez ces nouvelles formes verbales. Ce temps est le passé simple et son usage, sans être obligatoire, est traditionnel dans un conte. Est-ce qu'il y a une autre forme verbale que l'on pourrait utiliser à sa place ?

4. ÊTES-VOUS « POLYCHRONIQUE » ?

A. Une personne « polychronique » est quelqu'un qui fait souvent plusieurs choses en même temps. Posez les questions suivantes à un camarade afin de savoir s'il est « polychronique ».

	jamais	parfois
1. Est-ce que vous mangez en regardant la télévision ?		
2. Est-ce que vous parlez en mangeant ?		
3. Pouvez-vous rire en étant triste ?		
4. Pouvez-vous être avec votre petit/e ami/e et penser à un/e autre personne ?		
5. Est-ce que vous travaillez en écoutant la radio ?		
6. Est-ce que vous travaillez en fumant ?		
7. Est-ce que vous mangez en travaillant ?		
8. Est-ce que vous fumez en prenant un bain ?		
9. Est-ce que vous chantez en vous rasant / en vous maquillant ?		
10. Pouvez-vous envoyer un SMS à un/e ami/e en écoutant le professeur ?		
11. Pouvez-vous être aimable avec quelqu' un en pensant : « Quel imbécile ! »		
12. Est-ce que vous parlez en dormant ?		

B. Maintenant, présentez vos conclusions à votre camarade.

- Je crois que tu es assez polychronique, car tu manges en travaillant...

5. QU'EST-CE QUE C'EST ?

A. Lisez ces phrases et devinez de quoi on parle.

1. Sans lui, la vie est impossible mais il est tellement chaud qu'on ne peut pas l'approcher.
2. Il est tellement apprécié qu'on le boit pour toutes les grandes occasions.
3. Elle est tellement fidèle qu'elle ne nous quitte jamais, mais on ne la voit que lorsque le soleil brille.
4. Il est devenu tellement indispensable dans les pays développés que ces derniers sont prêts à tout pour l'obtenir.
5. Elle a l'aspect d'un meuble et elle a pris tellement d'importance dans notre vie qu'elle est souvent au centre de la salle à manger.
6. On ne l'aime pas beaucoup mais elle est tellement nécessaire que, sans elle, le monde deviendrait un immense désert.

B. Avec un camarade, écrivez d'autres devinettes sur ce modèle en utilisant **tellement ... que**.

6. CHAQUE PROBLÈME A UNE SOLUTION

A. Mettez-vous par groupes de quatre et faites une liste des problèmes les plus importants concernant l'écologie, l'éducation, la distribution des richesses, votre emploi du temps, la santé... Une seule condition : soyez concrets !

· Le trou dans la couche d'ozonne s'agrandit.

· Il n'y a pas de métro après 2 heures du matin pour rentrer chez moi.

B. Maintenant, vous allez passer votre liste à un autre groupe qui va écrire en face de chaque problème la solution qu'il propose.

LE PASSÉ SIMPLE

Le passé simple s'emploie seulement à l'écrit et essentiellement à la troisième personne du singulier et du pluriel. Il situe l'histoire racontée dans un temps séparé du nôtre.

*Ils se **marièrent** et **eurent** beaucoup d' enfants.*

*Pâris **lança** une flèche qui **traversa** le talon d' Achille.*

GARD**ER**	il gard**a**	ils gard**èrent**
FIN**IR**	il fin**it**	ils fin**irent**
CONNAÎTRE	il conn**ut**	ils conn**urent**

TELLEMENT/SI ... QUE

Tellement/si + adverbe ou adjectif exprime une très grande intensité.

*Je suis **si/tellement fatigué** !*

Tellement/si ... que annonce une conséquence.

*Je suis **si/tellement** fatigué **que** je m' endors absolument partout.*

CAR ET PUISQUE

Car introduit une cause que l'on suppose inconnue par l'interlocuteur.

*Un piercing avant 16 ans n' est pas recommandable **car** les adolescents peuvent encore grandir.*

Puisque introduit une cause que l'on suppose connue par l'interlocuteur.

- *Tu vas au cinéma ce soir ?*
- *Non, **puisque** tu m' as dit que tu ne m' accompagnais pas.*

AFIN DE ET POUR QUE

Afin de + INFINITIF introduit un objectif, un but à atteindre.

*Je dois étudier beaucoup **afin de** réussir tous les examens et partir tranquille en vacances.*

Pour que + SUBJONCTIF introduit un objectif, un but à atteindre. On l'emploie quand les sujets de la première et de la deuxième phrase sont différents.

*Montre tes pattes **pour que** nous puissions te voir.*

POURTANT

Pourtant met en évidence le fait que quelque chose nous semble paradoxal.

- *Tu t'es perdu ?*
- *Oui, complètement !*
- ***Pourtant,** ce n'est pas la première fois que tu viens chez moi !*

LORSQUE

Lorsque s'utilise de la même manière que **quand**.

***Lorsque** je suis sorti, il pleuvait.* (= **Quand** je suis sorti, il pleuvait.)

TANDIS QUE ET PENDANT QUE

- *Comment les enfants se sont-ils comportés ?*
- *Oh, très bien ! Paul a tout rangé **pendant que/tandis que** Judith a mis la table.*

LE GÉRONDIF

Quand le sujet fait deux actions simultanées, on peut utiliser le gérondif.

- *Moi, je lis toujours le journal **en prenant** mon petit-déjeuner.*

PRÉSENT	GÉRONDIF
nous **pren**ons	en pren**ant** [ã]
nous **buv**ons	en buv**ant** [ã]
nous **conduis**ons	en conduis**ant** [ã]

Attention :
ÊTRE → en **étant**
AVOIR → en **ayant**
SAVOIR → en **sachant**

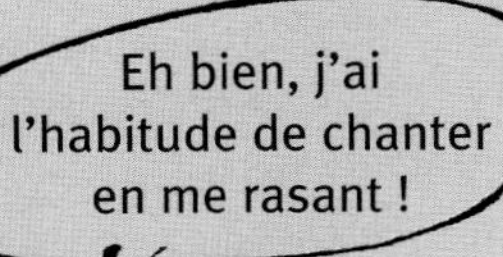

Eh bien, j'ai l'habitude de chanter en me rasant !

- *Pour réduire le trou dans la couche d'ozone, on devrait interdire la circulation des voitures.*
- *Pour qu'il y ait des métros après 2 heures du matin, on pourrait envoyer une lettre au maire ou bien au ministre des transports.*

7. CAR OU POURTANT ?

A. Complétez chaque phrase avec **car** ou **pourtant** en fonction de ce qui vous semble le plus adéquat. Comparez vos choix avec ceux d'un camarade.

	car ou pourtant	
1. Les pays développés détruisent leur surproduction d'aliments		des millions d'enfants dans le monde meurent de faim.
2. Tout le monde sait que l'alcool est dangereux au volant		beaucoup de conducteurs conduisent après avoir bu.
3. Les antivirus ne sont plus assez efficaces pour protéger les ordinateurs		les virus informatiques sont de plus en plus sophistiqués et rapides.
4. Les baleines sont en voie de disparition		on les a trop chassées dans le passé.
5. Les aliments des fast-foods sont très mauvais pour la santé		ils sont saturés de graisses.
6. L'équipe de France a été éliminée en quart de finale de l'Euro 2004		il y avait dans cette équipe de grands joueurs comme Zidane, Lizarazu ou Henry.

B. Est-ce que vous avez déjà été victime d'une injustice ? Est-ce que vous avez déjà observé quelque chose qui n'est pas logique ou qui n'était pas du tout prévisible ? Parlez-en avec deux camarades. Vous aurez sans doute besoin du connecteur **pourtant**.

- *Moi, par exemple, je pense que j'ai été victime d'une injustice : j'ai eu une très mauvaise note à l'examen, pourtant j'avais étudié...*

8. COURSE CONTRE LA MONTRE

Ce matin, Gilles et Marité ne se sont pas réveillés à temps et doivent tout faire en 30 minutes. Il est 7 h 00 et à 7 h 30 ils doivent partir. Regardez la liste des choses à faire et aidez-les à s'organiser :

- préparer le café : 5 mn
- faire les lits : 5 mn
- prendre le petit-déj' : 10 mn
- s'habiller : 5 mn chacun
- beurrer 4 tartines : 4 mn
- donner à manger au chat : 1 mn
- habiller les enfants : 5 mn
- prendre une douche : 5 mn chacun

	Gilles	Marité
7.00-		
7.20-7.30	*Prendre le petit-déjeuner*	*Prendre le petit-déjeuner*

- *Pendant que Marité prépare le café, Gilles peut...*

9. LA PIERRE PHILOSOPHALE

A. Lisez ce conte moderne. Vous voyez les fragments soulignés ? Pour que l'histoire soit un peu plus claire, essayez de les réécrire en utilisant les mots suivants.

- **afin de**
- **tellement ... que**
- **pourtant**
- **lorsque**
- **car**
- **puisque**

B. Est-ce que vous avez aimé cette histoire ? À votre avis, quel est le sens de cette histoire ? Quelle est sa moralité ? Parlez-en avec deux autres camarades.

- Moi, je pense que la morale de cette histoire, c'est qu'il ne faut pas...

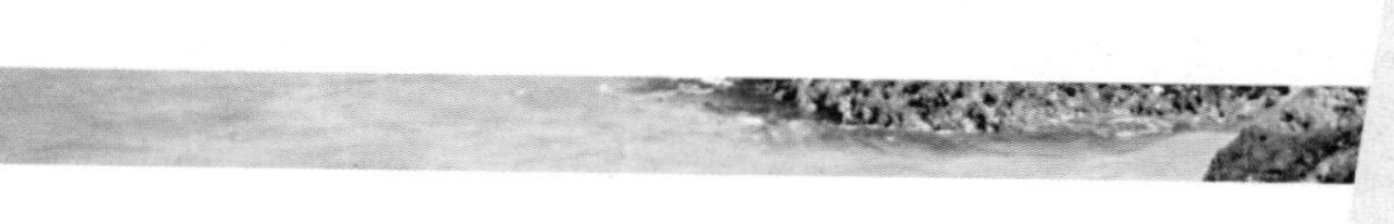

Il était une fois un homme d'affaires qui s'appelait Benjamin. Il voyageait beaucoup : il avait beaucoup de clients importants aux quatre coins du monde et sa femme et ses deux fils ne le voyaient presque jamais. Benjamin ne s'intéressait pas seulement aux affaires, il s'intéressait aussi aux vieux livres et, de temps en temps, il allait dans une petite librairie spécialisée. Un jour il tomba sur un livre dont le titre l'intriga : « La Pierre philosophale ». Il l'acheta puis rentra chez lui : il voulait le lire tranquillement. Ce livre était très intéressant : il le lut en deux heures. On y parlait d'une pierre philosophale qui donnait la sagesse et on expliquait que cette pierre était sur une petite île déserte en Océanie. Benjamin décida de partir à la recherche de cette pierre-là. Il divorça et laissa toute sa fortune à ses deux fils. Il n'avait pas besoin d'argent : il n'y avait rien sur l'île. Quelques jours plus tard, il était sur l'île. Il se mit aussitôt au travail. Le livre ne disait pas où se trouvait exactement la pierre, mais il expliquait qu'elle provoquait une certaine chaleur quand on la tenait dans la main. Alors, Benjamin commença à ramasser une à une les pierres de la plage : il gardait chaque pierre un certain temps dans le creux de sa main pour savoir si elle était chaude. Il ne sentait rien... Il la jetait à la mer. Des jours passèrent puis des semaines, puis des mois et au bout d'un an, Benjamin n'avait encore rien trouvé. Un jour, alors que Benjamin prenait les pierres et les jetait automatiquement à l'eau, il laissa passer la chance de sa vie : en effet, comme il commençait à être fatigué, son geste était devenu automatique. Il toucha alors une pierre qui était plus chaude que les autres, il la jeta à l'eau d'un geste machinal. Désespéré, il se mit à l'eau pour la rechercher mais ne la retrouva jamais. Benjamin avait jeté à l'eau ce qu'il cherchait depuis si longtemps.

10. À VOUS DE RACONTER !

A. Mettez-vous par petits groupes et faites une liste de contes que vous connaissez, puis décidez ensemble lequel de ces contes vous voulez raconter.

B. Maintenant, pensez à un intrus, c'est-à-dire, un personnage réel ou fictif qui normalement n'est pas dans ce conte mais que vous voulez introduire dans l'histoire. Puis écrivez votre version du conte.

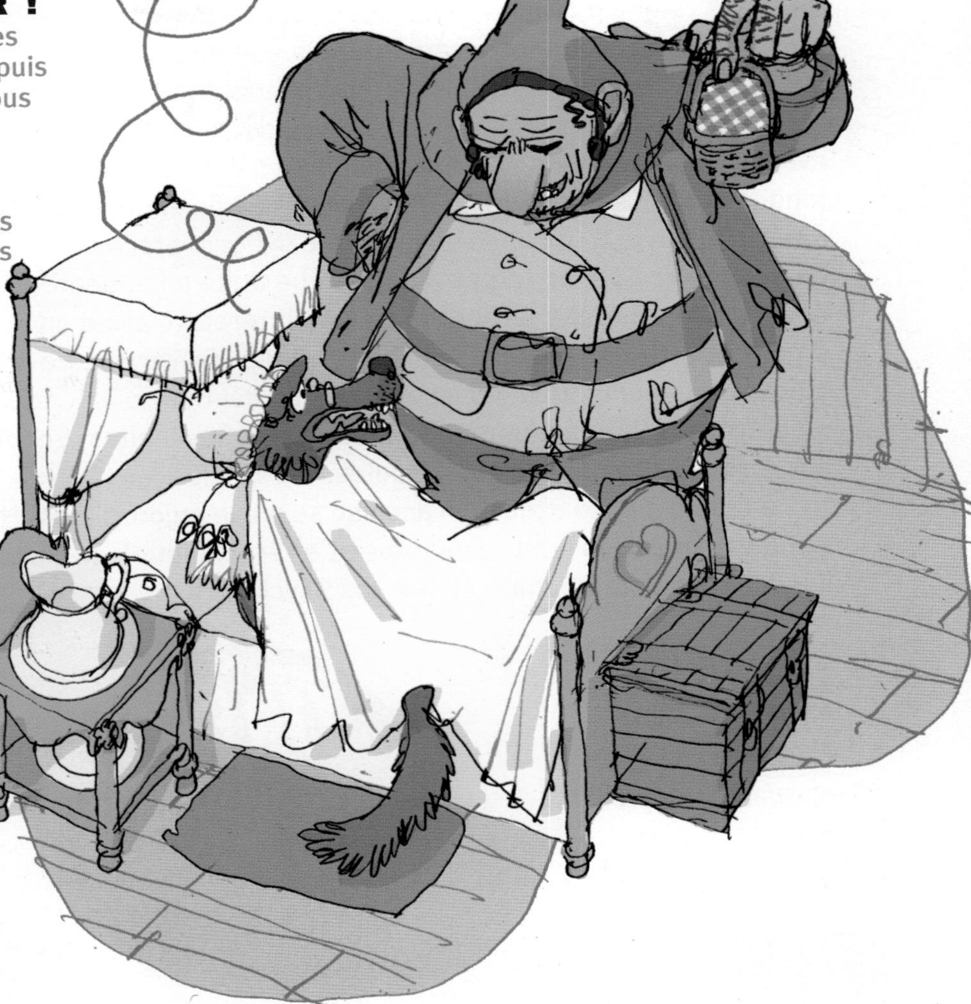

C. Une fois le texte corrigé par votre professeur, vous pourrez mettre au point une lecture dramatisée de votre conte et le présenter à la classe.

11. LE CRÉOLE

Lisez le texte suivant sur l'origine de la langue créole. Est-ce que vous connaissez l'histoire de votre propre langue ou d'autres langues de votre entourage ?

LA LANGUE CRÉOLE

Le créole à base lexicale française est né du métissage du vocabulaire français des XVIIe et XVIIIe siècles avec des expressions d'origine africaine. Capturés sur leur terre natale, les Africains déportés aux Antilles étaient répartis sur diverses îles pour éviter que les tribus se reconstituent et provoquent des révoltes. Ainsi, devant la nécessité de survivre et de communiquer avec des compagnons parlant des langues différentes, ils ont créé une langue commune, reprenant des mots français et quelques termes amérindiens, le tout construit avec une syntaxe proche de celle des langues d'Afrique. Le temps a donné une unité à l'ensemble et s'est développée progressivement toute une littérature orale en langue créole à partir des contes, des chants et des proverbes. Le créole est aujourd'hui une langue à part entière et il est même la langue officielle de deux pays indépendants : Haïti et les îles Seychelles. En fait, il n'existe pas un créole mais plusieurs créoles.

Le créole à base lexicale française se parle aujourd'hui à Haïti, aux Antilles françaises (en Guadeloupe et en Martinique), en Guyane, sur l'île Maurice, à la Réunion et aux îles Seychelles.

enmé :	aimer
chapo-a :	chapeau
appwan :	apprendre
ayen :	rien
gadé :	regarder

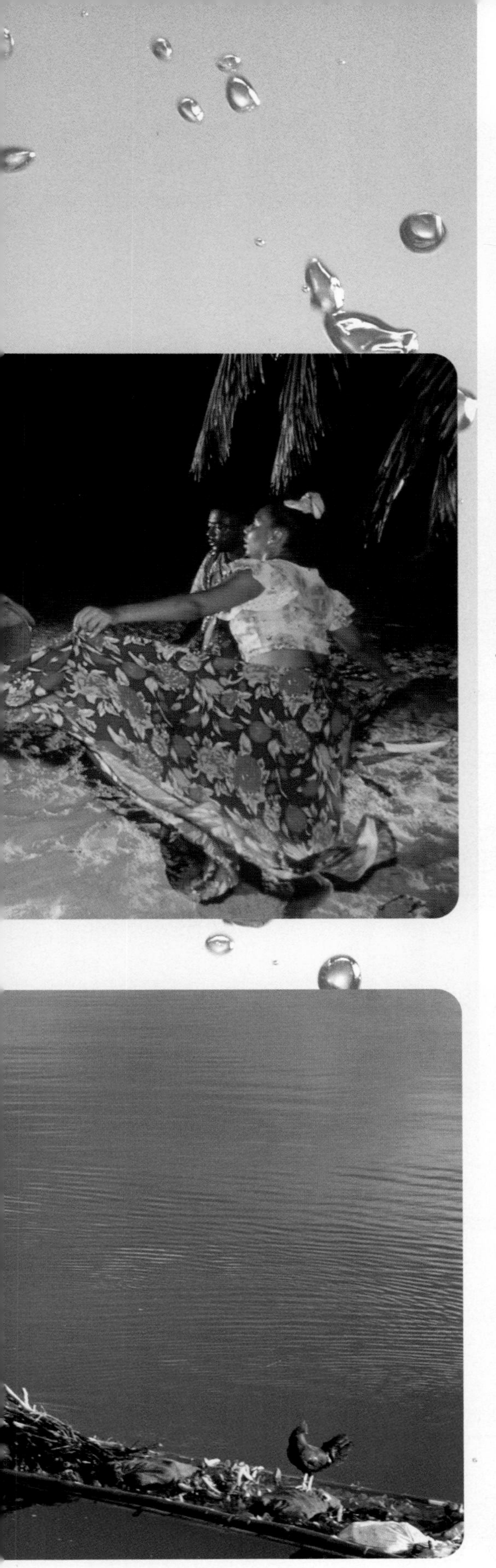

TI POCAME était un gentil petit garçon qui vivait chez sa tante car il était orphelin. Sa tante ne l'aimait pas du tout, elle préférait ses deux fils. Elle réservait à ses fils les plus beaux habits et pour Ti Pocame, les vieux habits ; pour ses deux fils, les bons morceaux de viande et pour Ti Pocame, les os. De même, Tipocame faisait toutes les corvées : aller chercher de l'eau à la rivière, nourrir le cochon et les poules, éplucher les légumes... Souvent, Ti Pocame était puni injustement et dans ses colères, sa tante menaçait de le donner au diable.

Mais Ti Pocame était courageux et il ne se plaignait jamais. Il songeait souvent à sa chère marraine chez qui il aimerait bien partir vivre un jour. Un soir, alors qu'ils étaient à table, la Tante ordonna à Ti Pocame d'aller cueillir un piment afin d'épicer le repas. Il faisait noir et, tout de suite, Ti Pocame pensa : « C'est ce soir que ma tante m'envoie au diable ! »

Avant de sortir, il prit soin de glisser dans sa poche les sept pépins d'orange qui portent chance et que sa marraine lui avait donnés pour son anniversaire. Arrivé dehors, la nuit l'enveloppa tout entier. Il prit garde à faire le moins de bruit possible afin que le diable ne le remarque pas. Soudain, il vit une petite lumière comme celle d'une luciole et celle-ci se mit à foncer sur lui, « le diable », pensa-t-il.

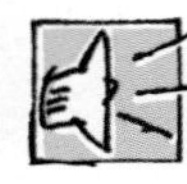

12. TI POCAME

A. Lisez le début de ce conte antillais. À quels autres contes vous fait-il penser ?

B. Comment croyez-vous que le conte continue ? Voulez-vous le savoir ? Écoutez !

JOUER, RÉVISER, GAGNER

s allons créer un
sur des thèmes
stoire, de géogra-
, de cultures fran-
nones et de langue
çaise, et nous
ns faire un bilan
al de notre appren-
age du français.

cela nous allons
endre à :

- rmuler des questions
mplexes
- oondre à des ques-
ns sur la France, la
ncophonie et la langue
nçaise
- primer des désirs et
s volontés

ous allons utiliser :

- subjonctif après les
rbes qui expriment un
sir ou une volonté
- expressions d'affir-
ation ou de négation.
- *puis* et *il y a... que*
- à la forme interro-
gative

9

1. COMBIEN Y A-T-IL D'HABITANTS EN FRANCE, ENVIRON 40, 50 OU 60 MILLIONS ?
2. QUELLE VILLE FRANÇAISE EST LE SIÈGE DU PARLEMENT EUROPÉEN ?
3. CONJUGUEZ, AU PASSÉ COMPOSÉ, LE VERBE **ÊTRE**
4. QUEL EST LE NOM DE L'INSPECTEUR HÉROS DES ROMANS DE G. SIMENON ?
5. DANS QUELS PAYS PARLE-T-ON FRANÇAIS EN EUROPE ?
6. CITEZ 2 CHANTEURS/EUSES OU GROUPES FRANCOPHONES
7. CONJUGUEZ LE VERBE **CONNAÎTRE** AU FUTUR
8. PARLEZ D'UN FILM, D'UN SPECTACLE OU CONCERT QUE VOUS AVEZ VU
9. CITEZ 3 PLATS FRANÇAIS
10. CONJUGUEZ LES VERBES **FAIRE** ET **VENIR** AU PRÉSENT
11. PRÉSENTEZ UN DES MEMBRES DE VOTRE GROUPE ET PARLEZ DE LUI
12. CITEZ 5 VILLES FRANÇAISES
13. PRÉSENTEZ VOTRE FAMILLE
14. PARLEZ DE VOS GOÛTS
15. — EST-CE QUE JEAN A OFFERT UNE ROSE À ALINE ? — NON, IL .... A OFFERT UNE ORCHIDÉE.
16. CITEZ 2 ÉCRIVAINS FRANCOPHONES
17. CITEZ 4 MONUMENTS DE PARIS
18. QUEL NOM, INSPIRÉ DE SA FORME, DONNE-T-ON À LA FRANCE ?
19. DANS QUELLE VILLE FRANÇAISE A LIEU UN IMPORTANT FESTIVAL DE THÉÂTRE ?
20. QUEL TYPE DE VACANCES PRÉFÉREZ-VOUS ?
21. QUEL EST LE PARTICIPE PASSÉ DE **PRENDRE**, DE **SAVOIR** ET DE **DIRE** ?
22.
23. QUELLES SONT LES LANGUES PARLÉES EN SUISSE ET EN BELGIQUE ?
24. EXPLIQUEZ COMMENT ALLER DE VOTRE DOMICILE À L'ÉCOLE/AU TRAVAIL
25. CHANTEZ AVEC TOUTE VOTRE ÉQUIPE : " SUR LE PONT D'AVIGNON "
26. CONJUGUEZ **SAVOIR** AU PRÉSENT
27. DONNEZ LE NOM DE 7 ALIMENTS SALÉS ET DE 4 ALIMENTS SUCRÉS
28. FAITES 6 RECOMMANDATIONS POUR RESTER EN FORME
29. QU'EST-CE QUE VOUS FAITES NORMALEMENT LE SAMEDI ?
30. CITEZ 3 PERSONNAGES DE BANDES DESSINÉES FRANCOPHONES
31. CONJUGUEZ **SE LEVER** AU PRÉSENT
32. FAITES FAIRE UNE MINI-SÉANCE DE GYMNASTIQUE À UNE ÉQUIPE ADVERSE
33. CITEZ UNE VILLE DU QUÉBEC
34. CITEZ 5 OBJETS QUI N'EXISTAIENT PAS DU TEMPS DE VOS GRANDS-PARENTS
35. CITEZ 10 NOMS PROFESSIONS

1. JEU DE L'OIE

A. Formez 4 groupes dans la classe et lisez les règles du jeu. Il vous faut un dé, un pion pour chaque groupe et des cartes joker.

B. Maintenant commencez à jouer. Chaque groupe lance son dé et effectue les épreuves. Le professeur corrige les réponses et joue le rôle d'arbitre en cas de litige.

RÈGLES DU JEU

À tour de rôle, chaque groupe lance le dé et répond à la question. Si la réponse est correcte, avancez d'une case (deux cases, si la suivante est la prison) avant de laisser jouer le groupe suivant. Vous avez 30 secondes pour vous mettre d'accord sur la réponse à donner.

- Si vous tombez sur une case VERTE, et que vous répondez correctement, vous pouvez avancer de trois cases.
- Si vous tombez sur une case **JOKER** et que vous répondez correctement à la question, vous gagnez un joker.
- Si vous tombez sur une case **PRISON**, vous devez passer deux tours, mais si vous avez un joker, vous pouvez sortir au tour suivant.
- Si vous êtes sur une case ROUGE, et que vous répondez correctement à la question, vous pouvez rejouer.
- Si vous tombez sur une case SABLIER, vous devez parler pendant 1 minute sur le thème proposé.

36 … T NÉ EN 1975 … ONDRES, IL … MARIÉ AVEC … SPICE GIRL, … EST-CE ?

37 QU'EST-CE QUE VOUS AVEZ FAIT AU MOIS D'AOÛT DERNIER ?

38 RACONTEZ CE QUE VOUS AVEZ FAIT HIER

39 … VOUS GAGNIEZ … U LOTO, OÙ … IEZ-VOUS EN … ACANCES ?

40

41 CONJUGUEZ CROIRE AU PRÉSENT

… FEREZ-VOUS … DANT VOS … CES ?

43 COMMENT SERAIT VOTRE COPAIN/COPINE IDÉAL(E) ?

44 CITEZ 6 MOYENS DE TRANSPORT

45 DITES CE QUE VOUS ALLEZ FAIRE CE WEEK-END

46 IL A FAIT DES FILMS DOCUMENTAIRES, IL A BEAUCOUP NAVIGUÉ. QUI EST-CE ?

47 EN FRANCE, L'ÉCOLE EST OBLIGATOIRE JUSQU'À...... ?

48 CITEZ 4 RÉGIONS FRANÇAISES

49 CITEZ 4 VILLES FRANCOPHONES NON FRANÇAISES

50 CONJUGUEZ LE VERBE FINIR AU PASSÉ COMPOSÉ.

51 LISEZ CES CHIFFRES : 3 792, 99 778

52 QUI A EU L'INITIATIVE DE L'ÉCOLE PUBLIQUE, LAÏQUE EN FRANCE ?

53

54 CONJUGUEZ LES VERBES ÊTRE ET AVOIR AU PRÉSENT DE L'INDICATIF ET DU SUBJONCTIF

55 CITEZ 8 VERBES NON PRONOMINAUX DONT LE PASSÉ COMPOSÉ SE CONJUGUE AVEC L'AUXILIAIRE ÊTRE

56 UNE MARQUE DE VOITURES FRANÇAISE QUI PARTICIPE AU CHAMPIONNAT MONDIAL DE FORMULE I

57 DÉCRIVEZ LES VÊTEMENTS QUE VOUS PORTEZ AUJOURD'HUI

58 QU'EST-CE QUE LES FRANÇAIS PRENNENT POUR LE PETIT-DÉJEUNER LE MATIN ?

59 QUELS SONT LES INGRÉDIENTS DE LA QUICHE LORRAINE

60 QU'EST-CE QUE VOUS AIMEZ DANS VOTRE VILLE ET POURQUOI ?

61 QU'EST-CE QUE VOUS N'AIMEZ PAS DANS VOTRE VILLE ET POURQUOI ?

62 DONNEZ 6 ADJECTIFS POUR DÉCRIRE VOTRE MEILLEUR/E AMI/E

63 « IL EST BIEN ARRIVÉ À L'AÉROPORT ? » POSEZ CETTE QUESTION DE DEUX AUTRES FAÇONS

64 CITEZ LES JOURS DE LA SEMAINE EN COMMENÇANT PAR LE DERNIER

65 CONJUGUEZ LE VERBE VOULOIR AU PRÉSENT

66 DITES CE QU'IL Y A COMME INGRÉDIENTS DANS UN DE VOS PLATS PRÉFÉRÉS

67 CONJUGUEZ LE VERBE DEVOIR AU PRÉSENT

68 PRÉFÉREZ-VOUS LA VILLE OU LA CAMPAGNE ? DONNEZ 3 ARGUMENTS

69 RÉCITEZ LES MOIS DE L'ANNÉE EN COMMENÇANT PAR LE DERNIER

70 FAITES UNE DESCRIPTION DÉTAILLÉE DE VOTRE SALLE DE CLASSE

2. LE QUÉBEC, VOUS CONNAISSEZ ?

A. Essayez de répondre individuellement à ce questionnaire sur le Québec. Si vous avez des doutes, mettez un point d'interrogation.

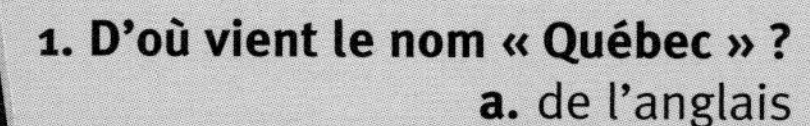

1. D'où vient le nom « Québec » ?
a. de l'anglais
b. du français
c. d'une langue indienne

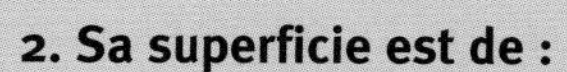

2. Sa superficie est de :
a. 7 fois la France
b. 5 fois la France
c. 3 fois la France

3. Au Québec il y a :
a. environ 7 millions d'habitants
b. environ 15 millions d'habitants
c. environ 25 millions d'habitants

4. Climat. La température à Montréal en été est de :
a. 9 à 18 degrés
b. 10 à 26 degrés
c. 17 à 32 degrés

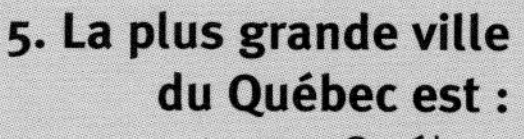

5. La plus grande ville du Québec est :
a. Québec
b. Montréal
c. Ottawa

6. La principale source d'énergie du Québec est :
a. l'énergie solaire
b. l'énergie hydraulique
c. l'énergie nucléaire

7. Au dernier référendum en 1995, les Québécois ont voté :
a. majoritairement pour l'indépendance du Québec
b. majoritairement contre l'indépendance du Québec

8. Le Québec est :
a. un état fédéral au sein du Canada
b. un état souverain
c. un pays indépendant

9. La population est à 82% francophone, 9% anglophone :
a. l'anglais est langue officielle, et les services fédéraux sont bilingues
b. l'anglais et le français sont langues officielles
c. le français est la langue officielle, mais les services fédéraux du Québec sont bilingues

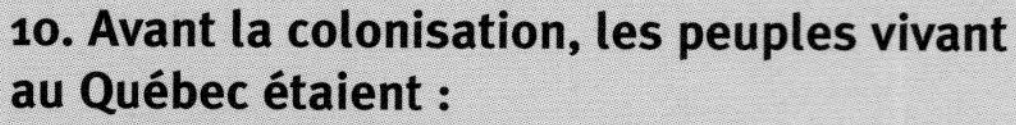

10. Avant la colonisation, les peuples vivant au Québec étaient :
a. les Algonquiens, les Iroquois, les Inuits et les Micmacs
b. les Aztèques, les Mayas, les Incas et les Huicholes
c. les Appaches, les Sioux , les Navajos, les Commanches

11. La population est majoritairement :
a. protestante
b. catholique
c. athée

12. Ces derniers temps, les groupes d'immigrants le plus importants du Québec sont :
a. les Anglais et les Français
b. les Italiens et les populations de l'Europe de l'Est
c. des immigrants francophones de l'Afrique, du monde arabe et des Antilles

B. Comparez vos réponses avec celles d'un camarade.

- Au Québec, je ne pense pas que la religion dominante soit le catholicisme.
- Moi, si. Il me semble...

C. Écoutez cette émission radiophonique sur le Québec et vérifiez vos réponses.

D. Lisez ce texte sur la revendication identitaire au Québec. Qu'en pensez-vous ? Avez-vous vécu, ou entendu parler de situations identiques dans votre pays ou dans d'autres pays ?

QUESTIONS D'IDENTITÉS

Les Québécois sont, en général, très fiers d'être francophones, même s'ils ne sont qu'un petit îlot dans le Canada anglophone.

Ils veulent qu'on leur parle en français et certains aimeraient que le Québec soit un état indépendant du Canada reconnu comme un pays francophone. Pourtant, il y a eu deux référendums à ce sujet, dont le premier en 1980 : le Gouvernement du Québec voulait que le pays obtienne le pouvoir exclusif de faire ses lois, de percevoir ses impôts et de gérer ses relations extérieures. En résumé, le projet de loi proposait que le Québec devienne un état souverain tout en maintenant une association économique avec le Canada. Mais la majorité des Québécois se sont opposés à l'indépendance avec 59,56% de « non ». En 1995, peu de temps après l'arrivée au pouvoir du parti québécois, le premier ministre a proposé un avant-projet de loi sur la souveraineté du Québec. L'option de la souveraineté du Québec a de nouveau échoué au référendum.

Depuis l'arrivée au pouvoir du Parti québécois, en 1976, la politique linguistique du Québec a pris un tournant décisif. La charte de la langue française, plus connue sous le nom de « loi 101 », adoptée en 1977, assure la prédominance de la langue française. Mais quelle est la réalité linguistique actuelle ?

Le français est la langue officielle du Québec, mais la situation réelle est un quasi-bilinguisme. Le français est devenu la langue de la législature et de la justice, de l'administration publique, du travail, du commerce, des affaires et de l'enseignement. La langue parlée dans 82% des foyers est le français. Mais, si un anglophone au Québec veut que ses enfants aillent dans une école anglophone ou qu'on lui parle en anglais dans les services publics, il a le droit de l'exiger.

À travers la loi 101, le gouvernement du Québec a aussi voulu que les populations autochtones puissent parler leur langue maternelle, c'est à dire l'algonquin, l'attikamek, le micmac, le montagnais et le mohwh. Cette loi insiste sur l'importance de l'utilisation des langues amérindiennes dans l'enseignement public dispensé aux Amérindiens.

- En Finlande, il y a une minorité qui parle suédois et, au Nord, il y a aussi les Lapons, qui...
- Oui, et en Belgique...

3. DEPUIS QUAND ?

Lisez les petites histoires suivantes et, avec un camarade, mettez-vous d'accord pour compléter les espaces vides. Vous devrez, d'abord, faire le calendrier des mois d'avril, mai et juin 2004.

Mars

L	M	M	J	V	S	D
1	2	3	4	5	6	7
8	9	10	11	12	13	14
15	16	17	18	19	20	21
22	23	24	25	25	27	28
29	30	31				

Marie et François

Nous sommes en mars 2004. Le 17 mars exactement.
Marie et François sont mariés depuis 3 ans (date de leur mariage : ....)
et chaque année ils fêtent leur anniversaire de mariage le 17 mars. Mais cette année, ils ne le fêtent pas car Marie est à l'hôpital depuis 3 jours. Elle a eu un accident en traversant la rue, (date de l'accident : ....).
Rien de grave heureusement ! Mais Marie et François ne pourront pas fêter leur anniversaire avant 15 jours, quand Marie sortira de l'hôpital (date de sortie de l'hôpital : ....).

Pierre

Nous sommes le 27 mai 2004.
Pierre est allé aux sports d'hiver il y a exactement 2 mois (date : ....).
C'était un samedi soir. Il est arrivé à Andorre tard dans la nuit.
Le lendemain matin (date : ....), il s'est précipité sur les pistes car il avait très envie de skier.
Il est tombé et s'est cassé la jambe.
Il est resté pendant 3 jours à l'hôpital (dimanche compris), puis il est rentré chez lui un mardi (date : ....).
Depuis sa sortie de l'hôpital, il ne travaille plus.
Il reprendra son travail dans 12 jours (date de reprise du travail : ....).

4. MOTS BIZARRES

Lisez ces mots puis, avec un camarade, mettez-vous d'accord sur ce qu'ils désignent.

un brochet	un objet, un poisson ou une plante ?
une pastèque	un objet, un fruit ou une profession ?
un cygne	une profession, un oiseau ou un objet ?
un plombier	une profession, une plante ou un oiseau ?
une hirondelle	un objet, une profession ou un oiseau ?
un jonc	un poisson, une plante ou une profession ?
un flacon	une plante, un fruit ou un objet ?

● Moi, je crois que brochet c'est une plante.
○ Non, pas du tout, c'est un...

LA QUESTION À LA FORME INTERRO-NÉGATIVE

La forme interrogative peut se combiner avec la forme négative.

*La Réunion **n'a-t-elle pas été** une prison ?*
*Est-ce que la Réunion **n'a pas été** une prison ?*

Dans ce cas, si la réponse est affirmative, la réponse n'est pas **oui** mais **si**.

***Si**, c'était une prison dès le XVIIe siècle.*

~~Oui~~, c'était une prison.

RÉPONDRE À UNE QUESTION AUTREMENT QUE PAR OUI OU NON

Si la question est affirmative et la réponse affirmative : **tout à fait, en effet.**

● *Vous connaissez le Québec, n'est-ce pas ?*
○ ***En effet**, je l'ai visité il y a deux ans.*

Si la question est affirmative et la réponse négative : **pas du tout**.

● *Vous avez habité là-bas ?*
○ ***Pas du tout**, j'y suis allé comme touriste.*

Si la question est négative et la réponse affirmative : **si, bien sûr** (**que si**).

● *Vous n'êtes pas sorti de Montréal, n'est-ce pas ?*
○ ***Si**, j'ai visité une grande partie du Québec.*

Si la question est négative et la réponse est négative : **absolument pas, vraiment pas.**

● *Vous n'avez pas eu peur ?*
○ ***Absolument pas.***

DEPUIS / IL Y A... QUE

Ces deux indicateurs temporels peuvent se construire avec une durée chiffrée.

*La Réunion est française **depuis** le XVIIe siècle.*
***Il y a** quatre siècles **que** la Réunion est française.*

Ils se construisent également avec des adverbes de temps : **longtemps**, **peu de temps**, etc.

***Il y a longtemps qu'** elle est française.*

*Elle travaille ici **depuis peu de temps**.*

DANS

Dans peut représenter un repère dans le futur du locuteur, il est suivi d'une durée chiffrée.

__Dans__ 10 jours, ils partiront pour la Réunion.

Je crois que nous arriverons __dans__ deux heures à Marseille.

LE SUBJONCTIF APRÈS LES VERBES QUI EXPRIMENT UNE VOLONTÉ OU UN DÉSIR

Tu **aimes**		
Il **veut**		
Elle **préfère**		+ SUBJONCTIF
On **exige**	**que**	je **fasse** les courses.
Vous **adorez**		
Elles **souhaitent**		

Cette construction apparaît quand le sujet des deux verbes n'est pas le même.

Nous voulions que nous fesions les courses.

5. NI OUI NI NON

À tour de rôle, vous allez répondre aux questions que la classe va vous poser, mais attention : vous ne pouvez répondre ni **oui** ni **non**. Si la question est mal formulée, votre professeur le signale et vous ne répondez pas. Si vous répondez **oui** ou **non**, vous avez perdu et vous laissez votre place à un camarade.

- ● Tu vas bien aujourd'hui ?
- ○ Parfaitement bien.
- ● Tu es sûr ?
- ○ Absolument.

6. MAIS SI !

Par groupes de trois. Chaque membre du groupe rédige quatre propositions à la forme négative concernant l'un des deux autres. Ensuite, vous lui demandez de confirmer vos suppositions. Celui qui se trompe plus de deux fois, laisse son tour au suivant.

- ● Mario, tu n'aimes pas les fruits de mer, n'est-ce pas ?
- ○ Mais si, j'aime les fruits de mer.
- ● Tu n'habites pas en ville ?
- ○ Non, j'habite à la campagne.

7. ILS VEULENT QUE...

Complétez ces textes comme dans l'exemple en choisissant le verbe correct.

Le stressé

Mon père veut que je sois un musicien célèbre.
Ma mère souhaite que je fasse le ménage.
Ma sœur exige que je lui .... un petit copain.
Mon prof de maths veut que je lui .... des devoirs tous les jours.

Le je-m'en-foutiste

Les gens veulent que je .... une cravate.
Mon prof de physique exige que je .... une blouse blanche pendant les cours.
Mon père aime que je .... les meilleures notes.
Les gens de mon quartier ne veulent pas que je .... de la batterie le soir.

Le dragueur

Fanny veut que je .... toujours avec elle.
Marianne préfère que je ne .... pas en moto, elle a peur que j'aie un accident.
Julie adore que je l'.................... en boîte.
Stéphanie aime que je .... en rocker.

emmener • s'habiller • porter • rouler • faire • jouer • mettre • sortir • trouver • avoir

8. LE QUIZ

A. Formez des équipes de trois ou quatre. Vous devez préparer six questions par groupes sous forme de fiches. Vous écrivez les réponses sur une feuille à part. Les questions peuvent se référer à la culture ou à la langue, mais on doit pouvoir trouver les réponses dans ce manuel.

- On pourrait demander la conjugaison d'un verbe au conditionnel, par exemple. C'est dans l'unité 1.
- D'accord, on pourrait prendre le verbe pouvoir.

B. Maintenant, lisez attentivement les règles du jeu avant de commencer.

- Moi, je crois que le conditionnel de pouvoir, à la première personne, ça s'écrit P, O, U, R, R, A, I.
- Non, il faut un s à la fin.
- D'accord. Donc, " je pourrais " s'écrit : P O U R R A I S.

Règles du jeu

1. Chaque équipe va donner son jeu de questions au professeur qui les distribuera à un autre groupe.
2. Les équipes ont 15 minutes pour décider des réponses qu'elles écriront sur cette même carte.
3. Le porte-parole de chaque équipe lit les questions et les réponses de son groupe.
4. Le professeur dit si la réponse est correcte ou non.
5. Si un groupe ne connaît pas la réponse ou donne une réponse incorrecte, les autres équipes ont alors le droit de donner la réponse.

Ponctuation

Bonne réponse : 3 points
Réponse donnée à la place d'une autre équipe : 5 points
Mauvaise réponse : –3 points

9. MOI ET LE FRANÇAIS : MON BILAN

Répondez individuellement à ces questions, puis commentez vos réponses par petits groupes.

1. Voilà deux ans que vous étudiez le français, et maintenant, que pensez-vous faire ?

- Je pense continuer à l'étudier comme je l'ai fait jusqu'à présent.
- J'en sais suffisamment pour continuer seul.
- Je vais arrêter.
- Autre :

2. Pendant les prochaines vacances, vous voudriez :

- Suivre un cours intensif de français.
- Vous inscrire à un cours de…
- Aller en France ou dans un pays francophone.
- Ne rien faire.
- Autre :

3. Avec ce manuel vous pensez que :

- Vous avez beaucoup appris.
- Vous n'avez pas suffisamment appris.
- Vous avez trop travaillé.
- Vous n'avez pas assez travaillé.
- Autre :

4. Dans ce manuel :

Vous avez aimé qu'il y ait…
Vous avez aimé qu'on fasse…
Vous n'avez pas aimé qu'il y ait…
Vous n'avez pas aimé qu'on fasse…

5. Vous voulez que votre prochain livre de français soit :

Plus…
Moins…
Aussi…

6. Pour vous, le français c'est…

7. Si vous deviez recommencer à apprendre le français, qu'est-ce que vous changeriez ? Qu'est-ce que vous ne changeriez pas ?

10. DEUX ÎLES : LA MARTINIQUE ET L'ÎLE DE LA RÉUNION

Vous connaissez ces îles ? Lisez ces dépliants touristiques.
Laquelle de ces deux îles vous attire le plus ? Pourquoi ?

LES D.O.M.-T.O.M. (Départements et territoires d'outre-mer)

La Guadeloupe, la Martinique, la Guyane et l'île de la Réunion sont des départements d'outre-mer français. Cela signifie qu'elles ont le même statut qu'un département français avec en plus quelques spécificités, notamment en ce qui concerne la fiscalité.

LA MARTINIQUE

La Martinique est une île de 64 km de long qui s'étale sur à peine 20 km de large, et est dominée par la montagne Pelée qui culmine à 1397 m. Elle présente une grande diversité de paysages. Le Sud est constitué de collines à la végétation peu abondante. Le Nord est montagneux. Ses premiers habitants, les indiens arawaks, l'appelaient Madinina, « l'île aux fleurs », en raison de sa végétation tropicale. Sa population est multiculturelle : européenne, africaine, hindoue, caraïbe, asiatique...

Un peu d'histoire

Avec l'arrivée de Christophe Colomb en 1502, commencent de nombreuses guerres pour la domination des Antilles. Anglais, Hollandais et Français se disputent la Martinique jusqu'à ce qu'elle devienne un département français d'outre-mer en 1946. C'est avec l'arrivée de Belain d'Esnambuc, en 1635, que s'ouvre une longue période de commerce entre les Indes Occidentales, l'Afrique et l'Europe. Commence alors la déportation de millions d'esclaves noirs vers les plantations de canne à sucre. Après l'abolition de l'esclavage en 1848, de nombreux indiens viennent remplacer la main-d'œuvre noire dans les champs de canne à sucre.

Le climat

Il y a trois saisons en Martinique. De fin décembre à mai, c'est la saison sèche pendant laquelle il fait très beau. De mi-juin à novembre c'est la saison humide, et de fin août à octobre, la période des cyclones. La température peut dépasser 28 degrés de juillet à octobre et ne descend pas au-dessous de 26 degrés durant la saison sèche. Les pluies peuvent être particulièrement abondantes. Toute l'année, le soleil se lève entre 5 h et 6 h et se couche entre 17 h 30 et 18 h.

Les plages

Elles surprennent par leur beauté et leur incroyable diversité, avec des couleurs qui vont du sable blanc lumineux au noir volcanique. L'eau est transparente et dans les fonds marins on trouve des bancs de poissons colorés. Il y a les plages tranquilles du Sud-caraïbe, bordées de cocotiers, et celles plus tumultueuses, de la côte atlantique.

Production

L'agriculture est dominée par le secteur bananier et plus de la moitié de la récolte de canne à sucre est destinée aux distilleries pour la fabrication de rhums. La Martinique a aussi une assez longue tradition de culture de l'ananas et de l'avocat. À part les fruits, on cultive aussi les fleurs qui sont exportées en Métropole et aux États-Unis ou vendues sur place aux touristes.

L'ÎLE DE LA RÉUNION

L'île est peuplée de 600 000 habitants d'origine africaine, chinoise, européenne, indienne, indonésienne et malgache.
Elle est située dans l'Océan Indien, à l'est de l'Afrique, et son littoral est constitué de 207 km de côtes au pied des montagnes, dont 30 km de plages. La Réunion est une petite île presque ronde. C'est une montagne posée sur la mer que dominent deux pics : le Piton des Neiges (3069 m) et le Piton de la Fournaise (2632 m), un volcan toujours en activité qui entre régulièrement en éruption.

Un peu d'histoire
Jusqu'au milieu du XVIIe siècle, l'île était inhabitée. C'est en 1638 que la petite île devint possession du roi de France, pour être utilisée comme prison. Les premiers colons s'y installent à partir de 1663, développant la culture du café et l'esclavage. Les esclaves étaient capturés par des négriers sur les côtes de Madagascar et d'Afrique de l'Est, puis transportés et vendus aux colons français de la Réunion. En 1848, l'esclavage est aboli. Pour cultiver la canne, on fait alors appel à une population issue des côtes sud-est de l'Inde. Ces Tamouls apportent alors leur mode de vie et leur religion, l'hindouisme. Plus tard, l'île connaîtra d'autres flux de migrations ; elle verra arriver les Indiens musulmans venus du Goujrat et des chinois. En 1946, l'île obtient le statut de Département français d'outre-mer.

Langue et culture métissées
Les Réunionnais d'aujourd'hui sont donc issus de ce métissage de cultures. Pour se comprendre, les habitants de la colonie ont forgé un créole épicé de mots d'origine malgache ou tamoule. Mais la grande majorité de la population s'exprime en français, qui est la langue officielle.
Les pratiques religieuses sont très présentes dans la vie quotidienne d'une majorité d'habitants. L'hindouisme est présent sur les façades des temples qui fleurissent dans toute l'île. Entre octobre et novembre, la fête de la lumière, le « Dipavali », réunit des milliers de fidèles. Les processions et les spectaculaires « marches sur le feu » sont organisées selon le rythme d'un calendrier ancestral.

La nature en fête
Le climat de l'île est tropical : la température sur la côte varie entre 18 et 31 degrés tandis que, en altitude, elle peut chuter à 4 degrés. Grâce à cette variété, une flore originale s'est développée sur le littoral comme dans les forêts de montagne. De nombreuses plantes, issues de tous les rivages tropicaux, ont été apportées par l'homme. On y trouve des palmiers de tous les continents et, dans les bois, des orchidées sauvages.

MÉMENTO GRAMMATICAL

PARLER DE NOS GOÛTS ET DE NOTRE MANIÈRE D'ÊTRE : LE PRÉSENT

Le présent de l'indicatif s'utilise pour parler de nos goûts et de notre manière d'être.

j'aime	la musique brésilienne / ces robes / ... (NOM)
j'adore	cuisiner / danser / ... (INFINITIF)
je déteste	

la musique brésilienne / cette robe / ... (NOM)	**me** **te** **lui** **nous** **vous** **leur** **plaît**
cuisiner / danser / ... (INFINITIF)	

Je (ne) supporte (pas)	le désordre / la musique techno / ... (NOM)

NOM SINGULIER Le bruit / la fumée de cigarette / ...	(ne) me pla**ît** (pas).
	(ne) me dérang**e** (pas).
	(ne) m'irrit**e** (pas).
	(ne) me gên**e** (pas).
NOM PLURIEL Les enfants / les chats / ...	(ne) me plais**ent** (pas).
	(ne) me dérang**ent** (pas).
	(ne) m'irrit**ent** (pas).
	(ne) me gên**ent** (pas).

Tous les verbes n'ont pas les mêmes désinences au présent.

Les verbes en **-er** comme **aimer** : **-e**/**-es**/**-e**/**-ons**/**-ez**/**-ent**
Les verbes en **-re** comme **prendre** : **-s**/**-s**/**Ø**/**-ons**/**-ez**/**-ent**
Les autres verbes : **-s**/**-s**/**-t**/**-ons**/**-ez**/**-ent**

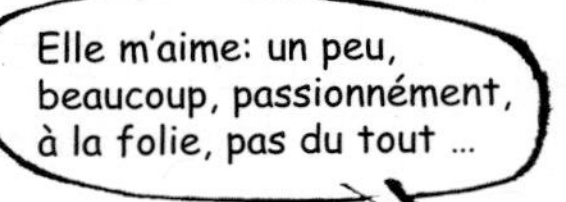

Certains verbes ont une seule racine ou base phonétique pour toutes les personnes.

AIMER

j'aim	-e	[ø]
tu **aim**	-es	[ø]
il/elle **aim**	-e	[ø]
nous **aim**	-ons	[ɔ̃]
vous **aim**	-ez	[e]
ils/elles **aim**	-ent	[ø]

Certains verbes ont deux ou trois racines ou bases phonétiques.

Les verbes avec deux bases phonétiques ont la même base pour **je**, **tu**, **il**/**elle**/**on** et **ils**/**elles**, et une autre pour **nous** et **vous**.

PRÉFÉRER (**préfèr-préfér**)	
je **préfèr**e	
tu **préfèr**es	
il/elle/on **préfèr**e	
	nous **préfér**ons
	vous **préfér**ez
ils/elles **préfèr**ent	

La même base pour les trois premières personnes et une autre base pour les personnes du pluriel.

DORMIR (**dor-dorm**)	
je **dor**s	
tu **dor**s	
il/elle/on **dor**t	
	nous **dorm**ons
	vous **dorm**ez
	ils/elles **dorm**ent

Une base pour les trois premières personnes du singulier, une base pour **nous** et **vous,** et une autre base pour **ils/elles**.

BOIRE (**boi-buv-boiv**)		
je **boi**s		
tu **boi**s		
il/elle/on **boi**t		
	nous **buv**ons	
	vous **buv**ez	
		ils/elles **boiv**ent

Au présent, certains verbes fréquemment utilisés ont des formes très différentes de leur forme à l'infinitif. C'est le cas des verbes irréguliers **être**, **avoir** et **faire**.

ÊTRE	AVOIR	FAIRE
je **suis**	**j'ai**	je **fais**
tu **es**	tu **as**	tu **fais**
il/elle/on **est**	il/elle/on **a**	il/elle/on **fait**
nous **sommes**	nous **avons**	nous **faisons**
vous **êtes**	vous **avez**	vous **faites**
ils/elles **sont**	ils/elles **ont**	ils/elles **font**

AVOIR L'AIR

Avoir l'air suivi d'un adjectif exprime une impression ou une ressemblance.

● *Quelle impression vous fait Nathalie Reine ?*
❍ *Elle **a l'air sympathique**.*

L'adjectif s'accorde avec **air**, au masculin singulier, quand l'impression vient d'un indice visuel (aspect, mine, ressemblance, physionomie de la personne).

*Tu as vu ses cernes ? Elle **a l'air** très fatigu**é** !*

Quand l'impression vient d'indices non visuels, lorsque l'expression a le sens de « doit être », l'adjectif s'accorde avec le sujet.

● *Comment tu trouves ta nouvelle collègue ?*
❍ *Eh bien, elle est très sympa, mais elle a l' air un peu désordonn**ée**.*
● *Ah oui ? Pourquoi tu dis ça ?*
❍ *Elle perd tout.*

La tendance actuelle est d'accorder avec le sujet dans tous les cas.

j'**ai l'air**	fatigué/e
tu **as l'air**	satisfait/e
il/elle/on **a l'air**	content/e
nous **avons l'air**	surpris/es
vous **avez l'air**	sympathique/s
ils/elles **ont l'air**	sérieux/ses

PLUTÔT

Plutôt indique un choix entre deux qualités opposées.

● *Il est petit ton appartement ?*
❍ *Non, il est **plutôt** grand. Il fait 100 m² et il y a un grand balcon.*

Plutôt permet aussi au locuteur de ne pas être catégorique et d'indiquer une tendance.

● *Il est comment le nouveau prof d' informatique ?*
❍ *Il a l' air sympa.*
■ *Je ne suis pas d' accord ! Il est **plutôt** bizarre !*

TELLEMENT, SI

Tellement et **si** se placent devant un adverbe ou un adjectif et expriment une très grande intensité pour le locuteur. (Pour les degrés d'intensité, voir ***Rond-point 1***, page 101.)

● *Tu aimes Brad Pitt ?*
❍ *Oui, j' adore. Il joue **si** bien et il est **tellement** beau !*

Tellement et **si** sont interchangeables.

*Il joue **tellement** bien et il est **si** beau !*

SITUER QUELQUE CHOSE ET GUIDER QUELQU'UN DANS L'ESPACE

À droite de, à gauche de, à côté de, en face de, devant, derrière, en haut de, en bas de, au bout de, au-dessus de, au-dessous de, ... servent à situer un objet ou un lieu par rapport à un autre.

● *Pardon Monsieur, pourriez-vous me dire où se trouve la poste ?*
❍ *Oui, elle est **en face de** l' église.*
● *Et où est l' église, s' il vous plaît ?*
❍ ***Derrière** l' hôtel de ville.*

Devant et **derrière** ne sont pas suivis de la préposition **de**.

Derrière ~~de~~ l' hôtel de ville.
Devant ~~de~~ la piscine municipale.

La préposition **de** se contracte devant les articles **le** et **les**. Ainsi, de + le devient **du**, et de + les, **des**.

● *Où est-ce que tu habites ?*
❍ *En face **du** cinéma Le Rex.* ❍ *En face ~~de le~~ cinéma.*

● *Où sont les toilettes, s' il vous plaît ?*
❍ *En bas **des** escaliers.* ❍ *En bas ~~de les~~ escaliers.*

À droite, à gauche, tout droit permettent de guider dans l'espace. **Jusqu'à** indique le point d'arrivée.

● *Pardon Monsieur, pour aller à la poste ?*
❍ *Allez **tout droit jusqu'à** la boulangerie puis tournez **à droite**.*

Jusqu'à se contracte devant les articles **le** et **les**. Jusqu'à + le devient jusqu'**au**, et jusqu'à + les, jusqu'**aux**.

*Allez jusqu' **au** carrefour et tournez à droite.*

*J' ai marché jusqu' **aux** galeries commerciales puis j' ai pris le métro.*

Avec un nom de pays ou de région au féminin singulier, **jusqu'à** devient **jusqu'en**.

*Les Vikings sont arrivés **jusqu'en** Amérique au VIIe siècle.*

L'IMPÉRATIF

L'impératif est un mode verbal qui permet d'exprimer un ordre, une recommandation, un conseil et pour cette raison, il s'utilise souvent pour indiquer le chemin à quelqu'un.

● *Pardon Monsieur, pour aller à la poste ?*
❍ ***Allez** tout droit jusqu' à la boulangerie puis **tournez** à droite.*

● *Comment on va chez toi ?*
❍ *C' est facile, **prends** le métro jusqu' à la Porte de Clignancourt et quand tu arrives, **passe-moi** un coup de fil.*

À l'impératif, il y a seulement trois personnes et on n'utilise pas les pronoms personnels sujets. Pour le former, on part des formes au présent.

Prends !	Ne prends pas !	Va !*	Ne va pas !	Tourne !*	Ne tourne pas !
Prenons !	Ne prenons pas !	Allons !	N'allons pas !	Tournons !	Ne tournons pas !
Prenez !	Ne prenez pas !	Allez !	N'allez pas !	Tournez !	Ne tournez pas !

* Pour les verbes en **-er,** le **-s** de la deuxième personne au présent disparaît à l'impératif.

● ***Va*** *chez ta grand-mère et surtout ne* ***parle*** *pas avec des inconnus !*

Les verbes pronominaux à l'impératif affirmatif sont suivis des pronoms toniques (**toi, nous, vous**).

SE LEVER	Lève-**toi** !	Ne te lève pas !
S'AMUSER	Amusez-**vous** !	Ne vous amusez pas !

Les verbes **être** et **avoir** ont une forme propre à l'impératif.

ÊTRE	**Sois** sage !	**Soyons** prêts !	**Soyez** aimables !
AVOIR	N'**aie** pas peur !	N'**ayons** pas peur !	**Ayez** l'air aimable !

EXPRIMER UN DÉSIR : LE CONDITIONNEL

Le conditionnel est un des modes du virtuel, c'est-à-dire que l'action est vue comme possible ou hypothétique. Le conditionnel sert donc à exprimer un désir.

● *Quelle personne célèbre est-ce que* ***tu aimerais*** *rencontrer ?*
❍ *Moi,* ***j'aimerais*** *bien rencontrer la reine d' Angleterre !*
● *Tu plaisantes ?*
❍ *Oui, bien sûr ! C' est absolument impossible !*

Le conditionnel sert aussi à demander ou à exprimer quelque chose avec prudence ou très poliment.

Est-ce que ***tu pourrais*** *me prêter ta voiture ce week-end ?*

● *Je me sens très fatigué ces derniers temps !*
❍ ***Tu devrais*** *prendre un peu de vacances et oublier le travail.*

AIMER			
j'aimer**ais**	[ɛ]	nous aimer**ions**	[iõ]
tu aimer**ais**	[ɛ]	vous aimer**iez**	[ie]
il/elle aimer**ait**	[ɛ]	ils/elles aimer**aient**	[ɛ]

Remarque : **ai** peut se prononcer [e] ou [ɛ], en fonction des mots et des aires linguistiques.

VERBES RÉGULIERS		
rencontrer	**rencontrer-**	-ais
inviter	**inviter-**	-ais
sortir	**sortir-**	-ait
préférer	**préférer-**	-ions
écrire	**écrir-**	-iez
prendre	**prendr-**	-aient

VERBES IRRÉGULIERS		
être	**ser-**	-ais -ais -ait -ions -iez -aient
avoir	**aur-**	
faire	**fer-**	
savoir	**saur-**	
aller	**ir-**	
pouvoir	**pourr-**	
devoir	**devr-**	
voir	**verr-**	
vouloir	**voudr-**	
venir	**viendr-**	
valoir	**vaudr-**	

POSER DES QUESTIONS

Il existe trois manières de poser une question.

À l'oral et dans un registre de langue familier, on exprime l'interrogation avec une **intonation montante**, en gardant la construction de la phrase affirmative.

Vous êtes français ?
Hélène est française ?

Dans un registre de langue standard, on exprime l'interrogation avec **est-ce que**, placé au début de l'interrogation.

***Est-ce que** vous êtes français ?*
***Est-ce qu'**Hélène est française ?*

Dans un registre de langue soutenu, on exprime l'interrogation avec une **inversion du verbe + pronom personnel sujet**.

***Êtes-vous** français ?*
*Hélène **est-elle** française ?*

Dans ce cas, il y a un trait d'union entre le verbe et le pronom.

Avez-vous déjà vécu en colocation ?
Aimes-tu le hip-hop ?

LES MOTS INTERROGATIFS : OÙ, QUAND, COMMENT, COMBIEN, POURQUOI

Normalement, à l'oral, l'intonation est légèrement montante sur la dernière syllabe.

Vous partez quand ? *Quand est-ce que vous partez ?* *Quand partez-vous ?*

À l'oral, dans un registre de langue familier, les mots interrogatifs se situent à la fin de la question.

***Comment** il s' appelle ?*	*Il s' appelle **comment** ?*
***Où** tu vas ?*	*Tu vas **où** ?*
***Combien** coûte le loyer ?*	*Le loyer coûte **combien** ?*

Quand la question est formulée avec **une inversion** (voir ci-dessus « Poser des questions »), on ajoute **-t-** entre deux voyelles pour faciliter la prononciation.

*Comment s' appelle-**t**-il ?* *Pourquoi étudie-**t**-elle autant ?*

LES MOTS INTERROGATIFS : QUI, QUE/QU'

Le mot interrogatif **qui** se réfère à une personne et ne s'apostrophe jamais.

● ***Qui** est-ce ?*
❍ *C' est Hervé, le frère de Céline.*

● *Avec **qui** est-ce que tu sors ce soir ?*
❍ *Avec Tim et Caroline.*

Quand le mot interrogatif **qui** est aussi sujet du verbe, il ne peut pas être à la fin de la question.

***Qui** a téléphoné ?*
***Qui** est là ?*

Le mot interrogatif **que** se réfère à une chose et s'apostrophe devant une voyelle.

● ***Qu'**est-ce que vous prenez ?*
❍ *Moi, un café.*
■ *Moi, de l' eau minérale.*

● ***Qu'**étudie-t-elle ?*
❍ *L' arabe.*

LES ADJECTIFS INTERROGATIFS : QUEL, QUELLE, QUELS, QUELLES

Les adjectifs interrogatifs s'accordent avec le nom auquel ils se rapportent.

MASCULIN SINGULIER	**Quel** âge avez-vous ?
FÉMININ SINGULIER	**Quelle** est votre formation ?
MASCULIN PLURIEL	**Quels** sont vos projets ?
FÉMININ PLURIEL	**Quelles** sont ses propositions ?

Quand **être** est le verbe, on ne peut pas formuler la question avec **quel/le/s/les + est-ce que**.

***Quel est** votre âge ?*
***Quel** âge **est-ce que** vous avez ?*

***Quelle est** votre formation ?*
***Quelle** formation **est-ce que** vous avez ?*

SI ON ALLAIT AU THÉÂTRE ?

DÉCRIRE ET ÉVALUER UN SPECTACLE, UN FILM

Pour parler d'un spectacle, l'approuver ou le critiquer, on peut utiliser plusieurs formules.

C'était + (très) ADJECTIF

● *Ce film, **c'était nul** !*
○ *Mais non ! **C'était génial** !*

*Ce spectacle de danse, **c'était très** original, tu ne trouves pas ?*

Pour insister sur l'appréciation, on peut répéter plusieurs fois **très** ou utiliser **vraiment**.

● *Ce concert de musique, c' était **vraiment** génial !*
○ *Ah, oui ? Tu trouves ?*
● *Oui, vraiment **très très** chouette.*

On peut aussi dire :

*C' était **très bien** hier en discothèque !*

Pour donner une opinion plus tempérée :

*C' était **pas mal**.*   *J' ai **bien aimé**.*

Il y avait plein de + SUBSTANTIF

Pour décrire des événements à l'oral, on peut aussi utiliser **plein de** qui est un synonyme de **beaucoup de**.

● *Tu as vu? Il y avait **plein de** monde au Festival de la Publicité !*
○ *Ouais, il y avait **plein de** gens sympas.*

PROPOSER, SUGGÉRER DE FAIRE QUELQUE CHOSE

On peut employer plusieurs structures pour proposer à quelqu'un de faire quelque chose.

Ça me/te/lui/nous/vous/leur dit de/d' + INFINITIF

● ***Ça te dit d'aller** prendre un verre ?*
○ *Oui, d' accord, à six heures au Ricot ?*

Ça **me** dit d'aller en boîte.
Ça **te** dit d'aller au ciné ?
Ça **lui** dit d'aller manger une pizza.
Ça **nous** dit de regarder la télé.
Ça **vous** dit de faire du ski ?
Ça **leur** dit de visiter un musée.

Une façon délicate de faire une proposition sans brusquer l'interlocuteur est d'utiliser le conditionnel.

__Ça te dirait de__ venir chez moi demain ?

Si on + IMPARFAIT ?

Pour inciter quelqu'un ou lui proposer de faire quelque chose avec vous.

● *Et __si on allait__ au cinéma ce soir ?*
❍ *Oui, d'accord, à quelle heure ?*

__Si on faisait__ des crêpes ?

__Si on organisait__ une fête ?

Le **on** a ici la signification de « nous », mais il se conjugue comme à la 3e personne du singulier.

Avoir envie de + INFINITIF / NOM

Cette structure sert à exprimer le désir de faire quelque chose ou le désir de quelque chose.

● ***Tu as envie de danser** ?*
❍ *Oui, allons en boîte !*

● ***J'ai envie d'une glace** au chocolat, et toi ?*
❍ *Moi, **j'ai envie d'un bon café au lait.***

ACCEPTER OU REFUSER UNE PROPOSITION

Vous acceptez sans prendre une décision complètement définitive.

● *On va au cinéma ce soir ?*
❍ *D'accord, **pourquoi pas** ?*

Vous refusez en donnant une excuse.

***Désolé, je ne peux pas,** je ne suis pas libre.*

Vous pouvez continuer en expliquant pourquoi vous refusez.

***Désolé, je ne peux pas,** je ne suis pas libre, je dois aller dîner chez...*

PRENDRE / DONNER / AVOIR RENDEZ-VOUS

Si vous prenez l'initiative du rendez-vous.

***J'ai pris rendez-vous chez le dentiste** mardi prochain à 17 heures.*

Si vous avez organisé un rendez-vous, seul ou avec quelqu'un d'autre.

***J'ai donné rendez-vous** à Martin devant le cinéma Rodin.*
***On s'est donné rendez-vous** chez lui à 8 heures.*

Vous n'êtes pas spécialement à l'origine de ce rendez-vous.

J'ai rendez-vous *avec Nadia, elle m' a téléphoné hier au soir, on se voit demain.*

Vous prenez rendez-vous pour bénéficier d'un service.

J'ai rendez-vous *chez le coiffeur la semaine prochaine.*

LES MOMENTS DE LA JOURNÉE

La journée est découpée en trois parties : **le matin, l'après-midi** et **le soir.**

Samedi matin, *on joue au basket et* ***samedi après-midi,*** *on va au festival de ciné.*

Quand on utilise **matinée** et **soirée,** on insiste sur la durée qui compose cet espace temporel. C'est la même chose avec **jour/journée, an/année.**

J' ai passé ***la matinée*** *à faire le ménage.* (J'ai fait la lessive, puis j'ai nettoyé les vitres, et enfin, j'ai repassé).

INDIQUER UN LIEU

à Berlin
au centre ville
dans le quartier chinois
pas loin du métro Saint Michel
à côté de la boutique de vêtements/**du** bar Les trois vents
(tout) près de la fac/**du** port
(juste) à côté de la gare/**du** stade
(juste) en face de la poissonnerie/**du** magasin de sport
devant le restaurant
au coin de la rue Cigor et **de** la rue Dumont
sur la place du marché
à la plage/**au** café des sports/**à l'**hôtel
au 3, rue Victor Hugo

● *On va où ce soir ?*
❍ *Chez Arsène,* ***au 10, rue*** *de la Loupette.*

● *Je connais une librairie géniale.*
❍ *Elle est où ?*
● *Place Wilson,* ***tout près de*** *la faculté !*

L'IMPARFAIT

L'imparfait situe une action au passé sans signaler ni le début ni la fin de l'action. L'imparfait sert aussi à parler de nos habitudes dans le passé, à décrire des personnes, des choses, une action en cours dans le passé ou à décrire les circonstances qui ont entouré un événement.

Pour obtenir la base de l'imparfait, il faut partir de la première personne du pluriel du présent.

La mode ce n'est plus ce que c'était.

PRÉSENT	IMPARFAIT	
nous **dorm**ons	je dorm**ais**	[ɛ]
	tu dorm**ais**	[ɛ]
	il/elle/on dorm**ait**	[ɛ]
	nous dorm**ions**	[iõ]
	vous dorm**iez**	[ie]
	ils/elles dorm**aient**	[ɛ]

Attention ! ÊTRE

j'**étais** | nous **étions**
tu **étais** | vous **étiez**
il/elle/on **était** | ils/elles **étaient**

Emploi

L'imparfait sert à décrire des habitudes dans le passé.

PRÉSENT	IMPARFAIT
Tous les matins, **elle se lève** à 6 heures.	À cette époque-là, **elle se levait** tous les matins à 6 heures.
À notre époque, les gens **travaillent** seulement 8 heures par jour.	Autrefois, les gens **travaillaient** 14 heures par jour.

Il sert aussi à décrire une action en cours dans le passé.

● *Que **faisiez**-vous hier à 17 heures ?*
❍ *Moi, je **regardais** la télévision.*
■ *Moi, je **prenais** un café avec un copain.*

Dans ce cas, on peut dire aussi :

*J'**étais en train** de regarder la télévision.*
*J'**étais en train** de prendre un café avec un copain.*

Il sert également à décrire des choses ou des personnes dans le passé.

● *Comment **était**-il ? Vous pouvez le décrire ?*
❍ *Il **était** grand. Il **portait** une veste marron et un pantalon noir.*

● *Décrivez-moi ce que vous avez vu.*
❍ *Et bien, **c'était** une sorte de sphère, très lumineuse. Il y **avait** une petite porte...*

Enfin, l'imparfait sert à décrire les circonstances qui situent ou expliquent un événement.

*J' **étais** fatigué alors je me suis couché tôt.*
*De la fumée **sortait** du moteur alors je me suis arrêté.*
*Je ne suis pas venu en cours parce que j' **étais** malade.*

LE PASSÉ COMPOSÉ

Le passé composé sert à raconter un événement du passé qui est déjà achevé.

*Lundi 5 juillet, deux hommes **ont attaqué** la superette de la rue des Rosiers. Ils **ont menacé** les clients et le personnel avec des armes à feu et ils **ont emporté** l' argent de la caisse. La police les **a arrêtés** le lendemain matin.*

Formation

Le passé composé est formé d'un auxiliaire (**avoir** ou **être**) au présent de l'indicatif, suivi du participe passé du verbe. La plupart des verbes se conjuguent avec l'auxiliaire **avoir**.

● *Qu' est-ce que vous **avez fait** samedi ?*
❍ *Moi, j' **ai étudié** toute la journée.*
■ *Moi, j' **ai fait** les courses et le ménage.*
❐ *Moi, je **suis allé** au cinéma.*

ÉTUDIER		ALLER	
j'**ai** tu **as** il/elle/on **a** nous **avons** vous **avez** ils/elles **ont**	**étudié**	je **suis** tu **es** il/elle/on **est** nous **sommes** vous **êtes** ils/elles **sont**	**allé/e/s/es**

Les verbes qui expriment la transformation du sujet d'un état à un autre ou d'un lieu à un autre, se conjuguent avec l'auxiliaire **être**. Les verbes pronominaux se construisent aussi avec le passé composé.

*Je **me suis réveillée** très tôt ce matin.*
*Ils **se sont mariés** à Amsterdam.*

SE RÉVEILLER	
je **me suis** tu **t'es** il/elle/on **s'est** nous **nous sommes** vous **vous êtes** ils/elles **se sont**	**réveillé/e/s/es**

Enfin, l'auxiliaire **être** s'utilise également dans le cas de certains verbes intransitifs, c'est-à-dire qui n'acceptent pas de complément d'objet, de leurs contraires sémantiques et, généralement, de leurs dérivés : **naître, mourir, venir, devenir, revenir, apparaître, arriver, partir, entrer, aller, rester, tomber, demeurer...**

● *Tu **es** finalement **parti** en vacances ?*
❍ *Non, je **suis resté** tranquillement chez moi.*

Attention ! Certains verbes, également intransitifs et qui signalent un déplacement, se conjuguent avec **avoir**.

*J' **ai couru** toute la journée !*
*Nous **avons** beaucoup voyagé l' été dernier.*
*Ils **ont marché** pendant des heures.*

Je ~~suis~~ couru toute la journée !
Nous ~~sommes~~ beaucoup voyagé.
Ils ~~sont~~ marchés pendant des heures.

Les verbes **monter, descendre, sortir, passer, retourner, rentrer** se conjuguent avec l'auxiliaire **être** quand ils sont intransitifs.

*L' ascenseur ne marche pas, je **suis monté** à pied.*
*Samedi soir, je **suis sorti** avec mes amis.*

Mais ils se conjuguent avec l'auxiliaire **avoir** quand ils sont transitifs.

*Xavier est vraiment en forme, il **a monté les escaliers** en courant.*

● *Est-ce que tu **as sorti** le chien ?*
❍ *Oui, je **l'ai sorti** il y a une demi-heure.*

À la forme négative

À la forme négative, les particules **ne** et **pas** encadrent l'auxiliaire.

*Je **ne** me suis **pas** réveillé ce matin. Mon réveil **n'** a **pas** sonné.*
*C' est un film horrible, nous **ne** sommes **pas** restés jusqu' à la fin.*

En langue orale, **ne** disparaît souvent.

● *Vous êtes sortis hier ?*
❍ *Non, on est **pas** sorti.*

LES PARTICIPES PASSÉS

À l'écrit, il y a huit terminaisons différentes de participe passé, mais à l'oral, il n'y en a que cinq.

-é	[e]	Lulu et moi, on s'est **rencontré** à Londres.
-i	[i]	Je n'ai pas **fini** mon travail.
-it	[i]	Julien a **conduit** toute la nuit.
-is	[i]	Ils ont **pris** le train de nuit.
-ert	[er]	Mes amis m'ont **offert** un super cadeau.
-u	[y]	Vous avez **lu** le dernier roman de Nothomb ?
-eint	[ɛ̃]	Qui a **peint** la Joconde ?
-aint	[ɛ̃]	Un client s'est **plaint** au directeur de la revue.

Attention ! Faites bien la différence entre le présent et le passé composé.

je [ə] *finis / **j' ai*** [e] *fini*   ***je*** [ə] *fais / **j' ai*** [e] *fait*   ***je*** [ə] *dis / **j' ai*** [e] *dit*

L'accord

Quand un verbe se conjugue avec l'auxiliaire **être,** le participe passé s'accorde normalement avec le sujet.

*Alain est **rentré** cette nuit à une heure du matin.*	(MASCULIN SINGULIER)
*Elle est **rentrée** à 8 heures chez elle hier soir.*	(FÉMININ SINGULIER)
*René et Thierry sont **rentrés** à 11 heures du soir.*	(MASCULIN PLURIEL)
*Estelle et Julie sont **rentrées** à 10 heures du soir.*	(FÉMININ PLURIEL)

Quand un verbe se conjugue avec l'auxiliaire **avoir,** le participe passé ne s'accorde pas avec le sujet. Mais quand l'objet direct est placé avant le verbe, le participe passé s'accorde avec cet objet direct.

● *Tu as sorti le chien ?*
❍ *Oui, je **l'** ai **sorti** il y a une demi-heure.* (MASCULIN SINGULIER)

● *Elle est jolie cette chemise !*
❍ *Oui, c' est **une chemise** en soie que j' ai **achetée** en Chine.* (FÉMININ SINGULIER)

● *Et les malfaiteurs ?*
❍ *Ce matin, la police **les** a **arrêtés**.* (MASCULIN PLURIEL)

● *Est-ce que tu as vu Hélène et sa sœur ?*
❍ *Non, je ne **les** ai pas **vues**.* (FÉMININ PLURIEL)

La place des adverbes

Les adverbes se placent normalement après le verbe conjugué. Par conséquent, au passé composé, les adverbes se placent après l'auxiliaire.

Il (n') **a** (pas)	**encore** **beaucoup** **trop** **assez** **bien** **mal**	travaillé dormi bu

SITUER DANS LE TEMPS

Hier, **Hier matin,** **Hier après-midi,** **Hier soir,** **Avant-hier,**	je suis allé au cinéma.

Pour préciser qu'il s'agit d'un moment de la journée en cours, on utilise les adjectifs démonstratifs **ce, cet** et **cette**.

ce matin
ce midi
cet après-midi
ce soir
cette nuit

● *Quand est-ce qu'elle est partie ?*
❍ ***Ce matin.***
● *Et quand est-ce qu'elle va rentrer ?*
❍ ***Cette nuit.***

***Dimanche / lundi / mardi / mercredi**... j'ai joué au football.*

Pour annoncer l'heure exacte à laquelle quelque chose a (eu) lieu, on utilise la préposition **à**.

● *À quelle heure commence le film ?*
❍ ***À** dix heures trente.*

Heures approximatives

environ = about

● *À quelle heure est-ce que vous êtes sorti hier soir ?*
❍ ***Vers** 19 heures.*
*À 20 heures **environ**.*
***Il était environ** minuit.*
***Il devait être** 5 heures et demi.*

LA SUCCESSION DES ÉVÉNEMENTS : D'ABORD, APRÈS, ENSUITE, PUIS...

Des mots comme **d'abord**, **ensuite**, **puis**, **après** et **enfin** indiquent la succession des événements dans un récit.

***D'abord**, j'ai pris mon petit déjeuner.*
***Ensuite**, je me suis douché.*
***Puis**, je me suis habillé.*
***Après**, je suis sorti.*
***Et puis**, j'ai pris l'autobus.*
***Enfin**, je suis arrivé au travail.*

Avant de parler, il faut tourner sept fois sa langue dans sa bouche.

Un moment antérieur

\+ NOM
***Avant** les examens, j'étais très nerveux.*

\+ INFINITIF
***Avant de** me coucher, je me suis douché.*

Un moment postérieur

\+ NOM
***Après** le déjeuner, ils ont joué aux cartes.*

\+ INFINITIF PASSÉ
***Après avoir déjeuné**, ils ont joué aux cartes.*

L'infinitif passé se forme avec l'auxiliaire **être** ou **avoir** à l'infinitif suivi du participe passé du verbe.

*J'ai décidé de devenir médecin **après avoir vu** le film « Johnny s'en va-t-en guerre » de Dalton Trumbo.*

***Après être montés** jusqu'au sommet du Mont-Blanc, à 4 807 mètres d'altitude, ils sont redescendus jusqu'à Chamonix.*

SE RAPPELER, SE SOUVENIR

Ces deux verbes sont synonymes.

● *Tu crois qu' elle a oublié notre rendez-vous ?*
❍ *Impossible, elle a une mémoire incroyable, elle* ***se rappelle/souvient*** *toujours de tout.*

Rappeler a deux racines ou bases phonétiques (**rappell-rappel**). Devant **e muet,** le verbe double le **l.**

SE RAPELLER (**rappell-rappel**)	
je me rappel**le** [ø]	
tu te rappel**les** [ø]	
il/elle/on se rappel**le** [ø]	
	nous nous rappel**ons** [ɔ̃]
	vous vous rappel**ez** [e]
ils/elles se rappel**lent** [ø]	

Souvenir a trois racines ou bases phonétiques (**souvien-souven-souvienn**) : une base pour les trois premières personnes du singulier, une base pour **nous** et **vous**, et une autre base pour **ils/elles**.

SE SOUVENIR (**souvien-souven-souvienn**)		
je me **souviens**		
tu te **souviens**		
il/elle/on se **souvient**		
	nous nous **souven**ons	
	vous vous **souven**ez	
		ils/elles se **souvienn**ent

ÇA SERT À TOUT !

LA MATIÈRE

Pour indiquer en quelle matière est fait un objet, on utilise la préposition **en**.

● *C' est **en** quoi ?*
❍ *C' est **en** plastique.*

un sac **en** **papier**
tissu
cuir
plastique

une boîte **en** **carton**
bois
porcelaine
fer
verre

TAILLES, FORMES ET QUALITÉS

C'est + ADJECTIF.

*C' est **petit**.*
***grand**.*
***plat**.*
***long**.*
***rond**.*
***carré**.*
***rectangulaire**.*
***triangulaire**.*

L'USAGE : ÇA SERT À..., C'EST UTILE POUR..., ÇA PERMET DE... + INFINITIF

● ***À quoi ça sert ?***
❍ ***À enlever** les tâches des vêtements.*

***Ça sert à** écrire.* ***C'est utile pour** ouvrir une bouteille.* ***Ça permet d'**écouter de la musique.*

LE MODE DE FONCTIONNEMENT

Ça marche avec de l'/de la/du/des + SUBSTANTIF.

Ça marche avec ***de l'**essence.*
***de la** vapeur.*
***du** gaz.*
***des** piles.*

On peut spécifier aussi le mode de fonctionnement en utilisant le singulier.

*Ça marche **à l'**essence / **à la** vapeur / **au** gaz.*
Ça marche ~~aux~~ piles.

LES PRONOMINAUX PASSIFS

Pour ne pas préciser qui fait l'action, on peut utiliser une forme pronominale.

Ça ***se lave*** *facilement / en machine.* (= on peut laver ça facilement / en machine.)
Ça ***se mange.*** (= on peut manger ça.)

Les pronominaux passifs s'emploient également pour décrire un processus qui peut se faire sans l'intervention d'une personne. C'est souvent une manière d'exprimer qu'un objet est très facile à utiliser.

● *C' est difficile à mettre en marche ?*
❍ *Non, ça* ***se met*** *en marche tout seul. Tu appuies sur ce bouton, c' est tout.*

● *Comment on appelle une porte qui* ***s'ouvre*** *toute seule ?*
❍ *Une porte automatique.*

PRONOMS RELATIFS : QUI ET QUE

Qui et **que** sont des pronoms relatifs. Ils introduisent des informations supplémentaires sur l'objet ou la personne placés devant eux.

● *Qu' est-ce que tu veux pour ton anniversaire ?*
❍ *Je veux une voiture* ***qui*** *se transforme en robot intergalactique.*

● *C' est quoi un baladeur ?*
❍ *C' est un petit appareil* ***qu'****on porte sur soi pour écouter de la musique.*

Qui représente le sujet grammatical.

C' est un objet ***qui*** *est rectangulaire,* ***qui*** *marche avec l' électricité et* ***qui*** *sert à griller le pain.*

Que représente le complément d'objet direct.

C' est un objet ***que*** *vous portez dans votre sac ou dans votre poche et* ***que*** *vous devez éteindre en classe, au cinéma ou dans un avion.*

LE FUTUR

Emploi

Le futur sert à formuler des prévisions ou à faire des prédictions.

Demain, il ***fera*** *soleil sur tout le pays.*
Dans 30 ans, nous ***marcherons*** *sur Mars.*
Au siècle prochain, tout le monde ***parlera*** *chinois.*
Bientôt, nous ***habiterons*** *sous la mer.*

Le futur sert aussi à faire une promesse.

Demain, je ***viendrai*** *te chercher à 16 heures 30.*
Cet appareil vous ***facilitera*** *la vie.*

Il s'utilise pour demander un service.

Tu ***pourras*** *acheter le pain, s' il te plaît ?*

Enfin, le futur sert aussi à donner un ordre ou une consigne.

Vous prendrez *un cachet trois fois par jour après chaque repas.*

Situer dans le temps : périodes et dates

ce soir
demain
après-demain
dans deux jours / une semaine / 5 ans / quelques années / le futur
lundi (prochain)
la semaine prochaine
le mois prochain
l'été prochain
l'année prochaine
le 24 juin

Bientôt, prochainement et **un jour** annoncent que quelque chose se réalisera dans le futur mais sans donner une indication temporelle précise.

*En vente **prochainement** dans votre supermarché, « l'essuie-tout magique » !*
***Bientôt**, il y aura des villes sous la mer.*

● *J'aimerais bien aller en vacances aux Antilles.*
❍ *On ira **un jour**. Je te le promets.*

Formation

Verbes réguliers.

MANGER	**manger-**	-ai -as -a -ons -ez -ont
ÉTUDIER	**étudier-**	
VOYAGER	**voyager-**	
SORTIR	**sortir-**	
DORMIR	**dormir-**	
FINIR	**finir-**	

Verbes qui se terminent par **-re**.

BOIRE	**boir-**	-ai -as -a -ons -ez -ont
ÉCRIRE	**écrir-**	
PRENDRE	**prendr-**	
ENTENDRE	**entendr-**	

Les verbes en **-eter, -eler, -ever, -ener** ou **-eser** redoublent la consonne ou prennent un accent grave devant un **e** muet.

je/j'	jett**er**- a**pp**eller- ach**è**ter- p**è**ser- l**è**ver- m**è**ner-	-ai
tu		-as
il/elle/on		-a
nous		-ons
vous		-ez
ils/elles		-ont

Les verbes en **-oyer, -uyer** changent l'**y** en **i** devant un **e** muet.

je/j'	nettoier- essuier-	-ai
tu		-as
il/elle/on		-a
nous		-ons
vous		-ez
ils/elles		-ont

Les verbes irréguliers ont un radical très différent de celui de leur infinitif.

<table>
<tr><td>je/j'</td><td rowspan="6">(ÊTRE) ser-
(AVOIR) aur-
(FAIRE) fer-
(SAVOIR) saur-
(ALLER) ir-
(DEVOIR) devr-
(POUVOIR) pourr-
(VOIR) verr-
(ENVOYER) enverr-
(MOURIR) mourr-
(VOULOIR) voudr-
(VENIR) viendr-
(VALOIR) vaudr-</td><td>-ai</td></tr>
<tr><td>tu</td><td>-as</td></tr>
<tr><td>il/elle/on</td><td>-a</td></tr>
<tr><td>nous</td><td>-ons</td></tr>
<tr><td>vous</td><td>-ez</td></tr>
<tr><td>ils/elles</td><td>-ont</td></tr>
</table>

EXPRIMER LA CAUSE : GRÂCE À

Grâce à + SUBSTANTIF exprime une cause considérée comme positive.

<table>
<tr><td rowspan="5">Grâce</td><td>à Internet,</td><td rowspan="5">nous ne nous sentons jamais seuls !</td></tr>
<tr><td>à la télévision,</td></tr>
<tr><td>à l'ordinateur personnel,</td></tr>
<tr><td>au téléphone,</td></tr>
<tr><td>aux satellites,</td></tr>
</table>

Grâce à Internet, je ne suis pas obligée d'aller au bureau tous les jours.

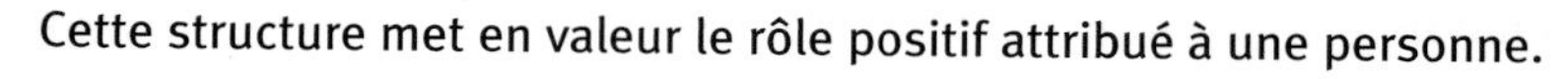

Cette structure met en valeur le rôle positif attribué à une personne.

__Grâce à__ mes parents, j'ai pu partir étudier aux États-Unis.

Grâce à peut aussi être suivi des pronoms toniques pour exprimer l'aide apportée **par quelqu'un.**

Grâce à __moi/toi/lui/nous/vous/eux/elles,__ Marie a réussi son examen.

EXPRIMER LE BUT

INFINITIF
***Pour ne pas vous fatiguer**, utilisez l'ascenseur !*

INFINITIF
***Pour ne plus penser** à vos problèmes, partez en vacances à la Réunion !*

EXPRIMER LA FACILITÉ OU LA SIMPLICITÉ D'UTILISATION : SUFFIR (DE)

*Pour obtenir une surface brillante, un simple geste **suffit**.* (= un simple geste est suffisant)

INFINITIF
*Pour ouvrir la porte, **il suffit d'**appuyer sur le bouton rouge.*

LES PRONOMS COD ET COI

Quand on parle de quelqu'un ou de quelque chose qui a déjà été mentionné ou bien est identifiable grâce au contexte, pour ne pas le répéter, on utilise les **pronoms compléments d'objet direct** (COD) et **compléments d'objet indirect** (COI).

Le complément d'objet direct (COD)

Le COD représente la chose ou la personne sur laquelle s'exerce l'action exprimée par le verbe.

● *Tu regardes beaucoup **la télévision** ?*
❍ *Non, je **la** regarde surtout le week-end.*

● *Tu écoutes **le professeur** quand il donne des explications ?*
❍ *Bien sûr que je **l'écoute** !*

● *Vous aimez **les huîtres** ?*
❍ *Euh non, je ne **les** digère pas très bien.*

*Un véritable ami, c' est quelqu' un qui **nous** écoute, **nous** comprend et **nous** aide.*

Il (ne)	**me/m'**	regarde (pas)
	te/t'	écoute (pas)
	le/l'	comprend (pas)
	la/l'	aide (pas)
	nous	aime (pas)
	vous	...
	les	

Afin d'identifier le COD dans une phrase, on peut poser des questions avec **que** ou **qui**.

● ***Qu'est-ce que** tu regardes ?*
❍ ***La télévision.***

● *Tu écoutes **qui** ?*
❍ ***Le professeur.***

Le complément d'objet indirect (COI)

Le COI est la personne ou la chose qui a un rôle de destinataire de l'action que fait le sujet. Il est introduit par une préposition.

● *Qu' est-ce que vous offrez **à Charlotte** pour son anniversaire ?*
❍ *On **lui** offre un pull-over.*

● *Alors, qu' est-ce qu' il **t'**a dit ?*
❍ *Il ne **m'**a rien dit.*

● *Est-ce que tu as téléphoné **à tes parents** ?*
❍ *Oui, je **leur** ai téléphoné ce matin.*

Afin d'identifier le COI dans une phrase, on peut poser des questions avec **à qui**.

● *Alors, qu' est-ce qu' il a dit **à qui** ?*
❍ ***À toi.***

● *Tu as téléphoné **à qui** ?*
❍ ***À tes parents.***

Il (ne)	**me/m'**	téléphone (pas)
	te/t'	offre (pas)
	lui	dit (pas)
	nous	explique (pas)
	vous	parle (pas)
	leur	...

Les pronoms COD et COI se placent devant le verbe dont ils sont compléments,

- *Et ton travail ?*
- ❍ *Je peux **le** faire demain.* — *Je ~~le~~ peux faire demain.*

- *Tu as parlé à Marie-Laure ?*
- ❍ *Je vais **lui** parler ce soir.* — *Je ~~lui~~ vais parler ce soir.*

sauf quand le verbe est à l'impératif affirmatif. Dans ce cas, le pronom complément est à la forme tonique (à exception de la troisième personne du singulier et du pluriel) et se place derrière le verbe.

*Regarde-**moi** quand je te parle !*

*Regarde-**la** bien ! Tu ne trouves pas qu' elle ressemble à mamie Marguerite ?*

- *J' ai invité Yvan et Juliette samedi soir.*
- ❍ *Ah ! Très bien. Explique-**leur** bien le chemin, parce que ce n' est pas facile d' arriver jusqu' ici.*

À l'oral, on utilise souvent les pronoms COD et COI avant même d'avoir mentionné l'élément auquel ils se réfèrent.

*Alors, tu **les** as faits **tes devoirs** ?*

*Qu' est-ce que tu **lui** as acheté **à maman** pour son anniversaire ?*

Avec certains verbes, les pronoms qui représentent une personne sont toujours à la forme tonique : **moi, toi, lui, elle, nous, vous, eux, elles**.

- *Tes parents te manquent beaucoup ?*
- ❍ *Oui, je pense souvent **à eux**.* — *Je ~~leur~~ pense souvent.*

- *J' ai rencontré Elisabeth au supermarché.*
- ❍ *Ah justement, je pensais **à elle** ce matin.* — *Je ~~lui~~ pensais ce matin.*

- *Je vais faire une course, tu veux bien t' occuper **de ton petit frère** ?*
- ❍ *D' accord, je m' occupe **de lui**.* — *Je me ~~lui~~ occupe.*

À l'oral, on trouve parfois les pronoms **y** et **en** pour reprendre des noms de personnes.

- *Tu penses souvent **à tes parents** ?*
- ❍ *Oui, j' **y** pense souvent.*

- *Je vais faire une course, tu veux bien t' occuper **de ton petit frère** ?*
- ❍ *D' accord, je m' **en** occupe.*

LES PRONOMS COMPLÉMENTS Y ET EN

Y et **en** représentent une chose ou une idée.

● *Tu as pensé à acheter un cadeau à papa ? C' est son anniversaire demain.*
❍ *Oui, j' **y** ai pensé.*

● *Est-ce que vous pourriez vous occuper de mes plantes pendant mon absence ?*
❍ *Bien sûr. Partez tranquille, je m' **en** occuperai.*

Y et **en** expriment souvent la situation dans l'espace, mais avec des nuances différentes.

Y reprend un nom de lieu où l'on va ou bien où l'on est.

● *J' irai **à Venise** pour le week-end.*
❍ *Venise ? J' aimerais bien **y** aller un jour !*

● *Tu habites **à Strasbourg** ?*
❍ *Oui, j' **y** habite depuis deux ans.*

En reprend le nom d'un lieu d'où l'on vient.

● *Tu vas **à la piscine** ?*
❍ *Non, j' **en** viens.*

En reprend la notion de quantité.

● *Est-ce qu' il reste **du fromage** dans le frigo ?*
❍ *Non, il n' y **en** a plus.*

● *Pour être dompteur, il faut avoir **du sang-froid**.*
❍ *Oui, il **en** faut beaucoup.*

S'il s'agit d'un nom dénombrable déterminé par **un/e**, la reprise par le pronom **en** demandera en écho l'emploi de **un/une**, du **nombre** approprié, ou d'un **quantificateur**.

● *Pardon Monsieur, est-ce qu' il y a **un parking** par ici ?*
❍ *Oui, il y **en** a **un** sur la place du marché et il y **en** a **deux** autres dans la rue Honoré de Balzac.*

● *Qu' est-ce qu' il pleut ! Tu peux me prêter **un parapluie** jusqu' à demain ?*
❍ *Oui, pas de problèmes. J' **en** ai **plusieurs**.*

FAIRE UNE HYPOTHÈSE DANS LE PRÉSENT

Pour exprimer une action hypothétique dans le présent, on utilise **si** + IMPARFAIT, CONDITIONNEL PRÉSENT.

● ***Si vous gagniez** beaucoup d' argent à la loterie, qu' est-ce que vous **feriez** ?*
❍ *Je **ferais** le tour du monde.*
● *Moi, **j'arrêterais** de travailler.*

● ***Si vous étiez** un animal, quel animal **seriez**-vous ?*
❍ *Moi, **je serais** un éléphant.*

LE CONDITIONNEL

Pour former le conditionnel, il suffit de prendre la base du futur simple et d'ajouter les désinences de l'imparfait.

Être

FUTUR	CONDITIONNEL		
ser-	je **ser-**	**-ais**	[ɛ]
	tu **ser-**	**-ais**	[ɛ]
	il/elle/on **ser-**	**-ait**	[ɛ]
	nous **ser-**	**-ions**	[iõ]
	vous **ser-**	**-iez**	[ie]
	ils/elles **ser-**	**-aient**	[ɛ]

Le conditionnel d'autres verbes

ÉTUDIER	**étudier-**	
AIMER	**aimer-**	
RENCONTRER	**rencontrer-**	
INVITER	**inviter-**	
SORTIR	**sortir-**	**-ais**
DORMIR	**dormir-**	
PRÉFÉRER	**préfèrer-**	**-ais**
ÉCRIRE	**écrir-**	
PRENDRE	**prendr-**	**-ait**
AVOIR	**aur-**	**-ions**
FAIRE	**fer-**	
SAVOIR	**saur-**	**-iez**
ALLER	**ir-**	
POUVOIR	**pourr-**	**-aient**
DEVOIR	**devr-**	
VOIR	**verr-**	
VOULOIR	**voudr-**	
VENIR	**viendr-**	

Pour d'autres usages du conditionnel, voir **Mémento grammatical** de l'Unité 1.

PARLER DE NOS QUALITÉS

Avoir de la/du/de l'

● *Est-ce que vous* ***avez de la patience*** *?*
○ *Moi oui, j' en ai beaucoup.*
● *Moi non, je n' en ai pas du tout. Je ne pourrais pas m' occuper d' enfants.*

● *Il* ***a de l'imagination****, cet enfant !*
○ *Oui, il en a même trop !*

Attention à la forme négative.

Je n' ai pas ***de*** *patience.*
Il n' a pas ***de*** *sang-froid.*
Nous n' avons pas ***d'****imagination.*

Je n' ai pas ~~de la~~ patience.
Il n' a pas ~~du~~ sang-froid.
Nous n' avons pas ~~de l'~~ imagination.

Manquer de/d'

L'absence d'une qualité peut s'exprimer aussi avec le verbe **manquer de/d'** + SUBSTANTIF.

● *Est-ce que tu as de la patience ?*
❍ *Non, je* ***manque de patience.***

● *Tu l' imagines écrivain ?*
❍ *Non, pas du tout. Il* ***manque d'*** *imagination.*
● *Oui, et il* ***manque de*** *constance aussi.*

AVOIR PEUR

Avoir peur du/de la/de l'/des + SUBSTANTIF.

Joana ***a peur du vide.***
Éric ***a peur de la pollution.***
Danièle ***a peur de l'eau.***
Alain ***a peur des chiens.***
Maman ***a peur de tout.***
Et moi, je n' ***ai peur de rien*** *!*

Avoir peur de + INFINITIF.

● *Pourquoi tu ne viens pas en voiture ? C' est beaucoup plus pratique.*
❍ *Oui, je sais mais* ***j'ai peur de me perdre*** *dans les rues de Paris.*

COMPARER

Comme

Comme établit des similitudes entre deux choses ou deux êtres.

● *Tu ne trouves pas que Rémy ressemble vraiment beaucoup à son père ?*
❍ *Totalement ! Il est exactement* ***comme*** *Christophe quand il était petit. La copie conforme de son père.*

Les sumotoris ont l' air obèse mais ils sont souples ***comme*** *des chats et forts* ***comme*** *des bœufs !*

Attention ! Ne confondez pas **comme** et **comment.**

● *Je ne sais pas* ***comment*** *ouvrir cette machine.* (= de quelle manière ?)
❍ *C' est facile ! Tu fais* ***comme*** *ça et hop, c' est ouvert !* (= de cette manière)

Comparer une qualité

L'adjectif qualificatif se place entre les deux marqueurs de la comparaison.

Lucien est ***plus*** *patient* ***que*** *Philippe.*
Kevin est ***aussi*** *sérieux* ***que*** *Sybille.*
Vincent est ***moins*** *dynamique* ***que*** *Nathalie.*

Les adjectifs **bon** et **mauvais** ont une forme particulière.

bon/ne/s → **meilleur/e/s**
mauvais/e/es → **pire/s**

À mon avis, Sybille est ***meilleure que*** *Vincent pour ce travail.*

On peut nuancer la comparaison avec **un peu, beaucoup, bien...**

*Vincent est **un peu/beaucoup/bien moins** dynamique **que** Nathalie.*

*Sybille est **un peu/bien meilleure que** Vincent pour ce travail.*
Sybille est ~~beaucoup~~ meilleure que Vincent.

Comparer une quantité

On peut comparer des quantités (exactes ou approximatives) pour démontrer la supériorité, l'égalité ou l'infériorité.

Paul a beaucoup de patience.
Yannick a très peu de patience.

	plus de
Paul a	***autant de** patience **que** Janick.*
	moins de

On peut préciser une comparaison avec **un peu**, **beaucoup**, **bien**, **six fois**, **mille fois**, etc.

*Paul a **un peu** / **beaucoup** / **bien** / **mille fois** plus de patience que Yannick.*

Comparer une action

- *Vincent travaille **moins que** Kevin.*
- *Non, je crois qu' ils travaillent **autant** l' un **que** l' autre.*

	plus
Je travaille	***autant que** Julien.*
	moins

Plus, autant et **moins** se placent normalement après le verbe.

*En été, on dort **moins** qu' en hiver.* *En été on ~~moins~~ dort...*

Avec un temps composé, il est possible de trouver le premier élément de la comparaison juste après l'auxiliaire ou bien après le participe passé.

*Vincent **a** plus / autant / moins **travaillé** que Kevin l' année dernière.*
*Vincent **a travaillé** plus / autant / moins que Kevin...*

TUTOYER ET VOUVOYER

En fonction de l'interlocuteur, les Français tutoient ou vouvoient. **Tu** exprime **une relation de familiarité** et s'utilise pour parler aux enfants, aux membres de la famille, aux amis et, dans certains secteurs professionnels, aux collègues de même niveau hiérarchique.

Vous s'utilise pour marquer **le respect** ou **la distance**.

Les statuts des interlocuteurs ne sont pas toujours égaux et souvent l'un des interlocuteurs tutoie tandis que l'autre vouvoie, c'est le cas des professeurs et des élèves au collège et ou lycée.

Quand des locuteurs francophones veulent passer au tutoiement dans une situation où les conventions linguistiques exigent normalement le vouvoiement, ils le proposent clairement.

On se tutoie ?

- *Tu peux me tutoyer, si tu veux.*
- *Bien, Monsieur le Directeur !*

L'EXPRESSION DE L'OPINION

Se positionner

Pour signaler sa position par rapport à un thème, on peut dire :

Personnellement, | je suis **pour le/la/les**
 je suis **en faveur du/de la/de l'/des** + NOM <br> je suis **contre le/la/les**

- ● *Qu'est-ce que vous pensez de l'interdiction de fumer dans les restaurants ?*
- ❍ ***Personnellement, je suis en faveur de l'****interdiction de fumer dans les restaurants.*
- ■ *Moi aussi,* ***je suis pour*** *(l'interdiction de fumer dans les restaurants).*
- ❒ *Moi,* ***je suis contre*** *(l'interdiction de fumer dans les restaurants).*

❒ *Je suis ~~en~~ contre de l'interdiction de fumer dans les restaurants.*

Présenter son opinion

Pour introduire une opinion, on peut utiliser différentes formules.

À mon avis,
Pour moi,
D'après moi,
Selon moi, — le français est plus facile que l'anglais.

Je pense que/qu'...
Je crois que/qu'... + INDICATIF

Je pense que le français **est** plus facile que l'anglais.

Je ne crois pas que/qu'...
Je ne pense pas que/qu'... + SUBJONCTIF

Je ne crois pas que le français **soit** plus facile que l'anglais.

Exprimer son accord ou son désaccord

Face aux opinions des autres, on peut exprimer ouvertement son accord ou son désaccord.

Je (ne) suis (pas) d'accord avec | ce que dit Marcos / ce que vous dites.
 toi/lui/elle/vous/eux/elles.
 cela/ça.

Je (ne) partage (pas) | l'opinion de Sandra.
 ton/votre/son/leur point de vue.
 l'avis de monsieur Delmat.

Certains adverbes comme **pas du tout, absolument, totalement, tout à fait** permettent de nuancer l'expression de l'accord ou du désaccord.

Si on souhaite exprimer son adhésion totale, on peut dire :

Oui, vous avez ***tout à fait*** *raison.*
Oui, tu as ***totalement*** *raison.*

Pour rejeter catégoriquement l'argument ou l'opinion de l'interlocuteur.

Je ne suis ***pas du tout*** *d' accord avec vous.*
Je ne partage ***absolument*** *pas votre point de vue / le point de vue de...*

Pour nuancer et exprimer avec courtoisie son désaccord, même si celui-ci est total.

Je ne suis pas ***tout à fait / complètement / totalement*** *d' accord avec vous quand vous dites que...*
Je ne partage pas ***tout à fait / complètement / totalement*** *votre opinion.*

Contredire avec courtoisie

Pour contredire d'une manière courtoise, on reprend les arguments ou les idées exposés par l'interlocuteur avant d'introduire sa propre opinion ou argument.

Oui, bien sûr, mais
C'est vrai, mais + OPINION
Il est vrai que ... mais

Il est vrai que *les parents doivent surveiller ce que leurs enfants regardent,* ***mais*** *la télévision est un service public et...*

AUTRES RESSOURCES POUR DÉBATTRE

On sait que *le tabac est mauvais pour la santé.*	On présente un fait que l'on considère admis par tout le monde.
En tant que *médecin, je dois dire que...*	On situe un point de vue depuis un domaine de connaissance ou d'expérience.
Par rapport à *l' interdiction de fumer dans les restaurants, je pense que...*	On signale le sujet ou le domaine dont on veut parler.
D'une part, *les jeunes ne sont pas assez informés sur les risques du tabac,* ***d'autre part...***	On présente deux aspects d'un sujet, d'un fait, ou d'un problème.
Interdire n' est pas la bonne solution. ***D'ailleurs,*** *l' histoire l' a très souvent démontré.*	On justifie, développe ou renforce l'argument ou le point de vue qui précèdent.
Une meilleure communication intergénérations serait souhaitable, **c'est-à-dire** que les parents parlent avec leurs enfants.	On introduit ou développe une explication.
Augmenter le prix du tabac pour réduire sa consommation ne sert à rien. ***En effet,*** *les ventes continuent d' augmenter régulièrement.*	On confirme et renforce l'idée qui vient d'être présentée. Dans un dialogue, son usage est aussi une marque d'accord avec l'idée énoncée par l'interlocuteur.

*Les gens continueront à fumer **même si** le prix du tabac augmente beaucoup.*	On introduit une probabilité que l'on rejette.
*Fumer est dangereux, **car** des particules de goudron se fixent dans les poumons et...*	On introduit une cause que l'on suppose inconnue par l'interlocuteur.
*Le tabac est en vente dans des distributeurs automatiques, **par conséquent,** il est très facile pour un mineur d' en acheter.*	Introduit la conséquence logique de quelque chose.
*La cigarette est mauvaise pour la santé, **par contre,** un bon cigare de temps en temps ne fait pas de mal.*	On introduit une idée ou un fait qui contraste avec ce qu'on a dit précédemment.

LE SUBJONCTIF

L'emploi du subjonctif signifie que le locuteur considère la réalisation d'un processus comme :

- **Nécessaire, souhaitable, possible.**

*Pour améliorer votre niveau de français, **il faudrait que vous fassiez** un séjour en France.*
***J'aimerais que Julio vienne** samedi à la fête que j' organise.*

- **Incertaine, douteuse, peu probable.**

*Les jeunes ne croient pas **que les tatouages et les piercings soient** si dangereux.*
*Je ne suis pas sûr **que tu puisses** faire ce travail.*

Quand le sujet de la première et de la deuxième phrase est le même, on met le verbe de la deuxième à l'infinitif.

*Je ne suis pas sûr **de pouvoir** venir samedi.* *Je ne suis pas sûr ~~**que je puisse**~~ venir samedi.*

Le présent du subjonctif est construit à partir de la troisième personne du pluriel (**ils**) du présent de l'indicatif et des formes **nous** et **vous** de l'imparfait.

DEVOIR	
Présent de l'indicatif	**Subjonctif**
ils **doiv**-ent	que je doiv-**e** que tu doiv-**es** qu'il/elle/on doiv-**e** qu' ils/elles doiv-**ent**
Imparfait	**Subjonctif**
nous **devions** vous **deviez**	que nous **devions** que vous **deviez**

Attention ! Les verbes **être**, **avoir**, **faire**, **aller**, **savoir**, **pouvoir**, **valoir**, **vouloir** et **falloir** sont irréguliers.

ÊTRE
que je **sois**
que tu **sois**
qu'il/elle/on **soit**
que nous **soyons**
que vous **soyez**
qu'ils/elles **soient**

AVOIR
que j'**aie**
que tu **aies**
qu'il/elle/on **ait**
que nous **ayons**
que vous **ayez**
qu'ils/elles **aient**

FAIRE
que je **fasse**
que tu **fasses**
qu'il/elle/on **fasse**
que nous **fassions**
que vous **fassiez**
qu'ils/elles **fassent**

ALLER
que je **j'aille**
que tu **ailles**
qu'il/elle/on **aille**
que nous **allions**
que vous **alliez**
qu'ils/elles **aillent**

SAVOIR
que je **sache**
que tu **saches**
qu'il/elle/on **sache**
que nous **sachions**
que vous **sachiez**
qu'ils/elles **sachent**

POUVOIR
que je **puisse**
que tu **puisses**
qu'il/elle/on **puisse**
que nous **puissions**
que vous **puissiez**
qu'ils/elles **puissent**

VOULOIR
que je **veuille**
que tu **veuilles**
qu'il/elle/on **veuille**
que nous **voulions**
que vous **vouliez**
qu'ils/elles **veuillent**

VALOIR
que je **vaille**
que tu **vailles**
qu'il/elle/on **vaille**
que nous **valions**
que vous **valiez**
qu'ils/elles **vaillent**

FALLOIR – il faut (verbe impersonnel)
qu'il fa**ille**

Attention ! Ne confondez pas la prononciation de **que j'aie** [ɛ] et **que j'aille** [aj].

CARACTÉRISER DES ÊTRES OU DES CHOSES : DONT

Dont remplace un groupe de mots introduits par la préposition **de/d'**.

Il peut être complément du nom.

Je connais un garçon ***dont*** *le père est animateur à la télé.*
(= le père de ce garçon est animateur)

S'il est complément du nom, **dont** est toujours suivi des articles définis **le, la, les**.

● *Mais de qui tu parles ?*
❍ *De la fille* ***dont les*** *parents ont un restaurant sur les Champs Élysées.*
❍ *De la fille ~~dont ses~~ parents...*

Il peut être complément prépositionnel d'un verbe accompagné de la préposition **de**.

C' est une chose ***dont*** *on parle souvent.* (= **on parle de** la télévision)

● *Et si on allait au Japon cet été ?*
❍ *Fantastique ! C' est un voyage* ***dont*** *je rêve depuis des années.*
(= **je rêve de** faire un voyage au Japon.)

INTRODUCTION D'UN THÈME EN FRANÇAIS

Pour introduire ou annoncer un thème à exposer ou à débattre en français, on évite l'introduction directe telle que :

Aujourd' hui, nous allons parler de l' influence de la télévision sur les enfants.

On amène généralement le sujet de façon progressive en utilisant de multiples techniques rhétoriques.

On peut par exemple amener le sujet sous forme de **devinette**, de **métaphore**, de **petit jeu**, ou bien commencer par **une question**, **une petite histoire**, **un proverbe** ou **un dicton**, ou encore utiliser **l'ironie**, **la répétition**, **l'antithèse**, **un cliché**, etc. Le but de tous ces recours est de susciter l'intérêt et de séduire l'audience ou les participants dès le début.

Elle est présente dans presque tous les foyers et elle a pris une telle ampleur qu' elle est devenue le premier des loisirs. Elle a une place prépondérante au centre du salon et les enfants la regardent en moyenne 12 heures par semaine. Ce soir, nous allons donc parler de la télévision.

RACONTER UNE HISTOIRE, UN SOUVENIR D'ENFANCE, UNE ANECDOTE

Une histoire, c'est une succession d'événements que nous mettons au **passé composé**.

*Il **a versé** le café dans la tasse.*
*Il **a mis** du sucre dans le café.*
*Avec la petite cuiller, **il l'a remué**.*
***Il a bu** le café.*

(Pour la formation du passé composé, voir **Mémento grammatical** de l'Unité 3.)

Pour chaque événement, nous pouvons faire une pause et **expliquer les circonstances qui l'entourent**. On utilise alors l'**imparfait**.

*Il a versé le café dans la tasse. Il **était** très fatigué et **avait** très envie de prendre quelque chose de chaud. Il a mis du sucre, c' **était** du sucre roux.*

(Pour les autres usages et la formation de l'imparfait, voir **Mémento grammatical** de l'Unité 3.)

Pour évoquer des circonstances qui se sont passées avant l'histoire ou l'événement qu'on raconte, on utilise le **plus-que-parfait**.

*Il **avait** mal **dormi** et avait sommeil, alors il a pris une bonne tasse de café. C' était du café brésilien qu' il **avait acheté** la veille.*

On utilise aussi le **plus-que-parfait** quand on n'a pas raconté dans l'ordre chronologique tous les événements d'un récit et que l'on fait un retour en arrière.

● *T' es contente de ta nouvelle voiture ?*
❍ *Oui très ! Y a plein de gadgets pour garantir la sécurité. Par exemple, ce matin, je m' assois au volant, je vérifie tous mes rétroviseurs, je démarre et, à ce moment-là, j' entends un bip sonore et une lumière rouge se met à clignoter. Je n' **avais** pas **mis** la ceinture de sécurité ! Alors, j' accroche ma ceinture, je passe en première puis j' appuie sur l' accélérateur et, de nouveau, j' entends un bip et une autre lumière rouge s' allume !*
● *C' était quoi ?*
❍ *Je n' **avais** pas **desserré** le frein à main !*

LE PLUS-QUE-PARFAIT

Pour former le **plus-que-parfait**, on met l'auxiliaire **avoir** ou **être** à l'**imparfait** et on ajoute le **participe passé**.

Comme au passé composé, la plupart des verbes se conjuguent avec l'auxiliaire **avoir**.

j'**avais**	**fait** **acheté** **dormi** **vu** **lu** **peint**
tu **avais**	
il/elle/on **avait**	
nous **avions**	
vous **aviez**	
ils/elles **avaient**	

Les verbes qui expriment une transformation du sujet qui passe d'un état à un autre ou d'un lieu à un autre se conjuguent avec l'auxiliaire **être**. C'est le cas des verbes pronominaux et de certains verbes intransitifs.

(Pour la liste des verbes intransitifs qui se conjuguent avec l'auxiliaire **être**, voir **Mémento grammatical** de l'Unité 3.)

j'**étais**	**allé/e/s** **arrivé/e/s** **sorti/e/s** **entré/e/s**
tu **étais**	
il/elle/on **était**	
nous **étions**	
vous **étiez**	
ils/elles **étaient**	

je **m'étais**	**réveillé/e/s** **perdu/e/s** **assis/e/s**
tu **t'étais**	
il/elle/on **s'était**	
nous **nous étions**	
vous **vous étiez**	
ils/elles **s'étaient**	

L'ACCORD DU PARTICIPE PASSÉ

Quand le verbe se conjugue au passé composé ou au plus-que-parfait avec l'auxiliaire **être**, le participe passé s'accorde avec le sujet.

Quand est-ce que Thierry est parti ? (MASCULIN SINGULIER)

L'autre jour, j'ai rencontré Catherine aux Galeries Lafayette.
*Elle était venu**e** avec sa mère pour acheter une robe de mariée.* (FÉMININ SINGULIER)

*L'été dernier, Yvan et Stéphane sont allé**s** aux Antilles.* (MASCULIN PLURIEL)

*À quelle heure Estelle et sa sœur sont-elles rentré**es** ?* (FÉMININ PLURIEL)

Quand le verbe se conjugue avec l'auxiliaire **avoir**, le participe passé ne s'accorde pas avec le sujet, mais avec le **COD** (complément d'objet direct) **s'il y en a un et si celui-ci est placé devant le verbe**.

● *Tu as vu les photos de nos vacances ?*
○ *Oui, Claudine me **les** a montré**es** l'autre jour.*

● *Qu'est-ce que tu nous prépares ?*
○ *Une omelette avec **les champignons** que j'ai cueilli**s** ce matin.*

SITUER DANS LE TEMPS

L'autre jour, lundi dernier, il y a un mois ...

Pour commencer à raconter un souvenir d'enfance ou une anecdote vécue, on utilise des **marqueurs temporels** qui permettent de situer dans le passé ce souvenir ou cette anecdote.

On peut raconter une anecdote en la situant dans un passé lointain, mais sans préciser quand.

***Un jour**, j'étais très petite, je suis allée à la plage avec mes cousins...*

On peut raconter une anecdote en la situant dans un passé récent, mais sans préciser quand.

***L'autre jour**, je suis allé au cinéma avec des amis.*

On peut aussi raconter une anecdote en la situant d'une manière plus précise dans le passé.

Il y a un mois environ, *j' ai fait un voyage en Italie.*
Lundi dernier, *je suis allée au restaurant avec mes parents.*
Samedi soir, *j' ai vu une chose étrange.*

hier
avant-hier
il y a deux jours
lundi / mardi / mercredi /... (dernier)
la semaine dernière
le mois dernier
l'été dernier
l'année dernière
le 24 juin

(Pour les moments de la journée, voir **Mémento grammatical** de l'Unité 2.)

Il y a

Il y a suivi d'une expression de durée situe l'action dans le passé.

● *Quand est-ce que vous vous êtes rencontrés ?*
❍ ***Il y a*** *deux ans.*

● *Tu as déjeuné ?*
❍ *Oui,* ***il y a*** *une demi-heure.*

il y a	cinq minutes
	une heure
	deux jours
	trois mois
	un siècle
	mille ans

Il y a peut être suivi d'une expression de durée imprécise.

● *Quand est-ce que vous vous êtes mariés ?*
❍ *Oh,* ***il y a*** *longtemps !*
■ *Oui,* ***il y a*** *une éternité !*

À cette époque-là, cette année-là, ce jour-là, ce soir-là, à ce moment-là, ...

Ces marqueurs temporels servent à introduire des circonstances qui entourent ou précèdent l'événement. En général, il s'agit d'une reprise de la donnée temporelle.

Après mon bac, *j' ai fait mes études à la fac de lettres, à Montpellier.* ***À cette époque-là,*** *j' habitais dans un petit appartement au centre-ville.*

Lundi dernier, *je suis allée au restaurant avec Daniel.* ***Ce jour-là,*** *il pleuvait des cordes.*

Samedi soir, *j' ai vu une chose étrange.* ***Ce soir-là,*** *j' avais décidé de faire une balade au bord de la rivière.*

« Quand la Grande Guerre a éclaté, Mata-Hari, a continué à voyager librement à travers l' Europe. (...) ***Cette année-là,*** *la tension était très forte en France. La guerre durait depuis trois ans et elle avait fait beaucoup de victimes. »*

La veille, deux jours avant, quelques jours auparavant, la semaine précédente, ...

Ces marqueurs temporels servent à introduire les circonstances qui ont précédé l'événement.

*La police a arrêté jeudi un homme qui, **quelques jours auparavant / deux jours avant / la veille** avait cambriolé la bijouterie de la place Thiers.*

Au bout de

Ce marqueur temporel indique la durée entre deux événements et il suppose une conclusion qui intervient au terme de la seconde action.

*J'ai attendu très longtemps puis **au bout de deux heures**, je suis parti.*
(= après avoir attendu deux heures)

*Ils se sont mariés en 2001 et, **au bout de quelques années**, Charline est née.*
(= après quelques années de mariage)

***Au bout de longues négociations**, les deux parties sont enfin tombées d'accord.*

Tout à coup, soudain

Ces marqueurs temporels signalent une rupture de temps et l'arrivée brusque d'un événement.

*J'étais tranquillement assis dans la bibliothèque, (quand) **tout à coup** il y a eu un grand bruit.*

*J'étais en train de regarder la télé, (quand) **soudain** la lumière s'est éteinte.*

Tout à coup et **soudain** sont souvent précédés par le marqueur **quand**.

Finalement

Pour conclure le récit et indiquer quelle est la conséquence de l'histoire.

*Je me suis levée tard, j'ai renversé mon café, le téléphone a sonné, ma voiture ne voulait pas démarrer. **Finalement**, je suis arrivée en retard au travail.*

LA VOIX PASSIVE

Dans une phrase **à la voix active, le sujet du verbe fait l'action.**

*André Le Nôtre **a dessiné** les jardins de Versailles.*

Dans une phrase à la **voix passive, le sujet du verbe ne fait pas l'action**.

*Les jardins de Versailles **ont été dessinés** par André Le Nôtre.*

Dans une phrase à la voix passive, l'agent (celui qui fait l'action) n'est pas toujours explicite.

*Le château de Versailles et ses jardins **ont été construits** au XVIIe siècle.*

À la voix passive, le temps verbal est indiqué par l'auxiliaire et le participe s'accorde toujours en genre et en nombre avec le sujet.

*Le cyclone **a été** très violent ; plusieurs maisons ont **été détruites** et beaucoup d'arbres **ont été déracinés**.*

IL ÉTAIT UNE FOIS...

LE PASSÉ SIMPLE

Le **passé simple** s'emploie seulement à l'écrit et essentiellement à la troisième personne du singulier et du pluriel. Il a les mêmes valeurs que le passé composé, mais il situe l'histoire racontée dans un temps séparé du nôtre. Voilà pourquoi le passé simple est traditionnellement d'usage dans les récits où le surnaturel est présent comme dans les contes, les mythes et les légendes.

*Ils **se marièrent** et **eurent** beaucoup d' enfants. (*Blanche Neige et les sept nains*)*

*Pâris **lança** une flèche qui traversa le talon d' Achille. (*Le cheval de Troie*)*

*Georges **demanda** l' aide d' un dieu inconnu de la princesse : le dieu des chrétiens. (*Saint Georges et le dragon*)*

On peut aussi trouver le passé simple dans des biographies ou des romans.

*Dans sa ronde, elle **se heurta** contre moi, **leva** les yeux. **Je vis** se succéder sur son visage plusieurs masques —peur, colère, sourire. (Andreï Makine,* Le testament français*)*

Dans un récit au passé simple, il peut y avoir aussi des passés composés, notamment lorsque les personnages dialoguent.

Une minute et quatre cent vingt mille lieues plus loin, le Petit Poucet arriva devant le palais royal. Il entra dans la salle où le roi tenait un conseil de guerre.
« Que viens-tu faire ici, mon garçon ? » demanda le roi d' un ton sévère. « Ce n' est pas un endroit pour un petit enfant ! »
*« **Je suis venu** pour gagner beaucoup d' argent », expliqua le Petit Poucet en saluant le roi.*

FORMATION DU PASSÉ SIMPLE

DANSER	FINIR	CONNAÎTRE	VENIR
je dans -**ai**	je fin -**is**	je conn -**us**	je v -**ins**
tu dans -**as**	tu fin -**is**	tu conn -**us**	tu v -**ins**
il/elle/on dans -**a**	il/elle/on fin -**it**	il/elle/on conn -**ut**	il/elle/on v -**int**
nous dans -**âmes**	nous fin -**îmes**	nous conn -**ûmes**	nous v -**înmes**
vous dans -**âtes**	Vous fin -**îtes**	vous conn -**ûtes**	vous v -**întes**
ils/elles dans -**èrent**	ils/elles fin -**irent**	ils/elles conn -**urent**	ils/elles v -**inrent**

TELLEMENT / SI ... QUE

Tellement et **si** se placent devant un adverbe ou un adjectif et expriment une très grande intensité pour le locuteur.

(Voir **Mémento grammatical** de l'Unité 1, « Si, tellement ».)

Tellement et **si** annoncent souvent une conséquence qui sera introduite par **que**.

*Je suis **si/tellement** fatiguée **que** je m' endors absolument partout.*

*Mange plus **lentement** ! Tu manges **si/tellement** vite **que** tu vas avoir mal à l' estomac !*

*Cendrillon descendit l' escalier **tellement** vite **qu'**elle perdit une de ses pantoufles de verre.*

TELLEMENT (DE) / TANT (DE) ... QUE

Tellement et **tant** modifient un verbe et expriment l'intensité.

● *Qu' est-ce qui lui arrive ? Il est aphone ?*
❍ *Oui, complètement. Il a **tellement/tant** chanté hier soir !*

Tellement et **tant** peuvent annoncer une conséquence.

● *Qu' est-ce qui lui arrive ? Il est aphone ?*
❍ *Oui, complètement. Il a **tellement/tant** chanté hier soir qu' il n' a plus de voix du tout !*

*Benjamin voyage **tellement/tant** qu' il ne voit presque jamais sa famille.*

Tellement/tant de modifient un nom.

● *Et si on allait à la plage ?*
❍ *Je suis désolé, mais j' ai **tellement/tant de** travail **que** je ne peux pas sortir ce week-end.*

*La télévision a **tellement/tant** pris **d'**importance dans notre vie, **qu'**elle est souvent au centre de la salle à manger.*

À l'oral et dans un registre familier, **tellement/tant (de)** sont parfois placés derrière le participe passé quand le verbe est conjugué à un temps composé (passé composé, plus-que-parfait).

*Il a **chanté tellement/tant** hier soir qu' il n' a plus de voix du tout !*

*La télévision a **pris tellement/tant d'**importance dans notre vie, qu' elle est souvent au centre de la salle à manger.*

Tellement (de) est un peu plus fréquent que **tant (de)**.

LA CAUSE : PARCE QUE, CAR ET PUISQUE

Parce que indique la cause de manière neutre, c'est-à-dire que le locuteur ne fait aucune supposition à propos de ce que l'interlocuteur sait. **Parce que** peut se placer au début d'un énoncé ou bien entre deux propositions.

*Il n' est pas allé en cours **parce qu'**il était malade.*

● *Pourquoi vous n' êtes pas venu en cours ?*
❍ ***Parce que** j' étais malade.*

Car sert à introduire une cause que l'on suppose inconnue par l'interlocuteur.

*Un piercing au nombril avant 16 ans n' est pas recommandable **car** les adolescents peuvent encore grandir et la peau peut éclater.*

(Voir **Mémento grammatical** de l'Unité 6.)

Normalement, **car** n'est pas placé en début de phrase.

*~~**Car**~~ les adolescents peuvent encore grandir, un piercing au nombril avant 16 ans n' est pas recommandable.*

Puisque sert à introduire une cause que l'on suppose connue par l'interlocuteur.

● *Tu vas au cinéma ce soir ?*
❍ *Non, **puisque** tu as dit que tu ne m' accompagnais pas.*

Puisque peut se placer au début d'un énoncé ou bien entre deux propositions.

*Tu pourrais faire les courses **puisque** tu finis à midi.*
***Puisque** tu finis à midi, tu pourrais faire les courses.*

AFIN / POUR QUE

Afin de introduit un objectif, un but à atteindre.

Afin de est suivi de l'infinitif.

*Je dois étudier beaucoup **afin de** réussir tous les examens et partir tranquille en vacances.*

Afin que est suivi du subjonctif, lorsque le sujet de la première phrase est différent de celui de la deuxième.

*Nous avons téléphoné à sa mère **afin que** Pierre **puisse** venir avec nous en vacances.*

Pour que introduit un objectif, un but à atteindre mais s'emploie quand il y a un changement de sujet entre la première phrase et la deuxième. **Pour que** est suivi du subjonctif.

*« Loup, montre tes pattes **pour que** nous **puissions** voir si tu es vraiment notre chère maman », dirent-ils en cœur. (*Le loup et les sept chevreaux*)*

POURTANT

Pourtant s'emploie pour mettre en évidence quelque chose qui nous semble paradoxal.

● *Tu t' es perdu ?*
❍ *Oui, complètement !*
● ***Pourtant** ce n' est pas la première fois que tu viens chez moi !*

Pourtant peut être renforcé par **et** ou **mais**, en exprimant ainsi la surprise, la contrariété.

● *Quel sale temps !*
❍ *Comme tu dis ! **Et pourtant** la météo avait annoncé du soleil !*

*Je ne retrouve plus mes clefs, **mais pourtant** je suis sûr de les avoir laissées ici !*

LORSQUE

Lorsque s'utilise de la même manière que **quand**.

Lorsque peut signifier **à l'époque où**.

***Lorsque** j' étais petite, j' habitais en banlieue parisienne.*
(= **Quand** j'étais petite, j'habitais en banlieue parisienne.)

Lorsque peut signifier **au moment où**.

***Lorsque** je suis sorti ce matin, il pleuvait.* (= **Quand** je suis sorti ce matin, il pleuvait.)

Lorsque peut signifier **chaque fois que**.

*Le week-end, **lorsqu'**il fait beau, nous allons à la plage.*
(Le week-end, **quand** il fait beau, nous allons à la plage.)

LA SIMULTANÉITÉ : TANDIS QUE ET PENDANT QUE

Tandis que et **pendant que** indiquent la simultanéité de deux actions ou de deux états.

● *Comment les enfants se sont-ils comportés ?*
❍ *Oh, très bien ! Paul a rangé tous les jouets **pendant que / tandis que** Judith mettait la table.* (= Paul a rangé tous les jouets **et pendant ce temps** Judith a mis la table.)

● *Claude n' a pas téléphoné ?*
❍ *Ah, peut-être... Le téléphone a sonné deux fois **pendant que/tandis que** je prenais une douche.*

***Tandis que** le Petit Chaperon rouge marchait dans les bois, le loup mangeait sa grand-mère.*

● *Qu' est-ce que tu feras **pendant que** je serai absente ?* (= pendant mon absence)

Mais si on veut insister sur le fait que ces deux actions ou ces deux états sont très différents on emploie **tandis que**.

*Parfois il pleut **tandis que** le soleil brille.*

LE GÉRONDIF

Quand le sujet fait deux actions simultanées, on peut utiliser le gérondif.

● *Je n' ai jamais le temps de lire le journal.*
❍ *Moi, je le lis toujours **en prenant** mon petit déjeuner.*
(= Je lis le journal **et en même temps** je prends mon petit déjeuner.)

Ainsi, on peut dire aussi :

*Je prends mon petit déjeuner **en lisant** le journal.*

Le gérondif sert aussi à exprimer de quelle manière les choses se passent.

● *Maman, je me suis tordu le genou.*
❍ *Comment tu t' es fait ça ?*
● ***En jouant** au football.*

Il peut aussi exprimer la cause ou la condition.

*Tu l' as contrarié **en refusant** de participer.*
(= Tu l'as contrarié **parce que** tu as refusé de participer.)

*De nos jours, c' est difficile de trouver un travail de commercial **en ne sachant pas** parler l' anglais.*
(= C'est difficile de trouver un travail de commercial **si on ne sait pas** parler l'anglais.)

Quand on exprime la manière, la cause ou la condition, le sens de la phrase peut changer radicalement selon le verbe que l'on met au gérondif.

*Je me suis tordu le genou **en jouant** au football.*
J' ai joué au football ~~en me tordant~~ le genou.

*Tu l' as contrarié **en refusant** de participer.*
Tu as refusé de participer ~~en le contrariant~~.

Formation du gérondif

Le gérondif se forme avec **en** + PARTICIPE PRÉSENT.

Pour former le participe présent, on prend comme base la 1re personne du pluriel du présent de l'indicatif et on lui ajoute **-ant**.

Présent	**Gérondif**	
nous **pren**ons	en **pren**-ant	[pʀənɑ̃]
nous **buv**ons	en **buv**-ant	[byvɑ̃]
nous **conduis**ons	en **conduis**-ant	[kɔ̃dɥizɑ̃]

Voilà trois gérondifs irréguliers.

ÊTRE → **en étant**
AVOIR → **en ayant**
SAVOIR → **en sachant**

À la forme négative

À la forme négative, les particules **ne** et **pas** encadrent le participe présent.

*Tu nous as déçus en **ne** réussissant **pas** l' examen.*

en **n'**étudiant **pas**
en **ne** sachant **pas**
en **ne** parlant **pas**
en **n'**écoutant **pas**
en **ne** faisant **pas**

DEPUIS (QUE), IL Y A ... QUE, ÇA FAIT ... QUE

Depuis suivi d'une expression de durée chiffrée indique qu'un état ou une action qui a commencé dans le passé dure encore dans le présent.

Marie et François sont mariés ***depuis trois ans****.*

On peut dire la même chose de deux autres manières.

Il y a *trois ans* ***que*** *Marie et François sont mariés.*
Ça fait *trois ans* ***que*** *Marie et François sont mariés.*

Depuis, ça fait ... que et **il y a ... que** peuvent également être suivis d'un adverbe de temps qui exprime une durée.

J' habite à Paris ***depuis longtemps****.*
Ça fait longtemps que *j' habite à Paris.*
Il y a longtemps que *j' habite à Paris.*

Depuis peut aussi être suivi d'une date fixe ou d'un nom qui exprime un événement.

● *Depuis quand est-ce que tu habites à Paris ?*
❍ *J' habite à Paris* ***depuis 1998****, c' est-à-dire* ***depuis mon mariage****.*

● *Comment tu vas ? Ça fait longtemps qu' on ne s' est pas vus !*
❍ *Oui, nous ne nous sommes pas vus* ***depuis l'été dernier*** *!*

Dans ces cas-là, la phrase ne peut pas se construire avec **il y a... que** ou bien **ça fait... que**.

Depuis que est suivi d'une phrase verbale.

J' ai arrêté de travailler ***depuis que ma fille est née****.* (= depuis la naissance de ma fille)

depuis 1998 / janvier 2004 / Noël / l'été dernier / lundi dernier / mon mariage / le départ de Gérard / la naissance de mes enfants / le jour où je t'ai vue / la première fois où... / toujours / **que** je suis venu en France / **que** ma fille est née / **que** j'ai rencontré Frédéric / ...

IL Y A

Il y a suivi d'une expression de durée situe l'action dans le passé.

● *Quand est-ce que vous vous êtes rencontrés ?*
❍ ***Il y a deux ans.***

● *Je t' offre un café ?*
❍ *Non merci, j' en ai pris un* ***il y a une demi-heure****.*

● *Tu sais où est Sarah ?*
❍ *Elle était là* ***il y a deux secondes à peine****.*

Il y a peut être suivi d'une expression de durée imprécise.

● *Quand est-ce que vous vous êtes mariés ?*
❍ *Oh,* ***il y a longtemps*** *!*
● *Oui,* ***il y a une éternité*** *!*

il y a deux secondes / cinq minutes / une heure / deux jours / trois mois / vingt ans / un siècle / mille ans / longtemps / quelque temps / une éternité / ...

DANS

Dans suivi d'une expression de durée chiffrée situe l'action dans le futur.

● *Tu es prête ?*
❍ *Pas tout à fait, **dans cinq minutes** !*

● *Quand est-ce que tu pars pour Atlanta ?*
❍ ***Dans deux semaines.***

Dans peut être suivi d'une expression de durée imprécise.

*Je suis fatiguée de vivre en ville. **Dans quelques années,** j'ai l'intention d'acheter une petite maison tranquille à la campagne et de m'y installer définitivement.*

***Dans quelque temps,** nous irons en vacances sur la Lune.*

dans deux secondes / cinq minutes / une heure / deux jours / trois mois / vingt ans / un siècle / mille ans / quelques années / quelque temps / le futur /...

Pour l'expression du futur, voir aussi le **Mémento grammatical** de l'Unité 4.

OUI, NON, SI

Papa, tu me prêtes la voiture, n'est-ce pas ?

Quand la question introduite par **est-ce que** ou bien par **une intonation montante** contient une négation, la réponse est **si** pour signifier « oui ».

● *Vous **ne** faites **jamais** de sport ?*
❍ ***Si,** de la natation trois fois par semaine.*

● *Vous **n'**avez pas **encore** visité le musée d'Orsay ?*
❍ ***Si,** je l'ai visité l'année dernière. Il est superbe !*

La réponse est **non** pour confirmer l'information demandée.

● ***(Est-ce que) vous n'**aimez **pas** danser ?*
❍ ***Non,** je n'aime pas danser.*

N'est-ce pas

N'est-ce pas à la fin d'une question est une demande de confirmation.

● *Vous aimez le fromage, **n'est-ce pas** ?*
❍ ***Oui,** beaucoup.*

● *Vous savez qui est Ronaldo, **n'est-ce pas** ?*
❍ ***Oui,** bien sûr, c'est ce joueur de foot si célèbre.*

À l'oral, dans un registre familier, on emploie souvent **hein** [ɛ̃] au lieu de **n'est-ce pas** pour demander une confirmation.

● *N'oublie pas de venir samedi. Tu me l'as promis, **hein** ?*
❍ ***Oui,** je viendrai.*

● *Maman, je peux aller à la piscine avec mes amis ?*
❍ *Demande à ton père !*
● *Papa, tu veux bien que j' aille à la piscine avec mes amis, **hein** ?*

Sa place est mobile dans la phrase.

*Papa, **hein que** tu veux bien que j' aille à la piscine avec mes amis ?*
*Papa, tu veux bien, **hein, que** j' aille à la piscine avec mes amis ?*

RÉPONDRE À UNE QUESTION AUTREMENT QUE PAR OUI OU NON

Pour répondre affirmativement, on peut dire : **tout à fait, en effet, effectivement, évidemment, absolument, bien sûr, bien entendu...**

● *Vous êtes donc convaincu de votre découverte ?*
❍ ***Tout à fait**, vous en doutez ?*

● *Vous connaissez le Québec, n' est-ce pas ?*
❍ *En effet, je l' ai visité il y a deux ans.*

Pour l'usage de **en effet**, voir aussi **Mémento grammatical** de l'Unité 6.

● *Vous êtes sûr de ce que vous dites ?*
❍ ***Évidemment**, j' en suis certain !*

● *Vous êtes un écologiste perspicace ?*
❍ ***Absolument.***

● *Vous pensez que l' exercice physique est bon pour la santé ?*
❍ ***Bien sûr**, c' est très sain !*

● *Vous êtes persuadé de ce que vous affirmez ?*
❍ ***Bien entendu**, totalement persuadé.*

Pour répondre négativement, on peut dire : **pas du tout, absolument pas, vraiment pas, pas vraiment, pas tout à fait...**

● *Vous avez habité en Martinique ?*
❍ ***Pas du tout**, j' y suis allé comme touriste.*

Vraiment pas et pas vraiment

Vraiment pas est une négation catégorique de même que **pas du tout** ou **absolument pas**.

● *Vous aimez la bière, n' est-ce pas ?*
❍ *Non, **vraiment pas**.*

Pas vraiment est une négation partielle. **Pas vraiment** signifie **pas beaucoup** et s'emploie parfois par courtoisie.

● *Vous aimez la bière ?*
❍ ***Pas vraiment**. Je préfère boire de l' eau si c' est possible.*

● *Vous avez compris, n' est-ce pas ?*
❍ *Euh, je suis désolé mais **pas vraiment**. Est-ce que vous pourriez répéter ?*

Table des matières

Unité 1 CHERCHE COLOCATAIRE 06

Nous allons chercher dans la classe une personne avec qui partager un appartement.

Pour cela nous allons apprendre à :
- ♦ parler de nos goûts, de notre manière d'être et de nos habitudes
- ♦ décrire l'endroit où nous habitons
- ♦ exprimer des ressemblances, des différences et des affinités
- ♦ nous orienter dans l'espace
- ♦ communiquer nos impressions et nos sentiments

Et nous allons utiliser :
- ♦ **adorer, détester, ne pas supporter**
- ♦ (m') **intéresser,** (m') **ennuyer,** (me) **déranger**...
- ♦ le conditionnel : **moi, je préférerais...**
- ♦ les prépositions de localisation dans l'espace : **à droite, à gauche, en face de...**
- ♦ les questions : **quand, où, à quelle heure...**
- ♦ **avoir l'air** + adjectif qualificatif
- ♦ les intensifs : **si, tellement**

À la fin de l'unité nous serons capables :
- ♦ de comprendre des questions et de fournir des informations précises concernant nos goûts et nos traits de caractère

Unité 2 SI ON ALLAIT AU THÉÂTRE ? 16

Nous allons organiser un week-end dans notre ville pour des amis français.

Pour cela nous allons apprendre à :
- ♦ exprimer nos préférences en matière de loisirs
- ♦ faire part des expériences
- ♦ faire des suggestions et à exprimer une envie
- ♦ inviter quelqu'un
- ♦ accepter ou refuser une invitation et à fixer un rendez-vous (lieu et date)

Et nous allons utiliser :
- ♦ **c'était (très)** + adjectif ; **il y avait plein de** + substantif
- ♦ **ça te dit de** + infinitif
- ♦ **si on** + imparfait
- ♦ **avoir envie de** + infinitif
- ♦ le futur proche
- ♦ les jours de la semaine, les moments de la journée et l'heure
- ♦ des prépositions et locutions de localisation : **au centre de, (pas) loin de, (tout) près de...**

À la fin de l'unité nous serons capables :
- ♦ de suggérer de faire quelque chose et de prendre rendez-vous
- ♦ d'accepter ou refuser cette suggestion
- ♦ de recommander et d'évaluer un spectacle ou une sortie

Nous allons mettre au point un alibi et justifier notre emploi du temps.

Pour cela nous allons apprendre à :
- ♦ raconter des événements en informant de leur succession dans le temps
- ♦ décrire un lieu, une personne, des circonstances
- ♦ demander et à donner des informations précises (**l'heure, le lieu**, etc.)

Et nous allons utiliser :
- ♦ l'imparfait et le passé composé
- ♦ **d'abord, ensuite, puis, après, enfin**
- ♦ **avant** + nom, **avant de** + infinitif
- ♦ **après** + nom/infinitif passé
- ♦ **être en train de** (à l'imparfait)
- ♦ **se rappeler** au présent
- ♦ **il me semble que**
- ♦ le lexique des vêtements (couleurs et matières)
- ♦ la description physique
- ♦ les marqueurs temporels : **hier soir, dimanche dernier, avant-hier vers 11 heures trente...**

À la fin de l'unité nous serons capables :
- ♦ de comprendre la description d'événements
- ♦ de raconter des événements
- ♦ de donner des détails sur les circonstances qui entourent des événements

Nous allons mettre au point un produit qui facilitera notre vie.

Pour cela nous allons apprendre à :
- ♦ nommer et à présenter des objets
- ♦ décrire et à expliquer le fonctionnement d'un objet
- ♦ caractériser des objets et à vanter leurs qualités
- ♦ convaincre

Et nous allons utiliser :
- ♦ le lexique des formes et des matières
- ♦ les pronominaux passifs : **ça se casse, ça se boit...**
- ♦ quelques expressions avec des prépositions : **être facile à/utile pour, servir à, permettre de...**
- ♦ les pronoms relatifs **qui** et **que**
- ♦ le futur simple
- ♦ **grâce à...**
- ♦ **si** + présent
- ♦ **pour/pour ne pas/pour ne plus** + infinitif

À la fin de l'unité nous serons capables :
- ♦ de comprendre un document publicitaire
- ♦ de définir les caractéristiques de quelque chose de concret dont nous ne connaissons pas le nom
- ♦ de décrire l'usage et le mode d'emploi d'un objet

Unité 5 JE SERAIS UN ÉLÉPHANT 46

Nous allons élaborer un test de personnalité et préparer un entretien de recrutement.

Pour cela nous allons apprendre à :
- évaluer des qualités personnelles
- formuler des hypothèses
- choisir des manières de s'adresser à quelqu'un selon la situation et les relations entre les personnes
- comparer et justifier nos choix

Et nous allons utiliser :
- **avoir du, de la, des** + substantif
- **manquer de** + substantif
- l'hypothèse **si** + imparfait, conditionnel
- les pronoms compléments directs et indirects : **me, te, le, la, nous, vous, les, lui, leur**
- le tutoiement et le vouvoiement
- la comparaison

À la fin de l'unité nous serons capables :
- de comprendre quelques expressions idiomatiques
- de comprendre et de répondre à un questionnaire
- d'utiliser différentes manières de s'adresser à quelqu'un selon la situation de communication

Unité 6 JE NE SUIS PAS D'ACCORD ! 56

Nous allons organiser un débat sur la télé et en tirer des conclusions afin d'améliorer la programmation de la télévision.

Pour cela nous allons apprendre à :
- exposer notre point de vue et à le défendre
- prendre la parole
- reformuler des arguments
- utiliser des ressources pour le débat

Et nous allons utiliser :
- le subjonctif après les expressions d'opinion à la forme négative: **je ne crois/pense pas que**
- **on sait que, il est vrai que... mais, par rapport à, c'est-à-dire, d'ailleurs, en effet, car, par conséquent, d'une part... d'autre part, même si, par contre**
- le pronom relatif **dont**

À la fin de l'unité nous serons capables :
- de suivre et de participer à un débat
- d'argumenter sur un thème

Unité 7 QUAND TOUT À COUP... 66

Nous allons raconter des anecdotes personnelles.

Pour cela nous allons apprendre à :
- raconter au passé
- organiser les événements dans un récit
- décrire les circonstances qui entourent le récit

Et nous allons utiliser :
- l'imparfait, le plus-que-parfait et le passé composé dans un récit
- quelques marqueurs temporels : **l'autre jour, il y a, ce jour-là, tout à coup, soudain, au bout de, à ce moment-là, la veille, le lendemain, finalement**
- la forme passive

À la fin de l'unité nous serons capables :
- de raconter en distinguant les différents temps du récit
- de décrire une atmosphère, une ambiance au passé
- de partir d'un souvenir pour rédiger une anecdote, à plusieurs, et de la raconter

Nous allons raconter un conte traditionnel en le modifiant.

Pour cela nous allons apprendre à :
- exprimer des relations logiques de temps, cause, finalité et conséquence
- raconter une histoire

Et nous allons utiliser :
- passé composé
- **pour que** + subjonctif
- **afin de** + infinitif
- **si/tellement... que**
- **lorsque**
- le gérondif
- **pendant que, tandis que**
- **pourtant**
- **puisque**

À la fin de l'unité nous serons capables :
- de raconter et d'écrire une histoire ou un conte

Nous allons créer un quiz sur des thèmes d'histoire, de géographie, de cultures francophones et de langue française, et nous allons faire un bilan global de notre apprentissage du français.

Pour cela nous allons apprendre à :
- formuler des questions complexes
- répondre à des questions sur la France, la francophonie et la langue française
- exprimer des désirs et des volontés

Et nous allons utiliser :
- le subjonctif après les verbes qui expriment un désir ou une volonté
- les expressions d'affirmation ou de négation
- **depuis** et **il y a... que**
- **si** à la forme interro-négative

À la fin de l'unité nous serons capables :
- de formuler des questions et d'y répondre
- de parler de ce que l'on a appris pendant le cours de français avec Rond-Point 1 et Rond-Point 2.

Cette méthode est basée sur une conception didactique et méthodologique de l'approche par les tâches en langues étrangères développée par Neus Sans Baulenas et Ernesto Martín Peris.

ROND-POINT 2
Livre de l'élève

Auteurs
Josiane Labascoule
Catherine Flumian
Corinne Royer

Édition
Agustín Garmendia Iglesias et Eulàlia Mata Burgarolas

Conseil pédagogique
Neus Sans et Virginie Tamborero

Correction
Katia Coppola et Christian Lause

Conception graphique
A2-Ivan Margot ; Cay Bertholdt

Couverture
A2-Ivan Margot ; illustration : Javier Andrada

Mise en page
Cay Bertholdt ; Ronin-David Mateu ; Carme Muntané

Illustrations
Javier Andrada et David Revilla

Photographies et images
Toutes les photographies ont été réalisées par Marc Javierre Kohan sauf : Frank Kalero : p. 7 (Fabienne), p. 9 (Lara), p. 15 (Steven), p. 24-25 (graffitis et trentenaires au travail), p. 64 (Des chiffres et des lettres). Photographies cedées par la Région de Bruxelles-Capitale : p. 17 (patinoire). © Cin & Scen 2004 : p. 25 (Gloubiboulga Night). Frank Micelotta / Getty Images: p. 26 (6). Cover, Agencia de Fotografía : p. 26 (4), p. 47 (C), p. 55, p. 65 (Dicos d'or). Le Livre de Poche, Librairie Générale Française : p. 34-35. Stock Exchange : p. 26-27 (1, 2, 3, 7, 8), p. 38 (avion), p. 65 (geïsha), p. 82, p. 94 (plage). Frank Micelotta / Getty Images : p. 26 (6). Europa Press : p. 26 (5). © Jacques Carelman, VEGAP, Barcelone 2004 : p. 44-45. © Marsu by Franquin, 2004. www.gastonlagaffe.com : p. 75. Bharath Ramamrutham / STMA: p. 84-85 (Moutia at Anse Takamaka). ACI Agencia de Fotografía : p. 84-85 (canoë à Haiti), p. 94 (Fort de France). Photographies cédées par Tourisme Québec : p. 88-89. Serge Sayn : p. 95 (temple, Cirque de Cilaos).

Enregistrements
Voix : Carine Bossuyt, Christian Lause (Belgique) ; Catherine Flumian, Lucile Herno, Josiane Labascoule, Yves Monboussin, Olivier Penela, Corinne Royer, Jean-Paul Sigé (France) ; Valérie Veilleux (Québec) ; Rebecca Rossi (Suisse).
Musique : p. 35, version de « Bonnie and Clyde », paroles et musique de Serge Gainsbourg. P. 55, version de « Le travail, c'est la santé », paroles de Boris Vian et musique de Henri Salvador.
Studio d'enregistrement : CYO Studios

6e réimpression : février 2007

ISBN édition internationale : 978-84-8443-173-2
D.L. : B-36.498-2004

Imprimé en Espagne par Tallers Gràfics Soler, S.A.

difusión
Français Langue Étrangère

C/ Trafalgar, 10, entlo. 1a
08010 Barcelone (Espagne)
Tél (+34) 93 268 03 00
Fax (+34) 93 310 33 40
fle@difusion.com

www.difusion.com